Sedimentation in Submarine Canyons, Fans, and Trenches

Sedimentation in Submarine Canyons, Fans, and Trenches

Edited by

Daniel Jean Stanley
Smithsonian Institution
Washington, D.C.

and

Gilbert Kelling
University of Keele
Staffordshire, England

Dowden, Hutchinson & Ross, Inc.
Stroudsburg Pennsylvania

This volume is dedicated to Philip H. Kuenen (1902–1976). He led the way.

Library of Congress Catalog Card Number: 77-19163
ISBN: 0-87933-313-8

80 79 78 1 2 3 4 5
Manufactured in the United States of America

Library of Congress Cataloging in Publication Data

Main entry under title:
Sedimentation in submarine canyons, fans, and trenches.
Includes index.
1. Marine sediments. 2. Submarine valleys. 3. Submarine fans. 4. Submarine trenches. 5. Island arcs. I. Stanley, Daniel J. II. Kelling, Gilbert.
GC380.15.S4 551.4'608 77-19163
ISBN 0-87933-313-8

Distributed world wide by Academic Press,
a subsidiary of Harcourt Brace Jovanovich,
Publishers

Foreword

The presence of the heads of some submarine canyons near the shore may have been known locally for several thousand years by Phoenician and other early fishermen and seamen who lived in Lebanon and northern Israel. The earliest ships of the Mediterranean peoples were drawn ashore at night, and the favorite landing beaches then as now probably included those at the heads of submarine canyons whose deep water refracted the waves and diminished the height of the surf. Moreover, fishermen of ancient as well as modern times must have been well aware of sharply deeper water, different bottom materials, frequently colder water, and presence of deepwater species of fish and bottom-living animals in the heads of the canyons. When systematic mapping of continental shelves began in earnest about two hundred years ago, many more canyons were discovered, particularly ones whose heads only indent the shelf break.

About one hundred years ago the rise of scientific inquiry about how the Earth's surface was formed and how it changes included speculations about the origins and ages of submarine canyons. These hypotheses reached their maximum number thirty or forty years ago. During this period, excessive local breakage of submarine telegraph cables was recognized to occur in places where the cables lay across the mouths of submarine canyons. Recently and even more importantly for the future, a thorough knowledge of submarine canyons is needed and is being sought in order to answer questions of slope stability for offshore oil drilling and production platforms and of suitability of canyon heads as sites for disposal of sewage, industrial wastes, and dredge debris with reasonable expectation of subsequent natural transportation into deeper waters. The canyons also may prove to be effective conduits for seaward migration of shelf waters cooled during autumns, particularly at high latitudes, to form and augment slope waters.

Early inferences about origins of submarine canyons had to be based only upon morphology derived from lead-line soundings supplemented by small bottom samples. Later, these scant data were much improved by echo soundings and cores and still later by measurements by using continuous seismic reflection profilers, current meters, water samplers, and transparency meters. All of these methods of data collection are indirect ones, but the development during the past fifteen years of both small and large highly mobile and well-equipped research submersibles has provided the means for direct observation and precisely positioned data analogous to those obtainable by walking an outcrop on land. As the methods and quality of data collection improve, the number of tenable origins of submarine canyons decreased, but the complexity of the processes is seen to increase.

Application of many of these techniques to deeper water at the mouths of submarine canyons began to reveal about thirty years ago the presence of large piles of stratified sediments—too large, in fact, to represent only the material that had been eroded from the canyon. These submarine fans then are the sites of deposition of sands and other sediments trapped by the canyon heads and transported along the lengths of the canyons chiefly by turbidity currents. This process allows coarse-grained sediments to bypass the continental slopes, which typically are floored only by fine silts and clays. Anastomosing submarine channels mark changing positions of the main routes of turbidity currents emitted from the canyon mouths, and their patterns are similar to those of stream channels that cross alluvial fans on land. Even though many of the submarine channels are bordered by natural levees, especially large turbidity currents may top the levees and form new channels particularly after the floors of the earlier channels had been thickly sedimented.

Probably the most irregular-shaped fans are the youngest ones, and thev are as varied as are the subaerial parts of deltas, with both having shapes that are controlled mostly by the relief of the region in which they are deposited. Subsequent growth of the submarine fans to maturity must modify their shapes according to the origin of their continental margins as a function of plate tectonics—divergent, translation, and convergent. Along divergent continental margins, where fan building and channel shifting has continued for a very long time, the fans have coalesced into huge continental rises (as off eastern North America). Further blurring of individual fans is attributable to erosion and redeposition by currents of bottom waters that flow counter to a major surface current such as the Gulf Stream. Off continental margins of translation, submarine canyons and their fans generally are small owing to frequent lateral disruption of drainage systems on land. However, some quite large deltas with submarine canyons occur off the St. Lawrence, Amazon, and Niger Rivers near the junction between continental margins of divergence and translation. The submarine fans of these deltas have grown and prograded along the adjacent continental margins, including those of translation.

Far more complex are the submarine fans at the mouths of submarine canyons that cross continental margins of convergence. There, the larger fans tend to be elongate parallel to the coast and within the adjacent deep ocean trench. Their sizes (areas and thicknesses) are functions of their rate of deposition by turbidity currents and other marine processes and their destruction by tectonic subduction. This subduction highly contorts the fan deposits and plasters some of them against the continental crustal plate and carries some of them downward to become assimilated with oceanic crusts, with the melt rising as diapirs of intermediate magma that forms andesitic volcanoes.

Knowledge of now-filled former submarine canyons and their fans has enormously expanded during the past twenty years through the use of continuous seismic reflection profiles along traverses parallel to the shelf break on existing continental shelves and slopes. Additionally, more and more submarine canyons of ancient origin are being discovered in the geological record, as geologists apply their knowledge of modern processes to interpretations of the origin and environment of ancient strata. Some of these ancient canyons have been recognized in outcrops on land, but many others have been identified from samples and logs of oil wells. Where sands and gravels have been found off the mouths of former submarine canyons, they can be potential petroleum reservoir beds. The varied lithologies of these sediments also provide clues to the composition and paleogeography of the coastal and continental hinterlands, and the organic materials yield information about paleobiology.

Modern submarine canyons and those that have been thinly covered by sediments of the continental shelf and slope can be attributed to subaerial (upper parts) and subaqueous (lower parts) erosion accompanied by submarine deposition of sediments as fans at the canyon mouths during Pleistocene glacial times of low sea level. Other canyons that have been completely filled with Cenozoic and Cretaceous sediments must have been due to processes unrelated to glaciation, and they may not even have required temporarily lowered sea levels. Perhaps, as our knowledge of these features increases, some former submarine canyons may even be attributable to erosion during Precambrian times. Thus the canyons and their fans may well be due to normal geological processes and not to special ones unique to glacially lowered sea levels.

All of the above discussion indicates that submarine canyons and their associated fan deposits at the base of slopes and in trenches offer not only sites for basic studies of Earth processes, but they also provide opportunities for practical and economic application of the results of these studies. The collection in this volume of twenty-five articles written by experienced marine geologists, stratigraphers, and engineers is an up-to-date summary of the state of knowledge of both modern and ancient submarine canyon, fan, slope, and trench sequences. It should prove to be especially useful during coming years as a basis for further progress in understanding the outer continental margin and its counterparts in the rock record.

K.O. Emery
Woods Hole Oceanographic Institution

Preface

. . . experiments might help to clear the field of speculation.–Ph. H. Kuenen, 1937, Leidsche Geologische Mededeelingen

Submarine canyons, fans, and trenches appear to represent the principal repositories of terrigenous sediment in modern world oceans and marginal seas beyond the shelf break, and sequences ascribed to deposition in these features are increasingly recognized in the geological record. The growth of the technological expertise needed to sustain research concerned with outer margin sedimentation and the impetus provided by increased socio-economic pressures have combined to promote rapid expansion in such research. During this phase of development a number of important specialized volumes have appeared that deal with certain aspects of the topic, such as tectonic influences, morphologic characteristics, or sedimentary processes.

It now appears appropriate to devote to this important trio of deeper marine environments a volume that is concerned with all aspects of sedimentation. Such a contribution must take note of the intimate relationship between depositional patterns and tectonic controls and recognize the broad spectrum of sedimentary processes that operate on slopes beyond the shelf edge. By incorporating data and ideas from studies of both modern and ancient marine sequences, we may better test prevailing hypotheses and assess the validity of the conceptual models being generated to synthesize and systematize new information.

The nucleus of this volume is provided by a series of invited contributions presented at a symposium entitled "Submarine Canyon and Fan Sedimentation in Time and Space" convened by D.J. Stanley and F.P. Shepard at the Annual Meeting of the Society of Economic Paleontologists and Mineralogists held in New Orleans, Louisiana, on May 26, 1976. About one-third of the chapters in this volume were solicited subsequent to this meeting in an effort to achieve coverage of this field of research that is as comprehensive as possible. The twenty-five chapters that follow involve a total of fifty contributors, approximately one-half of whom are academics, more than one-quarter are associated with governmental research organizations, and the remainder are almost equally divided between oceanographic research institutions and industry. These statistics demonstrate the growing involvement of government agencies in this field, and an analysis of the authorship of the chapters underlines the trend toward collaborative work.

No plan of organization is sacrosanct or unique. We have assigned the contributions to five main categories that we consider to be the most logical groupings: bottom currents and biological processes in submarine canyons are treated in the six chapters of Part I; gravity-induced processes in submarine canyons and fans are the concern of five chapters in Part II; turbidites (sand as well as mud) and hemipelagic processes in submarine fans are discussed in Part III (six chapters); the tectonic and stratigraphic setting of some submarine canyons and fans are discussed in the three chapters in Part IV; tectonics and sedimentation in trench basins and arc settings and the downslope transport of conglomerates are dealt with in Part V (four chapters). A final chapter attempts to appraise the development and present status of this area of research and offers some predictions of future growth. The scope of the topics cataloged above is indicative of the state of the art, both in terms of the areas of active investigation and, by inference, the gaps in our knowledge. Although the volume reflects parity between ancient and modern studies (twelve chapters in each category), some imbalance in environmental emphasis is apparent, particularly with respect to trenches and slopes. From the nature of the contributions, it is also possible to detect the developing trends in deep marine sediment investigations. In addition to the renewal of sedimentological interest in trenches, concerted and individual efforts are being directed towards elucidating the genesis of those grades of sediment (gravels and muds) that hitherto have been neglected.

We hope that this volume will prove useful not only to marine geologists, sedimentologists, and oceanographers, but also to those concerned with the stratigraphy, structural geology, and environmental sciences of continental margins. In preparing this volume, we, as editors, have gained better insight and a broader appreciation of the manner in which this field is developing, with consequent influence on the direction of our own research. We trust that the reader will benefit equally.

A number of colleagues were called upon to serve as outside reviewers and thus supplemented the pool of authors who assisted in the critical reading of chapters in this volume. We owe a particular debt to the following specialists who interrupted their own work to provide constructive criticism and valuable counsel: P.F. Ballance, J.D. Collinson, J.C. Crowell, W.R. Dickinson, D.S. Gorsline, M. Hampton, R. Hesse, R.J. Knight, E.F. McBride, G.V. Middleton, J.W. Pierce, O.H. Pilkey, H.G. Reading, R.W. Tillman, R.G. Walker, and J.D. McD. Whitaker.

The color photograph used on the jacket of this volume shows a diving geologist working on the seafloor in the head of the Salt River Submarine Canyon off St. Croix, U.S. Virgin Islands. It was generously provided by R.F. Dill of the West Indies Laboratory, Fairleigh Dickinson University, Christiansted, St. Croix.

Our parent organizations, the Smithsonian Institution and the University of Keele, have provided the lion's share of secretarial and other logistical support to alleviate our editorial burdens. Our appreciation to these organizations is expressed, more particularly, for the provision of the scholarly climate enabling us to undertake this task. We thank Ms. M. Forgione and Mr. L. Isham, both of the U.S. National Museum of Natural History, for their assistance with secretarial and drafting requirements, respectively.

Finally, in recording our thanks we would be remiss were we to neglect to acknowledge our indebtedness, and that of our contributors, to the foresight, energy and enthusiasm of the scientist who was primarily responsible for focusing attention on both modern and ancient sequences discussed here. This book is dedicated to Philip H. Kuenen.

Daniel Jean Stanley
Gilbert Kelling

Contents

Contributors

Alvin A. Almgren
Union Oil Company of California
Southern California District
P.O. Box 6176
Ventura, California 93003

Barton C. Birdsall
Gulf Oil Corporation
Houston, Texas 77001

Arnold H. Bouma
U.S. Geological Survey
Marine Geology Branch
345 Middlefield Road
Menlo Park, California 94025

David A. Cacchione
U.S. Geological Survey
Marine Geology Branch
345 Middlefield Road
Menlo Park, California 94025

Paul R. Carlson
U.S. Geological Survey
Marine Geology Branch
345 Middlefield Road
Menlo Park, California 94025

Robert M. Carter
Department of Geology
Otago University
Dunedin, New Zealand

Richard A. Cooper
National Oceanic and Atmospheric Administration
Woods Hole, Massachusetts 02543

Stephen P. Cossey
Department of Geology
University of South Carolina
Columbia, South Carolina 29208

Robert F. Dill
West Indies Laboratory
Christiansted, St. Croix
U.S. Virgin Islands 00820

Robert H. Dott, Jr.
Department of Geology and Geophysics
University of Wisconsin
Madison, Wisconsin 53706

David E. Drake
U.S. Geological Survey
Marine Geology Branch
345 Middlefield Road
Menlo Park, California 94025

Robert Ehrlich
Department of Geology
University of South Carolina
Columbia, South Carolina 29208

Kenneth O. Emery
Woods Hole Oceanographic Institution
Woods Hole, Massachusetts 02543

Patrick G. Hatcher
Atlantic Oceanographic and Meteorologic Laboratories
National Oceanic and Atmospheric Administration
Miami, Florida 33149

John Holroyd
Geology Department
University of Wales
Swansea SA28PP, Great Britain

Murray D. Hicks
Department of Geology
Otago University
Dunedin, New Zealand

George H. Keller
School of Oceanography
Oregon State University
Corvallis, Oregon 97331

Gilbert Kelling
Department of Geology
University of Keele
Keele, Staffordshire ST55BG, Great Britain

Roy Kepferle
U.S. Geological Survey
University of Cincinnati
Cincinnati, Ohio 45221

Donald E. Koelsch
Woods Hole Oceanographic Institution
Woods Hole, Massachusetts 02543

Laverne D. Kulm
School of Oceanography
Oregon State University
Corvallis, Oregon 97331

Alexander Malahoff
National Oceanic and Atmospheric Administration
Rockville, Maryland 20850

Andrés Maldonado
Instituto "Jaime Almera"
University of Barcelona
Avenida de José Antonio, 585
Barcelona 7, Spain

Neil F. Marshall
Geological Research Division
University of California, San Diego
Scripps Institution of Oceanography
La Jolla, California 92093

Peter J. McCabe
Department of Geology
University of Nebraska-Lincoln
Lincoln, Nebraska 68588

Emiliano Mutti
Istituto di Geologia
Universita di Torino
10123-Torino, Italy

C. Hans Nelson
U.S. Geological Survey
Marine Geology Branch
345 Middlefield Road
Menlo Park, California 94025

Tor H. Nilsen
U.S. Geological Survey
Marine Geology Branch
345 Middlefield Road
Menlo Park, California 94025

William R. Normark
U.S. Geological Survey
Marine Geology Branch
345 Middlefield Road
Menlo Park, California 94025

Richard J. Norris
Department of Geology
Otago University
Dunedin, New Zealand

Harold D. Palmer
Dames and Moore, Consultants
7101 Wisconsin Avenue
Washington, D.C. 20014

Tullio Pescatore
Istituto di Geologia e Geofisica
Università di Napoli
Largo S. Marcellino, 10
80138 – Naples, Italy

David J.W. Piper
Departments of Geology and Oceanography
Dalhousie University
Halifax, Nova Scotia, Canada

Franco Ricci Lucchi
Istituto di Geologia
Universita di Bologna
Bologna, Italy

David A. Ross
Woods Hole Oceanographic Institution
Woods Hole, Massachusetts 02543

Gilbert T. Rowe
Woods Hole Oceanographic Institution
Woods Hole, Massachusetts 02543

William J. Schweller
School of Oceanography
Oregon State University
Corvallis, Oregon 97331

Richard M. Scott
Cities Service Oil Company
Exploration and Production Research
Tulsa, Oklahoma 74150

E.M. El Shazly
Egyptian Academy of Science and Technology
Egyptian Atomic Energy Commission
Cairo, Egypt

Francis P. Shepard
Geological Research Division
Scripps Institution of Oceanography
University of California, San Diego
La Jolla, California 92093

Richard A. Slater
Department of Earth Sciences
University of Northern Colorado
Greeley, Colorado 80639

Daniel J. Stanley
Division of Sedimentology
Smithsonian Institution
Washington, D.C. 20560

Colin P. Summerhayes
Exxon Production Research Co.
Houston, Texas 77001

Ian M. Turnbull
New Zealand Geological Survey
Dunedin, New Zealand

Elazar Uchupi
Woods Hole Oceanographic Institution
Woods Hole, Massachusetts 02543

Stephen G. Vedros
Exploration Division
Northern Natural Gas Company
5155 East 51st Street
Tulsa, Oklahoma 74135

Glenn S. Visher
Department of Earth Sciences
University of Tulsa
Tulsa, Oklahoma 74104

Arthur van Vliet
Koninklijke/Shell
Exploratie en Produktie Laboratorium
Rijswijk, The Netherlands

John E. Warme
Department of Geology
Rice University
Houston, Texas 77001

Robert D. Winn, Jr.
Denver Research Center
Marathon Oil Co.
Littleton, Colorado 80120

Sedimentation in Submarine Canyons, Fans, and Trenches

Part I

Bottom Currents and Biological Processes in Submarine Canyons

Chapter 1

Currents in Submarine Canyons and Other Sea Valleys

FRANCIS P. SHEPARD
NEIL F. MARSHALL

Geological Research Division
Scripps Institution of Oceanography
La Jolla, California

ABSTRACT

The vast quantities of sediment carried down submarine canyons is demonstrated by the huge fans at canyon mouths. After eight years of studying currents in submarine canyons, we have acquired an extensive body of information and a considerable understanding of water movements in canyons and their relation to the transport of sediment to canyon mouths. We find that near-bottom currents (rarely exceeding 50 cm/sec) flow almost continuously up- and downcanyon, with net flow commonly in a downcanyon direction and with sufficient speed to transport large quantities of fine sediment down the canyons. In the deeper portions of canyons and in regions with large tidal ranges, the length of these periods of flow and times between reversals show a close relationship to the tides, whereas in shallow canyon heads and in regions with small tidal ranges, the unidirectional flow and time between reversals is much shorter. The more frequent alternations in the direction of flow and length of periods of flow may be caused by internal waves and internal tides as is shown by tracing their patterns along the canyon axes. Periods of crosscanyon flows are observed particularly in canyons with wide floors, and their perioαs are in some cases related to phases of the tide.

Much faster downcanyon flows are observed at infrequent intervals and are interpreted here as turbidity currents. Such flows have been inferred from the loss of some of our current meters during storms and on a few occasions were observed in current meter records with downcanyon speeds of up to about 200 cm/sec (Inman et al. 1976). These flows occur in association with onshore storms, high swell, and during periods of large discharge from rivers that have canyon heads at their mouths. That earthquakes cause large turbidity flows has yet to be documented by current meters.

INTRODUCTION

Our studies of currents in submarine canyons and fan valleys have yielded about 22,000 hours of records showing speed and direction. The current meter equipment used in these studies has been described elsewhere (Marshall 1975). Depths of the various current meter stations range from 46 m to 4,400 m. The records cover periods of from three to twenty-six days. Most of these records come from canyons off the California coast, but others have been collected in canyons and other types of valleys off Baja California, central western Mexico, British Columbia, Kauai in the Hawaiian Islands, the East Coast of the United States, Saint Croix in the Virgin Islands, northwest Luzon, and the Congo Canyon off west Africa. This wide variety of areas and of depths gives us some confidence in making a few generalizations about the nature of currents, both in canyons and in the small valleys that characterize the submarine fans and slopes off large subaerial deltas. We have learned that normal and exceptional currents exist in these sea valleys and that conditions sometimes change radically during a recording period.

NORMAL ALTERNATION OF CURRENT DIRECTIONS

By far the most consistent finding from our records has been that canyon currents alternate their direction of flow from upcanyon to downcanyon (Figure 1-1). The periodicity of the alternations may occur in as short a time as fifteen minutes or for as long as twenty-four hours. None of our records show continuous unidirectional up- or downcanyon flows. One record obtained off the Var River in southern France by Gennesseaux et al. (1971) showed five days of undirectional downcanyon flow with variable speeds. Another record, reported

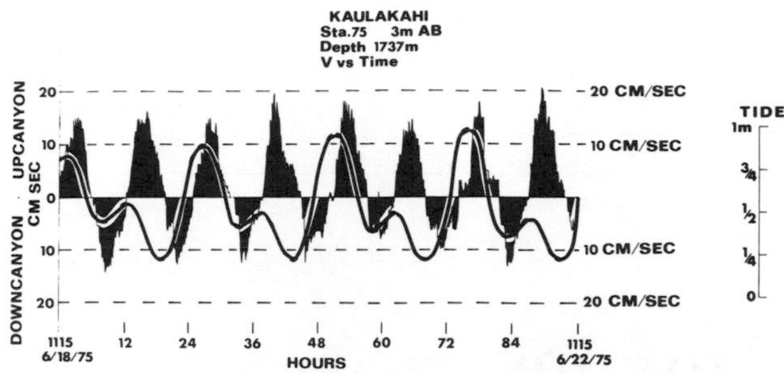

Figure 1-1. Example of the alternating up- and downcanyon currents observed in various types of submarine valleys. Note: Velocities are corrected for divergence from canyon axis by cosine of angle between flow direction and axis. The tide relation is obtained from the predicted tide of the nearest land station.

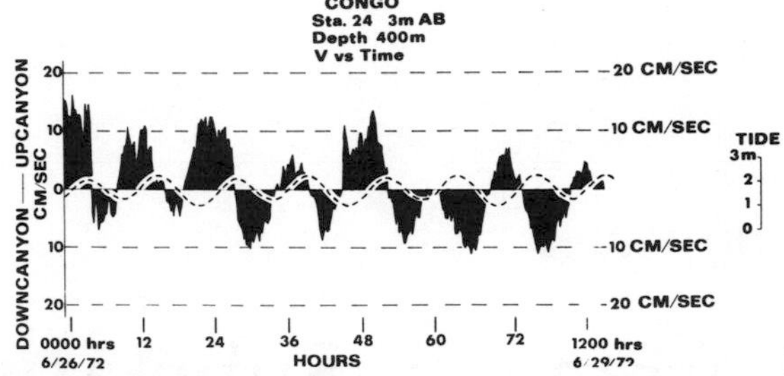

Figure 1-3. Example of canyon currents in Congo Canyon at 400 m depth that are closely related to the tides. Note: Flow is upcanyon during the flood tide and downcanyon during ebb tide, accelerating and decelerating as do typical tidal currents. For an example of exact opposite of flow direction and tide, see Figure 1-1.

by Cacchione et al. (1976) from the outer gorge of Hudson Canyon off New York, also registered three days of continuous, unidirectional downcanyon flow with variable speeds.

A progressive vector plot of the measured current data shows the net direction and approximate distance of travel of a flow. To date, our results show that net direction of flow is generally the same in successive runs at the same station. Current data collected in canyons off California and western Mexico show mostly a net downcanyon flow. The principal exceptions are at stations located near the heads of canyons. Our data and that of other investigators from the East Coast of the United States show approximately the same number of net upcanyon and downcanyon flows.

The currents in some canyons show frequent and very quick changes in speed (Figure 1-2), while in other canyons the period of change occurs slowly and is more similar to the familiar tidal flow that slowly reaches its peak flow and just as slowly decreases speed until reversal occurs (Figure 1-3).

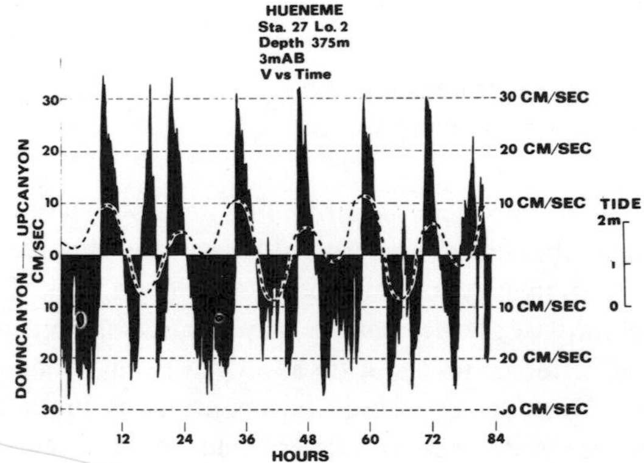

Figure 1-2. Example of the sudden temporal variations of current directions and speeds that typify most canyon currents (see also Figure 1-9).

REVERSAL CYCLES RELATED TO TIDE AND DEPTH

Almost all of our early records were from canyon depths of less than 200 m where we found little relation to tidal cycles (Shepard and Marshall 1969). As soon as records from deeper water were obtained, however, we discovered that the periodicity of current flows approximated the semidiurnal tidal cycles all along the California coasts with few exceptions (Shepard et al. 1974). We also found that in shoaler depths the periodicity of flow cycles (one up- and one downcanyon flow) increased in length with increasing depth until they matched the 12.5-hour cycle (Shepard 1976). In areas with small tidal ranges, as off the island of Kauai (Hawaii) and off the west coast of Mexico, our data indicate that the length of cycles does not approach the tidal frequency until much greater depths where the tides are relatively high. Canyons with depths as great as 1,000 m have relatively short flow cycles when the tidal range is less than or equal to a meter.

Combining all of our information on the average lengths of up- and downcanyon flow cycles versus the canyon depths where the measurements were made, and with respect to tidal range and periodicity, we find that some fairly consistent relationships emerge (Figure 1-4 and Table 1-1). Data from Congo Canyon off west Africa show a tidal relation at a shoaler depth than is found in other canyons (see Figure 1-3). Also, the currents measured at a station in the head of a valley off the Fraser River Delta at a depth of 85 m and in an area with a tidal range of up to 4 m show a tidal periodicity. Two other exceptions occur off the California coast where currents were recorded during neap tidal periods.

CROSSCANYON FLOWS

In general, the currents in canyons flow approximately in an up- or downcanyon direction as shown in our polar plots that display direction versus speed for each

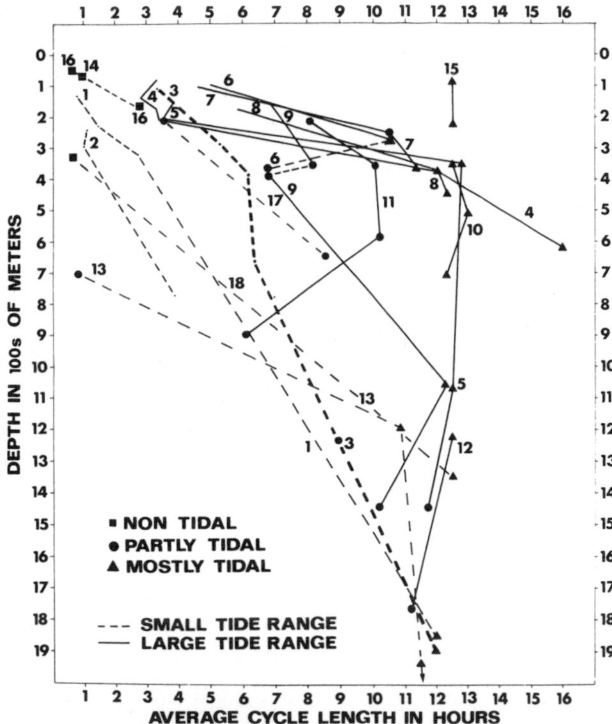

Figure 1-4. Comparison of data collected at different station depths with average cycle length versus tidal range. Note: Solid lines connect stations of the same canyon that had tidal ranges of more than 1.5 m; dashed lines connect stations where the tidal ranges were 1.5m or less (see Table 1-1). Numbers refer to the following canyons: (1) San Lucas (Baja California), (2) Kauai sea-valleys (Hawaii), (3) Rio Balsas (western Mexico), (4) La Jolla (California), (5) Carmel (California), (6) Redondo (California), (7) Newport (California), (8) Hueneme (California), (9) Monterey (California), (10) Hydrographer (Massachusetts), (11) Santa Cruz (California), (12) Hudson (New York), (13) Kaulakahi Channel (Kauai), (14) Christiansted (St. Croix Island), (15) slope valley off Fraser Delta (British Columbia), (16) Salt River (St. Croix Island), (17) Abra (off Abra Delta, West Luzon), (18) Pinget (West Luzon).

Table 1-1. *Tide Ranges*

Canyon	*Range (m)*
1. San Lucas	1.4
2. Kauai	0.7
3. Rio Balsas	0.8
4. La Jolla	2.6
5. Carmel	2.4
6. Redondo	2.5
7. Newport	2.4
8. Hueneme	2.5
9. Monterey	2.3
10. Hydrographer	1.5
11. Santa Cruz	2.3
12. Hudson	1.7
13. Kaulakahi	0.7
14. Christiansted	0.5
15. Fraser River	3.7
16. Salt River	0.5
17. Abra	1.5
18. Pinget	1.5

data increment (diagram A, Figure 1-5). Many exceptions to "normal" flow were found, and some of our data show little or no relation to the orientation of the canyon axes (diagrams B and C, Figure 1-5). One record from Hueneme Canyon (Santa Clara Delta, California) shows some concentration of points in the upcanyon direction, but a very dispersed pattern for downcanyon. This record also shows that the strongest currents occur almost at right angles to the canyon axis. The polar plot of the current data from our deepest station in Hudson Canyon has a random scatter of points around the compass and nothing to indicate the general northwest-southeast axial trend of the canyon. Figure 1-6 shows a plot of crosscanyon current vectors from Hueneme

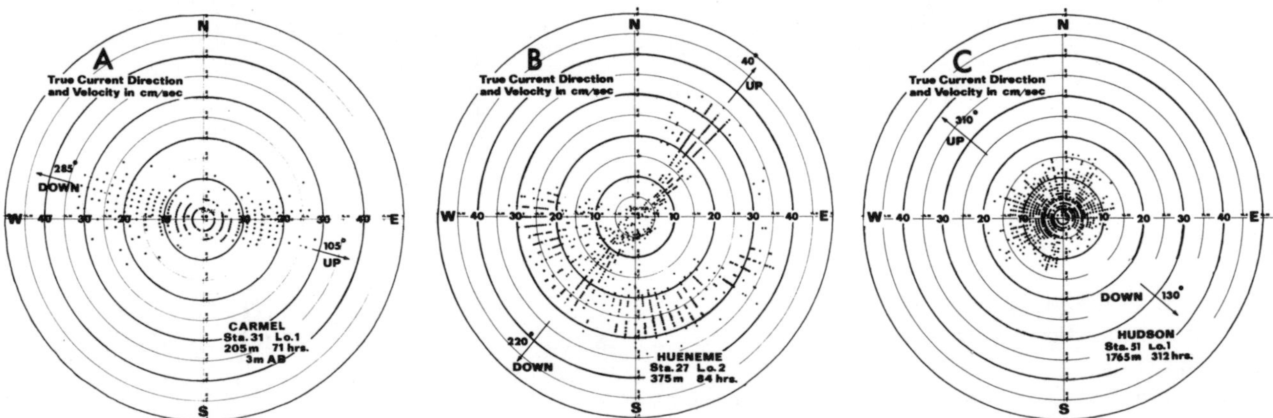

Figure 1-5. Polar plots of canyon currents. Note: Diagram A is an example of flow direction close to that of canyon axis. This polar plot gives the direction (285° downcanyon, 105° upcanyon) of flow and average velocity for each 7.5 minutes of a record in Carmel Canyon. Diagrams B and C show the polar plots of currents in Hueneme (040° upcanyon, 220° downcanyon) and Hudson (130° downcanyon, 310° upcanyon) Canyons. These two charts show little relation between the direction of current flow and the axial trend of the canyons, except for the upcanyon flow in Hueneme Canyon.

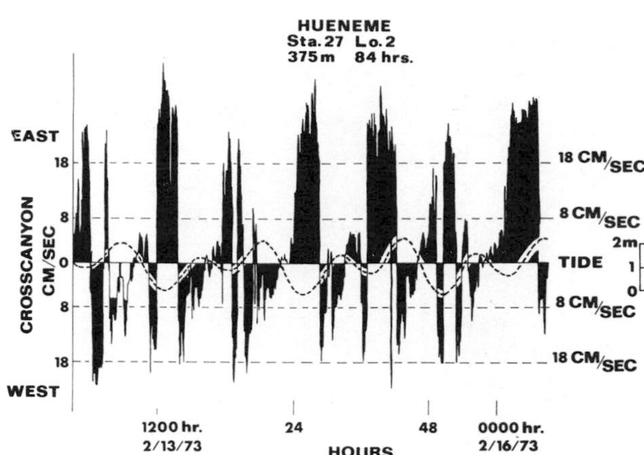

Figure 1-6. Record from Hueneme Canyon showing crosscanyon flow vectors with velocities corrected with the cosine of angle between flow direction and direction normal to the canyon axis. Note: Most of the strong currents of long duration occurred at low tide or shortly after it, thereby suggesting a relation between crosscanyon flows and the stage of the tides.

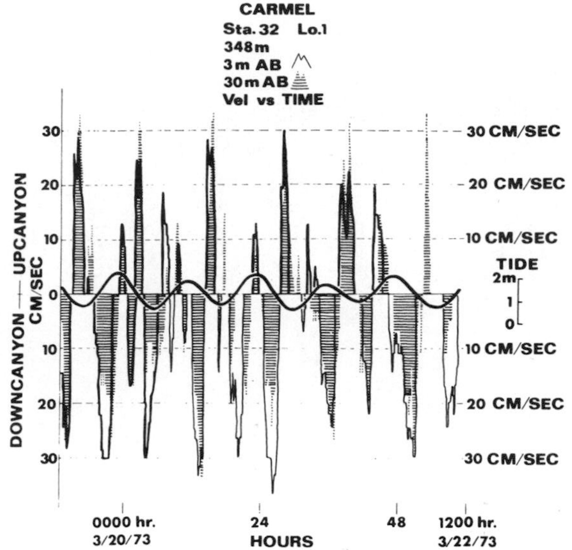

Figure 1-8. Measurements in Carmel Canyon illustrating the close similarity of current velocities and direction at 3 m and 30 m above the bottom (AB).

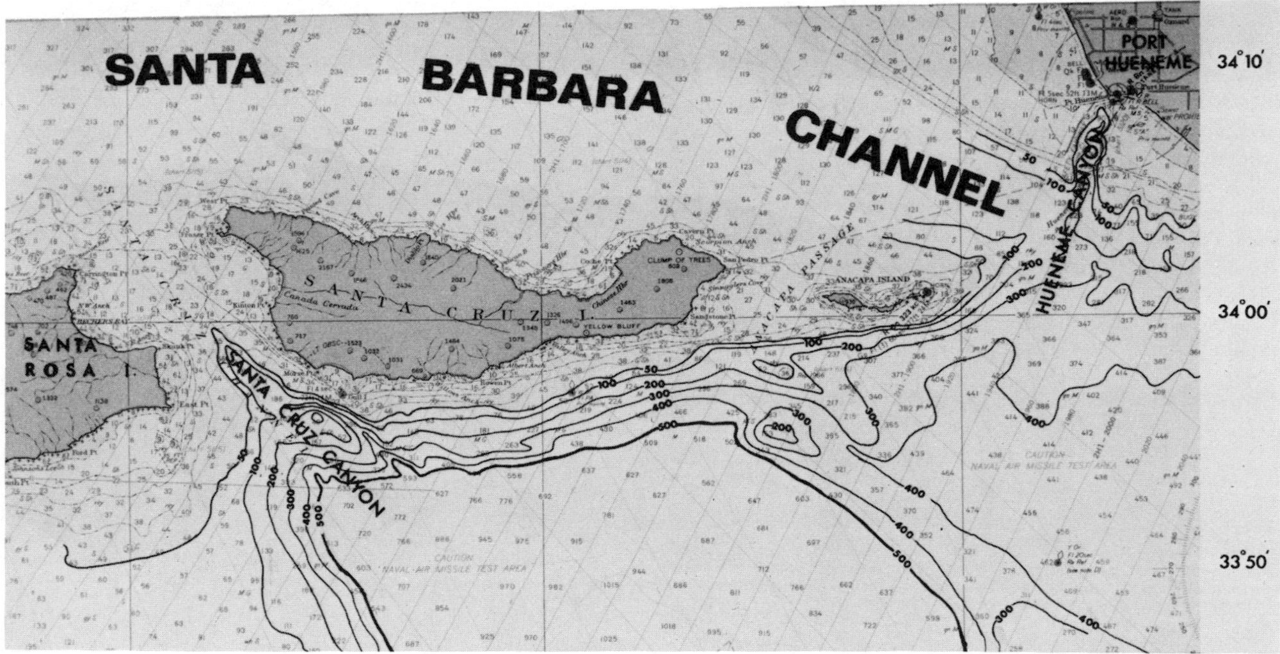

Figure 1-7. Map showing relations of Santa Cruz and Hueneme Canyons to the basins and passes between islands in the Santa Barbara area.

Canyon. All of the strong crosscanyon flows occur at approximately low tide. A reasonable expectation seems to be that the strong crosscanyon flows took place during the slack at low tide when strong wind-driven currents flow across the canyon from the Santa Barbara Channel to the west (Figure 1-7). The bottom flow seems to have the same predominant easterly flow as does the surface current in this area. Crosscanyon flows are especially common wherever a canyon or valley has a wide floor.

RELATIONSHIPS OF CURRENTS MEASURED SIMULTANEOUSLY AT VARIOUS HEIGHTS ABOVE A CANYON FLOOR

Where two current meters were used at different heights above the bottom at the same station, for example, 3 m and 30 m (Shepard and Marshall 1973a), some apprecia-tion of the nature of the vertical distribution of the flows in canyons was acquired. In most cases, directions and

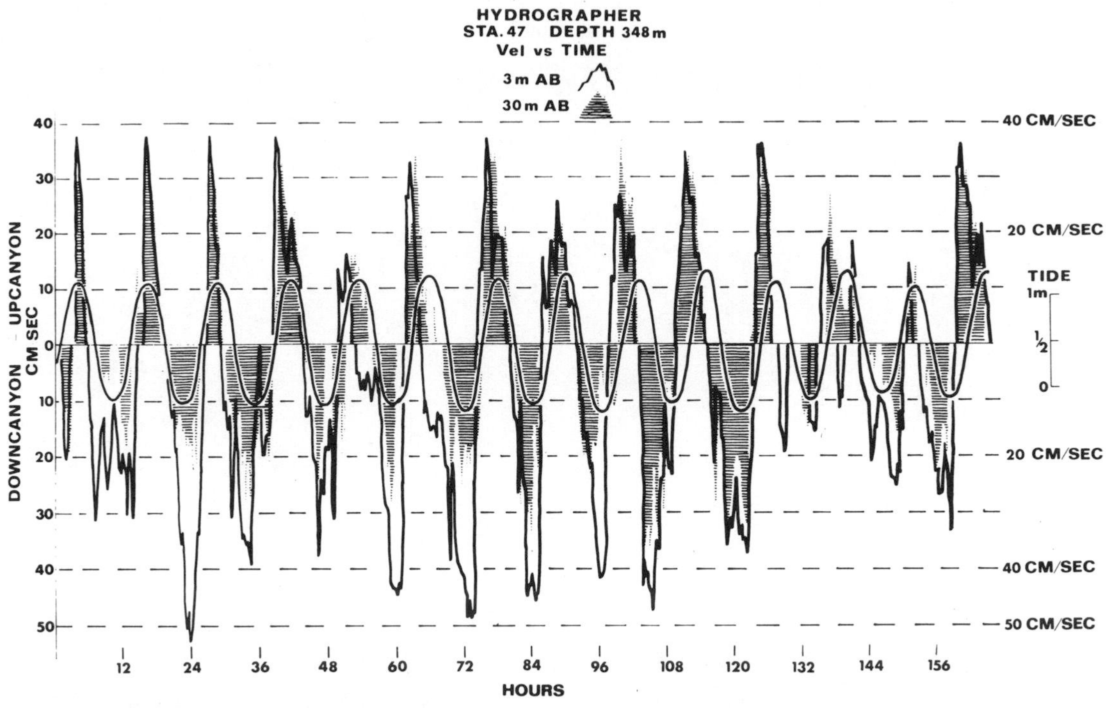

Figure 1-9. Measurements in Hydrographer Canyon illustrating current velocities at 3 m and 30 m above the bottom along the axis of the canyon. Note: The direction reversals and peak velocities are approximately coincident.

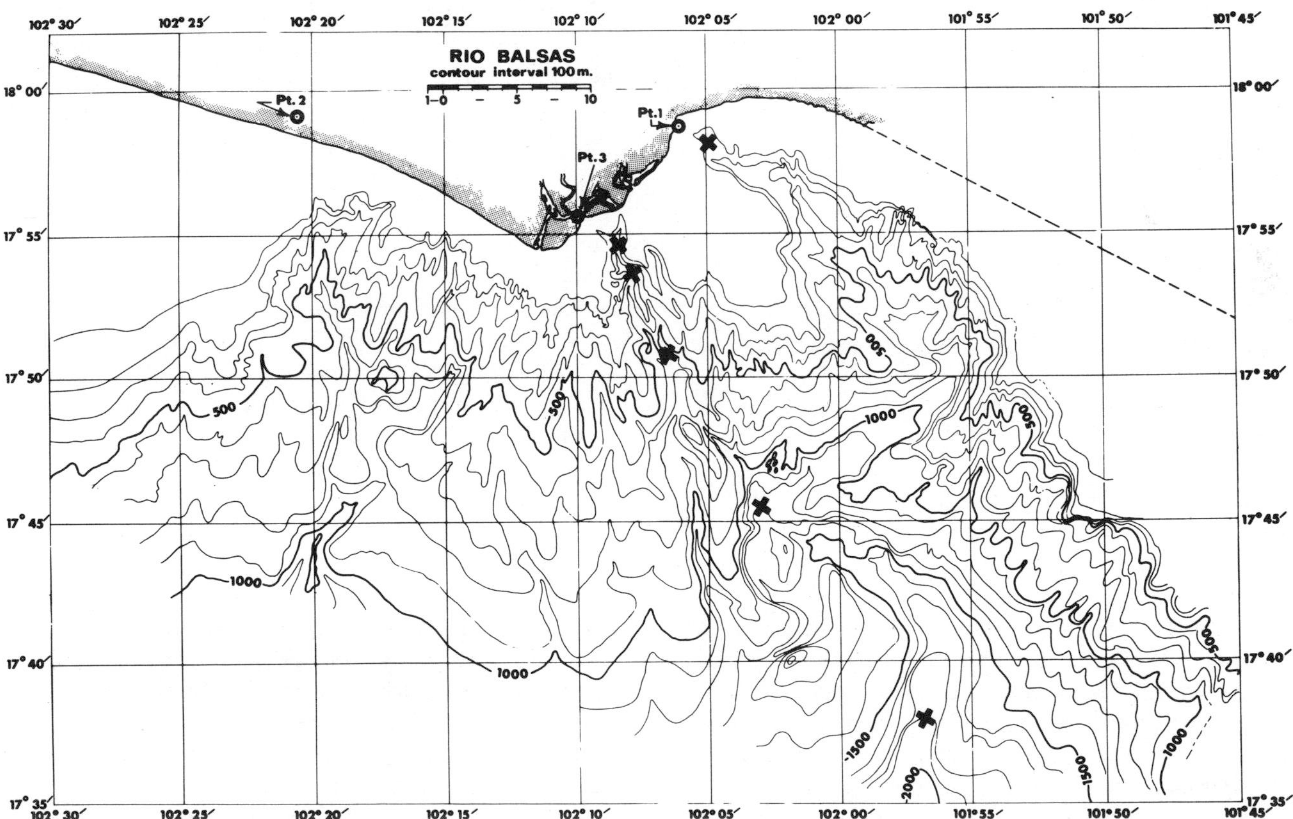

Figure 1-10. Chart of the area off Rio Balsas Delta in western Mexico showing the valleys on the adjacent sea floor and the location of current meter stations (crosses).

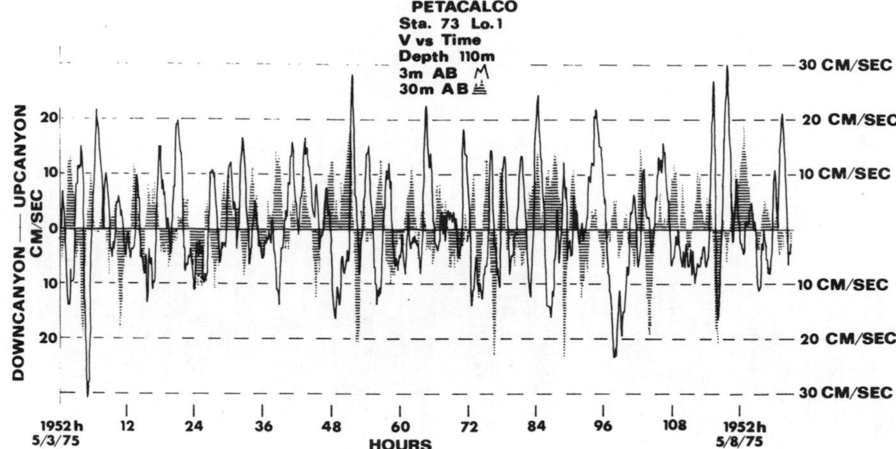

Figure 1-11. Measurements in Petacalco Canyon off the Rio Balsas area illustrating currents at 3 m and 30 m above the bottom. Note: There is almost no relation in the direction of currents at the two heights.

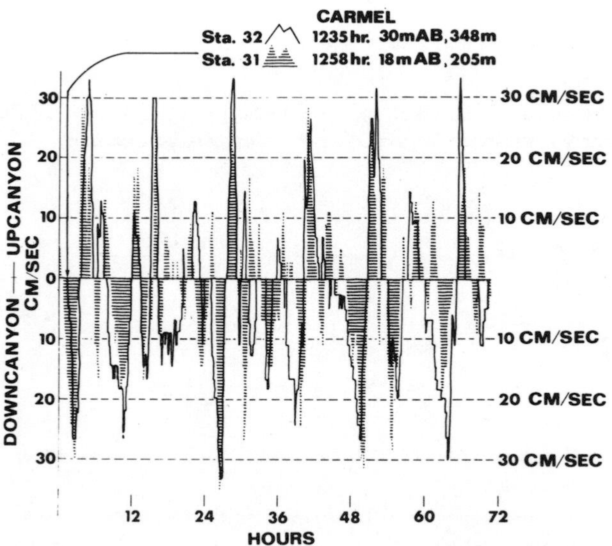

Figure 1-12. Example of the progressive advance of current patterns up a submarine canyon, as shown by time offset of records at two stations along Carmel Canyon. Note: These fits suggest upcanyon movement of internal waves.

speeds were found to be reasonably comparable, for example, in Carmel Canyon off California (Figure 1-8) and in Hydrographer Canyon south of the Gulf of Maine (Figure 1-9). On the other hand, a comparison of current measurements at 3 m and 30 m above the bottom in Petacalco Canyon off Rio Balsas, Mexico (Figure 1-10) shows that the currents at the two levels generally flowed in opposite directions (Figure 1-11). This pattern is unusual with respect to most of our current records. The only possible explanation for the difference in the Petacalco Canyon records from measurements made in other canyons is that they were made in shoaler depths (i.e., 110m).

ADVANCING INTERNAL WAVES ALONG THE CANYON AXES

The fact that currents were comparable at different heights above the canyon floors led us to compare data taken simultaneously at adjacent stations along canyon axes. Two stations in Carmel Canyon, California, separated only by 1.2 km showed an excellent agreement when the data was shifted twenty-three minutes so that flow patterns matched. This temporal offset is considered to be the difference in arrival time for an advancing internal wave (Figures 1-12 and 1-13). Comparisons of many other data sets have shown that agreements between current patterns at different locations along a canyon axis are not usually as good as that found at Carmel Canyon. Matching adjacent stations is somewhat difficult where cycles of alternating direction agree with the semidiurnal tides. In these cases the best fit of the curves was selected. For example, in Figure 1-13 a moderately good fit can be found by shifting the graph in either direction, but the fit is far better if the curve for station 46 is shifted to the right as has been done. In that case the relatively small currents in the middle of the curve show a far better agreement. If we compare the records at the three stations in Hydrographer Canyon (see Chapter 2), we can see that the tide shifts in relation to the curves progressively, which adds validity to the method. Furthermore, where we have many records of closely spaced stations in La Jolla Canyon, we find that almost all of them can be matched by a small shift in the same direction. Unfortunately, most of our stations are far apart, and the patterns do not match well, thereby creating some doubt about the method.

As seems reasonable, an attempt was made to get the best possible fit for all of the stations where two or more records were taken along the axis at the same

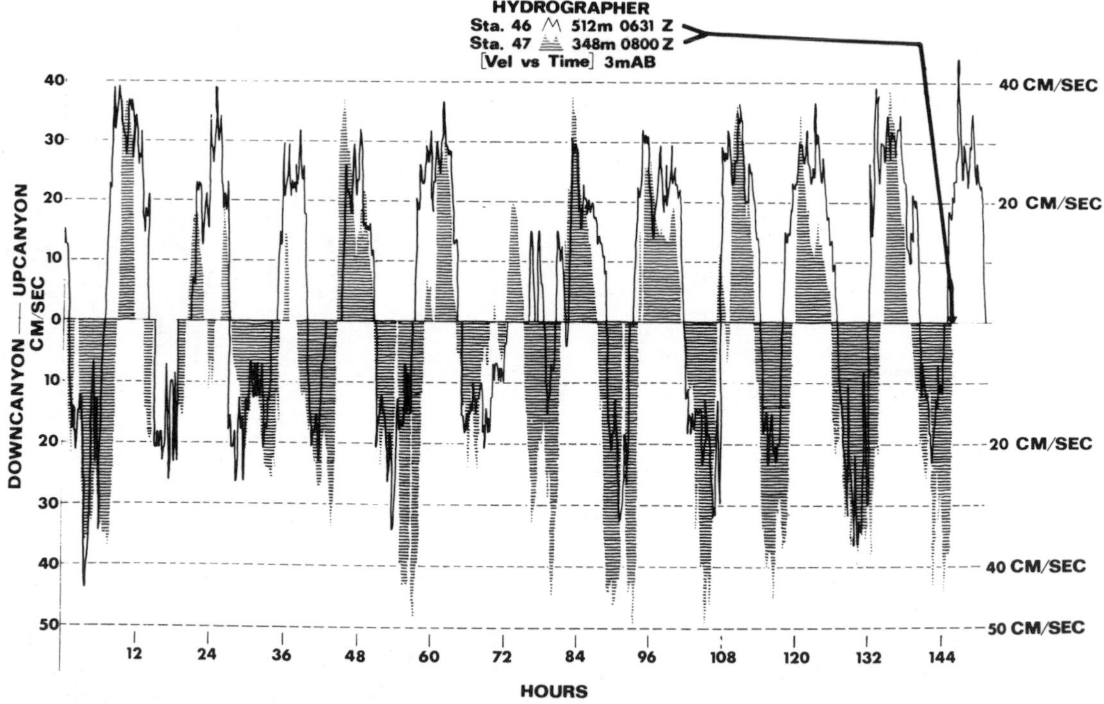

Figure 1-13. Example of the progressive advance of current patterns up a submarine canyon, as shown by time offset of records at two stations along Hydrographer Canyon. Note: The correlated times of the two stations are indicated by an arrow.

time (Figure 1-14). We have put question marks on all of those axes where the fit was not very good. We determined the speed of advance of these supposed internal waves by finding how much we had to shift the stations to get the best fit and by determining the distance between the stations.

In Figure 1-14 we see that in most of the canyons the supposed internal waves were advancing upcanyon. Also the curves can generally be fitted far better where the advance is in this direction. However, in Santa Cruz Canyon the best fit was found on two occasions in which the supposed internal waves seemed to be advancing downcanyon (Figure 1-15). Although these fits are not very good, they are far better than any other solution in both cases. Thus, we appear to have some evidence for downcanyon advance. In this case the canyon heads in the passageway between Santa Rosa and Santa Cruz Islands (see Figure 1-7), which might suggest that the internal waves were coming from Santa Barbara Basin and being carried through the passageway and thence down the canyon axis.

We can conclude that we have based our discussion on much speculation. Where the stations are close together, we have fairly good evidence of waves of some sort advancing along the canyon axes. We need much more information, however, before we can substantiate the advance of these waves, particularly in a downcanyon direction.

RELATION OF CURRENTS TO WINDS AND WAVES

Most of our records were collected during periods of rather low wind and swell conditions. Our early records gave the impression that current speeds showed little if any relation to surface conditions. In fact, one record indicated that current speed actually increased when wind strength and wave height decreased.

The first good opportunity to compare currents with storm conditions occurred during a cruise of the NOAA ship, *Researcher*, with George Keller as chief scientist. Current meters were deployed in both Hydrographer and Hudson Canyons prior to a storm that developed winds up to 90 km/hr (*Researcher* ships' log). The data from the current meters showed a relationship to the storm only at the shoalest station in Hydrographer Canyon at the 343 m depth (Figure 1-16) and even in this case, the relation was not clearly indicated as some of the fast currents developed at the very earliest buildup of the storm. These flows may be related to a pressure wave that had preceded the storm. Also, the even shoaler 223 m station in Hudson Canyon, with the current meter at 30 m above the bottom, showed no relationship to the storm (see Chapter 2). Another opportunity to look for relationships during a storm with winds up to 75 km/hr in two deep stations off Carmel, California, showed no relationship between sea surface conditions and current speeds. As will be explained in the next section, the apparent lack

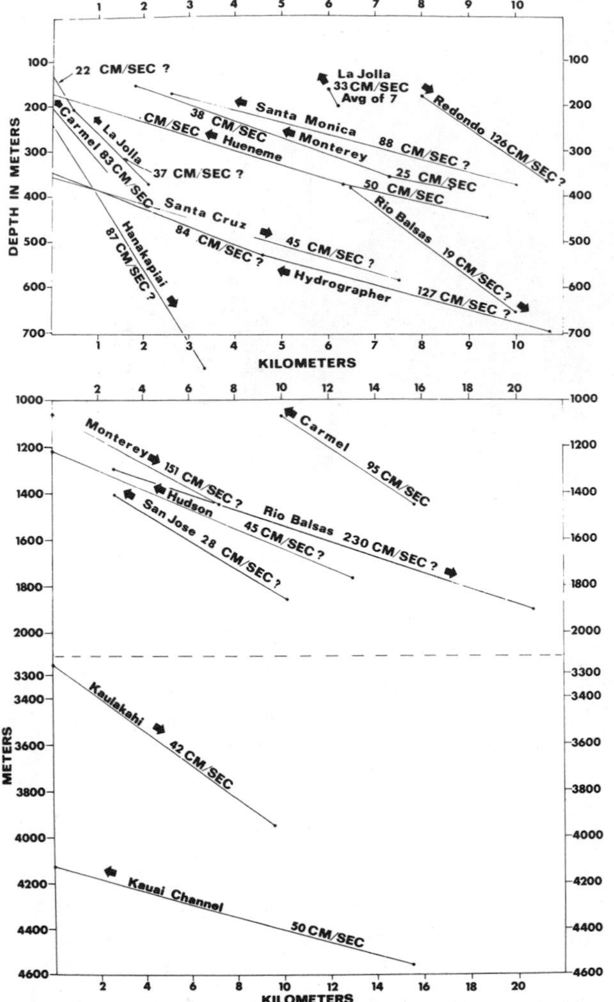

Figure 1-14. Diagrams showing the more probable direction and very approximate speed of advance of supposed internal waves up and (less frequently) down the axes of submarine canyons. Note: The speed of advance is approximate since the matching of most records is subject to much uncertainty, particularly where the advance appears to be downcanyon.

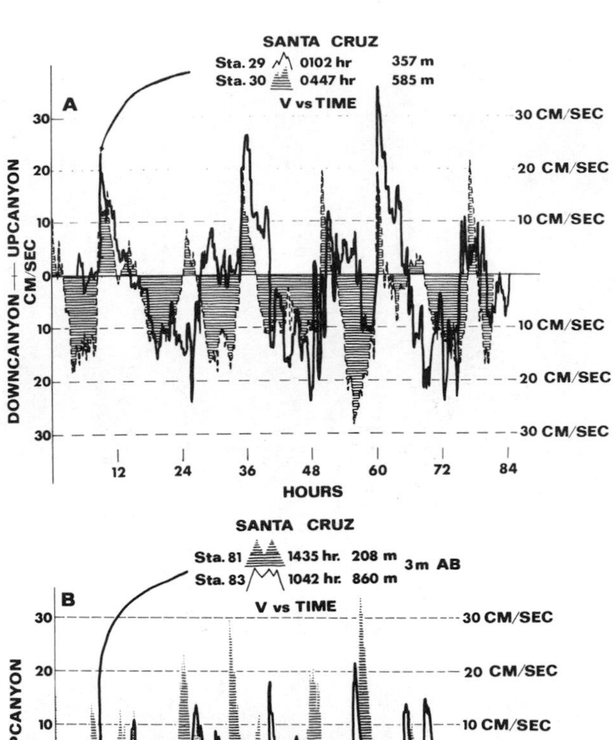

Figure 1-15. Examples of rather poorly matching patterns of currents at stations along the axis of a canyon that indicate the downcanyon advance of internal waves. Note: Both examples come from Santa Cruz Canyon where the canyon head is located in a pass between two islands (see Figure 1-7) where water is introduced from Santa Barbara Channel by the strong winds that normally blow to the southeast down the strait between the islands.

of correlation of the wind and waves to the currents in canyons only suggests that not all storms influence these currents. Exceptions are discussed below.

TURBIDITY CURRENTS

One of the risks of placing current meters in submarine canyons is the occasional occurrence of what we interpret as turbidity currents. If the high-speed currents like those reported from the Grand Banks earthquake (Heezen and Ewing 1952) were common, current meters would be lost so frequently that observations would become prohibitively expensive. Fortunately, such currents occur very infrequently. During a storm in La Jolla Canyon that had onshore winds blowing up to 65 km/hr, we lost two current meters that were subsequently recovered. They had been buried by a mass of kelp and sediment 0.5 km downcanyon from their deployment location as determined by horizontal sextant angles (Shepard and Marshall 1973b). The records (Figure 1-17) showed that the maximum current speeds increased as the wind speed rose, although they continued their periodic reversal up- and downcanyon. Finally, a large downcanyon surge reached a speed of about 50 cm/sec, and shortly afterwards the record stopped, the rotor apparently being tangled by kelp. How much faster the current became later we do not know, but during the dive in the *Nekton* submersible that recovered the lost current meters, a trough with walls more than half a meter high that had been cut into the silty sand fill of the canyon floor was observed, which suggested erosion by a powerful current.

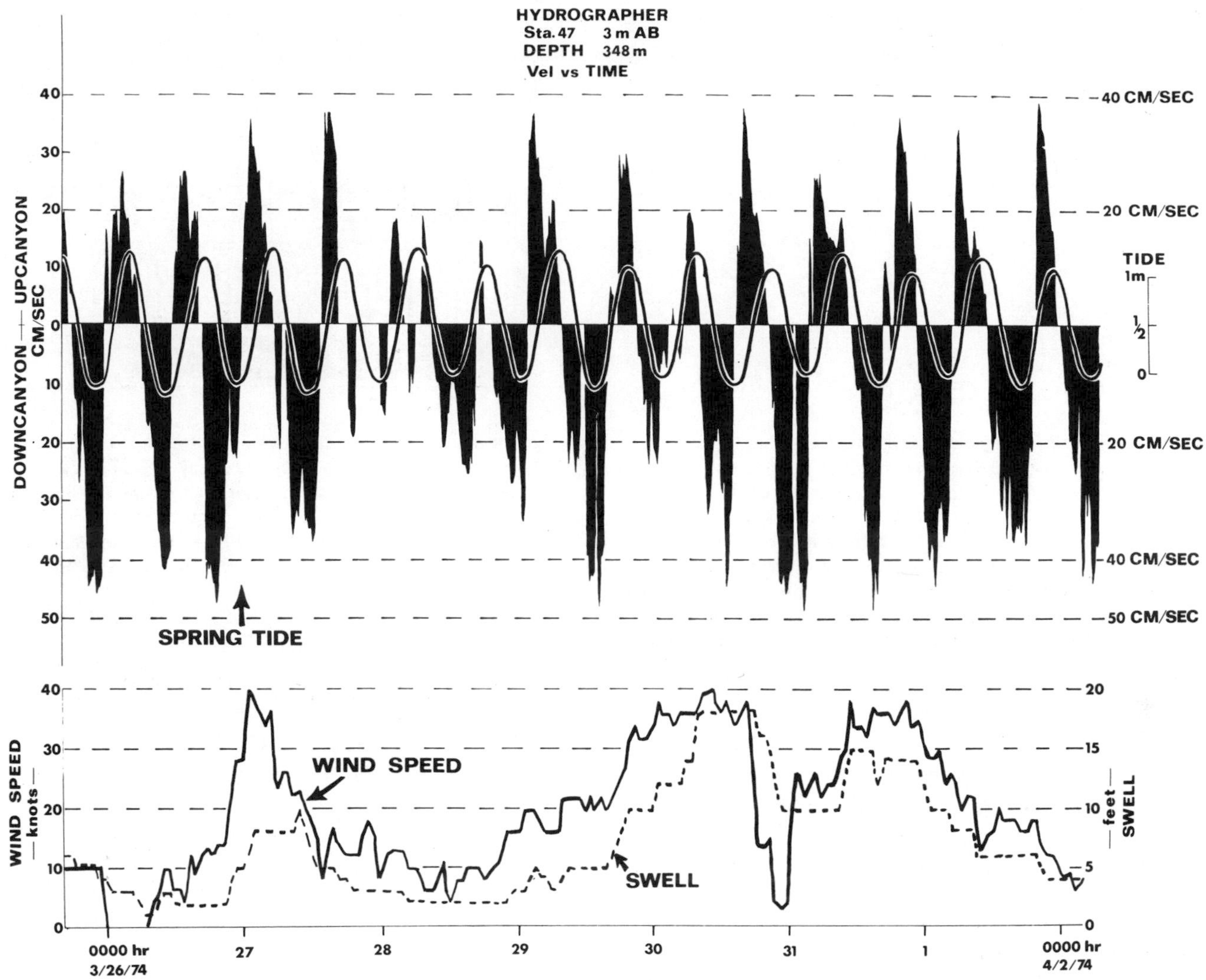

Figure 1-16. Relation of wind speed and swell height to the magnitude of up- and downcanyon flows during a storm period in Hydrographer Canyon. Note: The slowest currents appear during the time of low winds and swell.

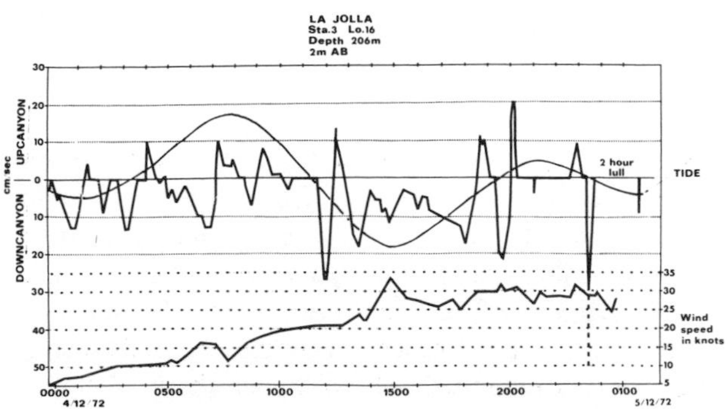

Rigure 1-17. Relation of wind velocity to up- and downcanyon flows prior to the ending of the record during an onshore wind storm when two current meters were carried down La Jolla Canyon for 0.5 km, but were later recovered during a dive in a *Nekton* submersible. Note: The dashed line at 2320 hrs. indicates current velocity for one minute at peak of flow.

Inman et al. (1976) reported that similar currents frequently erode all of the fill from the head of Scripps Canyon. They measured velocities up to 190 cm/sec during one onshore storm. After this speed was recorded, the current meter was carried away. They have measured other currents with speeds to 30 cm/sec that they attribute to the energy developed by edge waves carrying water along the coast to the head of the submarine canyon.

In a record obtained in Rio Balsas Canyon off western Mexico (Figure 1-18; also see Figure 1-10) current speeds up to 62 cm/sec were recorded (Shepard et al. 1975). Because these relatively high speeds occurred during a period of large swell and during high tides and because the meter was covered with silt and fine sand, we believe that the record represents the downcanyon passage of a turbidity current. The surges were similar to the current encountered by Reimnitz (1971) in a tributary of Rio

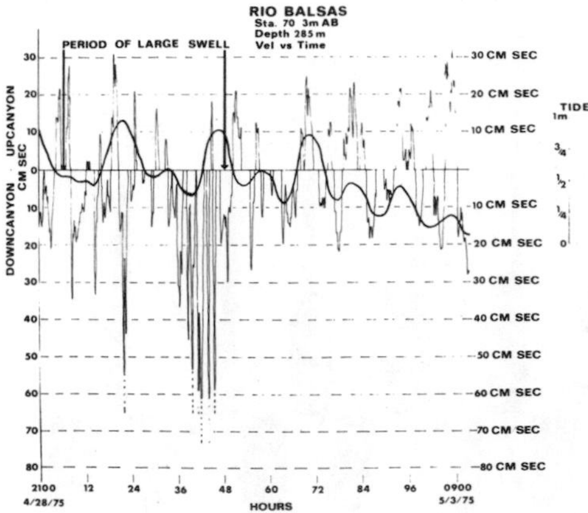

Figure 1-18. Record showing probable passage of turbidity current with down canyon pulses during a period of high swell in the canyon head off Rio Balsas, Mexico (for contour map, see Figure 1-10). Note: The pulses occurred close to the time of two succeeding high tides. The tidal peaks occurred somewhat later because the tide gauge was in an estuary where a tide lag would be expected. The bold arrows indicate arrival and termination of seiche waves at tide gauge. The dashed lines show maximum velocity measured for one minute during peak flow conditions.

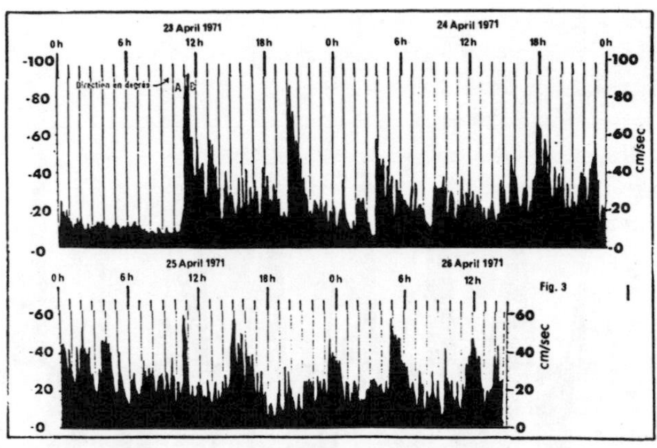

Figure 1-19. Current record showing the continuous downcanyon flow obtained by Gennesseaux et al. (1971) in Var Canyon off the French Mediterranean coast during a storm and flood. Note: Pulses up to 95 cm/sec occurred at intervals somewhat like those shown in Figure 1-18. (Source: Gennesseaux et al. 1971.) These are interpreted by Gennesseaux and others as turbidity currents.

Balsas Canyon during a period of high surface swell. Reimnitz thought that the strong current surge was associated with a rip current surge at the water surface.

In July 1976 we had current meters at 594 m in a canyon located directly off the Abra Delta and off the mouth of the Abra River in northwest Luzon. The records at both 3 m and 30 m above the bottom both showed a sudden change from the ordinary up- and downcanyon alternation to a strong downcanyon surge with a maximum velocity of 72 cm/sec at 3 m above the bottom and of 53 cm/sec at 30 m above the bottom. This example of a turbidity current occurred without a storm condition, but at a time when the river was in flood and was carrying large quantities of sediment to the sea as could be detected by the plume of muddy water that extended well out to sea off the river mouth. Also, heavy rain was reported by the Philippine weather bureau for the area. The seas were moderately high and the winds had occasional strong gusts coming from the southwest. All three of these examples of relatively weak turbidity currents that broke into the up- and downcanyon pattern had a considerable resemblance in their records (Shepard et al. 1977).

Gennesseaux et al. (1971) had a current meter at 800 m in Var Canyon off the Var River in southeastern France and measured a series of variable speed, downcanyon surges with velocities up to 90 cm/sec for a period of four days. These surges occurred during the passage of a storm and during a flood of the Var River (Figure 1-19).

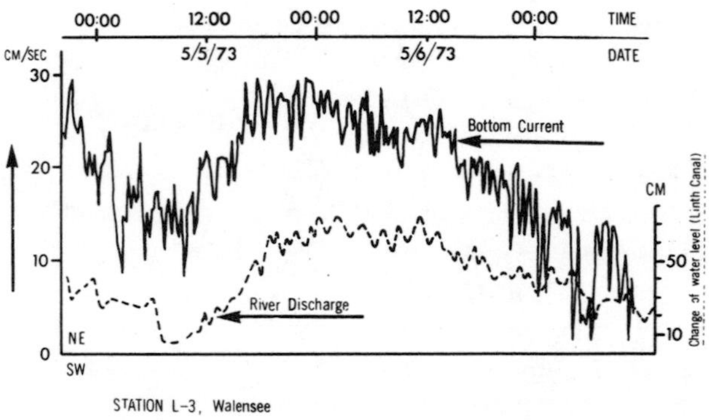

Figure 1-20. Pulsating downslope flow in Lake Walensee, Switzerland, due to relatively weak turbidity currents. Note: The strongest flows were during the period of peak river discharge during the spring flood. These data are from a record obtained by Marshall in 1973. (Source: Lambert et al. 1976.)

Currents suggestive of turbidity currents were measured in the Walensee of Switzerland (Figure 1-20) by Lambert et al. (1976). The currents flowed continually downslope, away from the river mouth with speeds as high as 30 cm/sec for periods of from one to two hours. These flows reached their highest speed during the largest discharge of the river, but the speed did not exceed about 30 cm/sec. Here again, the current meters were coated with a fine layer of sand.

Richard Slater (personal communication, 1976) who made a dive into Oceanographer Canyon in a *Nekton* submersible off the northeast coast of the United States encountered a turbidity current. His first indication that something unique was about to occur was a rapid rise in water temperature which suggested that warmer water from shallower depths was being carried down the canyon. Presently, a cloud of sediment arrived, and when the *Nekton* was turned into the current, the maximum two-knot speed (100 cm/sec) of the submersible could not keep up to the current. The current was estimated to be flowing at up to four or five knots (7.4 or 9.3 km/hr).

COMPARISON BETWEEN CURRENTS IN CANYONS AND FAN VALLEYS

There are various types of valleys on the sea floor (Shepard and Dill 1966, chapter 2). Most of those in which we measured currents are the narrow, deeply incised, steeply walled type to which the name *submarine canyon* should be applied. These differ in character from the slightly incised valleys with bordering natural levees that are found seaward of the true canyons and are also fairly common on the slopes bordering deltas. The latter, usually called *fan valleys*, are represented in our investigations by the deep portion of the Hudson, by the slightly incised deepwater valleys off northwest Kauai, and by the valleys off the Fraser Delta. Somewhat intermediate types are found off the Abra Delta in Luzon and off the Rio Balsas Delta in western Mexico. The currents in these fan valleys are usually slower than those in the true canyons, as might be expected because the fan valleys are usually wider floored so that the currents are less constricted. Elliptical patterns (Figure 1-21) comparable to those found where tidal currents cross Georges Bank off New England were observed in some of the wide-floored valleys off Kauai. Also, the deeper portions of the valleys off Rio Balsas and Abra Deltas have records with considerable gaps in the continuity of the currents.

The rapid surges that we interpret as turbidity currents are apparently more common in the rather shallow, incised valleys with natural levees found off large deltas. Two out of three of our recordings of these strong surges have come from such valleys, and we have about ten times as many records from the narrow, steep-walled canyons, but only one of these strong surges was recorded in the canyons.

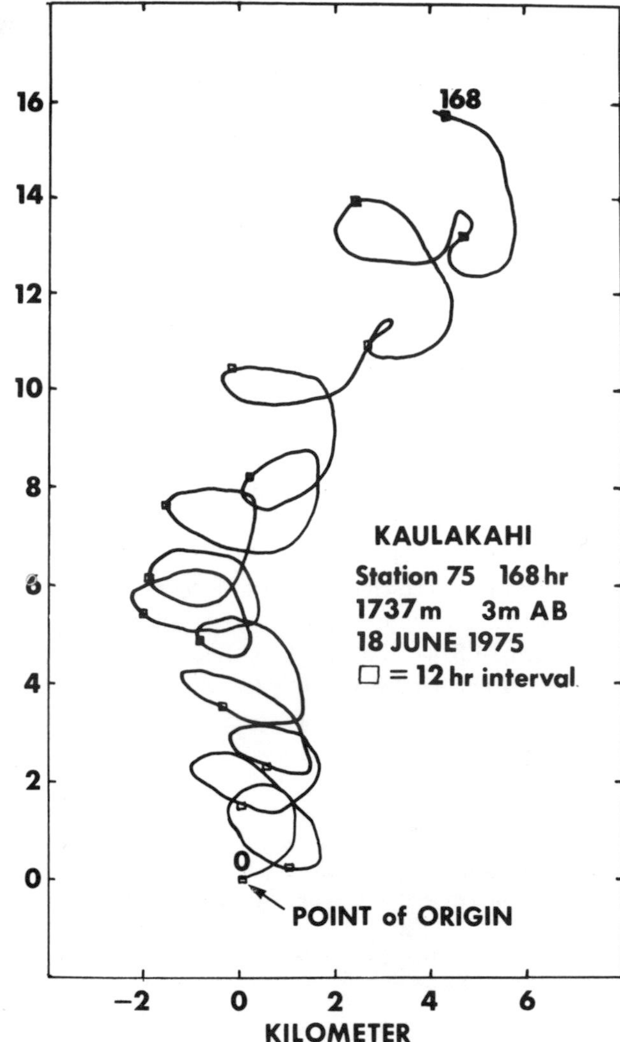

Figure 1-21. Graph showing the net flow along the floor of Kaulakahi Seavalley northwest of Kauai. Note: The ordinate represents the canyon axis (upcanyon at top). The elliptical movement of water particles is indicative of a tidal pattern superimposed on general upvalley flow.

CONCLUSIONS

There is considerable evidence to support the occurrence of current speeds of the order of 0.3 to perhaps 5 knots (0.6 to 9.3 km/hr) during the passage of turbidity currents down canyons. These are much weaker than the estimated speed of about 100 km/hr (Heezen and Ewing 1952) for the turbidity current associated with the Grand Banks earthquake and are also weaker than those estimates of Menard (1964, pp. 207-10) and Shepard (1963, pp. 339-41). Menard calculated 68 km/hr for the maximum, and Shepard obtained 29 km/hr for the average. We cannot help wonder whether most turbidity currents are not of relatively low velocity and perhaps fast currents occur only as a result of great earthquakes or exceptionally great storms and due to the large accompanying slides or slumps. Obtaining more information on this phenomenon will be interesting.

ACKNOWLEDGMENTS

The authors wish to acknowledge the long-continued help of Patrick A. McLaughlin and Gary G. Sullivan both in the field and in the laboratory. Richard Slater kindly provided unpublished information on currents noted during submersible operations. For arranging ship time for some of our operations, the authors express appreciation to D. S. Gorsline, G. H. Keller, R. E. Andrews, D. D. Drake, P. J. Fischer, and R. F. Dill. The ship time was given on *Velero IV* of Hancock Foundation; on the *Researcher* of NOAA; on the *Silas Bent* and the *Acania* of the U. S. Navy; on the *Sea Lion* of the Canadian Geological Survey; and on the *Thomas Washington*, the *Ellen B. Scripps*, the *Dolphin*, and the *Gianna* of the Scripps Institution of Oceanography. The authors appreciate the help and good will of the crews of these ships. The U. S. Geological Survey provided instruments for our Rio Balsas Expedition, and Erk Reimnitz was co-chief scientist of the expedition. The work was largely supported by National Science Foundation Grant DES74-22089 and by Office of Naval Research Contract N00014-69-A-0200-6049.

REFERENCES

Cacchione, D.A., Rowe, G.T., and Malahoff, A., 1976. Sediment processes controlled by bottom currents and faunal activity in lower Hudson submarine canyon. (Abst.) *Amer. Assoc. Petrol. Geol. Bull.*, 60: 654-55.

Gennesseaux, M., Guibout, P., Lacombe, H., 1971. Enregistrement de courants de turbidité dans la vallée sous-marine du Var (Alpes-Maritimes). *C.R. Acad. Sc. Paris*, 273, ser. D., pp. 2456-59.

Heezen, B.C., and Ewing, M., 1952. Turbidity currents and submarine slumps, and the Grand Banks earthquake. *Amer. Jour. Sci.*, 250: 849-73.

Inman, D.L., Nordstrom, C.E., and Flick, R.E., 1976. Currents in submarine canyons: An air-sea-land interaction. *Ann. Rev. Fluid Mechanics*, 8: 275-310.

Lambert, A.M., Kelts, D.R., and Marshall, N.F., 1976. Measurements of density underflows from Walensee, Switzerland. *Sediment.*, 23: 87-105.

Marshall, N.F., 1975. The measurement and analysis of water motion in submarine canyons. *IEEE Ocean '75:* 351-56.

Menard, H.W., 1964. *Marine Geology of the Pacific.* McGraw-Hill, New York, 271 pp.

Reimnitz, E., 1971. Surf-beat origin for pulsating bottom currents in the Rio Balsas Submarine Canyon, Mexico. *Geol. Soc. Amer. Bull.*, 82: 81-90.

Shepard, F.P., 1963. *Submarine Geology* (2nd Ed.). Harper & Row, New York, 577 pp.

——, 1976. Tidal components of currents in submarine canyons, *Jour. Geol.*, 84: 343-50.

——, and Dill, R.F., 1966. *Submarine Canyons and Other Sea Valleys.* Rand McNally, Chicago, 381 pp.

——, and Marshall, N.F., 1969. Currents in La Jolla and Scripps Submarine Canyons. *Science*, 165: 177-78.

——, and Marshall, N.F., 1973a. Currents along floors of submarine canyons. *Amer. Assoc. Petrol. Geol. Bull.*, 57: 244-64.

——, and Marshall, N.F., 1973b. Storm-generated current in La Jolla Submarine Canyon, California. *Mar. Geol.*, 15: M19-M24.

——, Marshall, N.F., and McLoughlin, P.A., 1974. Currents in submarine canyons. *Deep-Sea Res.*, 21: 691-706.

——, Marshall, N.F., and McLoughlin, P.A., 1975. Pulsating turbidity currents with relationship to high swell and high tides. *Nature*, 258: 704-06.

——, McLoughlin, P.A., Marshall, N.F., and Sullivan, G.G., 1977. Current-meter recordings of low speed turbidity currents. *Geology*, 5: 297-301.

Chapter 2

Currents and Sedimentary Processes in Submarine Canyons off the Northeast United States

GEORGE H. KELLER

School of Oceanography
Oregon State University, Corvallis

FRANCIS P. SHEPARD

Geological Research Division
Scripps Institution of Oceanography
La Jolla, California

ABSTRACT

Detailed current and sediment distribution observations in five of the major East Coast submarine canyons (Hydrographer, Hudson, Wilmington, Washington, and Norfolk) indicate that canyons south of Georges Bank are today relatively inactive in regard to the transport of coarse-grained sediments. They do, however, appear to be serving as conduits for the movement of fine-grained material out onto the continental rise and abyssal plain. The presence of sands and bare rock in the axes of the Georges Bank canyons versus the predominance of silt and clay in the canyons to the south further suggests that our short-term observations are indicative of the long-term processes taking place in these canyons.

Tidal and internal wave forces appear to be the major forcing functions responsible for the semidiurnal up- and downcanyon flow reversals that are characteristic of these canyons. Bottom current measurements in the canyons at depths ranging from 224 m to 1,767 m for periods of 11.5 to 19 days indicate that velocities and flow directions differ markedly both within a single canyon as well as from canyon to canyon. Canyons off Georges Bank, north of Veatch Canyon, display the highest currents of those measured in the East Coast canyons. Velocities as high as 70-75 cm/sec have been observed in Hydrographer Canyon as has the downcanyon migration of sand ripples at depths of 710 m. Maximum and mean near-bottom current velocities (cm/sec) in Hudson, Wilmington, Washington, and Norfolk Canyons were on the order of (33,10), (44,12) (20,6), and (32,7), respectively. Although net bottom transport in the canyons was generally downcanyon, notable exceptions were found. In Hydrographer canyon, for example, simultaneous current measurements revealed strong downcanyon net transport at the 348 m and 713 m stations, but a distinct upcanyon transport at the intermediate 512 m station. Of the canyons studied, only Wilmington Canyon displayed a consistent net transport upcanyon at the two (723 m and 915 m) recording stations. Examination of sediment threshold velocities at various depths in the five canyons revealed that in all but Washington and Norfolk Canyons there were periods when these velocities were exceeded and sediment transport occurred. The most pronounced sediment transport was noted in Hydrographer Canyon where the threshold velocity was exceeded during 26 percent of the 11.5-day observation period. Submarine canyons along the northeastern margin of the United States display sedimentary processes distinctly different from those in the well-studied canyons off Southern California that are very active in the seaward transport of coarse-grained sediments.

INTRODUCTION

The outer continental margin of northeastern North America is uniquely characterized by its abundance of submarine canyons and gullies (Figure 2-1). Between Laborador and Cape Hatteras at least 190 canyons dissect the continental slope (Emery and Uchupi 1972). In contrast to most canyons off Southern California, which have their heads very close to the beach, the East Coast canyons all head near the continental shelf break anywhere from 85 to 250 km from shore. This fact in itself leads to distinctly different dynamic conditions and transport processes between the canyons off the respective coasts (Shepard et al. 1974a).

The reason for the highly dissected nature of the northeastern continental margin has not been adequately

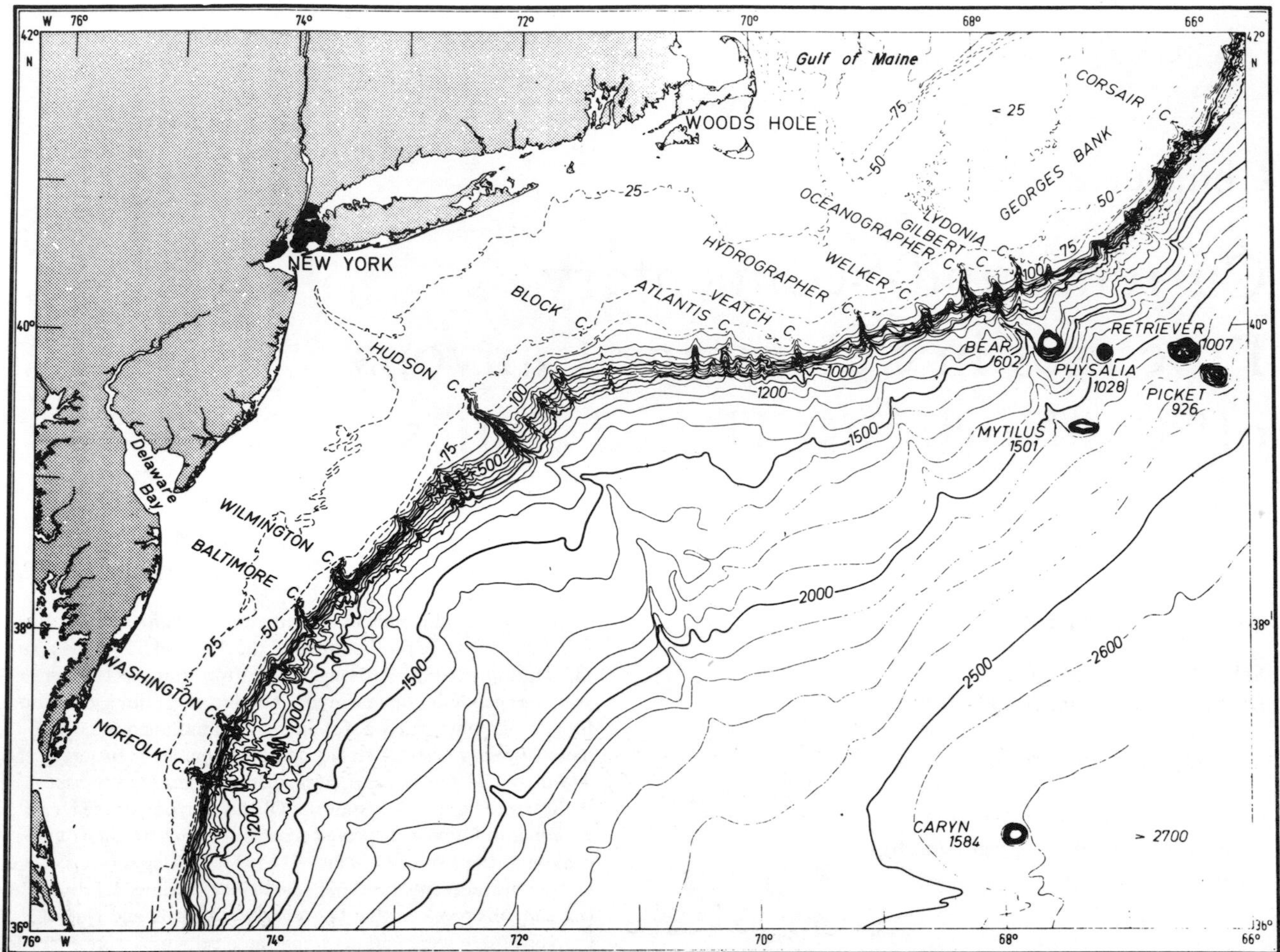

Figure 2-1. Submarine canyons along the continental margin off the northeastern United States. Note: Depths are in fathoms. (Source: Dietrich et al. 1975.)

researched, but it would seem logical to expect that the close proximity of the Wisconsin ice sheet played a significant role in the formation of many of these canyons (Emery and Uchupi 1972). There is, however, evidence that a number of these canyons existed prior to the Pleistocene (Roberson 1964).

The Hudson Canyon, although anomalous in that it extends much farther into the continental shelf than any other East Coast canyon, may be the type example for a number of the canyons on this margin. Its connection to the Hudson River via a shelf channel makes it apparent that the Hudson Canyon is the result of subaerial erosion during a lower stand of sea level. That it served as a conduit for the transport of coarse-grained shallow-water sediment to the deep sea during such periods of lowered sea level has been confirmed by the presence of sands, graded bedding, and shallow-water shells in the rise and abyssal plain deposits out beyond the canyon (Ericson et al. 1961; Horn et al. 1971).

Although a bathymetric map of the northeastern continental shelf does not reveal the presence of any other shelf channels, seismic reflection surveys have delimited buried channels extending shoreward from a number of the canyons (Roberson 1964; Knott and Hoskins 1968; McMaster and Ashraf 1973). Similar drainage ways for a number of canyons from Wilmington Canyon to the south have been postulated by Kelling et al. (1975) based on mineralogical analyses of sand-size material from the shelf, slope, and rise. Uchupi's (1968) seismic reflection profiling survey of Wilmington. Baltimore, Washington, and Norfolk Canyons as well as our surveys parallel to the continental margin from Cape Hatteras to Hydrographer Canyon (McGregor et al. 1975) clearlv support the concept that the East Coast canyons are erosional in origin. As is quite apparent, the canyons off the northeastern United States owe their formation to erosion, which was probably a much more forceful agent on the continental slope in the past than it appears to be today due to lower stands of sea level. In the case of Hudson Canyon, however, only some few hundreds of years may have passed since sand was actively transported through the canyon to the deep sea floor (Ericson et al. 1952).

Somewhere between Veatch and Hydrographer Canyons a notable transition appears to take place. The canyons to the west and south of Veatch appear to be relatively inactive and commonly are blanketed with fine-grained sediments, except in their uppermost reaches where headward erosion of Pleistocene sands and gravels is taking place. On the other hand, the major Georges Bank canyons from Hydrographer northeastward, with the exception of Lydonia, appear to be undergoing active erosion with either bare rock or coarse sands flooring the canyon axes. To the far north, on the Nova Scotia margin, gravels are present in the Gully Canyon out to depths of 1,100 m (Stanley 1967). Lydonia Canyon is relatively anomalous in that it is blanketed by fine-grained sediments and exhibits little or no indication of current activity (Ross 1969). Currents as high as 70 to 75 cm/sec may be relatively common in these northern canyons as recorded in Hydrographer Canyon (B. Heezen, personal communication, 1974), but to the south maximum velocities of 30 cm/sec are more frequently observed. Richard Slater (personal communication, 1976), during a dive in Oceanographer Canyon in the submersible *Nekton*, observed a turbidity current moving down the canyon with an estimated velocity of 100 to 200 cm/sec.

Mineralogical analyses of shelf, slope, and rise sediments off northeastern United States lead to an impression that a transition zone exists in the area between Hudson and Block Island Canyons (McMaster and Garrison 1966; Milliman et al. 1972). To the south of Hudson Canyon the heavy mineral assemblage consists of an abundance of garnet and amphibole (easily eroded minerals) whereas staurolite and other erosion-resistant minerals predominate in the high energy areas of Georges Bank and the Scotia Shelf (Milliman et al. 1972). Carbon-nitrogen ratios in slope deposits also reveal a similar transition zone with higher ratios south of Hudson Canyon than to the north (Milliman 1973). These findings suggest the influence of different source areas on the depositional environment.

Although the sediment and current data suggest that the outer shelf and slope environment to the north and south of the New York Bight differ markedly, an explanation is still awaited. Possibly this situation may be due to a combination of circumstances. The proximity of the active canyons to Georges Bank where the combination of currents, waves, and relatively shoal conditions result in higher near-bottom energy levels may be a factor in a "suitable" explanation of the behavioral difference. The lack of fine sediment in the upper and middle portions of the canyons may possibly be attributed to the entrapment of fines in the Gulf of Maine and the availability primarily of coarse-grained material from the adjacent shelf and slope. Insufficient information is available to explain the anomalous fine-grained sediments and low-energy regime observed by Ross (1968) in Lydonia Canyon. Until much more is known about bottom current mechanics and sediment transport in the canyons, or on the outer shelf, slope, and rise, a satisfactory explanation for the differences discussed above must be deferred.

Although the first current measurements in submarine canyons were those of Stetson (1937) in Lydonia and Gilbert Canyons, the majority of such observations have been made in the canyons off the California coast where Shepard has conducted numerous studies since 1938 (Shepard et al. 1939; Shepard and Marshall 1973a; Shepard et al. 1974b; Shepard 1975, 1976). Of particular note from all the studies made on currents within canyons is the unique characteristic of flow reversal. Currents predominantly move up and down the canyons, and in the case of many of the canyons off Southern California the downcanyon flow is by far the strongest and most pronounced (Shepard and Marshall 1973a). Although all canyons display an up and down flow, those off Southern California possess distinctly different characteristics when compared with those off the northeastern United States. Quite often the reversals are much more frequent than those in the East Coast canyons where currents appear to be much more closely allied to the tidal cycle with periods of reversal on the order of 12.5 hours. The net flow in the California canyons is almost always downcanyon, often at such speeds that large volumes of sand are very rapidly transported down the canyons (Shepard 1965; Shepard and Dill 1966; Shepard and Marshall 1973a). Coastal processes and their influence on the California canyons contribute significantly to these different characteristics. As opposed to the very "active" California canyons those off the northeastern United States, south of Hydrographer Canyon, can be considered relatively inactive today with only fine-grained sediment being transported in suspension out through these canyons (Shepard 1965; Keller 1975).

INSTRUMENTS AND DATA DISPLAY

During the course of this study, three types of current meter were used to investigate the characteristics of the currents within the canyons off northeastern North America. The Isaac-Schick (Isaac et al. 1966) and the Geodyne photo-recording type meters were used predominantly, while an Aanderaa meter was used less frequently in conjunction with a time-lapse bottom-mounted camera. All current meter arrays were deployed as free vehicles with subsurface floats and timed or acoustic release systems. With the exception of the Aanderaa meter, which was positioned 100 cm above the sea floor, the meters were positioned 3 m and 30 m above the bottom in dual arrays and 3 m above the bottom in single meter arrays. Current speed and direction was recorded at 7.5-to 15-minute intervals depending on the meter used.

The current data results from our study are plotted in two graphical forms: time-velocity curves (corrected for divergence from axis) and progressive vector diagrams (net flow). For each of the canyon stations, currents are plotted to show the frequency and velocity of flow up and down the canyon. The predominant axial flow direction along the canyons was determined from an initial analysis of the current data and the orientation of the canyon axis. All values occurring in a quadrant 45° to either side of the selected axial trend were then used in plotting the up- and downcanyon flow characteristics. In the case of the current stations outside the canyons on the adjacent continental shelf and slope, the velocities are presented in component (north-south, east-west) form. The progressive vector plots are particularly useful in a study such as this where net current transport is of major interest. They also serve to provide a graphic demonstration of the nature of the flow through the canyon in a relatively simple yet clear manner.

Current data presented herein were collected in 1974 except for those from Washington and Norfolk Canyons and the adjacent slope station, which were obtained in 1975. Complementary observations are based on seventeen *Alvin* dives by the senior author in Hudson and Veatch Canyons between 1972 and 1975.

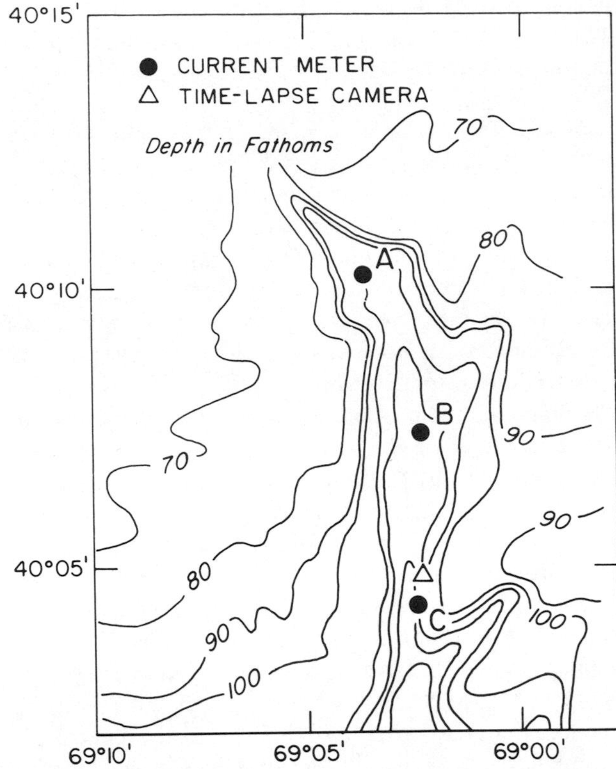

Figure 2-2. Location of current meter arrays and time-lapse camera observations in Hydrographer Canyon. Note: Stations A and B are dual meter arrays; station C is a single meter.

CURRENT CHARACTERISTICS

Hydrographer Canyon

Three current meter arrays at depths of 348 m, 512 m, and 713 m with three bottom meters and two meters positioned 30 m above the bottom recorded simultaneously for 11.5 days in Hydrographer Canyon (Figure 2-2). Hydrographer Canyon displays the most dynamic conditions of the canyons studied. Bottom velocities were as high as 53 cm/sec. Bruce Heezen (personal communication, 1973), however, observed current speeds on the order of 70-75 cm/sec in the upper reaches of the canyon.

Bottom photographs taken along the axis of the canyon indicate that the canyon floor consists primarily of bare rock and coarse-grained sediment down to a depth of approximately 860 m. Submersible observations in the same part of the canyon verify our impression that it is an actively eroding canyon to depths of at least 900 to 1,000 m (D. Cacchione and B. Heezen, personal communications, 1973).

The reversal, between up- and downcanyon flow directions, is clearly apparent at all of the Hydrographer Canyon stations (Figure 2-3). The periodicity of current

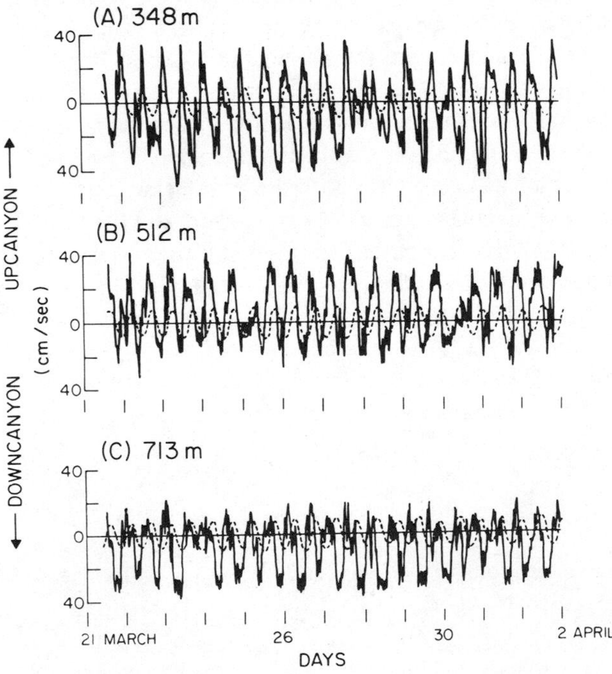

Figure 2-3. Time-velocity curves of currents at three stations in Hydrographer Canyon measured 3 m above the bottom. Note: Tidal curves (indicated by the dashed profile) are derived from Nantucket station.

reversals coincides with the period of the semidiurnal tides. Such a relationship has been reported from canyon investigations in other parts of the world (Shepard 1976), but it is dependent on measurement depth in the canyons and, with a few exceptions, on the tidal heights. When the tidal amplitude is larger than 1 m, and water depths exceed about 200 m, then the current reversals are regular and occur approximately every 6.25 hours. If, however, the tidal range is less than 1 m and the depths are less than 200 m, the current reversals are more frequent and less regular. Since we made no current measurements shoaler than 224 m, not surprisingly the periodicity of reversals is essentially tidal.

On the other hand, the importance of internal waves in developing the flow patterns may be indicated by comparing the flows at the various stations along Hydrographer Canyon (Figure 2-4). By shifting the time-velocity curve

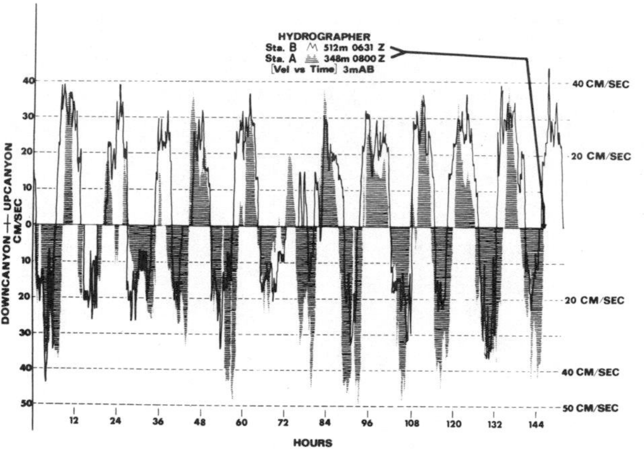

Figure 2-4. Superimposition of time-velocity curves from meters at 3 m above bottom (AB) at stations A and B in Hydrographer Canyon, with A profile shifted 1 hour and 29 minutes to the left. Note: Shifting the curves in the other direction would make a very poor fit for the period of small currents in the middle of the plot.

at the 348 m station by 1 hour and 29 minutes to the left, we find that the frequency of current reversal is similar to that at the 512 m station. Based on Shepard's (1975) study of this phenomenon in a number of canyons, our findings may possibly indicate that internal waves are advancing up the canyon and superimposing their effect on the tidal flows.

Considering that the measurements were made synoptically in Hydrographer Canyon, an interesting note is that the large differences in the axial currents were recorded approximately 4.5km apart. Currents were by far the strongest near the canyon head (station A, 348 m), with velocities of 35 to 40 cm/sec occurring frequently (see Figures 2-3 and 2-4). The strongest flows were downcanyon and, as shown in Figure 2-5, the net mass

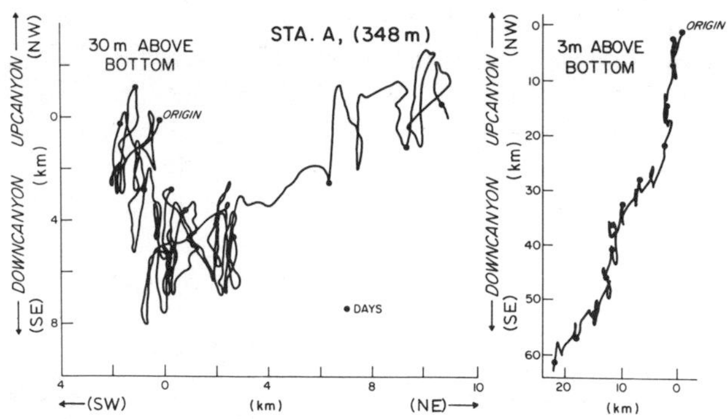

Figure 2-5. Progressive vector diagrams of currents at station A in Hydrographer Canyon.

transport was distinctly downcanyon at this site. Of particular interest is the current behavior at station B (512 m). Here, the bottom velocities were somewhat less than those recorded at the shallower station A, but the flow direction was predominantly upcanyon (see Figure 2-3). Although not presented, a progressive vector diagram of these measurements shows a plot basically similar (same slope) as that of the bottom meter at station A (see Figure 2-5) except that the origin is at the bottom of the plot rather than at the top. A similar upcanyon transport, but with a strong crosscanyon drift to the east, was recorded 30 m above the bottom at this site (station B). Farther down the canyon, at a depth of 713 m (station C), a pronounced downcanyon flow predominated during the period of observation (see Figure 2-3). Here, the downcanyon velocities were significantly greater than those at site B, yet were substantially lower than those recorded in the canyon head (station A).

As part of this study a time-lapse, camera-current meter unit was positioned on the bottom in the vicinity of station C at a depth of 710 m for a period of three days. The photographic observations documented what appeared to be ripples composed of sand-size material actually migrating down the canyon when velocities measured 100 cm above the floor exceeded 26 cm/sec. Unfortunately, no sediment sample was collected from this site. The direction and actual migration are readily documented, but the distance of ripple movement could only be approximated due to the lack of a fixed scale of reference. We estimated that the ripple crests moved at the rate of approximately 0.5 cm/min during the periods of higher velocities. Similar movement of coarse-grained sediments in the canyon has been observed from a submersible (B. Heezen, personal communication, 1973).

The anomalous upcanyon net flow observed at station B cannot readily be explained from the data at hand. The possibility that the array was positioned in such a way as to be influenced by local topography is a tempting explanation but one that cannot be verified.

Current measurements 30 m above the bottom at stations A and B revealed velocities somewhat lower than those recorded 3 m above the bottom. Superimposing the 3 m and 30 m records at station A shows that there is a close similarity between the times of reversals and maximum flow at the two levels of this station (Figure 2-6). This pattern has also been found in many of the records from other areas (Shepard et al. 1974c). Such agreements might be explained either by the tidal or the internal wave influence on the currents. There was, however, a notable difference in the net flow direction at these two levels (see Figure 2-5). At 30 m the influence of an eastward drift overshadowed the up- and downcanyon transport that was so pronounced 3 m above the bottom. Unfortunately, currents were not measured on the adjacent shelf due to the loss of an array to a trawler. However, current measurements 30 m above the bottom at the 348 m station reveal the strong influence of an eastward drift that possibly may have been the result of east-trending shelf currents. A similar eastward trend of bottom currents for this area at this time of the year (March) was reported by Bumpus et al. (1973).

Hudson Canyon

Three current meter arrays were positioned in Hudson Canyon at depths of 224 m, 1,222 m, and 1,765 m for a period of sixteen days (Figure 2-7). The shallow (224 m) and deep (1,765 m) arrays each consisted of two current meters, 3 m and 30 m above the bottom. A single meter mounted 3 m above the bottom comprised the array at 1,222 m.

As in Hydrographer Canyon, the strongest currents were found in the head of the canyon where a maximum velocity of 48 cm/sec was recorded. Current reversals in the canyon head occurred at intervals closely resembling those of the semidiurnal tide (Figure 2-8), but they were not as regular as those observed in Hydrographer Canyon. The irregularity may have resulted from near-bottom turbulence influencing the current measurements. Submersible observations in this part of Hudson Canyon revealed very turbulent flow that caused sporadic movement of gravel-size material. There is little variation in the current velocity at 3 m and 30 m above the bottom with only slightly higher speeds at 3 m, particularly in the downcanyon direction (see Figure 2-8). Although the duration of the measurements at

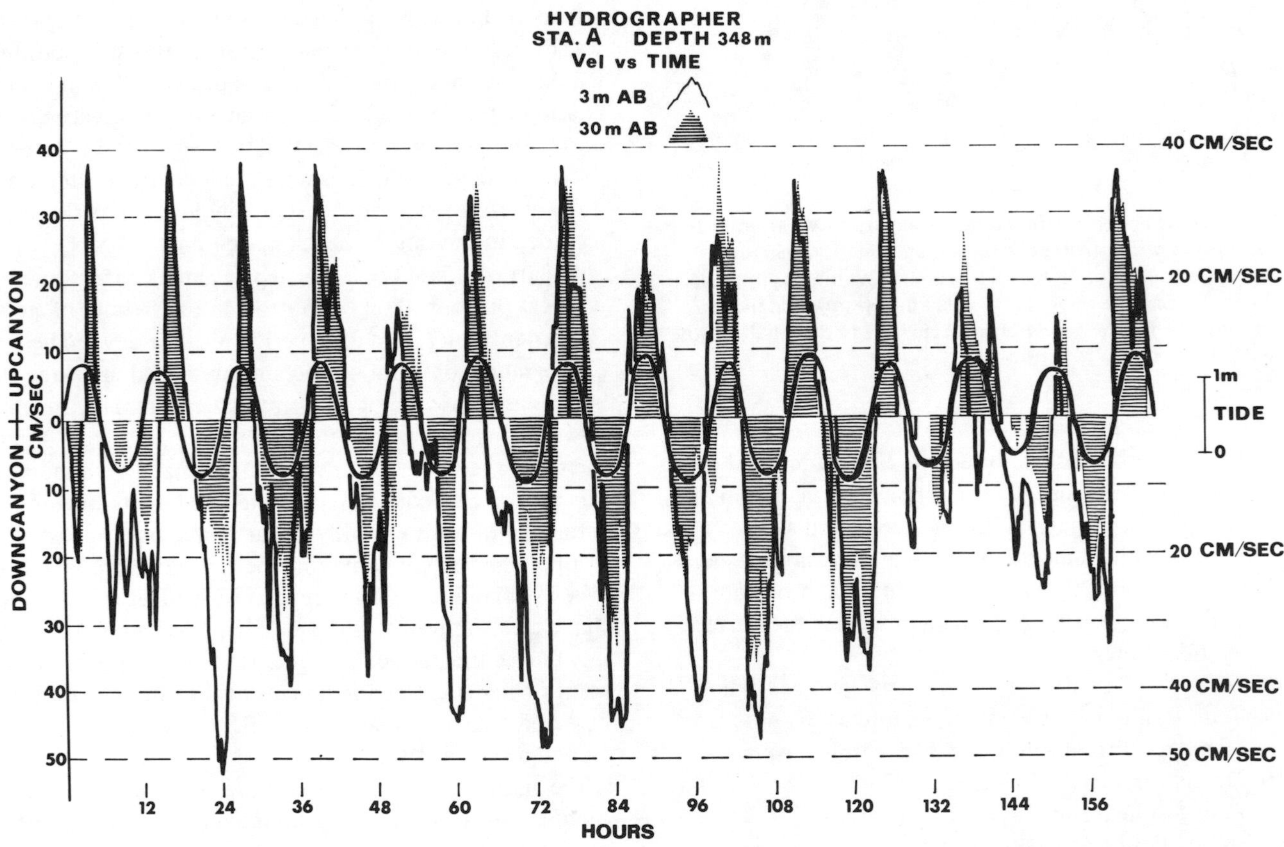

Figure 2-6. Comparison of current reversals and velocities at 3 m and 30 m above the bottom at the 348 m station in Hydrographer Canyon.

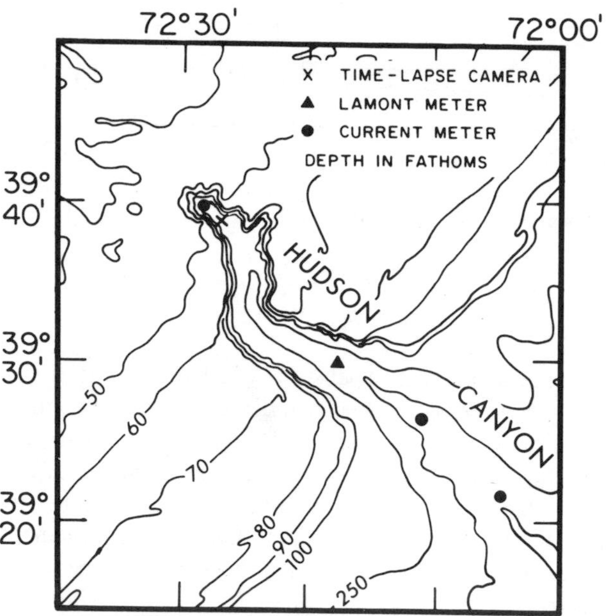

Figure 2-7. Location of current meter arrays and time-lapse camera in Hudson Canyon.

the two levels differed (the bottom meter failed after seven days), a comparison of the progressive vector diagrams reveals a considerable difference in the net flow directions. At 3 m above the bottom the net transport was strongly downcanyon (Figure 2-9), whereas at 30 m the downcanyon component was relatively weak compared to the southwest or crosscanyon drift (Figure 2-10). The predominant southwest bottom drift along the continental shelf in this area (Bumpus et al. 1973) may be a major factor contributing to the observed flow at 30 m. In conjunction with time-lapse photography at 224 m, three days of current measurements (March 27-29, 1974) at 100 cm above the sea floor recorded maximum velocities up to 26 cm/sec. The mean current velocity was approximately 15 cm/sec, and the net mass transport was predominantly downcanyon.

Maximum bottom velocities at the intermediate station (1,222 m) were in the order of 20 to 21 cm/sec, with a mean of about 10 cm/sec (chart A, Figure 2-11). A downcanyon drift component with a slight set to the southwest was characteristic of the flow at this site. At 1,765 m, the currents differed considerably from those at the two shallower stations (chart B, Figure 2-11). Maximum velocities of 17 to 18 cm/sec were recorded at 3 m and 30 m above the bottom. The mean speed at 3 m was about 6 cm/sec and was slightly higher than the mean speed recorded at 30 m (Table 2-1). At 1,222 m, the reversals are close to those of the semidiurnal tides, but less regular than those recorded in Hydrographer Canyon. At 1,765 m, the first part of the record shows an apparent relation to the tides similar to that of the 1,222 m station, but after the first eighty hours the irregularities increase considerably and only a slight correlation with the tides is indicated. A progressive vector plot of data from the 30 m and 3 m levels of the 1,765 m station revealed only a small upcanyon transport and a very distinct crosscanyon flow to the northeast for most of the sixteen days of observation. Such a direction of transport appears anomalous to earlier reports of currents along the continental slope. At comparable slope depths, 8 and 170 km northeast of Hudson Canyon, Oser (1969) and Emery and Ross (1968) respectively reported that the predominant current flow was to the southwest, parallel to the isobaths. This apparently anomalous behavior of the canyon currents at 1,765 m, both in regard to the other observations made in the canyon and to what may be a regional flow pattern along the adjacent slope, cannot be explained in the light of our data. The situation may well be one in which our short period of measurement is not representative with respect to the true characteristics of the flow.

Most recently, submersible and short-term (3.5 days) bottom current observations in the canyon at depths between 2,900 m and 3,450 m found a consistent downcanyon flow (Cacchione et al. 1976). Although the maximum recorded current reached 15 cm/sec, the

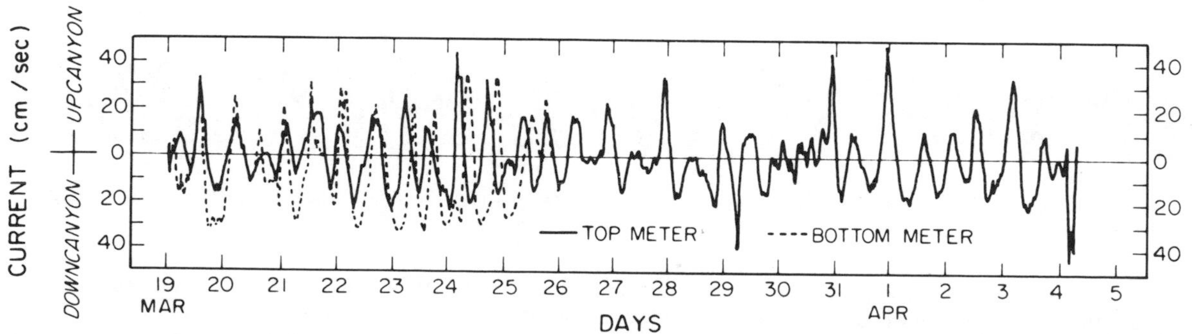

Figure 2-8. Time-velocity plot of currents at 3 m and 30 m above the bottom at the shoalest (224 m) station in Hudson Canyon.

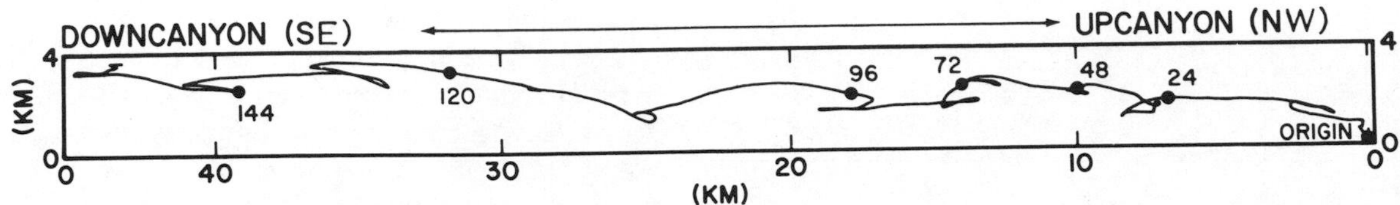

Figure 2-9. Progressive vector diagram of currents 3 m above the bottom at shoalest (224 m) station in Hudson Canyon. Note: Numerals alongside curve indicate hours.

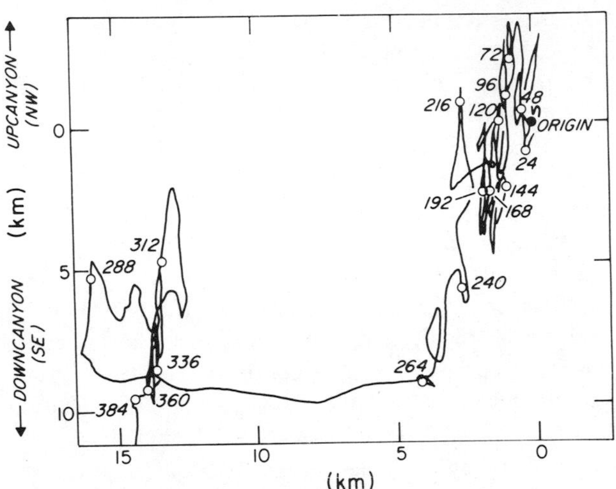

Figure 2-10. Progressive vector diagram of currents 30 m above the bottom at 224 m station. Note: Numerals alongside curve indicate hours. Data from Hudson Canyon study.

presence of a fine sand substrate and large boulders, thinly covered by sediment, indicates that bottom currents exceeded this velocity from time to time. The absence of thick accumulations of the soft, unconsolidated muds found farther upcanyon would tend to indicate that this deeper part of the canyon is not at grade or is influenced by currents quite different from those in the upper canyon.

In 1974 workers from Lamont Doherty Geological Observatory measured near-bottom currents (5 m above the bottom) in the axis of Hudson Canyon at a depth of 800 m (see Figure 2-7) for seven weeks (A. Amos, personal communication, 1976). A maximum current velocity of about 30 cm/sec was recorded in a downcanyon direction. In viewing the seven weeks of data, the higher velocities were commonly associated with the upcanyon flow. Analysis of these data showed that the net mass transport for the seven-week measurement period was upcanyon (A. Amos, personal communication, 1976). These are the only long-term current measurements in any canyon along

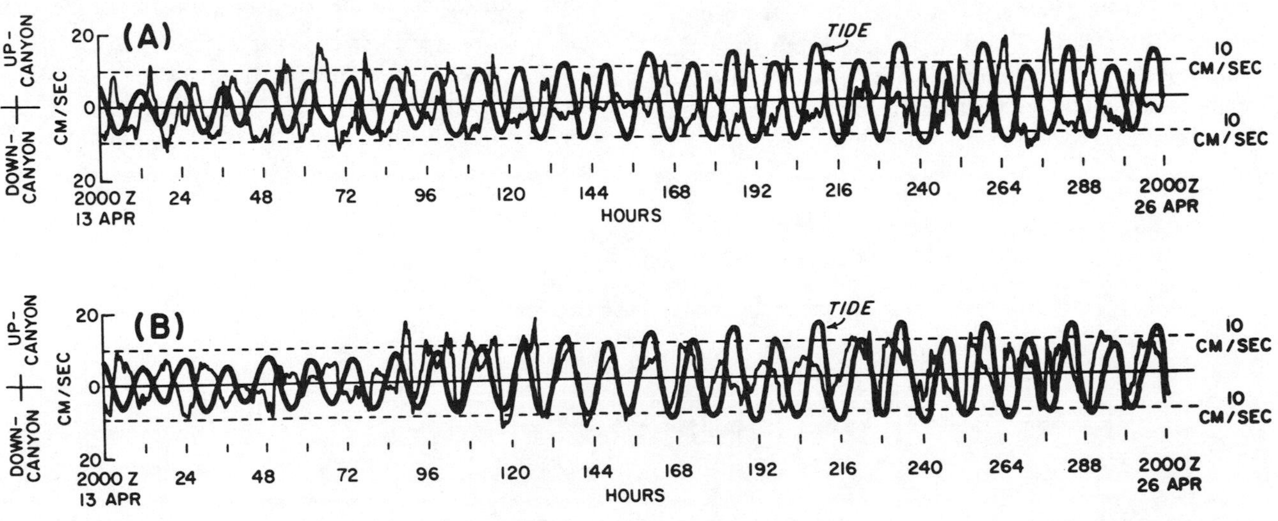

Figure 2-11. Time-velocity diagrams and tide curves for stations in Hudson Canyon. Note: Chart A shows the time-velocity curve and tide curve for the 1,222 m station where good agreement exists between current reversal and periodicities of the tidal curve. Chart B shows the time-velocity plot and tide curve for the 1,755 m station where the current direction alternations are more irregular than those shown in chart A and have a somewhat shorter average period than that of the tide.

Table 2-1. *Current and Sediment Characteristics*

Location	Station Depth (m)	Observ. Period (days)	Velocity (cm/sec) Maximum Top[a]	Maximum Bottom[b]	Mean Top[a]	Mean Bottom[b]	$\bar{U}_{300}$† (cm/sec)	$\bar{U}_{300}$ Exceeded (% of Obs. per.)	Net Flow Direction	Sediment Type	Sediment Median Diameter (mm)
Hydrographer Canyon	348	11.5	38	53[c]	16	18	41	5D	crosscanyon downcanyon	sand, gravel	.19
	512	11.5	39	44	10	14	35	4U	upcanyon[a] upcanyon[b]	sand	.11
	713	11.5		39		14	27	26D	downcanyon[b]	silty sand	.044
Hudson Canyon	224	16	48	33	17	17	44	0	downcanyon[a] downcanyon[b]	sand	.243
	1,222	16		21		10	18	1U	downcanyon[b]	clayey silt	.005
	1,765	16	17	18	5	6	18	0.6U	crosscanyon[a] crosscanyon[b]	silty clay	.004
Wilmington Canyon	723	19	40	44	9	12	27	2.6D 4.5U	downcanyon[a] upcanyon[b]	silty sand	.044
	915	19		22		8	26	0	upcanyon[b]	sandy silt	.38
Shelf Station	103	19		65		15	49	3W	west	sand	.35
Washington Canyon	600	15.5		20		6	22	0	downcanyon[b]	sandy silt	.21
Slope Station	896	12		12		4	18	0	southeast	clayey silt	.008
Norfolk Canyon	573	15.5	32		7		30		downcanyon[a]	silty sand	.068

[a]30 m above bottom.

[b]3 m above bottom.

[c]Velocities of 70-75 cm/sec have been observed by B. Heezen (personal communication).

† Threshold velocity.

U = upcanyon; D = downcanyon; W = West.

the East Coast and as such deserve serious consideration in the analysis of canyon current characteristics. Bottom photographs and visual observations of bottom characteristics made from submersibles along the axis of the canyon from 200 m to 3400 m indicate that the long-term sediment transport trend is downcanyon.

Wilmington Canyon

Two current meter arrays were placed in Wilmington Canyon at depths of 723 m and 915 m and one on the adjacent continental shelf at a depth of 103 m (Figure 2-12). The shelf and outer canyon arrays each consisted of one current meter positioned 3 m above the bottom. Two meters, 3 m and 30 m above the bottom, comprised the upper canyon array.

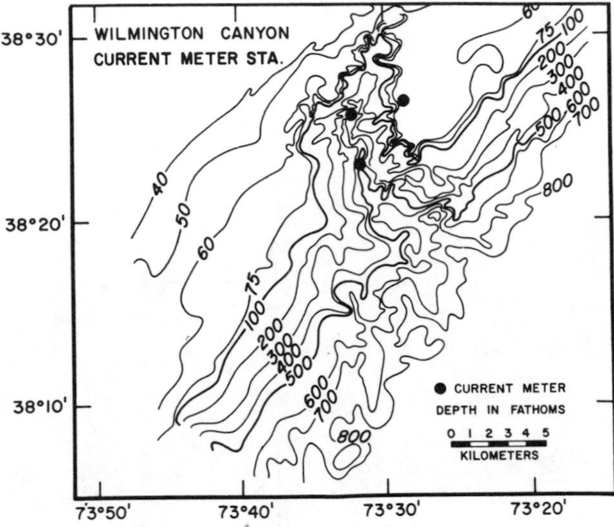

Figure 2-12. Location of current meter arrays in Wilmington Canyon.

Currents in the upper portion of Wilmington Canyon were generally higher than in Hudson Canyon. They averaged about 15-16 cm/sec, with a number of peaks at 37-40 cm/sec (Figure 2-13). Only a random relationship appears to exist between the peak velocities at the 3 m and 30 m levels. The short record at 30 m above the bottom shows current reversals of a period closely related to the semidiurnal tides, but the 3 m record is somewhat more irregular in reversal periods, yet still closely correlated with the tides. In a few instances the velocities at the two levels are similar but in others not only do the velocities differ markedly, but the flow directions are 180° out of phase. The current characteristics may be better seen in the progressive vector diagrams (Figures 2-14 and 2-15) where the upper meter indicates a slight downcanyon transport with a set to the west, whereas at 3 m the flow is distinctly upcanyon with a similar drift to the west. As these diagrams attest, the near-bottom currents are of an irregular character.

At 915 m, the bottom current was even more complicated than at 723 m. Velocities reached as high as 22 cm/sec, but averaged between 8-9 cm/sec. The rather quick flow reversals in the upper canyon (see Figure 2-13) occur much slower at the deeper station (915 m) similar to that found in canyons elsewhere. At any particular period of observation the net transport was found to be toward almost any point of the compass, however, over the nineteen-day period the net flow was upcanyon with a set to the west.

Short-term, intermittent current measurements in the canyon by others have recorded velocities not unlike those reported here (Fenner et al. 1971). Other studies in the canyon on such aspects as bottom features, sediment distribution, and the transport of suspended and bedload material suggest that the movement of material through the canyon is seaward (Stanley and Kelling 1968; Kelling and Stanley 1970; Lyall et al. 1971). If we assume that sedimentary features such as ripple marks and sediment dispersal patterns are more indicative of long-term current characteristics than current meter records, our measurements may possibly reflect a change in the character of the downcanyon flow.

Currents on the continental shelf adjacent to the canyon were much stronger and more undirectional than those measured in the canyon. Shelf currents as high as 65 cm/sec and trending primarily to the west (Figures 2-16 and 2-17)

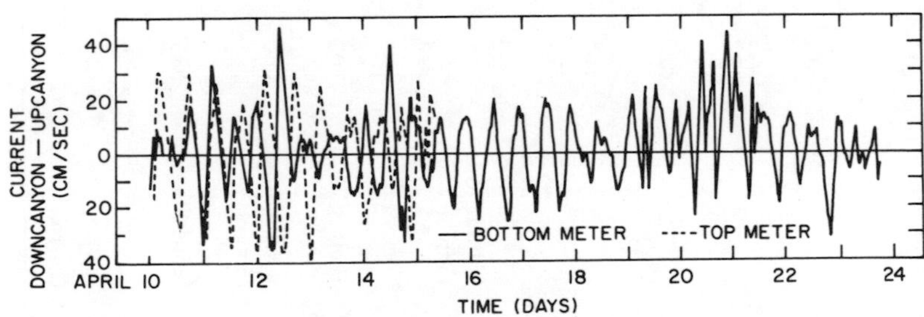

Figure 2-13. Time-velocity curves of currents at 3 m and 30 m above the bottom at 723 m station in Wilmington Canyon.

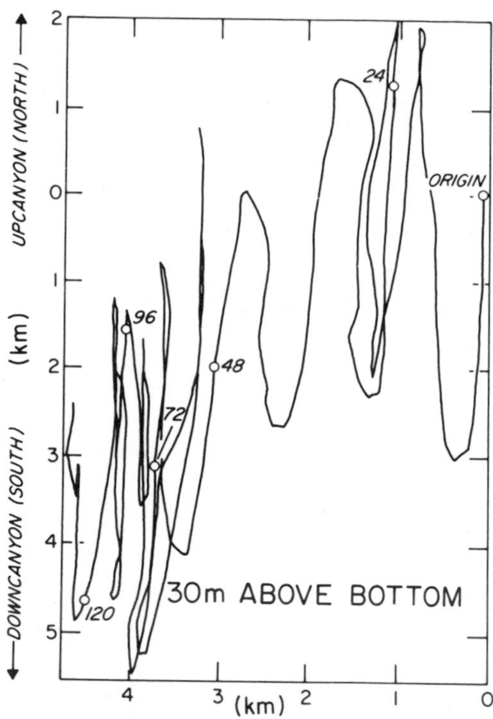

Figure 2-14. Progressive vector diagram of currents 30 m above the bottom at the 723 m station in Wilmington Canyon. Note: Numerals alongside curves indicate hours.

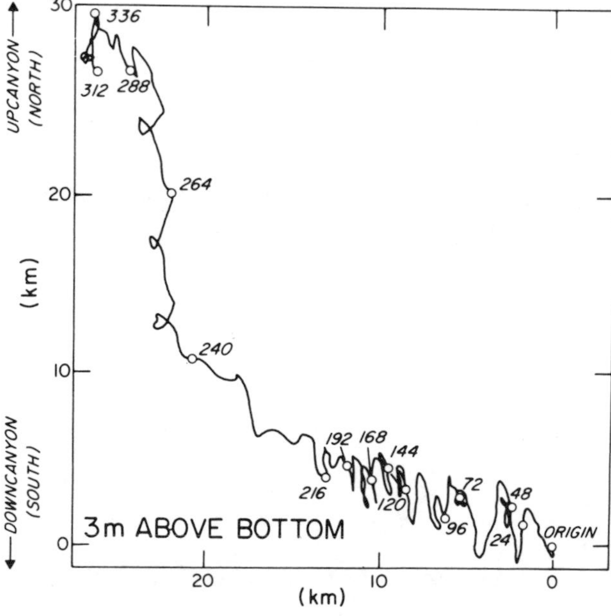

Figure 2-15. Progressive vector diagram of currents 3 m above the bottom at the 723 m station in Wilmington Canyon. Note: Numerals alongside curves indicate hours.

seem to dominate this area (Bumpus et al. 1973; McClennen 1973). Apparently, the strong westerly flow along the outer shelf is also reflected in the westward drift of the canyon currents.

Washington Canyon

A single meter array was positioned in Washington Canyon at a depth of 600 m for 15.5 days in March 1975 (Figure 2-18). Bottom currents at this depth trended primarily up- and downcanyon with a very small crosscanyon component (Figure 2-19). The relatively low velocity found in Washington Canyon appears anomalous in comparison to the currents reported in other east coast canyons. The highest flow was 20 cm/sec and the average over the 15.5-day period was about 6 cm/sec. Net transport during this period was distinctly downcanyon with a slight set to the northeast (Figure 2-20).

Bottom photographs indicate that the canyon head consists of medium-fine sand and shell hash, thereby suggesting that it is an area of active erosion. Ripples and sandy silt deposits were traced downcanyon to a depth of 740 m where clayey silt becomes the predominant sediment type. Except for the canyon head, the observations made during this study indicate the canyon to be relatively inactive.

Later that same year (September 1975) twelve days of current data were obtained from a depth of 896 m on the continental slope approximately 21 km northeast of Washington Canyon (37°48.8'N, 74°12.0'W). At that time the maximum recorded bottom current (3 m above the bottom) was 12 cm/sec with an average velocity of about 4 cm/sec (Figure 2-21). Net transport was to the southeast or down slope but not without notable deviations in the flow (Figure 2-22). Correlation between the canyon and adjacent slope currents is not feasible in this instance, but it is interesting to note that in both cases the current velocities were quite low.

Norfolk Canyon

Current measurements in Norfolk Canyon were carried out during the same period as those in Washington Canyon. An array consisting of two meters was placed at a depth of 573 m (See Figure 2-18). Because of flooding in the lower meter only data from the upper current meter (30 m above the bottom) were obtained. The time-velocity plot reveals an unusual current flow not seen in any of the other canyons (Figure 2-23). Although the current reversal period is similar to that found in other canyons along the East Coast, the regularity of the current characteristics are quite different from those recorded in these other canyons and tends to resemble that of a tidal current. The current velocities are consistently regular both up and down-

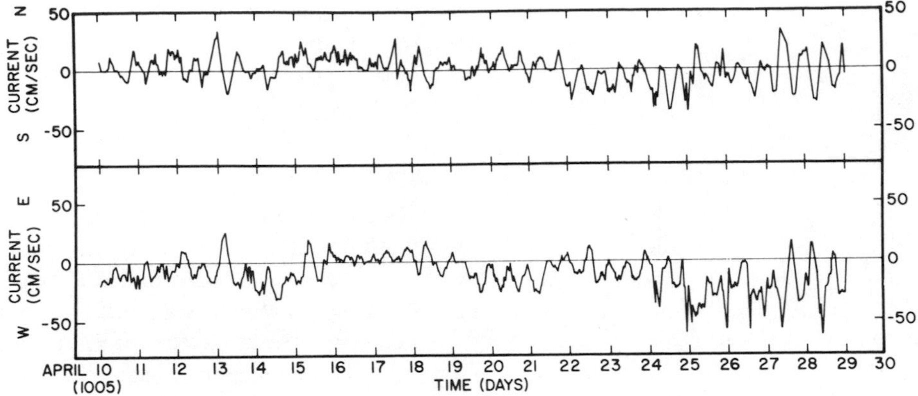

Figure 2-16. Time-velocity curve of north, south, east, and west current components measured 3 m above the bottom on the shelf adjacent to Wilmington Canyon.

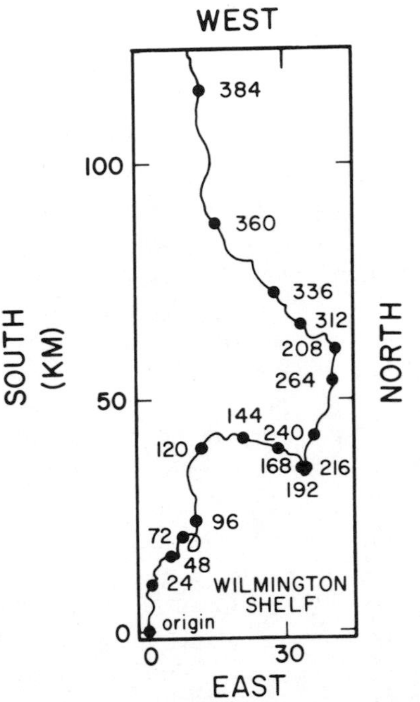

Figure 2-17. Progressive vector plot of shelf currents 3 m above the bottom near Wilmington Canyon. Note: Numerals alongside curve indicate hours.

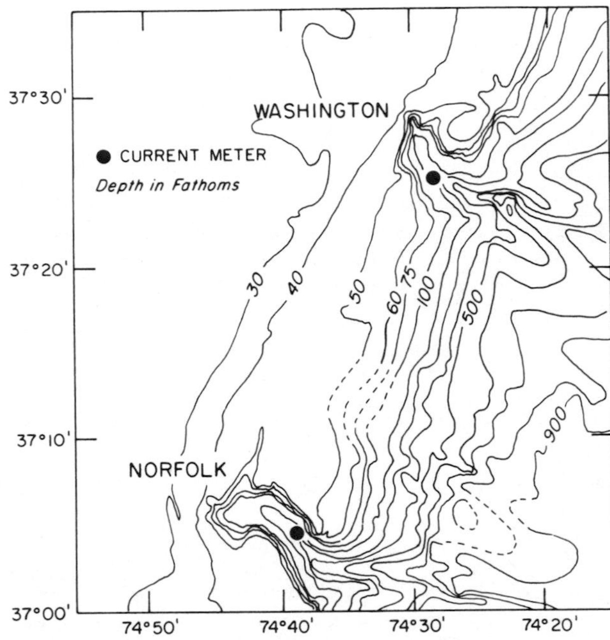

Figure 2-18. Location of current meter arrays in Washington and Norfolk Canyons.

canyon for days at a time which is another characteristic not commonly found in the East Coast canyons. These characteristic regularities plus the higher current speeds recorded in Norfolk Canyon may suggest a resonating influence on the currents. Unfortunately the available data only allow us to speculate as to the reason for the regularity of the current characteristics observed in Norfolk Canyon. Although there was a strong downcanyon transport, there appears to have been an equally strong drift to the northeast (Figure 2-24). In this case, having data from 3 m above the bottom would have been particularly valuable to determine whether this unusual flow pattern extended to the sea floor.

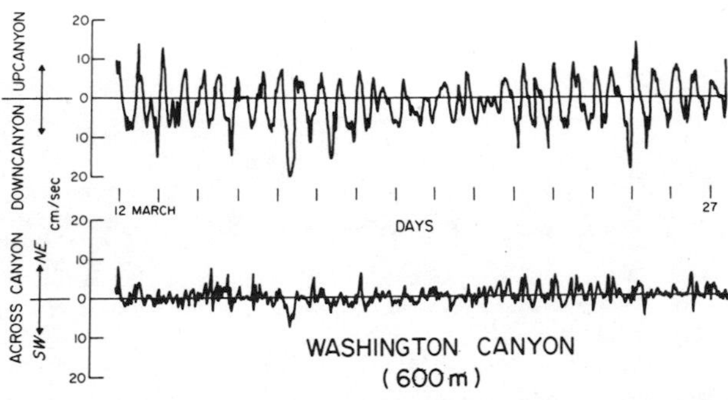

Figure 2-19. Time-velocity curve of currents 3 m above the bottom at 600 m depth in Washington Canyon.

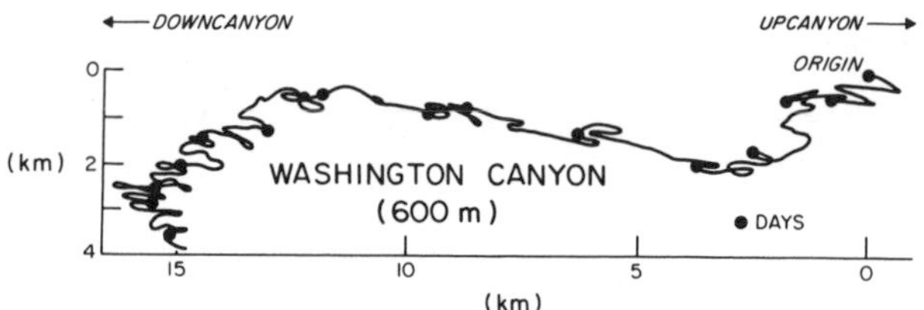

Figure 2-20. Progressive vector plot of currents 3 m above the bottom at 600 m depth in Washington Canyon.

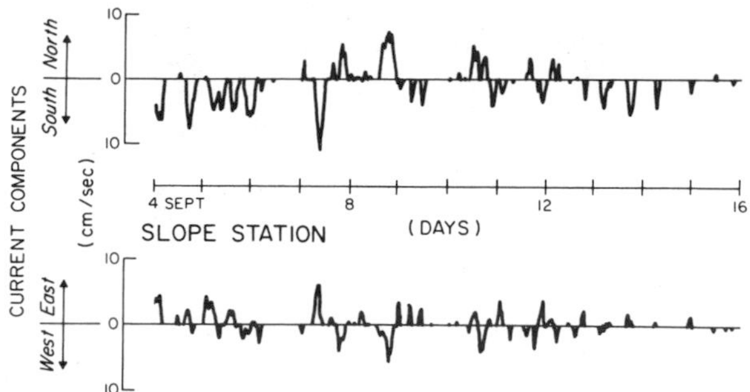

Figure 2-21. Time-velocity curve of the north, south, east, and west current components at 3 m above the bottom on the continental slope northeast of Washington Canyon.

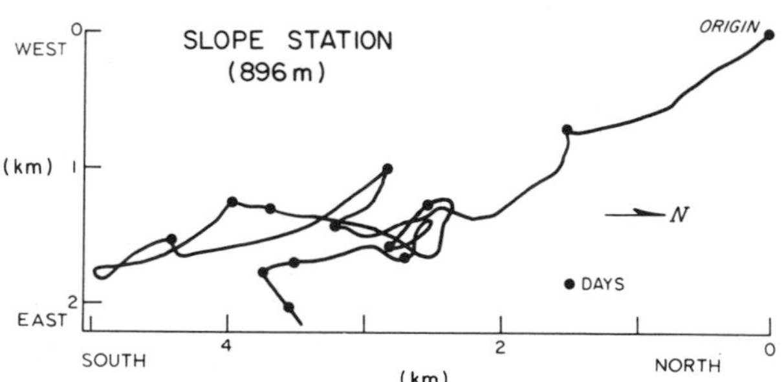

Figure 2-22. Progressive vector plot of currents 3 m above the bottom at the slope station (37°48.8'N, 74°12.0'W) near Washington Canyon.

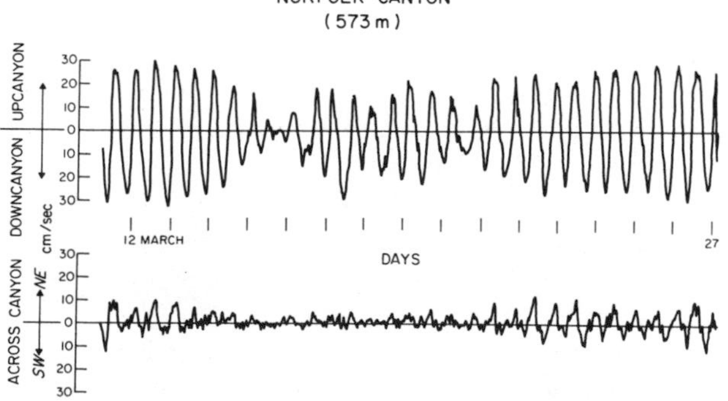

Figure 2-23. Time-velocity curve of currents 30 m above the bottom at 573 m depth in Norfolk Canyon.

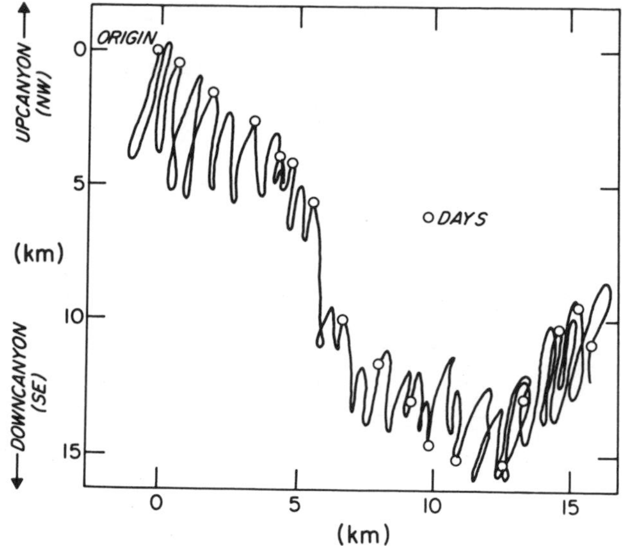

Figure 2-24. Progressive vector plot of currents 30 m above the bottom in Norfolk Canyon.

STORM EFFECTS

Current data presented in this study were usually collected during the winter months, which is commonly the period of severest storm activity. An attempt was made to determine what influence if any these storms may have had on the canyon currents. A comparison of the surface wind data recorded from our own vessel and from the Coast Guard weather station *Hotel* (38°N, 71°W) with the temporal variations of the currents in a number of canyons showed only one possible correlation (Figure 2-25). The hourly wind speed and swell height observed from the bridge of the *Researcher* during and preceding two storm periods is compared with the current velocities at the shoalest Hydrographer Canyon station during the same period (see Figure 2-25). As can be seen, the greatest downcanyon velocities occurred near the beginning of the first storm (short duration) and near the beginning of and during the second storm (moderate duration). The lowest velocities were during the calm period between storms. Possibly there is some connection between these sea conditions and the currents, but the data are not conclusive. An indication that wind might influence canyon processes came from an observation made with the submersible *Alvin* in Hudson Canyon. Some three years ago, three days after a hurricane passed over the inner part of the New York Bight a mass of highly turbid water was encountered at a bottom station (270 m) where dives four days earlier recorded good visibility. A dive returning to the site two days later again found good visibility. Whether the turbidity was the result of the storm or a coincidence cannot be ascertained, but the question does offer an intriguing

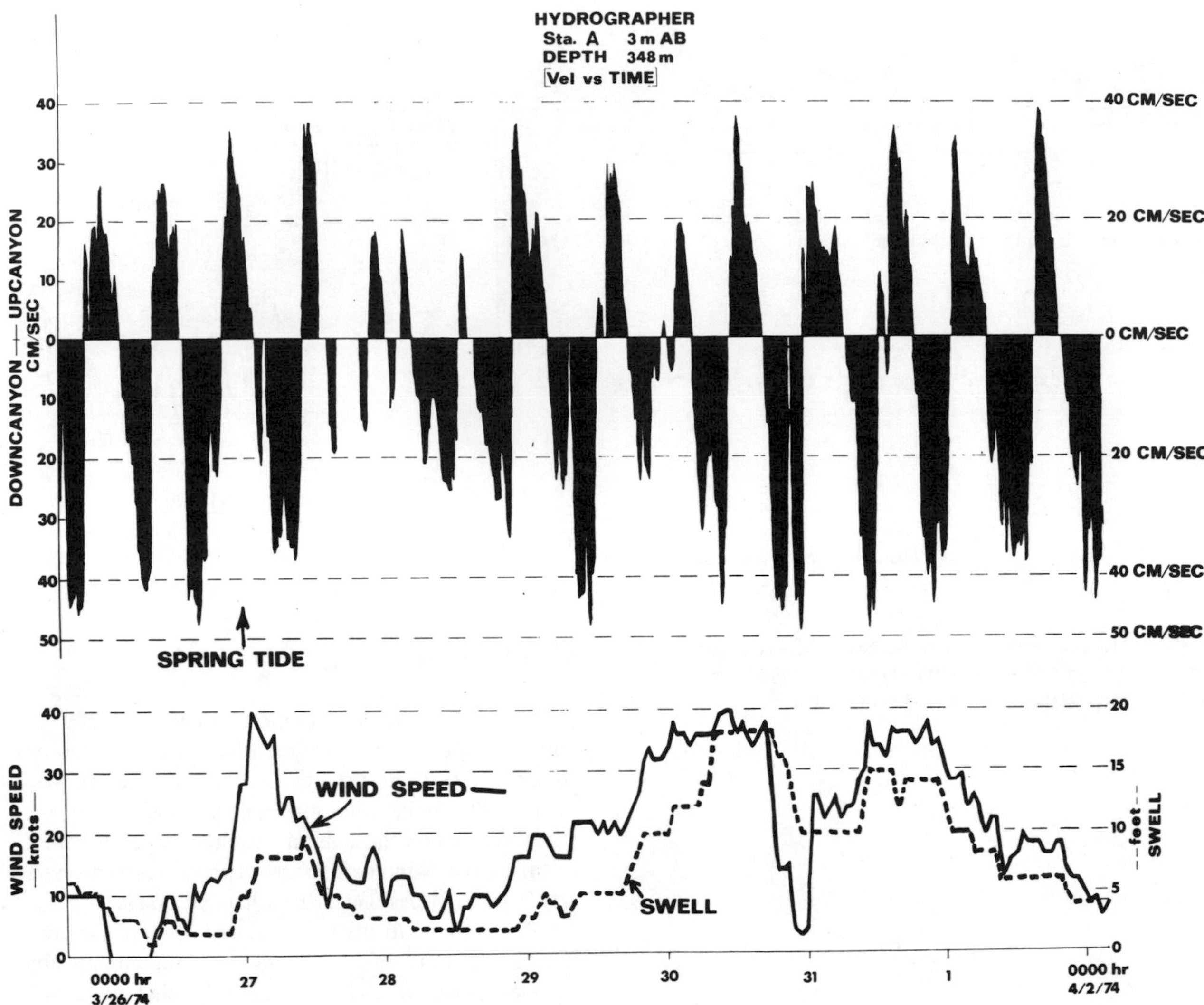

Figure 2-25. Possible relation of wind speed and swell to current velocities at the shoal station in Hydrographer Canyon. Note: The slowest currents appear to have been during the period between the two storms, but the fastest downcanyon currents do not center over the peak of the two storms.

concept that storms on the scale of a hurricane may be required to influence the processes in canyons this far from shore. The apparent lack of appreciable effects of a storm below hurricane velocities in the East Coast canyons is in contrast to the storm effects along the California coast where turbidity currents are initiated in the canyon heads by onshore winds of about 30 knots (Shepard and Marshall 1973b). Also, turbidity currents were indicated at the near-shore head of Rio Balsas Canyon in Mexico during a period of large swell (Shepard et al. 1975).

SEDIMENT TRANSPORT

In an attempt to gain some understanding of the actual sediment transport in the East Coast canyons, the measured current velocities were compared to the threshold velocities (the velocity slightly less than that necessary to initiate motion for a certain sediment grain size). Miller et al. (1977) in their study of various threshold velocity diagrams and the parameters upon which they are based developed a curve for $\bar{U}_{100}$ (water velocity 100 cm above the bottom) versus grain diameter for quartz density material. Assuming no change in the boundary layer characteristics from 100 to 300 cm above the bottom, the Miller et al. curve was modified to $\bar{U}_{300}$ versus grain diameter to provide the threshold velocities presented in Figure 2-26. $\bar{U}_{300}$ in place of $\bar{U}_{100}$ reflects the level, 300 cm, above the bottom at which our current measurements were made.

The limitations in using such a threshold diagram for conditions (fluid viscosity and grain shape and density) other than those upon which it is based must be realized.

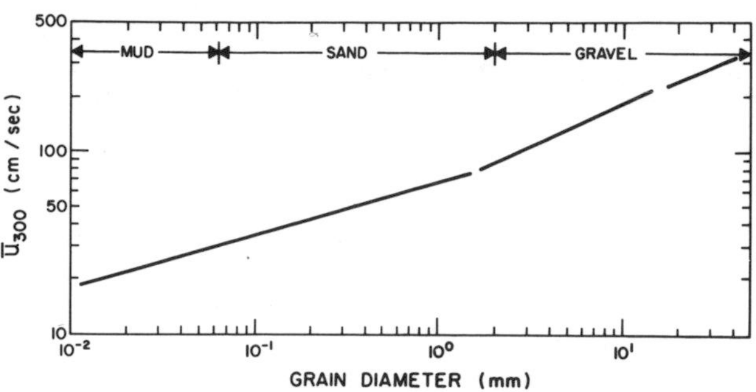

Figure 2-26. The grain diameter versus the flow velocity 300 cm above the bottom ($\bar{U}_{300}$) necessary for the threshold transport of quartz density material in water of temperature 20° C. (Source: Adapted from Miller et al., 1977.)

For this reason the threshold velocities discussed here are considered close approximates and serve only to indicate that bedload transport is an active process in the upper portions of a number of the East Coast canyons. As can be seen from the various current velocity versus time plots presented herein, the threshold velocity for the sediments in the vicinity of the respective current meters was often exceeded during even our relatively short periods of observation. As might be expected from the earlier discussion, sediment transport was found to take place in both the up- and downcanyon directions. During our periods of observation, threshold velocities were exceeded in some cases up to 26 percent of the time (see Table 2-1).

Threshold velocities were more commonly exceeded in Hydrographer Canyon than in any other canyon. As shown in Table 2-1, current velocities at each of the three stations were sufficient at times to transport bottom sediments. At the 512 m station the threshold velocity was only exceeded when the current flowed upcanyon.

Observations in Hudson Canyon indicate that even in the deeper portions of the canyon (1,200 m to 1,765 m), threshold velocities may be exceeded. In the case of such fine-grained sediment as is present in the outer canyon (median diameter 0.005 mm) the theory of threshold velocity and sediment transport is not well developed; thus whether transport does or does not occur at these depths is questionable. The data in Table 2-1 indicate that the threshold velocity was not exceeded in the head of the canyon (224 m) during our period of measurement. Observations by the senior author from the submersible *Alvin* at this depth, although at a different time, have documented bedload transport at times when bottom currents measured 27 cm/sec. The discrepency between these observations and the results shown in Table 2-1 may possibly be attributed to local turbulence and some form of hydraulic lifting. As clearly seen from *Alvin*, sediment particles were carried along the bottom, often in an erratic manner. In some instances the particles literally hopped about. Flow conditions in the canyon head as observed from *Alvin* appeared to resemble those that might be found in a "breaker zone" but at a markedly reduced scale. Such hydrodynamic conditions would not be accurately recorded by a current meter. We are confident from our firsthand observations that bedload transport does occur, despite the fact that the recorded currents did not exceed the threshold velocity as normally defined. Our submersible studies have documented that erosion and transport of even gravel-size material takes place in the canyon head.

In the upper part of Wilmington Canyon (723 m) the threshold velocity was found to be exceeded during both the up- and downcanyon flow (see Table 2-1). Within the period of current measurement, sediment transport in an upcanyon direction occurred about twice as often as did downcanyon transport. At the 915 m station the threshold velocity was not exceeded during the nineteen-day measurement period.

Although bottom photographs clearly revealed that bedload transport is a common process in the head of Washington Canyon (300 m) current measurements at 600 m indicate that the sediment threshold velocity was seldom even approached during the 15.5-day observation period.

Current velocities on the shelf (103 m) adjacent to Wilmington Canyon were frequently higher than those measured in the canyons; yet during the nineteen days of measurement the threshold velocity was exceeded only 3 percent of the time. No correlation was found between the period (April 25-28) of relatively high current velocity (see Figure 2-16) and the tidal and wind conditions for the general area of the station. Although much longer periods of measurement are required, these data tend to indicate that bedload transport is minimal during relatively "normal" conditions.

Bedload transport is, however, a significant process in this area of the shelf as shown by the presence of sediment ripples and lineations (McClennen 1973), sand waves (Knebel and Folger 1976) and sand tongues extending down into Wilmington Canyon along the eastern slope (Lyall et al. 1971). In light of our observations and the significance of bedload transport in this area, we can postulate that erosion and sediment transport occurs primarily during storm conditions. This concept tends to be supported by the findings of recent time-lapse photography studies of the bottom on the New Jersey shelf where sediment transport was clearly related to storm conditions (Butman et al. 1976).

At a depth of 896 m on the continental slope 21 km northeast of Washington Canyon, bottom currents during the relatively short period of observation did not appear to exceed the threshold velocity. Here, however, the fine-grained nature of the sediment does not permit more than a crude determination of threshold velocity, as was pointed out above.

Although our observations are rather limited in duration, they make evident that erosion and sediment transport is taking place in the canyons off the northeastern United States. Given considerably longer observation periods extending over the seasons a much more meaningful sediment flux rate could be established for the East Coast canyons.

SUMMARY

Some of the strongest bottom currents (70-75 cm/sec) in the canyons and on the slopes along the outer east coast margin are found off Georges Bank (see Table 2-1). Erosion and very active transport of coarse material downcanyon is most pronounced in Hydrographer Canyon where velocities as high as 70-75 cm/sec have been reported and where photographic evidence has revealed the downcanyon movement of sand ripples at a depth of 710 m. Submersible observations in the upper portion of Hydrographer canyon (shallower than 900 m) have revealed the presence of ripples superimposed on sand waves (B. Heezen, personal communication, 1973). A turbidity current observed in Oceanographer Canyon by Richard Slater (personal communication, 1976) from a submersible further suggests occasional large-scale downcanyon transport of sediment, or even erosion. Of the East Coast canyons studied, Hydrographer is perhaps the most active insofar as seaward transport of coarse sediment is concerned. Current velocities of 50-75 cm/sec appear to be relatively commonplace. Other Georges Bank canyons–Oceanographer and Corsair, as well as the Gully Canyon off Nova Scotia–display coarse-grained sediments extending considerable distances downcanyon (Trumbull and McCamis 1967; Stanley 1967; Ross 1968). Downcanyon flows appear to predominate in Gilbert Canyon (Stetson 1937), but the axial sediments are sandy silts at depths comparable to those where sands are found in the other canyons. A single submersible dive in Lydonia Canyon, 21 km from Gilbert Canyon, revealed no evidence of current activity and that fine-grained sediments blanketed the canyon floor (Ross 1969). Apparently, even in an area where many canyons are quite active, the processes are far from uniform.

Canyons west and south of Georges Bank appear to be relatively inactive in comparison to those off the Bank (see Table 2-1). Although a small number of current velocities as high as 40-48 cm/sec have been recorded, peak currents of 25-30 cm/sec appear to be more representative of the higher velocities encountered in these canyons. Transport of coarse-grained sediment is mainly confined to the heads and upper portions of these canyons. Coarse surface sediments in Hudson Canyon extend only a short distance downcanyon to a depth of about 275 m. In Wilmington Canyon, shelf sands are reported in the canyon at depths of about 740 m (Stanley and Kelling 1968). Coarse-grained sediments also are found extending to comparable depths in Washington and Norfolk Canyons.

Mapping the sediment distribution along the shelf break and the adjacent slope in the vicinity of the Middle Atlantic Bight makes evident that sand is being transported onto the slope and into the canyons (Pierce et al. 1974). A study of mica distribution in the same area indicates that fine-grained shelf deposits are being carried well down the slope (Doyle et al. 1975).

Tidal and internal wave forces seem to be major factors contributing to the characteristic up- and downcanyon flow noted in the East Coast canyons. The data presented here indicate that the net transport through these canyons is predominantly seaward although the available current meter records show considerable upcanyon flow in a number of cases (see Table 2-1). Analysis of progressive vector diagrams reveals that net transport in the canyons is very complex and not necessarily consistent even in the same canyon. Synoptic measurements in Hydrographer Canyon at three stations (348 m, 512 m, and 713 m) revealed bottom net flow to be down-, up- and downcanyon respectively for the 11.5-day observation period (see Table 2-1). Somewhat similar confusing transport regimes were noted in Hudson and Wilmington Canyons.

In contrast to the California canyons those off the northeastern United States appear to be considerably less active. Apparently, however, sediment threshold velocities in most of the East Coast canyons are often exceeded and sediments are moved through the canyons. Only in Hydrographer Canyon was coarse-grained sediment found being transported along the canyon floor. With the exception of this and other Georges Bank canyons, movement of fine-grained material to the middle and outer parts of the canyons appears to be the extent of sediment transport in the canyons off the East Coast today.

Long-term current measurements, on the order of months rather than days, are essential if the temporal properties of bottom currents in submarine canyons and the transport of sediment to the deep sea are to be understood. Such data are not yet available and at best we are now able to show only the complexity of canyon currents. As we have also demonstrated, the currents and the transport of sediment vary considerably not only from canyon to canyon, but within a canyon as well. Just as the genesis of submarine canyons appears to be complex, so the explanation of canyon currents also appears to involve several processes.

ACKNOWLEDGMENTS

The authors are indebted to the outstanding assistance provided to this study by George Lapiene, Gary Sullivan, and Patrick McLoughlin, without whose efforts the project could not have been conducted. The authors sincerely appreciate the assistance of colleagues Bonnie Stubblefield and Richard Bennett, who served as chief scientists during various cruises while this study was in progress. A critical review and discussion of an early draft of the manuscript by La Verne Kulm was particularly helpful to the authors as were the comments of Paul Komar and Martin Miller.

The authors particularly wish to thank the officers and crew of the NOAA ship *Researcher* for their fine efforts on behalf of this study. The authors acknowledge the many hours of effort from Sam Bush who assisted in the data processing. Funding support from the NOAA Atlantic Oceanographic and Meteorological Laboratories and National Science Foundation Grant DES74-22089, Office of Naval Research Contract N00014-69-A-0200-6049 made this study possible.

REFERENCES

Bumpus, D.F., Lynde, R.E., and Shaw, D.M., 1973. Physical oceanography. 1-1-1-72. *Coastal and Offshore Environmental Inventory Cape Hatteras to Nantucket Shoals.* Univ. of Rhode Island Mar. Pub. Ser. No. 2, 696 pp.

Butman, B., Folger, D.W., and Noble, M., 1976. Winter sediment mobility on the Atlantic outer continental shelf. (Abst.) *Geol. Soc. Amer. Progr.*, 8: 799.

Cacchione, D.A., Rowe, G.T., and Malahoff, A., 1976. Sediment processes controlled by bottom currents and faunal activity in lower Hudson submarine canyon. (Abst.) *Amer. Assoc. Petrol. Geol. Bull.*, 60: 654–55.

Dietrich, G., Kalle, K., Krauss, W., and Seidler, G., 1975. *Allgemeine Meereskunde, Eine Einführung in die Ozeanographie.* Gebrüder Borntraeger, Berlin, 593 pp.

Doyle, L.J., Pilkey, O.H., Hayward, G.L., and Arbogast, J.S., 1975. Sedimentation on the northeastern continental slope of the United States. *Proc. IX Inter. Cong. Sediment. Theme*, 6: 51–56.

Emery, K.O., and Ross, D.A., 1968. Topography and sediments of a small area of continental slope south of Martha's Vineyard. *Deep-Sea Res.*, 15: 415–22

——, and Uchupi, E., 1972. *Western North Atlantic Ocean: Topography, Rocks, Structure, Water, Life and Sediments.* Amer. Assoc. Petrol. Geol., Mem. 17, 532 pp.

Ericson, D.B., Ewing, M., and Heezen, B.C., 1952. Turbidity currents and sediments in North Atlantic. *Amer. Petrol. Geol. Bull.*, 36: 489–511.

——, Ewing, M., Wollin, G., and Heezen, B.C., 1961. Atlantic deepsea sediment cores. *Geol. Soc. Amer. Bull.*, 72: 193–286.

Fenner, P., Kelling, G. and Stanley, D.J., 1971. Bottom currents in Wilmington submarine canyon, *Nature*, 229: 52–54.

Horn, D.R., Ewing, M., Horn, B.M., and Delach, M.N., 1971. Turbidities of the Hatteras and Sohm abyssal plains, western north Atlantic. *Mar. Geol.*, 11: 287–323.

Isaac, J.D., Reid, J.L., Schick, G.B., and Schwartzlose, R.A., 1966. Near-bottom currents measured in 4 kilometers depth off Baja California coast. *J. Geoph. Res.*, 71: 4297–303.

Keller, G.H., 1975. Sedimentary processes in submarine canyons off northeastern United States. *Proc. IX Inter. Cong. Sediment. Theme*, 6: 77–86.

Kelling, G., and Stanley, D.J., 1970. Morphology and structure of Wilmington and Baltimore submarine canyons, eastern U.S.A. *J. Geol.*, 78: 637–60.

——, Sheng, H., and Stanley, D.J., 1975. Mineralogical composition of sand-size sediment on the outer margin off the Mid-Atlantic states: Assessment of the influence of the ancestral Hudson and other fluvial systems. *Geol. Soc. Amer. Bull.*, 86: 853–62.

Knebel, H.J., and Folger, D.W., 1976. Large sand waves on the Atlantic outer continental shelf around Wilmington Canyon, off eastern United States. *Mar. Geol.*, 22: M7–M15.

Knott, S.T., and Hoskins, H., 1968. Evidence of Pleistocene events in the structure of the continental shelf off the northeastern United States. *Mar. Geol.*, 6: 5–43.

Lyall, A.K., Stanley, D.J., Giles, H.N., and Fisher, A., Jr., 1971. Suspended sediment and transport at the shelf-break and on the slope, Wilmington canyon area, eastern U.S.A. *Mar. Tech. Soc.*, 5: 15–27.

McClennen, C.E., 1973. New Jersey continental shelf near bottom current meter records and recent sediment activity. *J. Sed. Petrol.*, 43: 371–80.

McGregor, B., Keller, G.H., and Bennett, R.H., 1975. Seismic profiles along the U.S. northeast coast continental margin. (Abst.) *Trans. Amer. Geoph. Union.*, 56: 382.

McMaster, R.L., and Garrison, L.E., 1966. Mineralogy and origin of southern New England shelf sediments. *J. Sed. Petrol.*, 36: 1131–42.

——, and Ashraf, A., 1973. Drowned and buried valleys on the southern New England continental shelf. *Mar. Geol.*, 15: 249–68.

Miller, M.C., McCave, I.N., and Komar, P.D., 1977. Threshold of sediment motion under unidirectional currents. *Sediment.* 24: 507–27.

Milliman, J.D., 1973. Marine Geology. 10-1-10-91. In: *Coastal and Offshore Environmental Inventory Cape Hatteras to Nantucket Shoals.* Univ. of Rhode Island Mar. Pub. Ser. No. 3, 393 pp.

——, Pilkey, O.H., and Ross, D.A., 1972. Sediments of the continental margin off the eastern United States. *Geol. Soc. Amer. Bull.*, 83: 1315–34.

Oser, R.K., 1969. Bottom environmental oceanographic data report Hudson canyon area, 1967. *Naval Ocean. Off. Informal Report*, 69-8, 43 pp.

Pierce, J.W., Southard, J.B. and Stanley, D.J., 1974. Shelfbreak processes and suspended sediment transport on the outer continental margin. In: D.J. Stanley and D.J.P. Swift (eds.), *The New Concepts of Continental Margin Sedimentation II: Process and Application.* Amer. Geol. Inst., Washington, D.C., pp. 639-740.

Roberson, M.I., 1964. Continuous seismic profiler survey of Oceanographer, Gilbert, and Lydonia submarine canyons, Georges Bank. *J. Geoph. Res.*, 69: 4779–89.

Ross, D.A., 1968. Current action in a submarine canyon. *Nature*, 218: 1242–44.

——, 1969. Geological observations from *Alvin*. *Woods Hole Inst. Ref.*, 69-13: 71–72.

Shepard, F.P., 1965. Importance of submarine valleys in funneling sediments to the deep sea. *Prog. in Oceanography*, 3: 321–32.

——, 1975. Progress of internal waves along submarine canyons.

Mar. Geol., 19: 131–38.
_____, 1976. Tidal components of currents in submarine canyons. *J. Geol.*, 84: 343–350
_____, Revelle, R., and Deitz, R.S., 1939. Ocean-bottom currents off the California coast. *Science*, 89: 488–89.
_____, and Dill, R.F., 1966. *Submarine Canyons and Other Sea Valleys*. Rand McNally and Co., Chicago. 381 pp.
_____, and Marshall, N.F., 1973a. Currents along floors of submarine canyons. *Amer. Assoc. Petrol. Geol. Bull.*, 57: 244–64.
_____, and Marshall, N.F., 1973b. Storm-generated current in a La Jolla submarine canyon, California. *Mar. Geol.*, 15: M19–M24.
_____, Cacchione, D.A., and Sullivan, G.G., 1974a. Submarine canyon currents: East and west coast compared. (Abst.) *Geol. Soc. Amer. Prog.*, pp. 951-52.
_____, Marshall, N.F., and McLoughlin, P.A., 1974b. Currents in submarine canyons. *Deep-Sea Res.*, 21: 691–706.
_____, Marshall, N.F., and McLoughlin, P.A., 1974c. "Internal Waves" advancing along submarine canyons. *Science*, 183: 195–98.
_____, Marshall, N.F., and McLoughlin, P.A., 1975. Pulsating turbidity currents with relationship to high swell and high tides. *Nature*, 258: 704–06.
Stanley, D.J., 1967. Comparing patterns of sedimentation in some modern and ancient submarine canyons. *Earth and Planet. Sci. Letters*, 3: 371–80.
_____, and Kelling, G., 1968. Sedimentation patterns in the Wilmington submarine canyon area. In: *Ocean Sciences and Engineering of the Atlantic Shelf*. Mar. Tech. Soc., Philadelphia, pp. 127–42.
Stetson, H.C., 1937. Current measurements in the Georges Bank canyons. *Trans. Amer. Geoph. Union,* 18: 216–19.
Trumbull, J.V.A., and McCamis, M.J., 1967. Geological exploration in an east coast canyon from a research submersible. *Science*, 518: 370–72.
Uchupi, E., 1968. Seismic reflection profiling survey of the east coast submarine canyons, Part 1: Wilmington, Baltimore, Washington, and Norfolk canyons. *Deep-Sea Res.*, 15: 613–16.

Chapter 3

Suspended Particulate Matter and Mud Deposition in Upper Hudson Submarine Canyon

DAVID E. DRAKE

U.S. Geological Survey
Menlo Park, California

PATRICK G. HATCHER

Atlantic Oceanographic and Meteorologic Laboratories
Miami, Florida

GEORGE H. KELLER

School of Oceanography
Oregon State University, Corvallis

ABSTRACT

Concentrations of suspended matter of 2,210 to 3,440 μg/liter were present near the bottom at depths of 200 m to 450 m in Hudson Canyon during March 1974. The suspended matter was mostly inorganic mineral grains and two samples contained significant amounts of coarse silt (10-20 percent) and very fine sand (5-10 percent). Suspended matter concentrations near the sea floor declined both shoreward and seaward from the head of the canyon; concentrations of about 600 μg/1 were measured at 893 m in the canyon and values of 400-600 μg/1 were present over the surrounding shelf.

Current measurements at 3 m above bottom in the canyon head (223 m) revealed reversing flows of tidal period with peak speeds of 25-35 cm/sec and a strong net downcanyon component which averaged about 8 cm/sec over a six-day period. The combined suspended matter and current velocity data indicate important resuspension of the fine-grained components of the canyon head sediments followed by transport to deeper water along the axis of the canyon.

A 5 m piston core recovered in the canyon axis at 430 m contained an apparently continuous record of silt and clay accumulation. Radiocarbon dates at three levels in the core suggest a mean sedimentation rate of about 80 cm/10^3 years over the past 6,300 years. If this single core is representative of the sedimentation regime in this part of Hudson Canyon, the implication is that much of the sediment resuspended in the head of the canyon (<300 m depths) is deposited before moving far downcanyon. Furthermore, the textural character of the core sediments implies that vigorous bottom scour has been infrequent over the past several thousand years except in the shallowest portions of the canyon. Sediment transport by turbidity currents of high velocity appears to be unimportant in this canyon at the present time.

INTRODUCTION

Recent observations in Hudson Canyon have indicated a large variability in suspended matter concentrations near the bottom (Keller et al. 1973). In order to investigate the suspended matter and bottom currents in the canyon, current meters were deployed and water samples were obtained during March 1974 aboard the NOAA ship *Researcher* (Figures 3-1 and 3-2). The following year a piston core was recovered from the axis of the canyon at a depth of 430 m. The results of this work provide an initial assessment of the recent sedimentary history and ongoing sedimentary processes in a large canyon that is at present relatively distant from major sources of terrigenous sediment. In addition, previous investigations of suspended matter concentrations in the vicinity of Hudson Canyon (Manheim et al. 1970; Meade et al. 1975; Biscaye and Olsen 1975) were performed during summer and fall seasons. As will be shown, the results of our late winter survey differ substantially from the previous findings.

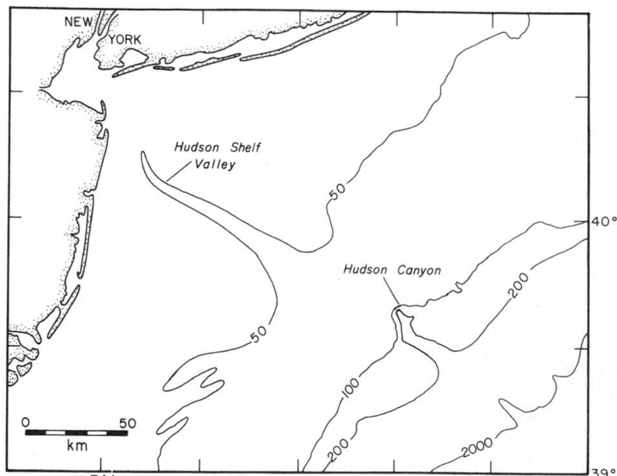

Figure 3-1. Location map showing Hudson Shelf Valley and Hudson Canyon. Note: Generalized depth contours are given in meters.

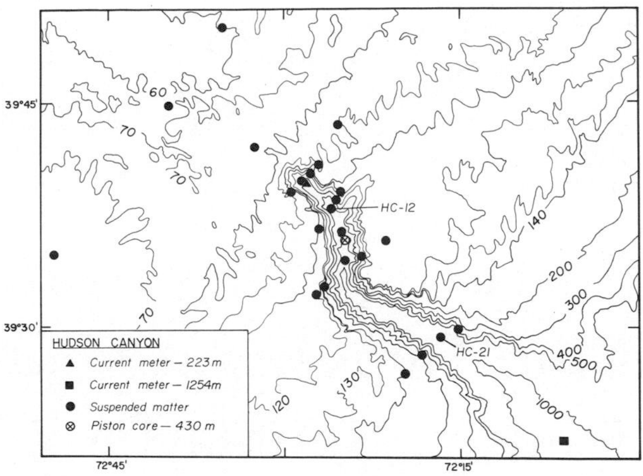

Figure 3-2. Bathymetric map of upper Hudson Canyon showing station locations and current meter sites.

INSTRUMENTS AND METHODS

F. P. Shepard and G. H. Keller deployed self-contained Savonius rotor current meters near the bottom (3 m and 30 m above bottom) at depths of 223 m, 514 m and 1,254 m in the axis of Hudson Canyon. Useful data were obtained from the sensors at 223 m and 1,254 m (see Figure 3-2). In the course of the present discussion important aspects of these current measurements as they relate to the suspended matter problem will be presented. The reader is directed to Chapter 2 for detailed discussion and interpretation of current meter data in this and other canyons.

Samples for suspended matter analysis were obtained at twenty three stations with a *General Oceanics*[a] Niskin rosette system mounted on a *Plessey* 9060 salinity-temperature-depth sensor. Samples within 5 m of the sea floor were collected with the aid of a pinger attached to the cable a known distance above the sampling package. Measured volumes of the water samples were vacuum filtered immediately aboard ship using preweighted 47 mm *Nuclepore* (0.4 μm pore size) polycarbonate membranes mounted in *Millipore* in-line filter holders. A minimum of 50 ml of filtered distilled water was used to free the entire surface of each filter of sea salts. The filters were then dried at 50° C for six hours and stored in self-sealing plastic petri dishes. At our shore laboratory the sample filters plus a series of control filters were placed in dessicators for several days and reweighed to a precision of ± 10 μg. None of our samples contained less than 300 μg of suspended matter and, consequently, analytical errors are considered negligible (<10 percent).

[a]The use of brand names in this chapter is for identification purposes only and does not imply endorsement by the U.S. Geological Survey.

All surface water and near-bottom samples and a selected number of samples from intermediate depths were examined with a polarizing microscope and scanning electron microscope (SEM). Small sections of each filter were cut for the SEM study and the remaining filter was reweighed and set aside for combustion analysis. Point counts for two hundred particles were made at magnifications of 50x and 500x (using SEM photo-micrographs) to estimate the percentages of detrital and biogenic minerals and amorphous organic matter. Several counts on the same filter indicate a reproducibility of about ± 10 percent.

The remaining large portion of each filter was combusted for six hours at 550° C in tared platinum foil crucibles according to the method of Manheim et al. (1970). Fresh *Nuclepore* filters were combusted along with each batch

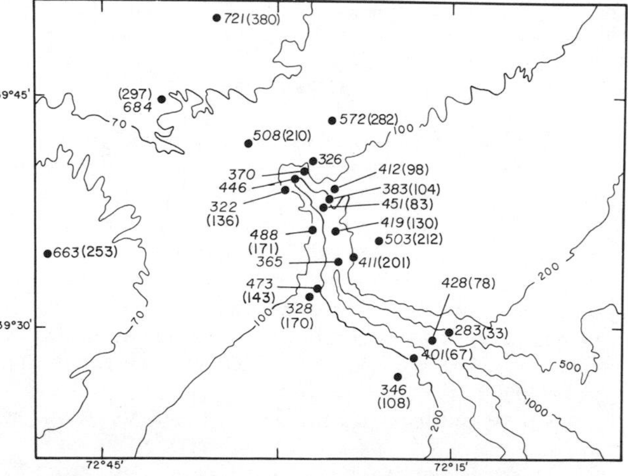

Figure 3-3. Concentrations of total suspended matter in the surface water, March 1974. Note: The noncombustible fractions are shown in parentheses. All values are in μg/1. The maximum and mean values are significantly above those measured by Manheim et al. (1970) and Meade et al. (1975).

of sample filters and our results show that these filters have an average ash residue of 0.12 percent or about 1 $\mu g/cm^2$. This correction was applied to our samples. As should be noted, the ash residue of suspended matter is composed of both biological and inorganic particles. Estimates of the relative amounts of these components were obtained using microscope analyses.

RESULTS

Suspended Matter

Concentrations of total suspended matter (TSM) in the surface waters ranged from 295 to 721 $\mu g/l$ (Figure 3-3). Values above 500 $\mu g/l$ were confined to our shallowest shelf stations (<80 m depth). In general, the concentrations of combustible organic matter did not change significantly over our survey area, but the ash residue values show a general seaward decrease that accounts for the southeastward decline in TSM (see Figure 3-3).

Fine to coarse silt-sized aggregates of amorphous organic matter, biogenic hard parts, and terrigenous minerals were present in all surface water samples. Estimates of the concentrations (by number) of aggregates using SEM and standard microscope point counts suggest a slight decrease from about 1,500/liter to 1,100/liter between our shallowest and deepest water stations. However, this trend may not be significant in view of the limited accuracy of this analytical technique.

At all stations TSM decreased at intermediate levels in the water column and the percentages of combustible matter declined to values in the range of 22 to 46 (for the minimum TSM sample at each station). Minimum TSM values ranged from a high of 328 $\mu g/l$ at our shallowest shelf station to a low of 84 $\mu g/l$ at our deepest canyon station (Figure 3-4). Thus, the regional TSM concentration pattern tended to mirror that present at the surface.

Concentrations of suspended matter and percentages of the terrigenous mineral fraction increased significantly near the sea floor (see Figure 3-4 and Figure 3-5). TSM values in excess of 3,000 $\mu g/l$ were present 5 m above bottom at three stations in the head of the canyon (see Figure 3-5), and significant, although less dramatic, increases occurred over the adjoining shelf. Ash percentages ranged between 66 and 82 at shelf stations shoaler than 200 m, and SEM examination shows that ~50-80 percent of this material was inorganic silt and

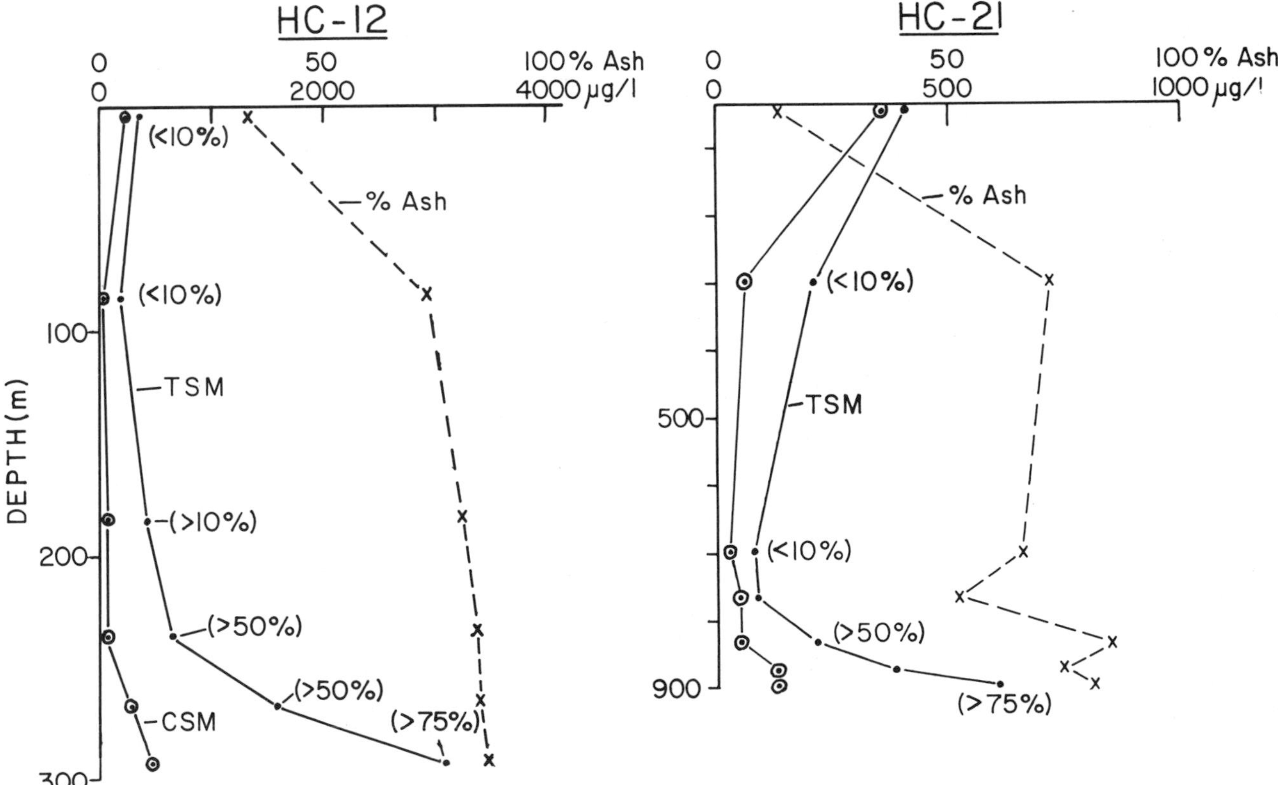

Figure 3-4. Suspended matter concentrations and compositions at 300 m and 900 m in the axis of Hudson Canyon (see Figure 3-2). Note: Percent ash and combustible suspended matter (CSM) were determined by combustion at 550° C. Estimates of amounts of inorganic detritus in TSM are shown in parentheses. Also note scale changes.

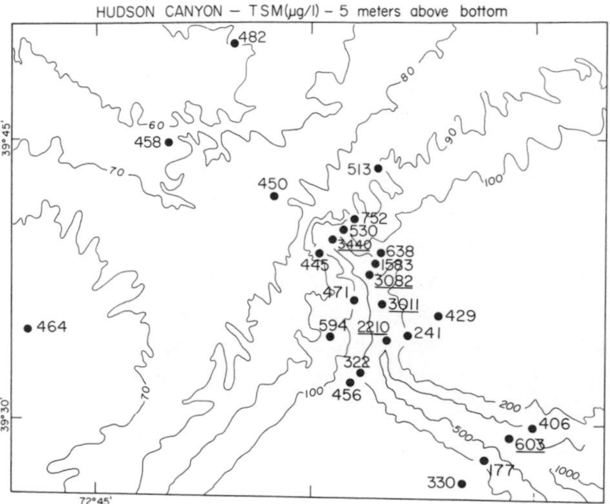

Figure 3-5. Total suspended matter in µg/1 at 5 m above the sea floor. Note: Underlined values denote canyon axis samples. Depths are in meters.

clay. In contrast to the spatial gradients in TSM concentrations at the sea surface and intermediate levels in the water column (see Figure 3-3), the near-bottom values show little meaningful correlation with depth except within the canyon.

We occupied five stations in the axis of the canyon: four between the depths of 200 m and 450 m and one at 893 m. Extremely high TSM concentrations were present in the bottom 50 m of the water column in the head of the canyon (see Figure 3-4). The highest concentration of 3,400 µg/1 was present at our shallowest canyon station and TSM decreased gradually to 2,210 µg/1 at 450 m. We have no data between this station and our deepest station (893 m) where the near-bottom concentration had declined to 603 µg/1.

Examination of our samples using scanning electron microscopy reveals significant changes in the quantities of coarse-grained mineral particles in suspension along the axis of the canyon. Our two shallowest near-bottom samples (TSM>3,000 µg/1) contain from 7 to 14 percent (by number) very fine sand (62 to 125 µm), and 12 to 24 percent coarse silt particles (31 to 62 µm). In contrast, the near-bottom suspended matter at 400 m, 450 m, and 893 m lacked the very fine sand components (<1 percent) and the coarse silt fraction (>31 µm) had decreased to 18, 16, and, 7 percent, respectively. The presence of the rapidly settling sand grains several meters above the floor of the canyon at depths of 200 m and 300 m indicates a significant intensification of bottom currents capable of resuspending and transporting the canyon sediments. The fact that the TSM concentrations near the bottom over the canyon walls and the shelf were substantially lower than those present above the canyon floor provides additional evidence for localized resuspension.

Although the bottom current meter at 223 m in the canyon axis was ideally located to examine the resuspension process, the sensor failed seven days after deployment and two days prior to our water sampling. The recovered data show that several days before we sampled the suspended matter, the head of the canyon was swept by reversing flows of tidal period with peak speeds of 25-35 cm/sec (Figure 3-6). To our knowledge there are no comparable data on bottom currents in the canyon head for other time periods. Consequently, we do not know whether the current speeds that we measured are normal or unusually high. Bottom grab samples taken during our survey recovered silty, noncohesive fine to medium sands near the current meter site. Available information on bed scour suggests that currents of at least 25-35 cm/sec (at 3 m above bottom) would be required to initiate bed movement and fine sediment resuspension (Sternberg 1972; White 1970). Our data suggest that currents of the proper velocity range occurred in March, and in light of the presence of very fine sand several meters above bottom, the possibility exists that the actual currents were substantially greater when we collected our water samples.

A progressive vector diagram of the flow at 223 m demonstrates a high degree of asymmetry in the oscillating canyon floor currents (Figure 3-7). The net flow was decidedly down the canyon, and if taken at face value, an advection averaging 7 cm/sec occurred over a period of six days. If we assume that this characteristic of the currents had not changed when we sampled the canyon water, the net downcanyon transport of suspended matter would be approximately 20 kg/m² /day. Attempting a more accurate calculation is meaningless because we cannot determine the spatial gradients in the bottom currents or suspended matter concentration. Nevertheless, substantial amounts of sediment were clearly resuspended and

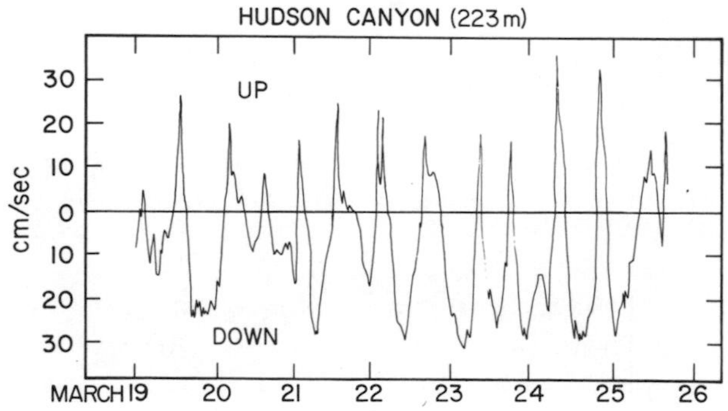

Figure 3-6. Current velocity 3 m above bottom at 223 m in upper Hudson Canyon. Note: Strong down- and upcanyon flows occurred for four days prior to sensor failure on March 26.

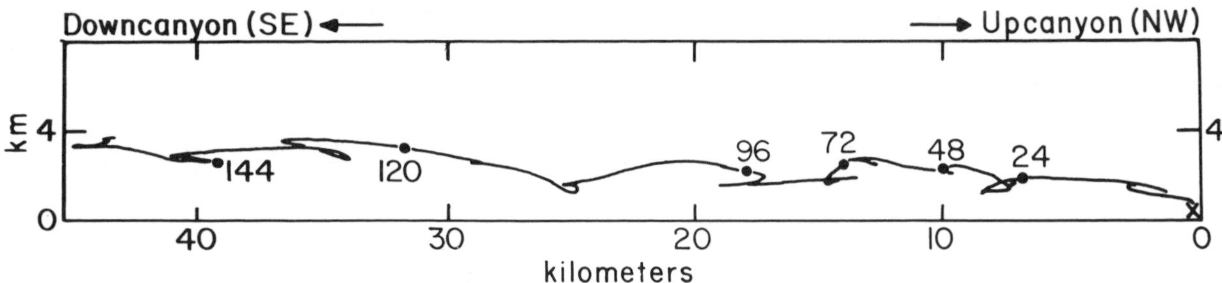

Figure 3-7. Progressive vector diagram based on data from 223 m, 3 m above bottom (see Figure 3-6). Note: Net flow averaged nearly 7 cm/sec downcanyon for almost seven days.

moved to deeper water by the canyon-controlled currents.

The gradual downcanyon decrease in TSM concentrations from 3,400 μg/1 at 200 m to 603 μg/1 at 893 m combined with the textural fining of the suspended terrigenous particles suggests that on the order of 80 percent of the sediment dropped from suspension within the upper Hudson Canyon.[b] Furthermore, these data demonstrate a significant reduction in the ability of the canyon currents below about 400 m to resuspend material or maintain particles in suspension.

We did not collect current meter data at deeper locations in Hudson Canyon during March 1974, but F. P. Shepard successfully measured the bottom currents at 1,254 m for twenty one days in April 1974 (see Chapter 2). Shepard's results show that peak velocities rarely exceeded 20 cm/sec, and net flow (Figure 3-8) was downcanyon (Shepard and Marshall 1976). As is noteworthy, the reversing currents at 1,254 m exhibit relatively long intervals during which speeds fall below 5 cm/sec in contrast to the abrupt current reversals typical of the canyon head. This aspect of the deeper currents would allow deposition of the more rapidly settling particles in the near-bottom water. Once deposited, subsequent current pulses of 20 cm/sec (3 m above bottom) probably would not achieve complete resuspension, and only the slowly settling components of the suspension would continue downcanyon.

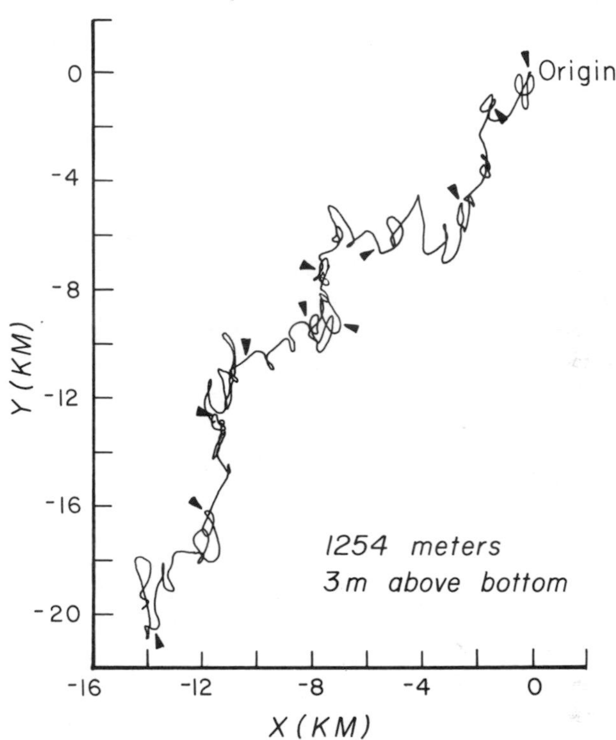

Figure 3-8. Progressive vector diagram for the current 3 m above bottom at 1,254 m in upper Hudson Canyon. Note: Arrows show twenty four-hour time marks beginning on April 11, 1974. The net drift was downcanyon (SE) at about 1-2 cm/sec. (Source: Data were kindly furnished by F.P. Shepard, Scripps Institution of Oceanography, and D.A. Cacchione, U.S. Geological Survey.)

Sedimentation Rates of Hudson Canyon Muds

The foregoing results along with earlier work reported by Keller et al. (1973) strongly suggest that sediments resuspended in the head of the canyon (<300 m) were transported seaward, but a significant portion of the material was deposited, beginning at canyon axis depths of 400-450 m. Keller and his associates made a unique set of direct observations and measurements in upper Hudson Canyon by using the submersible *Alvin*. Dives in the canyon axis at depths less than ~300 m revealed extensive exposures of bare rock along the canyon walls, evidence of headward erosion and rock sliding, and canyon floor sediments composed of fine to coarse terrigenous sands containing minor amounts of silt and clay. Below a canyon axis depth of about 400 m the character of the surficial bottom sediments changes dramatically to silty clays and clayey silts rich in organic

[b] The term *upper* Hudson Canyon is used herein to denote that portion of the canyon which is <1,000 m deep, and *canyon head* is arbitrarily defined to mean that part of the canyon floor less than 300 m deep.

matter. The shear strength of these muds proved to be exceptionally low, and visual appearances indicated that they would be especially susceptible to bottom current erosion. Surprisingly, during one dive, the bottom current reached 27 cm/sec with no discernible effect on the mud sediment. The sediments may have developed an *in situ* cohesiveness that was not accurately depicted by shear strength measurements. Alternatively, the difference in effective boundary shear stress produced by equivalent currents over the roughened canyon head sands and the relatively smooth muds may explain the lack of erosion.

In order to examine the time-dependent aspects of sedimentation in this zone of mud deposition, we collected a 5 m piston core at 430 m. Care was taken to insure that the ship was above the axis of the canyon and that this position was maintained during the coring operation. On the basis of our observations during the field sampling, we suspected that some portion of the surface sediment, which Keller et al. (1973) had described as "fluffy," had been lost. However, subsequent inspection of the core showed loose, highly organic, watery sediments from the surface to about 30 cm and a radiocarbon age of 665 ± 105 at 50 cm indicated a minimal loss of surface sediment (probably <20 cm).

Figure 3-9 presents a visual description of the core along with the results of radiocarbon dating of 5 cm sections centered at 50 cm, 250 cm, and 505 cm (the bottom of the core). Because the core underwent several other analyses that required relatively large amounts of sediment, using total carbon for the age dating was necessary.

We have not detected any lithologic changes in the core that might reflect major variations in sediment accumulation, slump-in, or loss of section. The core is predominantly composed of intensely bioturbated, olive gray and dark gray clayey silt with scattered 1 to 2 cm mollusc shells. Thin layers of silt and very fine sand are present at 186 cm, 423 cm, and 431 cm. Poorly developed laminations of silt and clay are present at 324 cm to 330 cm, and these are the only sedimentary structures in the core, as confirmed by radiographs. Although the activities of benthic burrowers undoubtedly destroyed some other laminated sections and thin sand beds, the overall textural uniformity and fineness of the sediment are indicative of persistent low-energy depositional conditions. However, silt and clay deposition was interrupted on at least three occasions by events of higher energy that resulted in deposition of thin (<1 cm) fine sand layers. The fact that medium- and coarse-grained terrigenous sands are not important components of the core strongly indicates that transport by high-velocity turbidity currents has been negligible in the Hudson Canyon over the time period represented by our core (~6,300 years). Nevertheless, our suspended matter samples at 400 m and 450 m contained substantial amounts of medium and coarse silt particles that were probably being advected down-canyon. We suspect that silt laminations (with a minor amount of very fine sand) would be common in our piston core if it were not for the intense mixing activities of burrowing organisms.

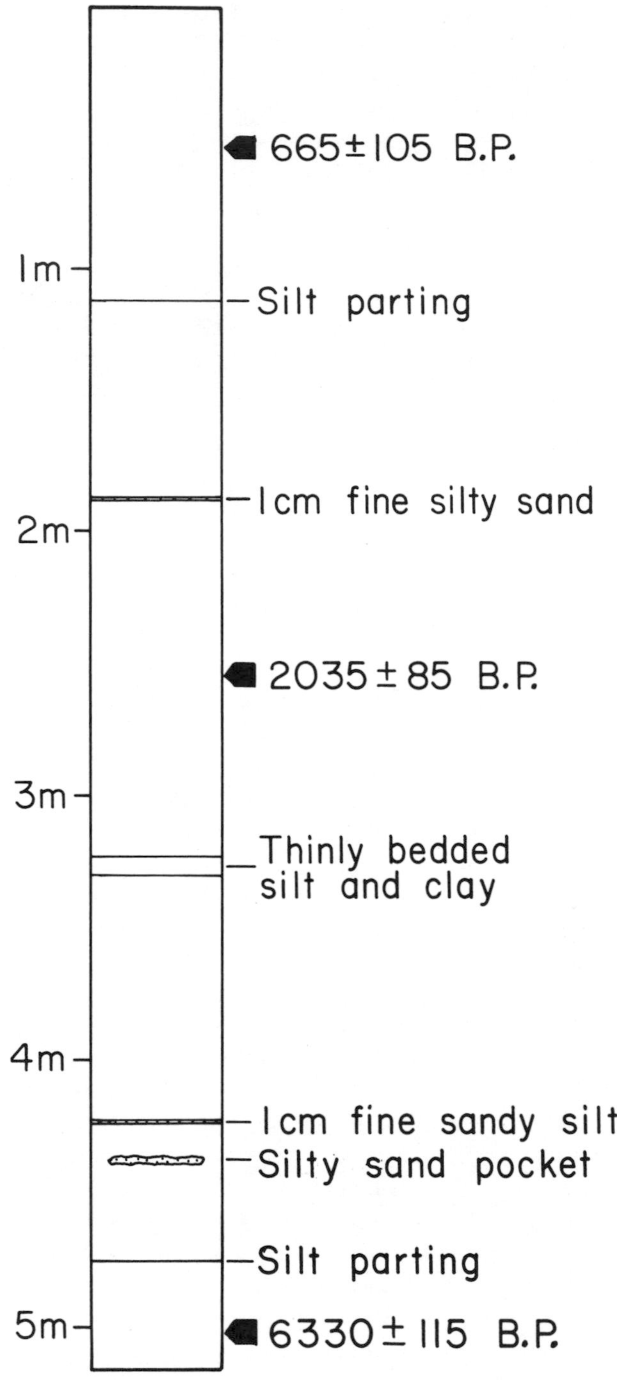

Figure 3-9. Lithology of piston core HM-3 recovered at 430 m in the axis of Hudson Canyon. Note: Intervals at 50 cm, 250 cm, and 505 cm were dated by using radiocarbon analysis of total organic carbon.

The radiocarbon dates show that an environment of silt and clay deposition has persisted just seaward of the canyon head for at least 6,300 years. The radiocarbon ages also suggest an approximately twofold increase in the sedimentation rate from 2,000 B.P. to the present (see Figure 3-9). Analyses are underway to determine whether this apparent change is real or simply a function of compaction. In any case, the average rate of sedimentation over the past 6,300 years has been on the order of 70-80 $cm/10^3$ years or five to ten times faster than rates measured on the nearby continental slope and rise (Turekian 1968; Doyle et al. 1976). The distribution of canyon axis sediments observed by Keller et al. (1973) suggests that our core is probably representative of the shallow end of a lens of mud extending from depths of about 400 m to at least 1,000 m.

DISCUSSION

A comparison of the basic characteristics of the suspended matter during our March survey with data presented by Manheim et al. (1970), Meade et al. (1975), and Biscaye and Olsen (1975) for the outer shelf and upper Hudson Canyon reveals substantial differences in both the surface and near bottom waters (Table 3-1). The sets of data of Manheim et al. (1970) and Meade et al. (1975) are based on samples collected during fair weather conditions in early and late summer. Manheim and his colleagues restricted their sampling to the surface waters where they found <125 μg/1 of TSM; combustion analyses showed that this material contained 80-90 percent amorphous organic matter. Meade et al. (1975) occupied four stations in Hudson Canyon and three stations within 100 km along the shelf break and upper slope. Their data for surface water TSM and an additional set of data collected in July 1975 by Biscaye and Olsen (1975) are in agreement with the results of Manheim et al. (1970).

In contrast, TSM concentrations in the surface waters during our survey ranged from 285 to 721 μg/1. The ash weight data (see Table 3-1) suggest the TSM difference was caused by a significant increase in noncombustible components in March 1974. We believe that this additional material, largely composed of grain aggregates, was resuspended from the bottom over the shelf by a storm that moved through the area on March 21 and produced moderately heavy seas (2-3 m) from March 21 to March 28 when conditions calmed sufficiently to allow water sampling. Relatively calm seas prevailed for about thirty six hours (March 28-30) before a second storm front forced curtailment of deck operations. The near-bottom currents that are associated with such winter storms in the Middle Atlantic Bight can exceed 50 cm/sec at depths of 50-100 m (Beardsley and Butman 1974), and Butman (personal communication, 1976) has obtained bottom photographs that provide direct evidence of shelf sediment resuspension by winter storm currents. The frequent storms that are a well-known characteristic of the North Atlantic from November through March must account for a significant proportion of the sediment transport on the outer shelf.

As expected, TSM and ash residue concentrations within 5 m of the shelf bottom exceeded previous measurements substantially (see Table 3-1). The most comparable sets of data (in terms of station density) were reported by Biscaye and Olsen (1975). These workers found 100-400 μg/1 within 10 m of the bottom over the shelf adjacent to Hudson Canyon during October 1974 and July 1975. Our "storm-influenced" TSM concentrations exceed this range by 50 to 100 percent.

Suspended matter samples have been collected in Hudson Canyon by Meade et al. (1975), Biscaye and Olsen (1975), and indirectly by Amos et al. (in press) by using a moored nephelometer and current meter at 803 m. Meade et al. (1975) reported TSM concentrations of 190 and 260 μg/1 at 320 m and 345 m in the canyon (within 5 m of bottom), and Biscaye and Olsen (1975) have shown that a lens of turbid water typically occupies the upper Hudson Canyon. However, the values observed by the latter researchers (see Table 3-1) are well below those present from the head of the canyon (>3,000 μg/1) to a depth of 900 m (603 μg/1) in March 1974. In fact, to our knowledge concentrations >2,000 μg/1 have not been measured previously seaward of the inner shelf (<40 km from shore) in the Middle Atlantic Bight. The results of the present study demonstrate that this situation is mainly a product of how and when we sample the shelf water column. Sediment entrainment and transport are highly nonlinear functions of the flow regime, and as is important to realize, fairweather data are likely to give a distorted picture of these processes.

Clearly, large variations occur in the amounts of material in suspension near the floor of upper Hudson Canyon. These changes probably reflect equally important variations in the velocity of the canyon currents and their ability to redistribute sediments. Unfortunately, while we believe that the storms that influenced the surrounding shelf also to some degree intensified the reversing currents in the head of the canyon, we have no current meter data from other time periods to form a basis for comparison. The present lack of long-term (one month or more) bottom current and suspended matter measurements at various positions in submarine canyons precludes an adequate understanding of the timing and intensities of sediment transport events.

On the other hand, the longer-term aspects of the sedimentary regime in upper Hudson Canyon are indicated by the distribution and accumulation rates of the postglacial sediments. The apparently abrupt change in sediment

Table 3-1. *Suspended Matter: Hudson Canyon Area*

Component	*Survey Time*	*Source*	*Surface Water*	*Near Bottom*	
				Shelf	*Canyon*
TSM[a] (μg/1)	6/65	1	<125		
	9/69	2	50-100	200-920[c]	190-260
	10/74	3		200-400	400-600
	7/75	3	<100	100-200	200-400
	3/74	4	285-721	241-752	603-3440
ASH (μg/1)	6/65	1	<20		
	9/69	2	<10	50-200	130-250
	3/74	4	33-380	210-680	520-3110
CSM[b] (%)	6/65	1	80-90		
	6/69	2	80-95	30-40	4-30
	3/74	4	50-90	18-34	12-23

[a]TSM = total suspended matter.

[b]CSM = combustible suspended matter.

[c]The value of 920 μg/1 was present 100 km NW of Hudson Canyon near the location of relict clayey silt bottom sediments. The bulk of the data (Meade et al. 1975) indicated a mean concentration of 300-400 μg/1 near Hudson Canyon.

Sources: 1, Manheim et al. (1970); 2, Meade et al. (1975); 3, Biscaye and Olsen (1975); and 4, This Study.

texture from relatively clean coarse sands in the head of the canyon to clayey silts at 350 m to 450 m coupled with the evidence from our piston core that suggests the minor role of high-energy turbidity currents at 430 m, leads to two conclusions. First, high-velocity turbidity currents have not been a significant transport agent for the past several thousand years; and secondly, the energy associated with canyon currents that oscillate in response to tides and internal waves (Shepard et al. 1974) seems to be stratified with respect to depth. The accumulation of mud in Hudson Canyon begins at approximately the same depth as the base of the permanent pyncnocline: 400 m to 500 m (Drake, unpublished data; Gordon et al. 1976). Although our interpretation must be regarded as speculative without longer current meter records at many depths in the canyon, a possibility is that the flow intensification caused by winter storms may extend to the pyncnocline but diminish rather rapidly below that level (400 m to 500 m). Simultaneous current measurements at 200-300 m and 500-600 m during the winter period would resolve this question.

The current meter record at 223 m reveals a strong net flow downcanyon. This type of current, superimposed on the up- and downcanyon oscillations, has been observed in many canyons (Shepard et al. 1974; Shepard and Marshall 1973) and remains one of the most important unexplained aspects of the flow regime. Recently, Inman et al. (1976) summarized many years of observations in the head of the La Jolla-Scripps Canyon systems. Their analysis suggests that periods of canyon current intensification are correlated to periods of intensified shelf water circulation during storms. They conclude that resuspension of sediment occurs in and around the head of the canyon as the velocities of the oscillating currents increase. If this process of flow intensification continues, enough sediment is placed in suspension to generate a turbidity current that can overcome the density stratification of the water column and produce strong downcanyon flow of up to several meters per second.

Important components in this model are wave and wind setup and topographically controlled circulations caused by edge waves. All of these factors diminish in importance as depth and distance from shore increase. Consequently, in the case of Hudson Canyon, the storm-related currents may be sufficient to stir the sediments in the head of the canyon, but incapable of producing the suspended sediment concentrations required for high-velocity, through-going turbidity flows.

Although this model appears to be compatible with the data we have presented, there are complications. For example, Cacchione et al. (Chapter 4) report a three-day observation of bottom currents in lower Hudson Canyon (~3,400 m) that shows essentially unidirectional downcanyon advection; visual observation from *Alvin* showed no unusual concentrations of suspended matter in the water column. As is quite apparent, the canyon current system is too complex to be properly understood on the basis of bits of data from different depths and different times.

While the exact cause of downcanyon near-bottom flow is not understood, there can be no doubt that this feature of the current system produces an important flux of sediment along the canyon axis. In particular, the distribution of organic carbon in surface sediments on the continental slope near Hudson Canyon (Figure 3-10) demonstrates the importance of the canyon as a supplier of organic matter to the deep sea (see Chapter 4).

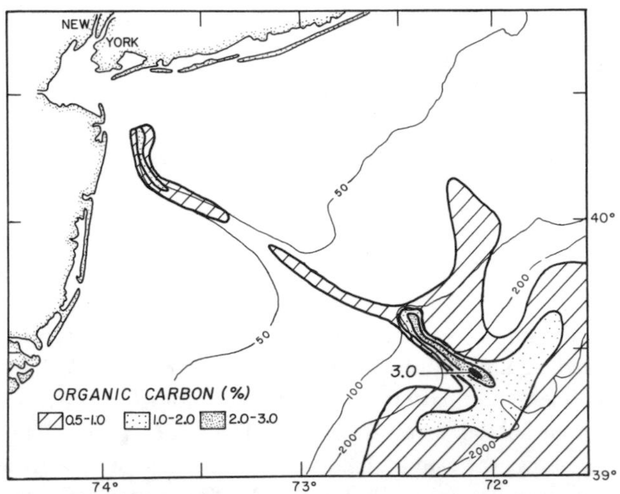

Figure 3-10. Organic carbon content of surface sediments in upper Hudson Canyon and on adjacent shelf and slope. Note: Depth contours are given in meters.

In summary, our results indicate that Hudson Canyon is both a sink and a transport conduit for sediment introduced to the head of the canyon. The fate of the suspended matter in the canyon appears to be a function of particle settling rates and depth stratification of current energy; rapidly settling particles (sand, silt and aggregates of clay and organic matter) are largely retained in the upper canyon (<1,000 m), whereas the more slowly settling particulate matter remains in suspension and is carried further through the system. Storm intensification of the normal reversing canyon currents may be an especially important factor in the transport regime. Finally, we believe that substantial progress in the study of present-day mechanisms of sediment transport in canyons will only be achieved when the proper long-term observations of currents and suspended matter concentrations become available.

ACKNOWLEDGMENTS

The authors express appreciation to F. P. Shepard, Scripps Institution of Oceanography, for making copies of his current meter results available; to Gary G. Sullivan, Scripps Institution of Oceanography, for help in the field; and to Otis Finley and Helen Gibbons for laboratory work and preparation of illustrations. The authors also thank Pierre Biscaye and Curtis Olsen, Lamont-Doherty Geological Observatory, for helpful discussions of their research in the Middle Atlantic Bight.

This research was supported by the NOAA COMSED project. The senior author participated in this investigation while he was supported by a NRC postdoctoral grant at the Atlantic Oceanographic and Meteorological Laboratories, NOAA, Miami, Florida.

REFERENCES

Amos, A.F., Baker, T., and Daubin, S., Jr. Near bottom currents and sediment transport in Hudson Canyon. *J. Geophys. Res.* (in press)

Beardsley, R.C., and Butman, B.,1974. Circulation on the New England continental shelf: Response to strong winter storms. *Geophys. Res. Lett.*, 1: 181-84.

Biscaye, P.E., and Olsen, C.R., 1975. Suspended particulate concentrations and compositions in the New York Bight. *Proc. Special Symposium on New York Bight.* Amer. Museum Nat. History, New York, Nov. 1975.

Doyle, L.J., Woo, C.C., and Pilkey, O.H., 1976. Sediment flux through intercanyon slope areas: U.S. Atlantic continental margin (Abst.) *Geol. Soc. Am. Prog.*, 8:843.

Gordon, A.L., Amos, A.F., and Gerard, R.D., 1976. New York Bight water stratification–October 1974, In: M.G. Gross (ed.), *Middle Atlantic Continental Shelf and the New York Bight.* Amer. Soc. Limn. Oceanogr., Lawrence, Kans., pp. 58–68.

Inman, D.L., Nordstrom, C.E., and Flick, R.E., 1976. Currents in submarine canyons: An air-sea-land interaction. *Ann. Rev. Fluid Mechanics*, 8: 275-310.

Keller, G.H., Lambert, D., Rowe, G. and Staresinic, N., 1973. Bottom currents in the Hudson Canyon. *Science*, 180:181–83.

Manheim, F.T., Meade, R.H. and Bond, G.C., 1970. Suspended matter in surface waters of the Atlantic Continental Margin from Cape Cod to the Florida Keys. *Science*, 167: 371-76.

Meade, R.H., Sachs, P.L., Manheim, F.T., Hathaway, J.C., and Spencer, D.W., 1975. Sources of suspended matter in waters of the Middle Atlantic Bight. *J. Sed. Petrol.*, 45: 171-88.

Shepard, F.P., and Marshall, N.F., 1973. Currents along floors of submarine canyons. *Amer. Assoc. Pet. Geol. Bull.*, 57: 244-64.

_____, and Marshall, N.F., 1976. Currents in submarine canyons and their relation to deep-sea fans (Abst.). *Amer. Assoc. Pet. Geol. Bull.*, 60: 721-22.

_____, Marshall, N.F., and McLaughlin, P.A., 1974. Currents in submarine canyons. *Deep-Sea Res.*, 21: 691-706.

Sternberg, R.W., 1972. Predicting initial motion and bedload transport of sediment particles in the shallow marine environment, In: D.J.P. Swift, D.B. Duane, and O.H. Pilkey (eds.), *Shelf Sediment Transport: Process and Pattern*. Dowden, Hutchinson & Ross, Stroudsburg, Pa., pp. 61–82.

Turekian, K.K., 1968. *Oceans.* Prentice-Hall, Englewood Cliffs, N.J., 120 pp.

White, S.J., 1970. Plane bed thresholds of fine grained sediments, *Nature*, 228: 152-53.

Chapter 4

Submersible Investigation of Outer Hudson Submarine Canyon

DAVID A. CACCHIONE

U.S. Geological Survey
Menlo Park, California

GILBERT T. ROWE

Woods Hole Oceanographic Institution
Woods Hole, Massachusetts

ALEXANDER MALAHOFF

National Oceanic and Atmospheric Administration
Rockville, Maryland

ABSTRACT

Data collected in the submersible *Alvin* during four dives into outer Hudson Canyon reveal that active erosion of the canyon walls and floor is occurring at water depths between 2,900 m and 3,600 m. Visual observations and sedimentary analyses of shallow cores obtained by *Alvin* indicate that strong bottom currents and benthic organisms have reworked the surface sediments of the canyon floor. Moats around the bases of erratic boulders and undercut bases of the rock walls provide additional evidence that bottom currents, flowing in a downcanyon direction, erode the canyon sediments and cut back the walls. This active erosion by strong bottom currents is probably an episodic process, because current measurements obtained on *Alvin* and with two moored current meters during the four-day dive period were consistently weak (speeds less than 15 cm/sec). Extensive burrowing, prevalent at each dive site, has probably contributed significantly to the erosion of the canyon walls.

INTRODUCTION

During August 13-16, 1975, four research dives by the deep submergence research vehicle (DSRV) *Alvin*, assisted by the surface support vessels R/V *Lulu* and R/V *Subsig*, provided unique information on the bottom morphology, sedimentation patterns, currents, and biological activities at four bottom localities in outer Hudson Canyon. Figure 4-1 shows the canyon location; the four dive sites (Figure 4-2) were between 2,900 m and 3,500 m water depth where the canyon incises the upper continental rise about 340 km southeast of New York City. One bottom mooring, instrumented with two vector-averaging current meters,

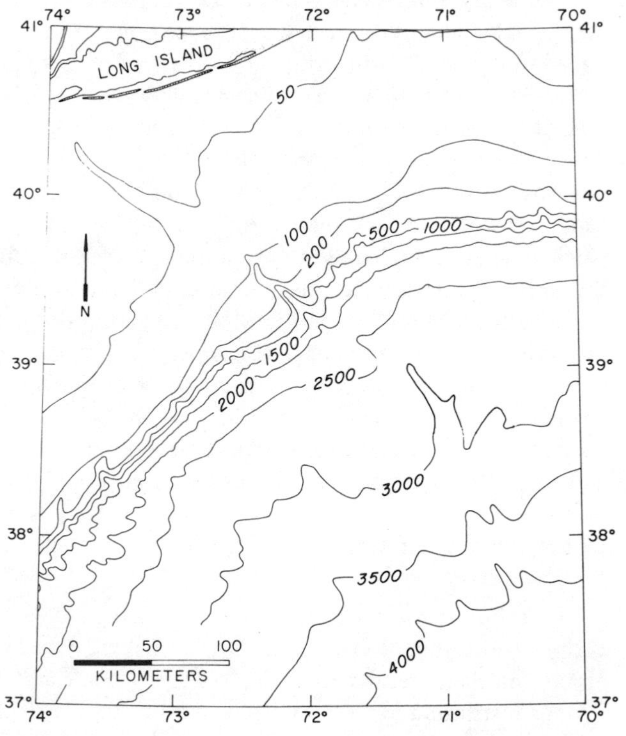

Figure 4-1. Bathymetric chart of the continental margin off eastern United States showing the geographic location of Hudson Canyon. Note: Depth contours are given in meters. (Source: Adapted from Uchupi 1968.)

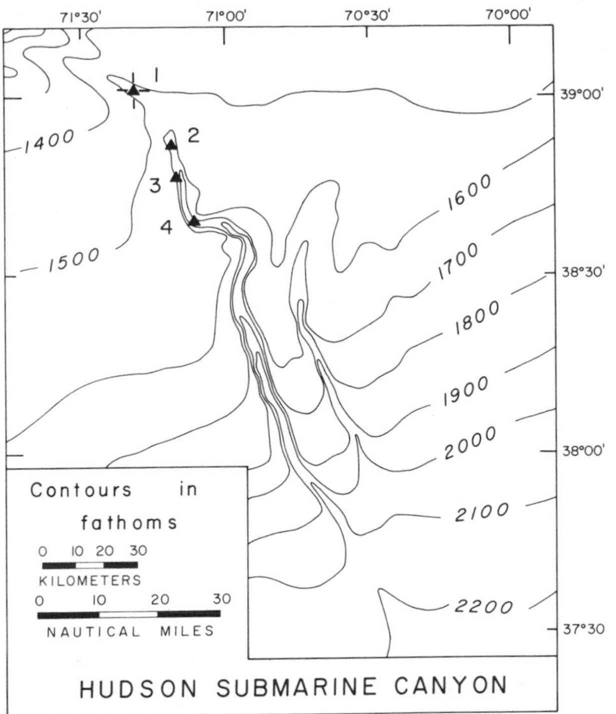

Figure 4-2. Dive site locations, numbered 1 through 4, in outer Hudson Canyon. Note: Dive site 1 was also the location of a single bottom mooring with two current meters at 7 m and 100 m above the canyon floor.

was deployed from R/V *Lulu* at dive site 1 and recovered after the last dive. The cruise was terminated after dive 4 on August 16, 1976, due to technical problems with *Alvin.*

Many submarine canyons extend from the coastal regions to the deep sea as continuous channels that are easily recognizable on bathymetric charts like that shown in Figure 4-1. The roles that these canyon pathways have in the exchange of sediments, nutrients, and pollutants between the shallow and deep waters of the oceans are not well understood.

The bathymetric charts in Figure 4-1 and 4-2 show that although Hudson Canyon extends seaward for over 500 km, the morphology of the canyon exhibits significant changes along its course. The steep walls and high relief that characterize the canyon topography on the continental shelf and upper continental slope diminish substantially beyond the base of the lower continental slope in water depths of 2,000 m to 3,000 m. However, at depths of about 3,000 m, the vertical relief of the canyon on the continental rise increases abruptly to about 400 m (see Figure 4-2), or a value that is comparable to those found on the outer continental shelf. The causes for these changes in canyon shape along its path are not known, but the topographic contrasts probably reflect distinct differences in the physical processes that control the sedimentation patterns at different depths within the canyon.

Recent investigations of bottom sediments in the shallow sections of Hudson Canyon (see Chapter 3) show that thick accumulations of organic-rich muds cover a large portion of the canyon floor. Submersible observations by Keller et al. (1973) corroborate the results. Although current meter data (Keller et al. 1973; see also Chapter 2 of this volume) suggest a net downcanyon flux at these relatively shallow sites, it is not known if the very fine-grained bottom and suspended materials are actively being transported downcanyon of if they are accumulating in the deeper sections. One purpose of the *Alvin* dives was to collect data that would help to answer these questions.

Ericson et al. (1961) and others suggest that turbidity currents have been a principal mechanism responsible for the origin of Hudson Canyon. The cores described by Ericson et al. (1961) taken from Hudson Canyon contained graded beds, displaced shallow-water organisms, and thin layers and lenses of gravel and pebbles, which are features characteristic of turbidite deposits. Several of these cores were obtained at locations near the *Alvin* dive sites (see Figure 4-2). These *Alvin* dives provided a unique opportunity to investigate the surface properties of turbidity current deposits and to examine their subsequent modification by more recent sedimentary and biological processes.

METHODS

The submersible *Alvin* has an elaborate array of special tools that can be manipulated by a mechanical arm to sample the local environment. During these dives, shallow cores (about 30 cm in length) of the near-surface sediments were obtained with tube and box coring devices. Photograph A in Figure 4-3 shows *Alvin's* arm extracting a tube core during one of the dives. Small sediment scoops were used to collect surface concentrations of pebbles and cobbles. Loose rocks and pieces of outcrops were recovered with the mechanical arm. A simple cone penetrometer was used to make *in situ* measurements of the bearing strength of the sediment surface. After the cone was released by *Alvin's* arm, the depth of penetration was observed and logged.

A specially constructed dye pellet device was used at several locations to measure the current velocity profile from the canyon floor to 1 m above the floor. The instrument consists of a series of dye pellets (made from Fluorescin dye) that are attached at regular intervals, typically 20 cm, to a thin nylon line (photograph B, Figure 4-3). This line is connected at the bottom to a weighted rubber tube and at the top to a buoyant syntactic float. Prior to each dive, the line with dye pellets was loaded into the rubber tube, and the tube was filled with kerosene to

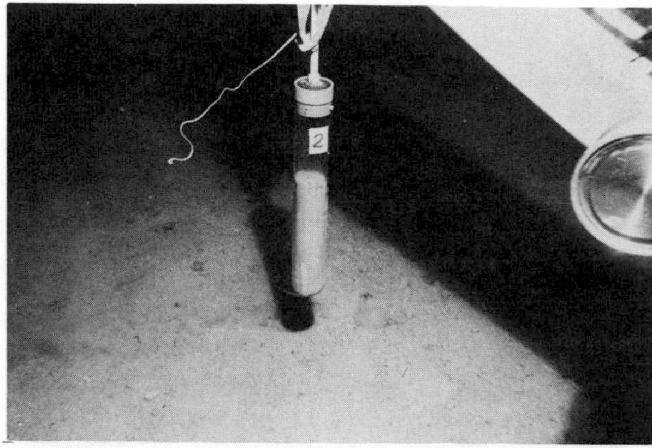

A

B

Figure 4-3. Data-collecting devices on the submersible *Alvin.* Note: Photograph A, taken from the port viewport on *Alvin,* shows *Alvin's* arm recovering a tube core at dive site 2 in 3,077 m water depth. The string-like material caught in the jaw section of the arm was not identified. Photograph B shows *Alvin's* dye pellet device fully extended. The pellets are located about 20 cm apart along the thin nylon line; the rubber tube at the bottom is attached to a metal cylinder with a hose clamp; the nylon line is attached to the metal inside the tube; and a syntactic foam float is attached to the nylon line at the top. *Alvin* is in the background resting in the skids of R/V *Lulu.*

compensate for the high ambient pressures at depth. The syntactic float was then used to plug the open end of the rubber tube.

On the bottom, *Alvin's* mechanical arm deployed the dye pellet device by squeezing the rubber tube and eventually freeing the syntactic float so that it could moor the dye pellet line vertically. After the device was deployed, *Alvin* was repositioned approximately 3 m distant and oriented such that the line of sight between *Alvin* and the dye pellets was normal to the local bottom current. The dye diffused from the pellet as a stream of small vortices that were advected downcurrent by the local flow (photograph A, Figure 4-4). The initial sizes of the vortices were determined by the dimensions of the pellets (all are about 0.5 cm). Each vortex retained its identity for distances of about 1 m to 2 m from the origin, depending on local flow speed and turbulence, before enlarging, mixing, and becoming indistinct. Fish often interfered with the dye streams as shown in photograph B in Figure 4-4. Distances travelled by individual vortices at each dye level over known time intervals were measured from 16 mm motion pictures and used to compute speed profiles over the sea floor.

A single current meter mooring deployed from R/V *Lulu* at dive site 1 on August 13, 1975, provided continuous records of temperature and current velocity at 7 m and 100 m above the sea floor for about four days. This mooring was visually inspected *in situ* to verify that it was in fact situated in the canyon axis, away from large local topographic irregularities. The onsite inspection also verified that the rotors on each current meter were turning at a rate commensurate with the local current speed.

RESULTS

Surface Sediments

The extremely thin cover of unconsolidated sediment that was observed on each dive is in marked contrast to the thick accumulations of fine-grained sediments in shallower sections of Hudson Canyon reported by Drake et al. (see Chapter 3). In many instances the bases of the canyon walls had been undercut by strong bottom currents, and large, irregular pieces of the overhanging wallrock had broken free and come to rest on the adjacent floor. The exposed boulders found in the canyon axis generally had moats along the downcanyon sides of their bases. Often the moats were littered with pebbles and cobbles of various shapes and colors (photograph A, Figure 4-5). Pebble composition is highly variable; many have probably been transported from continental sources.

The canyon floor generally consists of a silty sand that has been disturbed by strong bottom currents and benthic organisms (bioturbation consisting of tracks, burrows,

A

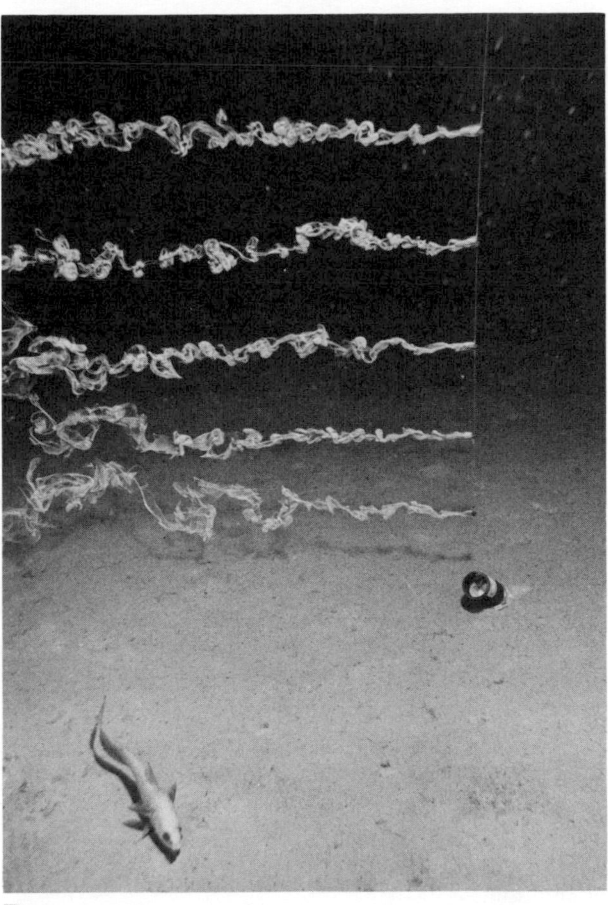

B

Figure 4-4. Dye development by the submersible *Alvin.* Note: Photograph A shows streams of Flourescin dye emanating from the dye pellets at dive site 2 in 3,050 m water depth. Photograph B shows unexpected disturbances to the dye streams by a fish (*Nematonurus*) of the rat-tail variety at dive site 1 in 2,940 m water depth. Both photographs were taken from the port viewport in *Alvin.*

mounds). Small, asymmetric sediment ripples having heights of about 1 cm to 2 cm and wave lengths of about 25 cm to 35 cm were observed during dives 2 and 3 in the narrower part of the canyon. The occurrences of sediment ripples were infrequent or "patchy" and were generally confined to areas downcanyon of larger topographic irregularities. The slip faces on the ripples are usually found on the downcanyon side, thereby indicating sediment motion in a downcanyon direction. Infauna and organisms at rest or moving across the sediment surface might account for the obliteration of older zones of sediment ripples.

In contrast to the graded beds that would be deposited by turbidity currents, the near-surface sedimentary units consistently had an inverse size grading. Except for core number 2 taken during dive 2, the size analyses shown in Table 4-1 indicate a coarse to fine trend from surface to bottom in the cores. The silt and sand fraction in these cores is subangular to subrounded and consists mainly of transparent quartz. Frosted and rose-colored quartz, feldspar minerals, and metamorphic fragments are also present in smaller quantities. Traces of mica, glauconite, and manganese casts are also found.

At dive sites 1 to 3 lineated patterns of pebbles and other coarse materials were found on the surface (photograph B, Figure 4-5). Generally these linear trails of pebbles oriented parallel to the canyon axis were located near the canyon walls or in the vicinity of the large rock exposures in the axis. These pebbles have a dark coating of manganese, less than 3 μ thick on several samples. Isolated pebbles without manganese coatings were also found at several depths in four of the cores.

Rock Outcrops

Exposures of rocks along the walls generally show beds that dip gently seaward ($<3°$). Wall slopes often exceed 30°; vertical cliff faces and overhangs were encountered on several occasions.

In general the canyon walls at each dive site consist of a series of narrow terraces of low gradient that are terminated by rough scarps whose slopes can be nearly vertical. The

Table 4-1. *Percent Sand in* Alvin *Cores Collected in Hudson Canyon*

Dive/Depth (m)	*Core*	*Depth in Core (cm)*	*% Sand (by weight)*
1/2940	1	0	66.6
	1	21	56.6
	2	0	77.9
	2	17	49.3
2/3160	1	0	57.3
	1	10	36.2
	1	18	20.7
	2	0	59.9
	2	2	80.0
	2	4	40.6
	3	0	62.3
	3	12	54.6
	4	0	38.9
	4	12	27.1
3/3240	1	0	65.3
	1	11	50.0
	1	20	33.4
	2	0	–
	2	12	44.2
	2	21	32.7
	3	0	61.1
	3	12	44.6
	4	0	71.8
	4	12	69.4

scarp faces are angular and often extend only for short distances in any one direction before abruptly changing orientation.

Two distinct sedimentary rock types were found during the traverses up the canyon walls. A dense, carbonate-poor, whitish-gray claystone of Paleocene to Eocene age (W. Berggren, personal communication, 1975) forms a considerable portion of the canyon wall. This sequence is probably the Eocene marl described by Heezen and Hollister (1971). The other major rock type is a moderately consolidated, gray-brown, sandy siltstone that was not as hard as the claystone. The siltstone appears to underlie the whiter claystone unit; however, on terraces at dive sites 1 and 3, a jumbled mixture of large blocks of each of these formations was found along the base of the scarp. Recent sediment cover and the complexity of the severely fractured rock sequences greatly obscured the stratigraphic relationships along the exposures.

Rock exposures that are either outcrops or large erratic boulders (up to 5 m in height) protrude above the canyon floor sediments at each of the dive sites. Large angular to subangular pieces of these rocks are strewn about the base of the exposures, which thus suggests their moderate resistance to erosion by currents.

At the base of the canyon walls and along the base of the terrace scarps, large fragments of the wall rock lie on the sediment surface, often with no sediment cover. Small talus piles of rock and other debris were observed along the wall base in many locations.

Currents

During the dives, current speeds measured with the rotor speed sensor on *Alvin* and estimated from movements of suspended materials over the bottom never exceeded 15 cm/sec. In the canyon axis near the bottom, current directions were downcanyon; weak, crosscanyon flows were measured at four locations on wall terraces.

Persistent downcanyon flow is also indicated in the current meter data. Figure 4-6 has hourly averages of the four-day record displayed as current vectors or "sticks." At no time is the flow in an upcanyon direction. The semidiurnal internal tide is only effective in modulating the current speed; it does not produce the flow reversals reported by others (Shepard et al. 1974). One-minute

A

B

Figure 4-5. Surface sediments in lower Hudson Canyon. Note: Photograph A shows a moat surrounding a boulder in the axis of the canyon at dive site 2. The boulder is a light-colored claystone probably derived from the Eocene formation forming a large section of the canyon walls. A beverage can is in the moat at right side. Water depth is 3,030 m. Photograph B shows linear rows or streams of pebbles (lower right) oriented downcanyon and lying near a rock outcrop at dive site 3. Pebbles are between 0.5 cm and 3 cm in diameter, subangular to subrounded, and often coated with a thin (<3 μ) layer of manganese. Their composition is highly variable; many are pieces of the canyon wall rock. Water depth is 3,210 m.

averages of the current measurements (not shown) also contain no occurrences of upcanyon flow. Mean currents at 7 m and 100 m above the canyon axis were 7.7 and 7.6 cm/sec, respectively; maximum currents at these levels were 14.1 and 19.7 cm/sec respectively.

Figure 4-7 is a progressive vector plot of the current meter data. A consistent downcanyon flow of about 7 km/day is readily apparent. The approximate cross-sectional area contained within the canyon walls at dive site 1 is about 2.3 km^2. Assuming an average rate of 7 km/day, a volume flux of 16.1 km^3/day or 0.19 x 10^6 m^3/sec in a downcanyon direction is derived for the duration of this experiment. This steady seaward flow, if representative of longer durations, would produce a very significant material flux to deeper zones. The relatively large quantity of macroscopic material observed in suspension within the canyon during the dives would be subject to downcanyon transport by this flux.

Representative examples of the dye pellet data at each location are shown in Figure 4-8. The points on each curve represent current speeds averaged over one-minute periods; in each case the flow direction was downcanyon. Corrections have been applied for refractive and angular errors; the solid lines represent 90 percent confidence intervals. Values of current speeds are all under 15 cm/sec, and the profiles are approximately logarithmic. Values of bed shear velocity, u*, were computed from the logarithmic velocity distribution, $u/u^* = 1/\kappa \ln z/K_S +$ (constant), where u is the velocity at a level, z, above the bottom; κ is von Karman's constant ($\kappa = 0.4$), K_S is the local bed roughness length scale, and u* is the bed shear velocity. The values of u* computed from these profiles are relatively small, or between 0.2 cm/sec and 0.7 cm/sec for all the measurements. These values probably are too low to transport sandy materials at the surface and certainly too low to generate the low amplitude ripples. However, we should note that these values are essentially short-term average measures of u*, and during periods of stronger currents, indicated by the scour and moats around outcrops and linear pebble trails, much higher values of u* would occur.

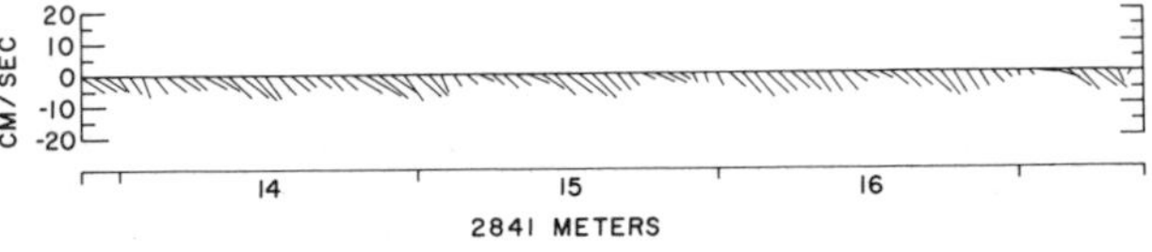

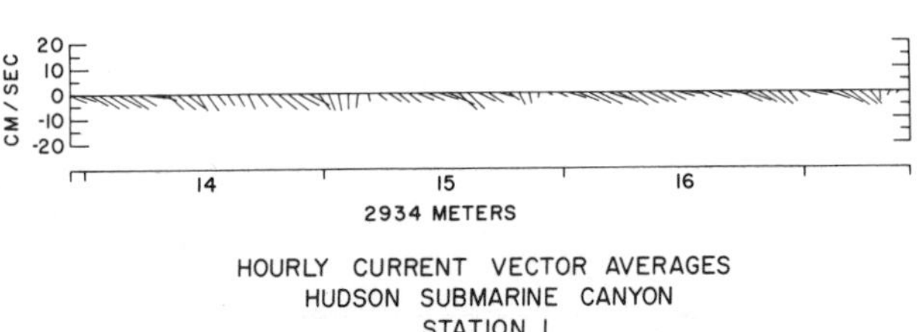

Figure 4-6. Current vectors or "sticks" plotted as one-hour averages of the current meter data taken at dive site 1. Note: The tidal modulation of the steady downcanyon flow is to the lower right. Water depths shown are for each current meter; total water depth is 2,941 m. North is toward the top of the page.

Figure 4-7. Progressive vector plot of the current meter data taken at dive site 1. Note: Average downcanyon flow (to the lower right) is about 7 km/day. The date is shown on each curve.

Biology

Shallow cores taken in the canyon floor have an assemblage of planktonic foraminifera that is normal for the latitude and depths of the samples. Fragmentation of the tests of the planktonic species is rare. Over one hundred different species of benthic foraminifera were found, mostly of Quaternary age common in depth zones greater than 1,000 m. However, rare occurrences of three species common in water depths less than 1,000 m were identified in the *Alvin* cores. The tests of these displaced benthic species were generally intact, which thus suggests that their mode of transport from shallower depths was gentle enough to prevent fragmentation. Ericson et al. (1961) noted that displaced benthic foraminifera present in turbidity current deposits commonly had low incidences of breakage. Upper Cretaceous foraminifera were found in two of the cores. This age correlates with the oldest date found in rock samples from submarine canyons off the northeastern United States (Gibson et al. 1968).

In contrast to the abundant epifauna observed in the shallower sections of Hudson Canyon (Rowe et al. 1974),

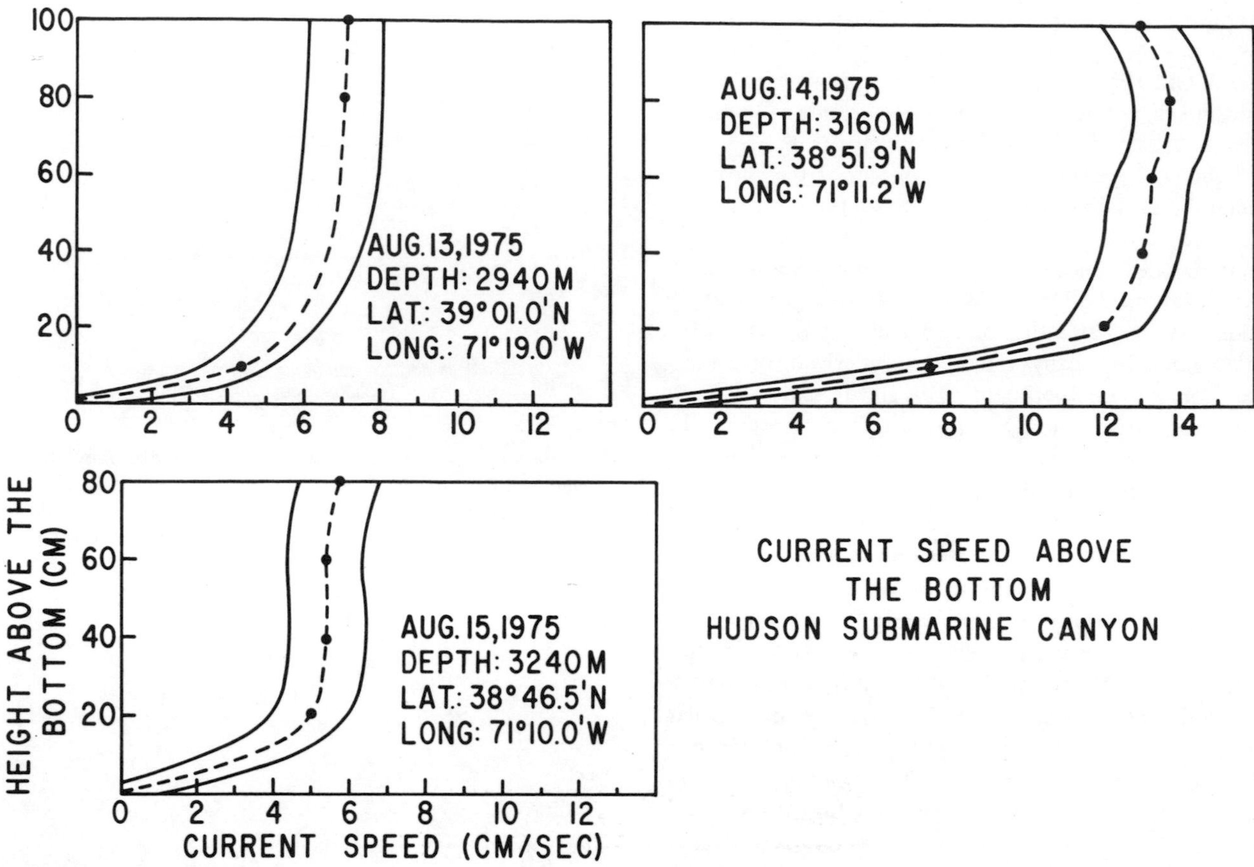

Figure 4-8. Current speed profiles measured from the dye streams at three dive sites. Note: Dashed curve is a least squares fit to the points; solid curves are the envelope of 90 percent confidence intervals estimated for the data reduction techniques.

there were generally few benthic organisms observed during the *Alvin* dives. Bottom traverses along the canyon floor showed that most organisms were grouped into zones of *Echinus affinus* (urchin) and *Dytaster insignus* (sea star). In the canyon axis at dive site 1, large numbers of *Echinus affinus* (about 2 cm to 3 cm in diameter) were observed to be moving slowly over the sediment surface. The motion of their tube feet and the spines dragging into the bottom left thin, shallow continuous trails along their paths. The sea stars, often seen partially buried in the surface sediment, commonly left star-shaped depressions in the surface. The net effect of these small disturbances to the sediment surface by the activities of organisms could contribute substantially to the redistribution of sediment within the canyon.

Irregular patches of the canyon floor were inhabited by small sea pens with stalks protruding about 1 cm to 2 cm above the bed. Stalks were usually oriented in a downcanyon direction, which thus suggests that strong currents recently flowed in that direction. The patches of sea pens most often occurred near the zones or trails of pebbles.

A circular pattern of holes, each hole about 1 cm to 3 cm diameter, was observed at many locations in the axis and on the terraces. These symmetrical patterns usually contain at least ten (often up to twenty) holes spread equidistantly along the perimeter. Often the area within the perimeter of holes was slightly raised in a dome-like shape above the general bed level. Similar structures have been observed by others (Heezen and Hollister 1971; Rowe 1971) during previous submersible dives in other areas.

In an effort to identify the organism responsible for these features, core tubes were driven 30 cm into the center and into the perimeter of two of these features. Both attempts failed to uncover a likely candidate. The animal causing these patterns is suspected to be a large asteroid (Rowe 1971), but this has yet to be demonstrated.

The principal fish seen near the bottom was the rat tail, *Nematonurus armatus,* which often swam very near *Alvin* during the bottom work. Photograph B in Figure 4-4 shows one of these fish.

During the descent to the bottom on dives 1 and 3, the water column appeared to contain numerous macro-

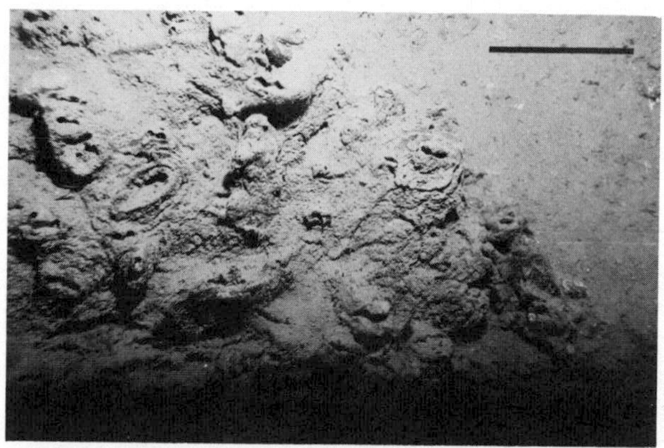

A

B

Figure 4-9. Effects of worm burrows at dive site 2. Note: Photograph A shows extensive burrows in a siltstone outcrop along the eastern canyon wall. The holes are paired in most cases and have a rosette surface appearance. No organisms were observed in the burrows, but polychaete worms are the suspected cause. Scale bar in top right is about 5 cm long; water depth is 2,940 m. Photograph B shows burrows broken off from rock outcrops in canyon axis. Weakening of the host rock and subsequent spalling or breaking of rock pieces was observed at several locations. Scale bar in lower right is about 25 cm long; water depth is 3,140 m.

scopic organisms, particularly salps and string-like aggregations of mucous materials. On the bottom at dive site 3 rows of abundant salp bodies rolling along the bottom downcanyon were organized in an axial direction and spaced about a meter apart. Continued downcanyon transport of these organisms, as was observed here, could contribute large quantities of organic matter as a food source and a cementing agent to the deep sea.

Many of the rock outcrops along the walls and in the axis have been extensively burrowed by worms, which has thus drastically altered the surface appearance and induced erosion. The burrows are generally confined to the rock surfaces. The photographs in Figure 4-9 illustrate the effects of worm burrows in the siltstone wall rock; worm burrowing has sufficiently weakened the host rock to cause collapse (photograph B).

The worm burrows often appear as double or paired tubes surrounded by overlapping or rosette-like structures of agglutinated materials (photograph A, Figure 4-9). The organisms responsible for these structures were not found; however, elaborate paired tubes are caused by several varieties of worms in the marine environment. Large onuphid polychaetes have been found to cause similar structures and are possible sources for the tubes described here.

Another less common feature found on the surfaces of the wall rock, usually most obvious on the underside of overhangs, are patterns of scallop-like depressions (Figure 4-10). The depressions could be caused by erosion during periods of intense bottom currents, but anemones occasionally seen near these features are another possible cause.

Figure 4-10. Unusual concave scallop-like scars in a siltstone along the eastern wall of Hudson Canyon. Note: Current-induced erosion or biological erosion by sea anemones are possible causes. Water depth is 3,205 m.

DISCUSSION

The bathymetric chart in Figure 4-1 shows that Hudson Canyon is characterized by steep walls and high relief on the upper and middle parts of the continental rise— that is, a profile more typical of canyons on the upper continental slope. The *Alvin* dives discussed in this chapter were undertaken to investigate the deep sections of Hudson Canyon; these dives mark the deepest excursion into a canyon system off the eastern United States that has been undertaken (as of this writing).

The unconsolidated surface sediment on the canyon floor and draping the terraces along the walls is mostly coarse silt to medium sand. This surface material often contains linear patches of small, subrounded pebbles whose composition suggests a continental origin. The thin manganese coating on many of these pebbles suggests a low depositional rate in this region. The low currents that were observed during the dives would suggest that deposition of fine-grained materials could occur if they were available. The finer grained clays and silts abundant in the shallower sections of the canyon (see Chapter 3) apparently have not been accumulating in the sections between 3,000 m and 3,600 m depth.

However, the sedimentary processes must be controlled in large part by the episodes of intense bottom currents and turbidity currents that undercut the base of the canyon walls and form moats around the exposed boulders. If finer materials are being deposited, the more powerful events would presumably winnow the surface material, entrain the fine fraction, and transport the resuspended materials downcanyon. Downcanyon transport by these higher-energy currents, similar to the steady downcanyon flow of the weaker currents represented by Figure 4-8, is indicated by the downcanyon orientation of the sea pens and by the preferred downcanyon position of moats. Although only rare occurrences of low amplitude sediment ripples were found, in each case the slip faces of the ripples were situated on the downcanyon side.

Erosion of the walls and exposed boulders in the axis is also caused by the activities of benthic organisms. Paired worm tubes have virtually destroyed the surfaces of the siltstone wall rock in most places. This burrowing apparently weakens the rock to the point where pieces break off and fall to the base, where currents can then more readily erode their surfaces. Various scars and depressions in the rock surfaces are probably the result of this type of process. The combination of burrowing by organisms and periods of intense current activity can be an effective mechanism producing widespread erosion of the canyon walls. Dillon and Zimmerman (1970) have reported similar findings for shallower sections of Block and Corsair Canyons.

Whereas the previous work of Ericson et al. (1951) had suggested that graded beds from turbidity currents should be found at depths near 3,200 meters in Hudson Canyon, the data presented here show that the sediments are coarser nearer the surface. This upward coarsening can be explained by winnowing and bioturbation of the surface and near-surface sediments, with subsequent removal of the finer materials by strong, episodic bottom currents. The upper sections of the turbidite sequences affected by the process of winnowing and bioturbation would eventually develop inversely graded beds. The depth to which this process is effective in modifying the turbidite deposits would be proportional to the current velocities and the intensity of local biological reworking of the sediments.

Finally, the steady transport of bottom water downcanyon over the four-day period, together with the frequent observation of large numbers of salp carcasses and other organisms being advected by this flow, suggests that the flux of material to the deeper water through Hudson Canyon is significant. If suspended and bed materials from the upper reaches of this canyon and the adjacent shelf are entering the dive site areas, these materials would probably be transported toward deeper water either by the relatively low, persistent downcanyon currents or by episodic pulses of stronger bottom currents. The evidence gathered here suggests that strong, episodic bottom currents combined with the action of benthic organisms have changed the surface sedimentary setting in the canyon axis and produced substantial erosion of the canyon walls.

ACKNOWLEDGMENTS

The authors express appreciation to C.D. Hollister for his discussions during cruise planning and to W. Berggren, J.V. Gardner, and P. Quinterno for their analyses of the microfauna in the core and rock samples. The authors also thank the *Alvin Lulu* group and the physical oceanography mooring group at Woods Hole Oceanographic Institution for making these measurements possible. This work was supported by the Ocean Science and Technology Division of the Office of Naval Research. D. A. Cacchione's contribution was supported by the U.S. Geological Survey.

REFERENCES

Dillon, W.P., and Zimmerman, H.B., 1970. Erosion by biological activity in two New England submarine canyons. *J. Sed. Petrol.*, 40: 524-47.

Ericson, D.B., Ewing, M., and Heezen, B.C., 1951. Deep-sea sands and submarine canyons. *Geol. Soc. Amer. Bull.*, 66: 961-66.

——, Ewing, M., Wollin, G., and Heezen, B.C., 1961. Atlantic deep-sea sediment cores. *Geol. Soc. Amer. Bull.*, 72: 193-285.

Gibson, T.G., Hazel, J.E., and Mello, J.F., 1968. Fossilferous rocks from submarine canyons off the northern United States. *U.S. Geol. Surv. Prof. Paper* 600-D, D222-D230.

Heezen, B.C., and Hollister, C.D., 1971. *The Face of the Deep.* Oxford University Press, New York, 659 pp.

Keller, G.H., Lambert, C., Rowe, G., and Staresinic, N., 1973. Bottom currents in the Hudson Canyon. *Science,* 180: 181-83.

Rowe, G.T., Keller, G.H., Edgerton, H., Staresinic, N., and MacIlvane, J., 1974. Time-lapse photography of the biological reworking of sediments in Hudson submarine canyon. *J. Sed. Petrol.* 44: 549-52.

Shepard, F.P., Marshall, N.F., and McLaughlin, P.A., 1974. Currents in submarine canyons. *Deep-Sea Res.*, 21: 691-706.

Uchupi, E., 1968. Atlantic Continental shelf and slope of the United States – Physiography. *U.S. Geol. Surv. Prof. Paper* 529-C, C1-C30.

Chapter 5

Physical and Biogenic Characteristics of Sediments from Hueneme Submarine Canyon, California Coast

RICHARD M. SCOTT

Skidaway Institute of Oceanography
Savannah, Georgia

BARTON C. BIRDSALL

Geology Department
University of South Florida
Tampa, Florida

ABSTRACT

Twelve box cores were collected along the axis of Hueneme Canyon in water depths between 14 m and 400 m. Analysis of sedimentary structures, mainly by x-ray radiography, and sediment textures permits characterization of three depositional environments within Hueneme Canyon: the canyon axis, the canyon floor (exclusive of the canyon axis), and the canyon wall.

Sediments from the canyon axis are characterized by primary physical structures including thin planar lamination, current-ripple lamination, and sometimes small-scale graded bedding; fine- to very fine-grained sands are dominant. Canyon floor sediments are highly bioturbated, fine-grained sandy muds and contain a wide variety of biogenic traces; remnant physical structures are sometimes present. Canyon wall sediments are thinly laminated, cohesive silty sands, with few biogenic structures.

Note: Richard M. Scott is currently with Cities Service Oil Company, Exploration and Production Research, Tulsa, Oklahoma.
Barton C. Birdsall is currently with Gulf Energy and Minerals Co. U.S., Houston, Texas.

Sediments from Hueneme Canyon consist predominantly of fine sands and silts with very little medium to coarse sand and no clay. All samples from the canyon axis and canyon floor environments contain small amounts of water insoluble hydrocarbon residue; canyon wall sediments contain no hydrocarbon residue.

This suite of cores indicates that transport of sand-sized sediment in the upper portion of Hueneme Canyon is channeled along the canyon axis. Primary physical structures indicate that sediments in the axis environment move by traction transport and/or small-scale mass sediment flow.

INTRODUCTION

Hueneme Canyon, defined by Shepard and Dill (1966) as a slope valley, is located 80 km northwest of Los Angeles, California, at the approximate center of the coast fronting the Oxnard Plain (Figure 5-1). The canyon heads within 200 m of the coast at Port Hueneme, extends 21 km south across the Santa Clara delta, and terminates in the Santa Monica Basin at a depth of 690 m. The average gradient along the slightly winding canyon axis is 30 m/km, although the upper, sampled portion of the canyon is typically steeper with an average gradient of 50 m/km. Hueneme Canyon differs from other submarine canyons along the California coast in the magnitude to which sediment input to the canyon has been affected by harbor maintenance programs during the past thirty-nine years.

A series of twelve box cores was collected in the upper portion of Hueneme Canyon during a cruise aboard the R/V *Velero IV* in July 1974. Box cores were collected along the canyon axis in water depths between 14 m, slightly seaward of the entrance to Port Hueneme, to

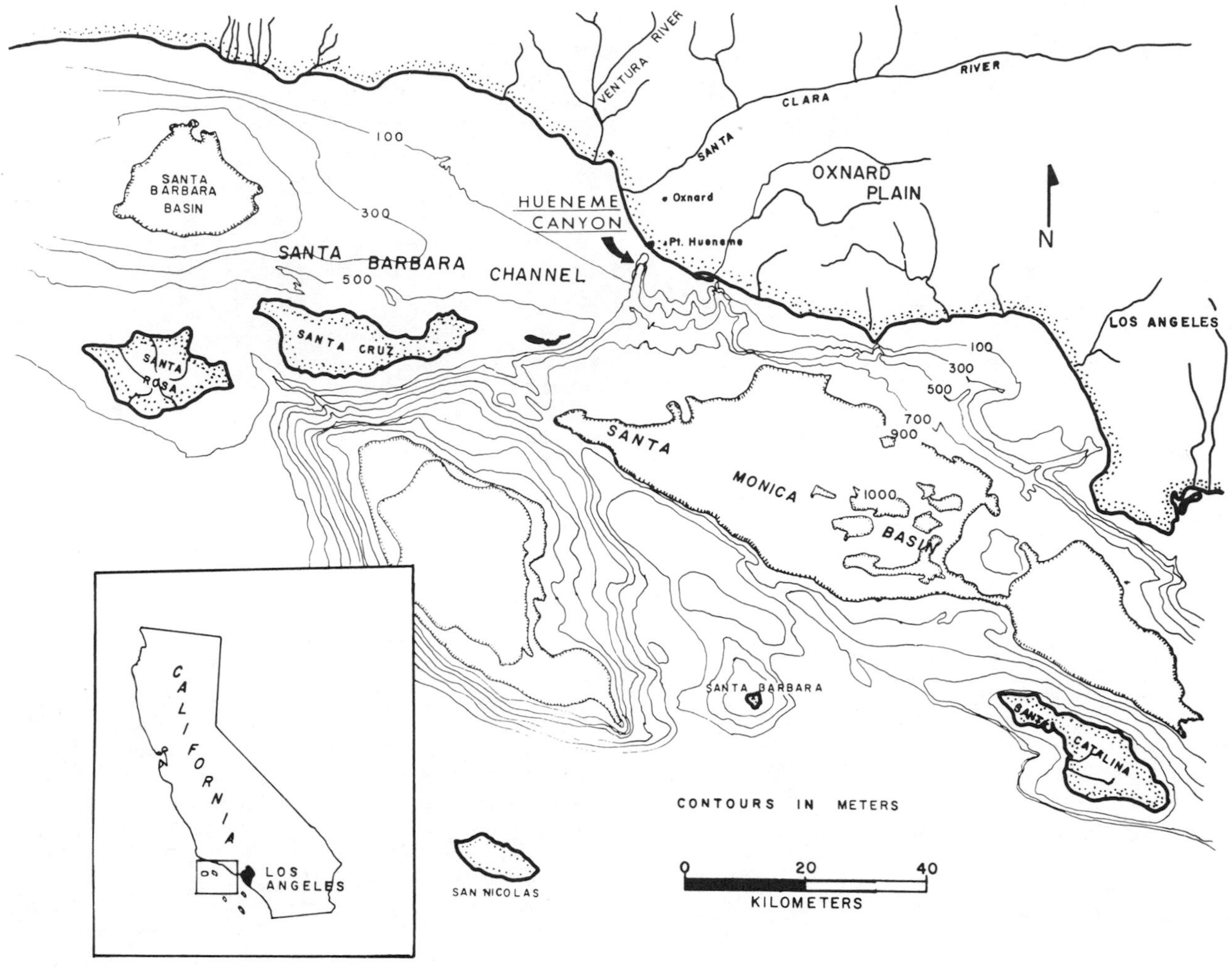

Figure 5-1. Bathymetric map showing the location of Hueneme Canyon and regional submarine morphology, in the vicinity of Hueneme Canyon. (Source: Adapted from Shepard and Emery 1941.)

400 m, approximately 8 km south of the harbor entrance (Figure 5-2). This investigation was undertaken to determine the nature of physical and biogenic sedimentary structures and textural characteristics of canyon sediments and to define depositional environments in Hueneme Canyon. Additional sediment samples from selected cores collected on two beach to shelf break transects, one east and one west of Hueneme Canyon (see Figure 5-2), were provided by J.D. Howard. Textures of these samples representing beach, shoreface, and shelf environments were analyzed to determine whether textures of canyon sediments were, or were not, similar to textures of sediments from adjacent depositional environments.

METHODS

Cores were collected with a N.E.L. spade corer, modified after Reineck (1958, 1963a) and Bouma (1969), with box dimensions of 60 cm x 30 cm x 21 cm. Cores were undisturbed and ranged in length from 16.5 cm to 56.5 cm, averaging 40 cm. Station locations were determined by radar, and sampling depths were determined by both precision depth recorder and metered wire length.

X-ray radiographs were made using Kodak *Industrex AA* type film (Bouma 1964, 1969; Howard and Frey 1975) of vertical sediment slices from each core, trimmed to a uniform thickness of 2 cm. Exposure times ranged from 1.2 to 2.2 minutes at a constant voltage of 55 kvp. Epoxy relief casts of each sediment slice were made (Bouma 1969). Bulk sediment samples and samples from discrete lithologic intervals were collected for textural analysis. The remainder of the core was carefully dissected to further examine biogenic traces and to collect organisms when possible.

Bulk sediment samples were freeze dried to minimize loss of fines (Malcolm 1968), and subsamples were sieved at 1/2 ϕ intervals (Ingram 1971). Pipette analysis (Folk

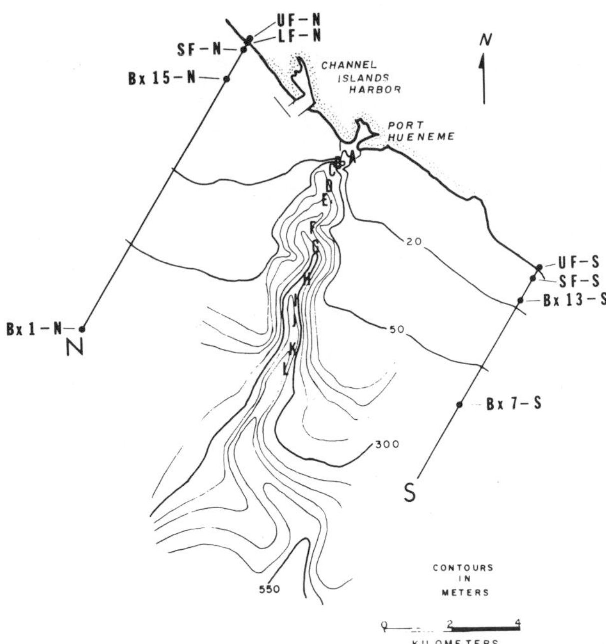

Figure 5-2. Bathymetric chart showing the location of samples HC-A through HC-L collected in Hueneme Canyon and samples collected on two beach to shelf break transects, N and S. Note: UF = upper foreshore; LF = lower foreshore; SF = shoreface; Bx = box core.

1974) was made using 1/2 ϕ divisions for samples with greater than 5 percent silt clay (pan fraction). Mean and standard deviation values were computed by the method of moments (McBride 1971).

During pipette analysis a water insoluble, oily residue was noted in several samples. Reagent grade petroleum ether was used to extract the residue from sediment subsamples, and a weight percent of residue was determined for each sample (method suggested by P. Meyers, personal communication, 1975). Extracted residues were analyzed by gas chromatography (Tissier and Oudin 1973).

DESCRIPTION OF STUDY AREA

Physical Setting of Coastal Area

Mean semidiurnal tidal range (MHH to MLL) in the vicinity of Port Hueneme is 1.65 m with a maximum range of 3.2 m. Local winds and waves approach primarily from the west and northwest, although wind effect is partially modified by Point Conception and the offshore islands; south and southwest winds sometimes occur, most frequently in association with winter storms (Savage 1957; Herron and Harris 1966). Breaker heights range between 1 m and 2.5 m. These wind and wave conditions produce dominant southward littoral and inner shelf (wind-driven) currents (Savage 1957; Drake and Gorsline 1973). In addition, Gorsline (1970) and Shepard et al. (1974a) have noted water circulation from the Santa Barbara Basin across Hueneme Canyon into the Santa Monica Basin, parallel to the Santa Barbara Channel.

Sources of Canyon Sediments

Hueneme Canyon sediments are derived primarily from the Santa Clara River, which discharges 10 km northwest of Port Hueneme and has been described as the largest river in Southern California in terms of both discharge and suspended load (Drake and Gorsline 1973). Herron and Harris (1966) indicate that significant amounts of sediment are also derived from the Ventura River and upcoast beaches and they estimate that the total southward littoral transport from these three source areas is approximately 1.2 million cubic yards (0.9 million cubic meters) annually (Figure 5-3).

In addition to the average annual sediment contribution from the Santa Clara and Ventura rivers, periodic heavy rainfalls and associated flooding, such as those recorded in January and February 1969, result in abnormally large influx of fluvial sediments into the near shore marine environment. Floods of similar magnitude occurred in 1938, 1909, 1884, 1861-2, 1832, and 1825 (D.S. Gorsline, personal communication, 1976). Drake et al. (1972) discussed the initial deposition and subsequent redistribution of 1969 flood sediments (an amount in excess of 35 x 10^6 metric tons) on the Santa Barbara-Oxnard Shelf. Through periodic sampling, they found that flood sediment deposits were still texturally discrete in mid-1971 and concluded that a minimum period of three years is required for sediments within the study area to return to "normal" following major introductions of river-borne material. Gorsline (personal communication, 1976) reports that box cores collected in 1976 in the lower fan and fan basin floor zones show large, recent sand flows and "turbidites" at distances of tens of kilometers from the mouth of Hueneme Canyon, and he tentatively correlates these with the 1969 rainfall event. These data indicate that the influence of flood events is extended throughout the region by continuous reworking and redistribution of flood sediments during the intervening years.

Drake and Gorsline (1973) attribute high turbidity (>5 mg/liter) within 1 km to 2 km of the coast to terrigenous particles derived from Santa Clara River and suggest southerly transport of suspended sediment by inner shelf currents. Shepard and Dill (1966) conclude that sediment transport known to take place over much of continental shelves intermittently provides sediments to submarine canyons. The dominance of west and northwest winds (Herron and Harris 1966) and indications of southward water circulation along the Santa Barbara

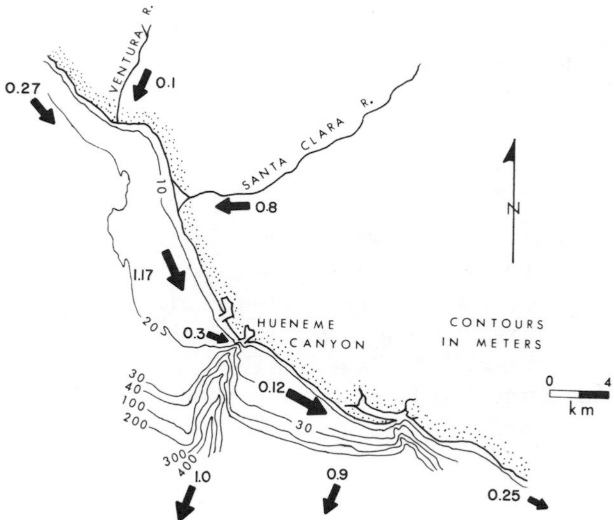

Figure 5-3. Topography and littoral sand movement in the vicinity of Hueneme Canyon. Note: Values are in million cubic yards per year. (Source: Adapted from Herron and Harris 1966.)

Channel (Gorsline 1970; Shepard et al. 1974a) suggest possible mechanisms for sediment transport across the Ventura Shelf toward Hueneme Canyon.

Harbor and river maintenance programs along the Southern California coast have affected sediment supply to Hueneme, Redondo, and Newport canyons. Redondo Canyon is used intermittently as a spoil site for dredging operations at King Harbor (D. S. Gorsline, personal communication). Felix and Gorsline (1971) noted that under normal runoff conditions only fine sands and mud reach the coast in the vicinity of Newport Canyon because all tributary rivers in the Newport Beach area have been dammed for flood control; the head of Newport Canyon is presently being filled with fine organic-rich sediment that consists largely of locally derived sewage effluent.

Hueneme Canyon, however, is unique in comparison, in that man's activities have largely disrupted natural longshore sediment transport in the vicinity of the canyon head. The affects of harbor construction and sediment bypassing operations on beaches in the vicinity of Hueneme Canyon are discussed in detail by Savage (1957) and Herron and Harris (1966). During the period from 1855 to 1938 the history of the coast segment from the Santa Clara River to Point Hueneme was one of accretion, while the downcoast segment was relatively stable. Jetty and harbor construction (1938 to 1940) initially resulted in rapid accretion of the beach upcoast of the west jetty and severe erosion of beaches downcoast of the east jetty at a rate of 1.2 million cubic yards (0.9 million cubic meters) annually (Savage 1957). Following initial rapid upcoast accretion to the west of the canyon, the bulk of the sediment necessary to maintain downcoast beaches (1.2 million cubic yards per year) was diverted into Hueneme Canyon (Herron and Harris 1966) in an amount probably in excess of 1 million cubic yards per year (0.75 million cubic meters). During the period from 1940 to 1958, approximately 2 million cubic yards (1.5 million cubic meters) of sand was dredged from Port Hueneme and upcoast beaches and pumped to downcoast beaches in an unsuccessful attempt to halt erosion.

Channel Islands Harbor, located 1.6 km west of Port Hueneme, was constructed as a littoral sediment trap in 1958 to 1960. The entrapment area landward of a breakwater 550 m offshore is dredged biennially, and sediments are pumped under the entrance of Port Hueneme to downcoast beaches. This procedure supplies 2 to 2.5 million cubic yards (1.5 to 1.8 million cubic meters) of sediment to downcoast beaches and has largely eliminated long-term coastal erosion (R. Bruno, personal communication, 1976). No recent estimates of input of littoral sediments into Hueneme Canyon are available.

Canyon Floor Currents

Canyon floor currents in Hueneme Canyon have been described by Drake and Gorsline (1973) and Shepard et al. (1974a, 1974b). Additionally, N.F. Marshall provided (unpublished) current data obtained at water depths of 201 m, 256 m, 362 m, and 406 m in Hueneme Canyon; current data were recorded simultaneously at 201 m, 256 m, and 406 m stations. Current data from these four stations are presented in Table 5-1.

Records show that current reversals are more frequent in the upper portion of the canyon and that the prevailing current flow direction is down the canyon. Downcanyon currents are typically of longer duration than upcanyon currents, although velocities of upcanyon currents are sometimes equivalent to or higher than velocities of downcanyon currents. Shepard et al. (1974a) noted that canyon floor currents in several submarine canyons are sufficiently strong (>18 cm/sec during peak flow) to transport fine sand and that more frequent and faster flows are downcanyon and result in a net transport of sediment seaward. Data from 1975 show a progressive downcanyon increase in the percent of time during which canyon floor current velocities (either up-, down-, or crosscanyon) exceed 10 cm/sec. (A current with a velocity of 10 cm/sec is capable of transporting sand-sized particles as large as 1 mm in diameter [Hjulstrom 1939].)

In addition to alternating upcanyon and downcanyon currents, Shepard et al. (1974a) observed strong, southeasterly crosscanyon currents (maximum velocities > 35 cm/sec at 3 m above the bottom) during a period of strong (40 km/hr) west winds. They consider winds to be particularly influential in relatively wide-floored canyons like Hueneme Canyon. Comparison of east-west flow records from 3 m and 30 m above the canyon floor at

Table 5-1. *Current Data from Hueneme Canyon*

Station	*Depth (m)*	*Starting Date and Duration (hr)*	*Height of Meter above Bottom*	*Average Duration Downcanyon Flow (hr)*	*Average Duration Upcanyon Flow (hr)*	*Average Duration and Direction of Dominant Crosscanyon Flow*	*% time velocity > 10 cm/sec.*
No. 84, lo. 1	201	9/10/75					
		96	3	4.8	2.4	1.4 (E)	30.6
No. 85, lo. 1	256	9/10/75					
		192	3	4.1	3.1	3.7 (E)	42.0
		232	30	4.1	3.8	N.D.D.	22.7
No. 78, lo. 1	362	8/7/75					
		520	3	6.0	3.1	N.D.D.	47.9
No. 86, lo. 1	406	9/10/75					
		232	3	6.0	4.1	N.D.D.	58.0

Note: Durations were computed using only flows with velocities exceeding 5 cm/sec; durations of crosscanyon flow were not determined if there was no dominant direction (N.D.D.).

Source: Data provided by Shepard, Marshall and McLoughlin (personal communication, 1976).

station 85 (see Table 5-1) reveals that during crosscanyon (eastward) flows of relatively long duration and high velocity (average velocity of 20 cm/sec) near the canyon floor, no corresponding crosscanyon flow occurred at 30 m above the floor. This finding indicates that in this instance the observed net southeast (crosscanyon) flow is not a result of wind direction but rather a result of local variation in the axial trend of the canyon (see Figure 5-2, station G for position). This conclusion is further supported by the absence of a similar crosscanyon flow at station 86.

Directions of flow, up or down the canyon, at 3 m and 30 m coincide consistently; the durations and maximum velocities of flows are also remarkably similar. Shepard et al. (1974a, 1974b) reported similar synchronous up- and downcanyon movements in several canyons off Southern California, and they concluded that these movements are caused by internal waves or tides—probably the former—that progress along canyon floors. Shepard (1976) found that along coasts with tidal ranges greater than 1.5 m tides exert a strong influence on current cycles in canyons at water depths between 250 m and 1,400 m. Current records from most depths in excess of 250 m in Hueneme Canyon (see Table 5-1, stations 78 and 86) show a strong relationship between reversal cycles of major up- and downcanyon flows and the semidiurnal tidal cycle. Tidal influence in Hueneme Canyon is further demonstrated by the occurrence of maximum crosscanyon velocities between times of slack peak high and peak low tide, recorded at a 374 m deep station (Shepard et al. 1974a).

In addition to up-, down-, and crosscanyon flows, occasional high-speed, downcanyon currents (turbidity currents) probably occur in Hueneme Canyon, but they have not been recorded. The existence of these currents in submarine canyons has been documented indirectly through the loss of current meters (Inman 1970; Shepard and Marshall 1973); such flows occurred during storms with strong onshore winds.

DESCRIPTION OF THE SAMPLES COLLECTED

Cores collected within Hueneme Canyon are divided into three groups representing the following submarine canyon environments: the canyon axis, the canyon floor exclusive of the canyon axis (herein referred to as the canyon floor), and the canyon wall. Cores in each group exhibit similar physical and biogenic sedimentary structures and sediment texture. Samples obtained during this study are assumed to be representative, and observed similarities within groups seem to justify the suggested environmental divisions in spite of the relatively low number of samples collected within the canyon. The reader is referred to an excellent discussion by Bouma (1965) of cores representing canyon

Table 5-2. *Textural Analysis of Sediments from Hueneme Canyon and Adjacent Depositional Environments.*

Depositional Environment	*No. of Samples*	*Mean Grain Size (Phi)*	*Phi Standard Deviation*	*Gravel (>-1φ) %>-1 Phi (2 mm)*	*Coarse Sand % between -1φ and 1φ (2 mm and 0.5 mm)*	*Medium Sand % between 1φ and 2φ (0.5 mm and 0.25 mm)*	*Fine Sand % between 2 and 4 (0.25 mm and 0.0625 mm)*	*Mud %<4φ (0.0625 mm)*	*Percent Hydro-Carbon*
Canyon Axis	5	3.33	0.76	0.00	0.56	4.15	76.83	18.45	2.68
Canyon Floor	5	3.88	0.95	0.00	1.27	2.18	48.85	47.72	1.94
Canyon Wall	2	2.87	1.28	0.00	12.05	13.21	60.72	14.02	0.00
Average (excluding canyon wall)	10	3.61	0.85	0.00	0.92	3.17	62.84	33.09	2.31
Beach	3	1.12	0.84	2.14	38.21	44.30	15.34	0.00	–
Shoreface	2	2.58	0.70	0.00	4.98	7.56	87.22	0.26	–
Inner Shelf	2	3.34	1.92	0.00	15.75	7.36	54.67	22.21	–
Outer Shelf	2	3.64	1.55	0.00	6.46	2.64	69.21	21.69	–
Average Shelf	4	3.49	1.74	0.00	11.11	5.00	61.94	21.95	–

$\phi = -\log_2 x$; x = diameter in mm.

axis and adjacent canyon sedimentary environments that were collected from several submarine canyons off Southern California and the southern tip of Baja California (see also Bouma 1964).

Textural analyses of canyon sediments and beach, shoreface, and shelf sediments are presented numerically in Table 5-2 and graphically in Figure 5-4. Physical and biogenic sedimentary structures observed in cores are indicated on contact prints of X-ray radiographs (see Figures 5-6 through 5-8) and are presented diagramatically, together with estimated degree of bioturbation, in Figure 5-5.

Canyon Sediments

Sediments from Hueneme Canyon, excluding canyon wall samples, consist predominantly of moderately-sorted very fine sand and silt with a mean grain size of 3.61 φ (0.08 mm) (see Table 5-2). Although coarse sand and gravel are abundant in some submarine canyons along Southern California and Baja California (Bouma 1965; Shepard and Dill 1966), sediments from Hueneme Canyon contain very little medium to coarse sand and no gravel. Medium and coarse sands are abundant in beach (mean = 1.12 φ; 0.46 mm) and shoreface (mean = 2.58 φ; 0.17 mm) sediments adjacent to Hueneme Canyon (see Table 5-2 and Chart A, Figure 5-4), and the paucity of coarser sediments in the canyon indicates that relatively little canyon sediment presently is derived from littoral sources. On one or two occasions gravel has been obtained in cores from Hueneme Canyon by researchers from the University of Southern California, and thick sequences of graded sand through gravel were present in piston cores collected in the lower canyon channels by Shell Oil Company (D. S. Gorsline, personal communication). These data indicate that coarse sand and gravel are present and perhaps abundant in Hueneme Canyon, but were not penetrated by the box corer (60 cm maximum penetration). The authors feel that entrapment of littoral sands at Channel Islands Harbor is largely responsible for the absence of coarse sediments in the cores collected in Hueneme Canyon during the present study.

Hueneme Canyon sediments are texturally similar to adjacent shelf sediments east and west of the canyon. Shelf sediments (mean = 3.46 φ; 0.09 mm) are typically poorly sorted and contain roughly three times as much medium to coarse sand (an average of 16 percent) as canyon sediments (see Chart B, Figure 5-4). Mud content (<4 φ, 0.0625 mm; see Table 5-2) of shelf sediments averages 22 percent, and less than 50 percent of the mud fraction is coarse silt (4 φ to 5 φ; 0.0625 mm to 0.031 mm); clay content (<8 φ, 0.0039 mm) averages 2.7 percent. In comparison, canyon sediments averaged 33 percent mud of which more than 50 percent is coarse silt. No clay was found in canyon sediments. The absence of fine silt and clay in canyon sediments is believed due to current turbulence that maintains fine sediment in suspension. Drake and Gorsline (1973) reported the presence of a nepheloid layer near the floor of Hueneme Canyon, with the highest particle concentrations (up to 4.6 mg/l) just above the canyon floor in the upper 150 m of the canyon. (Particle concentrations between 6.8 and 0.53 mg/l were observed within 1 m of the canyon floor following a period of

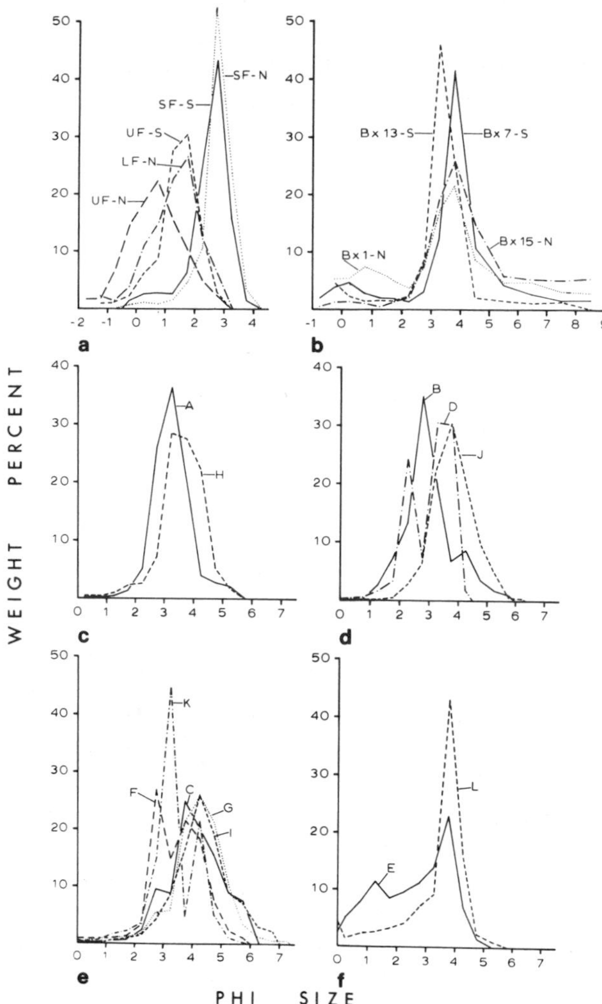

Figure 5-4. Grain-size frequency curves of bulk sediment samples collected in the vicinity of Hueneme Canyon. Note: Chart A describes the samples collected from the beach—that is, upper (UF) and lower (LF) foreshores—and the shoreface (SF); chart B, the shelf (Bx = box core samples); chart C, the canyon axis—samples with planar and ripple lamination; chart D, the canyon axis—samples with graded bedding; chart E, the canyon floor, exclusive of the canyon axis; and chart F, the canyon wall. Sample locations are shown in Figure 5-2.

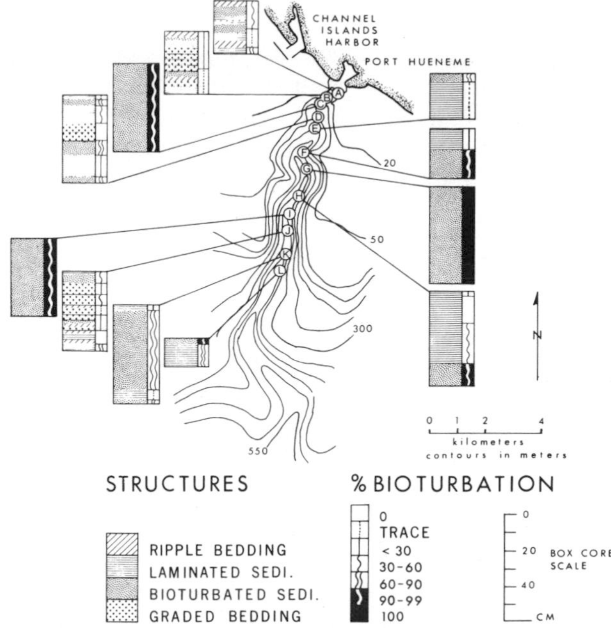

Figure 5-5. Graphic display of box cores. Note: The box core samples are drawn to scale, and station locations within Hueneme Canyon are indicated. Sedimentary structures are indicated diagramatically, and degree of bioturbation is indicated in the column to the right of each sample. Cores A, B, D, H, and J represent the canyon axis environment; cores C, F, G, I, and K represent the canyon floor environment; and cores E and L represent the canyon wall environment.

high discharge from the Santa Clara River in 1969.) Fine sand particles as large as 100 μ were present in nepheloid layers at Redondo and La Jolla canyons (Drake and Gorsline 1973), and velocities of bottom currents in these two canyons are of comparable magnitude to velocities observed in Hueneme Canyon (Drake and Gorsline 1973; Shepard et al. 1974a). Although no textural analysis of the nepheloid layer from Hueneme Canyon was reported, observed similarities between Hueneme Canyon and Redondo and La Jolla canyons suggest the presence of sufficient current turbulence in Hueneme Canyon to maintain silts and perhaps fine sands in suspension.

Organic material consisting of plant and wood fragments is present both in thin laminae interbedded with fine sand and silt and as isolated fragments, but is not a common constituent of canyon sediments. The thick accumulations of kelp and sea grass that form organic mat layers in other submarine canyons along Southern California (Bouma 1965; Shepard and Dill 1966) were not observed in cores recovered from Hueneme Canyon.

Analysis by gas chromatography revealed that the water insoluble residue was a hydrocarbon. Excluding canyon wall sediments that contain no hydrocarbons, sediments from Hueneme Canyon contain an average of 2.3 percent hydrocarbon residue by weight (see Table 5-2). A general decrease in hydrocarbon content with increasing water depth seaward along the canyon axis suggests downcanyon transport of hydrocarbons from a near-shore source. The most likely near-shore sources for the hydrocarbons are periodic reworking of residual hydrocarbons in beach (tar layers) and shelf sediments derived from the 1969 oil spill and/or minor incursion of material from Port Hueneme and the associated U.S. Naval Station. Oil seepages similar to those occurring in Redondo Canyon (Shepard and Dill 1966, p. 75) constitute another potential hydrocarbon source, although no seepages have been reported in Hueneme Canyon.

Canyon Axis

Samples from the canyon axis are characterized by a dominance of primary physical internal structures; biogenic traces are relatively scarce in this environment. Axis samples have been further subdivided into those characterized by planar and ripple lamination (print A, Figure 5-6) and those characterized by units of graded bedding in addition to lamination (print B, Figure 5-6). Textural analysis of samples in each of these two subdivisions is presented graphically in Charts C and D in Figure 5-4, respectively.

A total of five samples was collected from the canyon axis environment. These are cores: HC-A at 14 m water depth, HC-B at 64 m, HC-D at 159 m, HC-H at 343 m, and HC-J at 377 m. Of these, samples HC-B, HC-D, and HC-J contain graded beds.

Axis sediments consist predominantly of very fine sand (mean = 3.33 ϕ; 0.098 mm) and are moderately sorted. Two samples, HC-B and HC-D, exhibit bimodal distribution (see Chart D, Figure 5-4), thereby reflecting concentrations of fine sand in the graded beds.

Cores HC-A and HC-H consist predominantly of thinly laminated fine sands and silts. Laminations vary from less than 1 mm to a few centimeters in thickness; there is no consistency of lamination thickness, although laminations greater than 1 cm are relatively uncommon. The upper few centimeters of both cores (also HC-D and HC-J) are moderately bioturbated (see Figure 5-5), thereby indicating either decreased or intermittent sedimentation or increased biogenic activity (probably the former). Distinguishable burrows present in sample HC-H (see print A, Figure 5-6) include small (1 mm diameter) sinuous polychaete-type burrows (point 1) and a few lined tubes, typical of suspension feeding polychaetes (point 2). Planar lamination is the dominant sedimentary structure in the upper third (top 18 cm) of the sample, and as shown in Figure 5-6, irregularities and disruptions (point 3 on print A and point 6 on print B) are caused by burrowing organisms. Two thin ($<$1 cm) layers of clean, fine sand occur at depths of 15 cm and 18 cm (points 4 and 5, respectively, on print A, Figure 5-6) and the relatively sharp contact of these layers with underlying sediments suggests erosive scouring. Sediments from the lower two-thirds (below point 5 on print A, Figure 5-6) of the sample differ markedly from those of the upper third in that they are more highly bioturbated with only remnants of primary physical structures. Systems of small burrows (point 1 on print A, Figure 5-6) typical of foraging polychaetes are abundant in this portion of the sample, and this type of burrow pattern is representative of relatively low-energy environments. Fecal pellets are very abundant in sediments below the erosional contact at 18 cm and are extremely rare above the contact. Sediments from the lower two-thirds of the sample represent a period of slow deposition (or nondeposition) during which benthic organisms largely reworked the sediments. Subsequent renewal of deposition or increased rates of sedimentation (and current activity) produced a dominance of primary physical structures in the upper portion of the core; the vertical burrows (point 2 on print A, Figure 5-6) are related to the increased energy and higher rates of sedimentation. This sequence represents either a shift in the position of the canyon axis, or intermittent (periodic) sediment transport along the canyon axis. We should note that while the rate of sediment accumulation in the samples obtained is unknown, the authors feel that these sequences are the product of seasonal changes and most likely reflect variation in sediment supply. HC-A exhibits a wider variety of burrow types and is more densely burrowed than core HC-H, thereby suggesting either more prolific endobenthic life, greater substrate stability, or perhaps subdued current activity near the head of the canyon.

Samples HC-B, HC-D, and HC-J exhibit a dominance of primary physical structures and, unlike HC-A and HC-H, also contain distinct units of graded bedding (print B, Figure 5-6). Units range between 3.5 and 9.5 cm in thickness, and sharp contacts at the base of each unit are erosional. Sediments within a unit normally grade vertically upward from "clean" fine sand into silty sand or silt, and minor amounts of medium and coarse sand, subangular to subrounded shell fragments, and, in some cases, plant fragments are present at the base of some units. Sediments within a unit exhibit thin planar lamination, and in core HC-J (print B, Figure 5-6) current ripple lamination and wedging out of laminae (points 7 and 8) are also observed. This sequence of graded bedding with planar laminated and ripple laminated sediments is suggestive of the turbidite facies model (lower three intervals) described by Bouma (1962). However, this sequence lacks the upper two intervals of the Bouma model (indistinct parallel lamination and structureless pelitic interval), and the sharp contacts between intervals described by Bouma (1965, p. 308) are not present. The absence of fines in the lowest part of the graded bed indicates that the most probable origin of these graded beds is sediment transport and deposition by a current of gradually decreasing velocity and competency (Reineck and Singh 1973), which may or may not involve mass sediment flow. Frequent fluctuations in current velocity and periodic current reversals, which are well documented in many submarine canyons, including Hueneme Canyon, by Drake and Gorsline (1973) and Shepard et al. (1974a), provide a feasible mechanism for formation of graded beds without mass sediment flow.

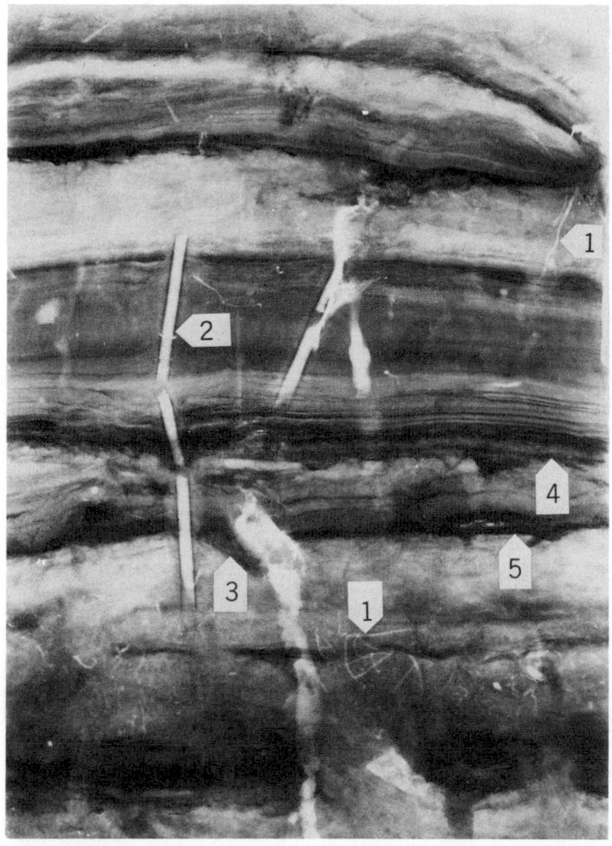

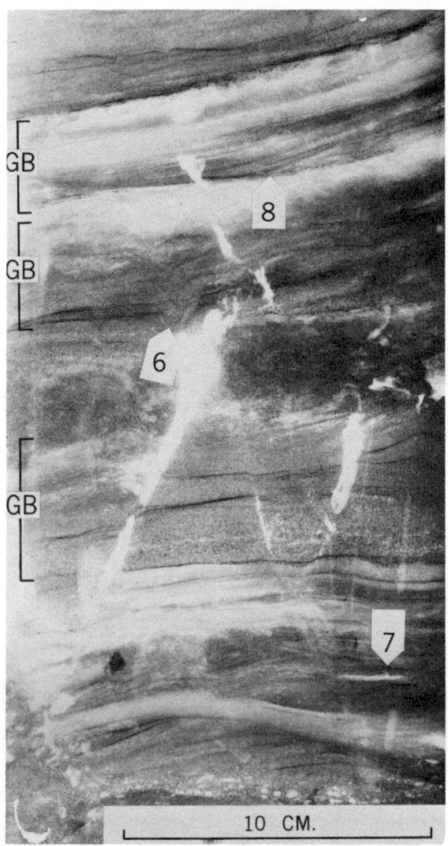

A **B**

Figure 5-6. Contact prints of X-ray radiographs made from vertical sediment slices of box samples collected in Hueneme Canyon, axis environment. Note: Print A shows upper 29 cm of sample HC-H from 343 m water depth. The relatively few biogenic traces present in laminated sediments include small sinuous burrow typical of foraging polychaetes (point 1). lined tubes of suspension-feeding polychaetes (point 2), and disruptions or irregularities in lamination (point 3) caused bv burrowing organisms (in this case a holothurian-type burrow extending downward at the right of point 3). Planar laminated fine sands and silts contain two thin sand layers with erosional lower contacts (points 4 and 5). Below point 5, sediments are more highly bioturbated (30-60 percent; see Figure 5-8), and biogenic structures are more numerous; complex burrow system and remnant primary lamination can be seen at lower point 1. Print B shows sample HC-J (bottom 29 cm of sample) from 377m water depth. Physical structures are dominant, and biogenic traces are rare (point 6 denotes lamination disrupted by mobile burrowing organism). Physical structures, in addition to planar lamination, include thin (<10 cm thick) graded beds (GB), current ripple lamination (point 7), and wedging out of laminae (point 8). Sharp lower contacts of graded beds are erosional. (The inclined white structure to the right of point 6 is artifical, due to partial collapse of core along one side during processing.)

Canyon Floor

A total of five samples were collected from the canyon floor: HC-C at 82 m water depth, HC-F at 25 m, HC-G at 254 m, HC-I at 339 m, and HC-K at 389 m. Sample HC-I collapsed during processing, and no X-ray radiographs or relief casts were obtained. A core description was obtained by carefully dissecting the sample, and a bulk sediment sample was collected and analyzed, thus permitting classification of the sample.

Samples from the central part of Hueneme Canyon outside the axis are characteristically highly bioturbated, and biogenic structures are very abundant compared to samples from the canyon axis. Examination of X-ray radiographs reveals isolated primary physical structures in some samples. Sediments from this group are finer grained and less well sorted than axis sediments. Frequency curves (see Chart E, Figure 5-4) typically exhibit bimodal distribution with primary and secondary modes in the 3 ϕ to 5 ϕ (0.125 mm to 0.031 mm) range (very fine sand to coarse silt).

As can be seen in Figure 5-7, cores HC-C (print A) and HC-G (prints B1 and B2) typify sediments outside the canyon axis. Both are highly bioturbated, although a few remnant primary physical structures are visible in core HC-C (point 9 on print A, Figure 5-7). Vertical open, lined dwelling tubes typical of suspension-feeding polychaetes are very abundant in the 2 cm thick slice of sample HC-C (point 10 on print A, Figure

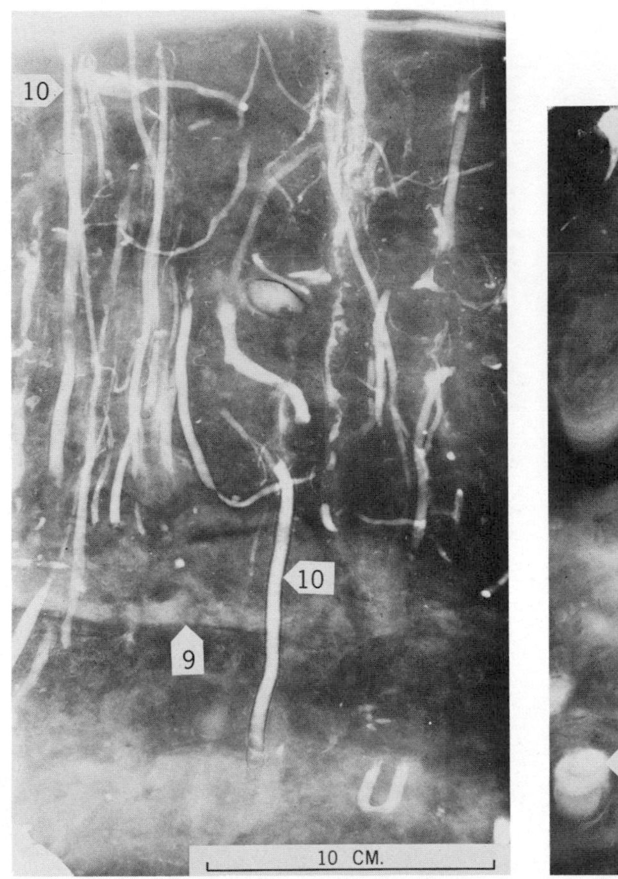

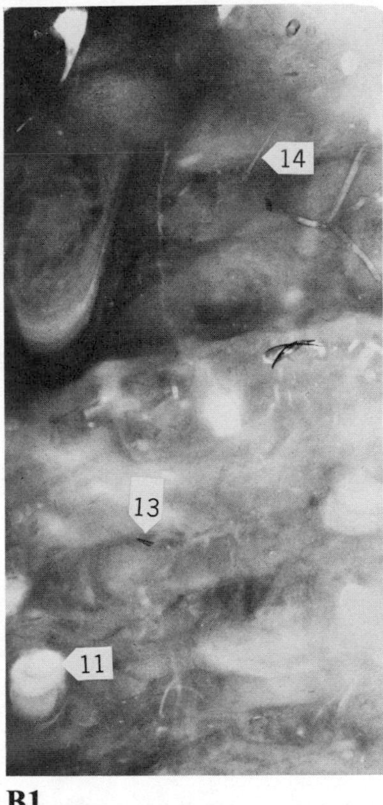

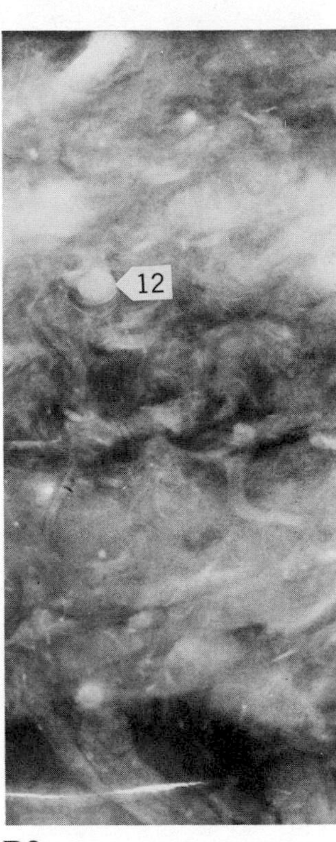

A **B1** **B2**

Figure 5-7. Contact prints of X-ray radiographs of samples representative of Hueneme Canyon floor environment. Note: Print A shows sample HC-C (upper 30 cm of sample) from 82 m water depth. Fine sandy muds are highly bioturbated (90-99 percent), and few remnant physical structures remain (point 9). Biogenic structures include numerous small polychaete burrows and a dense accumulation of lined tubes typical of suspension-feeding polychaetes (point 10); lower portions of longer tubes are sometimes sand-filled and uninhabited. Prints B1 and B2 show upper (B1) and lower (B2) portions of sample HC-G from water depth of 254 m. Fine sandy muds are highly bioturbated (100 percent, and no remnant physical structures are present. Abundant biogenic traces include open (point 11) and filled (point 12) burrows with concentric wall structures and numerous unidentifiable trails (especially in B2). The biogenic structure at point 13 resembles the trail made by a heart urchin moving laterally. Arenaceous tests ("trumpet tubes") of *Pectinaria* sp. (point 14, in living position) are found in both canyon axis and canyon floor environments.

5-7). A few of these tubes extend the entire length of the core (51 cm), although the lower portions were typically uninhabited and sand filled. Inhabited tubes more commonly extend to an average depth of 25 cm. (The apparent decrease in tube density at approximately 15 cm in print A, Figure 5-7 is due to the steeply inclined orientation of the tubes, which resulted in the lower portion of most tubes extending beyond the limits of the 2 cm thick sediment slice and thus necessitated their removal during processing.) We should note that this type of dwelling tube is a recent analogue of the trace fossil *Skolithos* that, where abundant, is considered indicative of a littoral or shallow sublittoral, relatively high-energy marine environment (Seilacher 1967; Frey 1975, Table 2.1). Similar tubes were observed, though not in the densities present in core HC-C, in several cores in Hueneme Canyon, the deepest being HC-K at a water depth of 388 m. This observation indicates that the distribution of endobenthic organisms within the canyon is not strictly controlled by water depth or sediment type and that organisms respond to other favorable stimuli, such as food supply and energy conditions, produced by down-canyon currents.

Sample HC-G (prints B1 and B2, Figure 5-7) lacks the dense accumulation of lined dwelling tubes like those found in core HC-C, although tubes are present in the core. As shown in Figure 5-7, abundant biogenic traces in both the upper (print B1) and lower (print B2) portions of the core include open and sand-filled burrows, some with concentric wall structures (points 11 and 12). A heart urchin (species unknown) was recovered at the surface of this sample, and lebensspuren patterns similar to those ascribed to this echinoderm by Reineck (1963b), Hertweck (1972), and Howard et al. (1974) are present, although not well

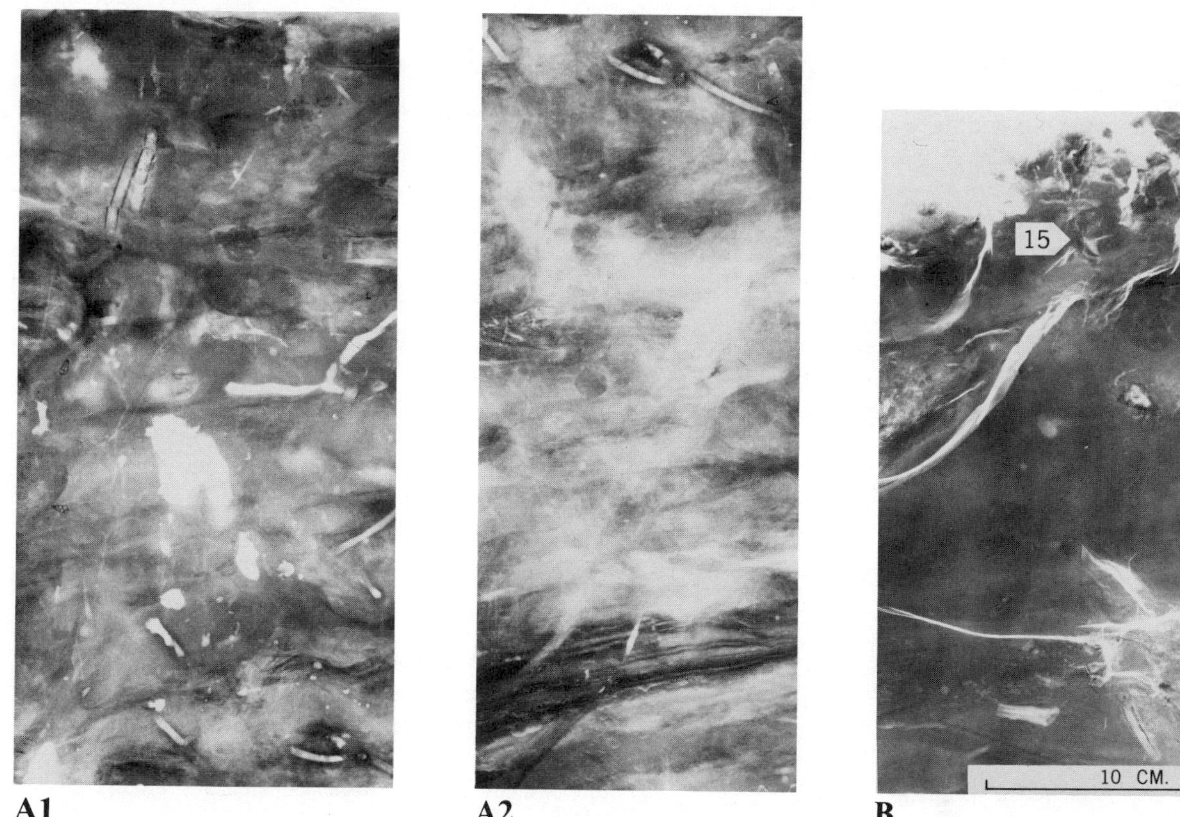

Figure 5-8. Comparison of contact prints of X-ray radiographs of samples collected from Hueneme Canyon floor and the canyon wall. Note: Prints A1 and A2 show the upper (A1) and lower (A2) portions of sample HC-K representing the canyon floor environment from water depth of 389 m. The muddy fine sands are highly bioturbated (60-90 percent), and biogenic traces are numerous. Remnant physical structures are present throughout most of the core, and the lower portion of the sample (bottom of print A2) is only slightly bioturbated. Print B shows sample HC-E (total core) representing the canyon wall environment from 189 m water depth. The surface layer (upward from point 15) consists of saturated muds with oxidized burrow fragments and clasts (point 15) as large as 2 cm in diameter. The remainder of the sample consists of thinly laminated, dense, cohesive muddy sands.

defined (point 13 on print B1, Figure 5-7). Heart urchins are generally considered to inhabit relatively coarse-grained sediments; however, Dörjes and Hertweck (1975, p. 475) report traces of *Echinocardium cordatum* in muddy, lower offshore sediments in the Gulf of Gaeta, Italy. Bouma (1965, Figure 14, p. 312) reported similar lebensspuren in muddy sediments from Santa Maria Canyon off the southern tip of Baja California.

Another characteristic of samples from the canyon floor environment is the presence of large quantities of fecal pellets in the sediments (except for sample HC-K). The pellets are typically elongated, with tapered, rounded ends, and are less than 1 mm in length. These pellets probably represent surficial accumulations that have been redistributed during subsequent bioturbation. The abundance of fecal pellets also suggests that suspension feeding organisms may contribute significant quantities of fine-grained sediment ($< 4\ \phi$, 0.0625 mm) that might normally be kept in suspension by bottom currents. This process, known as biogenic pelletization, is a well-known phenomenon in shallow-water marine environments (Pryor 1975). Isolated concentrations of fecal pellets present in several cores represent traces of deposit-feeding organisms such as *Pectinaria* sp. (point 14 on print B1, Figure 5-7) that were found in sediments from the canyon axis and canyon floor environments.

We consider samples HC-F and HC-K to be atypical of the canyon floor environment in that they contain higher percentages of sand (see Chart E, Figure 5-4) and are relatively less highly bioturbated (see Figure 5-5) when compared with the other three samples in this group. Remnant, thin planar lamination is present throughout the upper, bioturbated portion (80 percent) of core HC-K (print A1, Figure 5-8). And the lower portion of the core consists of lightly bioturbated, thinly laminated fine sand and silt (print A2, Figure 5-8). Textural characteristics and the presence of planar lamination suggest a close similarity to canyon axis sediments. Bouma (1965) observed that most cores collected outside the canyon axis were bedded, which thus suggests that sand is occasionally transported outside of the axis environment. Alternatively, this sample could represent either an inactive or abandoned axial channel, or intermittent sediment transport along this portion of the canyon axis. The latter condition would permit benthic organisms to keep pace with deposition and/or physical reworking of the sediments and would enable development of the thick sequence (44 cm) of bioturbated sediments. Combined moderate bioturbation and sparse remnant lamination throughout core HC-F suggest constant (uninterrupted) deposition and perhaps a relatively low abundance of benthic organisms.

A slow rate of continuous sediment accumulation in the canyon floor environment is evidenced by the degree to which sediments are bioturbated, thereby indicating that the rate of biogenic reworking is roughly equivalent to the rate of deposition. Also the infilled lower portions of inhabited, lined dwelling tubes indicates that the organisms are able to keep pace with sediment accumulation by extending the tube vertically. Therefore the authors feel that the sequence of fine bioturbated sediments probably represents sediment accumulation over a period of several seasons.

Canyon Wall

Two cores were collected from the canyon wall: HC-E at 189 m water depth and HC-L at 400 m. Canyon walls in the upper, sampled portion of Hueneme Canyon consist of Lower Pleistocene sediments of the San Pedro Formation that represent a shallow marine depositional environment (Greene 1976). Canyon wall sediments collected in this study are characteristically very dense, cohesive, poorly sorted muddy sand. The surface of canyon wall cores was typically covered with a saturated, soupy fluff layer, 3 to 5 cm thick, composed of mud, oxidized fragments of burrow linings, and subangular to subrounded clasts (up to 2 cm in diameter) of the underlying muddy sand (point 15 on print B, Figure 5-8). Canyon wall sediments contain an average of 25 percent medium to coarse sand compared to 5 percent and 3.5 percent medium to coarse sand in sediments from the canyon axis and canyon floor, respectively (see Table 5-2 and Chart F, Figure 5-4). Cores from the canyon wall averaged 21.25 cm in length compared to an average of 45.6 cm for cores from the other two environments, which thus reflects the relative resistance to penetration of canyon wall sediments.

Biogenic traces are not abundant in samples from the canyon wall, although a few open and sand-filled burrows, some as large as 1.5 cm in diameter, are present. Localized bioturbation is visible in X-ray radiographs, but samples are predominantly thinly laminated (print B, Figure 5-8). Convolute bedding observed in one of two sediment slices from sample HC-E may represent disturbance during coring operations or possibly *in situ* deformation. Pelecypod and gastropod shells and shell fragments recovered from sample HC-L have a bleached and yellowed "relict" appearance. Our two cores indicate that hydrocarbon residues, similar to those found in samples from the canyon axis and canyon floor, are not present in sediments from the canyon wall.

SUMMARY AND CONCLUSIONS

Cores collected in Hueneme Canyon represent three submarine canyon environments, each with a distinctive set of internal sedimentary structures and/or textural characteristics: the canyon axis, the canyon floor (exclusive of the axis), and the canyon wall.

Canyon axis sediments are characterized by abundant primary physical structures, and biogenic structures are relatively rare. The dominance of primary physical structures indicates that currents are of sufficient strength to transport and rework sediments in this portion of the canyon. Ripple lamination and wedging out of laminae suggest traction transport and erosion, and sequences of thin ($<$ 10 cm thick) graded beds suggest deposition from either small-scale mass sediment flow or a current of gradually decreasing velocity, or both.

Canyon floor sediments are typically highly bioturbated and contain a wide variety of biogenic traces. Although remnants of primary physical structures were observed in some samples, they are not common. This set of internal structures and typically high percentages of silt suggest that current activity is relatively lower when compared to the canyon axis environment.

Canyon wall sediments consist predominantly of thinly laminated, poorly sorted, cohesive, muddy sand; biogenic traces are not abundant. No hydrocarbon residue was present in canyon wall sediments.

Sediments from Hueneme Canyon consist predominantly of fine sands and silts (2 ϕ to 5 ϕ; 0.25 mm to 0.031 mm) with very little medium to coarse sand and no clays. The paucity of coarse sands indicates that relatively little canyon sediment is derived at present from littoral sources. Clays and fine silt are maintained in suspension by canyon floor currents that commonly exceed velocities of 10 cm per second.

With the exception of two cores collected from the canyon wall (HC-E and HC-L), all canyon sediment samples contained small amounts of water insoluble hydrocarbon residue, probably derived from a nearshore source.

Biogenic traces indicate that mobile benthic organisms are present in sediments from the canyon axis and abundant in canyon floor sediments. Tubes of suspension-feeding organisms were present in most samples from the canyon floor environment. Distribution of benthic organisms appears to be controlled by current activity and ability to adapt to periodic rapid sedimentation and erosion.

This suite of cores indicates that transport of sand-sized sediment is channeled along the canyon axis in the upper portion of Hueneme Canyon. Two samples, HC-H and HC-K, exhibit internal structures characteristic of both the canyon axis and canyon floor environments. This binding suggests either periodic shifts in the canyon axis position or intermittent sediment transport along portions of the axial channel.

We should emphasize that the sediments described herein (excluding the two canyon wall samples) represent deposition over a relatively short period of time, perhaps a few years at most. The authors consider that the described sedimentary structures reflect the "normal" day-to-day processes of sediment transport and deposition occurring within the submarine canyon environments rather than the less frequent catastrophic events such as large-scale sediment gravity flows and turbidites. During mass sediment flows, sedimentary deposits within the submarine canyon would presumably be eroded and/or reworked. Thus structures representative of such catastrophic events would be preferentially preserved in the sedimentary record.

ACKNOWLEDGMENTS

The authors wish to thank J. D. Howard and R. W. Frey for their counseling, encouragement, and editorial assistance. The authors also express sincere appreciation to D. S. Gorsline for his critical review and discussion of the paper and to the crew of the R/V *Velero IV* for their invaluable help in obtaining the samples. The field and laboratory assistance provided by C. Aenchbacher, J. Howard, Jr., and E. Warren is sincerely appreciated.

Discussions of problems with R. Bruno, H. E. Reineck, R. E. Carver, and P. Pinet have been helpful and are deeply appreciated. The authors gratefully acknowledge F. P. Shepard, N. Marshall, and P. McLoughlin for providing current data obtained in Hueneme Canyon.

Research was supported by Oceanographic Section, National Science Foundation, NSF Grant GA-39999X.

REFERENCES

Bouma, A. H., 1962. *Sedimentology of Some Flysch Deposits, a Graphic Approach to Facies Interpretation.* Elsevier, Amsterdam, 168 pp.

——, 1964. Notes on X-ray interpretation of marine sediments. *Mar. Geol.*, 2: 278–309.

——, 1965. Sedimentary characteristics of samples collected from some submarine canyons. *Mar. Geol.*, 3: 291–320.

——, 1969. *Methods for the Study of Sedimentary Structures.* Wiley-Interscience, New York, 458 pp.

Dörjes, J., and Hertweck, G., 1975. Recent biocoenoses and ichnocoenoses in shallow-water marine environments. In: R.W. Frey (ed.), *The Study of Trace Fossils.* Springer-Verlag, New York, pp. 459–91.

Drake, D. E., and Gorsline, D. S., 1973. Distribution and transport of suspended particulate matter in Hueneme, Redondo, Newport, and La Jolla Submarine Canyons, California. *Geol. Soc. Amer. Bull.*, 84: 3949–68.

——, Kolpack, R. L., and Fischer, P. J., 1972. Sediment transport on the Santa Barbara-Oxnard Shelf, Santa Barbara Channel, California In: D.J.P. Swift, D. B. Duane, and O. H. Pilkey (ed.), *Shelf Sediment Transport.* Dowden, Hutchinson & Ross, Stroudsburg, Pa., pp. 307-331.

Felix, D. W., and Gorsline, D. S., 1971. Newport submarine canyon, California: An example of the effects of shifting loci of sand supply upon canyon position. *Mar. Geol.*, 10: 177–98.

Folk, R. L., 1974. *Petrology of Sedimentary Rocks.* Hemphill Publ. Co., Austin, 182 pp.

Frey, R. W., 1975. The realm of ichnology, its strengths and limitations. In: R. W. Frey (ed.), *The Study of Trace Fossils.* Springer-Verlag, New York, pp. 13-38.

Gorsline, D. S., 1970. Report on a reconnaissance survey of hydrographic characteristics of the Hueneme-Mugu shelf to evaluate ground-water leakage from submarine exposures to coastal aquifers, June-July 1970. *Univ. Southern Calif., Geol. Rept.* 70-6, Los Angeles, 21 pp.

Greene, H. G., 1976. Late Cenozoic geology of the Ventura Basin, California. In: D. G. Howell (ed.), *Aspects of the Geologic History of the California Continental Borderland.* Amer. Assoc. Petrol. Geol., Pacific Section, Misc. Publ. 24, pp. 499-529.

Herron, W. J., and Harris, R. L., 1966. Littoral bypassing and beach restoration in the vicinity of Port Hueneme, California. *Proc. X International Conf. of Coastal Engineering,* Tokyo, pp. 651-75.

Hertweck, G., 1972. Georgia Coastal Region, Sapelo Island, U.S.A.: Sedimentology and biology. V: Distribution and environmental significance of lebensspuren and *in situ* skeletal remains. *Senckenberg. Marit.*, 4: 125–67.

Hjulstrom, F., 1939. Transportation of detritus by moving water. In: P. D. Trask (ed.), *Recent Marine Sediments.* Amer. Assoc. Petrol. Geol., Tulsa, pp. 5-31.

Howard, J. D., and Frey, R. W., 1975. Estuaries of the Georgia Coast, U.S.A.: Sedimentology and biology. I: Introduction. *Senckenberg. Marit.*, 7: 1–31.

Howard, J. D., Reineck, H. -E., and Rietschel, S., 1974. Biogenic sedimentary structures formed by heart urchins. *Senckenberg. Marit.*, 6: 185–201

Ingram, R. L., 1971. Sieve analysis. In: R. E. Carver (ed.), *Procedures in Sedimentary Petrology.* Wiley-Interscience, New York, pp. 49-67.

Inman, D. L., 1970. Strong current in submarine canyons. (Abst.), *Amer. Geophy. Union. Trans.*, 51:319.

Malcolm, R. L., 1968. Freeze-drying of organic matter, clays, and other earth materials. *U. S. Geol. Surv. Prof. Paper* 600-C, pp. C211-C216.

McBride, E. F., 1971. Mathematical treatment of size distribution data. In: R. E. Carver (ed.), *Procedures in Sedimentary Petrology.* Wiley-Interscience, New York, pp. 109-27.

Pryor, W. A., 1975. Biogenic sedimentation and alteration of argillaceous sediments in shallow marine environments. *Geol. Soc. Amer. Bull.*, 86: 1244–54.

Reineck, H. -E., 1958. Kastengreifer und Lotröhre "Schnepfe." *Senckenberg. Leth.*, 39: 42–48 and 54–56.

_____, 1963a. Der Kastengreifer. *Natur. u. Museum*, 93: 102–08.

_____, 1963b. Sedimentgefüge im Bereich der sudlichen Nordsee. *Abhandl Senckenberg Naturforsch Ges.*, 505: 1-138.

_____, and Singh, I. B., 1973. *Depositional Sedimentary Environments.* Springer-Verlag, New York, 439 pp.

Savage, R. P., 1957. Sand bypassing at Port Hueneme, California. *Beach Erosion Board, Army Corps of Engineers, Tech. Memo.* 92, pp. 1-34.

Seilacher, A., 1967. Bathymetry of trace fossils. *Mar. Geol.*, 5: 413–28.

Shepard, F. P., 1976. Tidal components of currents in submarine canyons. *Jour. Geol.*, 84: 343–50.

_____, and Dill, R. F., 1966. *Submarine Canyons and Other Sea Valleys.* Rand McNally, Chicago, 381 pp.

_____, and Emery, K. O., 1941. Submarine topography off the California coast: Canyons and tectonic interpretations. *Geol. Soc. Amer. Spec. Paper* 31, 171 pp.

_____, and Marshall, N. F., 1973. Currents along floors of submarine canyons. *Amer. Assoc. Petrol. Geol. Bull.*, 57: 244-64.

_____, Marshall, N. F. and McLoughlin, P. A., 1974a. Currents in submarine canyons. *Deep-Sea Res.*, 21: 691–706.

_____, Marshall, N. F., and McLoughlin, P. A., 1974b. "Internal waves" advancing along submarine canyons. *Science*, 183: 195–98.

Tissier, M., and Oudin, J. L., 1973. Characteristics of naturally occurring and pollutant hydrocarbons in marine sediments. *American Petroleum Institute, Environmental Protection Agency and U.S. Coast Guard, Joint Council on Prevention and Control of Oil Spills Conference, Washington, D.C.* pp. 205–14.

Chapter 6

Bioerosion in Submarine Canyons

JOHN E. WARME

Department of Geology
Rice University
Houston, Texas

RICHARD A. SLATER

Department of Earth Sciences
University of Northern Colorado, Greeley, Colorado

RICHARD A. COOPER

National Oceanic and Atmospheric Administration
Woods Hole, Massachusetts

ABSTRACT

Erosion by invertebrates and fishes has been documented in submarine canyons cut into the continental shelves off Georges Bank and off Southern California by using observations and collections made with scuba and manned submersibles. A high diversity of bioeroders is present, but they vary with geographic location, depth, and substrate. Bioerosion may presently be the dominant erosional process along the length of most canyons studied.

INTRODUCTION

Marine scientists have long known that many organisms bore by mechanical and chemical means into rocks cropping out in the shallow marine environment (e.g., Turner 1954; Yonge 1955; see reviews by Bromley 1970 and Warme 1975). With the advent of scuba (Limbaugh and Shepard 1957) and by observation from deep-diving submersibles (Moore 1965), this phenomenon was found to occur in deeper ocean depths. The process whereby borers erode rock outcrops, termed *bioerosion* by Neumann (1966), is in contrast with penetration and stirring of soft sediment by burrowers, or *bioturbation*, although the latter process is also important on the floors of submarine canyons (e.g., Trumbull and McCamis 1967; Shepard and Dill 1966).

Bioerosion of submarine canyon walls was first described by scuba-diving scientists exploring the heads of Scripps and La Jolla Canyons (Limbaugh and Shepard 1957; Dill 1964; Warme and Marshall 1969; Warme et al. 1971). Bioerosion of canyon walls below the scuba limit was documented more recently (Dillon and Zimmerman 1970; Palmer 1976), as well as bioerosion of large angular rock fragments lying on a canyon floor (Shepard and Marshall 1975). Most observers have omitted the names of the organisms responsible for the bioerosion, usually because they have only been able to observe and not sample them. Furthermore, a common difficulty is determining which animals are responsible for the primary borings, which ones secondarily modify abandoned borings, and which ones merely reside in preexisting holes (nestlers).

Bioerosion was not believed to be a major erosional process in submarine canyons until recently. A Georges Bank submarine canyons study (Cooper et al., in preparation) and numerous submersible dives in other submarine canyons by the authors as well as other recent studies (Dillon and Zimmerman 1970; Palmer 1976) substantiate the fact that on a small scale, such as that visible from a submersible or identifiable in a small outcrop sample, animals are the major eroders of most present submarine canyon walls. Rates of bioerosion are believed to be quite rapid, especially in some soft, semiconsolidated outcrops where over half of an initial rock volume is excavated to a depth of 10 cm from the surface in geologically short periods of time, probably in only a few years or decades (Warme and Marshall 1969). On a larger scale, landsliding, slumping, and other processes may be important, but they are more difficult to identify from submersibles, especially due to light limitations below the photic zone.

GEORGES BANK SUBMARINE CANYONS

Georges Bank submarine canyons of the New England continental shelf and slope (Figure 6-1) have been the subject of intensive investigations by the National Marine Fisheries Service from 1973 to 1977. Over sixty submersible dives have been made in Corsair, Lydonia, Gilbert, Oceanographer, Hydrographer and Veatch Canyons from depths of 150 m to 2,000 m. Ecology and geology of the canyon heads was documented on each of these dives.

Bioerosion, on a local scale, appears to be the major source of erosion in the Georges Bank submarine canyons, being observed on most of the submersible dives in the canyons. The bioeroders are (1) primary borers (mostly crustaceans), (2) secondary borers, crustaceans, and finfish enlarging abandoned borings, and (3) nestlers occupying preexisting holes with some degree of enlargement and shaping.

The most extensive bioerosion was observed in Veatch Canyon between depths of 150 m and 1,000 m, where the substrate is a semiconsolidated Pleistocene clay. In and near the head of the canyon the local bottom slopes range from about 1° to vertical. Bioerosion was evident throughout this gradient range and appeared most extensive between 20 and 50°, in areas we have termed "Pueblo Village" communities because of their likeness to the cliff dwellings of some western American Indian tribes. A "Pueblo Village" is described by Cooper et al. (in preparation) as a relatively localized area of a submarine canyon wall where megabenthic crustaceans and finfish have intensively bioeroded depressions and borings into the substrate and have occupied these sites.

Figure 6-2 portrays a 2.5 m by 2 m section of a "Pueblo Village" that was initially sketched by Cooper from the submersible *Nekton Gamma* at a depth of 185 m in the northeast corner of the head of Veatch Canyon. The slope there was approximately 45°. Borings in the substrate can be classified into three types on the basis of their size, shape, and degree of penetration. Progressing from the smallest, these types are as follows:

1. Relatively shallow depressions that measure up to approximately 10 cm across and that penetrate the substrate up to 6 cm and vary from circular to half-moon shapes. These depressions are most commonly occupied by galatheid crabs (*Munida* sp.) and juvenile jonah crabs (*Cancer borealis*). Six galatheids, noted by their long slender chelipeds, are shown in Figure 6-2.
2. Half-moon-shaped depressions that project horizontally into the substrate up to approximately 15 cm and up to 20 cm in width. This type of depression is generally occupied by juvenile and adult jonah crabs, as shown in the upper left and right portions of Figure 6-2.

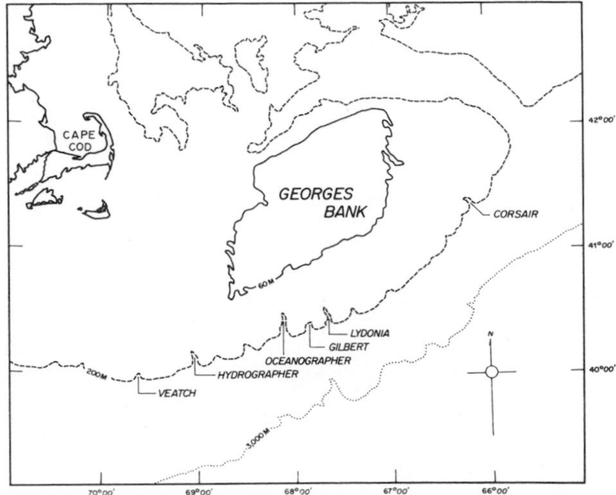

Figure 6-1. Distribution of Georges Bank submarine canyons investigated for bioerosion.

3. Borings that measure up to 1.5 m in width, 1 m in height, and 2 m or more in depth. These caverns are commonly occupied by juvenile and adult American lobsters (*Homarus americanus*), tilefish (*Lopholatilus chamaeleonticeps*), cusk (*Brosme brosme*), and cleaner shrimp (*Lysmata* sp.). Less frequent occupants are the octopus (*Octopus bairdii*), short big-eye (*Pseudopriacanthus altus*), and black-bellied rosefish (*Helicolenus dactylopterus*). Three of the latter are depicted in Figure 6-2. Cleaner shrimp were frequently observed evenly distributed along the outer roof and walls of these caves. Black-bellied rosefish are commonly oriented with respect to some of the cave entrances and appear to have a commensal relationship with the lobster. Between 5 and 10 percent of the Type 3 excavations have multiple openings; several caves were observed to have three openings.

The smallest excavations are occupied by juvenile galatheid and jonah crabs, which we believe are the primary borers or excavators (see Palmer 1976). Adult crabs and juvenile and adult lobsters occupy the next larger array of excavations. Jonah crabs and lobsters have been observed enlarging these depressions and pushing sediment to the foreground or entrance of the depressions with the chelipeds and the dactyl portions of the first pair of walking legs. Large adult lobsters (greater than 90 mm carapace length) and tilefish have been observed shaping and/or enlarging the larger excavations of the "Pueblo Village." Tilefish enlarge their caves by entering head first and going through a violent side-to-side stationary swimming motion that expels water from the excavation that is laden with suspended mud washed from the walls.

Figure 6-2. Artist's rendition of *in situ* sketches made of "Pueblo Village" community at the head of Veatch Canyon, 185 m depth, representing a 2 m by 2.5 m section of the canyon wall. Note: The bottom slope of Pleistocene clay substrate is about 45°. The community includes tilefish, American lobster, black-bellied rosefish, jonah crab, galatheid crab, and cleaner shrimp (see text).

Figure 6-2 illustrates the anterior view of a 3-4 kgm lobster and a 10-15 kgm tilefish at the entrance to their Type 3 excavations. The high density of depressions, caves, and organisms illustrated in Figure 6-2 is not representative of "Pueblo Village" communities in general, but the figure does illustrate the association of excavators and nestlers with the canyon substrate.

Bioerosion features similar to those illustrated in Figure 6-2 were also observed by us at depths of 300 m to 1,000 m, but were occupied primarily by the red crab (*Geryon quinquedens*) and the jonah crab. These excavated habitats are primarily of the Types 1 and 2 bioerosion categories. We believe that the juvenile jonah and red crabs are the primary borers with the adults enlarging the excavations. Type 3 excavations were rarely observed below 300 m.

Observations by Stanley et al. (1972), using underwater television in Wilmington Canyon (about 500 km southwest of Georges Bank), suggest that large fish and lobsters are agents of bioerosion there as well:

> . . . lobsters and colonies of ling cod were observed inhabiting hollows and crannies in outcrops. These, at the least, must maintain the hollows free of sediment and they may be capable of enlarging them at depths of 110 - 135 m [p. 627].

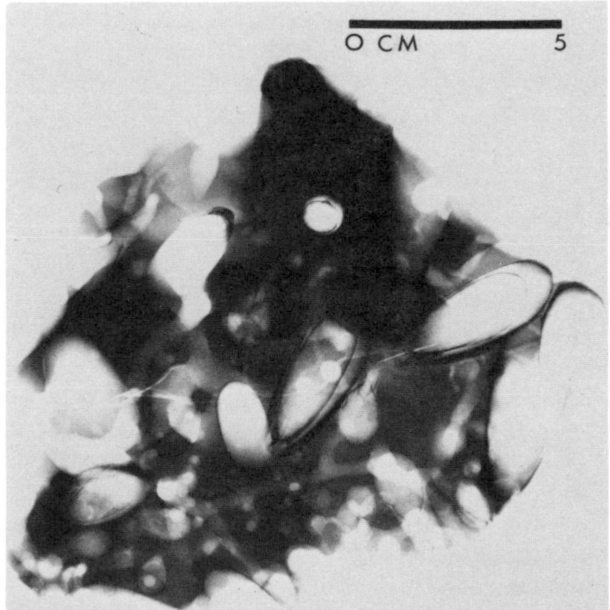

Figure 6-3. Radiograph of Eocene Ardath Shale from Scripps Canyon, 25 m depth. Note: Sample is approximately 3 cm thick, with longest dimension of 15.5 cm. Larger oval borings are of *Lithophaga* and the smaller ones are of *Nettastomella*; these are interconnected by small, branching borings with multiple openings made by polychaetes and/or crustaceans. (Radiograph, courtesy N.F. Marshall.)

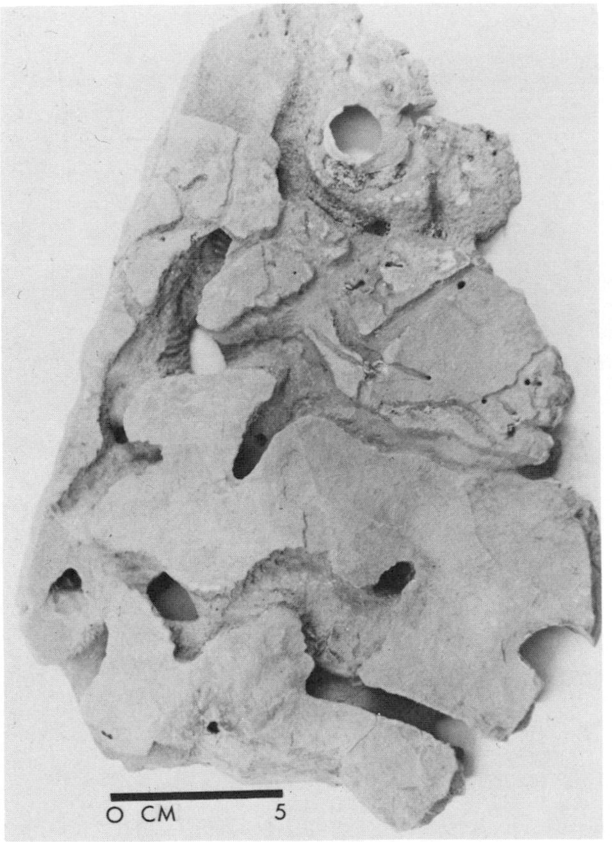

A

B

Figure 6-4. Slab of Eocene Ardath Shale from Scripps Canyon, 25 m depth. Note: In photograph A the sample is broken along bedding planes, approximately 1.5 cm thick, with longest dimension of 24 cm. Bivalve borings have been interconnected by crabs and larger worms (larger passageways) and smaller worms (smaller passageways). The radiograph (print B) of this sample shows connecting passageways within the rock. (Radiograph and photo, courtesy of N.F. Marshall.)

LA JOLLA AND SCRIPPS SUBMARINE CANYONS

La Jolla and Scripps canyons off Southern California serve to illustrate how different substrates attract different suites of borers. Shepard and Dill (1966) describe conditions at the head of La Jolla Canyon, which is semiconsolidated Pleistocene mud at 15 m to 30 m depths. We have documented the activities of pholadid bivalves (e.g., *Zirfaea*) and large gaper clams (*Panopea, Tresus*), as well as crabs, shrimp, ophiuroids, and other invertebrates that bore, burrow, and nestle in the stiff muds. In contrast, at similar depths nearby Scripps Canyon is cut into much harder outcrops of calcareous and noncalcareous mudstone and sandstone of the Eocene Ardath Shale (formerly Rose Canyon Shale; Shepard and Dill 1966). These harder rocks contain a largely different assemblage of borers that can be seen only by breaking away the substrate (Figures 6-3 and 6-4). Pholadids are of the genera *Penetella, Parapholas, Chacea* and *Nettastomella,* and the mytilid bivalves include *Lithophaga* and *Adula*. Several genera of crustaceans and polychaete worms are effective borers in these hard rocks (Warme and Marshall 1969; Warme 1970, 1975; Warme et al. 1971). Recently, McHuron (1976) has documented the many species of molluscs, polychaete annelids, sipunculids, crabs, and other borers that infest the canyon walls and has described the details of their borings. Initial larger penetrations usually are accomplished by the bivalves whose borings are then interconnected by the vagrant crustaceans and worms, thereby yielding a honeycombed substrate teeming with borers joined by a host of other residents (Figures 6-3 and 6-4).

Biological penetration is exceedingly important in erosion of these submarine canyons. Thoroughly lithified and cemented samples of the Ardath Shale from Scripps Canyon show weathering rinds only 1 cm to 2 cm deep next to outcrop/seawater interfaces and weathering halos only a few mm wide around borings. One sample from 50 m depth was tightly cemented and appeared dry when broken open; yet it was penetrated by bivalve borers for distances up to 8 cm. Clearly, within the local habitat of the megabenthos, processes of bioerosion are important and perhaps dominant in penetrating, excavating, and disintegrating exposed bedrocks or loose bedrock clasts in these West Coast canyons.

DISCUSSION AND CONCLUSIONS

The effects of boring and the other processes of bioerosion are demonstrably important in eroding the terrigenous sedimentary rocks exposed along the walls and floors of submarine canyons we have studied, just as they are in eroding shells, reefs, and other calcareous substrates as is well documented in intertidal and shallow water sites (Warme 1975). A diversity of animals bore, nestle, and otherwise occupy and modify hard substrates outcropping on the sea floor, and the walls of submarine canyons are especially susceptible to such bioerosion. Because the three examples we present here (the Georges Bank canyons and Scripps and La Jolla canyons) are each unique in the kinds of bioeroders present, the size and geometry of their excavations, and the substrates that they penetrate, there stands a strong likelihood that an even greater spectrum of bioerosion processes and products exists in other submarine canyons. Not only do the suites of hard substrate macrobenthos differ between localities, they also contrast markedly with faunas of the soft and level seabed that exist under otherwise similar ecological conditions such as temperature and depth. With the increased use of scuba and submersibles by biologists and geologists, a better understanding of the distribution, variation, and significance of these phenomena is sure to follow.

We do not wish to imply that borers are responsible for submarine canyons. Undoubtedly, during Pleistocene (and Tertiary?) lowered stands of sea level, sediment supply to canyon heads was more readily available, and sediment transport in most canyons was more vigorous than at present. These factors, as well as perhaps many others, probably account for the genesis and maintenance of submarine canyons, a topic not under consideration in this chapter. Furthermore, we realize that large-scale sliding, slumping, and creep are associated with submarine canyons (see Chapters 7 and 8). We wish to stress, however, that bioerosion is currently the most common small- to moderate-scale erosional phenomena (within scuba or submersible view) in the canyons and at the depths thus far investigated and that the erosional products of bioerosion are probably an important component of the sediment budget in some canyon systems.

Rates of bioerosion in submarine canyons have not been measured *in situ*. By a census of abundances of borers, and the sizes of their borings, and by assuming longevities for the animals, a model for bioerosion rates can be constructed (Warme 1975). Longevities may be greater than we are accustomed to consider, as evidenced by the large sizes of the fishes and crustaceans of the Georges Bank canyons and the integrated network of excavations in the California canyon walls. Even so, the sizes and densities of excavations suggest geologically significant bioerosion rates in the localities.

ACKNOWLEDGMENTS

The authors thank all those who have contributed to their studies. The manned submersibles and surface support ships used in the Georges Bank study were supplied by the National Oceanic and Atmospheric Administration's Manned Undersea Science and Technology Office, Rockville, Maryland. The authors wish to thank Mr. Frank Bailey, artist and draftsman at the Northeast Fisheries Center, Woods Hole, Massachusetts, for the "Pueblo Village" drawing and N.F. Marshall for providing radiographs and photograph of Scripps Canyon samples. Bioerosion studies in California were supported by grants to Warme from the National Science Foundation (GB 14321) and the Henry L. and Grace Doherty Charitable Foundation, Inc.

REFERENCES

Bromley, R. G., 1970. Borings as trace fossils and *Entobia cretacea* Portlock, as an example. In: T. P. Crimes and J. C. Harper (eds.), *Trace Fossils*. Geol. J. , Spec. Issue 3, pp. 49-90.

Dill, R. F., 1964. Sedimentation and erosion in Scripps Submarine Canyon head. In: R. L. Miller (ed.), *Papers in Marine Geology*, Shepard Commemorative Volume: McMillan, New York, pp. 23-41.

Dillon, W. P., and Zimmerman, H. B., 1970. Erosion by biological activity in two New England submarine canyons. *J. Sed. Petrol.* 40: 542-47.

Limbaugh, C., and Shepard, F. P., 1957. Submarine canyons. In: J. W. Hedgpeth (ed.), *Treatise on Marine Ecology and Paleoecology*. Geol. Soc. Amer. Mem., 67: 633-39.

McHuron, E. J., 1975. *Biology and paleobiology of marine boring organisms.* Unpublished Ph.D. Dissertation, Rice University, Houston, 280 pp.

Moore, D. G., 1965. Erosional channel wall in La Jolla Sea-fan Valley seen from bathyscaphe *Trieste II. Geol. Soc. Amer. Bull.*, 76: 385-92.

Neumann, A. C., 1966. Observations on coastal erosion in Bermuda and measurements on the boring rate of the sponge, *Cliona lampa. Limn. Oceanogr.*, 11: 92-108.

Palmer, H. D., 1976. Erosion of submarine outcrops, La Jolla Submarine Canyon, California *Geol. Soc. Amer. Bull.*, 87: 427-532.

Shepard, F. P., and Dill, R. F., 1966. *Submarine Canyons and Other Sea Valleys*. Rand McNalley, Chicago, 381 pp.

——, and Marshall, N. F., 1975. Dives into outer Coronado Canyon system. *Mar. Geol.*, 18: 313-23.

Stanley, D. J., Fenner, P., and Kelling, G., 1972. Currents and sediment transport at Wilmington Canyon Shelfbreak, as observed by underwater television. In: D. J. Swift, D. B. Duane, and O. H. Pilkey (eds.), *Shelf Sediment Transport.* Dowden, Hutchinson & Ross, Stroudsburg, Pa., 656 pp.

Trumbull, J. V. A., and McCamis, M. J., 1967. Geological exploration in an East Coast submarine canyon from a research submersible. *Science*, 158: 370-72.

Turner, R. D., 1954. The Family Pholadidae in the Western Atlantic and Eastern Pacific: Part I–Pholadinae. *Johnsonia,* 3: 1-64.

Warme, J. E., 1970. Traces and significance of marine rock borers. In: T. P. Crimes and J. C. Harper (eds.), *Trace Fossils*, Geol. J., Spec. Issue 3, pp. 515-26.

_____, 1975. Borings as trace fossils and the processes of marine bioerosion. In: R. W. Frey (ed.) *The Study of Trace Fossils*. Springer-Verlag, New York, pp. 181-227.

_____, and Marshall, N. F., 1969. Marine borers in calcareous terrigenous rocks of the Pacific Coast. *Amer. Zool.*, 9: 765-74.

_____, Scanland, T. B., and Marshall, N. F., 1971. Submarine canyon erosion: Contribution of marine rock burrowers. *Science,* 173: 1127-29.

Yonge, C. M., 1955. Adaptation to rock boring in *Botula* and *Lithophaga* (Lamellibranchia, Myt lidae), with a discussion of the evolution of this habit. *Quart. J. Microscop. Sci.*, 96: 383-410.

Part II

Gravity-Induced Processes in Submarine Canyons and Fans

Chapter 7

Large Storm-Induced Sediment Slump Reopens an Unknown Scripps Submarine Canyon Tributary

NEIL F. MARSHALL

Geological Research Division
Scripps Institution of Oceanography
La Jolla, California

ABSTRACT

The occurrence of a marine slump near Scripps Institution of Oceanography, and the fortuitous discovery of that slump, has provided the opportunity to develop concepts pertaining to the initiation of such slumps and their significance in the dynamic process of sediment transport to the deeper sea floor and to describe their particular morphology as it applies to the marine environment. Differential wave pressure, caused by large storm waves, is suggested as a mechanism for the initiation and headward advance of shallow water slumps on slopes of relatively low gradient.

INTRODUCTION

A recent slump, which developed during a storm, introduced approximately 10^5 m^3 of sediment into the axis of Scripps Canyon, La Jolla, California. This slump formed another tributary head for the canyon that extends approximately 400 m into unconsolidated shelf sediments to a depth of about 16 m. Exposed bedrock indicates that a tributary had previously existed, but had been filled and covered by recent deposits. Unique features were formed, such as secondary slumps perpendicular to the main slump trend; some were interspersed with sharply peaked ridges, while others were broadly arcuate.

After the event, sediments within the slump valley were supersaturated and had little cohesion. They remained in this state for some time and finally approached their previous state of stability after six months. The stability of the sediments on the shelf, even within 0.5 m of the scarps, remained unchanged from their preslump condition. A series of basin depressions observed along the valley floor suggest progressive slumping.

Even though slumping is a widely recognized mechanism for the progressive downslope movement of rocks and sediments to the deeper ocean, slumps in the marine environment have seldom been analyzed, and insight as to their significance in marine processes is limited. Shepard (1951), Chamberlain (1960), Dill (1964), and others have suggested that slumping and creep are important mechanisms for the movement of sediment down submarine canyons.

DESCRIPTION OF THE SLUMP

During the last week in May 1975, Nöel Davis, a marine ecology graduate student at Scripps Institution of Oceanography, noted during a regular underwater inspection of her research area northeast of Scripps pier, La Jolla, California (Figure 7-1) that a significant change had occurred in its depth. During a similar inspection of this area several days earlier, this feature was not in evidence. Be-

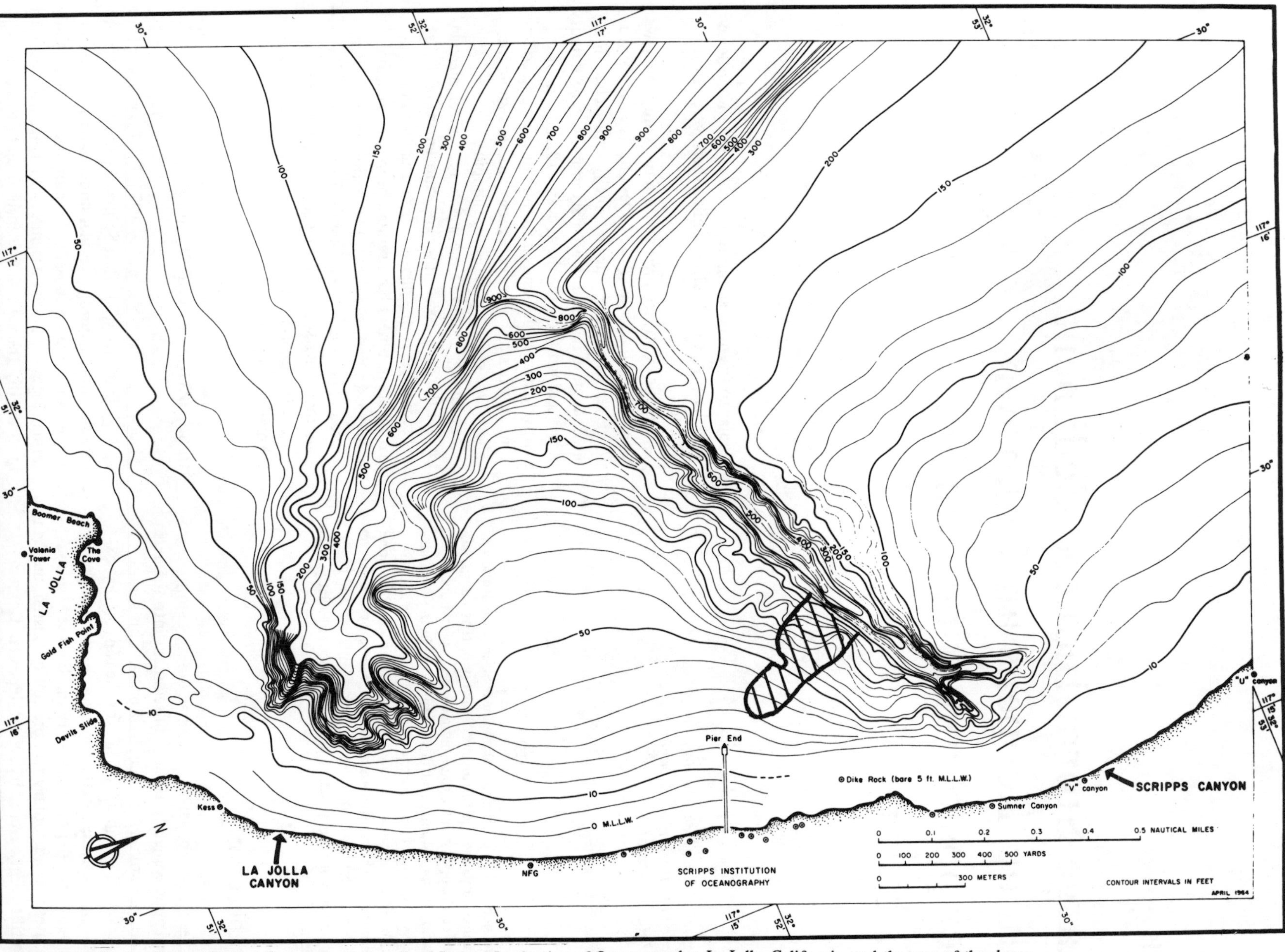

Figure 7-1. The location of Scripps Canyon to the north of Scripps Institution of Oceanography, La Jolla, California, and the area of the slump.

tween these two visits, a significant storm had occurred. The following week, an inspection of the area was made by the author.

At a depth of from 9 m to 11 m along the normally very gently sloping, sandy shelf, an abrupt scarp of about 1.8 m was observed. Many dead sea pens (*Stylatula elongata*) were noted strewn about the sea floor in depressions at the base of the scarp. The sediment surface below the scarp had an unusual pock-marked appearance, similar to a rain-pitted mud surface. These numerous 1 cm to 2 cm features are considered to reflect the escape of pore water from the sediment. The condition of the sediment above and within 0.5 m of the scarp was little changed from that of the surrounding shelf, but below the scarp it was manually oscillated with ease. Forcing one's arm into the sediment to a depth of nearly one meter—which is, according to this author's experience, an impossible feat in a normal shelf sediment—was quite feasible. Subsequent inspections indicated that the sediments remained highly saturated for about three to four weeks after which compaction effects became noticeable; however, almost six months had passed before they approached their previous state of stability. Although the cohesion of the sediment on the face of the scarp had been reduced during the slump, an angle of repose of around 30° was maintained. Numerous tiny secondary slumps had occurred on this scarp face. The sediments consist of fine sand, and the median grain size is approximately 177 microns.

Subsequent to the slump, a bathymetric survey was conducted and a chart of the area was prepared (Figure 7-2). The slump formed an irregular valley that is approximately 400 m in length, 8 m to 30 m deep, and as much as 100 m wide. Based on the apparent loss of sediment, it introduced approximately 100,000 m^3 of sediment into Scripps Canyon. The upper portion of the resulting valley is broadly rounded and bowl-shaped, and where the valley becomes narrow, the walls become steeper. At about mid-valley, the westerly wall turns to the west. Beyond this point, the wall becomes straighter and appears more like the headward section of the main slump. Farther to the west, another valley is shown by the new survey. Whether this valley is related to the new slump or to a previous episode is not known.

A chart of the same area (Figure 7-3), which is based on the most complete soundings available (Shepard 1951), indicates that other slumps to the left and right of the slump valley have left scars on the canyon slopes. The chart of the slump valley shows that some older slumps have been filled and are no longer recognizable. However, two other large tracts appear to have failed since the older ones were filled with sediments. For example, the depression east of the new slump indicates that between 1947 and 1975 another failure approximately equal in dimensions to the new slump valley had occurred. In addition, another slump valley in this general area was noted by Shepard in 1950 (Shepard 1951). These data suggest that slumps may occur frequently on the side slopes of canyon heads.

The volume of material removed by the slump is shown in the isopach map in Figure 7-4. It shows that the greatest volume of sediments was displaced near the steeper slopes close to the canyon, which is also shown by the profiles A-A' in chart B of Figure 7-4 that represent the shelf before and after the slump. The location of the profiles is shown in Figure 7-2.

Continued exploration of the slump valley by scuba diving has revealed some features of marine slumps that are perhaps unique to the marine environment. In the central portion of the valley the easterly wall is broadly scalloped (photograph A, Figure 7-5), while on the westerly wall there is a series of well defined sharp-peaked ridges and V-shaped cuts. Similar sharp ridges and cuts were observed by Reimnitz (personal communication) at Cedros Island off western Mexico. That slump had been accidentally initiated by pile driving at a construction project. These sharply defined features are not noted when slope failures occur on land and are probably related to the highly saturated condition of the sediment. Underwater slump scars observed in fine sediments and in deep water do not seem to display this structure, which suggests that they may be related to sediments of coarser grain size or perhaps to locations where liquefaction has been induced. In the Scripps Canyon slump valley numerous ripple marks on the upper valley walls show the effect of currents (photograph B, Figure 7-5). Farther down the walls, outcrops of shelf sediments occur (photograph C, Figure 7-5).

A unique feature of the sloping walls of the slump valley was the remnants of burrowing structures left protruding from the slope by as much as 12 cm (photo graph A, Figure 7-6). These were later destroyed, presumably by wave action, although they occurred in depths of 25 m. An explanation for these relatively fragile structures having been left intact after the slump is that winnowing from wave and current action, after liquefaction and slumping, gently removed loose saturated sediment from the slopes, thereby leaving these slightly indurated features exposed. Below the exposure of shelf sediments, exposed bedrock (photograph B, Figure 7-6) shows that at least the lower portion of the slump valley had been a small bedrock tributary to Scripps Canyon, and the attached remnant shell of an adult scallop, *Hinnites multirugosa*, that only lives on exposed rock surfaces indicates that this bedrock tributary has in the past been open for reasonably long periods.

The fact that sediment had substantially filled this tributary at the point where the steep slopes of the canyon begin, and where greater structural weakness versus sediment

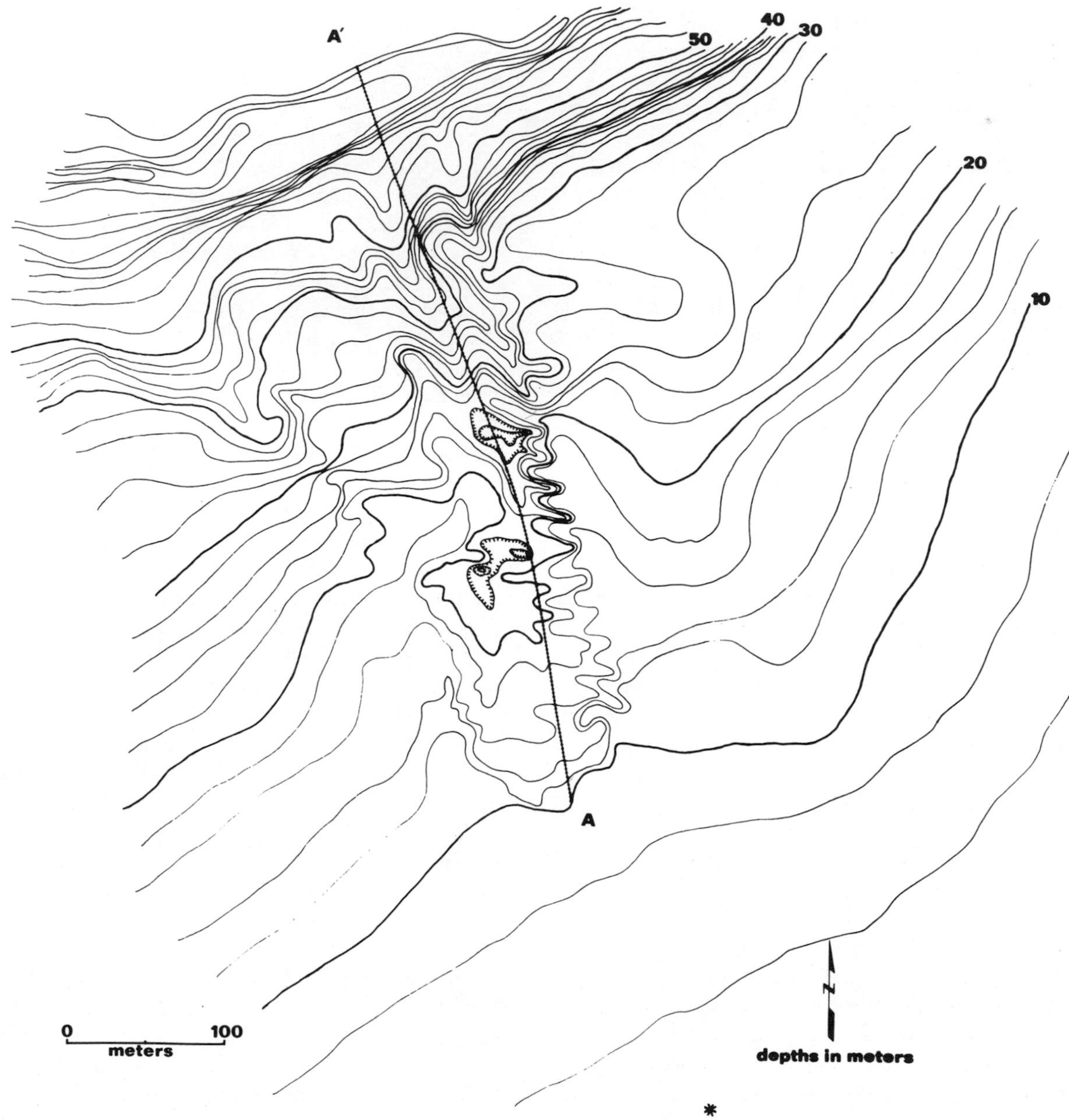

Figure 7-2. Bathymetric chart of the slump valley on the southerly flank of Scripps Canyon. Note: Contours are every 2 m to 50 m and every 10 m thereafter. The profile A-A' is represented in chart B, Figure 7-4. The small asterisk, lower center right, indicates the seaward end of Scripps pier.

depth and slope angle could occur, suggests that gravitational effects may have been significant. However, since the upper half of the valley was not associated with bedrock control and was on a more gently prograding slope, liquefaction appears to have been a significant factor in forming the valley head, which is substantiated by the condition of the sediments described earlier.

The process of gravitational slumping produces somewhat different results than liquefaction and sediment flow; however, both processes may cause a valley to be formed. Gravitationally induced slumps probably do not form true valleys. In all likelihood, liquefaction and flow aided by gravitationally induced failure are the conditions experienced in the formation of the headward and central section of the main valley of the slump in Scripps Canyon. These concepts will be further explored in the following sections along with a discussion on the causes of marine slumps on steep slopes and their contribution toward moving sediment to the deep sea.

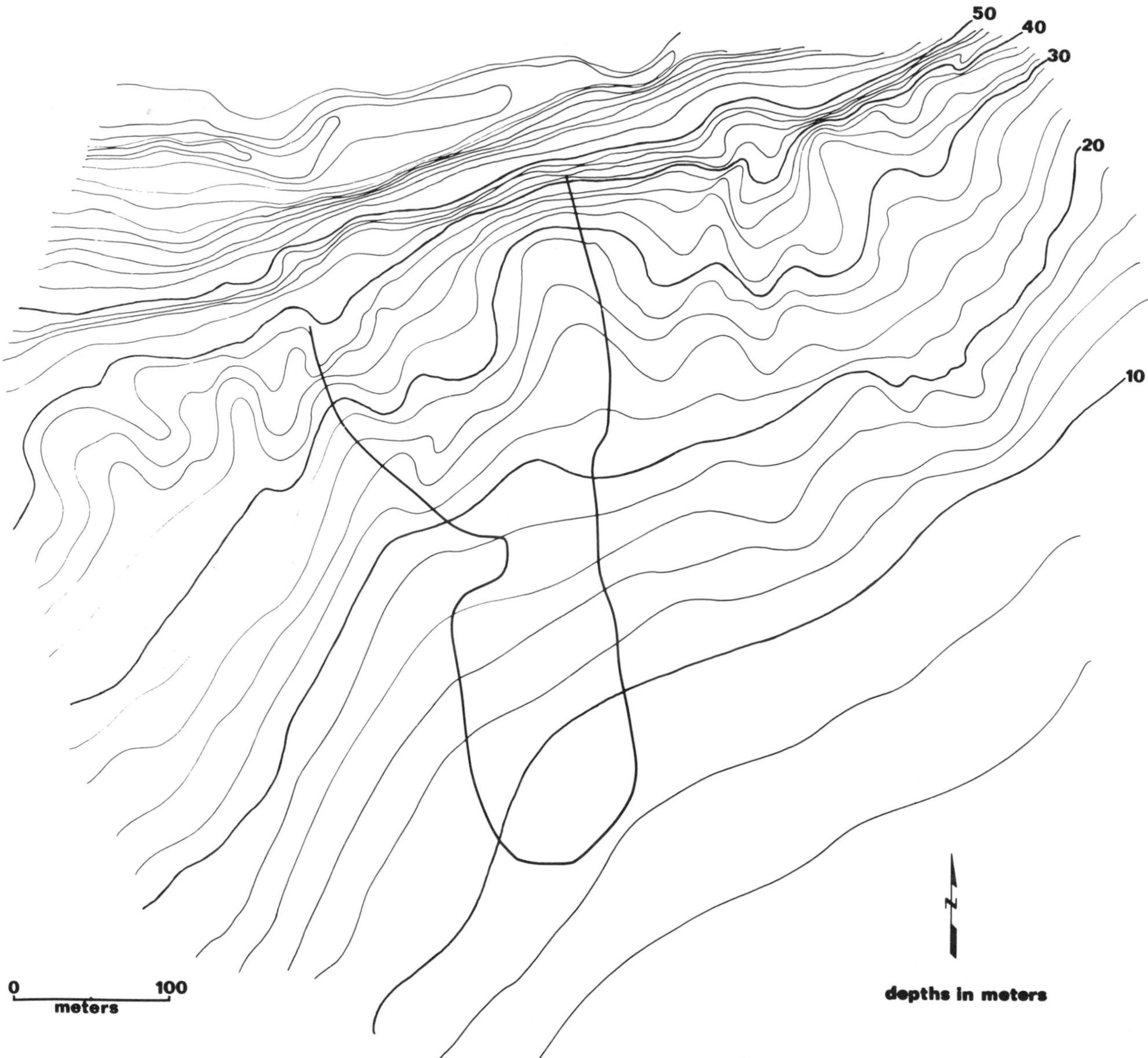

Figure 7-3. Bathymetry of the area of the slump valley with its outline superimposed on the chart. Note: Contour interval is every 2 m to 50 m and every 10 m thereafter. (Source: Based on soundings from 1947 by F.P. Shepard [1951].)

SLUMPS AND THEIR CHARACTERISTICS

As sediments accumulate on the continental shelf, they tend to modify the erosional cross-sections of narrow canyons by forming deposits with slopes that trend upward from the steeper bedrock walls. These slopes will, if undisturbed, lie at the angle of repose that is dictated by the particular internal cohesive forces of the sediment in question. Other factors, such as the binding effect created by bioturbate structures and organic growth, the variation of sediment type, and the deposition of salts, may also affect the angle of repose of sediments. As deposition continues, these slopes will become locally oversteepened, and slumps will occur.

Dill et al. (1975) have indicated that slumping on the slopes of canyons is an important process contributing shelf sediments to the floors of submarine canyons. These authors observed large numbers of slump scars on the upper canyon walls and slopes in the Rio Balsas Canyon, Mexico. Here, silty clay deposits at approximately 1,200 m to 1,500 m depth maintained an arcuate or scalloped appearance. The great spur ridges or buttresses formed between the scallops were separated by as much as 100 m to 300 m and are the remnants separating slumps. These buttresses are the sites most likely to experience future slumping. Similar features were observed along the upper walls in Pelekunu (Figure 7-7) and Amikopala Canyons off Molokai Island, Hawaii (Marshall and Hollister, in preparation).

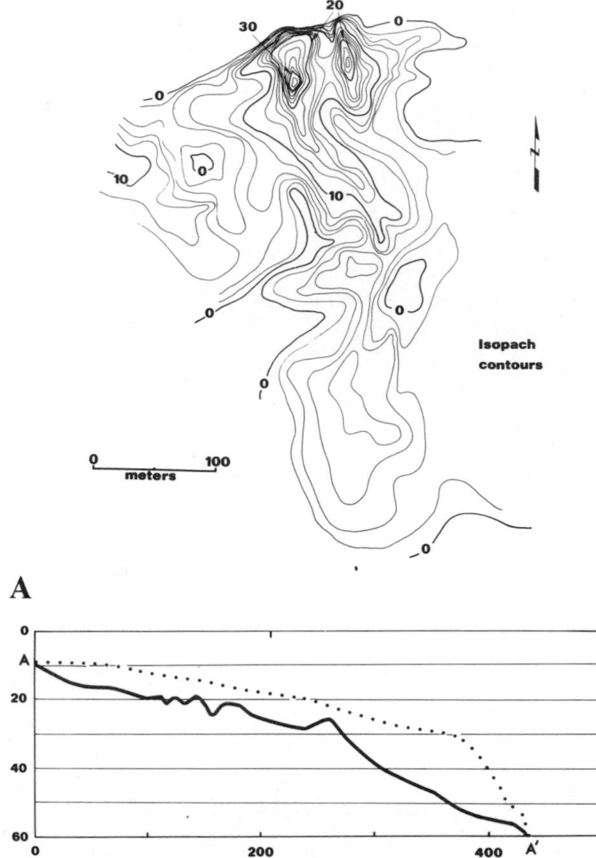

Figure 7-4. Volume and movement of sediment in the area of the slump. Note: Map A shows the isopach contours of the slump valley prepared as the difference between depths from Figures 7-2 and 7-3. Contour interval is 2 m. Chart B shows the before (dotted line) and after (solid line) profile sections through the slump valley. Location of the profile is indicated in Figure 7-2.

These slumps had not formed valleys like the one at La Jolla.

Slumps observed on the deeper slopes of canyons made relatively shallow cuts into adjacent deposits, and their disposition is directly related to very steep slopes. These slumps do not cut back deeply enough to form valleys, and we can assume that they are caused by gravitationally induced slope failure.

Large linear slumps, such as the one discussed here, may be the result of shallow water processes. However, since in all likelihood earthquakes induce cyclic loading (periodic application of stress) on marine sediments, they are no doubt a factor in generating slumps, and should slump valleys form in deep water, we suggest that earthquakes are the responsible mechanism. There were no significant earthquakes recorded in the La Jolla area during the period when the slump occurred.

SLUMPING AND DOWNSLOPE SEDIMENT MOVEMENT

Shepard et al. (1974) have shown that the normal current regime within submarine canyons is adequate to transport sediments and that net transport is downcanyon. Shepard and Marshall (1973) further suggest that movement of sediments during storm periods is a significant factor in downcanyon transport (also see Chapter 1).

Thus, this slump is likely to have formed a coherent turbidity or density flow that transported sediment for some distance down the canyon. This flow, when coupled with the energy of a very large storm, should be able to transport sediment over considerable distances, perhaps to the basins and troughs beyond the confines of a submarine canyon. In all likelihood, some turbidite sequences on submarine fans far from shore are related to turbidity flows caused by currents generated during exceptional storms (Shepard and Marshall 1973; Inman et al. 1976).

Any slump material remaining on the floor of a canyon after such an event will form a deposit blocking the axis of the narrow canyon. Behind this deposit, sediment will accumulate in a process leading to the more usual, evenly graded canyon floor. Normal canyon floor current activity is adequate to fill such a depression and even to cut the floor beyond the deposit or alternatively to erode and disperse the sediments.

Since lateral slumping introduces large quantities of sediment into submarine canyons at fairly frequent intervals, it is a very important process. Therefore, given the suggestions presented, consideration of the basic mechanisms of slumping within the marine environment is appropriate.

FAILURE IN MARINE SEDIMENTS

At present, the mechanical properties of noncohesive marine sediments are poorly understood. However, various authors have studied failure in saturated sediments, and some generalizations can be made. Concepts relating to liquefaction and cyclic stress are of importance to the interpretation of the factors pertaining to the new slump valley.

In general, if a saturated sand is subjected to cyclic loading, it will tend to compact and decrease in volume. If drainage is retarded, the tendency to decrease in volume results in an increase in pore water pressure. Should the pressure equal or exceed overburden pressure, the effective stress becomes zero and liquefaction occurs.

Lee and Seed (1967) state that the cyclic stress required to cause failure decreases as the number of stress cycles increases. They further indicate that failure in undrained, dense sands occurs under cyclic loading at levels from 1 to 8 kg/sq cm at around 100 loading cycles within sediments with relative densities from 38 to 78 percent. In addition,

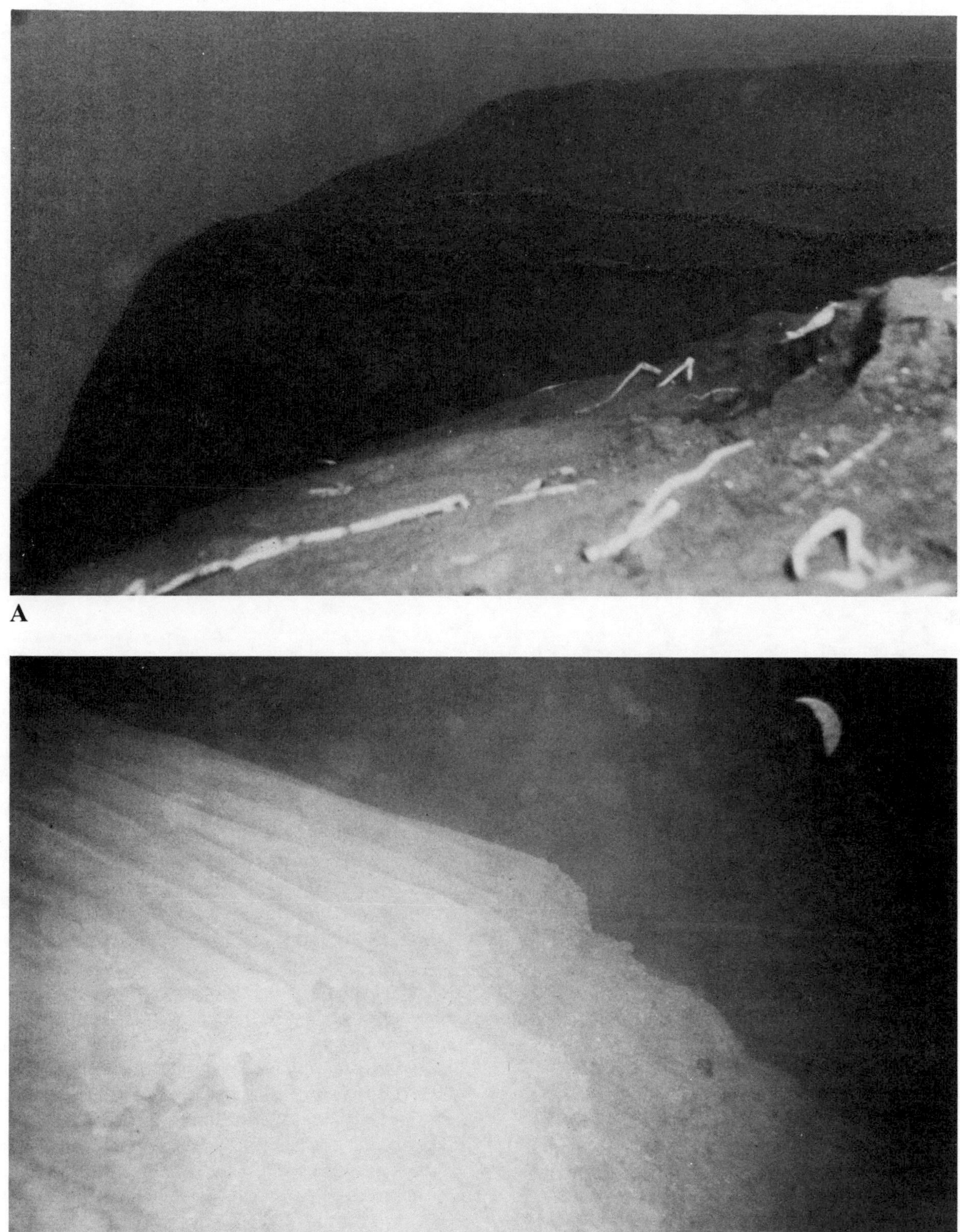

Figure 7-5. Features of marine slumps revealed by Scripps Canyon exploration. Note: Photograph A shows the broad arcuate scalloping of the easterly wall of the slump valley. One buttress is in the foreground with exhumed dead sea pens (*Stylatula elongata*), and another buttress is dimly visible in the background, approximately 20 m distant. Photograph B shows ripple marks along the upper edge of the central slump valley at around 20 m depth. Scale is approximately 10 m across the photograph.

C

Figure 7-5 Continued. Photograph C shows the outcropping layer of shelf sediment midway down the slump valley wall. Note the lack of sea pens here, compared to the upper slope in photograph A. Outcrop face is approximately 0.5 m high.

Finn et al. (1970) suggest that the previous strain history of sediments must be taken into consideration. They state, "Cyclic shear strains of more than 15 cycles of ±2 percent (of maximum loading) have a drastic effect on the resistance of sand to liquefaction in future cyclic loading. The threshold value at which cyclic shear strains weaken resistance to liquefaction probably depends on the number of times a shear strain is cycled" (Finn et al. 1970, p. 1932).

For the liquefaction of saturated sands, Seed and Lee (1966, p. 131) note, "If a saturated sand deposit of uniform density is subjected to uniform cyclic shear strains, . . . liquefaction will occur first in the zone where the overburden pressure is least—at the surface of the deposit."

The foregoing discusses the response of sediments to cyclic stress but does not indicate the methods by which failure or slumping occurs.

MECHANISMS OF SLUMP INITIATION

Slumping is induced in several ways, and in all likelihood more than one process may act to create this condition. Gravity is aided by deposition and oversteepening of slopes and accretion of material. Earthquakes, in conjunction with the above factors or by inducing liquefaction in deposits lying at low angles, are an important cause of slumps (Chamberlain 1960). Waves that have long periods, both storm and tsunami, are another very likely mechanism by which slumping is initiated. Additional factors in slump processes are joints in bedrock, bioerosion, erosive undercutting by currents, and other rock- and sediment-weakening factors. Wave-induced slumping will be explored in detail as it is considered to be the principal cause of the slump discussed herein.

WAVES AND THEIR EFFECT ON SEDIMENTS

Cyclic loading, taking the form of stress reversal, is generated in shallow water where surface water waves pass over the sea floor. Wave height and length become increasingly important as water depth decreases. Henkel (1970) has made calculations for failure on a given slope for fine-grained sediments off the Mississippi Delta. He suggests that sediment-shear failure resulting from differential wave pressure (alternating, loading and unloading, stresses on the underwater sediment surface) may occur in water depths to 125 m.

The depth to which disturbance within sediments occurs due to exceptional wave trains is further discussed by Wright and Dunham (1972). They suggest that for a wave producing a 2,930 kg/m^2 differential wave pressure within a soft clayey sediment, horizontal displacements in the upper 20 m of a deposit may approach 0.5 m to 1.2 m

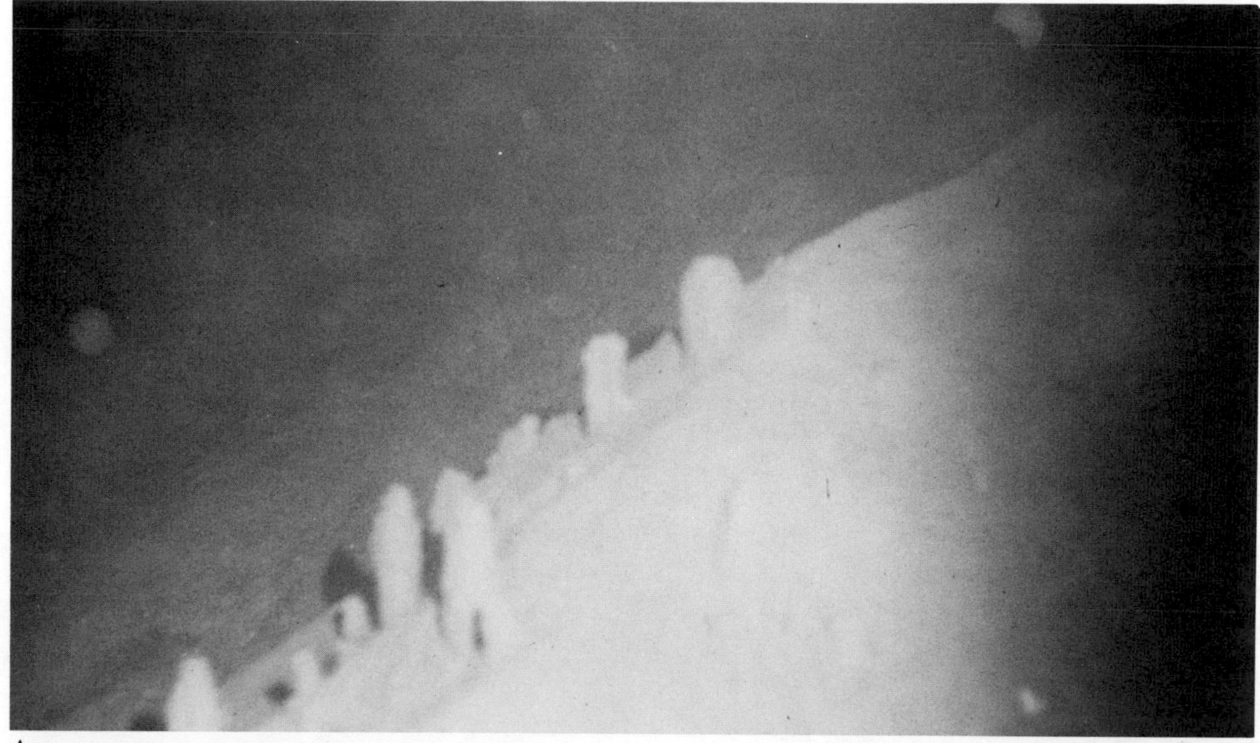

A

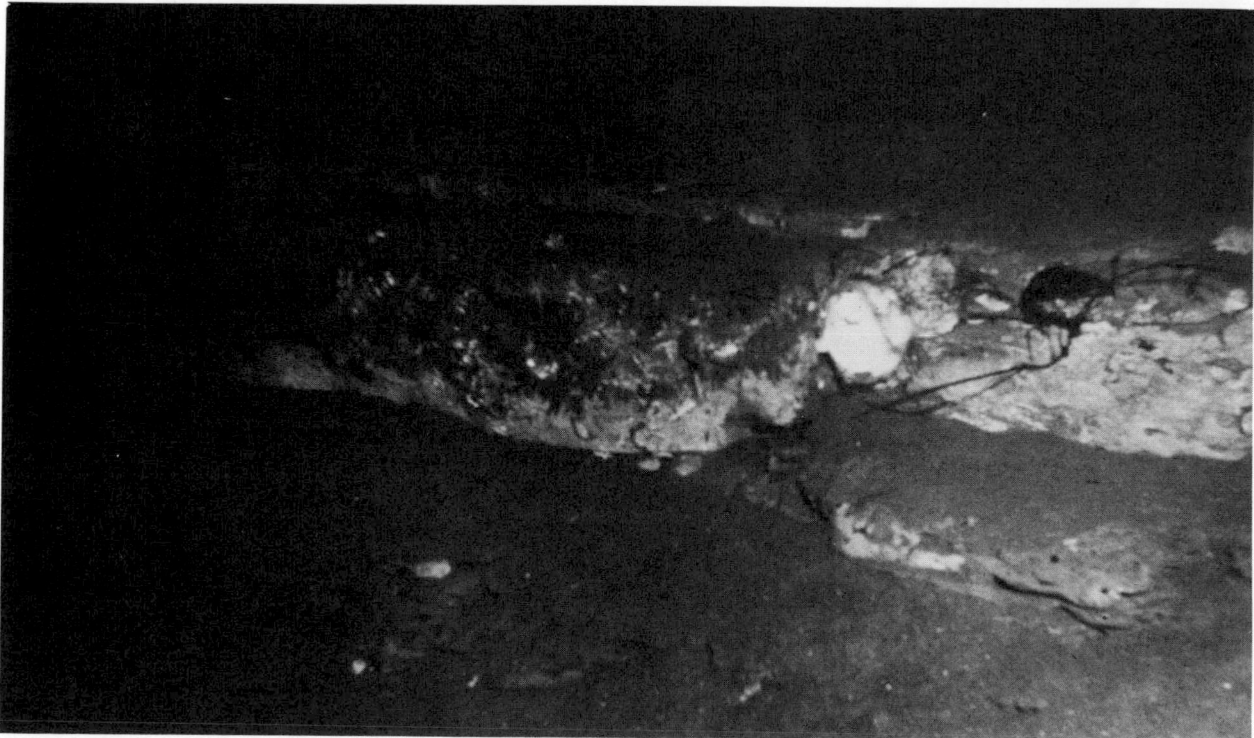

B

Figure 7-6. Unique features of the Scripps Canyon slump. Note: Photograph A shows remnant bioturbate structures on lower slopes of the slump valley. The tallest structure is approximately 15 cm high. Photograph B shows exposure of the bedrock tributary to Scripps Canyon that was exhumed when the slump occurred. The attached remnant adult shell of the purple hinge scallop *(Hinnites multirugosa)*, shown at center right, indicates that this tributary has at times been open for significant periods. Shell is approximately 8 cm across.

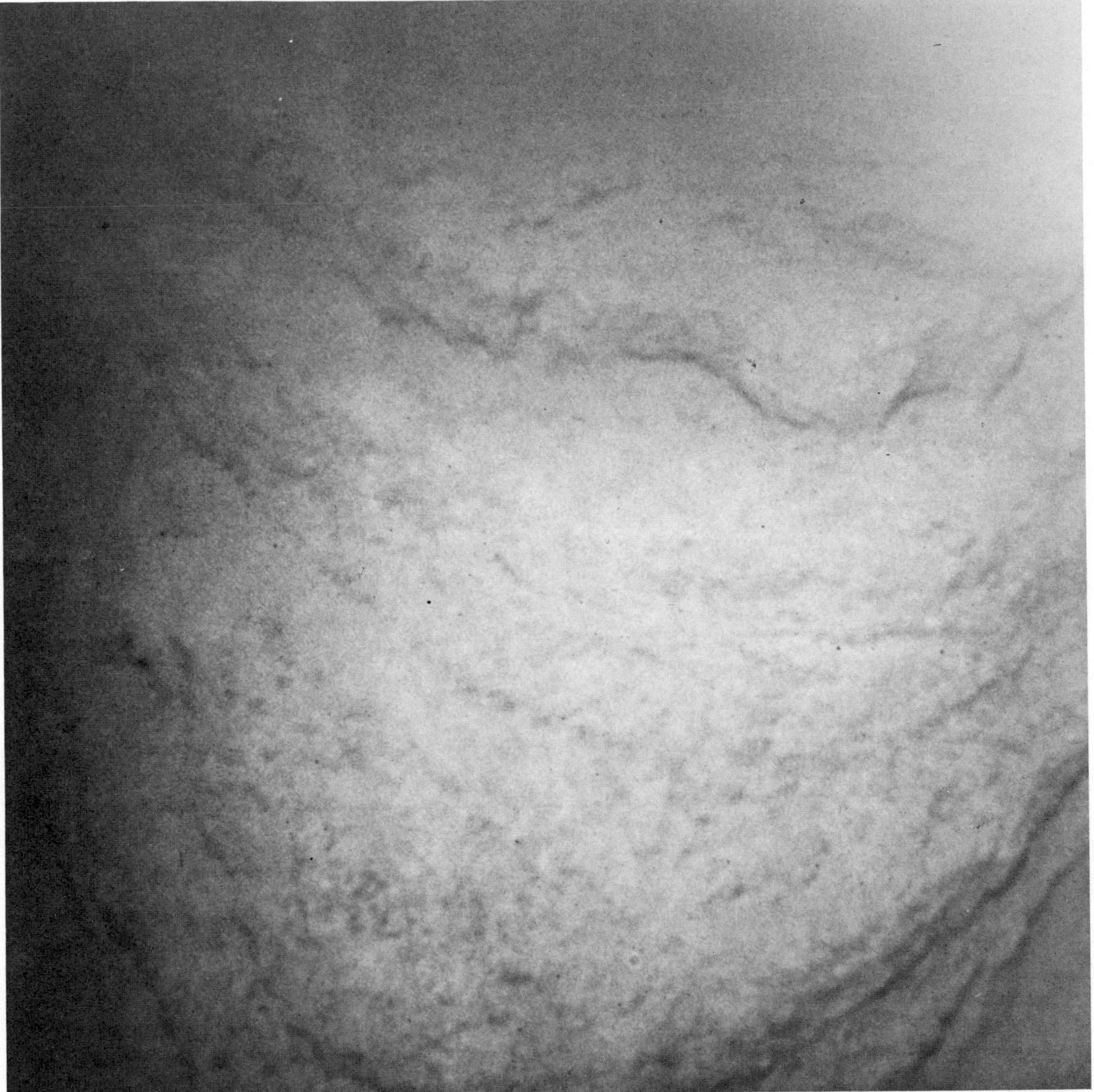

Figure 7-7. Large buttress of sediments between two arcuate slump scars at around 1,500 m depth in Pelekunu Canyon, Molokai Island, Hawaii. Scale is approximately 50 m across photo field. (Photo by C. Hollister.)

depending on the sediment and that displacements of 0.1 m to 0.3 m may likely occur to a depth of 60 m. With a wave pressure of 3,420 kg/m^2, the upper 30 m of sediment will experience 100 percent failure. They further state, "The magnitude of the displacements produced by wave pressure loadings indicate that severe interaction forces may be expected to considerable depths . . . in relatively soft-bottom sediments . . . . [and that] the development of large-bottom displacements and high stress levels may lead to a reduction in the shear strength of bottom sediments; . . . [possibly] leading to the initiation of extensive gravity slides" (Wright and Dunham 1972, p. I-858).

Bea (1971) reports a sediment-strength profile offshore of the Mississippi Delta that indicates a linear increase in strength with depth to 17 m with decreasing strength below that point, a factor which he suggests is related to differential wave pressure.

Although sediment grain sizes differ in the area under study, the preceding statements suggest that a possible mechanism for the formation of this large slump valley was sediment failure induced by differential wave pressure. The difference in the type of lateral infill scars from V- or inverted V-shape to the broad arcuate form of the slump valley itself is indicative that different mechanisms were involved. Although the conditions observed within the slump valley are consistent with the contention that differential wave pressure is responsible, an analysis of the May 1975 storm in comparison with the previously mentioned calculations suggests that the energy available was probably inadequate, by itself, to initiate a slump of this nature.

STORM CONDITIONS AT LA JOLLA

During the winter and spring seasons, La Jolla usually experiences two or three major storms. Figure 7-8 indicates the relationship of the storm suspected of initiating the slump to thirteen other stormy periods that occurred in the preceding two years (unpublished data from S. Pawka, Shore Processes Laboratory, Scripps Institution of Oceanography). Although the graph does not depict all storms for this period, the larger storms are shown. As can be seen, two major storms occurred during each of the two preceding winter seasons, and two out of those four, one in each season, were of higher energy than the storm in question. The maximum differential wave pressure for the storm was probably no greater than 1,000 kg/m^2. This value is somewhat low when compared to the calculations of Wright and Dunham. Since the relationship of the slump to the May 1975 storm is quite firmly established, the information available suggests that other conditions were necessary to set the stage for the final act of slumping. The probability exists, therefore, that if other factors, such as foreslope oversteepening, effects of strain history, and overburdening, are at an optimum, storm energy levels need not be as high as those calculated. Another factor besides those already mentioned might be the convergence by refraction at the location of the slump. Apparently, however, an energy level similar to that of the storm was all that was necessary to cause slumping and subsequent liquefaction.

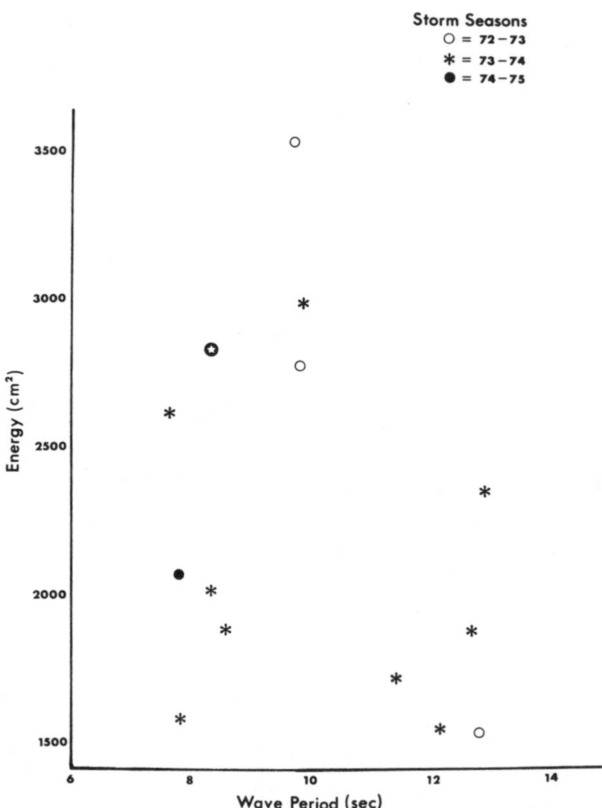

Figure 7-8. Relative energy of storms at the location of the slump, expressed as the sums of the surface area of the spectral peaks, in cm^2. Note: The storm related to the slumping is indicated as a black circle with a star in the center. Two other storms of greater relative energy occurred, one in each preceding year.

CONCLUSIONS

Slumping is one of the most active processes bringing sediment to submarine canyon floors, where the activities of ordinary canyon floor currents (Shepard et al. 1974) transport these sediments into deeper water.

The study of the formation of this slump valley has suggested that slumps on the slopes of canyons probably occur more frequently than was formerly believed. The findings of this and previous studies seem to indicate that at least two types of marine slumps are identifiable: those that are related to gravitationally induced slumping and those in which liquefaction becomes important. Those slumps that take the form of valleys are probably related to large wave- or earthquake-induced liquefaction slumping, while those that form large arcuate curves near slope breaks are probably related more specifically to gravitational factors, although earthquake shock could also be an important aid. The concept that large waves play an important role in moving sediment from the shelf by initiating slumping and then provide additional energy to help propel these sediments down submarine canyons onto deep sea fans is of considerable importance to the general understanding of sediment dynamics.

ACKNOWLEDGMENTS

The preceding work was accomplished under the auspices of the National Science Foundation Grant DES74-22089 and the Office of Naval Research Contract N000-69-A-0200-6049. The author is indebted to F.P. Shepard for his support and review of the manuscript; to E. Hamilton, R.F. Dill, and C. Nordstrom who also made helpful suggestions; to S. Pawka for his assistance with information on local storms; to M. Miller, N. North, G. Sullivan, P. McLoughlin, L. Holmes, and T. Harper for assistance in the field and office; and to N. Davis for reporting the occurrence of the slump.

REFERENCES

Bea, R.G., 1971. How sea-floor slides affect offshore structures. *The Oil and Gas Journal*, November 29: 88-92.

Chamberlain, T.K., 1960. *Mechanics of mass sediment transport in Scripps Submarine Canyon, California.* Unpublished Ph.D. Thesis, University of California, Scripps Institution of Oceanography, La Jolla, 200 pp.

Dill, R.F., 1964. *Contemporary submarine erosion in Scripps Submarine Canyon.* Unpublished Ph.D. Thesis, University of California, Scripps Institution of Oceanography, La Jolla, 269 pp.

——, Marshall, N.F., and Reimnitz, E., 1975. *In situ* submersible observations of sediment transport and erosive features in Rio Balsas Submarine Canyon, Mexico. (Abst.) *Geol. Soc. Amer.,* 7: 1052-3.

Finn, W.D.L., Bransby, P.L., and Pickering, D.J., 1970. Effect of strain history on liquefaction of sand, *Soil Mechanics and Foundations Division*, Amer. Soc. Civil Engin. Proc., pp. 1917-35.

Henkel, D.J., 1970. The role of waves in causing submarine landslides. *Géotechnique,* 20: 75-80.

Inman, D.L., Nordstrom, C.E., and Flick, R.E., 1976. Currents in submarine canyons: an air-sea-land interaction. *Ann. Rev. Fluid Mechanics,* 8: 275-310.

Lee, K.L., and Seed, H.B., 1967. Cyclic stress conditions causing liquefaction of sand. *J. Soil Mechanics and Foundations Division,* Amer. Soc. Civil Engin. Proc., pp. 47-70.

Seed, H.B., and Lee, K.L., 1966. Liquefaction of saturated sands during cyclic loading. *J. Soil Mechanics and Foundations Division,* Amer. Soc. Civil Engin. Proc., pp. 105-34.

Shepard, F.P., 1951. Mass movements in submarine canyon heads. *Trans. Amer. Geophys. Union,* 32: 405-18.

——, and Marshall, N.F., 1973. Storm-generated current in La Jolla Submarine Canyon, California. *Mar. Geol.,* 15: M19-M24.

——, Marshall, N.F., and McLoughlin, P.A., 1974. Currents in submarine canyons. *Deep-Sea Res.,* 21: 691-706.

Wright, S.G., and Dunham, R.S., 1972. Bottom stability under wave induced loading. *Offshore Technology Conference, paper no. OCT 1603*: I853-60.

Chapter 8

Coarse Sediment Transport by Mass Flow and Turbidity Current Processes and Downslope Transformations in Annot Sandstone Canyon–Fan Valley Systems

DANIEL JEAN STANLEY

Division of Sedimentology
Smithsonian Institution
Washington, D.C.

HAROLD D. PALMER

Dames & Moore
Washington, D.C.

ROBERT F. DILL

West Indies Laboratory,
Fairleigh Dickinson University
St. Croix, U.S. Virgin Islands

ABSTRACT

A series of submarine canyon axis, canyon wall, tributary canyon and fan valley, and interchannel facies of the Upper Eocene Annot Formation are well exposed in the French Maritime Alps. The dispersion of outcrops along major paleoslope trends allows definition of downslope, coarse sediment, depositional patterns in both channelized and unconfined outer margin environments. The geometry, stratification characteristics, internal structures, and fabric of the major sediment types reflect a broad spectrum of gravity-induced transport processes. This evaluation of downslope transport focuses on the different coarse facies sequentially, from those in proximal environments (which comprise a reduced proportion of obvious turbidites) to those in more distal parts of the paleobasin (characterized by enhanced proportions of turbidites).

The major depositional type in confined Annot Basin canyon and fan channel systems is a thick, amalgamated, nongraded or poorly graded, low-matrix pebbly sandstone that generally does not display typical turbidite structures. Most such massive units are attributed to high concentration sediment gravity flows, and associated with these are deposits from slumping, rockfalls, sandfalls, and a plexus of mass flow processes whose origin is less well defined. Classic turbidites account for only a minor portion of canyon-fan channel fills. The lithofacies assemblages that change downslope suggest a possible modification of grain support mechanisms and evolution from one type of flow to another. A scheme that involves down-axis segregation of material during transport is proposed to explain the distinction between sediment types in confined settings and those in more open environments further in the basin. Massive, low-matrix sands were deposited as "quick" beds from high-concentration underflows primarily in canyon and fan channels, whereas graded sands and finer fractions settled more progressively from lower-density turbidity current flows that bypassed canyon mouths and overtopped fan channel levees. The facies assemblages record significant changes in transport processes in base-of-slope environments, primarily in front of canyon mouths and on proximal fan apices.

INTRODUCTION

As is readily acknowledged, sediment entrapment, confinement, and transport in the submarine canyon and fan valley systems incised on continental margins are important factors in the emplacement of coarse sediment in deep ocean basins. There is ample evidence in both modern oceans and in the rock record that mud, sand, and pebble-size material of shallow origin can be moved through canyons to deeper environments well beyond the base of continental margins (reviews are provided by Shepard and Dill [1966] and Whitaker [1976]). The gravity-induced transport mechanism most commonly invoked in the long-distance transport of coarse sediment is turbidity current flow. As presently conceptualized, sediment set into motion by shelf edge spillover processes (particularly during eustatic low sea level stands) is introduced directly onto steep margins including the headward reaches of submarine valleys; much of this material, plus sediment provided by slump failure on slopes and in canyons, is subsequently transferred seaward by sediment gravity flows (Kuenen 1951).

The emphasis on turbidity currents and their role in downslope transport in modern canyons has diminished somewhat in recent years as a result of the increased attention paid to other gravity-induced mass flow processes (Dott 1963; Middleton and Hampton 1973; Carter 1975a). As has also been shown, traction transport by bottom currents induced by physical oceanographic phenomena such as storms, tides and internal waves can displace sediment downslope in canyons (see Chapter 1). Synthesis of available data indicates that a plexus of gravity-induced and traction processes are involved in the seaward movement of sand and pebbles beyond canyon mouths at the base of slopes (summary in Kelling and Stanley, 1976).

In spite of numerous studies of both modern and ancient outer continental margin deposits, knowledge of transport mechanisms in submarine canyons remains sketchy, and there has been some question as to the role of turbidity currents in some of the valleys studied. Scuba dives in the heads of canyons and direct visual observations made further downslope in the mid- and lower sectors of canyons and in fan valleys by sumersibles, as well as underwater photography and television records, indicate texturally variable, and often very coarse, terrigenous facies (Shepard and Dill 1966). However, dives generally reveal relatively tranquil conditions when, more probably, both short-term catastrophic events and slower, long-term movement such as creep prevail on slopes and in valleys. Data gathered during dives and with other presently available techniques do not effectively reveal the complexity of the submarine valley fill lithofacies distribution and geometry. Subbottom seismic systems are generally of insufficient resolution to detail sediment types or their configuration. Furthermore, coring programs in canyon-fan valley systems, with few exceptions (cf. Shepard et al. 1969; Nelson and Kulm 1973), are generally patchy, and core sections have been too short to provide sufficient information on the lithofacies variations in time and space.

Turbidity currents have on rare occasions been observed in canyons (Gennesseaux et al. 1971; R. Slater, personal communication), but in general, obviously graded sand turbidites do not appear to represent the predominant portion of surficial axial fills. The movement of sand and pebbles in the upper reaches of canyons is more commonly attributed to other mechanisms: bottom currents of diverse origin (see Chapter 1) and mass processes including slumping (see Chapter 8); muddy slurries and debris flow (Stanley 1974b); sandy slurries and grain flow; and glacier-like creep (Shepard and Dill 1966). However, cored sections in more distal modern environments such as fans, basin plains, and trenches reveal thick deposits that commonly include enhanced proportions of vertically and laterally graded sand sheets ranging from several millimeters to several meters in thickness; these units are of variable texture and structure and commonly comprise material of shallow water origin. In most instances, these laterally extensive graded sand layers display the sequences of internal structures as defined by Bouma (1962) and thus are confidently identified as turbidites (see the various chapters devoted to turbidites in Part III of this volume). The presence of such graded units in base-of-slope environments to date remains the strongest argument, albeit indirect, favoring the initiation of turbidity current flow in proximal slope sectors and in canyons.

These considerations have led to the broad acceptance of a simplified outer margin transportation-depositional model that depicts (a) moderate to rapid mass transport of sand and other sediment onto steep gradients of upper continental margins including canyon heads (off narrow shelves and/or during lower sea level stands); (b) the possible formation of metastable (collapsible) structure allowing periodic downslope displacement by some type of mass failure; and (c) subsequent transformation to some form of gravity sediment flow, including turbidity currents, somewhere between the upper reaches of canyons and less-confined environments beyond canyon mouths. The model is particularly applicable where canyons heads intercept moderate to large volumes of sediment—that is, during Pleistocene eustatic low stands and in modern ocean settings where canyons head close to land or have access to a significant sediment input.

We should note that in the La Jolla Canyon and Fan Valley off Southern California, which is probably the most thoroughly studied of the sedimentologically active submarine valley systems (and the study of which has lent credence to the above model), turbidity currents have not been recognized as being the prime mover of coarse sediment downslope at the present (Shepard and Dill 1966; Shepard and Marshall 1973). There is ample evidence that the fill in the upper La Jolla Canyon is periodically flushed downslope, that coarse materials, including sand and large blocks, occur along the length of the canyon, and that sandy turbidites and levee deposits occur on the La Jolla fan beyond the canyon mouth. However, the failure of explosives to start turbidity currents on the steeper gradients of the canyon head (Dill 1967) and the "nonturbiditic" nature of the sands (absence of graded beds, textural composition of sands, including low matrix content) in the canyon and fan valley (Shepard et al. 1969) has cast some doubt on the overall importance in the Holocene of turbidity currents, at least in the canyon proper.

Several well-exposed ancient submarine canyon-fan complexes in the French Maritime Alps provide some information on this matter. The Annot Sandstone exposures of Upper Eocene to the Lowermost Oligocene age are unusual in that although discontinuous, they present a relatively complete series of terrigenous facies dispersed along the major paleoslope trend (Figure 8-1). The fortuitous distribution of such sequences, rare in the rock record, makes possible examining the downslope sequential depositional patterns "frozen" in canyon-fan valley systems.

This chapter describes a broad spectrum of gravity-induced deposits within the ancient canyons and fan valleys and considers their origin in terms of possible transformation from various mass flow to turbidity current mechanisms along the paleoslope. In order to evaluate this latter aspect, attention is concentrated on Annot Sandstone lithofacies in (a) proximal canyon tributary and canyon wall environments close to the entry of coarse sand, pebbles, and mud, (b) within the canyon proper, and (c) on the upper apex of fan sections at and beyond the base of the slope.

GENERAL PALEOGEOGRAPHY AND FACIES DISTRIBUTION

The structural-paleogeographic framework of the late Eocene marine basin in the French Maritime Alps, termed the "Annot Basin," and the stratigraphy of the deposits that filled it have been detailed in earlier studies (Stanley 1961; Bouma 1962; Stanley and Bouma 1964). Facies analyses indicate that large volumes of medium- to coarse-grade sand and pebbles as well as mud were supplied by rivers draining a large emergent region south of the outcrop area. This Esterel-Corsica highland, much of which lay south of the present Mediterranean coast, provided materials that were transported northward across a relatively narrow shelf and into the structurally active, physiographically complex depression. The highland-bounded margin and mobile basins receiving the deposits have been compared with present settings off Southern California and to parts of the Mediterranean (Stanley and Unrug 1972).

In a recent synthesis of Annot Sandstone slope, submarine canyon, and fan deposits, Stanley (1975) described a variety of coarse-grained terrigenous facies, discussed their distribution in time and space, and outlined several possible mass gravity flow mechanisms responsible for their deposition. This regional petrologic survey of Annot Sandstone facies revealed a remarkable nonrandom distribution of sediment types that is in large part a result of the irregular northward sloping margin on which the sands and muds were transported. South- to north-trending submarine canyon-fan complexes are identified in three outcrop areas: Menton to Sospel, Contes to Peïra-Cava, and Annot to Le Ruch and Grand Coyer (respectively, a-b, c-d, and e-f-g in Figure 8-1). The major facies types described below and shown in Figure 8-2 are exposed in each of these three localities.

1. Coarse, massive sandstone, pebbly sandstone, and conglomerate layers—some to 30 m thick and for the most part amalgamated, nongraded to poorly graded, and moderately well-stratified—prevail in NNW-SSE-trending linear belts at Menton, Contes, and Annot (photograph A, Figure 8-2). These thick (> 300 m) belts trend down the paleoslope, and some of the more extensive outcrop belts can be followed continuously for distances of at least 8 km. The proportion of shale and fine-grained layers and of distinct, well-graded turbidites is remarkably low. These elongate tongues of massive units are interpreted as submarine canyon fills. Sandstone and shale units presenting typical slump structures such as disrupted and chaotic stratification are localized along, and merge with, the sandstone tongues; these are identified as canyon wall facies.

2. The thick sandstone belts are bounded to the east or west (or both) by sections of thinner and much better stratified, less-coarse strata presenting a classic flysch aspect (photograph B, Figure 8-2). This second facies, interpreted as intervalley slope deposits, includes thinner, moderately to well-graded, somewhat finer-textured sandstone turbidites alternating with laminated (and often graded) fine sandstone, siltstone, and shale units (photograph C, Figure 8-2). The latter may include a high proportion of mud turbidites.

3. Exposures near Sospel-Piena-San Michele, Peïra-Cava (photograph D, Figure 8-2), and Le Ruch-Grand Coyer—that is, north of the three relatively narrow sandstone and pebbly sandstone tongues listed as 1 above—cover greater surface areas and include thick (from 350 to over 650 m) alternating sequences of dark shale and sandstone beds of varied thicknesses, grain size, and lateral extent (cf. Stanley 1961; Bouma 1962). Medium- to coarse-grade sand turbidites of varied thicknesses prevail (usually < 2 m), but ungraded coarse sand layers ranging from 0.5 m to over 4 m in thickness are also common. These sandstone types occur both as laterally extensive sheets and as more restricted channelized deposits (see point FC in photograph D, Figure 8-2). Less well-stratified to chaotic pebbly mudstone layers are locally present. Distinct conglomeratic layers more than 1 m thick are uncommon;

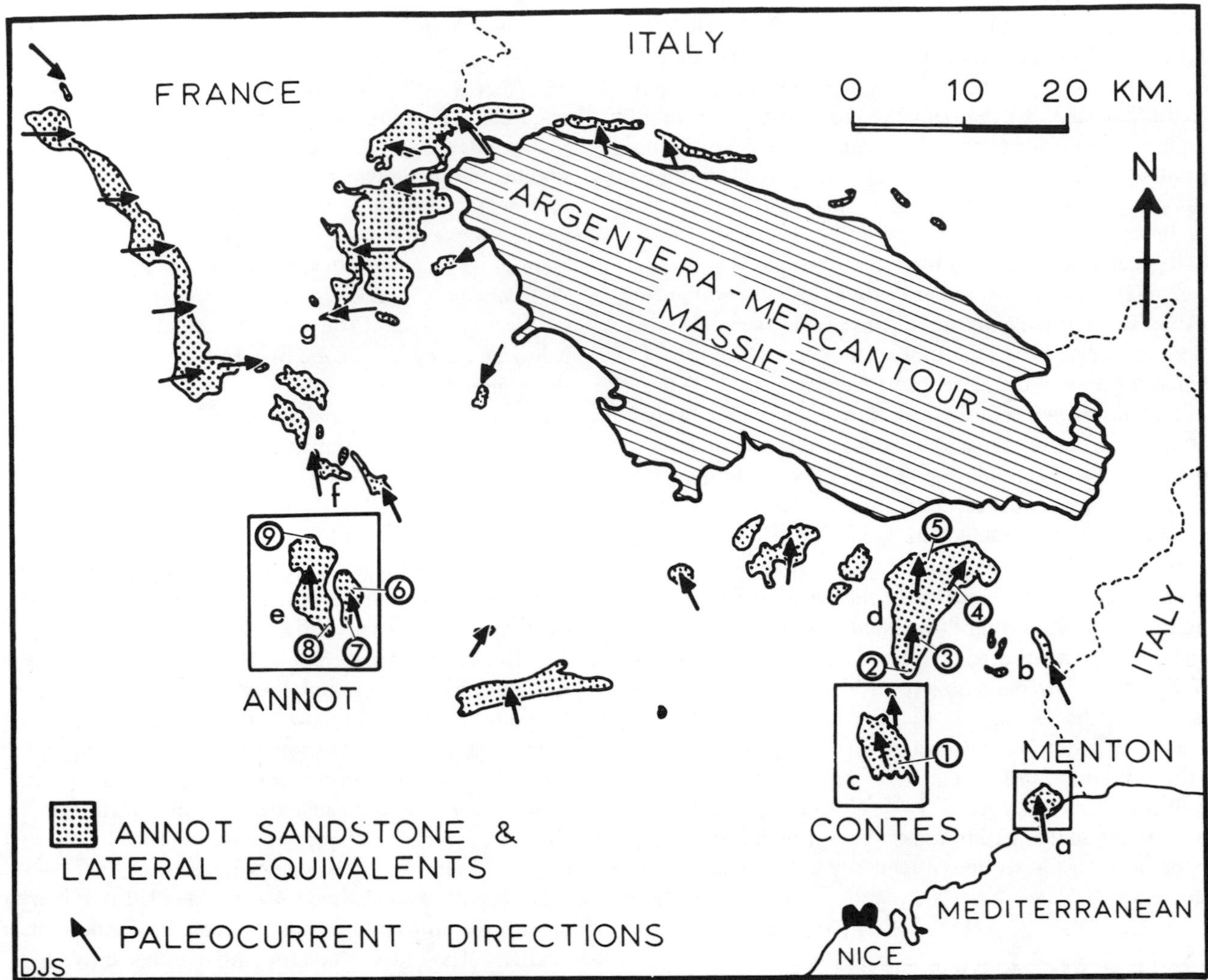

Figure 8-1. Map showing location of exposures of the Annot Sandstone Formation and lateral equivalents in the French Maritime Alps. Note: Arrows denote major paleocurrent trends; letters and numbers designate stratigraphic sections cited in the text and in other figures.

most often granules and pebbles (usually < 2 cm in diameter) of variable crystalline and metamorphic lithologies are distributed within sandstone units or are concentrated at their base. The thick turbidite-rich alternating sandstone-shale sequences of this type, which are the most representative Annot Sandstone facies in terms of volume and geographic distribution, are identified as submarine fan deposits; fan channel (FC in photograph D, Figure 8-2), levee, and interchannel overbank deposits are recognized (Stanley 1975).

The north-dipping paleoslopes are conservatively estimated at 6° to 1° on the basis of comparison with modern tectonically active margins (cf. Burk and Drake 1974) and analysis of the trace fossil assemblages (J. Wexler, personal communication, 1975). Slopes of this magnitude would imply that submarine valleys extended to depths of at least 1,000 m and that basin plains may have been as deep as 2,000 m. The predominant south to north dispersal in the three canyon-fan systems in this sector of the paleobasin has been established by detailed regional facies analysis. The spatial relationship between the submarine canyon, intervalley slope, and fan deposits is apparent on a series of maps depicting paleocurrents, sand:shale ratio, bed thickness, percentage and type conglomerate, mineralogy and other parameters (Stanley 1961, 1965, 1975; Stanley and Bouma 1964; Stanley and Unrug 1972). These earlier Annot Sandstone studies serve as the basis for interpreting the general facies-environment-process relationships discussed in this chapter. Herein, more specific attention is paid to the nature and origin of coarse-grade sandstone and pebbly sandstone types deposited on the canyon wall, in tributary canyon and canyon axis environments, and in deeper, more distal, and less confined paleoslope and fan settings. An evaluation of downslope transport processes is made by focusing on the different coarse facies sequentially—that is, from more proximal environments that com-

Figure 8–2. Major facies types in the Annot Basin. Note: Photograph A shows massive sandstone units forming thick tongue, which is interpreted as a submarine canyon fill, west of the Coulomp River near Annot (e-8 in Figure 8–1); deformed strata (arrows) record lateral infill processes—that is, large-scale slumping and slurry/sand flow processes. Photograph B shows well-stratified section of turbidite and probable hemipelagic deposits forming typical intervalley slope section; some complete $T_{a\text{-}e}$ turbidites are at top of photo (near e-6 in Figure 8–1). Photograph C shows thin, base cut-out turbidites in the section shown in photograph B. Photograph D shows thick sequence of well-stratified sandstone turbidites and mudstones of variable thickness and coarse, massive channelized unit (FC); the former are interpreted as fan interchannel and overbank deposits, and the latter as a fan channel fill. This sequence is located near the Col de l'Orme-Cabanette section near Peïra-Cava (d-3 in Figure 8–1).

prise a reduced proportion of obviously well-graded sandstone turbidites to those farther seaward (northward) in the paleobasin characterized by enhanced proportions of turbidite incursions.

DOWNSLOPE DEPOSITIONAL PATTERNS

Canyon Wall Facies

Sections displaying lithofacies characteristics interpreted as canyon wall depositional types are localized along the margins of the thick, NNW-SSE-trending, finger-like sandstone belts (Stanley 1975). Good exposures of canyon wall facies occur east and west of the city of Menton, at Contes (on the eastern part of the syncline, near Sclos de Contes [point c-1 in Figure 8-1]) and near Annot (east of the outcrops of massive units in the vicinity of the village of Braux [between e-8 and e-9 in Figure 8-1]). Particularly remarkable in sectors identified as canyon wall environments is the sharp discontinuity between the Annot Sandstone facies and the underlying Eocene Marnes Bleues Formation (grayish-blue shales and marls) that, folded and eroded, formed the paleoslope depressions in which the sands were transported seaward. A somewhat analogous situation is observed in the modern western Mediterranean submarine valleys where Quaternary to Recent pebbly sand and mud are transported in steep-walled valleys cut into stiff Pliocene clays (cf. Gennesseaux 1966; Got and Stanley 1974).

The most obvious attributes of canyon wall facies are the generally poor stratification and high proportion of chaotic (slump-deformed) coarse sandstone and shale layers (Figure 8-3). Bed thickness (2 m to 10 m) and geometric configuration of the strata are highly variable; the units forming this section are of limited areal extent. No two layers resemble each other. Contorted shale masses occur either as discontinuous layers that can be traced for tens or more meters between sandstones (photograph A, Figure 8-3) or as large, isolated chaotic masses within thick sandstone layers (photograph B, Figure 8-3). The generally massive sandstone strata consist of matrix-poor coarse sand as well as muddy sandstone, and both types often contain significant amounts of mica and plant fragments. Shale rip-up clasts of variable size and shape (photograph C and D, Figure 8-3) are abundant in most massive sandstone strata. Some disorganized conglomerates (cf. Walker and Mutti 1973) and pebbly mudstones are observed (photograph D, Figure 8-3).

The massive sandstones either are sharp based or display large erosional structures and load deformation markings (photograph A, Figure 8-3). Graded bedding is not generally apparent. Better stratified sequences of alternating thinner (1 m to 3 m) sandstone sheets and shale layers lie between the thicker, geometrically more restricted massive deposits. Some of these thinner sandstone units display graded bedding. Complete Bouma sequences are occasionally observed, but more often the units are truncated and display the basal *a* massive graded and *b* laminated divisions of Bouma. The number of T_{b-e} or T_{c-e} turbidites is markedly reduced in contrast to the adjacent intervalley slope sections. Paleocurrent directions are generally variable even within a single exposure and do not necessarily display the prevailing north-trending patterns of the units in the adjacent canyon axial fill and intercanyon slope sequences.

Canyon wall units are easily distinguished from the thinner, much better stratified slope intervalley sequences with which they interfinger. The latter include abundant sandstone turbidites (2 cm to 20 cm and occasionally to 1 m), some of which display complete Bouma sequences (top in photograph B, Figure 8-2), but more often are the T_{b-e} and T_{c-e} types (photograph C, Figure 8-2). The enhanced proportions of graded and laminated siltstone and shale probably include mud turbidites (Stanley 1970; Rupke and Stanley 1974). The sand:shale ratio of some Annot Basin intervalley slope sections is less than unity. In contrast to canyon wall facies, sole markings on the intervalley slope turbidites include a variety of flute as well as tool markings and display a generally more consistent northerly transport direction. Furthermore, bioturbation on sandstone soles is more extensive in slope deposits than in either canyon wall or canyon fill facies. Representative Annot Basin slope deposits are exposed near localities c-1 and e-6 in Figure 8-1.

Tributary Canyon Deposits: Crête de la Barre

A well-exposed section about 200 m to 300 m long and 65 m thick (photograph A, Figure 8-4) at the southern end of a cuesta-like crest called Crête de la Barre (about 2 km SSE of the village of Braux [e-7 in Figure 8-1]) is interpreted as the fill of a small tributary canyon (Stanley 1975, Figure 3). This interpretation is based on the geometry and nature of the units, position of the section relative to the thick north-northwest-trending tongue of massive sandstones to the west, and anomalous paleocurrent directions. The proportion of silt and shale layers (sand:shale ratio is about 2) displayed by this sequence is considerably lower than in the major NNW-SSE-trending canyon fill facies (sand:shale ratio to >32) exposed about 1 km to the west (arrow on photograph A, Figure 8-4).

The moderately stratified section includes the following stratal types in order of diminishing importance: (a) thick (to almost 4 m), massive sandstone beds that consist for the most part of moderately sorted coarse sand with a low matrix content, contain some dispersed granules and small pebbles (to 1 cm) of crystalline and metamorphic origin, and locally enclose large (to 1 m) shale rip-up clasts and discontinuous layers of shale (photograph B, Figure

Figure 8-3. Exposures interpreted as canyon wall deposits, near Sclos de Contes (c-1 in Figure 8-1). Note: Photograph A shows typical slump unit (arrows) between massive sandstone strata. Photograph B shows discontinuous, contorted shale horizon (arrow) within massive sandstone. Photographs C and D are views of large, shale rip-up clasts (hammer = 35 m; notebook = 14 cm). Photograph E shows sandstone block (arrow) that is more than 1 m in length in coarse, granule-rich sandstone.

Figure 8–4. Section interpreted as tributary canyon fill at the Crête de la Mauvaise Barre east of Annot (e-7 in Figure 8–1; see also photograph A, Figure 8–5). Note: Photograph A shows moderately well-stratified section pinching out (arrow) against older Marnes Bleues Formation that formed the tributary channel walls and illustrates the contrasting facies of massive sandstone forming major canyon fill in the background (close to section shown in photograph A, Figure 8–2). Photograph B shows thin turbidite sequence covered by massive sandstone unit enclosing discontinuous shale layer (book = 28 cm). Photograph C shows contorted sandstone-shale horizon within a sandstone stratum (book = 28 cm). Photograph D shows sandstone flow roll in silty shale near base of section (book = 20 cm).

Figure 8-5. Exposures of massive bedded sandstone in Annot area. Note: Illustration A shows the sole marking and slump structure paleocurrent trends on sandstone outcrop maps (e in Figure 8-1) and indicates the predominant NNW canyon axis trend. Photograph B shows typical massive sandstone sequence at the Chambre du Roi section (e-8 in Figure 8-1); amalgamated unit in foreground is more than 10 m thick. Photograph C shows pebbles in horizontally stratified layers and irregularly distributed within massive sandstone unit at the Chambre du Roi (scale = 20 cm). Photograph D shows soft sediment deformation and erosional structures at the base of a massive, granule-rich sandstone bed (key = 5 cm).

A

B

Figure 8-6. Internal structures in relatively clean, poorly cemented massive sandstone units forming typical canyon fill sections at the Chambre du Roi southeast of Annot (e-8 in Figure 8-1; see also photograph B, Figure 8-5). Note: Photograph A shows weathered surface that reveals horizontal laminated sandstone of medium grade eroded by graded granule and coarse sandstone layer; arrow indicates truncation of the convoluted laminations at contact between the two units. Photograph B shows flame structures (f) and pockets (p) of small pebbles that record the contact between two depositional episodes within a thick (> 20 m) sandstone bed.

C

Figure 8-6 Continued. Photograph C shows thin (1cm), discontinuous shale layer (dsl, near hand) truncated by a sandstone stratum displaying wavy base and inclined lamination; this unit and another discontinuous shale layer are truncated by a pebble sandstone (left of photo).

8-4); (b) irregularly stratified layered units (usually sandstone) that include chaotic shale horizons (photograph C, Figure 8-4); (c) poorly graded, coarse sandstone turbidites, about 30 cm to 100 cm thick, that commonly comprise the basal *a* or *a-b* Bouma divisions (upper part of photograph B, Figure 8-4); (d) alternations of fine grained sandstone and siltstone layers (some laminated T_{b-c} and T_{c-e} turbidite sequences) several to 20 cm thick that alternate with silty shales (see thin units below thick sandstone in photograph B, Figure 8-4); (e) a few coarse sandstone and, occasionally, muddy sandstone layers that include small dispersed pebbles to 1 cm; and (f) a discontinuous layer of large sandstone flow rolls.

The thick, coarse, nongraded sandstones and massive, poorly stratified sandstone including chaotic mud layers (units a and b above) account for somewhat more than half of the section. Some of these beds display moderately to well-defined groove marks (some are very large -that is, 30 cm wide and > 2 m long); load and slide marks and trace fossils are also common. The base of the section includes paleocurrent directions trending south to north– that is, roughly along the major slope trend as determined from the regional paleocurrent analysis (Stanley 1961). However, as one proceeds up section, paleocurrent directions change markedly –that is, trends vary progressively to northwest-southeast and subsequently to east-west and northeast-southwest. The east-southeast- and west-southwest- trending flute marks on the base of some beds indicate transport directions almost normal to the predominant north-northwest trend displayed by the massive sandstone units west of the Coulomp River.

Presumably material, initially moved from the south, was intercepted in a depression formed within the Marnes Bleues and in time displaced toward the west into the Annot Canyon. Most of the sandstone beds display remarkable lateral changes in thickness and internal structures along the exposure. The wedging out of strata is an obvious attribute of this sequence, and the section as a whole pinches out against the older Eocene Marnes Bleues facies (photograph A, Figure 8-4).

Canyon Fill Facies

The remarkable exposures of massive bedded sandstones west of the Coulomp River near Annot (photograph A, Figures 8-2 and 8-4; photograph B, Figure 8-5) and those at Contes can be traced 8 km and 6 km, respectively, along the paleoslope. Similar exposures at Menton, more generally covered by vegetation and dwellings, are less accessible. The thick (to 350 m) deposits in these three regions have been interpreted as canyon fill facies (Stanley and Unrug 1972), and their distribution patterns have been amply illustrated in earlier studies by Stanley (1961), Bouma (1962), Stanley and Bouma (1964), Stanley and Unrug (1972), and Stanley (1975). The elongate outcrop localities consisting of thick, coarse-grade sandstone units are little more than 1 km wide at the three localities. The gross geometry of the coarse, massive strata at these localities is remarkably similar. Isopach maps depict finger-like belts trending essentially NNW-SSE; within these belts are concentrated the thickest, massive sandstone strata, the largest number of pebble-bearing beds, and the largest pebbles. The paleocurrent directions measured in the massive sequence show a relatively consistent south-to-north dispersal pattern (illustration A, Figure 8-5). The vertical succession of *Nereites* and *Zoophycus* trace fossil assemblages at the three localities suggests a progressive shallowing from bathyal to possibly infraneritic depth with time (J. Wexler, personal communication, 1975).

East-west profiles across the thick bundle of Annot Sandstone strata show that the base of the formation is broadly concave-up; the basal sandstone units truncate the underlying Marnes Bleues Formation (Stanley 1975). The sequences west of the Coulomp River at Annot and near Contes consist of massive wedge-shaped sandstone strata with only minor intercalated shale units. In these areas there appears to be a progressive thickening, then thinning, of individual sandstone beds from south to north; a northward increase in the number and proportion of shale and fine-grade sandstone intercalations is also observed along the paleoscope trend. The diminution of sandstone thickness and increase of shale interbeds is even more abrupt in directions east and west away from the fill—that is, in sections exposing canyon wall and intervalley slope series described earlier.

There is a notable lack of lithologic uniformity from bed to bed in and near the axial fill, as in the case of canyon wall and tributary canyon facies. No two beds display the same sequence of structures and textures. Most sandstones are of the coarse grade, thick massive type. These are irregularly to moderately well stratified and range in thickness from about 1 m to almost 30 m (photograph B, Figure 8-5). Most strata thicker than 2 m are composite sedimentation units representing a fusion, or amalgamation, of successive sand and pebbly sand incursions. Successive events within an individual massive bed are recorded by marked changes in grain size (photographs C and D, Figure 8-5; photographs A and B, Figure 8-6). A few well-defined turbidites are observed between the massive sandstone beds (photographs A and B, Figure 8-7).

The stratal boundaries sometimes are separated by thin, usually discontinuous and deformed shale stringers (photograph C, Figure 8-6). In many cases only shale rip-up clasts are distributed along a horizon within a massive bed; these record what once must have been a more continuous mud layer. Mud layers of this type were eroded prior to consolidation by sand and pebbly sand flows as is attested by the abundance of rip-up shale clasts (photograph E, Figure 8-8) and armored mud balls in thick sandstones (photograph D, Figure 8-8). Occasionally, large sandstone blocks are observed within massive beds, thereby indicating that transport mechanisms in such confined environments also were able to erode and displace semiconsolidated sands. Slump deposits of various dimensions also are noted. In some instances, the entire sandstone or sandstone and shale unit appears to have moved en masse, as displayed in exposures west of the Coulomp River (photograph A, Figure 8-2); in other exposures only portions of a bed exhibit failure. The massive sandstones at the three localities do not commonly display features typical of traction sedimentation, although some thinner sandstone units show low to moderately inclined stratification (photograph C, Figure 8-2).

The base of some massive units reveal scour-and-fill and groove (photograph B, Figure 8-8) structures or large loads and other forms of soft-sediment deformation; other thick sandstone layers display a sharp base without notable structures (photograph A, Figure 8-8) and a sharp top. Graded bedding in massive sandstone units, when present, is subtle (photograph A, Figure 8-8) to well developed (photographs A and B, Figure 8-6) and coarse-tail graded bedding is sometimes observed at or near the base of, or within, thick units. Complete Bouma T_{a-e} sequences are uncommon. Granules or small pebbles (usually <1 cm diameter) concentrated at the base of sandstone strata are occasionally graded (photographs A and B, Figure 8-6) or inversely graded. In some instances, pebbles within the massive beds form some type of stratification (photograph C, Figure 8-5) and imbrication (photograph C, Figure 8-7); more often the small crystalline or metamorphic pebbles do not display distinct fabric (disorganized conglomerates; cf., Walker and Mutti 1973). Erosional structures such as channeling are commonly observed at the base of, as well as within, the thick pebbly sandstone strata. Pockets of gravel, shale clasts (photograph E, Figure 8-8), chaotic mixes of pebbles, and shale of the pebbly mudstone type (Crowell 1957), and chaotic sandstone-shale slump masses and large sandstone spheroids (photograph D, Figure 8-8; see also Stanley 1964) are irregularly distributed within the belts of massive sandstones. Vague to distinct horizontal and convolute lamination (see Figure 8-6), parting lineation, and dish structures (photograph C, Figure 8-8) are less frequently noted (Stanley 1974a). These internal structures are probably rather typical of many thick beds, but the highly weathered outcrop surfaces often preclude observation of these features.

In summary, the vertical sequence of some massive

Figure 8-7. Section near Chambre du Roi (e-8 in Figure 8-1) containing well-defined turbidites. Note: Photographs A and B show a 3 m thick sequence of well-defined turbidites between massive (> 10 m thick) sandstone layers. Photograph C shows imbricated pebbles in a graded layer of granules and coarse sand (hammer = 40 cm).

beds includes a basal structureless or graded sandstone or pebbly sandstone division, followed by a layer of deformed lamination and dish structures and topped by flat lamination. The zone of flat lamination may be truncated by an overlying sequence of structureless or poorly graded sandstones. The matrix content in the massive sandstone is generally lower (material <40μ, usually well under 10 percent) than in distinct sandstone turbidites in other Annot Basin environments (Stanley 1961).

Submarine Fan Facies

The paleocurrent directions recorded at outcrop localities north of Menton, Contes, and Annot indicate a divergent but predominantly northward transport trend (see Figure 8-1). These vectors and the composition of both pebbles and sandstones substantiate a provenance from southern source areas (Stanley 1961, 1965; Bouma 1962). The proportions of chaotic slump and pebbly mudstone strata and of conglomerate units (disorganized and organized types, generally less than 1 m thick) are somewhat lower, and the proportions of turbidites showing good graded bedding and Bouma sequences are considerably greater than at the Annot-Contes-Menton canyon localities to the south. The petrology, areal extent, and thickness of the well-stratified sandstone turbidite and shale alternations (>650 m north of Annot toward Grand Coyer and Lac d'Allos [photograph A, Figure 8-9]) record deposition in less confined environments at and beyond the base of slope in the Annot Basin. The stratigraphic sections display variable sand to shale ratios (generally high, from 4.0 to 8.0) and include some massive, coarse-grade sandstone and pebbly sandstone beds within thick sections of fine- to medium-grade sandstone turbidite-shale cycles (Figure 8-9; photograph D, Figure 8-2). The lobe-shaped areal configuration of these Annot Sandstone exposures, the alternating fining-upward and coarsening-upward cycles, stratification characteristics and general petrology, and paleocurrent patterns together suggest submarine fan deposition (Stanley 1975).

Thick (3 m to over 10 m), massive units of coarse-grade sandstone and pebbly sandstone within the Annot Basin proper have been interpreted as the coarse fills of fan

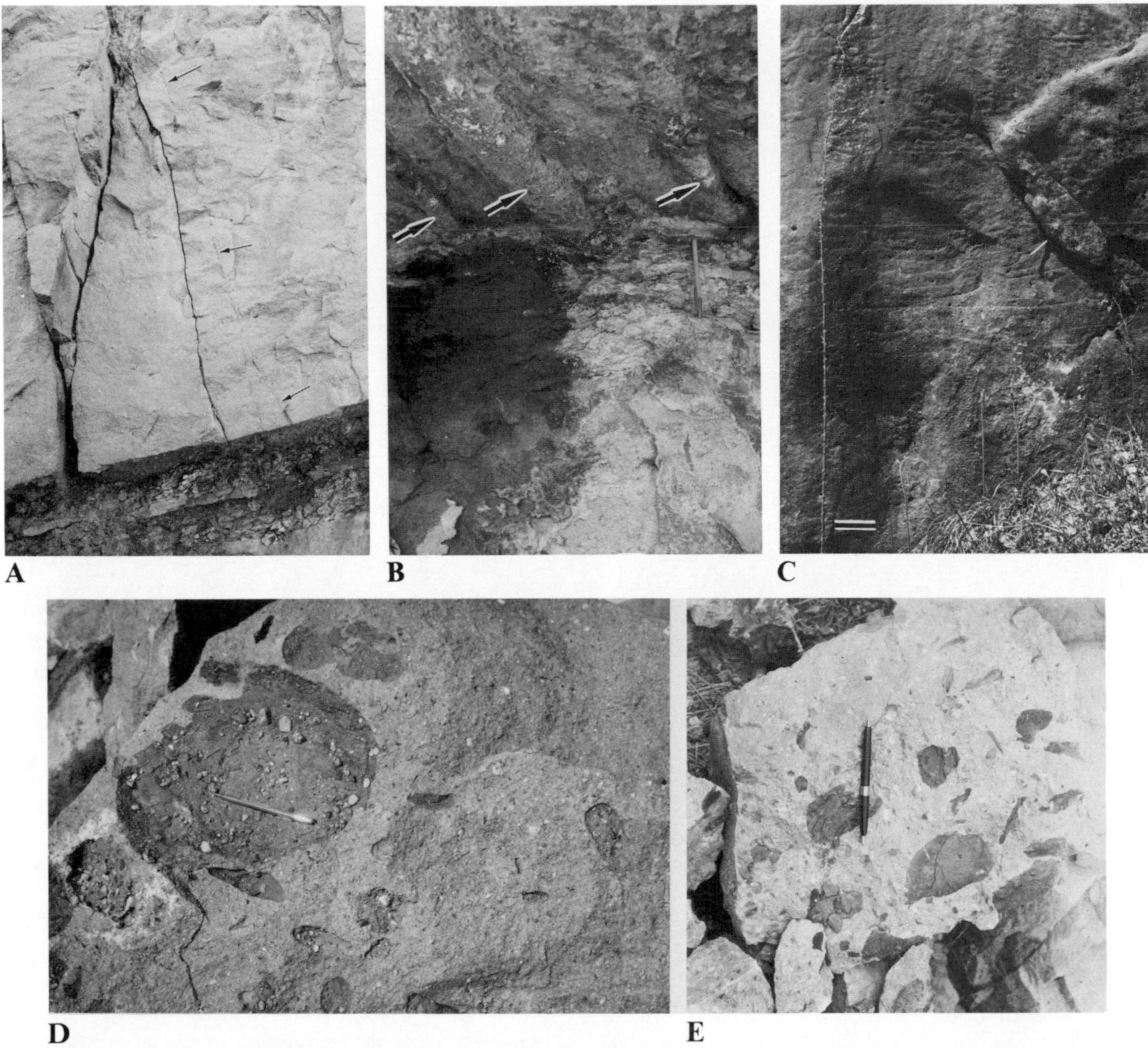

Figure 8-8. Some typical structures displayed by sandstone strata of the canyon fill type in the Annot region (e in Figure 8-1). Note: Photograph A shows sharp-based "structureless" unit with long (> 1 m) vertical burrow (arrows). Photograph B shows large grooves (arrows) on base of a thick, poorly graded layer (pencil = 14 cm). Photograph C shows a 150 cm thick section displaying dish structures in a massive (3 m) bed (bar scale = 20 cm). Photograph D shows an armored mud ball (pencil = 13 cm). Photograph E shows a shale rip-up clast layer in pebbly sandstone.

valley channels (Figures 8-10 and 8-11; photograph D, Figure 8-2), as defined in various submarine fan models (cf. Normark 1970; Mutti and Ricci Lucchi 1972; Nelson and Kulm 1973; Mutti 1974; Nelson and Nilsen 1974; and others). Many of these sandstone strata resemble those forming the more proximal canyon sequences to the south and described earlier —that is, they are usually composite amalgamated units. Some massive beds are graded and display horizontal lamination (photographs A and B, Figure 8-11; photograph A, Figure 8-12) and dish structures (photograph D, Figure 8-11) in their upper parts. The base of such sequences generally truncates underlying finer sandstone or shale sequences (photographs B and C, Figure 8-10). Shale clasts, some larger than 1 m in length, abound in such sandstone and pebbly sandstone strata (photograph D, Figure 8-10). The massive beds alternate with more numerous, thinner sandstone turbidite sheets and shale sequences (photograph B, Figure 8-9) of variable thicknesses (the latter probably includes mud turbidites as well as hemipelagic deposits). These alternations of finer-grade, well-stratified units are interpreted as interchannel deposits and include overbank

Figure 8-9. Thick turbidite-rich sections interpreted as fan deposits. Note: Photograph A shows the Col de la Cayolle region (> 600 m) in a view toward the east as seen from the Lac d'allos (g in Figure 8-1). Photograph B shows a section along the road between Moulinet and Turini in the Peïra-Cava region; thick sandstone units in background (arrows) are probably fan valley deposits, and the person is pointing to thin, fine grade sandstone, siltstone, and shale fan overbank interchannel layers that probably include both turbidites and hemipelagic deposits.

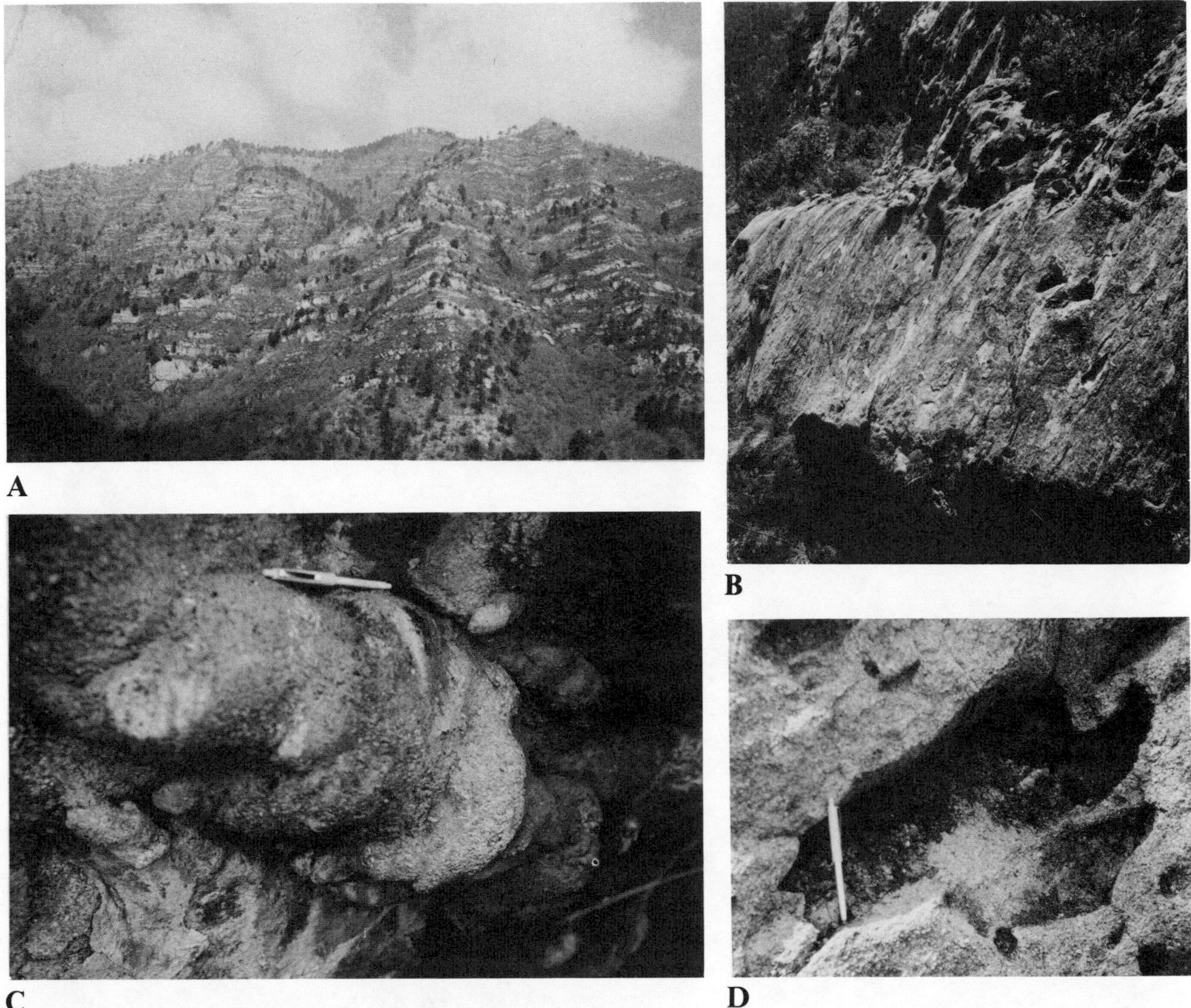

Figure 8-10. Section interpreted as fan channel deposits southwest of the Col de l'Orme in the Peïra-Cava syncline (near d-3 in Figure 8-1). Note: Photograph A shows massive sandstones particularly in the lower and middle part of the section. Photograph B shows lower part of the 10 m thick channelized massive sandstone body (shown as FC in photograph D, Figure 8-2). Photograph C shows large flute-like structure on base of bed in photograph B (pencil = 13 cm). Photograph D shows shale rip-up clasts distributed in "structureless" coarse sandstone forming bed shown in photograph B (pencil = 13 cm).

sand and mud flow deposits and possibly natural levee beds. The latter characteristically display small-scale wavy stratification and cross-bedding structures (photograph B, Figure 8-11).

The association of chaotically stratified sandstone strata (Figure 8-13) and disorganized conglomerate layers (photograph D, Figure 8-14) with above sequences indicates proximality to a zone of decreased gradient (i.e., near the mouth of a canyon or to a fan valley meander well on the fan. The pebbles in Annot Basin fan (and also canyon) sequences present attributes (shape and polymicitic nature) similar to those of some river gravels. The grain size analyses and regional paleogeographic interpretations suggest that these coarse elements had a fluvial origin. Pebbles, along with poorly to moderately sorted sands locally enriched with plant matter and mica, were transported directly from emerged source terrains to the south, through the coastal (possibly deltaic) zone, and into the deep basin with minor textural modification in the littoral and shelf environments. A possible modern analog are the fluvial and coastal pebbles mixed with sand and mud observed in Mediterranean canyons and rises such as those off the narrow coast of Provence (Bourcart 1964; Gennesseaux 1966) or Catalonia (Got and Stanley 1974). Kelling and Holroyd provide further discussion of the origin of pebbles in Chapter 11 of this volume.

The proportion of fine turbidite sandstone-shale interchannel sequences and coarse fan valley deposits

Figure 8-11. Sandstone structures observed along the La Cabanette-Col de l'Orme section near Peïra-Cava (d-3 in Figure 8-1). Note: Photograph A shows truncated (T_{a-b}) turbidite above section of shale and thin graded and laminated siltstone and sandstone sheets. Photograph B shows wavy-rippled and cross-laminated unit. Photograph C shows horizontal lamination and dish structures within a truncated turbidite (pencil = 17 cm). Photograph D shows a large plant fragment (about 25 cm long) in a sandstone that is rich in mica and disseminated plant matter.

vary vertically within the same section (i.e., alternating fining-upward and coarsening-upward sequences) as well as laterally between sections. The observed spatial and temporal variations in thickness and proportion of sandstone, siltstone and shale record the evolution of superposed upper to mid-fan lobe sequences as depicted by some fan models. Displacement of fan lobe deposition results from channel migration on a fan system developing in a closed basin (cf. Nelson and Kulm 1973).

The excellent exposures of fan deposits crop out in the region north of Menton and north of Annot, between Le Ruch, Grand Coyer, and the Lac d'Allos- Col de la Cayolle (respectively, e-9, f, and g in Figure 8-1; photograph A, Figure 8-9); these sections are detailed in Stanley (1961). The easily accessible Peïra-Cava exposures (d in Figure 8-1; photograph A, Figure 8-10) also display upper and suprafan to mid and possibly outer fan facies within a relatively restricted area and these are emphasized in this chapter. The outcrop locality in the syncline is fan shaped (about 14 km long from its southern apex at Col St. Roch [point 2 in Figure 8-1] to Turini-l'Aution [5 in Figure 8-1] in the north) and is oriented roughly parallel to the major paleocurrent trends (Bouma 1962, Figure 19). Thus an ideal disposition of sequences for evaluating downslope facies changes is provided.

Sandstone turbidites presenting Bouma sequences dominate the various exposures, although variable proportions of other strata types are observed. For instance, a thick

A

B

Figure 8-12. Thick (about 9 m) sandstone unit exposed on the Col de l'Orme-La Cabanette road in the Peïra-Cava syncline (d-3 in Figure 8-1). Note: The base of this section (photograph A) locally merges with the underlying slump (s1) and a disorganized granule layer (arrow) about 30 cm thick. This layer is followed upward by a 50 cm thick graded layer and then a 4.5 m thick "structureless" sequence (photograph B), which in turn is topped sequentially by a horizontally laminated section of 1.5 m (photograph C), a cross-laminated, convoluted laminae layer of 1.25 m, and a 20 m thick silty shale horizon. The hammer used as scale is 35 cm.

C

Figure 8-12 Continued.

(>6 m) slump unit covered by a thick (>20 m) massive sandstone sequence is exposed near the Col St. Roch at the southernmost sector of the syncline (photograph A, Figure 8-13). This type of deposit signals a zone of reduced gradient near the base of the slope, possibly an upper (inner) fan environment (cf. models of Normark 1970; Mutti and Ricci Lucchi 1972; Nelson and Kulm 1973; Mutti 1974; Nelson and Nilsen 1974; and others).

Massive beds of pebbly sandstone and coarse sandstone include both organized and disorganized conglomerate types (cf. Walker and Mutti 1973). As shown in Figure 8-14, examples of the former type include stratified and normally graded (photographs A through C) or reverse graded (photograph D) pebble-rich layers at the base, or within, massive beds. Most often, however, pebble fabric and structures are not well defined (photograph E, Figure 8-14). These disorganized conglomerates are commonly directly associated with slump or chaotic structures in the lower part of a sandstone unit (Figure 8-15; arrow on photograph A, Figure 8-12), and in other instances they appear below, and separate, from the base of sandstone layers or actually truncate sandstone beds (photograph E, Figure 8-14).

The base of coarse grade sandstone units often display large sole markings (photograph C, Figure 8-10), load structures (photograph C, Figure 8-14), and associated forms of soft sediment deformation (photograph F, Figure 8-13). The lower part of sandstone beds are structureless or poorly to moderately graded, whereas the upper parts occasionally show dish structures (photograph C, Figure 8-11) and, more commonly, horizontal lamination (photographs A and B, Figure 8-11). In southern and central part of the Peïra-Cava syncline such coarse, generally massive units are associated with slump deposits and conglomerates (Figure 8-13) and are interpreted as inner fan (suprafan) channel deposits. The finer-grade, better stratified, turbidite-shale sequences between them (details in photograph C, Figure 8-14) are of probable channel overbank origin. For additional descriptions on the stratigraphic section and lithofacies types cut by the road between the Col de l'Orme and La Cabanette (photograph D, Figure 8-2; photograph A, Figure 8-10), the reader is referred to the detailed profiles by Bouma (1962) and Lanteaume et al. (1967, Figure 3). Toward Turini (point 5 in Figure 8-1) further to the north in the Peïra-Cava syncline the proportion of shale and somewhat thinner, finer-grade laminated ($T_{b\text{-}e}$, $T_{c\text{-}e}$) turbidites increases (photograph C, Figure 8-9), and the sand to shale ratio is correspondingly lower than in sections to the south. These sequences are interpreted as mid-fan and more distal outer fan deposits. Coarse, massive sandstone, pebbly mudstone, and slump deposits decrease substantially in number and thickness. The massive sandstones are thinner, but as in the more proximal fan valley channels described above, they occasion-

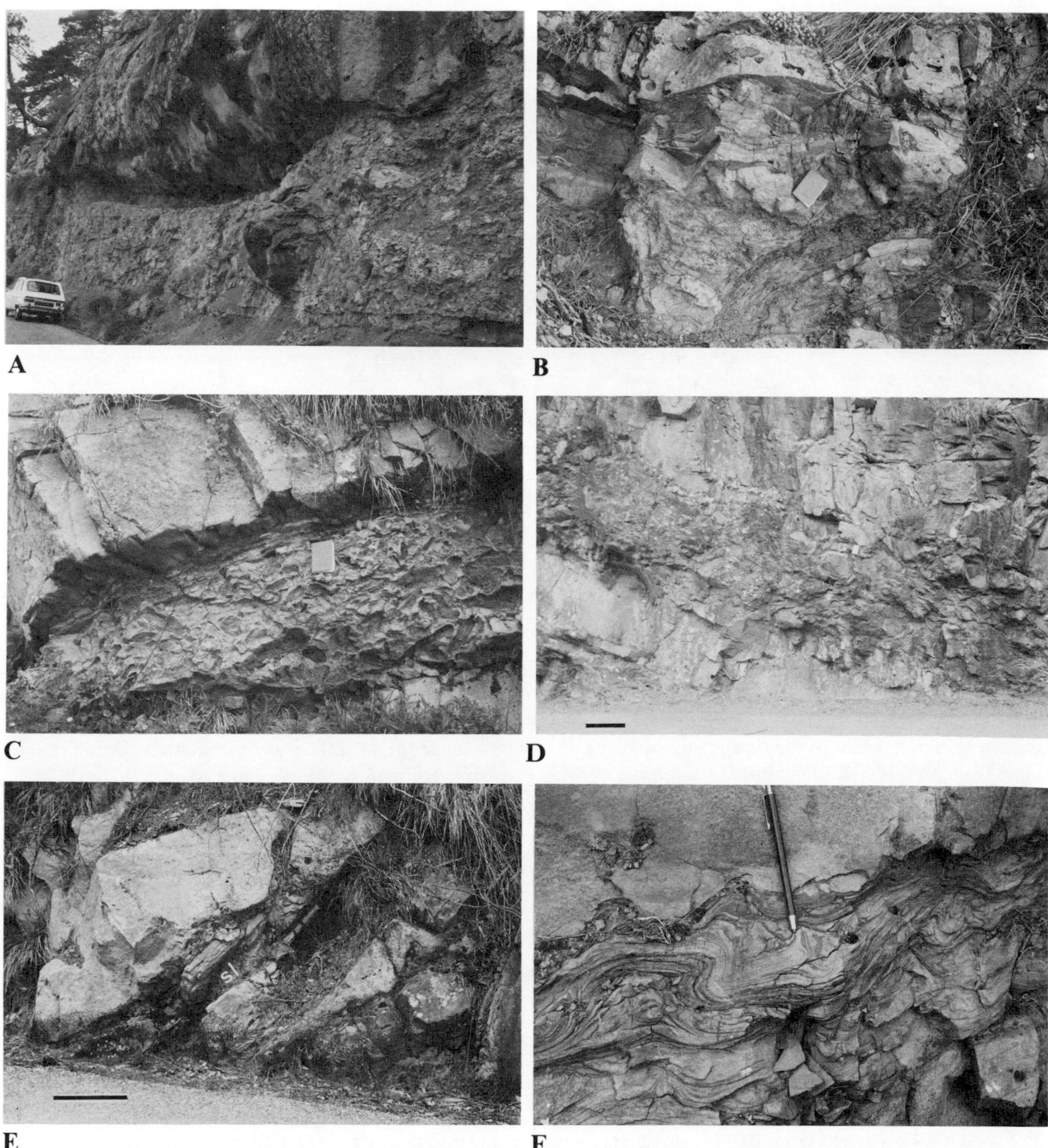

Figure 8-13. Exposures displaying various sandstone sequences discussed in text. Note: Photograph A shows a large slump deposit in the vicinity of the Col St. Roch (d-2 in Figure 8-1) that is covered by a massive sandstone unit about 20 m thick. Photographs B through E depict a selection of mass flow sequences and structures exposed on the Cabanette Road section in the Peïra-Cava region (d-3 in Figure 8-1) as follows: photograph B, slump-granule mixture (book = 20 cm); photograph C, muddy sandstone with abundant rip-up clasts interpreted as a slurry deposit; photograph D, deformed sandstone and shale horizon within a thick sandstone layer (scale = 1 m); photograph E, Slump (s1) - coarse sand flow sequence (scale = 50 cm); and photograph F, soft sediment deformation in fine-grained sandstone unit between coarser-grade layers (pencil = 15 cm).

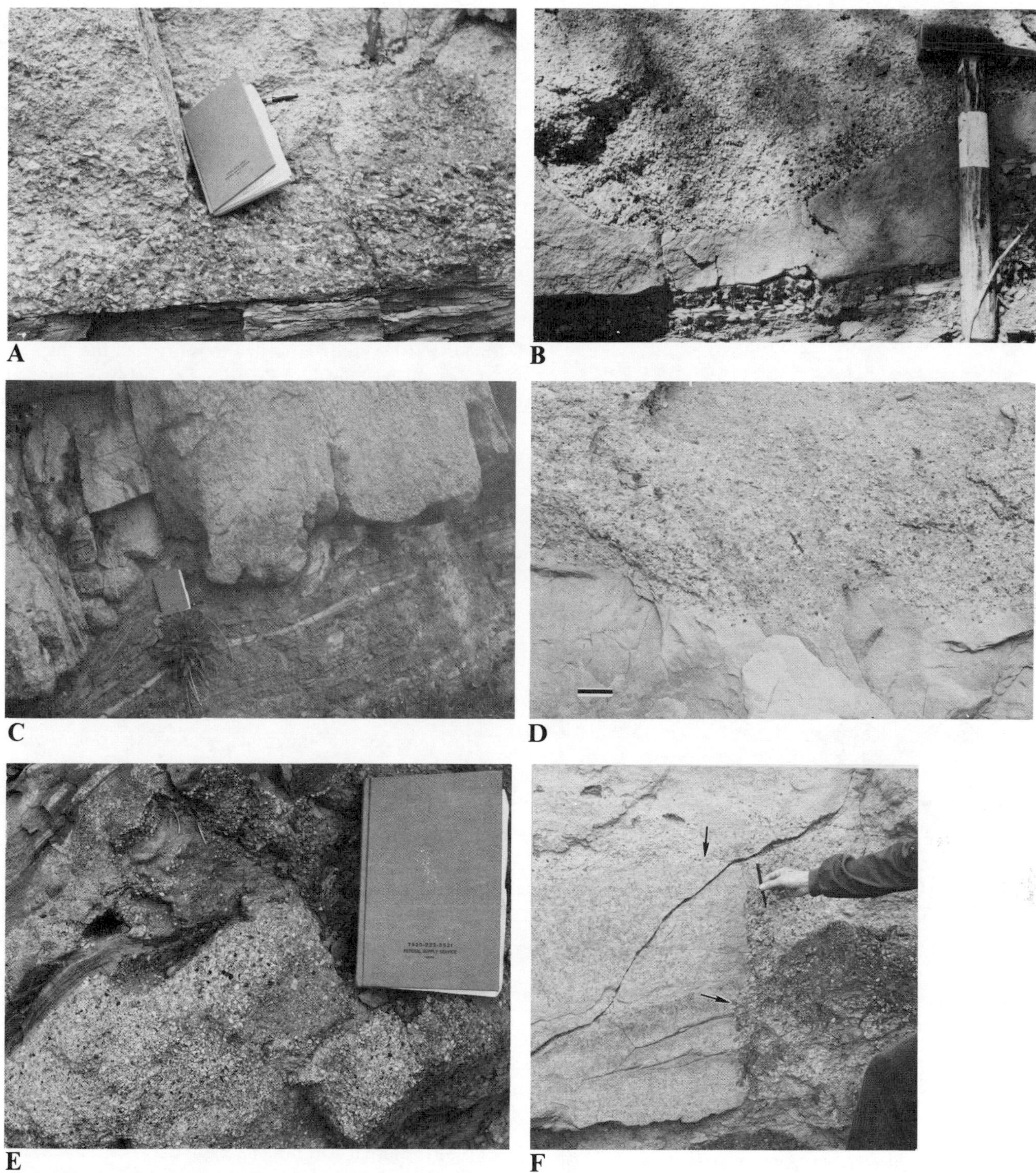

Figure 8-14. Pebble-rich layers observed along the Col de l'Orme-La Cabanette section, Peïra-Cava syncline (d-3 in Figure 8-1). Note: Photograph A shows a graded unit with some pebble imbrication (book = 20 cm). Photograph B shows a scour channel filled by a well-graded pebble layer (hammer = 35 cm). Photograph C shows large load structures formed by a graded granule-small pebble layer. Photograph D shows a pebble layer that fills scours in underlying medium-grade sandstone and that displays inverse graded, then normally graded, bedding (bar scale = 20 cm). Photograph E shows a disorganized small pebble-granule layer. Photograph F shows a sandstone layer sharply truncated (arrows) by a disorganized small pebble conglomerate.

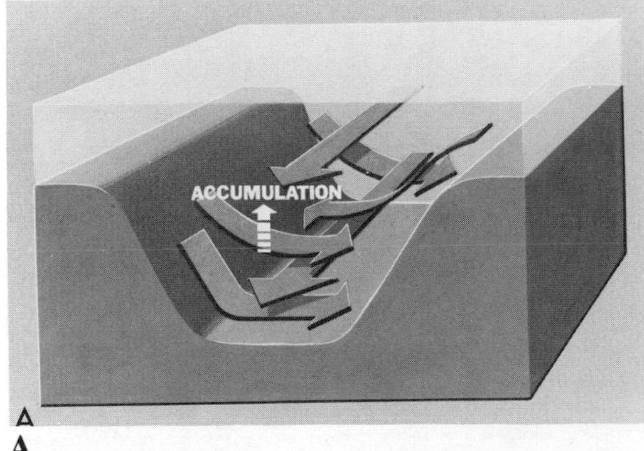

A

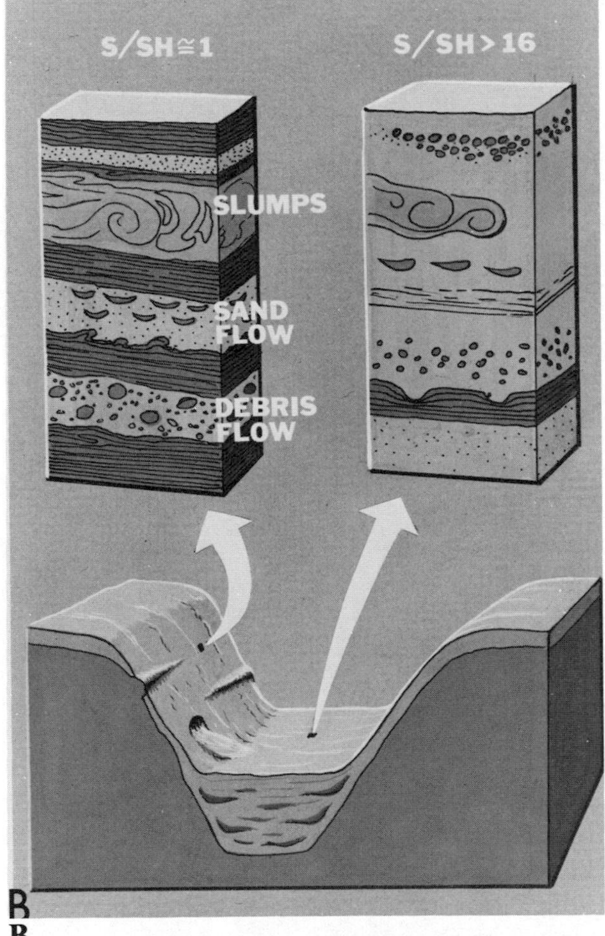

B

Figure 8-15. Schematic representation of lateral infill (diagram A), which is a major depositional factor in submarine canyons, and which combines with axial transport to result in different facies assemblages on canyon walls and axes (diagram B). Note: Drawings are not to scale.

ally display small pebble conglomerates at their base. A somewhat higher proportion of graded (organized conglomerate) units are observed in this sector, and these are less frequently associated with chaotic bedding at their base than in the southern part of the syncline.

The lower part of the section between the Col de Turini and Moulinet (d-4 in Figure 8-1) consists of thin turbidite sheets alternating with shale; this fine flysch sequence may represent an outer fan or distal interchannel facies. The number and thickness of sandstone beds varies vertically in stratigraphic sections here (as in the Sospel-Piena region north of Menton and Le Ruch-Grand Coyer sector north of Annot); these sandstone-shale cycles record a mid to outer fan evolution closely influenced by channel migration and overbank and hemipelagic deposition. This change in depositional pattern with time is accomplished by a change in paleocurrent directions (from northeast to north) as one proceeds up section and travels westward (Figure 8-1). Further study is needed to verify that the thicker shale sequences north of Peïra-Cava probably include mud turbidites of the T_et type of Rupke and Stanley (1974) as well as hemipelagic layers. In these more distal sectors of the basin, channelization of coarse sediment was less important a factor than progressive deposition from discrete, wide spread turbidity current incursions, associated overbank sheet flows of sand and silt (see Chapter 13) and hemipelagic deposition.

PROCESSES AND PRODUCTS IN ANNOT BASIN CANYONS AND FAN VALLEYS

Gravity-Induced Mechanisms in Channelized Systems

Interpretations of the principal processes responsible for the deposition of the Annot Sandstone Formation have changed progressively during the past two decades as concepts on coarse sediment transport have continued to evolve. In this section, the focus is on the plexus of sediment gravity flow mechanisms that appear to have operated in the Annot Basin submarine canyon and fan valley systems and that are responsible for the broad spectrum of lithofacies preserved in proximal to distal environments as outlined in the previous sections.

The emphasis initially on turbidites (Kuenen et al. 1957) rapidly expanded so as to include slumps and other flow types as well (Stanley 1961; Bouma 1962). Coarse-grained strata, which do not present the typical sequence of turbidite divisions, were initially termed fluxoturbidites (Stanley 1961) after the definition of Dzulynski et al. (1959). More recent investigations have considered an even broader spectrum of processes involved in downslope displacement of coarse sediment in the channelized environments of the Annot Basin. In consequence, coarse facies presenting internal structures distinct from idealized turbidites (Walker 1965) have been termed *sand flow deposits* (Stanley 1974a). Subsequently, grain flow (cf. Stauffer 1967) and/or fluidized flow (cf. Middleton and Hampton 1973) and debris flow (cf. Hampton 1972) along with the turbidity currents have been proposed as possible mechanisms for the emplacement of some massive, nongraded or poorly graded sandstone and pebble sandstone units (Stanley 1975).

The origin of this suite of coarse-grade stratal types is interpreted on the basis of internal structures and fabric. The next two sections identify the deposits of more readily identifiable gravitative transport process end-members: rockfall, sandfall, slumping, and turbidity current flows. A third section describes a series of sediment types whose specific origin is less well defined.

Slump, Rockfall, Sandfall, and Resulting Facies

Slumping, the en masse movement along slippage planes accompanied by variable internal deformation (Moore 1961; Dott 1963, Figure 7), has given rise to a series of diagnostic Annot Sandstone facies. The most remarkable slump deposits in terms of size and proportion are those comprising canyon wall, near-axial, and proximal base-of-slope sequences. Those laterally discontinuous, often chaotically bedded, mixes of sandstone, pebbles, and shale are irregular in geometry and occur in discontinuous fashion along the length of both canyon and fan valley sections. Particularly large slump masses include those exposed at the southern end of the Peïra-Cava syncline (d-2 in Figure 8-1; photograph A, Figure 8-13), in the Annot Canyon tongue west of the Coulomp River (best viewed from the Crête de la Barre [photograph A, Figure 8-2]), and east of Contes near the Sclos de Contes locality (Figure 8-3).

Slumps commonly incorporate contorted sequences of what originally may have been thin, moderately to well-stratified sand and shale sequences that accumulated on, or near, the channel walls. Such moderately deformed to highly chaotic units are frequently associated with, or are part of, thick, poorly stratified massive sandstones that display indistinct internal structure (photographs A and B, Figure 8-3). Some sandstones forming slump sequences are texturally clean, while others include muddy sands or muddy micaceous sands with abundant, finely disseminated, carbonized plant matter.

Locally associated with slump deposits are large clasts of shale that are either highly irregular in shape or are spherical and armored with coarse sandstone or pebbly sandstone. These allochthonous blocks are interpreted as rockfall deposits that were derived from Marnes Bleues sediments that formed the canyon walls and that at the time of failure were only partially lithified. Blocks of sandstone, some angular and more than 1 m in diameter (photograph E, Figure 8-3), are also identified as rockfall deposits; these blocks appear to have been partially lithified prior to failure and subsequent incorporation in slump and some sand flow deposits.

Although difficult to ascertain, some massive, irregular bedded sandstones (particularly those without distinct internal structure) associated with chaotic blocks could conceivably represent deposits of sand flows that were initiated as sand falls on very steep canyon walls (see, for example, units in photographs A and B, Figure 8-3). Sandfalls have been photographed in modern canyons off Baja California (Shepard and Dill 1966) and probably also occur on steep channel walls incised in more distal fan environments. Failure of sand is most readily explicable in such high-gradient settings, where slopes may approach the 30° natural angle of repose (Lowe 1976), although theoretical hydrodynamic and experimental studies suggest that failure can occur on considerably lower slopes (Carter 1975a,b).

A steepened surface (slump scar) resulting from slumping might generate subsequent failure and retrogressive flow sliding of the type discussed by Anderson and Bjerrum (1967), Carter (1975a), and Marshall (Chapter 7 in this volume). We propose herein that this type of process occurred not only in upper reaches of canyons but also in much deeper sectors, such as along the flanks of fan channels well within the Annot Basin.

In all likelihood, some coarse facies presenting "proximal"-type attributes in the sense of Walker (1967) observed in distal Annot Basin environments may be analogous to slump, rockfall (cf. Moore 1969), and sandfall deposits observed along the steep margins of some modern meandering fan valleys.

Turbidity Current Flow and Well-Defined Turbidites

At the other end of the gravity-induced transport spectrum, turbidity currents are the mechanisms that have been invoked most commonly in the case of the Annot Sandstone. As may be recalled, studies of this formation, particularly of sequences so well exposed at the La Cabanette section near Peïra-Cava (d-3 in Figure 8-1), resulted in the definition of the systematic arrangement of turbidite internal structures (Bouma 1962). The earlier parts of this chapter call attention to sandstone types in the different environments whose stratification and internal structures clearly identify them as turbidites. Apparently, then, turbidites comprise the single most important depositional type in terms of total numbers of sandstone beds and terrigenous volume in the Annot Basin.

Strata that are graded and display distinct Bouma divisions are by no means restricted to sequences now interpreted as submarine fan deposits (Le Ruch-Grand Coyer, Peïra-Cava, Sospel, and so forth). Well-developed turbidites form part of lithofacies attributed to the middle and lower reaches of canyons and adjacent inter-canyon slope sectors; they also are found along the length of canyon and fan valley sequences. The distribution of these so-called "typical" (or "ideal" or "mature") turbidites indicates that turbidity currents were one of the major sediment gravity flow types initiated well upslope, both in canyons and on intercanyon slope sectors. This point is emphasized inasmuch as some studies of modern outer margin settings have tended to downgrade the relative importance of turbidity current processes in proximal, relatively high-gradient environments such as canyons (Shepard and Dill 1966).

An approximation of the proportion of sandstone displaying well-defined turbidite characteristics, per Annot Basin environment, is as follows: generally less than 15 percent of the total number of strata in canyon wall and axial sequences; as high as 20 percent to 25 percent in the Crête de la Barre Canyon tributary fill section; less than 25 percent in channelized fan valley sections; and over 70 percent in the typical Annot Sandstone fan interchannel sequences. The substantial increase downslope in the proportion of sandstone strata showing turbidite features is in accord with the prevailing margin sedimentation mode that depicts a substantial increase of turbidite deposition at and beyond the canyon mouths (Komar 1972) and particularly in fan to basin plain environments.

A generalized distribution pattern of turbidite types is as follows: the middle to lower slope sectors between canyons include mostly thin, fine-to medium-grade sandstone units displaying base cut-out (T_{b-e}, T_{c-e}, T_{d-e}) and some complete T_{a-e} Bouma sequences (photographs B and C, Figure 8-2); in contrast, the adjacent (proximal) upper submarine valley sections include much higher proportions of coarser truncated (T_a, T_{a-b}, T_{a-c}) sandstone and pebbly sandstone sequences (photographs A and B, Figure 8-6) associated with massive nongraded sandstones of poorly defined origin. In the axial fill of both canyons and fan valleys, the most commonly encountered turbidites are poorly graded and of the T_a and T_{a-b} truncated types (photograph A, Figure 8-11), although complete T_{a-e} turbidites are locally recorded. For instance, a 4 m thick section of complete turbidites ranging in thickness from about 10 cm to 50 cm occurs between massive sandstone strata 20 m or more thick (photographs A and B, Figure 8-7) at the proximal Chambre du Roi Canyon section near Annot (e-8 in Figure 8-1). Higher proportions of fine-grade turbidites (coarse sandstone) are encountered in proximal Annot Basin fan sectors (upper fan, suprafan lobes); truncated series are commonly associated with fan channel sections while base cut-out sequences occur in interchannel environments (photographs A and B, Figure 8-11) and also in northern (more distal) fan localities such as in the vicinity of Turini north of Peïra-Cava (photograph B, Figure 8-9).

Facies (Including "Atypical Turbidites" and Problematica) Emplaced by Mass Flow Processes

A variety of depositional types, whose transport origin is much less well defined, occurs in association with the rockfall, sandfall, slump, and turbidite deposits cited above. This suite of deposits forms the largest proportion of sequences interpreted as canyon-fan valley fill facies and records a series of mass transport mechanisms that for the most part ranges between the plastic, high-concentration slump and fully turbulent flow spectrum (cf. Dott 1963; Middleton and Hampton 1973; Carter 1975a). In earlier studies of the Annot Sandstone most of the sediment types listed below were grouped in a general category termed fluxoturbidite (Stanley 1961; Stanley and Unrug 1972). The following sections provide brief descriptions of the more obvious sequences; a genetic terminology is applied to six of the identified types based on present knowledge of mass flow mechanisms.

*1. **Slump-debris flow sequences.*** Stratal types displaying both slump and debris flow structures (cf. Hampton 1972) are particularly characteristic; in these units, ranging from about 1 m to 3 m in thickness, a zone of contorted or chaotic bedding (slump) generally underlies a muddy pebbly sandstone or, in some cases, muddy, sandy, fine-grained conglomerates of the disorganized type. The two stratal types may merge (photograph B, Figure 8-13) or remain distinct (photograph E, Figure 8-14).

*2. **Slump-slurry sequences.*** These deposits display a basal contorted (slump) zone overlain by poorly stratified, muddy sand or very sandy mud or muddy sand with large numbers of small shale rip-up clasts (photograph C, Figure 8-13). The upper term, less coarse than debris flow deposits, is identified as a slurry deposit (cf. Carter 1975). The two stratal types may merge.

*3. **Slump-sand flow sequences.*** In this type, distorted and chaotically stratified units of varied thicknesses (to 3 m or 4 m) are covered by massive, sometimes pebbly, sandstone as much as 5 m thick that generally do not display marked graded bedding or other distinct turbidite structures. The contact between the two may be indistinct or well defined (photographs D and E, respectively, in Figure 8-13). The massive sandstone type with indistinct internal structures may be similar to that in sequences 4, 5, and 6 described below.

*4. **Debris flow-sand flow sequences.*** This type displays a basal pebbly muddy sandstone or pebbly mudstone (cf. Crowell 1957) facies or disorganized conglomerate (cf. Walker and Mutti 1973). This unit is covered by a massive sandstone division that may be either nongraded and without obvious internal structure or may display graded bedding, some horizontal lamination and, occasionally, vague dish structures (cf. sequence type 6 below). In the latter case (photographs A through C, Figure 8-12), the deposit would be termed a debris flow-incomplete (truncated) turbidite sequence.

*5. **Sand flow-truncated turbidite sequences.*** The largest proportion of thick, amalgamated deposits of coarse, poorly cemented sandstone and pebbly sandstone forming canyon and fan valley lithofacies are of this type. Small crystalline pebbles (usually < 2 cm in diameter) are commonly distributed irregularly throughout the sandstone; shale rip-up clasts (10 cm or more in diameter) as in the case of sequence 6 below, display some graded bedding, vague to well-defined horizontal lamination and, occasionally dish structures (photograph C, Figure 8-11); this upper part is interpreted as a truncated T_a or $T_{a\text{-}b}$ Bouma turbidite sequence.

6. Coarse, truncated (incomplete) turbidite sequences. This type, usually observed within thicker, massive, channelized sandstone series, displays poor to well-developed graded bedding followed by moderate to well-defined horizontal (photographs A and B, Figure 8-6; photograph A, Figure 8-11) and, occasionally, cross-lamination in its upper parts. Small rip-up shale clasts, sometimes imbricated (photograph C, Figure 8-7) occur along discrete layers. Coarse-tail grading prevails, although distribution grading (the whole size distribution tends to shift progressively upward to finer sizes) is occasionally encountered, particularly where pebbles are concentrated at poorly defined inverse grading, or inverse grading is followed by vague to moderate normal graded bedding (photograph D, Figure 8-14).

The six sequences described above, and gradational varieties thereof, commonly form coarse-grained channelized bodies on the slope and in the basin. None of the sequences occupies a unique or geographically restricted position in the canyon-fan valley systems. However, the proportion of sequences displaying some form of massive slumping (types 1, 2, and 3) is highest in (or in the proximity of) the sand tongue lithofacies interpreted as canyon deposits (i.e., near Annot, Contes and Menton). Unit 5 dominates the lower canyon fill sequences. In contrast, unit 6 interpreted as coarse, truncated turbidites forms the largest proportion of massive, channelized sequences in more distal sectors—that is, valleys beyond canyon mouths on the fan sectors well in the basin. Types 1 to 4 commonly form an important portion of base-of-slope deposits in proximity to outer canyon-upper fan sectors.

DOWNSLOPE EVOLUTION OF FLOWS IN THE CHANNELIZED SYSTEMS

Rationale

That a single gravity-induced transport episode responsible for the entrainment of sand and coarser material to deeper water may involve more than one specific flow mechanism has long been suspected. Kuenen (1951), for example, suggested the possible changes downslope from slumping to mud flow, and subsequently others also have proposed transformations from slumping and debris and related mass flow to turbidity current (Dott 1963; Morgenstern 1967; Komar 1971; Hampton 1972) primarily on the basis of theoretical hydrodynamic considerations.

In recent years there has been an attempt to systematize in more orderly fashion the possible grain interaction, pore fluid criteria, and responding evolution of grain support mechanisms during flow. Middleton and Hampton (1973, 1976) in a generalized analysis of sediment gravity flows distinguish between turbulence, the grain support type most characteristic of the turbidity current process end-member, and upward movement of pore fluids, dispersive pressure, and matrix strength that give rise, respectively, to fluidized flow, grain flow, and debris flow end-members. In a hypothetical scheme emphasizing changes of grain to fluid concentrations and grain support mechanisms in time and space Middleton and Hampton (1976, Figure 10) depict the transformation of one flow type to another and show a natural slump to debris-fluidized-grain-turbidity current flow evolution that may exist between initial failure and final deposit.

Carter (1975a) also presents a natural suite of granular, water-saturated sediment transport modes and outlines a somewhat broader spectrum of gravity-induced processes. Carter emphasizes, perhaps more than do Middleton and Hampton, the distinction between turbidity flow and inertia flow, with the latter referring to "transportation of sediment in which grains move above the bottom but are not supported by upward components of the turbulent movements within the fluid . . . the grains move differentially within the fluid and owing to their inertia, follow linear paths . . ." (Sanders 1965, p. 329). Inertia flow processes in Carter's scheme are broadly divided into those involving water as the interstitial fluid (or grain flow, with implication of transfer of momentum by intergranular impacts that also generate dispersive pressure) and those having an interstitial fluid of enhanced viscosity due to increased proportions of clay and silt (or slurry flow, an inertia flow likened to a viscous grain flow). Carter (1975a) also calls attention to slump creep, a process of glacier-like creep generally associated with slumping on all scales; this phenomenon, recorded in some modern submarine valleys, was earlier elucidated by Shepard and Dill (1966).

Conglomerates account for a relatively minor proportion (< 5 percent of total number of strata) of the Annot Sandstone sequences. Small crystalline pebbles most commonly occur as isolated clasts or as poorly defined horizons in a sand matrix (photograph C, Figure 8-5); more rarely they are concentrated as discretely graded, inversely graded, and pebbly mudstone-type strata (Figure 8-14). A number of pebble transport models are of some assistance in the interpretation of the pebbly sandstone and conglomeratic deposits observed in this formation. Various transport modes have been suggested: sliding (Natland and Kuenen 1951), creep and traction (Gennesseaux 1966), and dense mass flows (Crowell 1957; Dott 1963; Hampton 1972; Middleton and Hampton 1973; Stanley 1974b). Studies also suggest that pebble and cobble size material may be transported by turbidity currents, particularly in confined environments such as channels (Komar 1970; see also Chapter 24 of this volume). Several conglomerate transport models, based largely on observation of ancient deposits and emphasizing fabric and stratification, attempt to relate conglomerate type with environment (Walker and Mutti 1973; Walker 1975, Figure 10). These models depict a systematic downslope evolution of pebble deposition in response to hydrodynamic changes that occur between confined (canyon, fan valley) and more open sectors of outer margin environments.

Finally, we should recall that processes active during mobilization and active displacement of sand are not necessarily the same as those effective during the terminal, or "freezing," depositional stages, which is a factor emphasized by Walton (1967) and again more recently by Carter (1975a, Figure 7). We recognize that in many instances these "final stage" hydrodynamic conditions are what have given rise to the assemblage of internal structures (or their absence) in the coarse, channelized sequences exposed at the three Annot Basin canyon-fan valley localities.

Possible Transformations and Canyon Mouth and Fan Channel Bypassing

The field observations suggest the following generalized outer margin depositional patterns: mud, sand, granule, and pebble-size material, for the most part moderately to poorly sorted, fed by rivers draining emerged terrains in the south, bypassed a narrow, tectonically active shelf and were introduced into the heads of valleys cut on the shelf and north-dipping slopes. The high proportion of nongraded or poorly graded massive sandstone and pebbly sandstone of low matrix preserved in these channelized settings suggest transport by sediment gravity flows that differ from idealized low-density turbulent turbidity currents; the sand and granule fractions were sustained and displaced by grain interaction (dispersive pressure) or upward intergranular flow rather than by turbulence alone; and the geographic distribution and geometry of facies shows that a considerable proportion of coarser elements were deposited from high-concentration mass flows in confined settings, but that the bulk of sediment fed into this late Eocene-early Oligocene basin accumulated from lower-density turbidity currents as interchannel fan deposits beyond the canyon mouths. The lithofacies diversity records a broad spectrum of sediment gravity flow mechanisms effective in the downslope transfer of coarse-grade, terrigenous material in the channelized systems. The evidence that at various stages of their evolution such flows combine more than one theoretically distinct grain support mechanism is also suggested, but less readily explained.

The facies assemblages record large-scale slumping and associated failure on relatively steep slopes such as the canyon walls (Figure 8-15). These lateral infill processes in turn may have initiated movement of sandy, pebbly sandy, or muddy sand slurries that deposited the massive, moderately stratified, structureless, poorly sorted sandstone and pebbly sandstone facies identified as canyon fill deposits. Slumping also may have initiated debris flow and slump-creep mechanisms that resulted in strata enclosing randomly distributed pebbles supported by a sandy or sandy mud matrix (diagram B, Figure 8-15).

The superposition of massive sandstone strata (sand: shale ratios >32) and reduced number of silt and shale layers in both canyon and fan channel sequences suggest that sand flows of various types followed each other rapidly or, alternatively, that fine-grained layers that did accumulate were eroded by subsequent sand flows (Stanley 1975). The erosive nature of such flows in both canyons and fan valleys is shown by large groove markings on the base of beds, the scouring and truncation of successive sandstone layers within massive layers, and the frequent inclusion of large rip-up clasts. The vigorous entrainment of coarse material along the seafloor also is indicated by the abundant armored mud balls found in the canyon fill sections.

There are other explanations, albeit circumstantial, for the low proportion of fine-grained material, either as matrix or as discrete layers, in canyon and fan valley sections. Inasmuch as silt and clay size strata abound in intervalley slope, fan, and more distal Annot Sandstone sequences (sand:shale ratios of 1.0 to 3.0), both mud and sand were almost certainly fed to the canyon heads, and the low proportion of fines preserved in canyon and fan sequences is in large part a response to processes active in these confined settings. Several explanations may be proposed. The first, based on the observation of a few ideal turbidites in canyon-fan sequences, is that fines were displaced with sand primarily by low-density turbidity flows initiated in proximal sectors and transported to the fan and the basin plain beyond. This explanation, however, is not an entirely satisfactory argument. More probably, as will be shown subsequently, much of the mud and silt originally trapped and set in motion in the valleys was separated from the bulk of the coarse fraction as a result of sediment gravity mechanisms presenting grain support characteristics of inertia flow as well as turbidity current flow.

In order to obtain a reasonable solution, evaluating the

origin of the massive sandstone and pebbly sandstone units that comprise the major stratal types in valley systems is necessary. These facies have long been identified as the products of processes intermediate between slump and turbidity currents (i.e., the fluxoturbiditic flows of Kuenen 1958 and Stanley 1961) or as proximal turbidites (Walker 1967). More recently, Carter (1975a), who opts for the retention of the term *fluxoturbidite*, attributes these facies to deposits from the basal layer of "a turbidity current in which the immediately pre-depositional transport of the basal layer was by inertia-flow (p. 172)." Experimental flume study of high concentration flows provides information that may bear on this problem of origin of massive beds. Middleton (1967) summarizes as follows:

> In the high concentration flows the bed was observed to accumulate in four main stages, only the last of which was identical with that observed for the low concentration flows. The first three were: (i) deposition of sediment immediately behind the head, followed by extensive mass shearing of the bed (*not* movement by rolling or sliding); the top of the bed was not clearly defined; (ii) formation of an expanded, "quick" bed; Helmholtz waves formed at the upper surface of the bed and produced a circular shearing motion deep within the bed; and (iii) disappearance of waves and formation of a plane surface, accompanied by consolidation of the bed by as much as 20%. The behavior of high concentration flows and the formation of coarse-tail grading was a function of the pseudoplastic behavior of the particle suspensions at high concentrations, and was caused by particle interactions (dispersive pressure) within the suspension. The expanded bed was essentially the current itself after movement had almost ceased. The Helmholtz waves were caused by the passage over the bed of the dilute suspension constituting the mixing zone entrained by the current. This entrained zone continued to flow at high velocities, due to inertia, even after the concentrated suspension had ceased flowing [p. 503].

Although they differ considerably in appearance from turbidites displaying the typical Bouma divisions, the massive nongraded to poorly graded composite strata of low matrix may have originated in a manner somewhat analogous to the lower massive *a* division of turbidites. The presence of horizontal laminated layers, occasionally accompanied by dish structures, in the upper parts of some massive beds is compatible with this hypothesis. The rapid deposition and "freezing" of sediment accompanying dissipation of the underflow does not preclude the continuation of flow of an overlying mixing zone that entrains sediment farther downslope. Slower, more progressive settling of grains from this mixed zone of lower density probably gave rise to the more classic, matrix-rich, better-graded turbidites farther downslope. An attractive aspect of this process model is that it helps explain the marked increase of sandstone turbidites displaying Bouma divisions in the less-confined settings beyond the mouths of canyons in the Annot Basin.

The concentration of massive, nongraded, amalgamated units in some fan channel sequences suggests that the segregation process occurred well within the basin as well. As presently conceptualized, sand (particularly of medium and fine grade) and finer fractions bypassed fan channel margins by overbank flows of varying density and turbulence and were deposited as graded (base cut-out) sheets over extensive interchannel surfaces. The somewhat coarser sediment confined in fan valleys appear to have been moved by denser flows (debris flow, inertia, or high-concentration turbidity current underflow).

Fan channel bank overtopping and levee formation in modern settings may be related in part to a hydraulic jump phenomena in sectors of decreased gradient and reduced confinement—that is, usually at the base of slopes and seaward of canyon mouths (cf. Komar 1971). That some flows were thick enough to overtop channel banks cut on the inner fan in the Annot Basin may be recorded by some sandstone units that display intensely cross-laminated divisions. One area where such possible levee deposits are observed is the Cabanette-Pas de l'Escous road-cut section in the southeastern sector of the Peïra-Cava syncline (d-3 in Figure 8-1). In this same locality the proportion of thick, coarse grade, complete $T_{a\text{-}e}$ turbidites is remarkably high. Bouma (1962, Figure 25) has suggested that a complete turbidite sequence would most likely accumulate at the proximal apex of a lobate depositional zone covered by a turbidite sheet. Recent reexamination of the facies in this locality indicates that the sea floor at this point occupied a transitional zone between a steeper slope to the south (near Contes) where a large proportion of truncated turbidites were trapped in a channelized environment, and a less confined setting further north in the paleobasin (near Turini [d-4 and 5 in Figure 8-1]) where somewhat finer-grade base-cut turbiditic sequences accumulated on a fan (Stanley 1975).

There is additional evidence in support of this paleogeographic interpretation. The complete turbidites and possible levee deposits observed at la Cabanette are associated with a suite of strata types that record a variety of transitional sediment gravity flow mechanisms: slump and slurry to debris flow to high-density turbidity current flow; inertia flow to low-density turbidity current flow; and flows likely to deposit sand with structures attributable to both inertia and turbidity current mechanisms as discussed earlier. Furthermore, the presence of inversely and normally graded granule and small pebble conglomerates and some disorganized conglomerates would also identify this depositional site as one close to the base of a slope (Figure 8-16)— that is, near a canyon

Figure 8-16. Simplified diagram showing emerged highland and the complex submarine dispersal patterns in the Annot Basin that involve a broad spectrum of gravity-induced transport processes. Note: Numbers refer to figures in this chapter that illustrate a particular outer margin depositional sequence.

mouth and on a proximal fan apex according to the conglomerate depositional model of Walker (1975). Similar dispersal patterns also are recognized in other ancient outer margin deposits (Carter and Lindqvist 1975).

Base-of-slope facies assemblages near Le Ruch (e-9 in Figure 8-1) north of the channelized sandstone tongue at Annot and in the Sospel syncline north of the channelized tongue at Menton (b in Figure 8-1) are comparable to those at Peïra-Cava and may be interpreted in a similar manner. Recognition of a variety of mechanisms (slump, debris flow, slurry, and creep) in these two areas, as in the case of the Peïra-Cava syncline, enables the distribution and origin of the associated turbidite-rich fan facies to be evaluated in better perspective. On the one hand, the thick sequences of alternating sandstone turbidites and shales (including both mud turbidites and hemipelagic deposits) are interpreted as primarily interchannel fan deposits that were deposited by flows set into motion on the slope and in canyons and that continued seaward beyond canyon mouths and over fan channel banks. The generally good graded bedding and moderate matrix content of the sandstones indicate progressive settling from turbulent grain support and suspension mechanisms. The massive, generally amalgamated, poorly graded sandstone units defining fan channel sequences record the accumulation of somewhat coarser sizes from high concentration flows in the confined sectors. This segregation phenomenon occurred well in the basin on low (probably under 3°) slopes.

Somewhat comparable downslope mass flow transformations have been recognized in other ancient outer margin deposits (Carter and Lindqvist 1975). This phenomenon and associated depositional patterns that record flow segregation and by-passing likely prevail in some modern canyon-fan valley-basin settings, particularly those with access to significantly large supplies of sand-rich material. Examples at present might include canyons heading close to land and receiving sand from river or nearshore input (for instance, those on the California Borderland and in the Mediterranean). These coarse sediment transport patterns more commonly occurred in the Pleistocene when, as a result of eustatic sea level lowering, many more canyons trapped enhanced volumes of material. Theoretical hydrodynamic and experimental flume studies offer a means to evaluate transport processes, but there is also much to be gained by refining our knowledge of lithofacies distribution patterns in modern canyon-fan valley systems. This requires an advanced technology involving long, wide-diameter, sand-penetrating coring systems and methods to monitor artificially triggered sediment gravity flows in deep marine environments. Defining the transport of coarse sediment to deep water remains a challenging area of research.

CONCLUSIONS

Exposures of the Annot Sandstone Formation in the French Maritime Alps include a variety of coarse slope and basinal sediment facies deposited in both channelized and unconfined outer margin environments. The geometry, stratification, internal structures, and fabric of these lithofacies, examined sequentially from proximal (lower slope-inner fan) to distal (outer fan-basin) sectors, record a broad spectrum of gravity-induced transport processes.

Analysis of submarine canyon axis, canyon wall, tributary canyon and fan valley, and interchannel deposits dispersed along major paleoslope trends reveals downslope facies changes that are most readily explicable in terms of grain support modification and evolution from one type of flow to another during transport. These proximal to distal lithofacies changes serve to identify downslope mass flow transformations involving down-axis segregation of material. Massive, low-matrix sands were deposited in canyons and fan channels primarily as "quick" beds from high-concentration underflows, whereas graded sands and finer fractions that bypassed canyon mouths and overtopped fan channel levees settled more progressively from lower-density turbidity currents over broad areas. The most significant changes in transport processes are recorded in base-of-slope sectors seaward of canyon mouths and on proximal fan apices.

The thick, amalgamated, nongraded or poorly graded, low-matrix pebbly sandstone are attributed to high-concentration sediment gravity flows. Although such massive canyon-fan valley fills generally do not display typical turbiditic structures and do not resemble classic turbiditic sequences, we postulate that they were emplaced by processes involving turbidity current flow in association with a plexus of mass flow mechanisms whose origin is less well defined. Incomplete and truncated turbidites are common, particularly in channelized settings. Many of these coarse units, particularly in proximal base-of-slope (lower canyon axis, inner fan) sectors, include lower terms emplaced by slump, debris flow, and slurry flow processes and upper divisions of sand transported by turbidity current mechanisms.

Depositional patterns reflecting flow segregation, downslope transformation and by-passing likely prevail in some modern canyon-fan valley-basin settings, particularly those with access to significantly large supplies of sand-rich material. Examples at present include canyons heading close to land and receiving sand from river or nearshore input (for instance, those on the California Borderland and in the Mediterranean). These coarse sediment transport patterns more commonly occurred in the Pleistocene when, as a result of eustatic sea level lowering, canyons incised on subaerially exposed shelves trapped enhanced volumes of material.

Defining the modes of transport involved in the dispersal of coarse sediment to deep water remains a challenging area of research. More thorough investigation of modern environments such as canyons and fan valleys depends on the development of sophisticated technology involving long, wide-diameter, sand-penetrating coring systems and methods to monitor artificially triggered sediment gravity flows. This approach, undertaken with theoretical hydrodynamic and experimental flume studies and coupled with detailed analysis of ancient canyon-fan lithofacies is needed to advance knowledge on gravity-induced transport processes.

ACKNOWLEDGMENTS

Funding for this field study, part of the Mediterranean Basin (MEDIBA) Project, was provided by Smithsonian Research Foundation grants 71500532 and 71702120. Thanks are expressed to R. Slater for documentation on turbidity current flow in a modern canyon and to J. Wexler for data on trace fossil assemblages in the Annot Sandstone Formation. The manuscript was reviewed by A. H. Bouma and M. A. Hampton, U.S.G.S., Menlo Park, California, and G. Kelling, University of Keele, England.

REFERENCES

Andresen, A., and Bjerrum, L., 1967. Slides in subaqueous slopes of loose sand and silt. In: A. F. Richards (ed.), *Marine Geotechnique.* University of Illinois Press, Urbana, pp. 221-239.

Bouma, A. H., 1962. *Sedimentology of Some Flysch Deposits.* Elsevier, Amsterdam, 168 pp.

Bourcart, J., 1964. Les sables profonds de la Méditerranée occidentale. In: A. H. Bouma and A. Brouwer (eds.), *Turbidites.* Developments in Sedimentology, 3. Elsevier, Amsterdam, pp. 148-55.

Burk, C. A., and Drake, C. L. (eds.), 1974. *The Geology of Continental Margins.* Springer-Verlag, New York, 1009 pp.

Carter, R. M., 1975a. A discussion and classification of subaqueous mass transport with particular application to grain-flow, slurry-flow, and fluxoturbidites. *Earth Sci. Reviews,* 11: 146-77.

_____, 1975b. Mass-emplaced sand-fingers at Mararoa construction site, southern New Zealand. *Sediment.*, 22: 275-88.

_____, and Lindqvist, J. K., 1975. Sealers Bay submarine fan complex, Oligocene, southern New Zealand. *Sediment.*, 22: 465-83.

Crowell, J. C., 1957. Origin of pebbly mudstones. *Geol Soc. Amer. Bull.*, 68: 993-1010.

Dill, R. F., 1967. Effects of explosive loading on the strength of sea-floor sands. In: A. F. Richards (ed.), *Marine Geotechnique.* University of Illinois Press, Urbana, pp. 291-303.

Dott, R. H., Jr., 1963. Dynamics of subaqueous gravity depositional processes. *Amer. Assoc. Petrol. Geol. Bull.*, 47: 104-28.

Dzulynski, S., Ksiazkiewicz, M., and Kuenen, Ph. H., 1959. Turbidites in flysch of the Polish Carpathian Mountains. *Geol. Soc. Amer. Bull.*, 70: 1089-118.

Gennesseaux, M., 1966. Prospection photographique des canyons sous-marins du Var et du Paillon (Alpes-Maritimes) au moyen de la Troïka. *Rev. Géogr. Phys. Géol. Dynam.*, 8: 3-38.

_____, Guibout, P., and Lacombe, H., 1971. Enrégistrement de courants de turbidité dans la vallée sous-marine du Var (Alpes-Maritimes). *C. R. Acad. Sc. Paris*, 273: 2456-59.

Got, H., and Stanley, D. J., 1974. Sedimentation in two Catalonian canyons, northwestern Mediterranean. *Mar. Geol.*, 16: M91-M100.

Hampton, M. A., 1972. The role of subaqueous debris flow in generating turbidity currents. *J. Sed. Petrol.*, 42: 775-93.

Kelling, G., and Stanley, D. J., 1976. Sedimentation in canyon, slope, and base-of-slope environments. In: D. J. Stanley and D. J. P. Swift (eds.), *Marine Sediment Transport and Environmental Management.* Wiley, New York, pp. 379-435.

Komar, P. D., 1970. The competence of turbidity current flow. *Geol Soc. Amer. Bull.*, 81: 1555-62.

_____, 1971. Hydraulic jumps in turbidity currents. *Geol. Soc. Amer. Bull.*, 82: 1477-88.

_____, 1972. Relative significance of head and body spill from a channelized turbidity current. *Geol. Soc. Amer. Bull.*, 83: 1151-56.

Kuenen, Ph. H., 1951. Properties of turbidity currents of high density. In: R. D. Russell (ed.), *Turbidity Currents and the Transportation of Coarse Sediments of Deep Water.* Soc. Econ. Paleont. Miner., Sp. Publ. 2, pp. 14-33.

_____, 1958. Problems concerning source and transportation of flysch sediments. *Geol. Mijnb.*, 20: 329-39.

_____, Faure-Muret, A., Lanteaume, M., and Fallot, P., 1957. Observations sur les flyschs des Alpes Maritimes francaises et italiennes. *Bull. Soc. Géol. France*, 7: 11-26.

Lanteaume, M., Beaudoin, B., and Campredon, R., 1967. *Figures Sédimentaires du Flysch "grès d'Annot" du Synclinal de Peira-Cava.* Cent. Nat. Rech. Scient., Paris, 97 pp.

Lowe, D. R., 1976. Grain flow and grain flow deposits. *J. Sed. Petrol.*, 46: 188-99.

Middleton, G. V., 1967. Experiments on density and turbidity currents, III. Deposition of sediment. *Canad. Jour. Earth Sc.*, 4: 475-505.

_____, and Hampton, M. A., 1973. Mechanics of flow and deposition. In: G. V. Middleton and A. H. Bouma (eds.), *Turbidites and Deep-Water Sedimentation.* Soc. Econ. Paleont. Mineral., Pacific Section, Short Course, Anaheim, pp. 1-38.

_____, and Hampton, M. A., 1976. Subaqueous sediment transport end deposition by sediment gravity flows. In: D. J. Stanley and D. J. P. Swift (eds.), *Marine Sediment Transport and Environmental Management.* Wiley, New York pp. 197-218.

Moore, D. G., 1961. Submarine slumps. *J. Sed. Petrol.*, 31: 343-57.

_____, 1969. *Reflection Profiling Studies of the California Continental Borderland: Structure and Quaternary Turbidite Basins.* Geol. Soc. Amer., Spec. Paper 107, 142 pp.

Morgenstern, N., 1967. Submarine slumping and the initiation of turbidity currents. In: A. F. Richards (ed.), *Marine Geotechnique.* University of Illinois Press, Urbana, pp. 189-220.

Mutti, E., 1974. Examples of ancient deep-sea fan deposits from circum-Mediterranean geosynclines. In: R. H. Dott, Jr., and R. H. Shaver (eds.), *Modern and Ancient Geosynclinal Sedimentation* Soc. Econ. Paleont. Mineral., Sp. Publ. 19, pp. 92-105.

_____, and Ricci Lucchi, F., 1972. Le torbiditi dell-Appennino settentrionale: introduzione all'analisi di facies. *Mem. Soc. Geol. Italiana*, 11: 161-99.

Natland, M. L., and Kuenen, Ph. H., 1951. Sedimentary history of the Ventura Basin, California, and the action of turbidity current. In: R. D. Russell (ed.), *Turbidity Currents and the Transportation of Coarse Sediments to Deep Water.* Soc. Econ. Paleont. Mineral., Sp. Publ. 2, pp. 76-107.

Nelson, C. H., and Kulm, V., 1973. Submarine fans and channels. In: G. V. Middleton and A. H. Bouma (eds.), *Turbidites and Deep-Water Sedimentation.* Soc. Econ. Paleont. Mineral., Pacific Section, Short Course, Anaheim, pp. 39-78.

_____, and Nilsen, T. H., 1974. Depositional trends of modern and ancient deep-sea fans. In: R. H. Dott, Jr., and R. H. Shaver (eds.), *Modern and Ancient Geosynclinal Sedimentation.* Soc. Econ. Paleont. Mineral., Sp. Publ. 19, pp. 69-91.

Normark, W. R., 1970. Growth patterns of deep-sea fans. *Amer. Assoc. Petrol. Geol. Bull.*, 54: 2170-95.

Rupke, N. A., and Stanley, D. J., 1974. Distinctive properties of turbiditic and hemipelagic mud layers in Algéro-Balearic Basin, Western Mediterranian Sea. *Smithsonian Contr. Earth Sci.*, 13, 40pp.

Sanders, J. E., 1965. Primary sedimentary structures formed by turbidity currents and related resedimentation mechanisms. In: G. V. Middleton (ed.), *Primary Sedimentary Structures and their Hydrodynamic Interpretation.* Soc. Econ. Paleont. Mineral., Sp. Publ. 12, pp. 192-219.

Shepard, F. P., and Dill, R. F., 1966. *Submarine Canyons and other Sea Valleys.* Rand McNally, Chicago, 381 pp.

_____, Dill, R. F., and von Rad, U., 1969. Physiography and sedimentary processes of La Jolla Submarine Fan and Fan-valley, California. *Amer. Assoc. Petrol. Geol. Bull.*, 53: 390-420.

_____, and Marshall, N. F., 1973. Currents along floors of submarine canyons. *Amer. Assoc. Petrol. Geol. Bull.*, 57: 244-64.

Stanley, D. J., 1961. *Etudes sédimentologiques des grès d'Annot et de leurs équivalents latéraux.* Inst. Franc. Pétrole Ref. 6821, Société des Editions Technip, Paris, 158 pp.

_____, 1964. Large mudstone-nucleus sandstone spheroids in submarine channel deposits. *J. Sed. Petrol.*, 34: 672-76.

_____, 1965. Heavy minerals and provenance of sands in flysch of central and southern French Alps. *Amer. Assoc. Petrol. Geol. Bull.*, 49: 22-40.

_____, 1970. Flyschoid sedimentation on the outer Atlantic margin off northeast North America. In: J. Lajoie (ed.), *Flysch Sedimentology in North America.* Geol. Assoc. Canada, Spec. Paper 7, pp. 179-210.

_____, 1974a. Dish structures and sand flow in ancient submarine valleys, French Maritime Alps. *Bull. Centre Rech. Pau-SNPA*, 8: 351-71.

_____, 1974b. Pebbly mud transport in the head of Wilmington Canyon. *Mar. Geol.*, 16: M1-M8.

_____, 1975. Submarine Canyon and Slope Sedimentation (Grès d'Annot) in the French Maritime Alps. *Proc. IX Intern. Cong. Sediment.*, Nice, 129 pp.

———, and Bouma, A. H., 1964. Methodology and paleogeographic interpretation of flysch formations: a summary of studies in the Maritime Alps. In: A. H. Bouma and A. Brouwer (eds.), *Turbidites.* Developments in Sedimentology., 3. Elsevier, Amsterdam, pp. 34-64.

———, and Unrug, R., 1972. Submarine channel deposits, fluxoturbidites and other indicators of slope and base-of-slope environments in modern and ancient marine basins. In: J. K. Rigby and W. K. Hamblin (eds.), *Recognition of Ancient Sedimentary Environments.* Soc. Econ. Paleont. Mineral., Sp. Publ. 16, 287-340.

Stauffer, P. H., 1967. Grain-flow deposits and their implications, Santa Ynes Mountains, California. *J. Sed. Petrol.,* 37: 487-508.

Walker, R. G., 1965. The origin and significance of internal sedimentary structures of turbidites. *Proc. Yorks. Geol. Soc.,* 35: 1-32.

———, 1967. Turbidite sedimentary structures and their relationship to proximal and distal depositional environments. *J. Sed. Petrol.,* 37: 25-43.

———, 1975. Generalized facies models for resedimented conglomerates of turbidite association. *Geol. Soc. Amer. Bull.,* 86: 737-48.

———, and Mutti, E., 1973. Turbidite facies and facies associations. In: G. V. Middleton and A. H. Bouma (eds.), *Turbidites and Deep-Water Sedimentation.* Soc. Econ. Paleont. Mineral., Pacific Section, Short Course, Anaheim, pp. 119-58.

Walton, E. K., 1967. The sequence of internal structures in turbidites. *Scott. J. Geol.,* 3: 306-17.

Whitaker, J. H. McD. (ed.), 1976. *Submarine Canyons and Deep-Sea Fans, Modern and Ancient.* Dowden, Hutchinson & Ross, Stroudsburg, Pa., 460 pp.

Winn, R. D., and Dott, R. H., Jr., 1977. Large-scale traction-produced structures in deep-water fan-channel conglomerates in southern Chile. *Geology,* 5: 41-44.

Chapter 9

The Kinderscoutian Delta (Carboniferous) of Northern England: A Slope Influenced by Density Currents

PETER J. McCABE

Department of Geology
McMaster University
Hamilton, Ontario, Canada

ABSTRACT

A coarsening upward sequence from mudrock to fine- and medium-grade sandstone occurs in the Kinderscoutian (Namurian, Carboniferous) of the Central Pennine Basin of northern England. The sequence is underlain by submarine fan deposits and is overlain by fluvial sediments. The coarsening-upward sequence is interpreted as a prograding delta slope sequence. While most of the sediment appears to have settled directly from suspension, the uppermost sandstones were reworked by currents and waves, and some of the lower sandstones were apparently reworked by river-generated density currents.

Some of the sediment drapes over inclined truncation surfaces, at angles up to 18°. Units of such inclined beds are up to 16 m thick. The smooth nature of the truncation surfaces and their association with syn-sedimentary deformation structures suggest that they are slump scars.

Channels, up to 30 m deep, are found throughout the coarsening upward sequence. The channels are infilled by an upward sequence of unlaminated – horizontally laminated – cross-bedded coarse sandstones. Slump-generated turbidity currents probably cut and infilled these channels. Thin sandstone beds, resembling distal turbidites, are associated with the channels and are interpreted as overbank deposits.

Note: Peter J. McCabe is currently at the Department of Geology, University of Nebraska, Lincoln, Nebraska.

INTRODUCTION

This chapter describes a thick delta slope sequence in the Kinderscoutian (Namurian, upper Carboniferous) of the Central Pennine Basin of northern England (diagram A, Figure 9-1), which coarsens upwards in a complex manner. The sediments deposited from suspension have been modified by slumping and by density currents that carried little or no suspended sediment. Turbidity currents, bearing coarse sediment, have also cut and filled channels on the slope. The delta slope sediments differ from most present-day equivalents by their being coarse-grained, overlying a prograding submarine fan sequence, and being deposited in a basin of apparently fluctuating salinity.

The Kinderscoutian sediments have been extensively studied in the southern part of the basin (Allen 1960; Walker 1966a, 1966b, 1976; Collinson 1968, 1969, 1970), where a regressive sequence of submarine fan, slope, and delta top deposits has been recognized. This chapter is based on sediments studied in the northern part of the basin between Wharfedale and Longdendale (diagram B, Figure 9-1). Detailed sections were measured and the relationship between exposures was established by field mapping (McCabe 1975). The regressive sequence was found to be similar throughout the basin (diagram C, Figure 9-1). Examination of the sediments led to a reinterpretation of parts of the sequence. Previously unrecognized channels up to 40 m deep and in the order of 0.5 km to 1.5 km wide (diagram C, Figure 9-1) are interpreted as distributary channels (McCabe 1977).

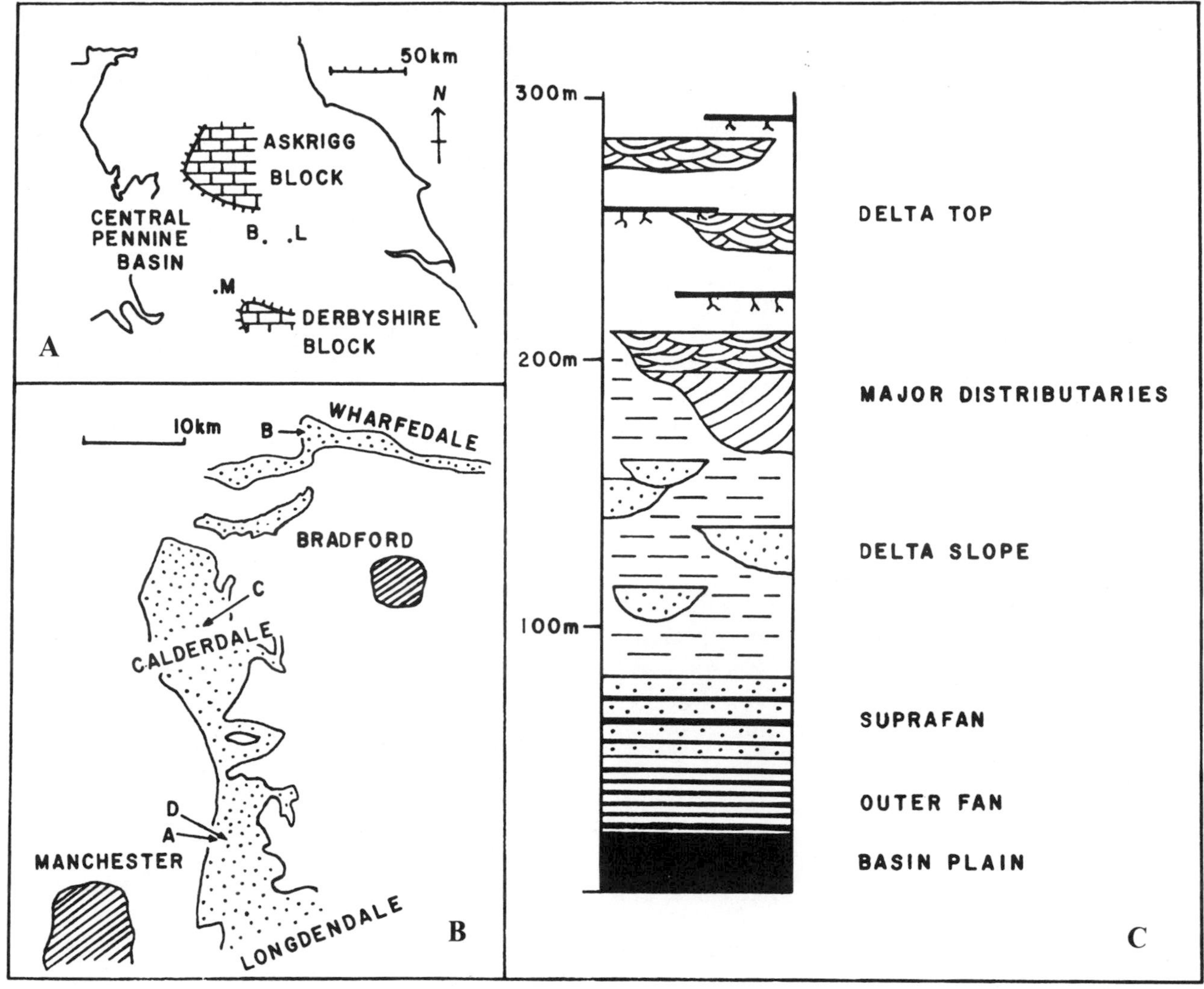

Figure 9-1. The study area and interpretive section of slope sediments. Note: Diagram A shows the location of the Central Pennine Basin, northern England (B = Bradford, L = Leeds, M = Manchester). Diagram B shows the location of the study area; the stippled area indicates the outcrop of the Kinderscoutian, and letters show location of sections in Figure 9-4. Diagram C shows idealized section through the Kinderscoutian with environmental interpretations; thicknesses are based on the Calderdale area.

Giant cross-beds within the channels that were interpreted by Collinson (1968) as deltaic sedimentation units are now interpreted as the internal bedding of large bedforms (McCabe 1977). Evidence is presented in this chapter that all the sediment between the distributary channels and the underlying submarine fan deposits was deposited on a slope built by a prograding delta. Previously only the lower half of this sequence was interpreted as a slope deposit, while the upper half (i.e., the upper Grindslow Shales and lowest Lower Kinderscout Grit in Derbyshire) was thought to be predominantly a delta top sequence (Collinson 1969). The deep distributary channels apparently were cut into the top of the delta slope.

While the total deltaic sequence varies between 200 m and 500 m in thickness thoughout the basin, the slope deposits are relatively constant and are about 100 m thick.

THE CENTRAL PENNINE BASIN

The Central Pennine Basin is bounded to the north by the Askrigg Block and to the south by the Derbyshire Block, both of which are sites of thick lower Carboniferous limestone deposition (see diagram A, Figure 9-1). During the Namurian, deltas infilled the basin from the northeast (Reading 1964). The Kinderscoutian delta was the most extensive of these features.

Marine fauna, including goniatites, bivalves, sponges, crinoids, and brachiopods, is restricted to thin beds that comprise a small percentage of the total Kinderscoutian sediment. While fossils are virtually absent from the remainder of the sequence, trace fossils are common. The most abundant is *Pelecypodichnus* Seilacher, 1953, which in the Carboniferous has been shown to be the resting trace of the bivalves *Carbonicola* and *Anthraconaia* (Hardy 1970; Eagar 1974). Two Anthraconaiad bivalves were found in association with the trace fossils in the Kinderscoutian. Belt (1975) has suggested that these bivalves existed in brackish water conditions. In all probability, therefore, the basin experienced salinity variations and only rarely was full marine.

No evidence has been found of significant tidal influence in any part of the Namurian in the north of England.

FACIES ANALYSIS

Introduction

Collinson (1969) first described in detail the facies of the delta slope sequence. Because of the reinterpretation of many facies and the recognition of facies previously undescribed, a new facies analysis is given here.

Most of the slope sediments can be divided (see Figure 9-3) into nine facies: groups of rocks distinguished by their grain size and sedimentary structures.

Facies 1: Homogeneous Mudrock

Description. Unlaminated mudrocks (mudrocks is used here in the sense of Blatt et al. [1972] for rocks finer than sandstone grade) vary from clay to coarse silt grade (scale of Wentworth [1972] is used throughout this chapter) and occurs as a shale. They have a high carbon content and are dark grey or black in color. Weathering of microscopic pyrite gives exposures a rusty brown coating of limonite. This facies is equivalent to Collinson's (1969) Facies 1.

Interpretation. The facies was probably deposited slowly in quiet water. The abundant carbon and pyrite suggests a lack of oxygen.

Facies 2: Laminated Siltstone

Description. This facies consists of laminae— of varying size grade— that are gradationally based. Laminae have an average thickness of 1.4 mm. Finer-grained laminae contain more carbon and pyrite, which thus results in a light/dark grey banding. The finer part of Facies 2 of Collinson (1969, Figure 5) belongs to this facies.

Interpretation. The nature of the lamination and lack of any sedimentary structures produced by traction currents suggests deposition from suspension. Variations in grain size indicate fluctuation in the sediment supply.

Facies 3: Gradationally Laminated Sandstone

Description. This facies is similar to the laminated siltstones, but differs in grain size. The coarser laminae are of fine to medium sand grade, while the finer are of coarse silt to very fine sand grade. The thickness of the laminae increases with grain size; laminae in the fine to medium sand grades average 25 mm in contrast to 10 mm for fine to very fine sand grades. The grading between finer and coarser laminae may be gradual (diagram A, Figure 9-2) or abrupt.

The finer units of the facies contain trace fossils: *Scolicia* de Quatrefegas, 1849 (an endichnial wandering trail); *Planolites* Nicholsen, 1873 (burrows that are vertical in the coarser laminae and horizontal in the finer laminae); and *Pelecypodichnus.* Parts of Collinson's (1969) Facies 2 and 3 are grouped into this facies.

Interpretation. Deposition of these sandstones appear to have been from suspension. The grain size indicates relatively strong currents above the bed but there is no evidence of tractional reworking. The laminae presumably reflect changes in the strength or sediment load of the current.

The laminated siltstones are assumed to be distal equivalents of this facies. The fluctuations in sediment supply were probably related to the changing water stage of the delta distributaries and may have been seasonal.

Facies 4: Cross-Laminated Sandstone

Description. This facies consists of beds of fine- to medium-grade sandstone that are entirely cross-laminated on a small scale. Beds are up to 2 m thick. Their bases are mostly sharp, sometimes with prod marks. Individual cross-laminated sets are up to 15 mm thick and have trough shaped bases. Foresets are concave upward, and bedding planes show a rib-and-furrow pattern. The upper parts of the foresets are truncated at varying levels by the overlying set and stoss-sides are never preserved. If the truncation occurs at a low level in the set, the sandstone appears to be parallel laminated. *Pelecypodichnus* is common in this facies (diagram B, Figure 9-2), and its escape traces are up to 0.13 m long. This facies is equivalent to Facies 5 of Collinson (1969, Figures 7, 8, and 9).

Interpretation. The bases of the sandstone beds indicate slight predepositional erosion. The cross-lamination is interpreted as having been produced by the migration of small-scale ripples. Parallel lamination can be produced by the migration of very low amplitude ripples (Smith 1971; McBride et al. 1975). Since the pronounced grain size segregation noted by Smith and McBride et al. is absent, the parallel nature of the lamination in parts of this facies is more probably the result of a low rate of sedimentation, which is a process discussed by Jopling (1966).

A

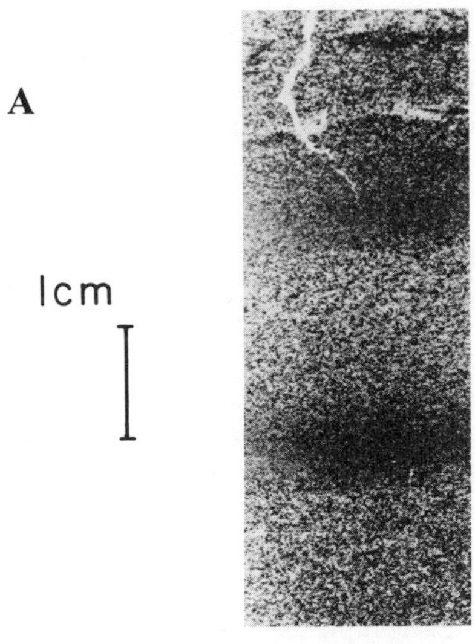

B

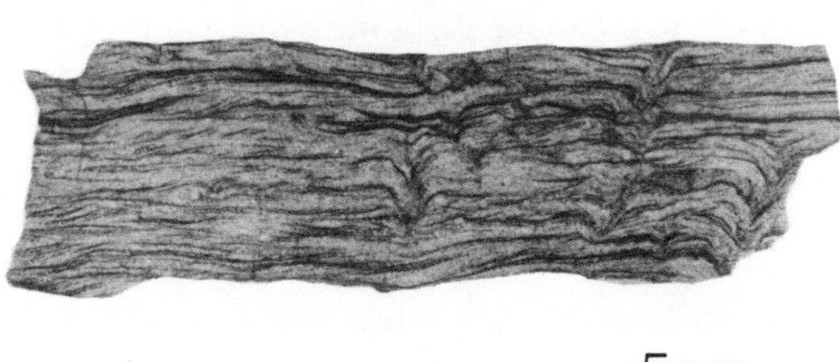

Figure 9-2. Laminated sandstone facies of the delta slope sequence. Note: Photograph A shows a thin section of gradationally laminated sandstone. Photograph B shows a cross-laminated sandstone section with V-shaped *Pelecypodichnus* escape traces.

Facies 5: Unlaminated Sandstone

Description. The sandstones of this facies are coarse- to very coarse-grained with granule grade grains up to 3 cm long. Mud clasts up to 0.45 m long are common and are concentrated at the bottom of a bed or scattered throughout. These are the "massive beds" of Collinson (1970). X-radiographs of slices of the rock show no laminae. There is also no apparent orientation of the granule grains.

The bases of most massive beds are strongly erosional, with large flutes and grooves. Bedding is difficult to delimit because many beds are amalgamated. This facies is part of Facies 7 of Collinson (1969, Figures 10 and 11).

Interpretation. Many suggestions have been made as to the origin of unlaminated beds and any interpretation must be based on their relationships to other facies. Flutes, however, necessitate a turbulent current.

Facies 6: Horizontally Laminated Sandstone

Description. Horizontal laminae occur in sandstones of medium to very coarse sand grade with scattered granule grains up to 2 cm long. The thickness of the laminae varies between 5 mm and 40 mm and is generally dependent on grain size. In the finer sandstones, a parting lineation (Allen 1964) is common. Beds of this facies are up to 2.5 m thick. Mud clasts are generally absent but occur in abundance on a few bedding planes. Facies 8 of Collinson (1969, Figure 12) is equivalent to this facies.

Interpretation. The thickness of the beds and the parting lineation suggest they were deposited under plane bed conditions of the upper flow regime. A lower flow regime plane bed, which is restricted to sediment coarser than 0.7 mm (Simons et al. 1965), can be discounted for at least the finer sandstones.

Facies 7: Cross-Bedded Sandstone

Description. Cosets of cross-bedding occur in medium- to coarse-grained sandstone. Individual sets range in thickness between 0.07 m and 0.6 m, and most have trough shaped bases. The bases of the cosets are generally flat, but where they overlie silts, they may be erosional. Mud clasts up to 50 mm in diameter occur on a few cross-bed foresets. Mud clasts, especially where they are very abundant, predominate in the lower part of the foresets. This facies is part of Collinson's (1969) Facies 9.

Interpretation. The trough cross-beds are interpreted as the deposits of sinuous crested or linguoid dunes.

Facies 8: Parallel-Sided Sandstones

Description. The sandstones of this facies are between 1 mm and 1.5 m thick. Bases are sharp and may show sole marks. Prod and groove casts are abundant. Flutes are less common but a few are well developed; for example, a 12 cm thick bed displays flutes of 12 cm width and 3.5 cm depth.

Four internal sedimentary structures are recognized: (1) massive medium- or fine-grained sandstone; (2) parallel lamination defined by mica plates up to 1.5 mm in diameter and showing a parting lineation (this lamination occurs in fine-grained sandstones); (3) parallel lamination defined by large mica plates, plant material, and in some cases small mudflakes (this lamination occurs in units that grade upward from fine-grained sandstone to siltstone); and (4) ripple form sets occurring in fine-grained sandstone.

Two basic types of structure sequence can be distinguished. In the first type, parallel laminae of Type 2 and/or ripples are present. There is a sharp break at the top of the sequence. In the second type, by contrast, the parallel laminae are of Type 3 and they grade upward into mudrock without a ripple phase. Many beds, without Type 3 parallel laminae, contain trace fossils, including *Sinusites* Demonet and van Straelen, 1938 (a sinuous surface trail); *Bergaueria* Prantl, 1946 (a vertical burrow); and *Pelecypodichnus.* Sandstones of this facies are included in the Facies 11 of Collinson (1969, Figure 14).

Interpretation. Type 2 parallel laminae and ripples are typical of the lower part of the upper and lower flow regimes, respectively (Simons et al. 1965). Type 3 plane lamination is more difficult to explain. While plane beds of the upper flow regime could feasibly form graded units under waning flow conditions, the possibility that the parallel lamination would grade perfectly into siltstones seems unlikely. The sandstone, however, is too fine grained to belong to the plane bed field of the lower flow regime. The large amount of plant material may have suppressed a ripple phase by depressing turbulence, thus extending the field of upper flow regime plane beds into lower stream powers.

The sharp bases, the graded nature of the beds, and the evidence of waning flows strongly suggest that these sandstones are turbidites.

Facies 9: Complexly Bedded Sandstone

Description. Four lithotypes occur in this facies:

1. Fine-grained sandstone displaying symmetrical ripples, which have sharp crests, chord lengths of about 80 mm, and vertical form indices of 7.
2. Cross-laminated fine- to medium-grained sandstone. Sets have trough-shaped bases, and bedding planes show a rib-and-furrow pattern. The tops of a few beds show asymmetrical ripples with an average height of 8 cm. Mica, carbonaceous material, and silt drapes emphasize the lamination of both ripple types. The orientation of both types is variable.
3. Parallel laminated, medium-grained sandstone that shows a good parting lineation.
4. Unlaminated coarse- to very coarse-grained sandstone, which occurs in beds between 1 cm and 20 cm thick, but mostly less than 4 cm. Some of the thicker beds are graded. Most bases exhibit load structures and only one groove cast has been observed. A few sandstones rest on an underlying ripple morphology without erosion.

No sequence of structures has been determined from this facies. A few of the thicker beds are distorted and recumbent folds occur. *Pelecypodichnus* occurs in the finer parts of the facies. This facies has not previously been distinguished in the Kinderscoutian.

Interpretation. The symmetrical ripples indicate that the association was deposited above wave base. Asymmetrical ripples and parallel lamination indicate deposition in the lower and upper flow regimes, respectively. The unlaminated sandstones were probably deposited from suspension with little or no prior erosion. The variety of structures and the unsorted nature of the sediment indicate a highly variable environment with rapid influx of sediment.

THE COARSENING UPWARD SEQUENCE

Description

The homogeneous mudrock, laminated siltstone, gradationally laminated sandstone, cross-laminated sandstone, and the complexly bedded sandstone facies form a coarsening upward sequence.

The homogeneous mudrocks and laminated siltstones occur in interbedded units up to 26 m thick, though commonly less than 10 m thick. In the lower part of the sequence, homogeneous mudrocks are dominant, while laminated siltstones become more common upwards. These two facies comprise a high percentage of the total sediment in the lower part of the coarsening upward sequence (Figure 9-3). Higher in the sequence the two facies become less common, although a reversal in the trend is seen at the top (an example of homogeneous mudrock occurring at the top of the sequence is shown in section B, Figure 9-4).

The gradationally laminated sandstones and cross-laminated sandstones also occur in close association, although the former is much more common (sections B and D, Figure 9-4). The two facies occur in units up to 20 m thick. Their abundance within the coarsening upward sequence is inversely related to that of the finer sediment, and they are most abundant in the middle of the sequence (see Figure 9-3).

Of the five facies, the complexly bedded sandstones are the least common. They are restricted to the uppermost part of the sequence (see Figure 9-3), and in several localities the facies is cut into by the succeeding distributary channels.

Interpretation

The coarsening upward sequence produced by the five facies is interpreted as the product of a prograding delta slope. Homogeneous mudrock was probably deposited in areas where the influence of distributaries was small. Deposition may have been near the base of the slope, on the slope lateral to lobes of coarser, distributary-derived sediment, or even lateral to the distributaries (see Figure 9-7). Laminated siltstone was probably only a slightly more proximal facies.

The gradationally laminated and cross-laminated sandstones are interpreted as sediment, deposited mainly from suspension, as lobes in front of distributary mouths. The gradationally laminated sandstones are similar to the

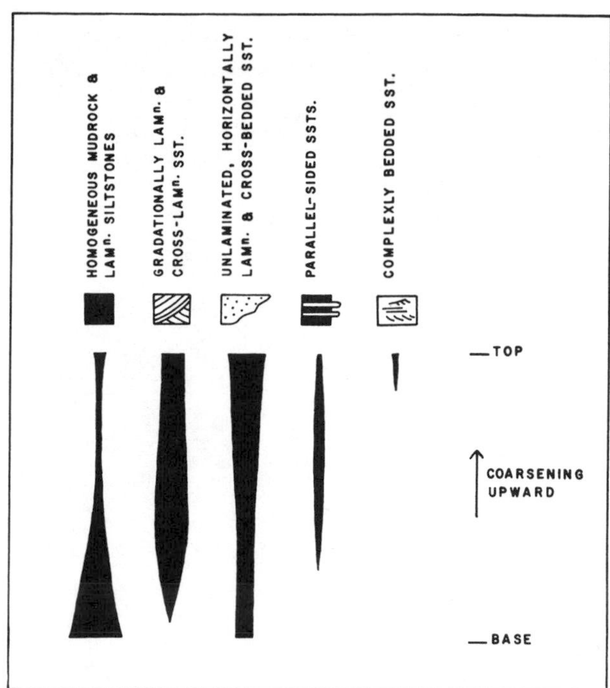

Figure 9-3. Estimated proportionate distribution of the facies throughout the coarsening upward sequence. Note: The estimates are based on detailed measurements of sections and field mapping over a wide area but are somewhat subjective due to the quality of the exposure.

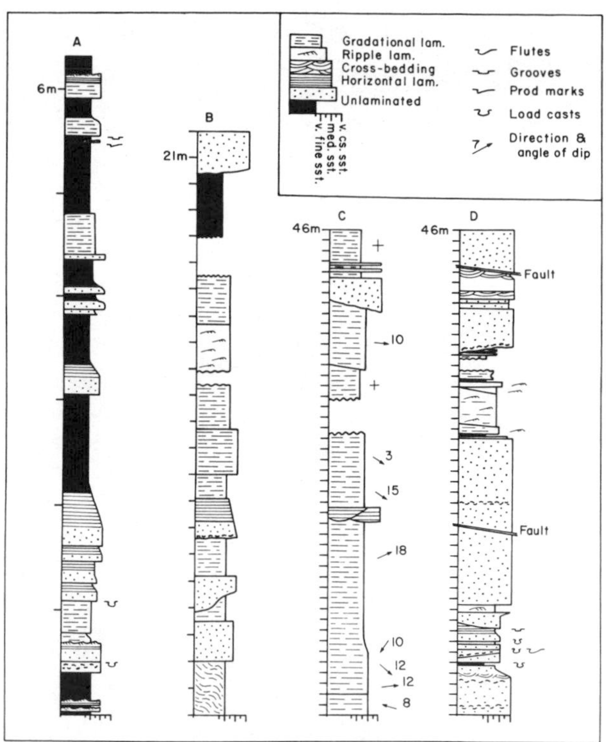

Figure 9-4. Representative logged sections of Kinderscoutian slope sediments. Note: Section A shows parallel-sided sandstones at Rams Clough, northeast of Manchester (British Grid Ref. SE018026). Section B shows predominantly finer-grained sediment with four small channels from the middle of the slope sequence at Holden Brook, Wharfedale (SE064455). The section also shows a slumped unit at the base (0-2 m). Section C is an upper slope sequence at Dale Clough, Calderdale (SD972266); orientation of inclined units is indicated (+ = horizontal): no tectonic dip. Section D shows upper part of slope sequence at Great Gruff, northeast of Manchester (SE028016); succession is mainly through slope channels. Note the differences in scale between sections.

parallel laminae of differing texture that occur in the Mississippi mouth bars (Coleman and Gagliano 1965).

The variable environment suggested by the complexly bedded sandstones was probably close to the river mouth. Coleman et al. (1964) and Wright and Coleman (1974) note that the proximal mouth bar is an area of rapid deposition and that the interaction of varying currents and waves results in multidirectional trough cross-lamination and occasional wave ripples. Coleman et al. (1964) also record concentrations of wood fragments. Prograding distributaries rework much of the upper axial mouth bar, and therefore, that this facies is uncommon is not surprising.

INCLINED UNITS

Description

In most exposures the bedding of the gradationally and cross-laminated sandstones is inclined to the regional dip, at angles up to 18°. These inclined beds occur in units up to 16 m thick (section C, Figure 9-4). Inclined bases of these units cut into other inclined units or into thick beds of unlaminated sandstone. The basal surfaces are gently curved and concave upwards. Surfaces are smooth and show no irregularities even when traced over a vertical distance of 14 m (Figure 9-5). Syn-sedimentary faults, with displacements up to 40 cm, occur within the sediment underlying some basal surfaces. In one outcrop four faults occur en echelon. Disturbed beds also underlie a few basal surfaces. Faults and disturbances do not extend into the overlying sediment.

The inclined beds of a unit are almost parallel to its lower bounding surface (Figure 9-5), although the dip may change upwards. The dip of one unit decreases from 14° to 0° over a lateral distance of 200 m. Individual beds are nearly parallel sided and sedimentary structures persist in beds traced laterally. Dip directions of the inclined beds display a wide range but are predominantly to the south and east (see diagram C, Figure 9-6).

Figure 9-5. Two units of gradationally laminated and cross-laminated sandstone at Hebden Dale (British Grid Ref. SD971302). Note: The nature of the sediment is heterogenous. The lower unit is virtually horizontal, while the upper unit, overlying the truncation surface A-B, is inclined at 8°; the outcrop is approximately 10 m high.

In a few localities the sediment shows syn-depositional folding of the sediment that is not associated with the base of an inclined unit (see section B, Figure 9-4). Such deformed units have a sharp, erosional top.

Interpretation

The basal surfaces of the inclined units are interpreted as slump scars. The smoothly curved nature of the surfaces with no apparent erosional structures is in marked contrast to the stepped, irregular erosional surfaces of the channels infilled with coarse sandstone (see next section). The disturbed beds below the basal surfaces could be due to drag by the overlying sliding sediment, and the penecontemporaneous faults are probably remnants of slump fault systems. The basal surfaces show all the features typical of a slump scar as described by Laird (1968, Table 1).

The units with syn-depositional folds are interpreted as slumps. However, that so few slumped units have been recognized is surprising, because the abundance of inclined units suggests slump scars were common. Poor exposure of the lower part of the sequence may have precluded the recognition of slumps. Large slumps may also be difficult to recognize, because only the slump toe shows compressional folding and thrusting (Lewis 1971).

Although slumping occurs in a variety of environments the apparent existence of 14 m high slump scars enhances the interpretation of the sequence as a delta front slope deposit and detracts from Collinson's (1969) interpretation of the upper half of the sequence as an interdistributary complex. Apparent slumps have been recorded from the upper slope of the Mississippi delta (Watkins and Kraft 1976) and from the Fraser River delta (Mathews and Shepard 1962).

The draped nature of the slump scar fills is to be expected if, as is postulated here, the gradationally laminated sandstones settled from suspension without passing through a tractional phase. The parallel nature of the rippled beds within an inclined unit requires that the current forming the ripples flowed with virtually equal strength over the entire inclined surface. Paleocurrent data (diagrams A and B, Figure 9-6) indicates that the ripple forming currents were in a downslope direction. In the absence of tides, such currents were probably the result of a river-generated

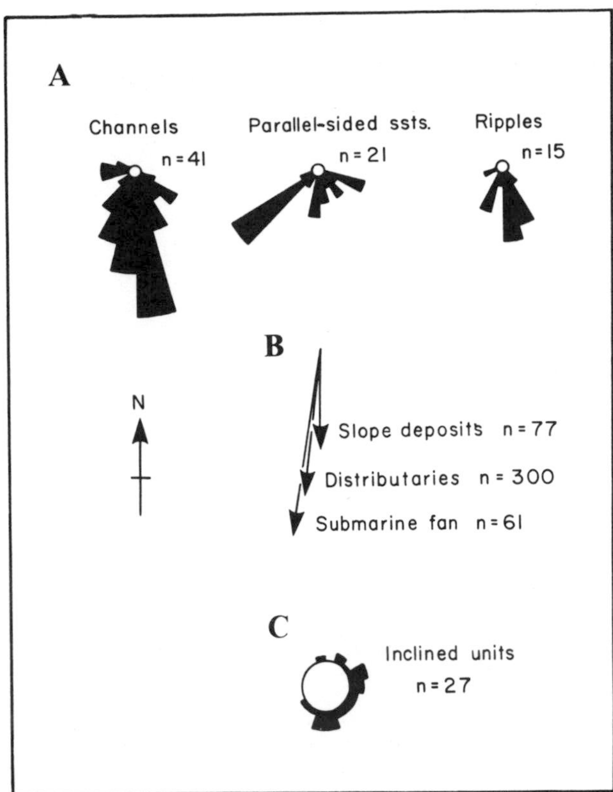

Figure 9-6. Paleocurrent data from the area between Wharfedale and Longdendale. Note: Diagram A shows histograms for various facies of the slope sequence. Diagram B shows vector mean for the grouped slope paleocurrent data compared to vector means for the underlying submarine fan and the overlying distributaries. Diagram C shows dip direction of inclined beds of gradationally laminated sandstone.

density flow. While their generation requires exceptional conditions in a marine environment, density currents are well documented from lakes (Gould 1960; Lambert et al. 1976). In the subaqueous channel of the Rhone River delta, in Lake Geneva, Shepard and Dill (1966) record active current ripples formed by a density flow. Density currents may have existed in the low salinity environment suggested for the Kinderscoutian.

River-generated density currents, whether resulting from contrasts in temperature or content of suspended material, are likely to persist longer than a classical slump-generated turbidity current. Such long-lived currents might deposit the sharp-based and sole-marked, rippled beds. The downcutting of ripples into one another may reflect the relatively small amount of sediment coming from suspension compared to the high fallout of sediment that results in the climbing ripples of a typical Bouma (1962) turbidite sequence. The possibility exists that the rippled sandstones are reworked and winnowed, gradationally laminated sandstones.

CHANNELS ON THE SLOPE

Description

Unlaminated, horizontally laminated, and cross-bedded sandstones occur within channels, which are cut into finer sediment or into other channels (see sections B, C, and D, Figure 9-4). The channels occur throughout the coarsening upward sequence but are more common in the upper part (see Figure 9-3).

Channel sides are irregular and stepped surfaces are common. The sides of many channels are steep with slopes up to 40°, and Collinson (1970) records vertical and overhanging sides. The axes of channels and the flutes on their base indicate the same current direction as all the Kinderscoutian sediments (see diagram A, Figure 9-6). The maximum depth of erosion observed is over 9 m. However, many channels are undoubtedly larger, and Collinson (1970) records infill thicknesses of 30 m. The limited exposure makes measuring the width of large channels impossible, but at one locality thick massive beds are continuous over 200 m laterally. Collinson (1970) records one well-exposed channel more than 530 m wide.

Unlaminated sandstones predominate in the lower parts of channels. Higher in the channels, horizontally laminated sandstones are dominant, but cross-bedded sandstones overlie horizontally laminated sandstones in some channels. Cross-bed cosets are up to 2 m thick. A few small channels in the order of 8 m wide and 1 m deep with sides dipping at up to 16° are infilled by a coarse sandstone similar to the horizontally laminated sandstones, but the laminae are parallel to the channel sides. Fine-grained, micaceous sediment occurs on the terraces of some channels.

Interpretation

The size of the channels and the presence of flute marks indicate that they were cut by strongly erosive turbulent currents. The fine-grained, micaceous sediments suggest a marked decrease in current strength between the erosion of some surfaces and the deposition of the unlaminated sandstones. The restriction of micaceous sediment to terraces suggests that infilling was periodic, although the general absence of fine-grained deposits indicates that erosion and deposition resulted from the same process. The upward transition – unlaminated → horizontal lamination → cross-bedding–suggests a waning of the depositional current.

Collinson (1970), envisaging a shallow water environment, suggested the channels resulted from fluviatile currents of varying discharge, which were extending distributary channels into the basin. The sequence of channel erosion, followed by a waning flow infilling the channels, was interpreted in terms of a flood cycle. Collinson (1970) suggested that the lack of lamination in

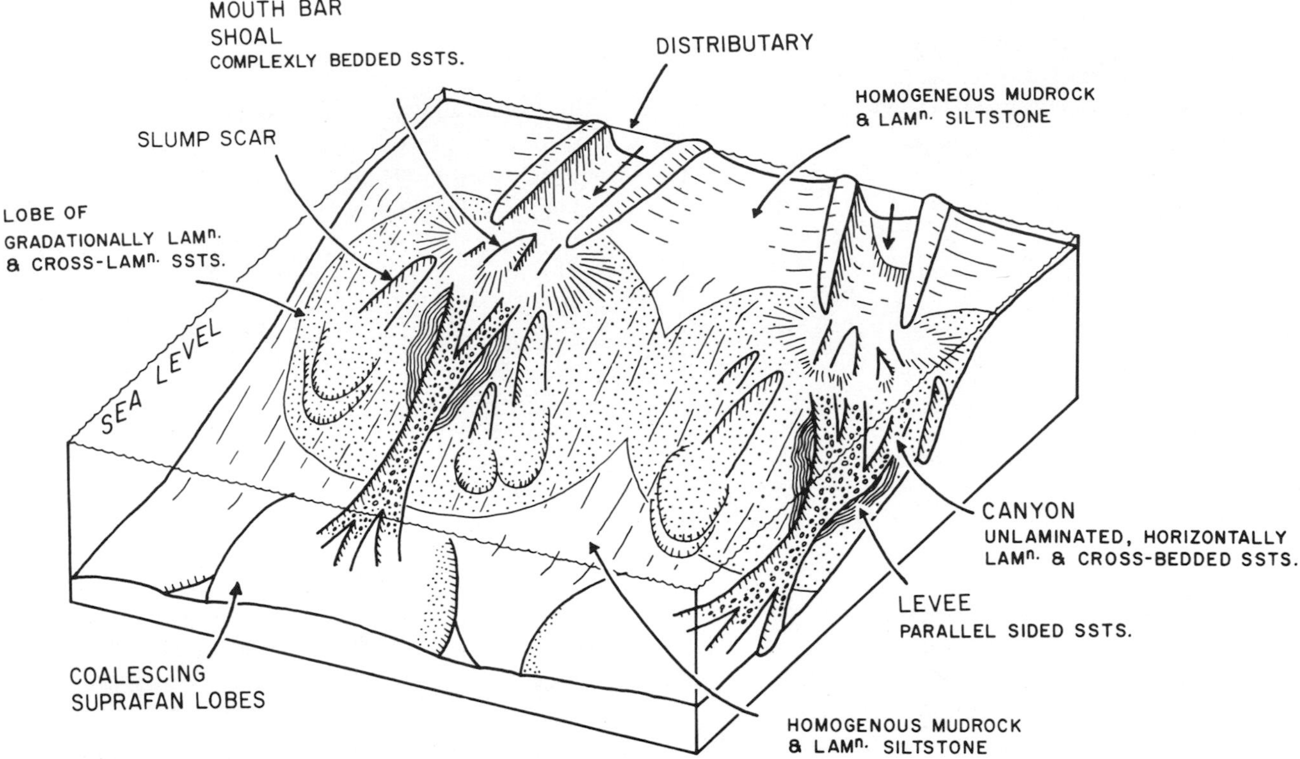

Figure 9-7. Model for deposition of the Kinderscoutian slope sediments.

the lower parts of the channels might be due to rapid fallout of sediment from suspension into a traction carpet, thereby swamping any sorting process.

Since evidence presented earlier indicates that the channels were cut into a delta slope, on which slumping took place, we can suggest that the channels were cut and infilled by turbidity currents. The maximum grain size of the sandstones is comparable to the coarsest part of the underlying turbidite sequence (Walker, 1966a, 1966b), and we can assume that these channels were feeders to the submarine fan complex (see Figure 9-7). In fact Walker (1966b) and Collinson (1969, 1970) report channels in the lowest part of the slope sequence that are infilled with amalgamated, thick and structureless turbidites.

Gradational contacts between facies suggests that some channels were infilled by one turbidity current. In contrast, the presence of fine sediment within a few channels and amalgamation between the unlaminated sandstone beds suggests that many channels have composite infills.

Since there are no fluid escape structures, the origin of the unlaminated beds may be explained by several processes: (a) deposition under upper flow regime (Walker 1965); (b) shearing of loosely compacted sediment (Middleton 1967), or (c) deposition by a rapidly decelerating current that failed to produce an equilibrium bedform (Walton 1967). Cross-bedding, though not typical of turbidity current deposition, has been observed in many coarse-grained turbidite deposits (e.g., Dzulynski et al. 1959; Thomson and Thomasson 1969; Hendry 1972; Davies and Walker 1974).

PARALLEL-SIDED SANDSTONES

Description

Parallel-sided sandstone beds are always interbedded with homogeneous mudrock and form distinctive units (see section A, Figure 9-4). These units are up to 20 m thick and are intercalated with units of gradationally laminated and cross-laminated sandstones or the coarse infilled channels. These parallel-sided sandstone/mudrock units are a minor constituent of the slope complex but occur throughout the upper two-thirds of the sequence (see Figure 9-3).

Paleocurrents from the parallel-sided sandstones are given in diagram A, Figure 9-6. In one exposure the sandstones are distorted into folds with horizontal axes, large ball structures, and boudins. The deformed unit has a sharp top that is overlain erosionally by a thin sandstone bed.

Interpretation

Although interpreted as turbidites, the interbedding with sediments interpreted as slope deposits makes unlikely that the parallel-sided sandstone units were deposited on a submarine fan. The deformed beds are interpreted as slump deposits with evidence of plastic deformation and movement before erosion and deposition of overlying beds. They provide further evidence of slumping on the slope.

Two possible origins for the-parallel sided sandstone beds are proposed. First, the sandstones may have been deposited by a density current unassociated with the formation of either the channels or the cross-laminated sandstones. Such currents may have been generated by storms sweeping sediment away from the shore. Second, the sandstones may have been deposited by the turbidity currents responsible for the cutting and infilling of the channels. As can be seen in Figure 9-7, the finer, upper part of these currents may have spilled over the channel banks, thereby building up levee deposits (cf. Walker 1966a). This theory is favored since in all likelihood, overbank deposition would occur if the turbidity currents were of a size capable of infilling a channel by one flow.

Why are the parallel-sided sandstone beds interbedded with homogeneous mudrock rather than the gradationally laminated sandstone? We can suggest that the mudrock may have been deposited by the fine tail of the turbidity currents and that the rapid deposition near channels may have swamped the slower process of delta front deposition.

MORPHOLOGY OF THE DELTA SLOPE AND INITIATION OF TURBIDITY CURRENTS

Individual facies occur preferentially with one or two other facies and general trends occur through the succession (see Figures 9-3 and 9-4). The sequence, however, shows great variability, and a facies can pass upward into virtually any other facies.

A suggested model for the deposition of the nine facies is shown in Figure 9-7. The homogenous mudrock and laminated siltstones were deposited in those areas where the influence of the distributaries was small or absent. The lobes of sediment in front of each distributary mouth were deposited predominantly from suspension and modified by slumping and reworking by density currents. In the fluctuating environment of the upper mouth bar, the complexly bedded sandstones were deposited. Channels, cut down the delta slope, fed turbidity currents to the underlying submarine fan. Overbank flows from the channels deposited the parallel-sided sandstone units.

Delta progradation, initiation and eventual abandonment of distributary channels, and switching of the main delta lobes all may have contributed to the complex relationships now seen in the Kinderscoutian.

A possibility is that at high discharges flows emanating from the distributaries as heavily laden suspension currents proceeded onto the slope. Once on the slope, such currents could develop into turbidity currents. The Congo river and its canyon (Heezen et al. 1964) may provide a modern analogue for this process (see Collinson 1970). A direct fluvial origin for the turbidity currents seems less likely, however, since the deposits of long-lived currents of fluctuating strength might produce reversals in the Bouma (1962) sequence. A slump origin is more probable, but despite the evidence of slumping, no slump source for the very coarse sandstones has been found. Rapid deposition of bed load occurs at distributary mouth bar crests (Wright and Coleman 1974), and the accumulation of sediment at the mouth bar crest may have been so rapid that oversteepening of the bar front caused slumping. The absence of bar crest deposits in the Kinderscoutian is hardly surprising in the light of their low preservation potential.

In conclusion, this study demonstrates that a prograding delta slope may be considerably more complex than the classical coarsening upward deltaic sequence. Such complexity is promoted by conditions that are conducive to density current development. In such a situation, discriminating between an ancient slope sequence with small submarine canyons and a shallow water shelf sequence may be difficult.

ACKNOWLEDGMENTS

The author wishes to thank John Collinson for his many helpful suggestions during the course of this study. Thanks also go to Colin Jones and John Baines for useful discussions of other Carboniferous deltaic deposits. The manuscript has been greatly improved by the constructive criticism of J. D. Collinson, C. M. Jones, G. V. Middleton, R. G. Walker, and the reviewers. The research was carried out during the tenure of a N.E.R.C. (U.K.) research scholarship at Keele University, and the manuscript was prepared during tenure of a postdoctoral fellowship at McMaster University, financed by N.R.C. (Canada) through a research grant to Dr. G. V. Middleton.

REFERENCES

Allen, J. R. L., 1960. The Mam Tor Sandstones: A turbidite facies of the Namurian deltas of Derbyshire, England. *J. Sed. Petrol.*, 30: 193-208.

———, 1964. Primary current lineation in the Lower Old Red Sandstone, Anglo Welsh Basin. *Sediment.*, 3: 89-108.

Belt, E. S., 1975. Scottish Carboniferous cyclothem patterns and their palaeoenvironmental significance. In: M. L. S. Broussard, (ed.), *Deltas, Models for Exploration.* Houston Geol. Soc., Texas, pp. 427-49.

Blatt, H., Middleton, G. V., and Murray, R., 1972. *Origin of Sedimentary Rocks.* Prentice-Hall, Englewood Cliffs, N. J., 634 pp.

Bouma, A. H., 1962. *Sedimentology of Some Flysch Deposits.* Elsevier, Amsterdam, 168 pp.

Coleman, J. M., and Gagliano, S. M., 1965. Sedimentary structures; Mississippi deltaic plain. In: G. V. Middleton (ed.), *Primary Sedimentary Structures and Their Hydrodynamic Interpretation.* Soc. Econ. Paleontol. Mineral., Sp. Publ. 12, pp. 133-48.

——, Gagliano, S. M., and Webb, J. E., 1964. Minor sedimentary structures in a prograding distributary. *Mar. Geol.*, 1: 240-58.

Collinson, J. D., 1968. Deltaic sedimentation units in the upper Carboniferous of northern England. *Sediment.* 10: 233-54.

——, 1969. The sedimentology of the Grindslow Shales and the Kinderscout Grit: A deltaic complex in the Namurian of northern England. *J. Sed. Petrol.*, 39: 194-221.

——, 1970. Deep channels, massive beds and turbidity current genesis in the Central Pennine Basin. *Proc. Yorks. Geol. Soc.*, 37: 495-520.

Davies, I. C., and Walker, R. G., 1974. Transport and deposition of resedimented conglomerates: The Cap Enragé Formation, Cambro-Ordovician, Gaspe, Quebec. *J. Sed. Petrol.*, 44: 1200-16.

Dzulynski, S., Ksiazkiewicz, M., and Keunen, P. H., 1959. Turbidites in flysch of the Polish Carpathian Mountains. *Geol. Soc. Amer. Bull.*, 70: 1089-118.

Eagar, R. M. C., 1974. Shape of shell of *Carbonicola* in relation to burrowing. *Lethaia*, 7: 173-256.

Gould, H. R., 1960. Turbidity currents, *Comprehensive Survey of Sedimentation in Lake Mead 1948-1949*, U.S. Geol. Surv., Prof. Paper 295, pp. 201-07.

Hardy, P. G., 1970. *Aspects of palaeoecology in the arenaceous sediments of upper Carboniferous age in the Manchester area.* Unpublished Ph.D. thesis, University of Manchester, England.

Heezen, B. C., Menzies, R. J., Scheider, E. D., Ewing, W. M., and Granelli N. C. L., 1964. Congo submarine canyon. *Amer. Assoc. Petrol. Geol. Bull.*, 48: 1126-49.

Hendry, H. E., 1972. Breccias deposited by massflow in the Breccia Nappe of the French pre-Alps. *Sediment.*, 18: 277-92.

Jopling, A. V., 1966. Some applications of theory and experiment to the study of bedding genesis. *Sediment.*, 7: 71-102.

Laird, M. G., 1968. Rotational slumps and slump scars in Silurian rocks, western Ireland. *Sediment.*, 10: 111-20.

Lambert, A. M., Kelts, K. R., and Marshall, N. F., 1976. Measurements of density underflows from Walensee, Switzerland. *Sediment.*, 23: 87-105.

Lewis, K. B., 1971. Slumping on a continental slope inclined at 1° – 4°. *Sediment.*, 16: 97-110.

Mathews, W. H., and Shepard, F. P., 1962. Sedimentation of Fraser River Delta, British Columbia. *Geol. Soc. Amer. Bull.*, 46: 1416-42.

McBride, E. F., Shepherd, R. G., and Crawley, R. A., 1975. Origin of parallel, near-horizontal laminae by migration of bedforms in a small flume. *J. Sed. Petrol.*, 45: 132-39.

McCabe, P. J., 1975. *The sedimentology and stratigraphy of the Kinderscout Grit Group (Namurian, R_1) between Wharfedale and Longdendale.* Unpublished Ph.D. thesis, University of Keele, Staffordshire, England, 172 pp.

——, 1977. Deep distributary channels and giant bedforms in the Upper Carboniferous of the Central Pennines, northern England. *Sediment.*, 24: 271-290.

Middleton, G. V., 1967. Experiments on density and turbidity currents, III: Deposition of sediment. *Can. J. Earth Sci.*, 4: 475-505.

Reading, H. G., 1964. A review of the factors affecting the sedimentation of the Millstone Grit (Namurian) in the Central Pennines. In: L. M. J. U. van Straaten (ed.), *Deltaic and Shallow Marine Deposits.* Elsevier, Amsterdam, pp. 340-46.

Shepard, F. P., and Dill, R. F., 1966. *Submarine Canyons and Other Sea Valleys.* Rand McNally, New York, 381 pp.

Simons, D. B., Richardson, E. V., and Nordin, C. F., 1965. Sedimentary structures generated by flow in alluvial channels: In: G. V. Middleton (ed.), *Primary Sedimentary Structures and Their Hydrodynamic Interpretation.* Soc. Econ. Paleontol. Mineral., Sp. Publ. 12, pp. 34-52.

Smith, N. D., 1971. Pseudo-planar stratification produced by very low amplitude sand waves. *J. Sed. Petrol.*, 41: 69-73.

Thomson, A. F., and Thomasson, M. R., 1969. Shallow to deep water facies development in the Dimple Limestone (Lower Pennsylvania), Marathon Region, Texas, In: G. M. Friedman (ed.), *Depositional Environments in Carbonate Rocks.* Soc. Econ. Paleontol. Mineral., Sp. Publ. 14, pp. 57-78.

Walker, R. G., 1965. The origin and significance of the internal sedimentary structures of turbidites. *Proc. York. Geol. Soc.*, 35: 1-32.

——, 1966a. Shale Grit and Grindslow Shales: Transition from turbidite to shallow-water sediments in the upper Carboniferous of northern England. *J. Sed. Petrol.*, 36: 90-114.

——, 1966b. Deep channels in turbidite-bearing formations. *Amer. Assoc. Petrol. Geol. Bull.*, 50: 1899-917.

——, 1976. Ancient submarine fans. In: R. S. Saxena (ed.), *Sedimentary Environments and Hydrocarbons.* New Orleans Geol. Soc., Amer. Assoc. Petrol. Geol., Short Course Notes, New Orleans, pp. 187-217.

Walton, E. K., 1967. The sequence of internal structures in turbidites. *Scott. J. Geol.*, 3: 306-17.

Watkins, D. J., and Kraft, L. M., 1976. Stability of continental shelf and slope off Louisiana and Texas. In: A. H., Bouma, G. T. Moore, and J. M. Coleman, (eds.), *Beyond the Shelf Break.* Amer. Assoc. Petrol. Geol., Marine Geology Committee, Short Course, vol. 2, New Orleans, pp. B1-B34.

Wentworth, C. K., 1922. A scale of grade and class terms for clastic sediments. *J. Geol.*, 30: 377-92.

Wright, L. D., and Coleman, J. M., 1974. Mississippi River mouth processes: Effluent dynamics and morphologic development. *J. Geol.*, 82: 751-78.

Chapter 10

Growth Fault–Controlled Submarine Carbonate Debris Flow and Turbidite Deposits from the Jurassic of Northern Tunisia: Possible Canyon Fill Sequences

STEPHEN P. COSSEY
ROBERT EHRLICH

Department of Geology
University of South Carolina
Columbia, South Carolina

ABSTRACT

Two thick, Jurassic (Bathonian) carbonate, mass-movement debris flow deposits are spectacularly exposed on the eastern flanks of Djebel Bou Kornine in northern Tunisia. Both deposits consist of micrite-supported pebbles, cobbles, and boulders in their upper portions, but contain clast-supported boulders in their lower portions. Most of the clasts were derived from the south or southeast from Liassic (Pliensbachian and Domerian) carbonate platform deposits and were lithified prior to the emplacement of the flows. However, some clasts show plastic deformation in outcrop. The first major deposit (up to 75 m thick) shows good "coarse tail" grading (indicating single instantaneous deposition). Both deposits are succeeded and separated by radiolarian-rich carbonate turbidites of both proximal and distal type.

Emplacement of the flow deposits was closely linked to movements on growth faults that controlled the pattern of drastic northerly thickening of six facies units spanning the Bajocian to early Oxfordian. This interval increases in thickness from 140 m to 350 m over a distance of less than 4 km. Facies within the Bajocian reflect the initial movements of these growth faults. Lack of thickness changes in the early Oxfordian reflect the dying phases of growth fault activity.

The mass-movement deposits were probably confined to a north-south-trending canyon on the southern edge of Tethys in northern Tunisia during the Jurassic. Each episode of debris flow activity was followed by a much longer period of turbidite activity, some of which was confined between growth fault blocks as shown by paleocurrents perpendicular to the canyon axis. Other turbidites were directed parallel to the canyon axis. The great volume and coarseness of the two debris flow deposits suggest a combination of earthquakes and tsunamis as a cause for their genesis.

INTRODUCTION

Bou Kornine is the northernmost representative of a line of djebels (mountains) called the "Jurassic Range" in northern Tunisia (Figure 10-1) whose trend follows a line of regional faulting (Castany 1955). Situated 20 km southeast of Tunis, Bou Kornine differs from the other djebels in many respects. The others are composed dominantly of Jurassic shallow water carbonate deposits (Castany 1955). Bou Kornine records a sequence of carbonates that indicates a submarine slope environment was prevalent during the middle Jurassic and part of the late Jurassic.

In a reconnaissance of the mountain, slope deposits were first seen in the form of a carbonate submarine debris flow deposit, well exposed in road cuts on the trail that winds to the summit of Bou Kornine. A first field season during the summer of 1975 was spent studying the deposit, and a second season was spent studying the enclosing facies of

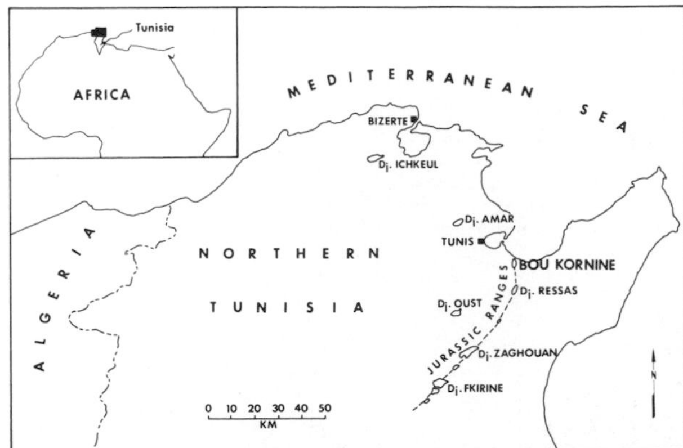

Figure 10-1. The Jurassic outcrops in northern Tunisia exposed in the cores of uplifted djebels.

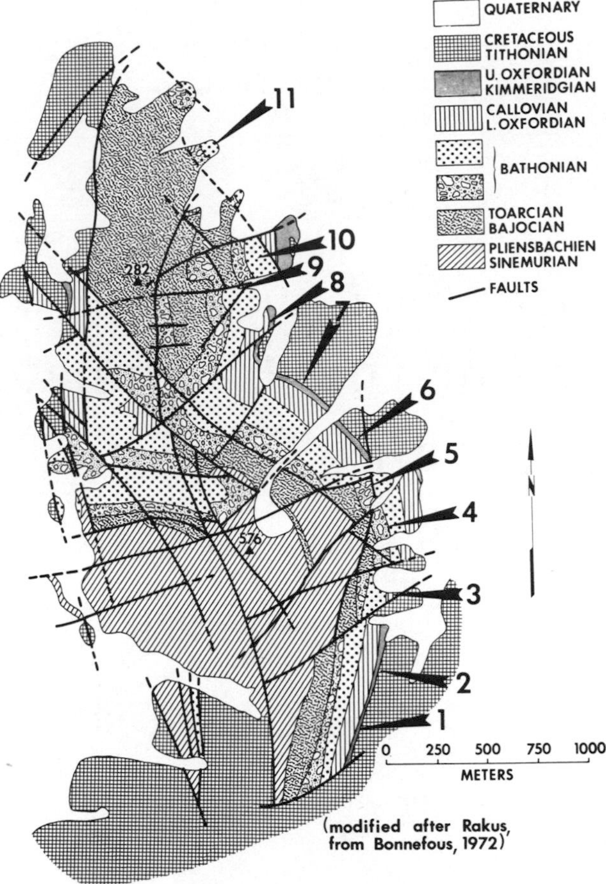

Figure 10-2. Geologic map of Bou Kornine showing section positions and outcrop of studied units (Bajocian to Oxfordian). (Source: Modified after Rakus [Bonnefous 1972].)

this deposit in order to determine the basinal setting associated with the emplacement of such a deposit. The densely forested slopes of Bou Kornine are not ideal for section measurement, but a total of eleven complete and partial sections were measured (Figure 10-2). Cross-sections based on these measurements will be discussed below.

Previous work has been limited to general stratigraphic studies and mapping. Studies have been carried out by Solignac (1927), Castany (1952), and more recently by Rakus (see Figure 12 in Bonnefous 1972) whose accurate map forms the basis for Figure 10-2. Bonnefous and Rakus (1965) were able to assign precise ages to the succession by using ammonite zones.

SLOPE FACIES OF THE MIDDLE AND LATE JURASSIC

Within the measured sections, six facies units spanning the Bajocian to the Oxfordian of Bou Kornine were defined (Figure 10-3). Carbonates predominate throughout the sequence.

BAJOCIAN FACIES

Unit 1

The Bajocian consists of an interstratified and complexly interfingering succession of pebbly mudstones, "rhythmites," and turbidite-like beds. Four different subfacies can be recognized, and although correlation of each between sections is possible, the resultant pattern tells us little.

1. The top of the oldest subfacies was chosen as the base for section measurements because it forms a generally continuous outcrop and is relatively homogeneous throughout. Subfacies 1 consists of monotonously, medium-bedded (20 cm) limestones containing calcareous spicules, radiolarians, calcispheres, and ammonoids. These finely laminated strata ("rhythmites" of Wilson 1969) are occasionally interbedded with chert. Sliding or slumping features are sometimes present (Figure 10-4). This evidence suggests a deep water origin in the form of an apron downslope from a typical carbonate shelf and in close proximity to the shelf margin (Wilson 1969).
2. Overlying the rhythmites, but limited to the sections north of section 5 (see Figure 10-2) is a second subfacies consisting of hard, thick-bedded (up to 3 m), matrix-supported pebbly micrites. These are slumped and plastically deformed in the region near section 5 and are interpreted as minor debris flow deposits in which the clasts are lithologically similar to the matrix. The significance of the areal limitation and slumping of this subfacies will be discussed in the second part of this chapter.
3. Succeeding these minor debris flow deposits are soft, matrix-supported, pebbly carbonate mudstones. The size of the clasts within this subfacies

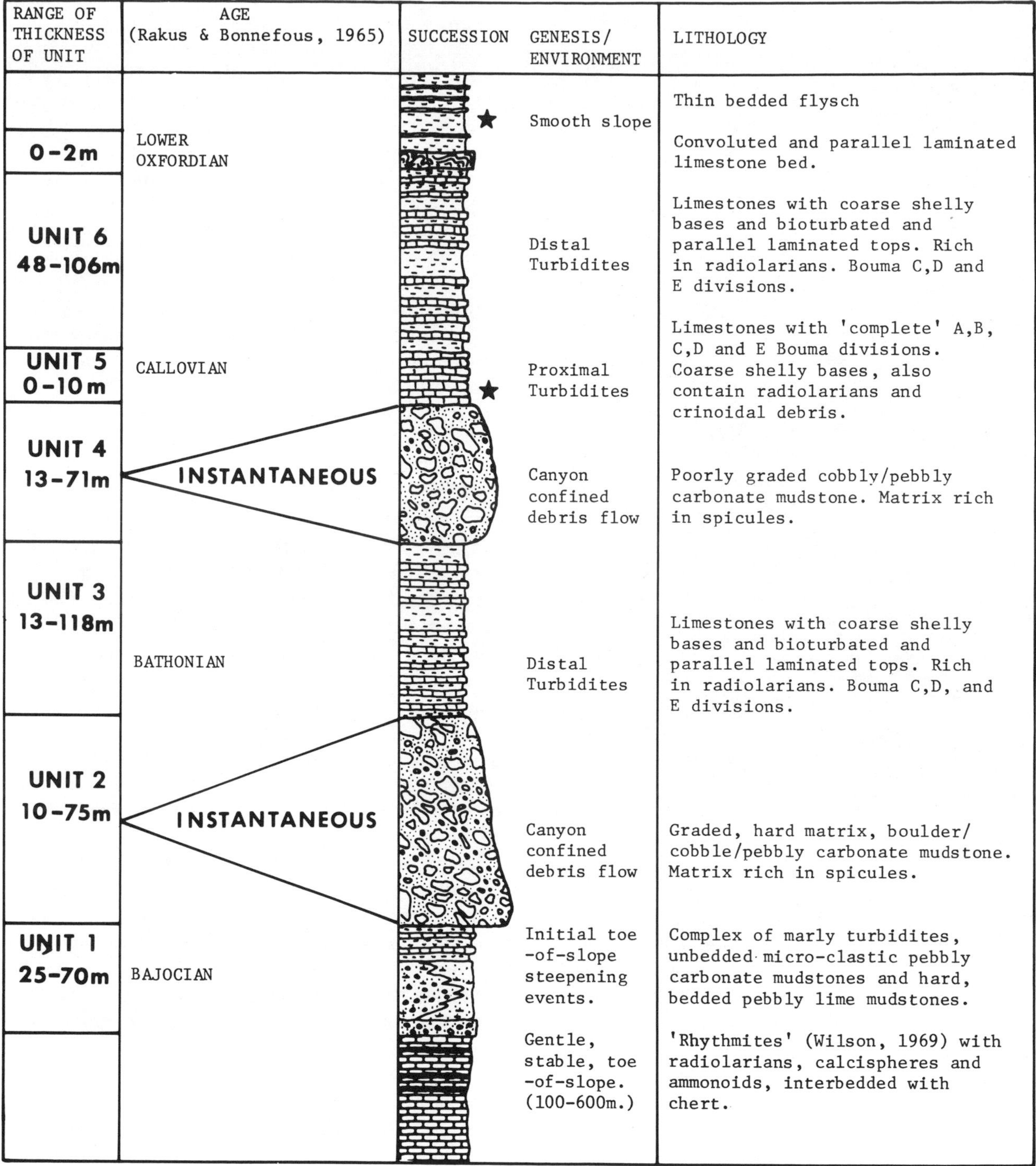

Figure 10-3. Generalized stratigraphic sequence for the middle Jurassic and part of the upper Jurassic of Bou Kornine.

varies from pebble size to millimeter size, and the unit is always massive and unbedded and weathers into small pillow shapes. This subfacies is also interpreted as consisting of minor debris flow deposits, but of a different character to those immediately below.

4. The fourth and youngest subfacies of the Bajocian is only seen in sections 9, 10, and 11. It directly underlies the base of the first Bathonian debris flow deposit and grades both laterally and vertically from the third subfacies. It consists of thin-bedded (approximately 10 cm) and bioturbated, coarse marly

Figure 10-4. Slump feature in medium (20 cm) bedded rhythmites. Note: Top of section is to the left; hammer is 35 cm long.

limestones that contain plastically deformed shale chips and are penetrated by escape burrows, thereby suggesting instantaneous deposition. These are interpreted as being turbidite deposits, possibly originating from local instability of the underlying Bajocian subfacies 1, 2, and 3. Subfacies 3 and 4 are the only two units seen to underlie the first Bathonian debris flow deposit.

The thickness of the measured subfacies within the Bajocian ranges from 25 m to 70 m from the top of the rhythmites to the base of the first Bathonian debris flow deposit.

BATHONIAN FACIES

Unit 2

The Bathonian is the stage in which most mass-movement activity took place. Two large mass-movement units can be distinguished. The first is a major carbonate debris flow deposit, which crops out fairly continuously for 4 km on the east side of Bou Kornine (see Figure 10-2) and which ranges from 10 m to 75 m in thickness. The upper portion of the deposit is poorly sorted (silt to boulders) and matrix supported (Figure 10-5), and the lower portion (Figure 10-6) is coarse, bouldery, and clast supported throughout.

Qualitative observations suggest that the decrease in clast size and increase in matrix proportion varies smoothly from the base to the top of the deposit. This evidence indicates that the entire deposit is a single unit rather than several smaller ones that have accumulated through time. A sampling program was devised to estimate quantatively the grain size and matrix proportion through the deposit at several localities in order to document and model the depositional processes.

At the four localities where an almost complete section through the deposit was completely exposed (see Sections 1, 4, 10, and 11, Figure 10-2), grain size and matrix proportions were recorded at intervals by drawing a variable-sized square on the outcrop at up to ten levels through the sequence at each locality. The side of the square varied from 10 cm to over 1 m in order to include approximately the same number of clasts within each of the squares. The long axis of every clast within the square (minimum diameter measured = ¼ cm) was determined, and a record was made of the different types of clast lithologies. Each square was then point counted to obtain a percentage value for the clast/matrix ratio. Obtaining exceptionally accurate long axis measurements was possible, because the clasts are resistant to weathering and were projecting from the outcrop.

A summary of this data from section 11, where the deposit is typically developed, is shown in Figure 10-7. The base of the deposit is coarse and bouldery throughout, with the basal portion being clast supported (chart C, Figure 10-7). In the remainder of the deposit, clasts are matrix supported. In general, mean clast size decreases in a logarithmic fashion upwards from the base (charts A and B, Figure 10-7).

Clast diameter measurements were taken and divided into half-phi size classes, and the relative percentage of each grouping was calculated on the basis of volume frequency. The volume proportion of clasts in each size class was plotted on a graph of percentage of total volume of clasts versus position in the deposit (chart D, Figure 10-7). This graph shows that each finer half-phi group in the intermediate sizes has a peak progressively nearer the top of the deposit. The finer size groups increase in percentage towards the top and the coarser size groups increase in percentage towards the base. There exists, for the intermediate sizes, a point within the deposit above which they are rarely found (chart D, Figure 10-7). Just below that point, clasts of that size occur in maximum abundance,

Figure 10-5. Poorly sorted texture typical of the middle of the early Bathonian debris flow deposit. Note: The clasts are slightly more resistant than the matrix. Knife is 9 cm long.

but become gradually less abundant downwards. These size characteristics are strong evidence that this particular deposit is a single catastrophic event and not the result of the accumulation of smaller debris flow deposits through time.

Seemingly, during downslope movement of the flow, matrix strength decreased, thereby allowing the clasts to settle through the matrix at varying rates dependent on the particle sizes until movement stopped, as the matrix "froze." The smaller pebble sizes acted as part of the matrix and were displaced relatively upwards (chart D, Figure 10-7) due to the downward motion of the larger particles or by fluid pressures unable to support the larger particles. Hampton (1975) suggested that the matrix of a debris flow has less shear strength when the flow is moving than when it has stopped, therefore allowing particles to settle due to density differences when the flow is in motion, but freezing the particles in their settling positions when the flow is arrested. The bouldery, clast-supported base of the deposit is significant in that it may reflect traction processes.

A typical debris flow has a positively skewed size distribution, with pebble-to boulder-size clasts set in a fine-grained matrix (Crowell 1957). Fluidized flows, in which the grains are supported by intergranular dispersive pressures, commonly show "coarse tail grading" (Middleton and Hampton 1973). The deposit on Bou Kornine has both carbonate clasts and matrix and combines the characteristics of both fluidized and debris flow mechanisms. It probably represents a hybrid deposit formed by a combination of these two mechanisms.

Clasts within the deposit are Domerian (Liassic) in age (Bonnefous and Rakus 1965). Although almost all of the clasts are limestone, chert clasts are also present. Most of the clasts were lithified prior to the emplacement of the deposit, but some show evidence of soft deformation. Contorted pieces of shaly limestone up to 10 m long occur near the middle of the flow at several localities. The deposit itself is dated as Bathonian by ammonites found within the micritic, spicule-rich matrix (Bonnefous and Rakus 1965). Assemblages within the matrix consist of shallow water pectinids (whole shells) with deeper water ammonite and belemnite faunas.

Unit 3

A thick sequence (13 m to 118 m), often of monotonous limestone beds of variable thickness (1 m to

Figure 10-6. Upturned, clast-supported bouldery base of the early Bathonian debris flow deposit. Note: Subfacies 3 of the Bajocian is in the foreground.

5 cm) and lutites, overlies the debris flow deposit. The limestones typically have erosive shelly basal portions that are almost devoid of radiolarians and calcispheres. Upwards from the base, radiolarian and calcisphere content increases, and structures in the middle of a typical bed include bioturbation and parallel lamination. Shale fragments are also found within the bed, and these have been subsequently cut through by burrows. Subfacies 4 of the Bajocian has many features similar to these beds. These limestones have many of the characteristics of Recent abyssal carbonate turbidites adjacent to the Campeche shelf in the Gulf of Mexico (Davies 1968).

No "complete" turbidites (Bouma 1962) were found with *a, b, c, d,* and *e* divisions. However, the typical sequence within each bed may represent the *c, d,* and *e* divisions (Bouma 1962) of the lower flow regime of a turbidity flow typical of more distal depositional sites. Hence, these limestones will be termed *distal turbidites* (Walker 1967).

Unit 4

The remainder of the Bathonian is represented by a second major carbonate debris flow deposit that lies unconformably on the underlying turbidites at sections 1, 3, and 7 (Figure 10-8). Clast size measurements were taken from this deposit at sections 1 and 7, where almost complete exposure was available. At section 1, this deposit shows poor peaking of size group curves when compared to the underlying lower Bathonian deposit. At section 7, grading is absent and apart from a bouldery base (basal 0.5 m), the deposit is essentially a pebbly carbonate mudstone. The matrix is very soft and marly and occasionally contains large (1.5 m) rafted boulders at the top of the deposit.

Clasts in this deposit have been derived from both Domerian and Pliensbachian (Liassic) sources (Bujalka et al. 1972). Clast types include biomicrites and shallow water algal limestones.

CALLOVIAN AND OXFORDIAN FACIES

Unit 5

The first facies of the Callovian is seen overlying the upper Bathonian debris flow deposit only at sections 5, 6, and 7 (Figures 10-2 and 10-8). This facies consists of up to 10 m of coarse, well-bedded, proximal turbidites characterized by complete Bouma *a-e* sequences. The beds have fluted bases and contain filamentous pelecypods, radiolarians, and crinoidal debris.

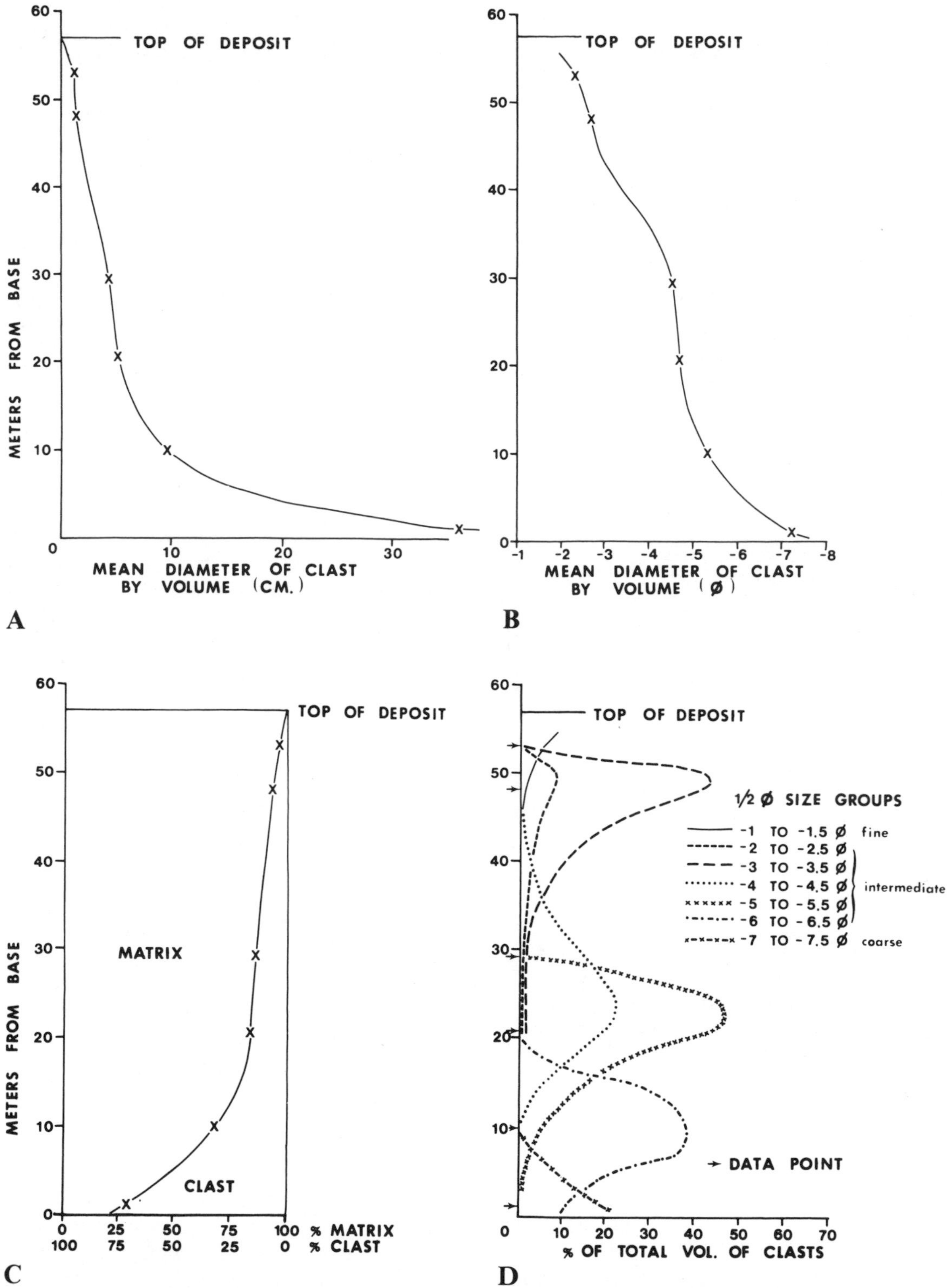

Figure 10-7. Clast size characteristics of the early Bathonian debris flow deposit, from measurements taken through the deposit at section 11. Note: Charts A and B show mean clast size variation through the deposit; Chart C shows clast/matrix ratio through the deposit; Chart D shows percentage variation of 1/2 phi size groupings through the deposit.

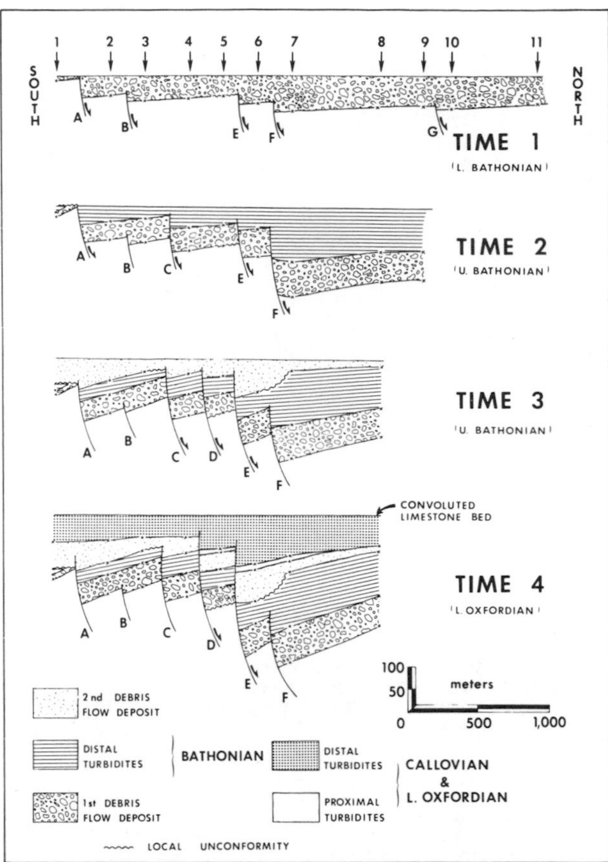

Figure 10-8. Thickness variations of the six measured units as explained by growth faults. Time datums are: Time 1, top of the early Bathonian debris flow deposit; Times 2 and 3, base and top of the late Bathonian debris flow deposit, respectively; and Time 4, Oxfordian convoluted limestone bed.

Unit 6

Facies unit 5 passes laterally and vertically into a thick sequence of well-bedded, laterally persistent limestones, which show parallel lamination, minor convolutions, grading, and bioturbation. Many of the beds contain deformed muddy clasts at their bases and tops. These beds are similar to those of unit 3 and are likewise interpreted as distal turbidites.

An upper datum for all complete measured sections is a persistent, parallel-laminated, and convoluted sandy limestone up to 2 m thick.

CHRONOLOGIC DEVELOPMENT

Up to one half of the middle Jurassic sequence at Bou Kornine can be assigned to two instantaneous single-event deposits. Assuming such single-event emplacement, the tops and bases of the two Bathonian debris flow deposits effectively represent time surfaces. The character of the upper datum—the convoluted limestone bed—changes little over the mountain, and it too can be used as a time horizon.

From the measured sections, four cross-sections, each representing sediment accumulation within a time interval, were constructed (see Figure 10-8). Distances between the measured sections vary from 250 m to 650 m. At any section where total thickness increased (total of units 1-6), most units increased in thickness. In many instances when one unit increased in thickness, overlying units thickened at the same location, which in addition to the great changes in thickness of units in such short distances, suggests a growth faulting mechanism. Bou Kornine is now intensely faulted (see Figure 10-2), and some of the presently mapped faults may be reactivated Jurassic growth faults. By careful reconstructions, the approximate movement on each growth fault, at each stage in its development through time, can be estimated. The measured thickness changes are extensive along the north-south cross-sections; this strongly suggests that the cross-sections are oriented almost perpendicular to paleostrike.

Time 1 (Early Bathonian)

If the top of the first debris flow deposit is used as a time datum, thickness changes in the debris flow can be explained by five small, down-to-basin growth faults (A, B, E, F, and G in Figure 10-8), which would have formed prior to this particular mass-movement event. The deposit is only 10 m thick and consists of clast supported boulders at section 1, but thickens to a maximum of 75 m at section 7. A small basal unconformity observed near section 5 (see Figure 10-8) is consistent with slight erosion at the edge of an upturned faulted block.

Time 2 (Middle Bathonian)

The interval between the early and late Bathonian debris flows was dominated by turbidite deposition, which resulted in much greater thicknesses in the north. The changes in thickness can be attributed to continuing movement on four growth faults. Fault B had ceased moving and another small growth fault, C, had developed (see Figure 10-8).

Time 3 (Late Bathonian)

The late Bathonian debris flow deposit is much more variable in thickness than the early Bathonian deposit (see Figure 10-8). The base is erosional in at least three places as is indicated by slight unconformities. The edges of small upturned fault blocks have been eroded at sections 1 and 3, and the underlying turbidites have been deeply channelled by the debris flow between sections 6 and 7 to give a 20° unconformity in that particular area. Faults A and F have ceased moving, and another small growth fault, D, has developed.

Time 4 (Late Oxfordian)

The fact that the late Oxfordian laminated limestone bed is areally persistent in thickness and character indicates a regional smoothing of the slope profile and also that growth faulting had essentially ceased. Prior to the late Oxfordian, the only two active faults were D and E (see Figure 10-8).

There are essentially no thickness changes in the turbidites overlying the upper Bathonian debris flow deposit at sections 1, 2, 3, and 4, but north of this, thickening does occur due to movements on faults D and E (see Figure 10-8). The proximal turbidites are limited to a structural/physiographic low in the region of most pronounced growth fault activity—that is, between sections 5 and 7. This area coincides with the region where the second debris flow deposit was deeply channeled into the Bathonian turbidites.

TECTONIC MODEL AND GROWTH FAULTS

The reconstruction of the sequence at four stages in its development reveals a history of episodic activity on at least six growth faults. Table 10-1 shows the amount of movement inferred for each of the growth faults. Inspection of this table, coupled with the field data suggests:

1. The emplacement of the late Bathonian debris flow coincided with the time when the sum of the movements on all the faults was at a maximum.
2. The total number of active faults decreased between the Bajocian and the Oxfordian.
3. The sum of the movements on all the faults had decreased drastically by the early Oxfordian, thereby indicating that growth fault activity was declining and may well have ceased by the late Oxfordian. Stratigraphic units younger than the early Oxfordian laminated limestone bed show no noticeable thickness changes on Bou Kornine.
4. Fault E was active throughout the whole time span from Bajocian to Oxfordian, and more movement occurred along it than along any other inferred fault. Fault E was thus probably one of the major middle Jurassic growth faults in the region. The geologic map of Bou Kornine (see Figure 10-2) shows a fault between sections 5 and 6 that coincides with the position of our postulated growth fault E. Tectonic reactivation probably occurred along the position of growth fault E, and percolating solutions through this zone have completely silicified the lower Bathonian debris flow deposit between sections 5 and 6.

Table 10-1. *Approximate Movement through Time of Postulated Growth Faults A through F*

AGE	DATUM	GROWTH FAULT MOVEMENT [meters]						
		A	B	C	D	E	F	Σ_F^A
EARLY BATHONIAN	TIME 1	45	15	NA	NA	20	25	105
LATE BATHONIAN	TIME 2	30	0	30	NA	10	55	195
	TIME 3	0	0	20	10	40	0	
LATE OXFORDIAN	TIME 4	0	0	0	30	40	0	70
	Σ_4^1	75	15	50	40	110	80	370

NA = NOT ACTIVE

5. If fault E is the major growth fault, it may well have controlled the distribution of facies during the Bajocian. Slumping of Bajocian pebbly carbonate mudstone occurs directly adjacent to fault E at section 6 (see Figure 10-8). Also, Bajocian activity of Fault E is suggested by the limitation of subfacies 2 of the Bajocian (thick pebbly micrites, representing minor mass movements) to the area down paleoslope from fault E. These two clues suggest early activity of fault E prior to the emplacement of the early Bathonian debris flow.
6. The active life span of the faults was in excess of twenty million years (Bajocian to early Oxfordian).

The spacing of these postulated growth faults is not unusually close. An ancient example of well-exposed growth faults in the Triassic of Svalbard (Edwards 1976) shows growth faults of comparable size to be as close as 260 m apart. Edwards observes a total of nine growth faults in a distance of 4 km, which is again comparable with the situation on Bou Kornine.

Many examples of detrital lithofacies texturally similar to the carbonates of Bou Kornine have been described from ancient formations (Lowe 1972; Peterson 1965). Paleogeographic reconstructions by various authors show that such examples were closely related to the existence of submarine canyons marginal to ancient continental shelves. Mutti (1969) has described an Oligocene example on the island of Rhodes where clastic lithofacies show vertical sequences of mass movement and turbidite units similar to those seen on Bou Kornine. According to Mutti (1969), these units record a sequence of depositional events at the base of a tectonically unstable and active slope in the Oligocene. In that example, the position of part of the actual canyon wall has been preserved in outcrop. A similar situation is envisaged for the middle Jurassic carbonates of Bou Kornine where the

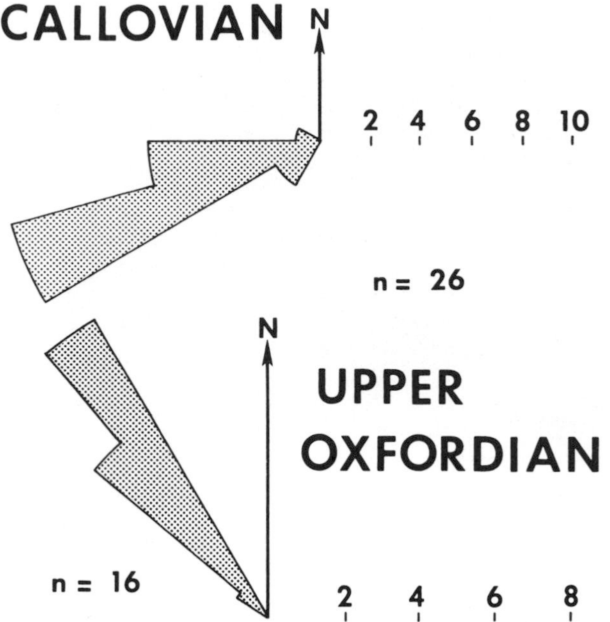

Figure 10-9. Paleocurrent data from flute marks in proximal turbidites of the Callovian and thin-bedded late Oxfordian carbonate flysch.

base-of-slope deposits were confined to generally low areas and the debris flow deposits were at least partially confined to a low at the distal end of a submarine canyon.

DIRECTIONAL DATA

Directional sedimentary structures were not found in the mass movement deposits themselves, but transport from the south or southeast is indicated by the presence of shallow water carbonates throughout the Jurassic at Djebel Ressas (Rakus 1973) 8km to the south of Bou Kornine (see Figure 10-1).

Two reliable sets of paleocurrent data were obtained from the turbidites that enclose the mass-movement deposits of Bou Kornine. These provide a more direct means for estimation of paleoslope directions. Flute marks on the bases of the Callovian proximal turbidites overlying the second Bathonian debris flow deposit indicate a southwest paleocurrent direction (Figure 10-9).

However, flutes from thin-bedded late Oxfordian flysch (above the laminated limestone bed) conform with the postulated regional paleoslope gradient and indicate a northwest paleocurrent direction (Figure 10-9). The slope profile was essentially smoothed by the late Oxfordian, so that we can assume that these late Oxfordian paleocurrents do indeed conform with the regional paleoslope. Therefore, the Callovian paleocurrents flowed at right angles to the regional paleoslope and canyon axis, channeled into a trough trending parallel to the basin edge, and ponded in the region of most growth fault activity—that is, between adjacent growth fault blocks. This trough coincides exactly in position with the channel cut by the late Bathonian debris flow into the underlying turbidites. This explanation may be one answer to the problem of the perpendicular discrepancy of turbidite paleocurrents that has often been encountered in ancient turbidite deposits (Potter and Pettijohn 1963, pp. 133 and 242-43).

In summary, if we assume that the mass movement deposits were confined to an essentially north-south axial canyon, then each episode of debris flow activity was followed by an interval of turbidite activity, some of which was in a ENE-WSW direction, across the axial trend of the canyon.

SUMMARY AND CONCLUSIONS

During the late Triassic and Jurassic, sedimentation patterns in northern Tunisia were under the influence of a westward opening, east-west-trending gulf, which formed part of the Tethys sea to the north (Nairn et al. 1977). An area of carbonate platform deposition dominated Jurassic sedimentation to the south of Bou Kornine. As the gulf developed, a steep southern margin formed and was accompanied by growth fault activity that started as early as the Bajocian. One or more canyons developed as the southern edge of the gulf steepened and early growth fault activity in the Bajocian triggered local debris flows that flowed northwards down the axis of a canyon close to the present-day position of Bou Kornine. Further growth fault activity coincided with the initiation of two major debris flows during the Bathonian. Most of the clasts were derived from already lithified carbonate formations, but the process of erosion of such large quantities of rock presents a problem. Some of the material may have been derived from submarine cliffs cropping out along the canyon walls as seen in recent canyons in detrital environments (Stanley 1974). The large volumes of sediment involved in both Tunisian deposits suggests catastrophic triggering and breakup of the source rocks to the south by strong earth tremors or tsunamis (Coleman, 1968).

Both episodes were followed by a much longer interval of turbidity current activity. Such currents may have been generated from a carpet of unconsolidated carbonate debris and shells on the shelf that lay a few kilometers to the south. Some of these turbidity currents may have utilized the canyons for their main axis of transport, but others were channeled for part of their course between active growth fault blocks in areas of maximum subsidence (Figure 10-10). This explanation is

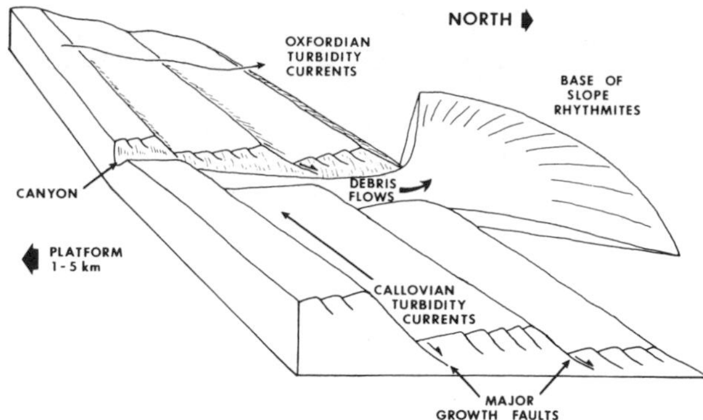

Figure 10-10. Paleogeographic model showing the interrelationship of sedimentation and structure in formation of the deposits on Bou Kornine.

supported by the fact that some of the turbidites show paleocurrents in a west-southwest direction, almost perpendicular to the north-south canyon axis and northerly regional paleoslope. Growth fault movements had decreased drastically by the early Oxfordian, and by the late Oxfordian, thin-bedded carbonate flysch shows paleocurrents that conform with the regional paleoslope (Figure 10-10).

ACKNOWLEDGMENTS

The authors thank Ahmed Azzouz, Director of the Service Géologique of Tunisia, for his helpful cooperation and also Nolan Gomm who was administrator in Tunis at the time the work was carried out. The authors also thank Dr. James Lee Wilson of Rice University for helpful discussions in the field and Robert Przygocki of the University of South Carolina for assistance with field measurements.

This research was carried out under the auspices of the University of South Carolina International Geological Programs, funded by National Science Foundation Grant GF-39074X1, which includes a grant from the Office of International Programs and a grant from the Earth Science Section of the Division of Environmental Sciences.

REFERENCES

Bonnefous, J., 1972. Contribution à l' étude stratigraphique et micropaléontologique du Jurassique de Tunisie. Unpublished Doctoral thesis, University of Paris, Tome 3.

———, and Rakus, M., 1965. Précisions nouvelles sur le Jurassique du Djebel Bou Kornine d'Hammam-Lif (Tunisie). *Bull. Soc. Géol. France,* 7: 855-59.

Bouma, A. H., 1962. *Sedimentology of Some Flysch Deposits.* Elsevier, Amsterdam, 168 pp.

Bujalka, P., Rakus, M., and Vacek, J., 1972. Carte géologique de la Tunisie, feuille #21 La Goulette, notice explicative. *Service Géol. Tunis.*

Castany, G., 1952. Le Jurassique du Djebel Bou Kornine d'Hammam Lif. *Bull. Soc. Nat. Tunisie,* 5: 195-204.

———, 1955. Les extrusions jurassique en Tunisie. *Ann. Mines et Géologie, Tunis,* 14.

Coleman, P. J., 1968. Tsunamis as geological agents. *J. Geol. Soc. Aust.,* 15: 267-73.

Crowell, J. C., 1957. Origin of pebbly mudstones. *Geol. Soc. Amer. Bull.,* 68: 993-1010.

Davies, D. K., 1968. Carbonate Turbidites, Gulf of Mexico. *Jour. Sed. Petrol.,* 38: 1100-09.

Edwards, B., 1976. Growth faults in Upper Triassic Deltaic Sediments, Svalbard. *Amer. Assoc. Pet. Geol. Bull.,* 60: 341-55.

Hampton, M. A., 1975. Competence of fine-grained debris flows. *J. Sed. Petrol.,* 45: 834-44.

Lowe, D. R., 1972. Implications of three submarine mass movement deposits, Cretaceous, Sacramento valley, California. *J. Sed. Petrol.,* 42: 89-101.

Middleton, G. V., and Hampton, M. A., 1973. Sediment gravity flows: Mechanics of flow and deposition. In: G. V. Middleton and A. H. Bouma (eds.), *Turbidites and Deep Water Sedimentation.* Soc. Econ. Paleont. Mineral., Pacific Section, Short Course, Anaheim, pp. 1-38.

Mutti, E., 1969. Sedimentologia delle arenarie de messanagros (Oligocene-Aquitaniano) nell'isola di Rodi. *Mem. Soc. Geol. Italiana,* (Eng. Abstract), 8: 1027-70.

Nairn, A. E. M., Stehli, F. G., and Kanes, W. H. (eds.), 1977. *Ocean Basins and Margins. IV: The Mediterranean.* Plenum, New York (in press).

Peterson, G. L., 1965. Implications of two Cretaceous Mass trasnport deposits, Sacramento valley, California. *J. Sed. Petrol.,* 35: 401-07.

Potter, P. E., Pettijohn, F. J., 1963. *Paleocurrents and Basin Analysis.* Springer-Verlag, Berlin, 269 pp.

Rakus, M., 1973. Le Jurassique au Djebel Ressas (Tunisie septentrionale). *Livre Jubilaire M. Solignac.* Annales des Mines et de la Géologie, Tunis, pp. 137-47.

Solignac, M., 1927. Etude géologique de la Tunisie septentrionale. Unpublished thesis, Tunis, Barlier edit.

Stanley, D. J., 1974. Pebbly mud transport in the head of Wilmington Canyon. *Mar. Geol.,* 16: M1-M8.

Walker, R. G., 1967. Turbidite sedimentary structures and their relationship to proximal and distal depositional environments. *J. Sed. Petrol.,* 37: 25-43.

Wilson, J. L., 1969. Microfacies and sedimentary structures in "deeper water" lime mudstones. In: G. M. Friedman (ed.), *Depositional Environments in Carbonate Rocks* Soc. Econ. Paleontol. Mineral., Sp. Publ. 14, pp. 4-19.

Chapter 11

Clast Size, Shape, and Composition in Some Ancient and Modern Fan Gravels

GILBERT KELLING
JOHN HOLROYD

Geology Department
University of Wales
Swansea, Great Britain

ABSTRACT

On the basis of external geometry and internal organization, rudite units in submarine canyon-fan complexes of Lower Paleozoic age in Scotland and Wales may be assigned to four intergrading megascopic categories (Type 1A: channeled, disorganized; Type 1B: channeled, organized; Type 2A: nonchanneled, disorganized; Type 2B: nonchanneled, organized). Type 1A units characterize more proximal regions of the canyon-fan complexes; Type 2B beds dominate the more distal sequences; and Types 1B and 2A rudites appear in intervening and overlapping situations.

Regional factors exert an overriding influence on the scalar, textural, and compositional attributes of the rudite units. However, individual rudite bodies display consistent relationships between parameters, such as unit thickness, maximum clast size, and maximum depth of basal erosion, which are consistent with the operation of debris flow mechanisms in Type 1 rudites and more tractional processes in the Type 2 beds. Moreover, Type 1 and Type A rudites within individual canyon-fan systems are generally coarser, more poorly sorted, and more positively skewed than the associated Type 2 and Type B units.

The shape attributes for extrabasinal large clasts in these submarine gravels are "fluvial" in character as attested by Zingg and Sneed and Folk shape classes, Krumbein and Folk sphericity measures, and Cailleux flatness indices. Internal evidence, reinforced by data from neritic, basinal, and canyon head gravels from the Annot Sandstone (Eocene) and the Holocene sediments of the Wilmington and Baltimore submarine canyons in the northwest Atlantic indicates that this character is attributable to shape sorting that is affected by the resedimenting mechanisms, rather than a result of inheritance through direct fluvial supply to the basin.

Three broad categories of clast composition are distinguished: extrabasinal, allochthonous intrabasinal, and autochthonous intrabasinal. Within individual canyon-fan systems the degree of polygenetic character in the extrabasinal element of the clast population tends to decrease distally, thereby leading to a relative enrichment in the content of siliceous clasts, which are especially common in Type 2B units. Large autochthonous intrabasinal clasts are most abundant in Type 1 rudites, thereby suggesting that the abundance of this element is closely related to the erosive capabilities of the gravel-bearing flows. A consistently high content of allochthonous intrabasinal clasts suggests either the existence of a wide shelf or platform adjacent to the submarine trough or the prolonged residence time for the extrabasinal components on this shallow platform. Roundness data facilitate distinction between these alternative modes of origin.

INTRODUCTION

Considerable effort has been expended in the past few years on the identification and description of modern and ancient sediments deposited in submarine fans and canyons, and the major features of sedimentary sequences formed in these environments have been integrated within a generalized model of sedimentation (Stanley and Unrug 1972; Mutti and Ricci Lucchi 1972; Walker and Mutti 1973; Normark 1970, 1974; Nelson and Kulm 1973; Kelling and Stanley 1976). Concurrently, several classifications of the rudaceous units encountered in such sequences have been devised (Walker 1970; Davies and Walker 1974) and have been linked with the spectrum of gravity-controlled

Note: Gilbert Kelling is currently at the Department of Geology, University of Keele, Keele, Staffordshire, England.

processes believed responsible for sedimentation of the coarser components in canyon and fan environments (Middleton 1970; Middleton and Hampton 1973, 1976; Carter 1975; Walker 1975).

The present study focuses on some hitherto neglected aspects of the fabric, texture, composition, and scalar attributes of rudites associated with ancient submarine canyon and fan sequences and with some modern counterparts, and it briefly assesses the genetic significance of these parameters. Data from geological formations representing a considerable stratigraphic and geographic range have been utilized in order to minimize regional constraints and to identify broadly applicable principles. Nevertheless, the conclusions reached should be regarded as indicative rather than definitive and as providing some refinement and amplification of existing ideas concerning the genesis of canyon-fan sequences.

Most of the data presented here are derived from rudite bodies of late Ordovician and early Silurian ages that occur in marginal zones of the Caledonian turbidite flysch troughs of southern Scotland and central Wales, Great Britain (Figure 11-1). The possible association of these rocks with ancient submarine canyons and fans has been recognized for some time (Kelling 1964; Walton 1965; Kelling and Woollands 1969), but definition of the subenvironments represented in these sequences has not been undertaken previously. A point of geotectonic significance is that these Scottish and Welsh flysch rudites are located on opposite borders of the paratectonic zone of the Caledonian orogen and, if current views are accepted, were emplaced on opposite margins of a Proto-Atlantic or Iapetus ocean, which was still relatively wide during this geological time interval (Dewey 1974; McKerrow and Ziegler 1972).

Additional clast shape and size data have been obtained mainly from rudites within the Annot Sandstone formation (s.l.), a late Eocene flysch sequence deposited in a relatively restricted marine basin in the Alpes Maritimes of southeast France (Stanley 1961; Stanley and Bouma 1964; Stanley 1975), and also from Holocene gravels sampled by dredge and grab in the heads of Wilmington and Baltimore submarine canyons off the eastern seaboard of the United States (Kelling and Stanley 1970).

A semantic note is required here to clarify our subsequent use of the terms *proximal* and *distal* in view of the variable application of these terms in the turbidite literature. Our use refers to distance from the source, or more accurately, from the point of input to the deep basin, but we recognize that channelized deposits, including debris flows, may extend for substantial distances from this point (cf. Embley 1976; Rodine and Johnson 1976) and may also coexist laterally, even in near-source situations, with finer thin-bedded sediments of overbank character that may appear "distal" on any lithologic criterion (see Chapter 8).

LITHOLOGICAL FRAMEWORK AND SEDIMENT GEOMETRY

In terms of geometry and organization, the following two major types of fan rudite units (Figure 11-2) may be recognized.

Type 1 (Channeled)

These are characterized by broad lensing of individual beds that results in rapid lateral expansion accompanied by deep channeling (up to 5 m) to give an eye-shaped pod of rudite. Such beds are highly impersistent laterally, relative to associated sediments (Kelling and Woollands 1969), and are commonly of amalgamated character. Dimensions are variable, but the general range is from 20 m wide and 4 m thick (S1 and S2 rudite bodies; see Table 11-1 and Figure 11-1) to 100 m wide and 12 m thick (W1 and S5 bodies). A few of the Lower Paleozoic rudite bodies are composed solely of such channeled units but most are composite, including both Type 1 and Type 2 units.

Type 2 (Nonchanneled)

This type comprises laterally persistent units of rudite with basal surfaces that may be erosional and scoured, but that display a broadly planar attitude lacking the abrupt, deep channels characteristic of the Type 1 units. The maximum thickness of individual beds ranges from less than 1 m to more than 10 m. Traced laterally, they show slight variations in thickness and may display shallow basal scours. Ultimately (over distances of 100 m to 500 m), the units wedge out or pass gradationally into the enclosing facies. We concede here that most, if not all, such units were originally deposited within submarine channels and gullies; however, the distinction we make is that on the scale of the outcrop, Type 2 units appear to have been formed by processes generally incapable of the localized scouring achieved in the case of the Type 1 rudites.

Mappable rudite bodies are commonly formed entirely by a number of Type 2 rudites, but units of this type also occur in combination with Type 1 beds, and in such bodies the Type 2 rudites are generally subordinate in number, are of finer grade, and occur beneath and lateral to the Type 1 rudites. Type 1 rudite beds are generally thicker than Type 2 units but the latter are more extensive laterally.

The most common vertical transition observed in the rudite bodies involves a gradual upwards coarsening

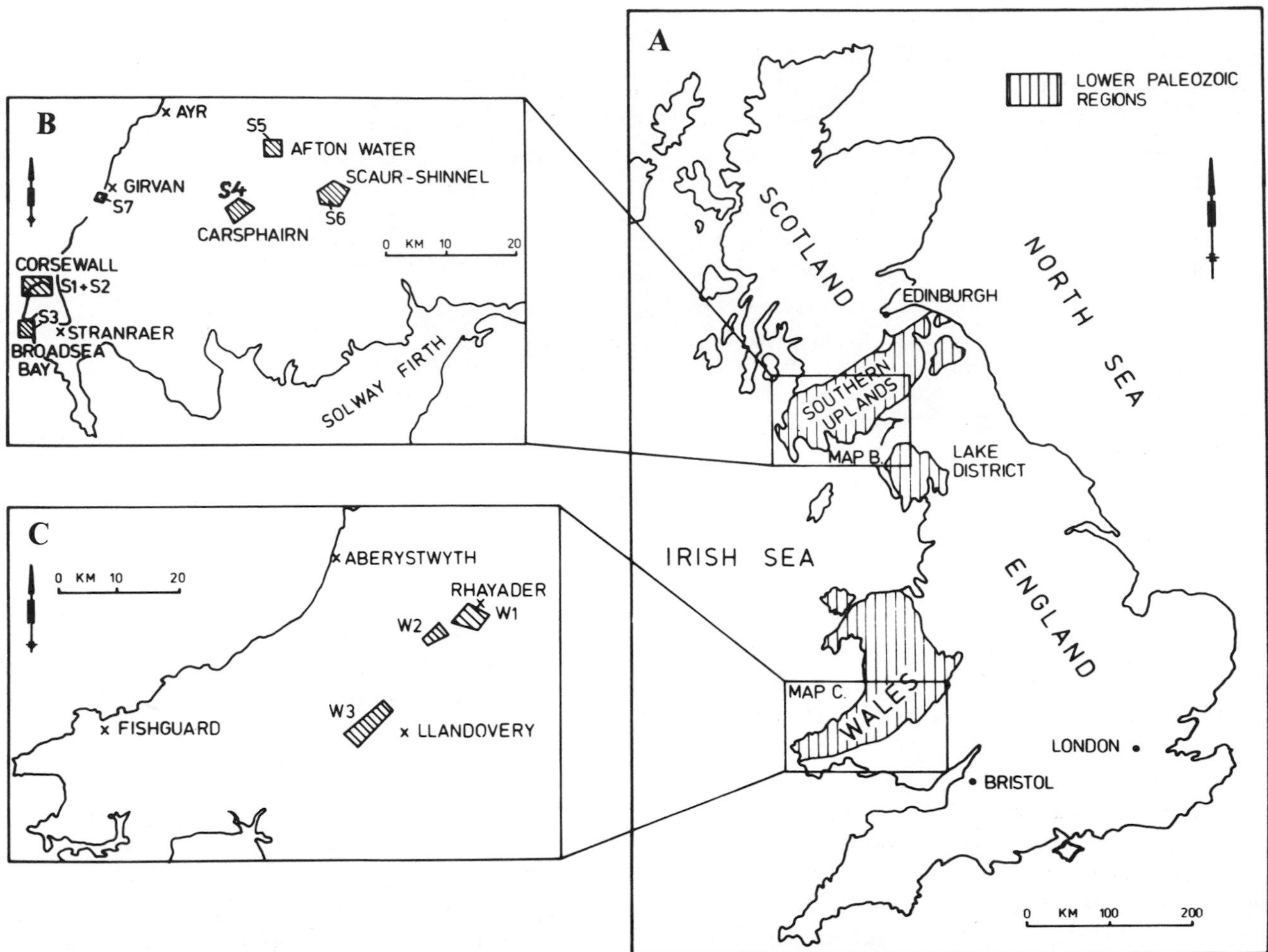

Figure 11-1. Location maps for the British Lower Paleozoic rudite complexes sampled. Note: Map A shows general location of regions containing Lower Paleozoic deep marine sediments in mainland Britain. Map B shows location of sampled rudite complexes in the western part of the Southern Uplands region, Scotland. (Codes – S1 and so forth – relate to Table 11-1.) Map C shows location of sampled rudite complexes in central Wales.

(within a few tens of meters) from lutites to flaggy arenites, followed by development of Type 2 rudites. These rudites gradually become interbedded with an increasing proportion of Type 1 units, which are dominant towards the central part of the complex. Above, there occurs a general and relatively rapid reversal to the prerudite lithology.

Internal Organization

Type 1 and 2 rudite units show a similar range of internal features, but these are developed in a variable manner and a further division thus may be made into Types A and B on the extent and nature of internal organization displayed (cf. Walker 1975). Alternatively, subdivision is possible on the basis of the proportionate abundance of large clasts with respect to the finer, groundmass component, which thus enables the distinction of *clast-supported* and *matrix-supported* rudite units.

Disorganized (Types 1A and 2A). These "chaotic" units possess relatively little or no internal structure; the entire bed exhibits random distribution of clasts. Some units (most common in the W1, S1, and S2 complexes) are of "bipartite" character, with a lower portion composed of chaotic rudite and an upper interval that is usually thinner and comprises poorly graded or ungraded arenite. The junction between the contrasting grades is invariably abrupt but not erosive (Photograph A, Figure 11-3).

Type 1A units occur predominantly in the coarse, central part of the rudite complexes (W1, W3, and other bodies of central Wales and most complexes in Scotland). Disorganized beds displaying Type 2 geometry involve finer rudites (W1, W2, and S4 complexes). The presence of abundant quartz and mudstone pebbles and granules imparts a speckled appearance to some Type 2A rudites, which have been termed "Haggis Rock" and which are appropriately widespread in Scotland.

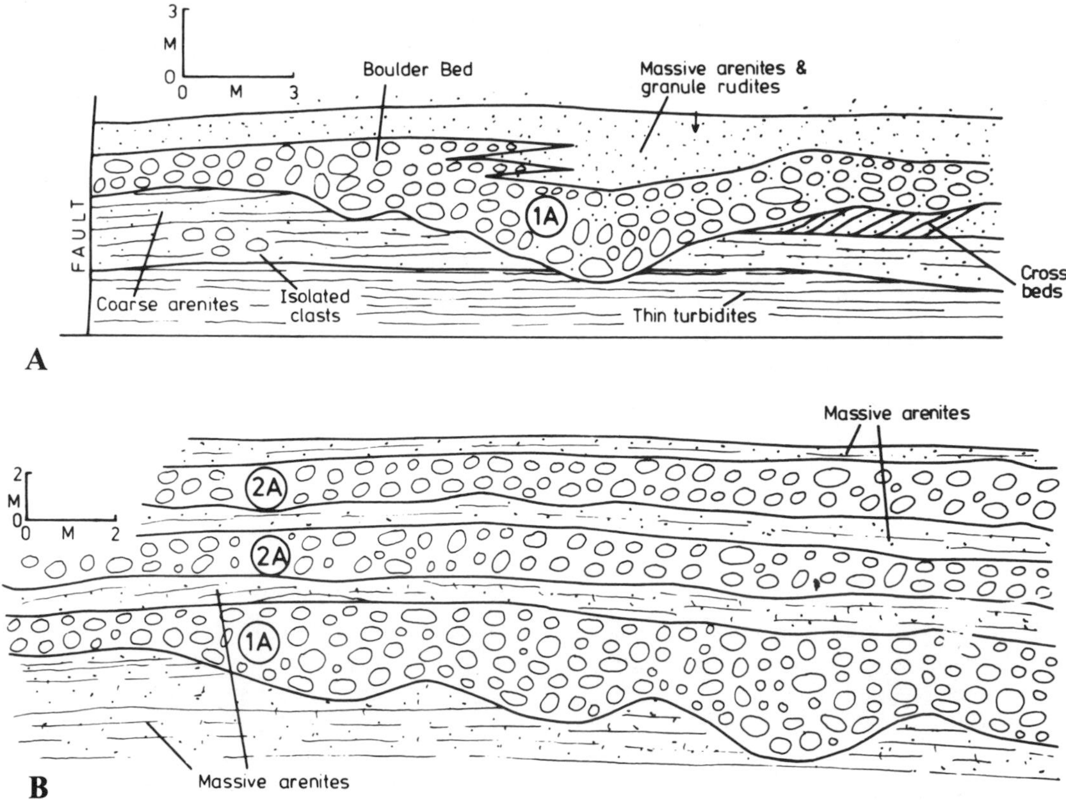

Figure 11-2. Field sketches (slightly schematized) illustrating the geometry of Type 1 and Type 2 rudite units. Note: Sketch A shows Type 1A compound unit, Boak Port rudite complex (S1), Rhinns of Galloway, SW Scotland. Sketch B shows Type 1A and Type 2A units, Corsewall Point rudite complex (S2), Rhinns of Galloway, SW Scotland.

Organized (Types 1B and 2B). Units of this type display one or more of the following features: grading (normal and inverse), imbrication, and stratification. Such features are most common in Type 2B rudites.

Grading is encountered in planar Type 2 rudites and in thinner beds of finer grade. A few graded rudites display internal sequences analogous to Bouma intervals *a-e*, but most units display only T*ab*(*c*?) sequences. Examples of grading are especially abundant within the quartz pebble rudites of mid Wales and the S5 body in Scotland. In the latter sequence the Bouma equivalent intervals may be up to one meter thick. Inverse grading is more prevalent in cobble-rich rudites and is therefore common in the Welsh complexes.

Original *imbrication* is most commonly developed in Type 2B rudites, especially the Kilranny and Craigskelly (S7) rudites of Girvan, Scotland, and those units of the S4 complex that contain only intrabasinal clasts, and is generally lacking in Type 1 rudites.

Stratification involves a succession of layers of differing grade and generally chaotic internal structure that constitute units up to several meters in thickness. The layers tend to display gradational mutual contacts and have a crude size relationship (layers dominated by large clasts tend to be followed by bands of clasts of moderate size, rather than by finer grades). The relationship of maximum size to layer thickness is irregular. Stratification is most common and conspicuous within the Type 1B rudites of the S1, S2, S5, S6, and W1 bodies. Only a few Type 2 rudites exhibit this feature, and these stratiform Type 2 units form a transitional class between Types 1B and 2B.

Clast Arrangement and Abundance

The allocation of rudite units to clast- and matrix-supported categories was achieved on a largely arbitrary, qualitative basis. However, the proportion of subpebble groundmass in the clast-supported units is estimated to range approximately from 5 to 35 percent, with the "matrix" content tending to increase with decreasing mean size of the rudite. Vertical transitions from clast to matrix-supported portions occur in many of the thick-graded, stratified, and bipartite units.

Clast-supported rudites. This feature occurs in Type 1 and 2 rudites, irrespective of grade, but is more common in cobble and boulder grade units. Close packing of clasts results in a low content of granule and sand grade "matrix" and a high degree of clast-to-clast contact.

Table 11-1. *Lower Paleozoic Rudite Complexes*

Designation	*Location*	*Stratigraphic Horizon*
S1	Boak Port-Milleur Point, Rhinns of Galloway, SW Scotland	Lower Caradoc (Ordovician)
S2	Corsewall Point, Rhinns of Galloway, SW Scotland	Lower Caradoc
S3	Broadsea Bay, Rhinns of Galloway, SW Scotland	Middle Caradoc
S4	Carsphairn area, southern Scotland	Lower/Middle Caradoc
S5	Afton Water Reservoir, New Cumnock, southern Scotland	Lower/Middle Caradoc
S6	Scaur and Shinnel Waters, Thornhill, southern Scotland	Middle/Upper Caradoc
S7	Craigskelly Rocks, shore section at Girvan, SW Scotland	Lower Llandovery (Silurian)
W1	Caban Coch area, Rhayader, central Wales	Lower Llandovery
W2	Drygarn area, Abergwesyn, central Wales	Middle Llandovery
W3a	Bwlch Trebanau, Llanwrda, central Wales	Lower Llandovery
W3c	Caio area, Llandovery, central Wales	Middle Llandovery

Matrix-supported rudites. Units exhibiting this texture consist of a relatively smaller number of clasts, often of pebble grade, "floating" in a finer matrix, producing a diluted rudite (the pebbly sandstone facies of Walker and Mutti 1973). However, in some rudite bodies (e.g., S1 and S2), the units of coarser material pass laterally into clast-supported units that occupy a central position in the body. Matrix-supported rudites are common in all the investigated complexes but especially in the central Wales bodies and the S4 complex, Scotland.

In general, clast-supported rudites display disorganized fabrics (Type A), whereas most of the matrix-supported units fall into the well-stratified or graded categories of Type B, organized rudites.

SCALAR AND TEXTURAL ATTRIBUTES OF RUDITE UNITS

Rudite units display a variety of scalar features that are amenable to quantitative analysis. Some dimensional attributes relate to the unit as a whole, while others are concerned with the size distribution of the contained clasts. Both these types of parameter may be linked, in varying degrees of correlation, with the megascopically defined rudite classes described in the previous section.

Scalar Aspects of Rudite Units

Attributes of this type considered here are (i) maximum unit thickness (largest visible dimension of a rudite bed measured perpendicular to bedding); (ii) maximum depth of basal erosion (the greatest dimension of a distinct and localized erosional scour or hollow, measured below the general level of the basal surface of a rudite unit); (iii) maximum clast size (strictly, a textural attribute; derived from the mean of the ten largest clasts encountered in a given unit).

These parameters display mutual relationships which are regionally variable but are, with only a few exceptions, consistent for individual rudite bodies (diagrams A through E, Figure 11-4), which yield regression lines possessing narrowly defined confidence limits. The aggregate plots of unit thickness versus maximum clast size reveal a high degree of correlation, as determined from the correlation coefficients for individual bodies (Figure 11-4), which range from 0.56 to 0.95. Most of the regression lines displayed in Figure 11-4 are derived from bodies that include more than one of the four main rudite categories, although one type tends to be dominant, as indicated in the diagram. In a few instances, differentiating the rudite categories and examining the relationship between unit thickness and maximum clast size in more detail has proved possible. Thus, data from the Type 1B and Type 2B rudites of the Scaur-Shinnel complex (S6) of southern Scotland share the same general trend and are separable only through the extended size range of the nonchanneled rudites, although the few data from Type 2B quartz pebble rudites provide a trend of much higher gradient (diagram E, Figure 11-4). There is some indication that on a regional basis, rudite bodies predominantly composed of Type 1 units tend to yield regression lines of low gradient whereas data from those sequences characterized by Type 2 units provide steeper regression lines (diagram F, Figure 11-4).

In general, the distinction between Type A and Type B rudites is even less marked, although graded and ungraded units from the Caban (W1) body of central Wales provide two contrasting regression lines (diagram F, Figure 4-11): one of low gradient (Type A) and another of steep slope (Type B).

A **B**

Figure 11-3. Internal features of "organized" rudites. Note: Photograph A shows "bipartite" unit in the Caban Coch quarry, rudite body W1, central Wales. Dark hemipelagic mudstone at base is overlain by pebbly gravel of probable debris flow origin, succeeded by poorly graded graywacke. The overlying gravel unit abruptly follows the arenite. Hammer (35 cm long) provides scale. Photograph B shows the upper part of a mega-graded unit (no. 16 in chart A, Figure 11-12) from the Afton Water rudite complex (S5). Parallel-laminated alternations of coarse and medium sand of the Bouma *b* division show an upward decrease in grade that passes up into a single set of steeply inclined cross-strata (Bouma *c* division) forming the top of the photograph. Scale in center is 15 cm long. Part of the basal portion of this unit is shown in photograph A, Figure 11-10.

The relationship of unit thickness to maximum basal erosion is only calculable for the Type 1 (channeled) rudite units, and the data on this aspect are therefore restricted. However, these parameters also yield regression lines that are consistent for individual rudite bodies and provide correlation coefficients (0.56 to 0.98) that are slightly superior in aggregate to those for unit thickness versus maximum clast size. On a regional basis there appears to be no distinction between Type A and Type B rudites but the range in gradient of the regression lines is small (diagram G, Figure 11-4).

Textural Aspects of Rudite Units

The size distribution has been determined on a number frequency basis from field counts of all clasts with a diameter exceeding 1 cm (pebble, cobble, and boulder grades) encountered in a channel sample through the entire thickness of a limited number of rudite units. The largest visible dimension of each clast was measured, and the number of clasts measured in samples varies from 50 to 250 with an average of 110 clasts.

Size parameters for these large clast populations have been calculated, following the standard procedures (e.g.,

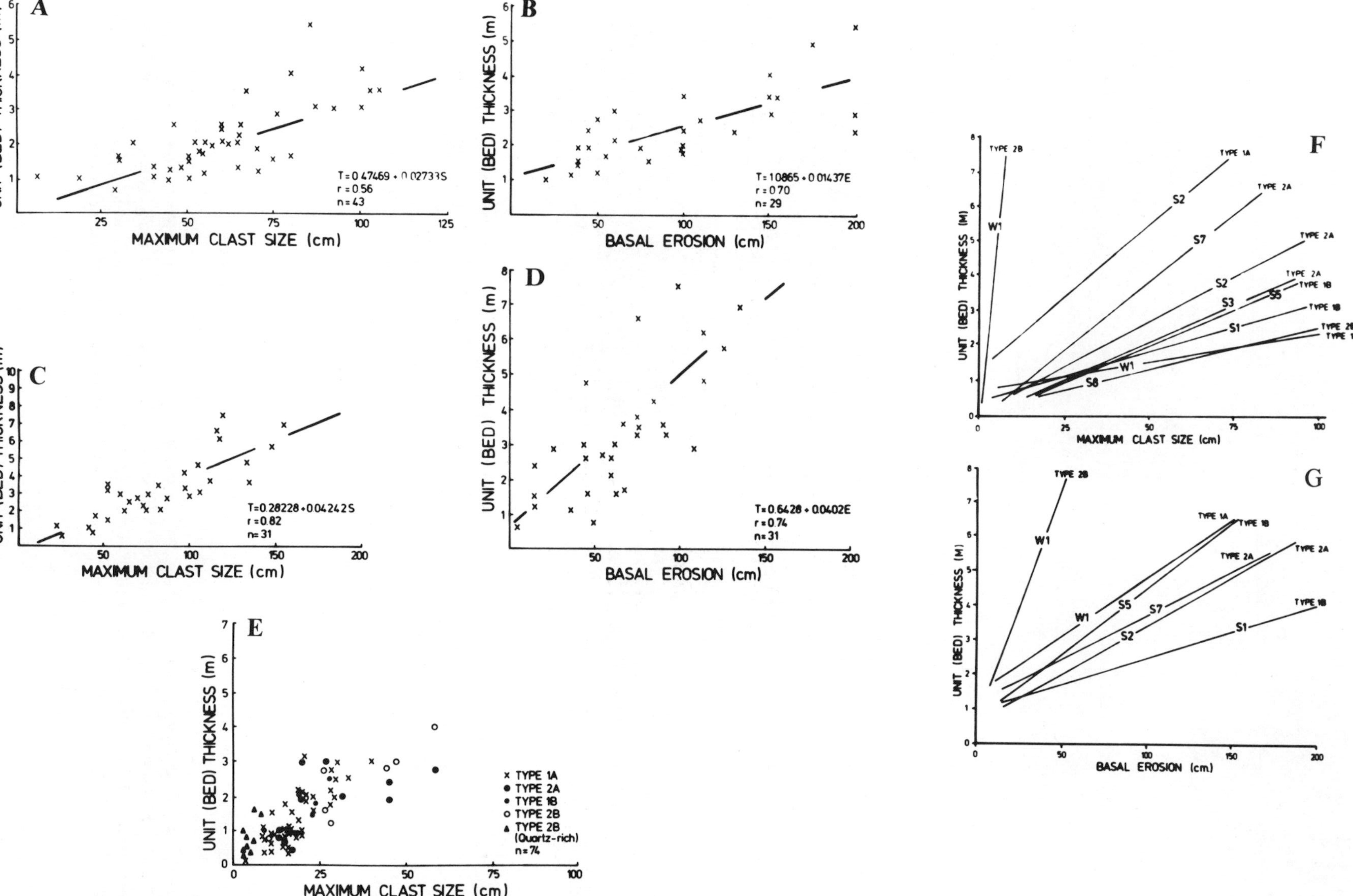

Figure 11-4. Scatter diagrams illustrating relationships between maximum clast size, unit thickness, and degree of basal erosion for rudite units in several complexes. Note: dashed lines on all diagrams indicate calculated regression lines. Diagrams A and B show Boak Port body (S1), Milleur Point, Rhinns of Galloway, Scotland (mainly Type 1A units). Diagrams C and D show Afton Water body (S5), New Cumnock, Scotland (mainly Type 1B units). Diagram E shows Scaur-Shinnel complex, with all rudite types differentiated. Diagrams F and G show calculated regression lines (relating unit thickness with maximum clast size and degree of basal erosion) from selected rudite bodies (see Table 11-1). The bodies are differentiated according to dominant rudite types. (See text for explanation.)

Table 11-2. *Textural Data for Lower Paleozoic Rudites*

Aggregated Mean Textural Data

		No. of Units	Mean Size (φ)	Size Range (φ)	Sorting (σ_I)	Skewness (Sk_I)
Type 1	Scotland	9	-7.6	4.3	0.93	+0.23
	Wales	3	-5.5	3.8	0.67	-0.07
	Total	12	-7.0	4.2	0.86	+0.15
Type 2	Scotland	10	-7.4	3.4	0.78	+0.06
	Wales	5	-4.8	2.3	0.49	-0.21
	Total	15	-6.6	3.0	0.69	-0.05
Type A	Scotland	16	-7.6	4.3	0.85	+0.08
	Wales	4	-5.4	3.8	0.67	-0.09
	Total	20	-7.2	4.2	0.81	+0.04
Type B	Scotland	3	-6.7	3.7	0.88	-0.12
	Wales	4	-4.8	2.1	0.43	-0.20
	Total	7	-5.7	2.8	0.63	-0.17
Undifferentiated	Scotland	19	-7.5	3.8	0.85	+0.16
	Wales	8	-5.1	2.9	0.56	-0.16
	Total	27	-6.8	2.5	0.76	+0.02

Textural Data for Selected Lower Paleozoic Rudite Units

Code[a]	Rudite Type	Mean Size (φ)	Size Range (φ)	Standard Deviation	Sorting (σ_I)	Skewness (Sk_I)
Sld	2A	-6.7	4.1	1.25	0.96	-0.21
Slc	1A	-6.9	4.3	1.28	1.03	-0.29
Slb	1A	-8.3	4.1	1.11	0.93	+0.54
Sla	1A	-8.4	4.1	1.26	0.98	+0.17
S2e	1A	-8.0	3.9	1.40	1.01	+0.03
S2d	2A	-7.8	4.2	1.36	0.98	-0.16
S2c	1A	-7.8	4.2	1.29	0.96	+0.12
S2b	2A	-8.1	4.1	1.27	0.95	+0.22
S2a	2A	-8.3	4.3	1.12	0.83	+0.34
S3e	2A	-7.6	2.2	1.04	0.51	+0.06
S3d	2A	-7.5	3.3	1.13	0.66	+0.24
S3c	2A	-7.4	3.4	1.14	0.86	+0.08
S3b	2A	-7.4	3.0	1.11	0.76	+0.17
S3a	2A	-7.5	3.9	0.79	0.62	+0.24

[a] Codes refer to designations of the rudite bodies as shown in Table 11-1. Alphabetical suffixes in each code number refer to successively higher units within each rudite complex selected.

Folk 1968), and some of these are listed in Table 11-2. The mean values of parameters such as size range, mean size, sorting, and skewness differ significantly from region to region, as indicated by the aggregated data for the Scottish and Welsh complexes (Table 11-2). These suggest that the Scottish rudites are, in general, coarser and have a wider size range, poorer sorting, and a slight positive skew compared with the rudites of central Wales. Such differences may be attributed to local provenance factors and ensure that the most significant within-region comparisons (between rudite bodies) are achieved from relative rather than absolute values. For this purpose, Table 11-2 lists rudite size parameters for both the Scottish and Welsh regions in terms of the categories (Types 1 and 2; Types A and B) delineated above in an effort to identify process-related factors.

Although the sample size for certain of these categories is inadequate to permit firm statistical appraisal, reasonably consistent differences are apparent between the average parameters for rudite types in both regions (Table 11-2); the Type 1 and Type A units are coarser, with a larger size range, poorer sorting, and more positive skew

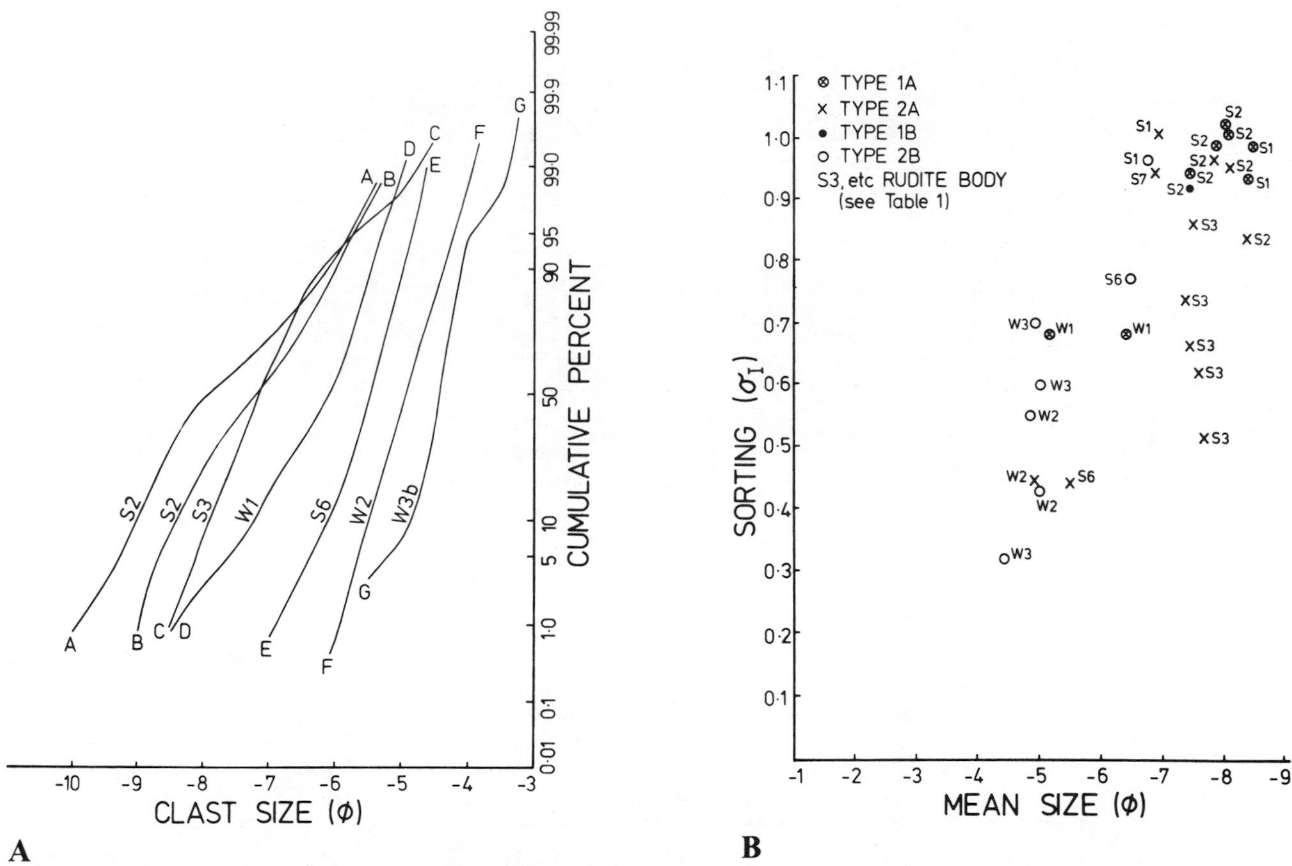

Figure 11-5. Size data and sorting index for large clasts in selected units. Note: Chart A shows the cumulative size distributions of large clasts in selected units derived from several rudite bodies. The coarse-component bimodality of curves A and B and fine-component bimodality of curve C and possibly G are noteworthy. Chart B shows relationship of mean size and inclusive sorting index for large clasts in selected derived units from various rudite bodies. An anomalous trend is displayed by the units from the S3 (Broadsea Bay) body. (See text for explanation.)

than the Type 2 and Type B rudites, respectively. Moreover, the differences in the average textural values between Types A and B are slightly greater than the contrasts between Types 1 and 2 parameters, thereby suggesting that internal organization of the rudites may have more genetic significance than unit geometry in textural terms. As is tentatively suggested by the evidence, within a given canyon fan system the Type 1A rudites possess the highest mean sizes and the poorest sorting, whereas the best sorting and smallest mean sizes are exhibited by the Type 2B units (see Walker 1975). The Type 1B and Type 2A rudites display less clear-cut, but intermediate, textural characters. These relationships are illustrated in chart A, Figure 11-5. However, both regional and local factors may intervene to modify this pattern. For example, the Type 2A rudites of the Broadsea Bay body (S3) occupy a field that is distinct from the general trend shown in chart B, Figure 11-5; they are better sorted, but of comparable mean size to the more proximal Corsewall bodies. This anomaly is explained by the nature of the clasts, most of which are calcareous graywackes of allochthonous intrabasinal type, presumably supplied from a local source but subjected to the enhanced sorting effects associated with deposition of the Type 2A units.

Vertical variations in size parameters appear to be significant in some sequences. Thus, successively higher units within the Corsewall complex (samples S2a to S2e in Table 11-2) reveal a general decrease in mean size that is paralleled by a tendency to poorer sorting and to more negative skewness. The Type 2A units of the more distal Broadsea Bay body (samples S3a to S3e in Table 11-2) display a more complex pattern. Here the smallest mean size and the poorest sorting is encountered in the central unit of the sampled sequence, which also exhibits a particularly low skewness value. These vertical trends, derived from similar categories of rudite unit, appear to represent a sensitive measure of relative proximality that is perhaps related to fan progradation or abandonment.

Although the great majority of the rudites yield unimodal size distributions, a few are bimodal or even polymodal (see chart A, Figure 11-5). The bimodality is generally not pronounced and may result from the addition of an excess of finer components, such as pebbles and small cobbles (see curves C and G in chart A, Figure 11-5). or, less commonly, an increment of very coarse (boulder grade) clasts usually of intrabasinal nature.

Matrix- and clast-supported rudites also display certain size-related differences. Quartz pebble matrix-supported rudites, together with intraclast-rich rudites and pebbly mudstones, generally exhibit a smaller maximum size and a more restricted size range and also display better sorting of the larger clasts (and higher roundness and sphericity) than the clast-supported rudites.

The above discussion, we should emphasize, is concerned solely with the coarse elements of the rudite units (boulders, cobbles, and pebbles) and has excluded the textural character of the groundmass or "matrix." Qualitatively (despite many exceptions), rudites of large mean size generally appear to possess a groundmass of granule or coarse sand grade, while many of the pebble rudites contain a silty or muddy matrix.

SHAPE ATTRIBUTES OF LARGE CLASTS

This section summarizes shape data derived from pebble- and cobble-grade populations in selected ancient rudites and modern deep marine gravels. This textural aspect is one that for practical reasons has not been emphasized in studies of ancient rudites (see Blatt et al. 1972, p. 70) and therefore is poorly documented for comparative purposes. Thus, most of the conclusions drawn from our data are based on a priori assumptions and must be treated with an appropriate degree of circumspection. We should also emphasize that the selection of the Lower Paleozoic rudites chosen for this shape study was governed largely by their degree of incoherence, which thus introduces a sample bias of unknown extent. Parameters utilized in the present analysis are: roundness, sphericity index (Krumbein 1941; Wadell 1934), the shape classes of Zingg (1935) and Sneed and Folk (1958), and the flatness index (aplatissement) of Cailleux (1945).

Roundness was determined by comparison with the standard silhouettes of Krumbein (1941) and is consequently a two-dimensional measure, while sphericity, shape classes, and the flatness index were assigned from direct measurement of the three principal diameters of each clast. For the Lower Paleozoic rudites, the sample size of extracted clasts averages 105, with a minimum of 55. In all the ancient and modern samples, clasts of "intrabasinal" nature (i.e., derived from contemporaneous sediments during transportation and deposition) have been excluded from the analysis. Figure 11-6 indicates the range of size and composition for each Lower Paleozoic sample, and the various shape parameter distributions are summated in Figures 11-6 to 11-9 and Table 11-3.

Some significant similarities and contrasts are revealed by the summated data. In particular, the Zingg and Sneed and Folk shape class histograms for the Lower Paleozoic flysch rudites are remarkably similar (Figure 11-6) in view of the stratigraphic and geographic intervals separating these samples. They reveal a dominance of spherical or compact forms with subordinate discoidal or oblate shapes and lesser amounts of rod-like or elongate pebbles and cobbles. This resemblance is also apparent in the frequency distributions of Wadell/Krumbein sphericity indices for these samples (Figure 11-6). The sample from the Caio (W3c) body of central Wales displays the same general shape characteristics, but it differs slightly from the others and the probable cause of this anomaly is discussed below.

Pebble lithology might be expected to exert an influence upon the shape populations, although the nearly identical form distributions in these flysch rudites have been derived from clasts of greatly differing aggregate composition (Figure 11-6). However, to assess the lithological control, Zingg shape data have been plotted by utilizing only the vein quartz pebbles (Figure 11-6). These reveal shape distributions that are broadly comparable to those obtained from the total clasts for each sample, but usually the quartz pebble populations are enriched in spheres and deficient in discs. The slightly anomalous aggregate shape distribution for the Caio sample, noted earlier, is probably attributable to the high proportion (nearly 60 percent) of quartz pebbles present in this rudite. Thus lithological control, while undoubtedly present, does not appear to exert a fundamental influence on the shape populations in these sediments.

This point is reinforced by consideration of data from other geological formations. Thus the Zingg shapes plotted as item E in Figure 11-6 are derived from data published by Van Loon (1970) and obtained from siliceous pebbles and small cobbles, in the lower, framework-supported portions of a graded muddy rudite in the Westphalian (Carboniferous) flysch of Cantabria, northern Spain. The resemblance to the British Lower Paleozoic examples is self-evident. Even more remarkable is the aggregate shape population displayed by clasts from four rudite bodies at widely separated localities in the basinal flysch facies of the Eocene Annot Sandstone formation in the Alpes Maritimes, southern France (chart B, Figure 11-7), which again reveals a strong similarity to the Paleozoic examples. Since the dominant clast types in these Annot rudites are schists, gneisses, and limestones, which are lithologies poorly represented in the older samples, the influence of pebble composition again can be considered negligible.

The effects of size on the relative abundance of different pebble shapes can be determined from Figure 11-8. The most consistent relationship is a general increase in the proportion of "rods" or prolate forms with increasing mean

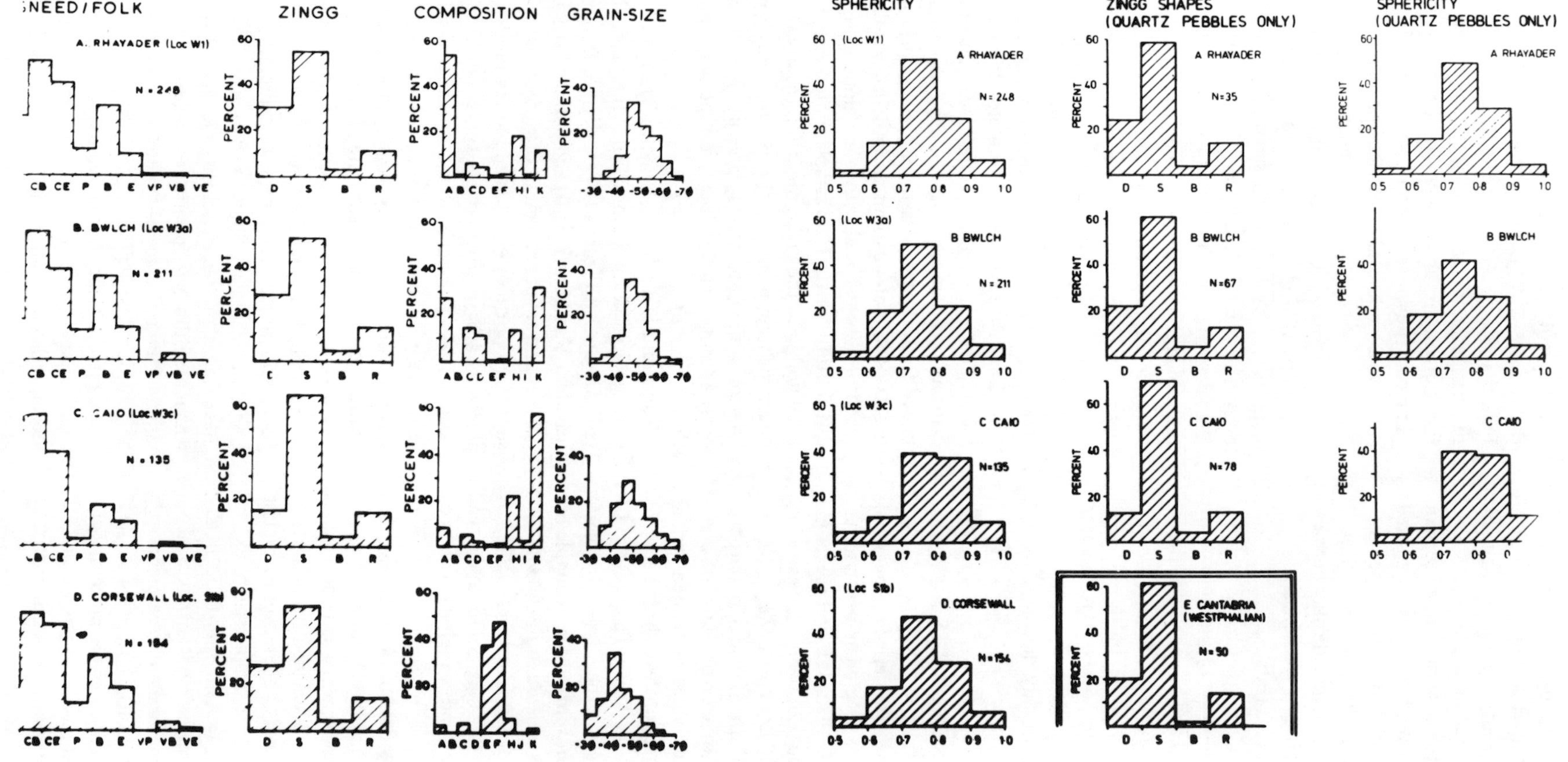

Figure 11–6. Histograms displaying shape, composition, and textural attributes of total large clast populations from units in selected rudite bodies, together with Zingg shapes and sphericity values for vein quartz pebbles and small cobbles in some of these samples. Note: Sample E is derived from a Westphalian flysch rudite unit in Cantabria, Spain (data in Van Loon 1970). For Zingg shape categories, D = discs, S = spheres, B = blades, and R = rods. For Sneed and Folk shape categories, C = compact, E = elongate, P = platy, B = bladed, CP = compact platy, CB = compact bladed, CE = compact elongate, VP = very platy, VB = very bladed, and VE = very elongate. For Composition column, rock type categories are: A, Arenites; B, Lutites; C, Acid volcanics; D, Tuffs and agglomerates; E, Basic volcanics and instrusives; F, Acid intrusives; H, Metaquartzites; I, Foliates; and K, Vein quartz.

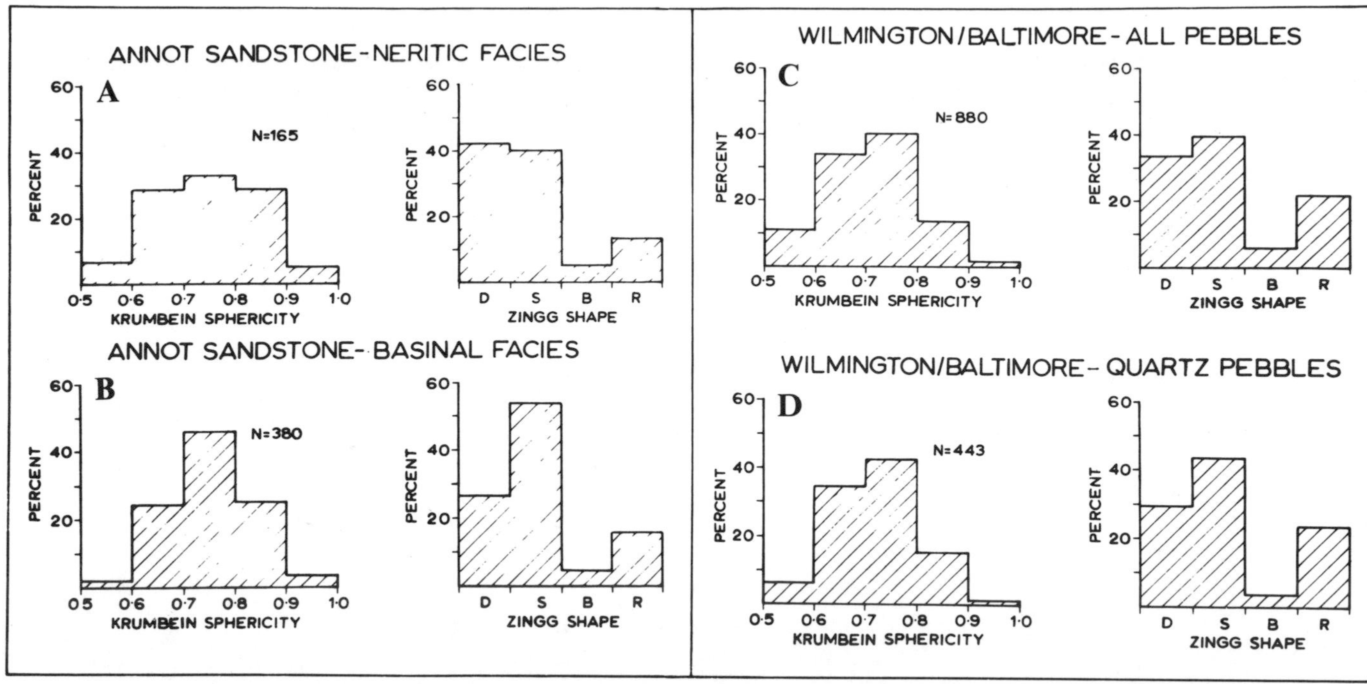

Figure 11-7. Shape data from Eocene and Holocene marine gravels. Note: Chart A shows pebbles and small cobbles combined from two localities in the neritic facies of the Annot Sandstone of the Alpes Maritimes, southern France (see Stanley 1975). Chart B shows pebbles and small cobbles combined from four localities in the basinal (submarine channel) facies of the Annot Sandstone. Chart C shows pebbles and cobbles combined from six grab and dredge samples of Holocene gravels obtained from axial regions of the heads of Wilmington and Baltimore submarine canyons, northwest Atlantic. Chart D shows shape of quartz pebbles from the six grab and dredge samples obtained from Wilmington and Baltimore canyon heads.

clast size (chart D, Figure 11-8). A corresponding decline in the percentage of spheres is negated only by the Corsewall samples (chart A, Figure 11-8), but neither discs nor blades reveal any uniform pattern, although individual samples display more consistent changes (charts B and C, Figure 11-8).

The frequency distributions for the Wadell-Krumbein sphericity index are broadly comparable in all the flysch rudite samples analyzed (note the similarity in the mean values listed in Table 11-3), but this parameter appears to be more variable than the shape class distributions. This enhanced variability is partly an artifact of the method of calculating sphericity, which assigns the same index to clasts of greatly differing shape, but may be ascribed in part to differing size ranges in the assemblages, since the mean sphericity is clearly size dependent (chart E, Figure 11-8). Again, lithology does not appear to exert the major control on sphericity, since the frequency distributions for quartz pebbles closely mimic those derived from the total compositional spectrum, even in those samples where quartz clasts constitute a minor fraction of the entire assemblage (see Figure 11-6).

The Cailleux flatness indices for the various flysch rudite samples reveal broadly similar cumulative frequency distributions (chart A, Figure 11-9) and the means are also comparable (Table 11-3). The values are generally low, as indicated by the 85 percentile values, both for the total samples and for quartz pebbles, which consistently display indices less than the value of 2.1, regarded by Cailleux (1945, Tables 4 and 5) as critical for distinction of littoral from fluvial gravels (see Table 11-3).

Roundness values are generally more variable than the other form parameters for the samples studied here (chart B, Figure 11-9), and the means display a correspondingly increased range (Table 11-3). Although size clearly influences the mean roundness values (chart F, Figure 11-8), this function cannot account for the variable data in Table 11-3 since these relate to pebbles from one size class (32 mm to 64 mm). Some lithological control may be involved, but quartz pebbles yield roundness frequency curves and mean values that are closely comparable with the whole sample equivalents, thereby suggesting that the observed variability either may be inherited or produced by dynamic factors operating during transport and deposition of the gravels. The standard deviations demonstrate that roundness sorting varies from poor to good (adjusted limits of Folk 1968, p. 11) and differs in a statistically significant manner from sample to sample (Table 11-3), which tends to favor a process control, as suggested by Bluck (1969a) for beach gravels.

The roundness of intrabasinal clasts is illustrated by some of the Afton Water (S5) graded rudites in which peb-

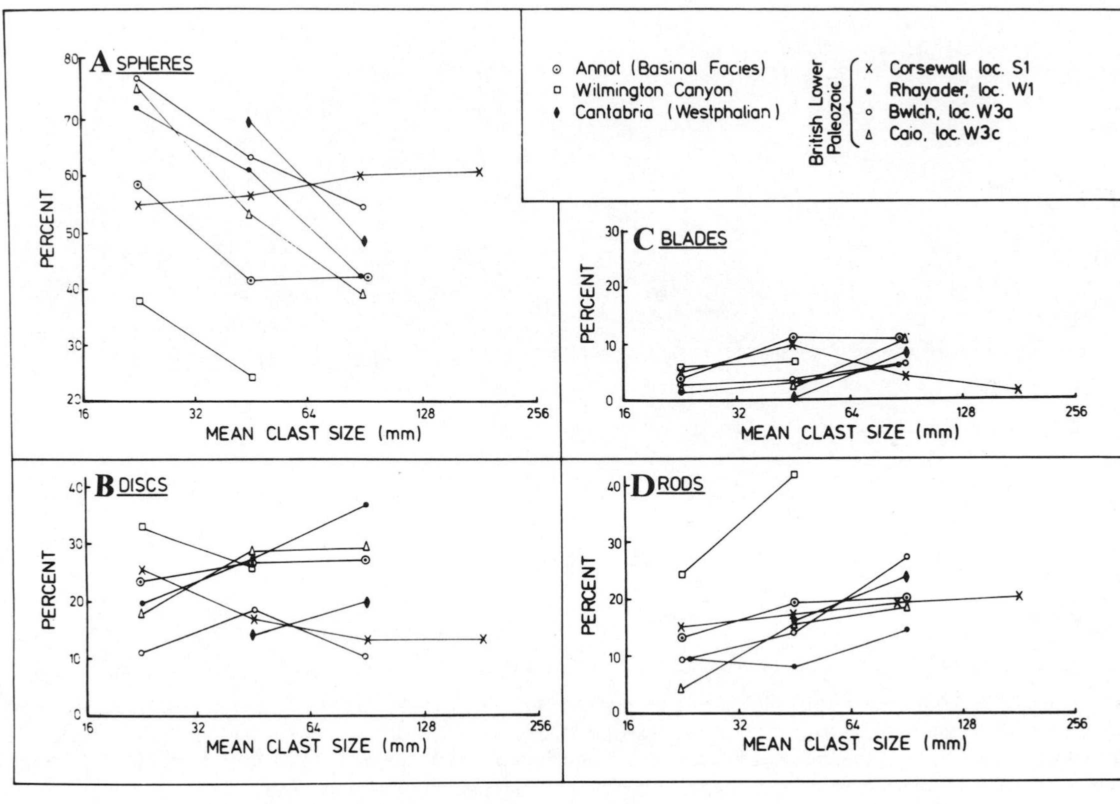

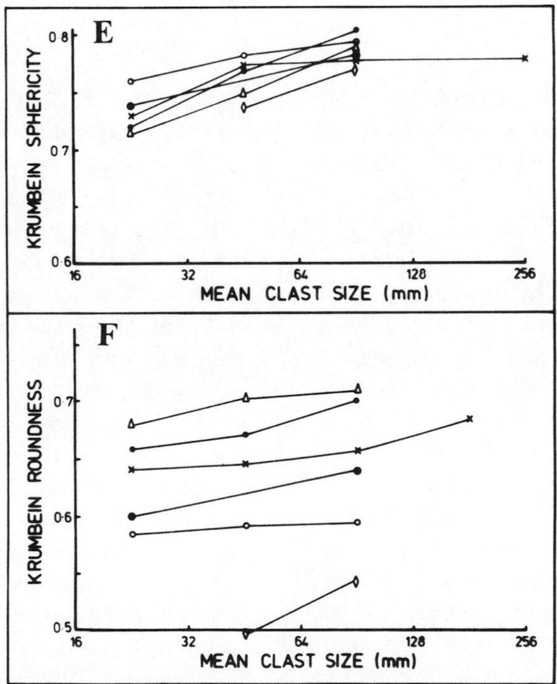

Figure 11-8. Relation between size and pebble shape. Note: Charts A through D show effects of size on the proportionate distribution of the four Zingg shape categories in pebble and cobble grades from the deep marine gravels described here. Chart E shows the effects of size on values of Krumbein sphericity. Chart F shows effects of size on values of Krumbein silhouette roundness in deep marine gravels described here.

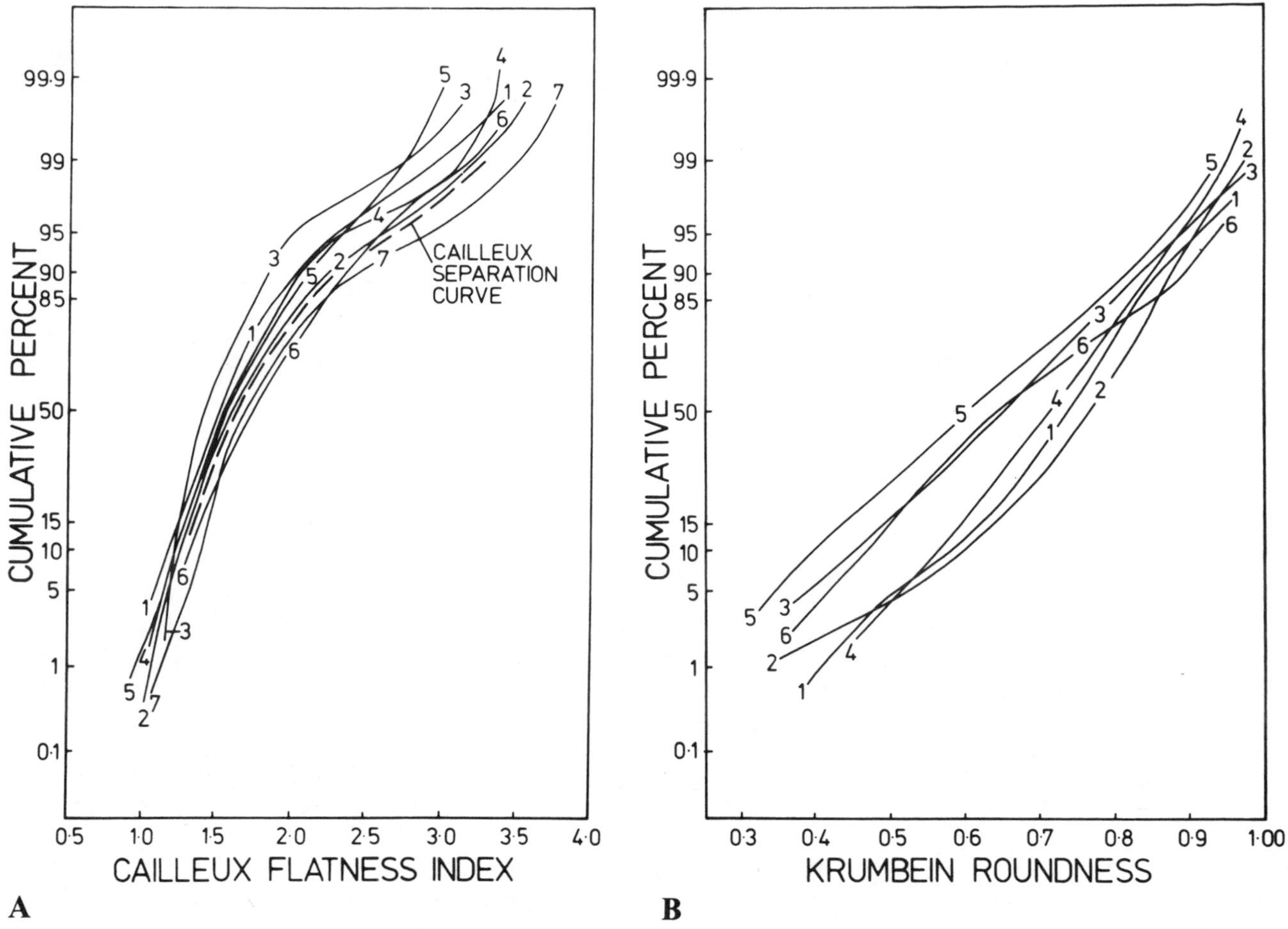

Figure 11-9. Cumulative curves of Cailleux flatness index (chart A) and Krumbein silhouette roundness values (chart B) for deep marine gravel samples. Note: 1 = Rhayader, central Wales (W1 complex); 2 = Bwlch, central Wales (W3a complex); 3 = Caio, central Wales (W3c complex); 4 = Milleur Point, southwest Scotland (S1 complex); 5 = Eocene Annot Sandstone (basinal facies), southern France; 6 = Eocene Annot Sandstone (neritic facies), southern France; and 7 = Holocene gravels, Wilmington and Baltimore submarine canyons. Dashed line in chart A is the curve suggested by Cailleux (1945) as separating "littoral" (to the right) from "fluvial" (to the left) gravels.

bles, cobbles, and even boulders of graywacke and lutite, clearly derived from contemporaneous erosion of canyon and channel walls, constitute a high proportion of the large clasts (Figure 11-10). Clasts of arenite are appreciably better rounded than those of lutite grade (photograph A, Figure 11-10), and this distinction is also displayed in several boulders composed of alternating graywacke and lutite layers (photograph B, Figure 11-10). Thus, the contrasting roundness of the two intrabasinal lithologies is presumably a function of the differing degrees of cohesion of these rock types and has been acquired during ephemeral transport from the local erosional site, thereby indicating that the transporting mechanism was capable of significant abrasion.

COMPOSITIONAL ATTRIBUTES OF RUDITE UNITS

The Lower Paleozoic flysch rudites considered here display substantial variations in clast composition. Some of these differences are regionally defined whereas others appear to be linked to the megascopic categories of rudite described earlier and thus may be related to process dynamics within individual canyon fan systems.

For convenience, two major groupings of clast rock types are recognized: extrabasinal and intrabasinal. Extrabasinal clasts include a wide variety of igneous, metamorphic, and (older, indurated) sedimentary rock types that originated from sources external to the depositional basin. Intrabasinal components are recognized mainly through their lithologic resemblance to rock types (most of which are sedimentary, but including occasional volcanic varieties) that are known to be broadly contemporaneous with the rudites. The intrabasinal clasts also differ from associated extrabasinal elements in their angular form and larger size. This intrabasinal grouping may be further subdivided into two classes: (a) allochthonous clasts, which comprise rock types (such as limestones, lutites, and well-sorted quartz

Table 11-3. *Shape Parameters*

Rudite Body	No. of Units Sampled	Unit Types Sampled	Mean Clast Size (CM)	No. of Clasts	All Clasts: Mean Krumbein Sphericity	All Clasts: Mean Eff. Settling Sphericity	All Clasts: Cailleux Flatness P_{50}	All Clasts: Cailleux Flatness P_{85}	All Clasts: Roundness[a] Mean	All Clasts: Roundness[a] S.D.	Quartz Only: Mean Krumbein Sphericity	Quartz Only: Mean Eff. Settling Sphericity	Quartz Only: Cailleux Flatness P_{50}	Quartz Only: Cailleux Flatness P_{85}	Quartz Only: Roundness[a] Mean	Quartz Only: Roundness[a] S.D.	Quartz Only: No. of Quartz Clasts
W1 (Rhayader)	2	1A	6.85	248	0.766	0.668	1.59	1.92	0.730	0.188	0.781	0.757	1.50	1.95	0.735	0.141	35
W3a (Bwlch)	1	1A	4.75	211	0.758	0.778	1.69	2.08	0.650	0.150	0.781	0.840	1.50	1.92	0.745	0.114	79
W3c (Caio)	1	2B	3.66	135	0.778	0.743	1.52	1.67	0.744	0.121	0.789	0.776	1.43	1.77	0.625	0.148	65
S2 (Corsewall)	3	1B+1A	7.85	154	0.762	0.792	1.63	1.96	0.704	0.118	–	–	–		–		–
Annot (Neritic)	2	–	6.88	165	0.741	0.701	1.78	2.25	0.685	0.156	–	–	–		–		–
Annot (Basinal)	4	1A+2A	8.42	380	0.762	0.785	1.64	2.00	0.610	0.186	–	–	–		–		–
Westphalian (Cantabria)	1	2B	6.30	50	–	–	–	–	–	–	0.782	0.812	1.58	1.86	0.530	0.155	50
Wilmington Baltimore (Recent)	5	–	4.20	880	0.721	0.695	1.85	2.25	0.585	0.124	0.772	0.713	1.81	2.15	0.617	0.130	449

[a]Determined from pebble grades only.

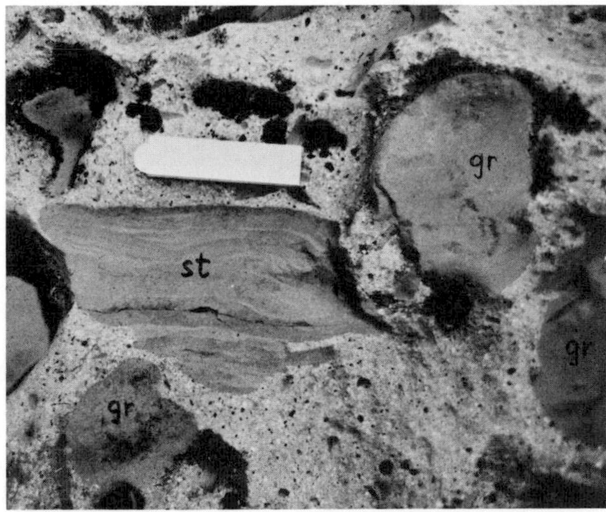

A

B

Figure 11-10. Photographs illustrating roundness of intrabasinal cobbles in mega-graded rudite units from the Afton Water (S5) complex, Scotland. Note: Photograph A shows moderately rounded clasts of graywacke (*gr*) that contrast with subangular clast of laminated siltstone (*st*). Scale is 15 cm long. (This section is from basal part of the same unit as in photograph B, Figure 11-3.) Photograph B shows large cobble in center that comprises two layers of graywacke (*gr*), with rounded periphery, separated by a more resistant laminated lutite (*st*) displaying angular boundaries. Scale is 15 cm long. (The section shown is from basal part of unit 20 in chart A, Figure 11-12.)

arenites) that are representative of marginal shelf environments, and (b) autochthonous clasts, which are lithologically and petrographically identical to sediments interbedded with the rudites. Graywacke arenites and dark gray to black lutites are typical of the latter class and these commonly display features indicative of their poorly consolidated condition during transportation (see Figure 8-10).

The *extrabasinal* clasts have been assigned to ten generalized categories of rock type (see Figure 11-6), and in aggregate the relative proportions of these differ on an essentially regional basis. Thus, there is a preponderance of acid extrusives, quartz arenites, and vein quartz in the Welsh rudites, while acid and basic intrusives and lavas dominate the Scottish clast assemblages. Moreover, the aggregate abundance of allochthonous intrabasinal clasts is conspicuously greater in the Welsh rocks.

When due account has been taken of these presumably provenance-determined regional differences, there remain additional variations in composition that appear within individual canyon fan systems and relate directly to sedimentologic attributes of the rudites. Such changes are most readily expressed in terms of the relative range of clast types (the degree of "polymicticity") and a compositional continuum can be recognized from Type 1A rudite sequences (highly polymict) through mixed Type 1B plus Type 2A sequences (less polymict, with increased vein quartz and intrabasinal components) to Type 2B rudites (dominated by allochthonous intrabasinal clasts with or without abundant vein quartz and quartzite pebbles/granules). The more "distal" part of this continuum is illustrated in Figure 11-11, which displays the lateral changes in *extrabasinal* clast constitution in the Type 2 rudites of the Drygarn complex (W2) of central Wales. Both the range and the aggregate abundance of nonsiliceous rock types decrease in the direction of transport, as indicated by independent current vector data.

The relationships between rudite categories and composition outlined above should be regarded as generalizations subject to important qualifications and exceptions. Thus, the S5 rudite sequence, which is dominated throughout by Type 1B units, exhibits a general and consistent upwards decrease in the proportion of extrabasinal clasts and a concomitant increase in the abundance of intrabasinal components (see chart A, Figure 11-12). Compositional changes of this type involve factors of both provenance and depositional dynamics (see later discussion, p. 157).

GENETIC IMPLICATIONS OF CLAST AND BED PROPERTIES

Aspects of flysch-associated rudites considered in this study may be categorized as follows: (a) geometry, dimensions, and facies relations of rudite units; (b) internal arrangement (fabric and structure); (c) textural attributes (size and shape of clasts); and (d) compositional attributes.

Each of these features reflects in varying degree the nature of the depositional environment, the sedimentation processes, and the provenance of the clasts. This section

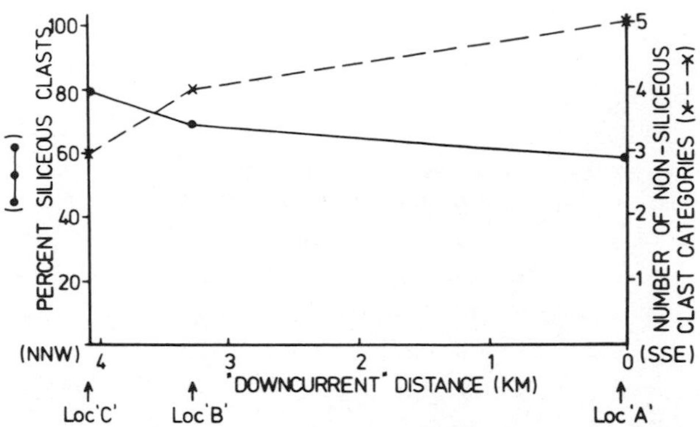

Figure 11-11. Change in overall composition of extrabasinal large clasts with "downcurrent" distance within the Drygarn rudite complex (W2) of central Wales. Note: Crosses joined by dashed lines represent the number of nonsiliceous compositional categories present in each channel sample through the complex at three localities. Dots joined by solid lines represent the percentage of siliceous clasts in each sample.

reviews the relative role of these factors in generating the features observed in these rudites.

With respect to external geometry and internal arrangement of the individual beds, few of the Lower Paleozoic rudite bodies display complete uniformity, and the sequences are best assessed in terms of the relative dominance of the various rudite Types and their recurring associations. Thus, Type 1 (channeled) units are most frequent in lenticular bodies that display abrupt lateral contacts with argillaceous or silty sediments. Moreover, Type 1 rudites typically possess disorganized (Type A) internal structure and are commonly framework supported. Conversely, Type 2 units are associated with wider, thinner rudite bodies, are typically organized (Type B) in character, and are matrix supported. In the context of existing canyon-fan models (e.g., Walker and Mutti 1973; Nelson and Kulm 1973; Walker 1975; Carter 1975; Kelling and Stanley 1976), these Type 1 and Type 2 associations probably represent the canyon-fan apex and suprafan regions, respectively.

The high statistical correlation that exists between bed thickness, maximum clast size, and maximum depth of basal erosion within most of the rudite bodies examined is a significant genetic feature. Most of the plots of bed thickness versus maximum clast size for individual bodies display the low slopes characteristic of mud flow or debris flow sediments (cf. Bluck 1969b, Figure 5), but the somewhat higher gradients for regression lines derived from bodies dominated by Type 2 units suggests that other (possibly tractional) processes may be important in the genesis of such rudites. The high gradient trend revealed by quartz pebble Type 2B rudites in the Scaur-Shinnel (S6) body indicates the operation of similar processes, and contrasts with the "mudflow-type" scatter plot derived from coarser, polymict Type 1 and Type 2 units in the same complex (see diagram E, Figure 11-4).

The uniqueness of each canyon fan system is well illustrated by a comparison of this Rhinns of Galloway sequence with the regression lines for the Caban Coch (W1) rudite body of central Wales (see diagram F, Figure 11-4), which, on the basis of geometry and facies relationships, was formed in a highly proximal situation (lower canyon or fan apex; Kelling and Woollands 1969). The logged section (chart B, Figure 11-12) demonstrates that this body is dominated by Type 1A rudite units that provide a very low gradient regression line (diagram F, Figure 11-4), that is substantially lower than the most "distal" of the Scottish bodies. However, the associated graded units (channeled and nonchanneled) that generally include smaller clasts, yield a steep regression line (diagram F, Figure 11-4) that is comparable to that of some stream-deposited fanglomerates (see Bluck 1969b, Figure 5; Bluck 1967, Figure 4). This evidence indicates a reduction in the relative competency of the formative mechanisms compared with the debris flows believed responsible for deposition of the disorganized rudites. Such a conclusion is compatible with the high values of competency claimed for both subaerial (Johnson 1970) and submarine (Hampton 1972; Middleton and Hampton 1976) debris flows. This Caban Coch sequence (together with the Corsewall body) includes several "bipartite" units, which are very similar to deposits formed by recent (subaerial) mudflows (Fahnestock 1963; Sharp and Nobles 1953). The lower, rudite portion of such units may be attributed to a debris flow origin while the upper, sandy, member represents suspension fallout from the superjacent turbulent layer (see Kelling and Stanley 1976, p. 414).

Bed thickness also displays marked statistical correlation with the depth of basal erosion within individual (channeled) rudite bodies and the regression lines exhibit a comparatively small range of gradients (diagram G, Figure 11-4). This evidence demonstrates that the channels were not preexisting features passively filled by gravels, but that the transporting agency was actively responsible for the deep basal scouring. Moreover, flows of higher capacity apparently possessed enhanced erosive capabilities. The composite diagram (diagram G, Figure 11-4) indicates that there is no overall distinction on this basis between organized and disorganized rudite units. However, in the Caban Coch (W1) sequence, the graded and nongraded (disorganized) units maintain their distinctive trends, and the processes responsible for the graded rudites apparently possessed relatively lower erosive energies.

The generally wider size range, increased mean clast size, and poorer sorting of channeled (and disorganized) rudites are presumably further functions of debris flow dynamics. Such units are commonly framework supported, which

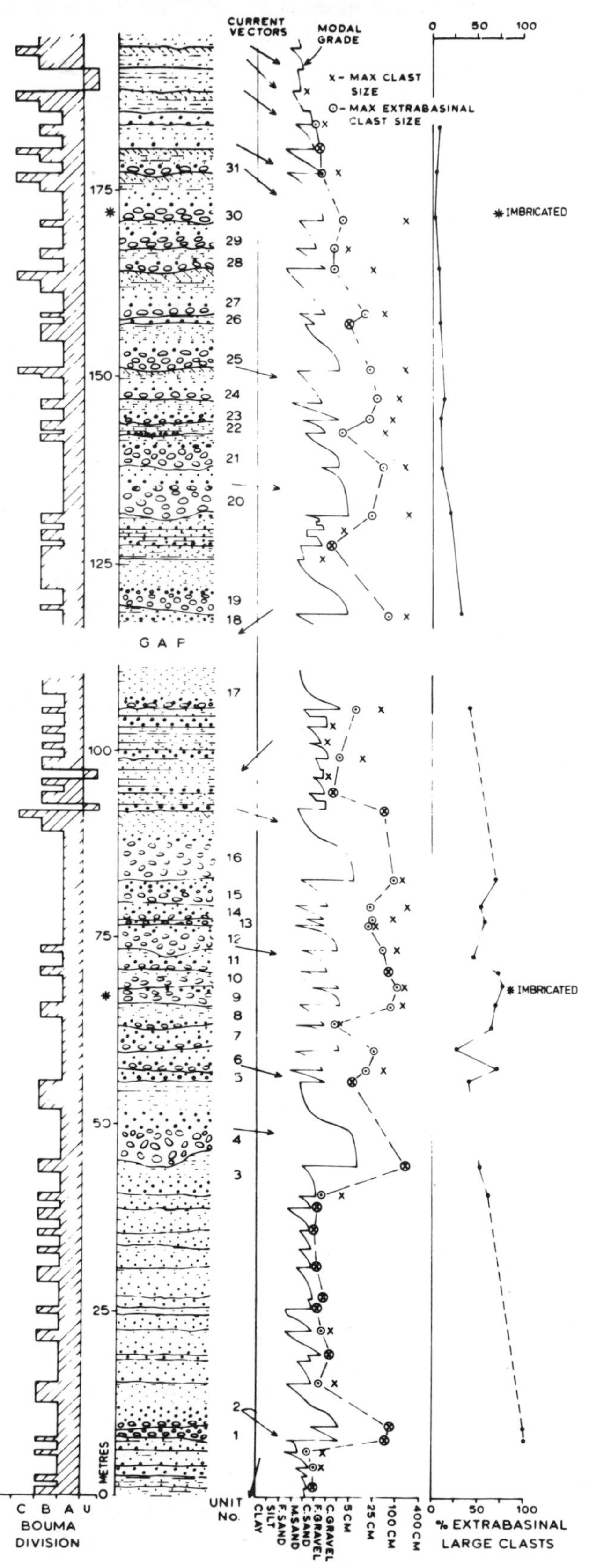

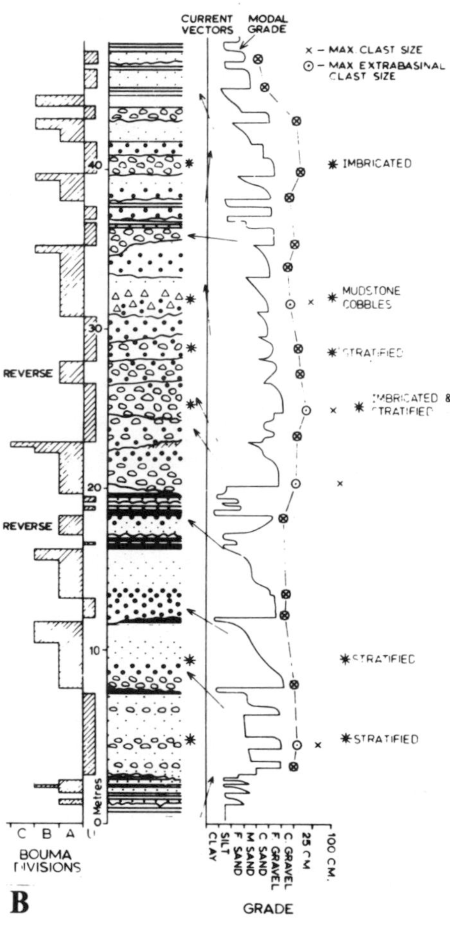

Figure 11-12. Sequence logs of the Afton Water rudite complex (S5), southern Scotland, and the Caban Coch quarry, Rhayader, central Wales. Note: In the Bouma Division column of both logs, "U" represents ungraded units. Chart A illustrates vertical variations in lithology, grade, and composition of exposures in the Afton Water immediately north of the reservoir. Type 1B units are dominant. Chart B illustrates verical variations in lithology and grade within the lower rudite body of the W1 complex. Type 1A and 1B units are dominant.

suggests that substantial winnowing and removal of finer grades was achieved during or shortly after deposition. Conversely, many of the graded and stratified rudites display a dispersed texture, large clasts being supported in a moderately sorted matrix, the modal grade of which varies in different units from granules to fine silt, thereby suggesting that deposition of these organized rudites occurred by some process akin to "freezing" out of suspension. Nevertheless, the pebbles and larger clasts in such units display enhanced sorting and a negative skew (see Table 11-2) that is probably the result of traction and "lagging." We thus suggest that a carpet of rolling pebbles and cobbles may have formed at the base of the more rapidly moving suspension of finer sediment.

The bimodality observed in some units almost invariably reflects the content of large, intrabasinal (wall rock) clasts composed of poorly cohesive sediment. Despite their larger mean size, these clasts may have been in hydraulic equilibrium with the extrabasinal material in view of the lower effective density of the water-saturated intrabasinal boulders.

Within individual bodies, the vertical variations in maximum clast size are generally more significant than changes in the modal grade. The sequences in the Afton Water (S5 body) and Caban Coch quarry (W1 body) are instructive in this respect (charts A and B, Figure 11-12). In the former, the upwards declining proportion of extrabasinal clasts is matched by a decrease in the maximum size of this component, but other parameters, such as the maximum size of intrabasinal clasts, the thickness of the rudite units, and the maximum depth of basal erosion, remain broadly uniform through most of the vertical succession and decrease rapidly only in the uppermost few meters (chart A, Figure 11-12). These relationships suggest that while this canyon fan system gradually became starved of coarse extrabasinal detritus, the resedimenting mechanisms responsible for the downslope transfer of sediment from the headward region maintained essentially uniform character and energy levels throughout the time interval required to accumulate nearly 150 meters of fill sediment. In contrast, the Caban Coch quarry sequence (chart B, Figure 11-12) is characterized by a high and relatively constant proportion of extrabasinal clasts. However, the maximum clast size increases towards the central part of the succession and then declines gradually in the upper part of the body. Moreover, the frequency, thickness, and internal organization of the rudite units and the depth of basal erosion reveal parallel changes. If similar changes recorded in mud flows on alluvial fans (Bull 1964, pp. 23-24) are utilized, the variations in the Caban Coch sequence are ascribed to changes in position on the submarine fan, with the central portion of this succession representing the most proximal part of the system.

The limited data presented here do not permit a clear distinction to be made between the various rudite types on the basis of shape criteria, but there is some indication that both in general and within individual systems the Type 2B units carry proportionately more spherical and fewer discoidal clasts than Type 1A rudites. Such changes are at least partly attributable to the decrease in mean size and increase in siliceous clast content in the Type 2B units.

As has been suggested elsewhere (Ricci Lucchi 1969), shape populations comparable to those described here are "fluvial" in character and, where found in deepsea gravels, may denote direct derivation from a river, without the intervention of a littoral or shallow marine phase of sedimentation. However, the preceding arguments indicate that the closely similar character of the shape populations in the flysch rudites is best attributed to a transporting mechanism capable of selective shape sorting, rather than considered an inherited feature or a function of source lithology. This point is reinforced when shape data from the Annot Sandstone (Eocene) and from the Holocene-to-modern gravels of Wilmington and Baltimore canyon heads are considered (Table 11-3; Figures 11-7 through 11-9). Pebble samples were obtained from both the "neritic" and the "basinal" facies of the Annot Formation (Stanley 1961, 1975; Stanley and Bouma 1964), and the shape data obtained from the submarine canyon and fan sequences of the Annot basin accord with those from older flysch rudites. However, the shallow marine facies yields a pebble assemblage that is rich in discs, with a broader spread of sphericity values and a significantly lower mean sphericity, together with substantially higher Cailleux flatness indices and slightly enhanced mean roundness value (see Figures 11-7 and 11-9; Table 11-3). These results clearly indicate a strong littoral influence in the formation of the neritic gravels (cf. Cailleux 1945; Dobkins and Folk 1970; Bluck 1967). Although bypassing of the shelf and direct fluvial input to form the deep marine rudites cannot be excluded (see Stanley 1974a, p. 354), these basinal gravels more probably have been introduced by resedimentation of shelf deposits, in which case the shape sorting results from mechanisms operating during downslope transportation.

The Holocene-to-modern canyon head and upper slope gravels display a composite shape population in which discs are almost as abundant as equant clasts, while prolate (rod-like) forms also bulk large (Figure 11-7). The mean Krumbein sphericity value is lower than that for the ancient flysch rudites (Table 11-3), while the Cailleux flatness indices are consistently higher and are comparable with those recorded from the Annot neritic facies (Table 11-3; chart A, Figure 11-9). However, roundness values are relatively low (Table 11-3). These clasts are reworked from

Plio-Pleistocene deltaic gravels deposited as the present shelf edge during sea level lowstands (Kelling and Stanley 1970; Stanley and Kelling 1968; Stanley 1974b), and this origin doubtless accounts for their low roundness. However, the other shape attributes clearly attest to significant littoral influence, presumably during interglacial and postglacial elevations of sea level, and suggest that the slumping and debris flow mechanisms that have emplaced these gravels in the deeper parts of the canyon heads have generated a shape population that is intermediate in character between the ancient "neritic" (cf. Annot samples, chart A, Figure 11-7) and the "basinal" or deep submarine fan gravels (chart B, Figure 11-7).

The absolute abundance of the various extrabasinal rock types in the rudites is ultimately dependent on the nature of the source rocks and thus is controlled essentially by geotectonic factors. However, as discussed earlier, there exists within individual canyon fan systems a relationship between the megascopic rudite category and the relative degree of polygenetic character displayed by the large clast population. Within individual systems the strongly channeled, disorganized Type 1A rudites exhibit the greatest variety of clast rock types, whereas many of the widespread sheets of Type 2B pebble rudite are dominated by vein quartz clasts. Thus, if arguments offered above are accepted, this feature affords a further crude measure of relative proximality within any given system. This effect is attributed to compositional sorting during transportation through the canyon and across the fan and is presumably related to the density, size, and shape characteristics of the extrabasinal components. Thus, the relative abundance of vein quartz in the finer and more distal Type 2 rudites is related both to the high proportion of this component in the pebble size grade in most of the sampled rudites and to the enhanced numbers of readily transported shapes (spheres) present in the vein quartz population as a whole (see Figure 11-6). These observations are also consistent with the conclusions outlined earlier concerning the influence of traction and rolling during deposition of the more distal, graded gravels.

Two categories of intrabasinal constituent have been distinguished (allochthonous or shelf derived, and autochthonous or subjacent). The character and varying porportions of these components within the deposits of individual fan systems provide further evidence relative to the genesis of the rudites. Units, such as those in the complexes of central Wales that carry high amounts of allochthonous intrabasinal material, such as neritic limestone clasts, probably record substantial submarine erosion of neighboring shallower regions and imply either the existence of a relatively wide shelf or a prolonged residence time for the gravels prior to downslope transfer. On the other hand, the distal decrease in the proportion of autochthonous intraclasts is a direct reflection of the declining erosive capability and competency of the flows.

CONCLUSIONS

The observations outlined above enable a number of conclusions to be advanced that are considered to be of general application.

First, rudite units ascribed to basin-margin canyon-fan complexes may be described in terms of external geometry as well as internal organization, thereby enabling the distinction of four intergrading megascopic categories (Type 1A: channeled, disorganized; Type 1B: channeled, organized; Type 2A: nonchanneled, disorganized; Type 2B: nonchanneled, organized). Sequences dominated by Type 1A rudite units tend to occur in the most proximal regions of the canyon-fan complex, while more distal portions are characterized by Type 2B rudites (cf. Walker 1975, Figure 10; Walker 1976, Figures 10 and 14). The relative location of the Type 1B and 2A units within individual systems is more equivocal, but they appear to occupy intervening positions.

The scalar and textural attributes of the rudites vary greatly in absolute terms, and regional factors exert an overriding influence. However, within each region the individual rudite bodies display consistent relationships between several of the dimensional and textural parameters; these relationships may be linked, albeit crudely, with the megascopic categories of rudite unit defined above.

Significant scalar relationships are unit thickness versus maximum clast size, which generally provides low-gradient regression lines in bodies dominated by Type 1 rudite beds, thus substantiating field evidence for the operation of debris flow mechanisms in these units. Plots of unit thickness versus maximum depth of basal erosion for individual bodies yield a set of regression lines displaying a narrow range of gradients, which is indicative of a fine balance between the capacity and the erosive capabilities of the flows responsible for deposition of both organized and disorganized rudites.

Texturally, within a given region, the Type 1 and Type A rudites are generally coarser, exhibit a wider size range, poorer sorting, and more positive skew than the associated Type 2 and Type B units. Moreover, vertical trends in size parameters may be detected in sequences consisting of one megascopic rudite category and may then provide a sensitive measure of increasing or decreasing proximality.

The shape populations for extrabasinal large clasts are generally of "fluvial" character, dominated by spherical and prolate forms, while the values of the Krumbein and Folk sphericity measures and the Cailleux flatness index also fall in the "fluvial" field. Although both mean size and clast lithology exert a considerable influence on the nature of the shape assemblage, shape sorting induced by the transport mechanism appears to be the most significant factor in determining the character of the population. There is some evidence that within individual canyon-fan systems, there is a general distal decrease in the proportion of discs and a corresponding enhancement in spherical shapes, whereas prolate forms achieve maximum abundance in an intermediate position in the system. While the increased proportion of spheres in more distal rudites might be a function of declining size, as indicated by chart A, Figure 11-8, the reciprocal decrease in the percentage of discs cannot be accounted for in this manner in view of the more equivocal role of the size factor in determining the relative abundance of this shape category (see chart B, Figure 11-8). Such changes are more crudely detectable in the shape attributes of the megascopic categories of rudite. Specifically, Type 1A rudites contain proportionately higher numbers of discoidal and prolate clasts, while Type 2B conglomerates are richer in spherical forms. These relationships accord with the size data in implying the more pronounced role of tractional processes in the organized rudites. The possibility of the observed shape populations being inherited directly from a fluvial source, as suggested by some authors (Ricci Lucchi 1969; Marschalko 1973), is not sustained by these observations, and supplementary data from the Annot Sandstone (Eocene) and from modern canyon head gravels also support the concept of within-system shape sorting.

The Lower Paleozoic rudites described here are compositionally diverse, and detailed differences in aggregate clast composition are functions of the ultimate or immediate sources. However, three broad categories of component have been recognized: extrabasinal, allochthonous intrabasinal, and autochthonous intrabasinal. With respect to the extrabasinal elements, within a given system the degree of polygenetic character decreases distally while the relative importance of siliceous (and especially vein quartz) clasts increases. Such changes are related to the size and shape of different types of clast and are attributable to the effects of sorting by the emplacement mechanisms. The proportion of intrabasinal clasts is related in some degree to the erosive capabilities of the gravel-bearing flows and more conspicuously so in the case of rudites that are rich in large autochthonous clasts. A consistently high content of allochthonous intraclasts (e.g., calcareous arenites or limestones) may indicate the existence of a laterally adjacent shelf region of some width contributing clasts through submarine erosion, or it may indicate prolonged residence time of the extrabasinal constituents on the marginal platform.

ACKNOWLEDGMENTS

The authors gratefully acknowledge the generosity of Daniel J. Stanley, who supplied shape data from the Annot Sandstone, and C. Michael Wear and Harrison Sheng, who assisted in compilation of data from the Holocene gravels from Wilmington and Baltimore submarine canyons. Financial assistance was provided by the United Kingdom Natural Environment Research Council Research Grant GT3/899 and by the Research Fund of the University College of Swansea.

REFERENCES

Blatt, H., Middleton, G. V., and Murray, R. 1972. *Origin of Sedimentary Rocks.* Prentice-Hall, Englewood Cliffs, N. J., 634 pp.

Bluck, B. J. 1967. Sedimentation of beach gravels: Examples from South Wales. *J. Sed. Petrol.,* 37: 128–56.

———, 1969a. Particle rounding in beach gravels. *Geol. Mag.,* 106: 1-14.

———, 1969b. Old Red Sandstone and other Paleozoic conglomerates of Scotland. In: M. Kay (ed.), *North Atlantic – Geology and Continental Drift.* Amer. Assoc. Petrol. Geol., Mem. 12, pp. 711-23.

Bull, W. B., 1964. Alluvial fans and near surface subsidence in Western Fresno County, California. *U.S. Geol. Surv. Prof. Paper* 437-A, 71 pp.

Cailleux, A., 1945. Distinction des galets marins et fluviatiles. *Bull. Soc. Géol. France,* ser. 5, 15: 375–404.

Carter, R. M., 1975. A discussion and classification of subaqueous mass-transport with particular application to grain-flow, slurry-flow and fluxoturbidites. *Earth Sci. Rev.,* 11: 145-77.

Davies, I. C., and Walker, R. G., 1974. Transport and deposition of resedimented conglomerates: The Cap Enragé Formation, Cambro-Ordovician, Gaspé, Quebec. *J. Sed. Petrol.,* 44: 1200-16.

Dewey, J. E., 1974. The geology of the southern termination of the Caledonides. In: A.E.M. Nairn and F. Stehli (eds.), *The Ocean Basins and Margins, II: The North Atlantic.* Plenum Press, New York, pp. 205-31.

Dobkins, J. E., Jr., and Folk, R. L., 1970. Shape development on Tahiti-Nui. *J. Sed. Petrol.,* 40: 1167-203.

Embley, R. W., 1976. New evidence for occurrence of debris flow deposits in the deep sea. *Geology,* 4: 371-74.

Fahnestock, R. K., 1963. Morphology and hydrology of a glacial stream – White River, Mt. Rainer, Washington. *U.S. Geol. Surv. Prof. Paper* 422-A, pp. 1-70.

Folk, R. L., 1968. *Petrology of Sedimentary Rocks.* Hemphills, Austin, 170 pp.

Hampton, M. A., 1972. The role of subaqueous debris flow in generating turbidity currents. *J. Sed. Petrol.,* 42: 775-93.

Johnson, A. M., 1970. *Physical Processes in Geology.* Freeman, San Francisco, 577 pp.

Kelling, G., 1964. The turbidite concept in Britain. In: A. H. Bouma and A. Brouwer (eds.), *Turbidites.* Developments in Sedimentology, 3. Elsevier, Amsterdam, pp. 75-92.

———, and Stanley, D. J., 1970. The morphology and structure of Wilmington and Baltimore submarine canyons, eastern U.S.A. *J. Geol.,* 78: 637-60.

———, and Stanley, D. J., 1976. Sedimentation in canyon, slope and base-of-slope environments. In: D. J. Stanley and D. J. P. Swift (eds.), *Marine Sediment Transport and Environmental Management.* Wiley, New York, pp. 379–435.

———, and Woollands, M. A., 1969. The stratigraphy and sedimentation of the Llandoverian rocks of the Rhayader district.

In: A. Wood (ed.), *The Pre-Cambrian and Lower Palaeozoic Rocks of Wales.* University of Wales Press, Cardiff, pp. 255-82.

Krumbein, W. C., 1941. Measurement and geological significance of shape and roundness of sedimentary particles. *J. Sed. Petrol.,* 11: 64-72.

Marschalko, R., 1973. Boulder beds in flysch and reconstruction of their transport mechanism on the basis of sedimentary structures. *Proc. X Cong. Carpathian-Balkan Geol. Assoc.,* Section II, Sedimentology, pp. 113-22.

McKerrow, W. S., and Ziegler, A. M., 1972. Silurian paleogeographic development of the Proto-Atlantic Ocean. *Proc. 24 ème Cong. Géol. Internat.,* Section 6, Montreal, pp. 4-10.

Middleton, G. V., 1970. Experimental studies related to problems of flysch sedimentation. In: J. Lajoie (ed.), *Flysch Sedimentology in North America.* Geol. Assoc. Canada, Spec. Paper 7, pp. 253-72.

——, and Hampton, M. A., 1973. Sediment gravity flows: Mechanics of flow and deposition. In: G. V. Middleton and A. H. Bouma (eds.), *Turbidites and Deep Water Sedimentation.* Soc. Econ. Paleont. Mineral., Pacific Section, Short Course, Anaheim, pp. 1-38.

——, and Hampton, M. A., 1976. Subaqueous sediment transport and deposition by sediment gravity flows. In: D. J. Stanley and D. J.P. Swift (eds.), *Marine Sediment Transport and Environmental Management.* Wiley, New York, pp. 197-218.

Mutti, E., and Ricci Lucchi, F., 1972. Le torbiditi dell' Appenino settentrionale. introduzione all'analisi di facies. *Mem. Soc. Geol. Italiano,* 11: 161-99.

Nelson, C. H., and Kulm, L. D., 1973. Submarine fans and deep-sea channels. In: G. V. Middleton and A. H. Bouma (eds.), *Turbidites and Deep Water Sedimentation.* Soc. Econ. Paleont. Mineral., Pacific Section, Short-Course, Anaheim, pp. 38-78.

Normark, W. R., 1970. Growth patterns of deep sea fans. *Amer. Assoc. Petrol. Geol. Bull.,* 54: 2170-95.

——, 1974. Submarine canyons and fan valleys: Factors affecting growth patterns of deep-sea fans. In: R. H. Dott, Jr., and R. H. Shaver (eds.), *Modern and Ancient Geosynclinal Sedimentation.* Soc. Econ. Paleont. Mineral., Sp. Publ. 19, pp. 56-68.

Ricci Lucchi, F., 1969. Composizione e morfometria di un conglomerato risedimentato nel Flysch miocenico romagnolo (Fontanelice, Bologna). *Giorn. Geol.,* Ser. 2, 36: 1-47.

Rodine, J. D., and Johnson, A. M., 1976. The ability of debris, heavily freighted with coarse clastic materials, to flow on gentle slopes. *Sediment.,* 23: 213-34.

Sharp, R. P., and Nobles, L. M., 1953. Mudflow of 1941 at Wrightwood, southern California. *Geol. Soc. Amer. Bull.,* 64: 547-60.

Sneed, E. D., and Folk, R. L., 1958. Pebbles in the Lower Colorado River, Texas: A study in particle morphogenesis. *J. Geol.,* 66: 114-50.

Stanley, D. J., 1961. *Etudes sédimentologiques des Grès d'Annot et leurs équivalents latéraux.* Editions Technip, Paris, 158 pp.

——, 1974a. Dish structure and sand flow in ancient submarine valleys, French Maritime Alps. *Bull. Centre Rech. Pau – SNPA,* 8: 351-71.

——, 1974b. Pebbly mud transport in the head of Wilmington Canyon. *Mar. Geol.,* 16: M1-M8.

——, 1975. Submarine canyon and slope sedimentation (Grès d'Annot) in the French Maritime Alps. *Excursion Guides IX Cong. Internat. Sedim.,* Nice, 129 pp.

——, and Bouma, A. H., 1964. Methodology and paleogeographic interpretation of flysch formations: A summary of studies in the Maritime Alps. In: A. H. Bouma and A. Brouwer (eds.), *Turbidites.* Developments in Sedimentology, 3. Elsevier, Amsterdam, pp. 34-64.

——, and Kelling, G., 1968. Sedimentation patterns in the Wilmington submarine canyon area. *Ocean Sciences and Engineering of the Atlantic Shelf.* Mar. Technol. Soc., Philadelphia, pp. 127-42.

——, and Unrug, R., 1972. Submarine channel deposits, fluxoturbidites and other indicators of slope and base-of-slope environments in modern and ancient marine basins. In: J. K. Rigby and W. K. Hamblin (eds.), *Recognition of Ancient Sedimentary Environments.* Soc. Econ. Paleont. Mineral., Sp. Publ. 16, pp. 287-340.

Van Loon, A. J., 1970. Grading of matrix and pebble characteristics in syntectonic pebbly mudstones and associated conglomerates with examples from the Carboniferous of northern Spain. *Geol. Mijnbouw,* 49: 41-56.

Wadell, H., 1932. Volume, shape and roundness of rock-particles. *J. Geol.,* 40: 443-51.

Walker, R. G., 1970. Review of the geometry and facies organization of turbidites and turbidite-bearing basins. In: J. Lajoie (ed.), *Flysch Sedimentology in North America.* Geol. Assoc. Canada, Spec. Paper 7, pp. 219-52.

——1975. Generalized facies models for resedimented conglomerates of turbidite association. *Geol. Soc. Amer. Bull.,* 86: 737-48.

——, 1976. Facies models 2: Turbidites and associated coarse clastic deposits. *Geoscience Canada,* 3: 25-36.

——, and Mutti, E., 1973. Turbidite facies and facies associations. In: G. V. Middleton and A. H. Bouma (eds.), *Turbidites and Deep Water Sedimentation,* Soc. Econ. Paleont. Mineral., Pacific Section, Short Course, Anaheim, pp. 119-57.

Walton, E. K., 1965. Lower Palaeozoic Rocks. In: G. Y. Craig (ed.), *The Geology of Scotland.* Oliver and Boyd, Edinburgh, pp. 161-229.

Zingg, T., 1935. Beiträge zur Schotteranalyse. *Schweiz min. pet. Mitt.,* 15: 39-140.

Part III

Turbidite and Hemipelagic Processes in Submarine Fans

Chapter 12

Turbidite Muds and Silts on Deepsea Fans and Abyssal Plains

DAVID J.W. PIPER

Departments of Geology and Oceanography
Dalhousie University
Halifax, Nova Scotia, Canada

ABSTRACT

Silts and muds deposited by turbidity currents are a major component of deepsea terrigenous sediment sequences. Mud turbidites have three divisions that are analogous to Bouma's divisions for sand turbidites: laminated mud, graded mud, and ungraded mud. Disorganized laminated muds are most common on levees and graded muds in deepsea fan valleys; ungraded muds predominate in distal environments. Floc breakup and reformation in a turbidity current results in a progressive segregation of silt and clay, so that sorted silt beds are characteristic of distal environments. Three types of turbidite silt are distinguished. The deposits of contour currents and lutite flows may often resemble mud turbidites, and there is no simple set of criteria to distinguish them.

INTRODUCTION

Deepwater continental margins are one of the most important sites of sediment accumulation. Turbidity currents are the principal mechanism of transporting sediment to the deep continental margins. The transport of sand by turbidity currents is the subject of a large body of literature; the transport of finer sediment by turbidity currents has been neglected in comparison.

Many authors (see reviews by Kuenen 1964; Hesse 1975) have demonstrated conclusively that fine sediment is transported and deposited by turbidity currents. This fine sediment may overlie coarse material deposited by the same turbidity current or in distal environments may rest directly on nonturbidite sediment.

This review deals with terrigenous mud turbidites; fine-grained carbonate turbidites have been recently reviewed by Hesse (1975). Carbonate turbidites are probably only important on low-latitude continental margins receiving little terrigenous sediment input.

The volumetric importance of turbidite muds to the world sediment accumulation budget can be crudely estimated in a number of ways. The data in Table 12-1 (when considered with information on petrography and rates of sedimentation) suggest at least half of the mud delivered to the deep ocean beyond the continental slope is carried by turbidity currents. Seismic reflection profiles and drill sites on the outer continental margin (e.g., Moore 1969; Kulm et al. 1973) suggests muds are two to ten times more important than sands. As is shown below, much of this mud is of turbidite origin. The shales and schists that make up ancient mountain belts are probably principally turbidites.

Many diverse mechanisms have been proposed for the transport of fine sediment into deep water. Many of the processes overlap with one another conceptually. They include settling of pelagic material, deposition from nepheloid layers, deposition from (nonturbidity) bottom (or "contour") currents, and deposition from low-density, low-velocity turbidity currents (lutite flows). All these processes result in slow rates of deposition that rarely exceed 0.1 mm/year. Those that are episodic result in beds less than 10 mm thick.

Muds deposited from large turbidity currents accumulate rapidly but episodically and form beds 10 to 10^3 mm thick in periods ranging perhaps from hours to months. The twin characteristics of thickness of beds and speed of deposition allow *large* turbidity current deposits to be distinguished from those of other processes.

This chapter is concerned only with the deposition of such thick mud and silt beds (say more than 10 mm thick). Thinner beds are not considered. An understanding of thicker beds may eventually help us to interpret thinner ones.

Table 12-1. *Some Examples of Significance of Fine-Grained Turbidites in Deepsea Environments*

Location and reference	*Age*	*% Hemipelagic Sediment*	*% Turbidite Sediment* *Sand*	*Silt*	*Mud*
Wilkes Abyssal Plain (Piper and Brisco 1975)	Miocene	18	7	15	60
Balearic Abyssal Plain (Rupke and Stanley 1974)	Late Quaternary	35	12		53
Navy Fan (Normark and Piper 1972)	Holocene				
distal		38	0	2	60
suprafan		53	25	2	20
levees on upper fan		98	1	1	0
Zementmergel (Hesse 1975)	Cretaceous	10	42	–	58

Evidence of Turbidity Current Origin

Sedimentary structures such as the restriction of bioturbation to the tops of beds and escape burrows may be used as evidence of rapid deposition. Petrologic characteristics, including microfossils, high organic carbon, or preservation of carbonate below the CCD, may indicate either a particular source area or rapid deposition and preservation by autoburial. Sequence analysis or petrologic comparison may relate muds to coarser recognizable turbidites. However, general criteria for the recognition of mud turbidites are difficult to establish; each case needs individual consideration.

The differences between large turbidity currents and some of the other processes mentioned may not be clear-cut. The tail of a turbidity current, from which mud is deposited, may be deflected by bottom currents. Turbidity currents may feed nepheloid layers. There may be a continuum between lutite flows and larger turbidity currents. Nevertheless, a class of fine-grained sediments can be recognized that have been deposited from large turbidity currents.

Determining whether a bed is the deposit of but a single turbidity current or whether it was laid down by a number of depositional events (which themselves may produce quite rapid sedimentation, such as small turbidity currents or lutite flows) is particularly difficult. There are several criteria that can be used to suggest deposition by a single event:

1. Is there a regular sequence to the bed (e.g., a partial or complete Bouma sequence), which because of its regularity is difficult to account for by a series of separate depositional events?
2. Are there criteria of rapid deposition present?
3. Is the thickness and extent of a bed consistent with the distance over which it has been transported. If rare beds of apparent turbidite sand or mud are found in a distal environment, it is a reasonable assumption that each bed could result from several depositional events that succeeded in reaching this distal location? (In proximal environments, episodic multiple depositional events other than turbidity currents can result in mud beds superficially resembling mud turbidites, as shown by Fleischer [1972] for Santa Barbara Basin.)

Nomenclature

In Bouma's (1962) structural scheme for turbidites, all fine material was classed as *e* division. Kuenen (1964) proposed distinction of e_t and e_p divisions (turbidite and pelagic). Van der Lingen (1969) and Hesse (1975) used e and f divisions. This latter usage is followed here (Figure 12-1).

In this chapter, Shepard's (1954) grain size nomenclature is used. In addition, the term *mud* is used for cohesive silt-clay sediment mixtures (thus including silty clay, clayey silt, and clay). *Silt* is used for granular sediment with >75 percent silt-sized particles.

In lithified sediments, examined in standard thin section, 30 μ thick, grains finer than 6 ϕ are difficult to distinguish. If studies of modern sediments are to be applied to ancient rocks, distinguishing coarse and medium silt grains from fine and very fine silt is important.

TURBIDITE SILTS

In proximal turbidite environments, muds directly overlie sands or very coarse silts. However, in more distal

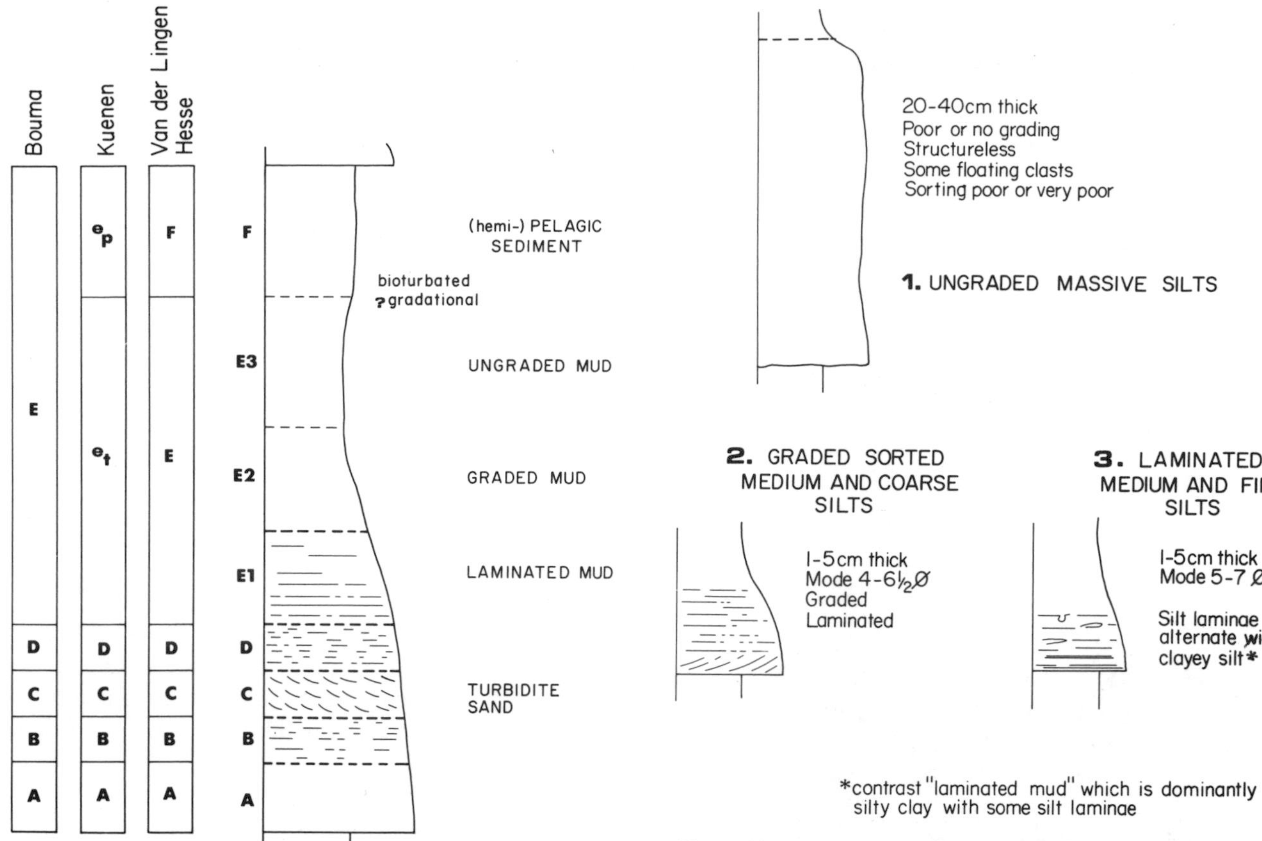

Figure 12-1. Models of the vertical sequence of sedimentary structures in turbidite sands and muds.

Figure 12-2. Three types of turbidite silt. (Source: Based on Piper 1973; Piper and Brisco 1975; Jipa and Kidd 1974; Horn et al. 1971.)

environments, distinct silt beds are common. Authors such as Fruth (1965), Horn et al. (1971), and Normark and Piper (1972) have demonstrated a change from sand to silt beds moving distally from deepsea fans to abyssal plains.

There have been few studies of the detailed sedimentology of turbidite silts. The tentative classification into three types developed below (Figure 12-2) is based on studies of partly lithified abyssal plain sediments in DSDP holes. All three types of silt beds have sharp and sometimes irregularly erosional bases.

Ungraded massive silts. These tend to be rather poorly sorted and occur in thick beds, often around 40 cm thick. They are generally structureless, and some contain "floating" clasts. They thus show resemblances to *a* division turbidite sands. On some abyssal plains, this lithology occurs only with modal grain size in the sand range (Horn et al. 1971).

Graded sorted medium and coarse silts. These are similar to very fine sands. They typically occur in beds 1 cm to 5 cm thick. Most are horizontally laminated: some show a Bouma *bcd* or *cd* sequence. Grading is usually clearly developed. Modal grain sizes may be as fine as 10 μ. Sorting is better than in type 1 silts.

Laminated medium and fine silts. These occur in beds several centimeters thick and consist of medium and fine silt laminae alternating with laminae of clayey silt. Individual laminae are 0.2 mm to 2 mm thick. They differ from graded sorted silts in not having any regular grading, but rather an alternation of laminae of differing grain sizes. Horizontal lamination is prominent, sometimes with load casting and other syn-sedimentary deformational structures. Small lenses of silt may be related to loading or ripple formation.

There is a widespread tendency for irregular lenticular bedding in turbidite silt beds, and this feature is much more common than in turbidite sands. It is a striking feature in many ancient turbidite silts (Figure 12-3).

TURBIDITE MUD

Turbidites in which sands grade up into turbidite mud, which is quite distinct from hemipelagic sediment, have been described by many authors (e.g., Kuenen 1964; van Straaten 1967; Enos 1969; and others reviewed by Hesse 1975, p. 388). Generally, the mud rests on *d* division sand and silt, and there is a gradation in both grain size and petrology from sand to mud. This characteristic,

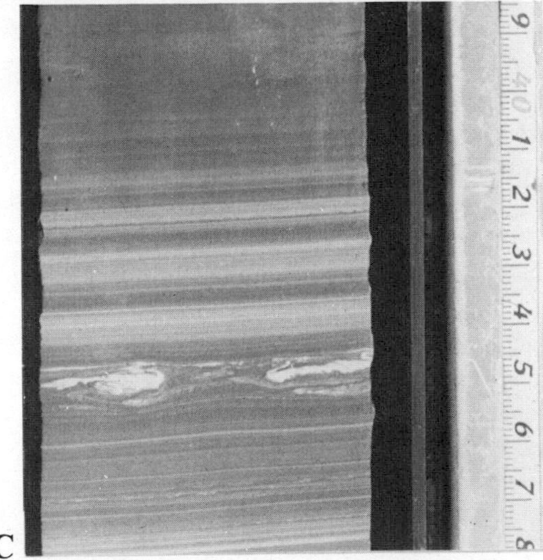

Figure 12-3. Turbidite silt beds. Note: Photograph A shows type 2 silt, Miocene, Wilkes Abyssal Plain (DSDP 269A-12-5-34/45). Photograph B shows type 3 silt, Minocene, Wilkes Abyssal Plain (DSDP 269A-10-5-108/117). Photograph C shows type 3 silt with syn-sedimentary deformation, Miocene, Wilkes Abyssal Plain (DSDP 269-A-10-4-39/48).

together with absence of erosion or bioturbation between sand and mud, indicates deposition from the same turbidity current. Rupke (1969) used bed thickness analysis to demonstrate continuity of disposition.

The work of van Straaten (1967), Griggs (1969), and Rupke and Stanley (1974) on modern sediments and that of Piper (1972a, b) on ancient rocks can be synthesized to establish a threefold division or sequence of turbidite muds (see Figure 12-1). Like the Bouma sequence for turbidite sands, not every division of the sequence can always be recognized. The divisions are:

F – hemipelagic sediment
E3 – ungraded mud
E2 – graded mud
E1 – laminated mud
D – laminated sand and silt.

Just as there are many turbidite sands that lack the lower part of the Bouma sequence, there are also turbidite muds that rest directly on hemipelagic or other non-turbidite sediment, rather than on sand. Some have a graded laminated (E1) division at the base; others rest on a single silty laminae; and some are mud throughout.

Some typical examples of mud turbidites are shown in Figures 12-4 and 12-5.

Division E1: Laminated Mud

This division immediately overlies the sand bed. (It may also rest directly in earlier turbidite or nonturbidite sediment.) The laminae are 0.2 mm to 2.0 mm thick and consist of alternating silt and silty clay. The transition from an underlying sandy *d* division may be quite gradual in grain size distribution and petrology (e.g., Rupke and Stanley 1974, Figure 11). This division has previously been called a graded laminated bed (Piper 1972a), since there is often an upward decrease in the thickness, frequency, and grain size of the silt laminae. This pattern is not, however, always found, and exceptionally thick and coarse silt laminae are often found high up in the E1 division (e.g., Piper 1972a, Plate 1C). Individual laminae are often lenticular (photograph A, Figure 12-6). Some silt laminae are sharp based; others grade up from the mud laminae, with coarse flat particles concentrated at the tops of laminae (Piper 1972b). The few observations on silt grain fabric suggest there is a preferred orientation of silt grains parallel to flow direction in the bedding plane of the silt laminae, but none in the mud laminae. In sections perpendicular to bedding, grains are sub-parallel to bedding, with no apparent imbrication (Piper 1972a and unpublished data).

Division E2: Graded Mud

This division is found either above the laminated mud or directly overlying sand. It may also rest directly on

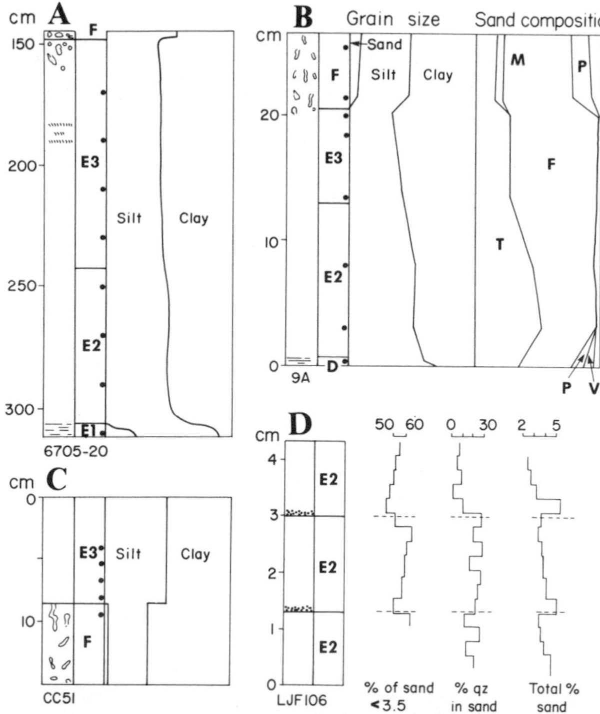

Figure 12-4. Grain size and petrologic variations within turbidite muds resting on hemipelagic sediments. Note: In Charts A through C, dots indicate samples; in Chart D, sampling was continuous. Chart A shows samples from Cascadia Channel (Griggs 1969, Figures 10 and 22; summarized in Griggs and Kulm 1970). Beds of green silty mud are up to 4 m thick, and turbidite origin is indicated by their geographic extent, high content of organic carbon, shallow benthic foraminifera, and restriction of bioturbation. Some beds extend from levees to channel floor, and deposition from several small turbidity currents or lutite flows is thus most improbable. Most muds overlie sands; 9 percent rest directly on (hemi)pelagic sediments. Chart B shows samples from Balearic Abyssal Plain (Rupke and Stanley 1974). Occasional turbidites are interbedded with hemipelagic sediments; most turbidites are E1-3 muds overlying sand. The distal environment, infrequency of turbidites, and grading all suggest a single depositional event. In the example shown there is only a thin silt lamina (D or E1) overlain by an E2 graded mud. In the sand composition column of the chart, T = terrigenous, M = mica, F = planktonic foraminifera, P = pteropods, and V = plants. Chart C shows samples from distal Navy Fan (Normark and Piper 1972). Micaceous mud beds 5 cm to 20 cm thick overlie distal turbidite sands; E3 beds up to 30 cm thick rest directly on hemipelagic mud in ponded basins. Microfossil content, petrology, and color distinguish turbidite from hemipelagic muds. The infrequency of the beds in a distal environment and the uniform texture within beds (despite considerable variations between beds) argue for a single depositional event. The example figured is not bioturbated (see Normark and Piper 1972, Figure 17 for photograph), but rests on highly bioturbated hemipelagic mud with open burrows. Chart D shows samples from La Jolla Fan Valley (Piper 1970). Graded E2 mud beds 1 cm to 2 cm thick overlie either a thick sand bed or a 1 mm thick lamina of silty mud. Grading is indicated by both petrology and grain size. The frequency, association of some beds with sand, and regular grading of these beds within an active turbidity current environment argue for a turbidite origin.

nonturbidite sediment. It shows an upward decrease in grain size, that is often accompanied by petrographic changes (e.g., Rupke and Stanley 1974; Griggs 1969). The mud usually lacks visible primary structure or shows very indistinct bedding in X-radiographs or lithified rock. Grading is often pronounced in beds with substantial amounts of very fine carbonate, which is often concentrated at the top of the bed (e.g., Hesse 1975, calcilutite over marlstone; Aksu personal communication, 1976, graded clay → carbonate turbidites in Baffin Bay). Graded muds can reach thicknesses of 10 cm to 20 cm. Thicker mud beds tend to show no grading in their upper part.

Some graded muds that rest directly on earlier turbidite or nonturbidite sediment have a distinct basal zone (1 mm to 3 mm thick) of very silty mud visible in X-radiographs, although sorted silt laminae are absent (print C, Figure 12-5).

Division E3: Ungraded Mud

This division overlies graded mud. There is no detectable, systematic primary gradation in grain size or petrology. Subtle trends may be masked by later bioturbational mixing with hemipelagic sediments. Some published analyses (e.g., Griggs 1969; Normark and Piper 1972) suggest these ungraded muds are not completely uniform. Van Straaten (1967, Plate 18) noted indistinct laminae, and some X-radiographs of recent ungraded muds show indistinct bedding.

Relationship to Underlying Sand

Many turbidite muds overlying sand rest gradationally on *d* division sand or coarse silt, and continuity of deposition is easily demonstrated. Frequently, although there is a petrologic gradation and a lack of erosional or bioturbational structures, there is a marked break in grain size between the *d* and *e* divisions (e.g., Bouma 1962; Griggs 1969, p. 43; Weiler 1970).

Griggs (1969) found about 25 percent of the turbidite muds in Cascadia Channel rested on *a* division sands, and he believed he could demonstrate continuity between the sands and muds. Graded beds showing *a-e* sequences are well known in ancient turbidites (Walker 1967). In the Silurian Aberystwyth Grits of Wales the type of laminated mud overlying sand varies from proximal to distal environments (Piper 1972a). Proximally, laminae of coarse silt alternating with silty mud are quite thick and well defined; distally, there is more of a gradation through finer silt, and laminae are more diffuse.

LAMINATED MUDS

Muds with silty laminae are an extremely common lithology on deepsea fans, continental rises, and the prox-

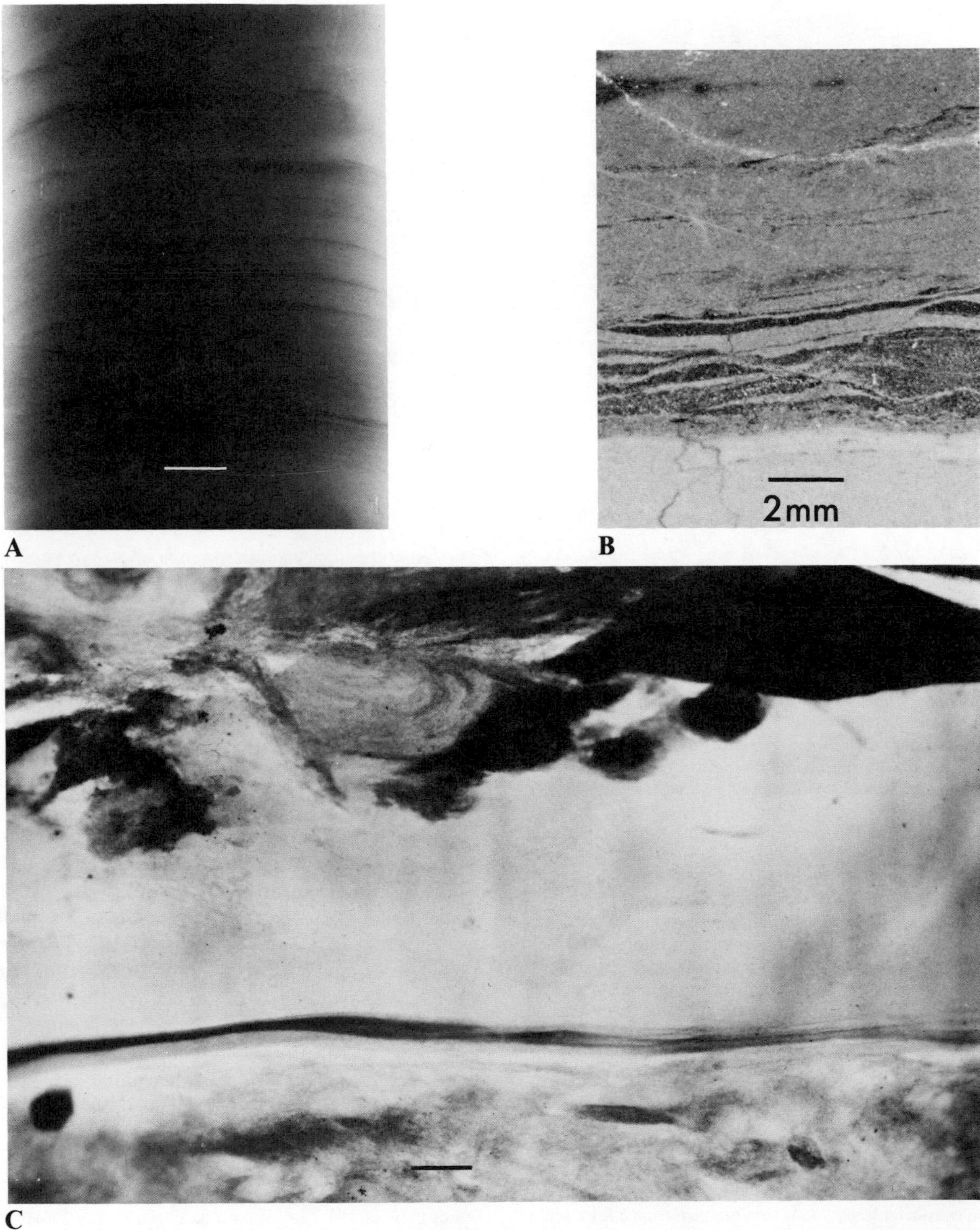

Figure 12-5. Sedimentary structures within turbidite muds resting on hemipelagic or earlier turbidite sediment. Note: In the X-radiograph prints (A and C), scale bar = 1 cm. Print A is an X-radiograph of sample from Newfoundland Abyssal Plain (Hemsley and Piper, in preparation). Rare, thin mud beds are over- and underlain by a coccolith ooze sequence on an abyssal hill elevated 30 m above the Newfoundland Abyssal Plain. Photograph B shows a thin section of sample from Gowlaun Formation, Silurian, Western Ireland (Piper 1972a). Mud sequences hundreds of meters thick were deposited on deepsea fan levees within a few kilometers of major conglomerate-filled channels. Many muds have graded laminated beds overlain by mud without silt laminae. Some graded laminated muds are affected by convolute bedding. Piper (1972a) presents a statistical analysis of the grading and shows a decrease in both proportion of silt and maximum size of silt grains in any laminated mud bed. Print C is an X-radiograph of sample from San Lucas Fan (Shepard and Dill 1966). Setting and evidence for turbidite origin is similar to La Jolla Fan (Chart D, Figure 12-4). The X-radiograph shows unbioturbated mud bed with laminated silty mud at base, over- and underlain by bioturbated sediment.

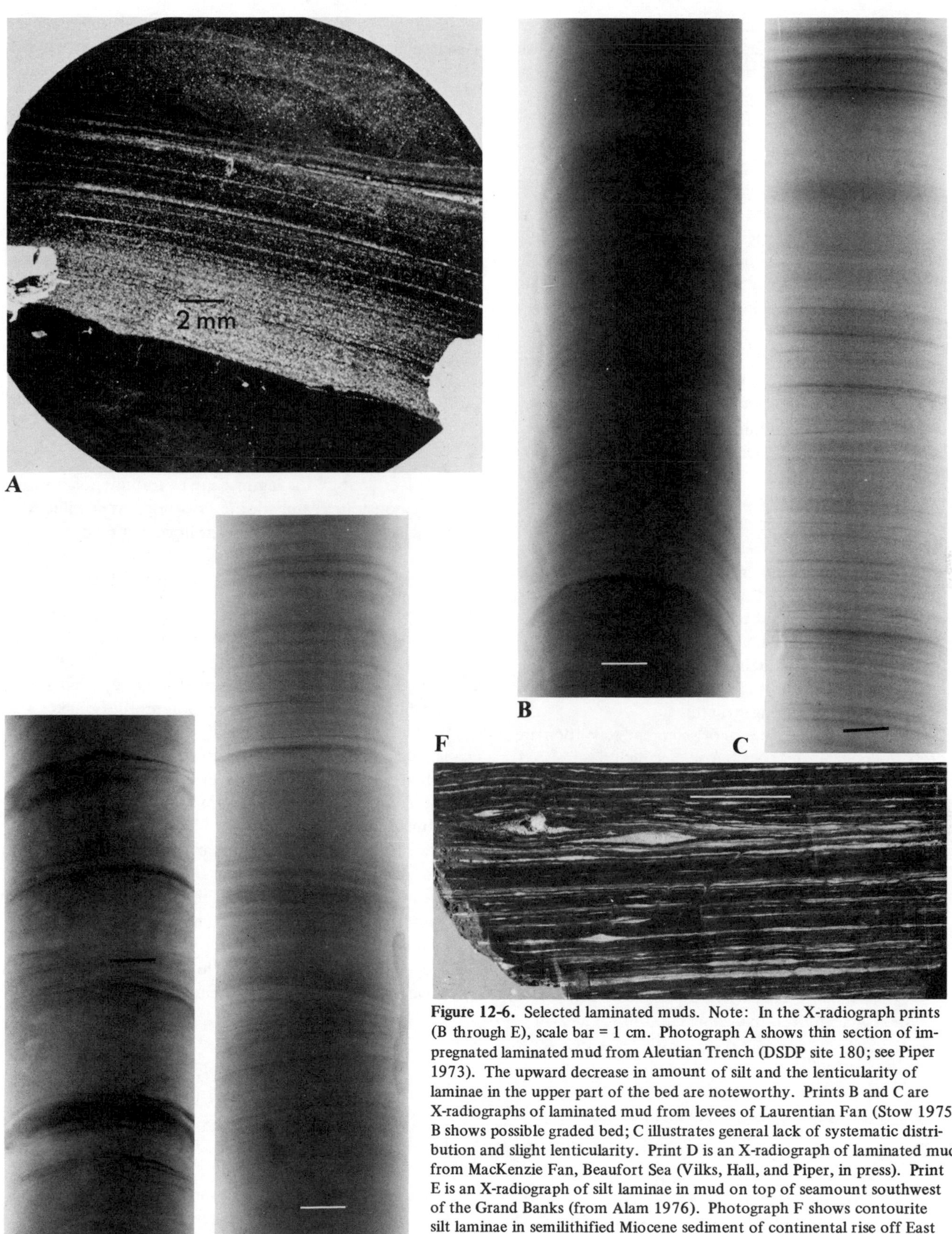

Figure 12-6. Selected laminated muds. Note: In the X-radiograph prints (B through E), scale bar = 1 cm. Photograph A shows thin section of impregnated laminated mud from Aleutian Trench (DSDP site 180; see Piper 1973). The upward decrease in amount of silt and the lenticularity of laminae in the upper part of the bed are noteworthy. Prints B and C are X-radiographs of laminated mud from levees of Laurentian Fan (Stow 1975). B shows possible graded bed; C illustrates general lack of systematic distribution and slight lenticularity. Print D is an X-radiograph of laminated mud from MacKenzie Fan, Beaufort Sea (Vilks, Hall, and Piper, in press). Print E is an X-radiograph of silt laminae in mud on top of seamount southwest of the Grand Banks (from Alam 1976). Photograph F shows contourite silt laminae in semilithified Miocene sediment of continental rise off East Antarctica (DSDP site 268, core 17-3; see Piper and Brisco 1975). The lenticularity of laminae, erosional scours not filled with coarser silt, and slight bioturbation are noteworthy.

imal parts of abyssal plains. We have seen that laminated muds can occur as the lowest part of a sequence of lithologies in a turbidite mud bed. Such beds are most recognizable where they alternate with hemipelagic deposits. However, in many sequences, such independent evidence of turbidite origin is lacking.

Many authors have interpreted silt laminae in deepwater muds as resulting from either the winnowing action of, or irregular deposition from, bottom currents. Some deepwater laminated muds are demonstrably of this origin (Hollister 1967; Davies and Laughton, 1972). Such muds may be referred to as contourites, although not all the bottom currents producing them follow contours. However, many laminated muds, without obvious grading or organization, are found on deepsea fans and abyssal plains in situations where a turbidite origin appears more probable than deposition by contour currents.

Criteria for distinguishing turbidite and contourite laminated muds are difficult to establish. Bouma and Hollister (1973) list criteria applicable to fine sands, but few of these criteria can be extended to thin silt laminae. Piper and Brisco (1975) point out that turbidity currents differ from contour currents primarily in their much greater suspended sediment load that is necessary to keep them in motion. They thus derive the following criteria for distinguishing turbidites and contourites.

1. Deposits with evidence of rapid deposition are probably turbidites. Such evidence would include climbing ripples with no stoss side erosion, animal escape burrows, and dewatering structures.

2. Deposits with evidence of sediment starvation are probably contourites. Such evidence includes isolated ("starved") ripples and perhaps scours not filled with unusually coarse sediments.

3. Evidence of erosion by highly turbulent flows is probably restricted to turbidity currents.

4. The petrographic composition of contourites should follow that of the coarser components of the interbedded sediment, which may also occur in turbidites, but does not appear common.

A number of examples, discussed below, illustrate the point that in many cases there are as yet no satisfactory criteria for clearly distinguishing contourites and turbidites. They are in order of decreasing evidence for a turbidite origin.

Quaternary Sediments of the Aleutian Trench

The Quaternary sequence of the eastern Aleutian Trench (Piper 1973) comprises alternating silts and muds. The overall rate of sedimentation is very high (1,800 m/million years), and although the muds contain some thin, coarse ice-rafted debris layers, there is little other distinguishable pelagic sediment. Silt laminae, both isolate and grouped, are very common in the mud sequences and frequently overlie silt beds. About 45 percent of the silt laminae overlie silt beds and frequently form a graded sequence, while 5 percent of the laminae are isolate. The remaining laminae occur in groups in the muds. Some 15 percent form a graded sequence (photograph A, Figure 12-6), while the other 35 percent show no noticeable systematic distribution. There are no recognizable differences between these several occurrences of laminae. A "contourite" origin of the laminae is unlikely, since:

1. Even where (? pelagic) diatomaceous muds occur, there is a lack of winnowed biogenic debris in the laminae.
2. Residual winnowed coarse sand and gravels are not found, despite the input of glacial erratics.
3. In the case of the graded sets of laminae, either a remarkable pattern of bottom current flow is required, or the bottom current must transport very large amounts of sediment.

However, that all the laminae are of turbidite origin cannot be demonstrated, and a lutite flow origin is possible for those laminae that do not occur in graded sequences.

Levees of the Late Pleistocene Laurentian Fan

The late Pleistocene sediments of prominent levees on the Laurentian Fan (Stow 1975) consist predominantly of laminated muds. The setting of these muds—adjacent to major turbidity current channels on levees that are recognizable from seismic reflection profiles—makes a contourite origin improbable. Only about 10 percent of the sequence consists of recognizable graded laminated beds. Otherwise, little regularity is visible (print C, Figure 12-6). Silt laminae range from 0.2 mm to 5 mm thick. The laminae are only slightly lenticular, and load casting is rare. Thicker silt laminae that can be correlated between cores become finer away from the channel. The median size of the coarser laminae is about 4.5 ϕ and 5.5 ϕ for the finer laminae (Stow 1976). Laminated muds are most common around 4000 m water depth and are rarer in both deeper and shallower water. Hollister (1967) interpreted apparently similar laminated muds to the southwest as contourites.

Continental Rise, Late Pleistocene, Beaufort Sea

A core described by Vilks, Hall, and Piper (in press) from the MacKenzie Fan in the Beaufort Sea has a very long sequence consisting of mud with silt laminae up to 1 cm thick. The silts tend to be grouped and show as much (or as little) systematic distribution as silts from the Gulf of Alaska or the Laurentian Fan (print D, Figure 12-6). This core illustrates the problem of determining, in the absence of other data, whether such silts are of turbidite or contourite origin.

Pleistocene Muds on Seamounts SW of Grand Banks

Alam (1976) has studied Pleistocene muds accumulated at 3 km water depth on top of seamounts 65 km from the shelf edge of the Grand Banks. The Western Boundary Undercurrent was probably active at times when mud was deposited. This interpretation is supported by evidence of winnowing at other stratigraphic horizons (Alam 1976). However, there is little in the laminae themselves (print E, Figure 12-6) to support or deny a "contourite" origin. Laminae are 0.2 mm to 2 mm thick and consist of silt with a modal size of around 10 μ. They often occur in groups, but no systematic distribution can be seen. Individual laminae are often lenticular. Some are sharp based; others are gradual. Since the muds contain almost no microfossils, the petrology of the laminae is no guide to origin. But for their setting and the lack of a systematic pattern, these laminae cannot be distinguished from those of turbidite origin.

DSDP Site 268, Miocene, Continental Rise of East Antarctica

Silt laminae, 0.2 mm to 3 mm thick, alternate with diatom-rich mud (photograph F, Figure 12-6; also see Piper and Brisco 1975, p. 729). About 25 percent of the total sequence consists of silt laminae. Laminae are quite well sorted and contain common size-sorted diatoms and radiolaria, thereby suggesting winnowing of bottom sediment. Many laminae are lenticular, and thinner laminae are discontinuous. Some lenses appear like small ripples, but are never more than 5 mm thick. Load structures are absent, but there is some disturbance by bioturbation. Some shallow erosional scours are filled with mud, rather than silt. Some individual laminae have sharp bases; others have gradational bases. No pronounced fabric has been observed.

The general stratigraphic setting (continental rise, lack of sand and silt beds) makes a strong a priori case for these silts being contourites, which is strongly supported by their petrology (discussed further by Piper and Brisco 1975, p. 729). Structures visible in most single laminae in thin section are little different from those seen in graded laminated turbidite muds. However, the local extreme lenticularity is distinctive; possibly comparable lenticularity is found only in turbidite silts showing considerable syn-sedimentary deformation.

Graded Laminated Mud Beds with Gradational Bases

In about 5 percent of well-graded laminated mud beds, there are two or three slightly siltier indistinct laminae near the top of the mud immediately below the thick sharp based lowest silt laminae. Examples have been previously figured (Piper 1973, Figure 11c; Piper 1975, Figure 2). The extreme base of some graded mud beds also shows reverse grading (or at least a zone of finer sediment) in the basal few millimeters. This phenomenon could be explained by a current of gradually increasing intensity, which then waned systematically. However, the location of many of the examples (e.g., the Gulf of Alaska sediments discussed above) makes a contourite origin unlikely. Griggs (1969, p. 86) found a thin turbidite mud layer beneath some turbidite silts in Cascadia Channel and compared them with the "prephase" of Meischner (1964). He interpreted them as depositied by a fast-moving body of the turbidity current moving more rapidly than the head. Alternatively, the head may be greatly diluted by mud eroded from the sea floor.

FACIES DISTRIBUTION

Systematic description of mud and silt facies have been prevented by a lack of a conceptual framework of classification, the thinness of beds, and the difficulty of observing structures and grain size distribution in fine-grained sediments. The paucity of data makes generalizing on facies distributions hazardous. (The discussion below is based on articles indicated by an asterisk in the list of references at the end of this chapter.)

Turbidite Silts

Of the three silt facies recognized, Type 1 appears to be most proximal and Type 3 most distal. There is insufficient data available to relate these facies to any particular bathymetric setting.

Type 2 silts have modal sizes in the 4 to 5 ϕ range on deep sea fans and channel levees. Modal sizes in the 6 ϕ range are found only on abyssal plains hundreds of kilometers from the sources of turbidity currents. Such fine silt is always deposited within muds, along with clay, in more proximal environments (Figure 12-7).

Turbidite Muds

Laminated muds, both overlying and distinct from sands, are most common on levees on deepsea fans, where they are often the most common facies (Figure 12-8). Proximally, little regularity is visible, and individual silt laminae tend to be thin. Distally, more graded laminated mud beds are recognizable, and some thicker laminae and even thin silt beds are visible. They are much less common on the more distal abyssal plains and make up only a few percent of the sediment thickness.

Graded muds are found in deepsea channels and fan valleys, but in more distal environments, grading becomes less pronounced. Good examples of ungraded muds occur on distal abyssal plains and in ponded basins.

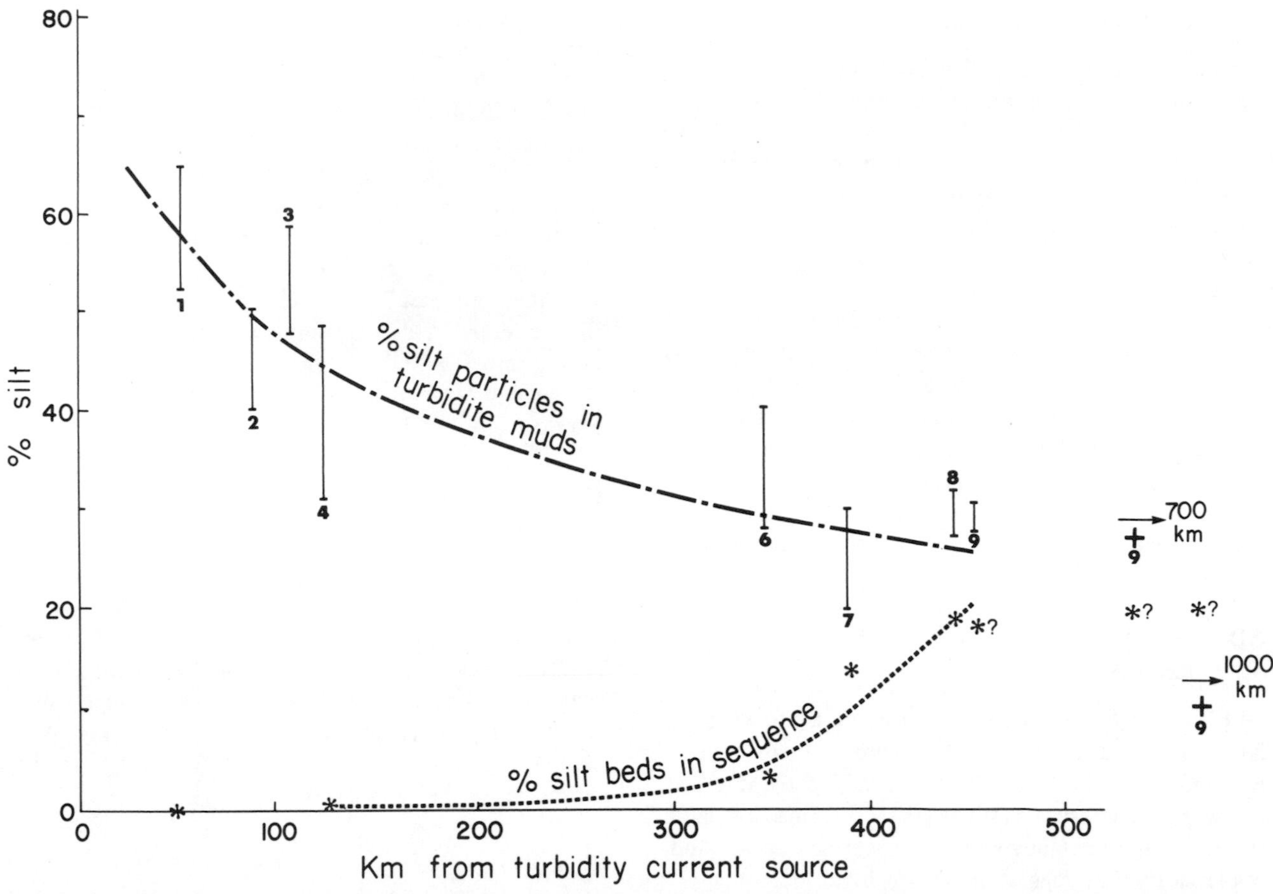

Figure 12-7. Proximal to distal changes in silt content of mud turbidites and proportion of silt beds in the sediment sequence. Note: Locations (and sources of data) are as follows: 1 = La Jolla Fan (Piper 1970 and unpublished data); 2 = Balearic Abyssal Plain (Rupke and Stanley 1974, cores 8 and 9); 3 = Scotian Slope, core 72-021-3; 4 = Distal Navy Fan (Normark and Piper 1972); 5 = Laurentian Fan (Stow 1975, 1976); 6 = Cascadia Channel (Griggs 1969); 7 = DSDP site 269 (Piper and Brisco 1975); 8 = DSDP site 178 (Piper 1973 and unpublished data); 9 = Sohm Abyssal Plain (Horn et al. 1971; percentage of silt beds estimated from figures). Mud turbidites overlying sand have not been considered.

Turbidite mud beds on deepsea fans commonly contain 30 to 60 percent silt and on abyssal plains, 10 to 25 percent silt. There is a general relationship between distance from turbidity current source, and percentage of silt in turbidite muds (see Figure 12-7).

DISCUSSION

Processes within the Flow

Dzulynski et al. (1959), Enos (1969), and other authors have pointed out that if Chezy or Kuelegan equations (Komar 1972) are applied to muddy turbidity currents, they predict that only very thin mud beds can be deposited from turbidity currents moving down slopes. Accounting for beds more than 10 mm thick is difficult. Hesse (1975) suggested that the sediment concentration in the flow could be increased if the water in the turbidity current, derived from much lesser depths, was substantially less dense than ambient seawater. Ponding in basins also provides a mechanism whereby thick mud beds can accumulate. The front part of a turbidity current will fill the basin, and the tail will mix with it or flow on top of it. In such cases, the rate of addition of sediment would be comparable with the rate of settling, and only the extreme top and base of the bed would be graded.

The proximal to distal sorting and separation of silt and clay fractions results from processes acting within the turbidity current, rather than at the bed, and is not likely a function of the state of flow near the bed, since the trend appears independent of depositional environment. There appears, rather, to be a sorting of material delivered to the bed. This sorting is a result of suspension by fluid turbulence, rather than autosuspension, which produces no grading (Bagnold 1966).

Fine sediment in a marine turbidity current will tend to flocculate as a result of high-sediment concentration. Frequent floc collisions in turbulent suspension will result in flocs becoming stronger and richer in clay and thereby releasing the less-cohesive silt fraction. A concentration

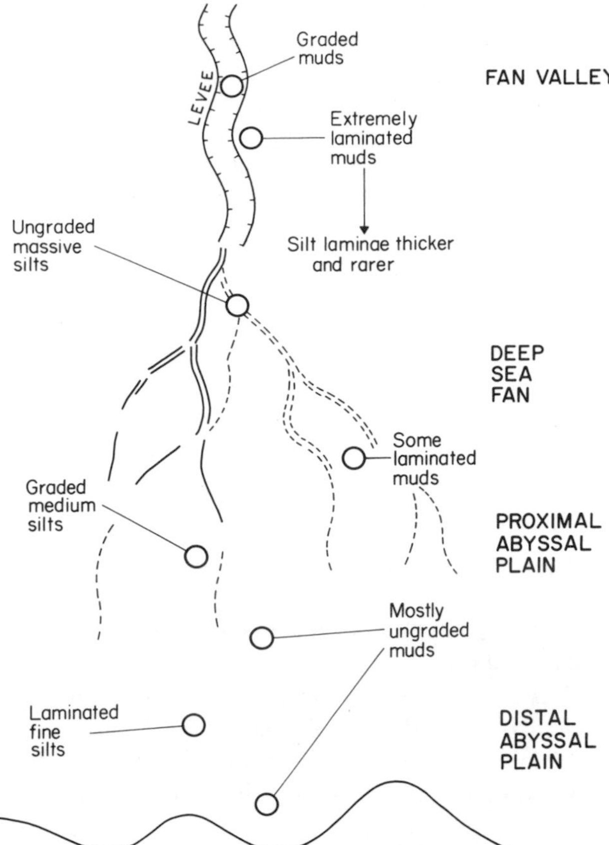

Figure 12-8. Speculative model of facies distribution for silt and mud turbidites.

gradient will develop, with larger flocs and silt near the base of the flow. If deceleration is slow, then the coarsest material will gradually be removed by sedimentation, and a graded bed will result. If deceleration is rapid, deposition will exceed the rate at which flocs break up, re-form, and sort out silt material. As a consequence, ungraded mud may be deposited. Ungraded muds may also deposit in distal environments, where the flocs consist predominantly of clay and can release little silt by breaking up and re-forming.

This process of floc collision and sorting, along with tractional deposition of coarser grains, leads to a distal separation of fine silt. In proximal environments, medium and coarse silts are deposited in a manner similar to fine sands. Thick, ungraded massive silts appear similar to Bouma *a* division sands; graded sorted silts are similar to *bcd* sands. The deposition of *a* division silts in the absence of sands appears rare and may reflect a lack of sand in the source material, rather than sorting by the current.

When a large amount of fine carbonate is present, the same process may lead to its concentration near the top of the turbidity current, since it has a lesser tendency to form large flocs. It will thus occur at the top of graded mud beds.

Processes at the Bed

Most of the distinctive sedimentary features of sand turbidites appear to be the result of near-bed processes: They reveal less about the behavior of sediment in the main body of the turbidity current. Interpretation of mud turbidites is hindered by our lack of understanding of processes on cohesive mud beds.

In at least the forward part of a large turbidity current, the concentration of suspended fine sediment exceeds 40 gm/liter (Grover and Howard 1938; Kuenen 1966; van Straaten 1967). Partheniades and his coworkers (Partheniades and Kennedy 1966; Partheniades 1972) found that at these high densities deposition of fines could take place quite rapidly at velocities of 10 to 30 cm/sec through adhesion of flocs to a mud bed. At suspended sediment concentrations less than 1 gm/l such as might occur in the tails of turbidity currents, rates of deposition are less (Mehta and Partheniades 1973), and deposition is primarily controlled by settling velocity (Hesse 1975).

As noted above, steady high-density flows with low velocities may be difficult to obtain. High densities are perhaps most easily obtained by overflow of channelized high velocity (> 500 cm/sec) flows.

The distinctive "break" in grading between a sand and an overlying mud is apparently the result of both flocculation and adhesion of muddy sediment in a turbulent flow onto a cohesive bed. Weiler's (1970) suggestion that it is solely due to flocculation is inadequate, since flocs rarely reach an adequate size to form a graded bed with fine sand (Migniot 1968); neither does it adequately explain the distal occurrence of fine silts.

Alternating silt and mud laminae apparently represent deposition of silt in bedload followed by deposition of suspended mud. An analogous alternation between bedload granules and suspended sand has been analyzed by Davies and Walker (1974) and Walker (1975). If velocity fluctuates, a decrease in velocity may result in the suspension capacity of the flow being exceeded, and deposition of suspended sediment will occur. With an increase in velocity, the flow again has sufficient capacity, and only bedload is deposited. Knowing the size of lenses of bedload sediment and using standard bedload equations, Walker was able to estimate the period of the velocity fluctuations. An analogous calculation suggests silt laminae similar to those in (photograph A, Figure 12-6) will require only a few or, at the most, tens of minutes to be emplaced in a current with a mean flow of around 20 cm/sec. Although cohesive bed deposition of mud could occur without decrease in velocity (Piper 1972a), decreases in velocity probably trigger mud deposition. Whatever the exact process of mud deposition, it frequently takes place rapidly enough to form an unstable lamina, and silt deposition above it forms load casts.

Variations in Flow

Laminated muds of turbidite origin suggest fluctuation in turbidity current flow. Lambert et al. (1976) and Normark and Dickson (1976) found fluctuations with a period of 15 to 30 minutes in turbidity currents flowing into lakes, and Davies and Walker (1975) inferred similar fluctuations in marine turbidites. Although silty laminae in a graded laminated bed might be the consequence of velocity fluctuations associated with turbulent eddies, the thickness of many suggests fluctuations of a longer period.

Levees are the sites of most irregular deposition of silt laminae, probably related to overflow of a turbidity current from the channel. Since flow within the head is of an upwelling nature (Middleton 1966) and the head passes any one point rapidly, a thin poorly sorted bed will likely result. Fine lamination in thick beds will result from body overflow and will tend to take place at gradients less than about 0.002, when body height exceeds head height, with a densiometric Froude number of about 0.75 (Komar 1972, 1973). Initial body overflow may be supercritical, as a result of continuity of momentum and decrease in flow thickness. The resulting shooting flow may be responsible for the very abundant laminated muds on levees on the middle parts of the continental rise, with gradients around 0.002 (Stow 1976). Deceleration and hence deposition will be rapid.

Reflection of a turbidity current at the edge of a ponded basin can result in a complex bed made up of several small graded beds (van Andel and Komar 1969). Van Straaten (1967) suggested "slow lateral movements of the suspension" in a ponded basin were responsible for subtle variations in texture. A more careful examination of ungraded beds in ponded basins might reveal the apparent lack of a systematic distribution is illusory.

The Tail of a Turbidity Current

What role is played by the dilute tail of a turbidity current in deposition of mud beds? As a turbidity current deposits sediment, it loses density and momentum. Autosuspension will take place only on steep slopes; distally, only progressively dampened fluid turbulence will maintain sediment in suspension. The turbidity current further dissipates by mixing with seawater. Bottom currents can play an increasing role in distributing the suspended sediment.

In submarine canyons and fan valleys, the tail of the turbidity current will probably be quite abrupt, as autosuspension maintains the flow down the steep slope. Normal tidally induced bottom currents will tend to reduce deposition of any residual cloud of suspended sediment. In more distal, less-constricted environments, a residual sediment cloud is presumably much more important. After a few months, it probably has a concentration similar to nepheloid layers.

Explaining the presence of ungraded muds in ponded basins by deposition from a residual sediment cloud is difficult. Simple settling from a sediment cloud is more likely to produce graded bedding than is deposition from a turbulent fluid. Order-of-magnitude calculations suggest deposition of more than 5 cm of mud from a residual cloud is most improbable, and more likely, less than 1 cm is deposited in this way. Residual sediment clouds deposit only thin layers that are easily mixed by bioturbation with overlying pelagic sediment.

One possible origin for laminated muds is interaction of fluctuating bottom currents with the tail of the turbidity current or reworking of turbidites by those bottom currents. The lack of laminated muds in fan valleys, where tidally induced bottom currents tend to be concentrated (Shepard and Marshall 1973), argues against this hypothesis.

SUMMARY

1. Mud turbidites are a major component of deepsea fans and abyssal plains.
2. Mud turbidites may overlie turbidite sands or rest directly on nonturbidite sediment.
3. Normal steady flow will result in a mud turbidite showing a threefold division overlying sand:

   F–hemipelagic sediment
   E3–ungraded mud
   E2–graded mud
   E1–laminated mud
   D–laminated sand.

4. Spillover on levees of deepsea fan valleys results in muds with silt laminae showing little systematic distribution. These disorganized laminated muds are most common in regions where the turbidity current body first exceeds the height. They may result from supercritical flow.
5. Laminated muds showing grading are common on deepsea fans, but become rarer distally on abyssal plains.
6. Graded muds are common in deepsea fan valleys and on the lower parts of deepsea fans. Distally, on abyssal plains grading becomes less pronounced.
7. Ungraded muds are found in many environments, but are most common on distal abyssal plains and in ponded basins. Lack of grading may result from: (a) development of a uniform stable floc composition within the flow; (b) continuous addition of sediment to a settling system in a ponded basin; or (c) superposition of many small graded beds due to repeated reflection of a turbidity current off the edge of a ponded basin.
8. Cohesive bed deposition of mud flocs explains the commonly observed break in grading between sands and muds. Alternating silt and mud laminae in laminated muds are a consequence of alternating bedload transport of silt and deposition of mud carried in suspension. Mud deposi-

tion is enhanced by cohesion. Unstable bed conditions, resulting in load casting of the overlying silt, may develop.

9. The breakup and re-forming of flocs within the flow leads to a progressive segregation of finer and finer silt. As a consequence, silt beds become increasingly common distally and have a decreasing modal size. Turbidite muds have a progressively lower content of silt-sized particles distally.

10. Three types of turbidite silt beds are distinguished. Ungraded massive silts, similar to *a* division sands, are commonest proximally. Graded sorted silts, similar to *bcd* division sands, are common on deepsea fans and abyssal plains.

11. The deposits of contour currents and lutite flows may often resemble mud turbidites, and there is no simple set of criteria to distinguish them. Mud turbidites are most easily recognized when they occur as thick beds in distal environments. Some beds identified as mud turbidites in this review may subsequently prove to be contourites or deposits of lutite flows.

ACKNOWLEDGMENTS

This work was begun at the Department of Geology, University of Cambridge, England, during the tenure of a Research Fellowship from Jesus College, Cambridge. Later work has been supported by Canadian National Research Council operating grants and a University Research Grant from Imperial Oil Ltd. Important data and ideas have resulted from participation in the Deep Sea Drilling Project, which is funded by the National Science Foundation.

REFERENCES

Alam, M., 1976. *Quaternary paleoclimates and sedimentation southwest of the Grand Banks.* Unpublished M.Sc. thesis, Dalhousie University, Halifax, Nova Scotia, 251 pp.

Bagnold, R. A., 1966. An approach to the sediment transport problem from general physics. *U.S. Geol. Surv. Prof. Paper* 422-I, 37 pp.

*Bouma, A. H., 1962. *Sedimentology of Some Flysch Deposits.* Elsevier, Amsterdam, 168 pp.

*_____, 1972a. Recent and ancient turbidites and contourites. *Trans. Gulf. Coast Assoc. Geol. Soc.,* 22: 205-21.

*_____, 1972b. Distribution of sediments and sedimentary structures in the Gulf of Mexico. In: R. Rezak, and V. J. Henry (eds.), *Contributions to the Geological Oceanography of the Gulf of Mexico.* Texas A & M University Oceanographic Studies, 3, pp. 35-65.

_____, and Hollister, C. D., 1973. Deep ocean basin sedimentation. In: G.V. Middleton, and A. H. Bouma (eds.), *Turbidites and Deep Water Sedimentation.* Soc. Econ. Paleont. Mineral., Pacific Section, Short Course, Anaheim, pp. 79-118.

*Cleary, W. J., and Conolly, J. R., 1974. Hatteras deep sea fan. *J. Sed. Petrol.,* 44: 1140-54.

Davies, I. C., and Walker, R. G., 1974. Transport and deposition of resedimented conglomerates: The Cap Enragé Formation, Cambro-Ordovician, Gaspé, Quebec. *J. Sed. Petrol.,* 44: 1200-16.

Davies, T. A., and Laughton, A. S., 1972. Sedimentary processes in the North Atlantic. In: A. S. Laughton, W. A. Berggren et al., *Initial Reports of the Deep Sea Drilling Project,* 12: 905-34.

Dzulynski, S., Ksiazkiewicz, M., and Kuenen, P. H., 1959. Turbidites in flysch of the Polish Carpathian Mountains. *Geol. Soc. Amer. Bull.,* 70: 1089-118.

Enos, P., 1969. Anatomy of a flysch. *J. Sed. Petrol.,* 39: 680-723.

Fleischer, P., 1972. Mineralogy and sedimentation history, Santa Barbara Basin, California. *J. Sed. Petrol.,* 42: 49-58.

*Fruth, L. S., 1965. *The 1929 Grand Banks turbidite and the sediments of the Sohm Abyssal Plain.* Unpublished M.A. thesis, Columbia University, New York, 257 pp.

*Griggs, G. B., 1969. *Cascadia Channel: The anatomy of a deepsea channel.* Unpublished Ph.D. thesis, Oregon State University, Corvallis, 183 pp.

*_____, and Kulm, L. D., 1970. Sedimentation in Cascadia deepsea channel. *Geol. Soc. Amer. Bull.* , 81: 1361-84.

Grover, N. C., and Howard, C. S., 1938. The passage of turbid water through Lake Mead. *Trans. Amer. Soc. Civil. Engrs.,* 103: 720-90.

*Hall, B. A., and Stanley, D. J., 1973. Levee bounded base of slope channels in the lower Devonian Seboomook Formation, Northern Maine. *Geol. Soc. Amer. Bull.,* 84: 2101-10.

Hesse, R., 1975. Turbiditic and non-turbiditic mudstone of Cretaceous flysch sections of the East Alps and other basins. *Sediment.,* 22: 387-416.

*Hollister, C. D. 1967. Sediment distribution and deep circulation in the western North Atlantic. Unpublished Ph.D. thesis, Columbia University, New York, 470 pp.

*Horn, D., Ewing, M., Horn, B. M., and Delach, M. N., 1971. Turbidites of the Hatteras and Sohm Abyssal Plains, Western North Atlantic. *Mar. Geol.,* 11: 287-323.

*Huang, T-C., and Stanley, D. J., 1972. Western Alboran Sea: Sediment dispersal, ponding and reversal of currents. In: D. J. Stanley (ed.), *The Mediterranean Sea.* Dowden, Hutchison & Ross, Stroudsburg, Pa., pp. 521-59

*Jipa, D., and Kidd, R. B., 1974. Sedimentation of coarser grained interbeds in the Arabian Sea, and sedimentation processes of the Indus Core. In: R. B. Whitmarsh, O. E. Weser, D. A. Ross, et al., *Initial Reports of the Deep Sea Drilling Project,* 23: 471-95.

Komar, P. D., 1972. Relative significance of head and body spill from a channelized turbidity current. *Geol. Soc. Amer. Bull.,* 83: 1151-56.

_____, 1973. Continuity of turbidity current flow and systematic variations in deep-sea channel morphology. *Geol. Soc. Amer. Bull.,* 84: 3329-38.

Kuenen, P. H., 1964. The shell pavement below oceanic turbidites. *Mar. Geol.,* 2: 236-46.

_____, 1966. Matrix of turbidites: Experimental approach. *Sediment.,* 7: 267-98.

Kulm, L. D., von Huene, R., et al., 1973. *Initial Reports of the Deep Sea Drilling Project,* 18. U.S. Govt. Print. Off., Washington, D.C., 1077 pp.

*LaJoie, J., and Chagnon, A., 1973. Origin of red beds in a Cambrian flysch sequence, Canadian Appalachians, Quebec. *Sediment.,* 20: 91-103.

Lambert, A. M., Kelts, K. R., and Marshall, N. F., 1976. Measurements of density underflows from Walensee, Switzerland. *Sediment.,* 23: 87-105.

*Article used in the synthesis of depositional environments of mud and silt turbidite facies.

*Maldonado, A., and Stanley, D. J., 1976. Late Quarternary sedimentation and stratigraphy in the Strait of Sicily. *Smiths. Contrib. Earth Sci.,* 16, 73 pp.

Mehta, A. J., and Partheniades, E., 1973. *Depositional Behavior of Cohesive Sediment.* Tech. Rept. 16. Coastal & Oceanographic Engr. Lab., University of Florida, Gainesville, 275 pp.

Meischner, K. D., 1964. Allodapische Kalke, Turbidite in riffnahen Sedimentationsbecken. In: A. H. Bouma and A. Brouwer (eds.) *Turbidites.* Elsevier, Amsterdam, pp. 93-105.

Middleton, G. V., 1966. Experiments on density and turbidity currents, I: Motion of the head. *Can. J. Earth Sci.,* 3: 523-46.

Migniot, C., 1968. Etude des propriétés physiques de différents sédiments très fins et de leur comportement sous des actions hydrodynamiques. *La Houille Blanche,* 7: 591-620.

Moore, D. G., 1969. Reflection profiling studies of the California Continental Borderland: Structure and Quaternary turbidite basins. *Geol. Soc. Amer. Spec. Paper* 107, 142 pp.

*Mutti, E., and Ricci Lucchi, F., 1972. Le torbiditi dell'Appennino settentrionale: Introduzione all'analisi di facies. *Mem. Soc. Geol. Ital.,* 11: 161-99.

*Normark, W. R., and Piper, D. J. W., 1972. Sediments and growth patterns of Navy deep sea fan, San Clemente Basin, California Continental Borderland. *J. Geol.,* 80: 198-223.

_____, and Dickson, F. H., 1976. Man-made turbidity currents in Lake Superior. *Sediment.,* 23: 815-31.

Partheniades, E., 1972. Results of recent investigations on erosion and deposition of cohesive sediment. In: H. W. Shen (ed.), *Sedimentation.* Water Res. Publ., Fort Collins, pp. 201 – 20.

Partheniades, E., and Kennedy, J. F., 1966. Depositional behavior of fine sediment in turbulent fluid motion. *Proc. X Intern. Conf. Coast. Eng.,* Tokyo, pp. 707-29.

Piper, D. J. W., 1970. Transport and deposition of Holocene sediment of La Jolla deep sea fan, California. Mar. Geol., 8: 211–27.

*_____, 1972a. Turbidite origin of some laminated mudstones. *Geol. Mag.,* 109: 115–26.

*_____, 1972b. Sediments of middle Cambrian Burgess Shale, Canada, *Lethaia,* 5: 169–75.

*_____, 1973. The sedimentology of silt turbidites from the Gulf of Alaska. In: L. D. Kulm, R. von Huene et al. (eds.), *Initial Reports of the Deep Sea Drilling Project,* 18: 847–67.

*_____, 1975. A reconnaissance of the sedimentology of lower Silurian mudstones, English Lake District. *Sediment.,* 22: 623–30.

*_____, and Brisco, C. D., 1975. Deep water continental margin sedimentation, D.S.D.P. Leg 28, Antarctica. In: D. E. Hayes, L. A. Frakes et al. (eds.), *Initial Reports of the Deep Sea Drilling Project,* 28: 727–55.

Rupke, N. A., 1969. Aspects of bed thickness in some Eocene turbidite sequences. *J. Geol.,* 77: 482–84.

*_____, and Stanley, D. J., 1974. Distinctive properties of turbidite and hemipelagic mudlayers in the Algéro-Balearic Basin, Western Mediterranean Sea. *Smithson. Contrib. Ear. Sci.,* 13, 40 pp.

*Scholle, P. A., 1971. Sedimentology of fine grained deep water carbonate turbidites, Monte Antola Flysch (Upper Cretaceous), Northern Appennines, Italy. *Geol. Soc. Amer. Bull.* 82: 629-58.

Shepard, F. P., 1954. Nomenclature based on sand-silt-clay ratios. *J. Sed. Petrol.* 24: 151–58.

_____, and Dill, R. F., 1966. *Submarine Canyons and Other Sea Valleys.* Rand McNally, Chicago, 381 pp.

_____, and Marshall, N. F., 1973. Currents along floors of submarine canyons. *Amer. Assoc. Petrol. Geol. Bull.,* 57: 244-64.

*Stow, D. A. V., 1975. *The Laurentian-Fan: late Quaternary stratigraphy.* Unpublished M.Sc. thesis, Dalhousie University, Halifax, Nova Scotia, 82 pp.

_____, 1976. Deep water sands and silts on the Nova Scotian continental margin (Abst.). *Geol. Assoc. Can. Prog.,* 61 pp.

Van Andel, T. H., and Komar, P. D., 1969. Ponded sediments of the mid-Atlantic ridge between 22° and 23° north latitude. *Geol. Soc. Amer. Bull.,* 80: 1163-90.

Van der Lingen, G. J., 1969. The turbidite problem. *N.Z. J. Geol. Geophys.,* 12: 7-50.

*Van Straaten, L.M.U.J., 1967. Turbidites, ash layers and shell beds in the bathyal zone of the southeastern Adriatic Sea. *Rev. Géogr. Phys. Géol. Dyn.,* 9: 219–39.

*Vilks, G., Hall, J. M., and Piper, D. J. W. The natural remanent magnetization of sediment cores from the Beaufort Sea. *Can J. Earth Sci.* (in press).

Walker, R. G., 1967. Turbidite sedimentary structures and their relationship to proximal and distal depositional environments. *J. Sed. Petrol.,* 37: 24–43.

_____, 1975. Generalized facies models for resedimented conglomerates of turbidite association. *Geol. Soc. Amer. Bull,* 86: 737-48.

Weiler, Y., 1970. Mode of occurrence of pelites in the Kythrea flysch basin (Cyprus). *J. Sed. Petrol.,* 40: 1255-61.

*Article used in the synthesis of depositional environments of mud and silt turbidite facies.

Chapter 13

Thin-Bedded Turbidites in Modern Submarine Canyons and Fans

C. HANS NELSON
WILLIAM R. NORMARK
ARNOLD H. BOUMA
PAUL R. CARLSON

U.S. Geological Survey
Menlo Park, California

ABSTRACT

Thin-bedded sand and silt turbidites are present in nearly all physiographic environments of modern submarine canyons and fans. In Holocene time, they are common within channel deposits because high stands of sea level restrict coarse-grained sediment supply in many canyon-fan systems, and typical channel fill of thick-bedded turbidites is not developed. During the late Pleistocene period, most canyon-fan valley systems were active pathways for the transport of coarse-grained sediment, and nonchannelized parts of fans were built by overbank deposition of thin-bedded turbidites.

Thin-bedded Pleistocene turbidite facies of presently existing canyons and fans exhibit sufficient differences in geophysical and sedimentological characteristics to define distinct depositional environments. In channelized areas of upper and middle fans, thin-bedded turbidites are associated with thick-bedded thalweg sands and gravels; they are found in thin sequences that pinch out laterally and are subject to cutting, filling, and slumping from channel walls. Individual turbidites contain Bouma $T_{c\text{-}e}$ sequences, cross-lamination, ripple drift, starved ripples, and lenticular sands. Sand:shale ratios are high because the fine-grained interbeds of turbidite and/or hemipelagic mud are thin.

Thin-bedded Pleistocene turbidite facies from interchannel, interlobe, and levee crests (on some fans) are finer grained and more continuous laterally than those from adjacent intrachannel or mid-fan lobe (suprafan) areas. They exhibit thicker stratigraphic sequences with lower sand:shale ratios and are less dominated by cross-lamination. Distal turbidites of basin plains occur in sequences hundreds of meters thick that are laterally continuous for tens of kilometers. Parallel lamination is the most characteristic sedimentary structure, and low sand:shale ratios prevail because of well-developed pelagic mud interbeds.

INTRODUCTION

Studies of modern marine turbidite environments have resulted in varied depositional schemes or models for modern submarine canyon and fans. These models typically are based on (1) the overall surface morphology as determined by echo sounding; (2) the gross structure of the deposit as determined by high-resolution seismic profiling; and (3) the distribution, lithology, and internal structure of the near-surface sediments observed mainly in gravity and piston cores (1 m to 10 m in length).

Various physiographic subenvironments of the submarine canyon and fan continuum contain thin-bedded sands and silts that alternate with silty clay or clay. These deposits are outwardly similar and usually are classified by their general appearance either as thin-bedded turbidites or as distal turbidites. Although no single criterion is sufficient to identify the specific subenvironment in which a thin-bedded lithologic unit was deposited, a combination of criteria allows a reasonable distinction to be made.

Distribution patterns of thin-bedded turbidites have changed between glacial times of the late Pleistocene and high eustatic sea levels of the Holocene. Holocene deposits of channelized areas in most present-day submarine canyon-fan systems reveal a dominance of thin-bedded sand and silt turbidites, compared to coarser-grained and thicker underlying late Pleistocene turbidites; this results because these channel areas have received little or no coarser-grained sediment during high sea level stands of the Holocene period (Nelson and Kulm 1973). Interchannel areas of these modern fans are now isolated from sources of terrigenous sediment and consequently are blanketed with Holocene hemipelagic mud. During the late Pleistocene, however, the canyon-fan valley systems formed active pathways for the transport of coarse-grained sediments, and fan growth in interchannel areas was largely by overbank deposition of predominantly thin-bedded turbidites (Nelson and Kulm 1973; Damuth and Kumar 1975).

Thin-bedded turbidites of late Pleistocene age have been found in cores from most depositional subenvironments of present day canyon-fan systems, and they predominate everywhere except in channel thalwegs and on suprafan lobes where coarser-grained, thicker beds occur. Valley walls, levees, and interchannel subenvironments of the upper fan consist mainly of thin-bedded units deposited during late Pleistocene time (Nelson and Kulm 1973). Late Pleistocene deposits of interlobe and adjacent lower fan or basin plain areas are mainly thin-bedded turbidites with some mud turbidites and hemipelagic deposits (Horn et al. 1971; Normark and Piper 1972; Davies 1972; Bouma 1973). Such thin-bedded deposits also occur sporadically in late Pleistocene canyon wall, fan valley floor, deepsea channel, and interchannel areas of suprafan lobes (Carlson and Nelson 1969; Nelson and Kulm 1973).

The purposes of this chapter are (1) to describe the sedimentologic and geophysical characteristics of thin-bedded turbidites in all environments, proximal to distal; (2) to indicate some differences between the different groups of thin-bedded and truly distal turbidites; (3) to suggest mechanisms for the deposition of thin-bedded tubidites; and (4) to delineate facies characteristics of thin-bedded turbidites to aid in paleogeographic reconstruction of deepsea fans.

Terminology

Since "thin-bedded" and "thick-bedded," as well as "proximal" and "distal," are commonly used adjectives for describing turbidite sediments, defining our usage is pertinent. Terms used to describe bed thickness follow the system proposed by McKee and Weir (1953): *Thick-bedded* is applied to layers more than 20 cm thick; *medium-bedded,* to layers 5 to 20 cm thick; and *thin-bedded,* to layers 0.5 cm to 5 cm thick. The term *lamination* is applied to layers less than 0.5 cm thick (Bouma 1962). Where bed thickness is variable, the term describing the predominant thickness is used.

The terms *proximal* and *distal* in turbidite literature initially were used in relation to distance from source (Walker 1967) on the assumption that high sand:shale ratios and thick beds characterize proximal areas and low sand:shale ratios and thin beds characterize distal regions. The presence of thin-bedded turbidites in areas inferred to be "proximal" subsequently led authors to use these terms in a more general sense to describe the relative thickness of beds and completeness of Bouma sequences. We believe that such distinctions are artificial and to minimize confusion have avoided the use of proximal and distal wherever possible.

Disagreements also exist concerning submarine canyon and fan terminology. Studies of modern fan systems depend heavily on morphologic characteristics; on the basis of these studies, the authors have distinguished four morphologic divisions (Figure 13-1) of the submarine canyon-fan surface: (1) the submarine canyon, steep walled and incised into underlying formations; (2) the upper fan, adjacent to the canyon and characterized by one or a few leveed fan valleys; (3) the middle fan, formed by rapid deposition associated with distributary channels at the end of leveed valleys and in some places characterized by a convex-upward depositional bulge termed the *suprafan* by Normark (1970); and (4) the lower fan, characterized by a depositional surface of low relief. Sediments of the lower fan differ from those of the abyssal plain only by their higher content of clastics, and where a fan is building into a restricted basin, it may merge with the ponded basin plain environment.

Physiographic Subenvironments

Submarine canyon. The submarine canyon represents the erosive section of the canyon-fan system. Although submarine canyons are extremely varied in such aspects as their dimensions, shapes, compositions of wall rocks, and types and amounts of sediment accumulating on the canyon floor, all can be divided into three readily delineated morphologic subdivisions: floor, walls, and tributaries. Canyons are morphologically distinguished from fan valleys by their greater wall relief, more V-shaped cross-sections, more sinuous, steeper axial profiles, and by the presence of tributary, rather than distributary, channels (Shepard and Dill 1966).

Upper fan. The upper fan has the steepest gradients of the fan surface (Figure 13-1). If the sediment supply is restricted to a single submarine canyon, the upper fan generally has one active fan valley that is continuous with the canyon. Upper fan valleys are characterized by prominent levees and valley floors that commonly are elevated above the surrounding level of the fan.

Middle fan or mid-fan. The mid-fan is commonly distinguished by a depositional bulge or lobe that appears as a convex-upward segment on a radial profile. This morphologic feature has been termed a *suprafan* (Normark 1970) and is located at the termination of the major active valley on the upper fan. The suprafan is characterized by numerous unleveed distributary channels or channel remnants that result from rapid channel migration due to deposition within the distributary system. On conventional seismic reflection profiles, the suprafan surface appears hummocky with numerous hyperbolic "side echoes" and discontinuous subbottom reflectors.

Several studies of modern submarine fans have noted the presence of a convex-upward segment on most radial profiles (Nelson et al. 1970; Normark 1970, 1974; Haner 1971; Damuth and Kumar 1975). Shifting of the valley(s) on the upper fan will lead to growth of new suprafan lobes. Abandoned suprafans become blanketed by muds that eventually smooth over the characteristic local relief, although the convex-upward profile normally remains.

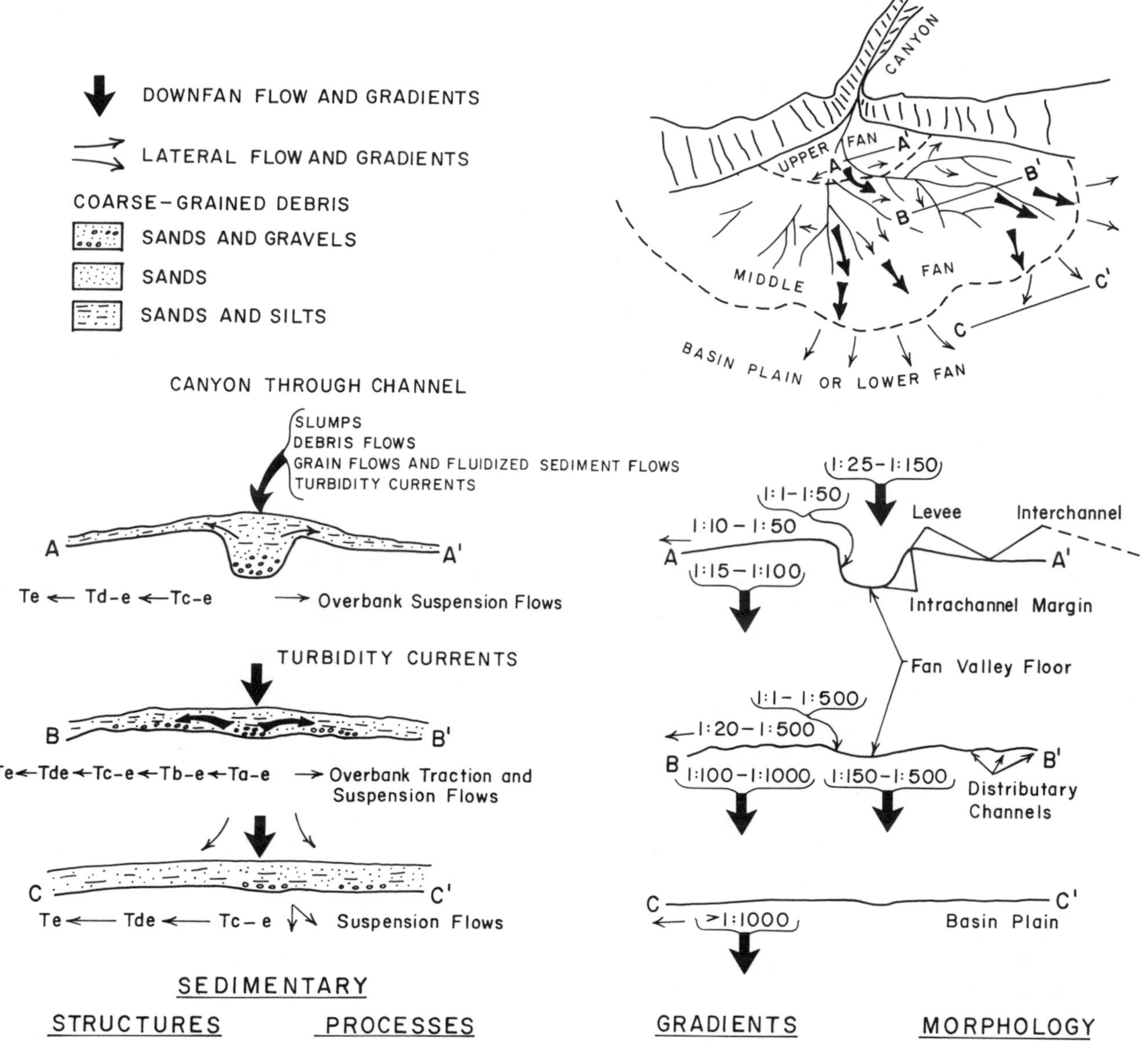

Figure 13-1. Morphologic and sedimentologic characteristics of submarine canyon-fan systems. (Source: Gradients from Nelson et al. 1970; Normark 1970; Haner 1971.)

Lower fan. The lower fan is the least known morphologic area of the canyon-fan system. Delineation of both the upper and lower boundaries of the lower fan is controversial. For example, Nelson et al. (1970) in describing Astoria Fan use "lower fan" for the more distal part of the channeled area that is characterized by numerous broad, shallow, and braided fan valleys and a concave-upward profile. Normark (1970) refers to this segment as the lower half of the suprafan or of the "middle fan." The nearly flat, nonchanneled region extending distally beyond the fan-shaped morphology has been termed *lower fan, fan fringe, basin plain,* or *abyssal plain* by various workers. The terminology employed in this chapter is that of Normark (1970) who refers to this morphologic unit as the "lower" fan. Features that characterize the lower fan portions of San Lucas and Navy Fans include nearly horizontal surfaces, uniform parallel bedding on seismic-reflection profiles, and no evidence of channels. Distinct sand beds generally are absent and the grain size of the interbedded silts and silty clays decreases distally (Normark and Piper 1972).

THIN-BEDDED TURBIDITES

Sedimentologic Characteristics

Bed thickness. Dominance of thin-bedded turbidites in levee and interchannel areas of upper fans (Figure 13-2) and on walls of canyons has been noted by many investigators (Carlson and Nelson 1969; Piper 1970; Haner 1971; Normark and Piper 1972; Bouma 1973; Nelson and Kulm 1973; Maldonado and Stanley 1976). Although little detail is available concerning the variation in thickness of beds in different interchannel environments because of the difficulty of correlating individual beds from core to core, the general impression that beds decrease in thickness and number away from channels (Piper 1970; Nelson and Kulm 1973) appears to be confirmed by data from a series of tuffaceous turbidites in Astoria Fan. These beds can be correlated from location to location because of their content of Mazama ash (Nelson et al. 1968). A transect laterally across the upper fan revealed that one such bed thinned gradually from a thickness of 25 cm in the thalweg of a channel to 5 cm over a distance of 25 km; 50 km from the channel this ash bed is 2 cm thick (Nelson 1976). Deposition on the Ascension Fan valley floor during the last glacial interval produced sandy horizons that are thicker, more numerous, and contain less clay-sized material than in the correlative section on the adjacent levee (Hess and Normark 1976). This evidence suggests the occurrence of a rather abrupt change from thick-bedded sandy turbidites in channel thalwegs to thin-bedded, finer-grained turbidites in interchannel locations and that the thickness and number of turbidites decreases laterally away from fan valleys. Nelson (1976) further suggests that thin-bedded turbidites may be better developed (e.g., coarser and thicker) on right levees of channels in the northern hemisphere, as is also indicated by morphologic evidence (Menard 1955; Hamilton 1967).

Levee and interchannel turbidites become fewer and thinner laterally away from the channel in the mid-fan region (Figure 13-2), although there appears to be downfan increase in the thickness of thin-bedded turbidites in this part of the fan (Piper 1970; Nelson and Kulm 1973). For example, late Pleistocene thin-bedded turbidites in interchannel areas on upper Astoria Fan average 2 cm thick; those on the upper mid-fan are 6 cm thick, and those on the lower mid-fan are 10 cm thick and nearly equal to the average thickness of adjacent early Holocene channel beds (Nelson 1976).

Thin-bedded turbidites are predominant throughout the lower fan subenvironment (Figure 13-2). Thick-bedded channel deposits apparently occur in this setting only if the fan is being actively dissected (Stanley et al. 1974; Horn et al. 1971; Normark 1970; Normark and Piper 1972). The transition from late Pleistocene turbidite regimes to Holocene hemipelagic sedimentation (Carlson

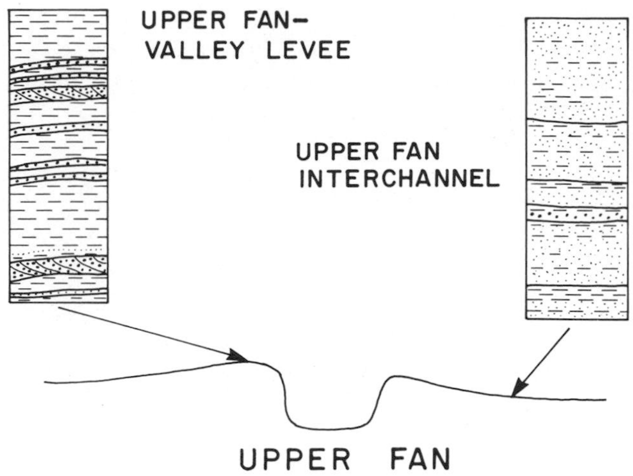

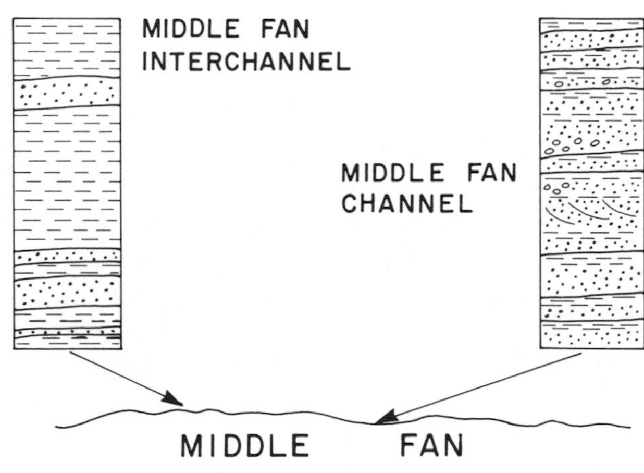

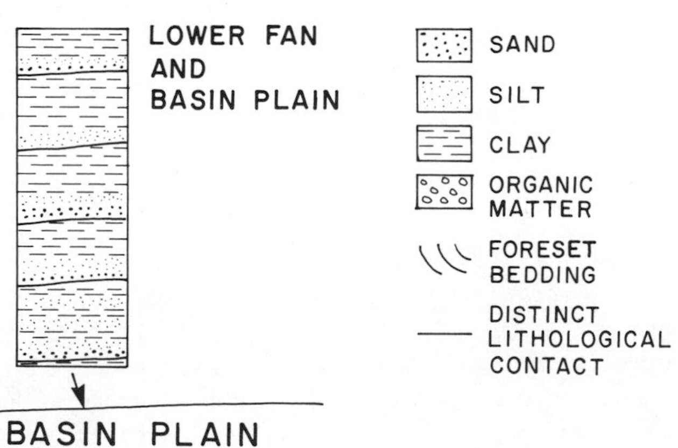

Figure 13-2. Diagrammatic sketches of thin-bedded turbidites in piston cores from deepsea fan environments, based mainly on late Pleistocene sections from Astoria Fan and Mississippi Fan. Note: Length of core segments is 15 cm.

and Nelson 1969; Nelson and Kulm 1973; Damuth and Kumar 1975) was accompanied by fining-upward sequences of thin-bedded turbidites over the entire fan as turbidity currents became progressively fewer and smaller (Nelson and Nilsen 1974).

Sedimentary structures. Major sedimentary structures that can be found in association with thin-bedded turbidites are known from canyon wall, fan valley floor, and levee areas. Locally steep walls and slump blocks of semiconsolidated sediment have been seen in La Jolla Fan Valley in submersible dives (Moore 1965). Slump deposits have been observed from deep-tow side-looking sonar records (Figure 13-3) and in cores from Astoria Canyon and adjacent fan valley floors (Carlson and Nelson 1969). Extensive scour depressions and meandering thalwegs have been found in valleys on Monterey and Navy Fans (Normark et al. 1976; Hess and Normark 1976). Lateral migrations of a thalweg may undercut the adjacent valley wall, thereby causing blocks of wall material and levee sediments – probably composed of thin-bedded turbidites – to slump into the channels (Normark 1976).

Internal sedimentary structures are prominent in thin-bedded turbidites of upper fan levees (see Figure 13-2). The thicker beds typically exhibit Bouma (1962) $T_{c\text{-}e}$ sequences; they contain abundant low-angle, fine-scale cross-lamination and climbing ripple drift lamination (Carlson and Nelson 1969; Bouma 1973). The thinner beds show abrupt changes in thickness along strike, thereby suggesting the presence of ripples evident in thicker beds; poor core preservation may obliterate cross-laminations. $T_{d\text{-}e}$ sequences containing flat laminations are characteristic of the very thin beds that predominate in interchannel areas far from valleys (Carlson and Nelson 1969). Such beds are found locally in the upper fan and canyon wall environments.

As can be seen in Figure 13-2, in the middle fan interchannel areas, the thin-bedded turbidites become thicker and consequently contain more complete Bouma sequences ($T_{b\text{-}e}$ and $T_{c\text{-}e}$) than the thin beds on the upper fan (Haner 1971). As also shown in Figure 13-2 thin-bedded turbidites of lower fan, basin plain, and abyssal plain environments are typically thinner-bedded than those of the middle and upper fan areas, and parallel lamination is the main internal sedimentary structure apparent in these very thin beds (Bouma 1973). The thin sand and silt beds of basin plains generally are found only in the lower parts of cores that penetrate through Holocene and into late Pleistocene deposits (Normark and Piper 1972; Bouma 1973; Nelson and Kulm 1973). Hemipelagic and turbidite muds are the principal constituents in the upper parts of these cores. Apparently, sand:shale ratios are very low in the near-surface basin and abyssal plain deposits exanined from piston cores.

Texture and composition. Lateral trends in the texture and composition of thin-bedded turbidites are similar to those noted in bed thickness and help to distinguish between proximal and distal depositional environments. At the lower end of Astoria Canyon, the average mean grain size of thin-bedded turbidites is 0 022 mm. near the base of the canyon wall, and 0.015 mm, high on the wall (Carlson 1967). On the right side of upper Astoria channel (looking downstream), the average mean diameter is 0.041 mm, at the base of the upper fan valley wall 2 to 3 km from the thalweg; 0.019 mm at the levee crest 10 km from the thalweg; and 0.014 mm in interchannel beds at the base of the levee 20 km from the thalweg. A similar grain-size probably exists on the left side of the channel, although these beds can be traced only a few kilometers to the base of the continental slope and are so thin that they cannot be analyzed individually (Nelson 1976).

These same trends appear to be present from the smallest to the largest fans. The average mean grain size of thin-bedded turbidites from the small Reserve Fan in Lake Superior is 0.053 mm in the channel axis, 0.031 mm on the levee crest, 0.022 mm on the mid-levee, and 0.017 mm at the edge of the right levee (Normark and Dickson 1976). A similar trend of diminishing grain size away from channels can be observed on the Mississippi Fan (Bouma 1973). A lateral decrease in grain size with distance from leveed valleys on the Bengal Fan (Bay of Bengal) is reflected by well-developed sorting of the biogenous component of sediments derived from turbidity currents passing through the valleys (J. Yount, U. S. Geological Survey, personal communication, 1976).

Sand-sized sediments in the thin-bedded turbidites of the Astoria Canyon-fan system (Figure 13-4) contain large amounts of mica and plant fragments or carbonaceous material (Carlson and Nelson 1969; Nelson and Kulm 1973). In addition both Wilde (1965) and Normark and Piper (1972) found that the largest quantity of mica in Monterey and Navy Fans occurs in the most distal thin-bedded turbidites.

Geophysical Characteristics

The acoustic character of thin-bedded turbidites in subsurface late Pleistocene and older deposits can be recognized on high-resolution reflection profiles to provide a useful means of mapping turbidite packages and their associated depositional environments over wide areas. We must point out that individual thin-bedded turbidites cannot be resolved on 3 to 4 kHz reflection records; however, coring of the near-surface packages in conjunction with high-resolution profiling permits extrapolation based on their acoustic character, which is a very useful technique in reconstructing fan depositional systems. Deep-tow and conventional surface ship 3.5 kHz records have been used for this purpose.

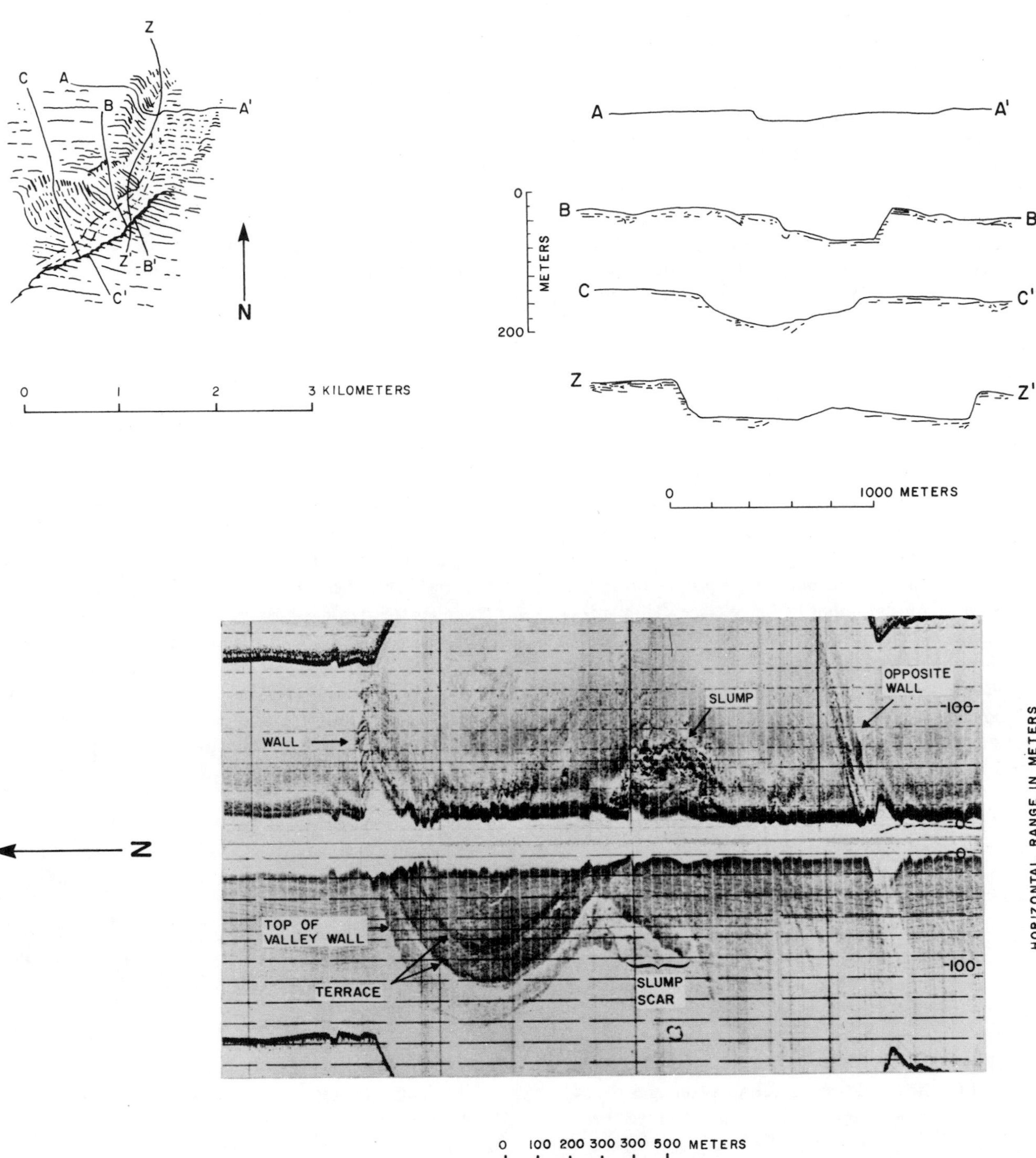

Figure 13-3. Side scan (bottom) and deep-tow (top right) profiles of valley meander on La Jolla submarine fan. Note: Physiographic sketch (upper left) by Tau Rho Alpha shows lines of profiles across the valley. Profile A-A' is a line drawing of narrow-beam echo-sounding record with no vertical exaggeration; B-B', C-C', and Z-Z' are line drawings of 3.8 kHz reflection profiles with vertical exaggeration of X3.5. Slump from fan-valley wall is seen in profile B-B' and on left hand (top) side-looking sonar record; side-looking sonar records are from profile Z-Z'.

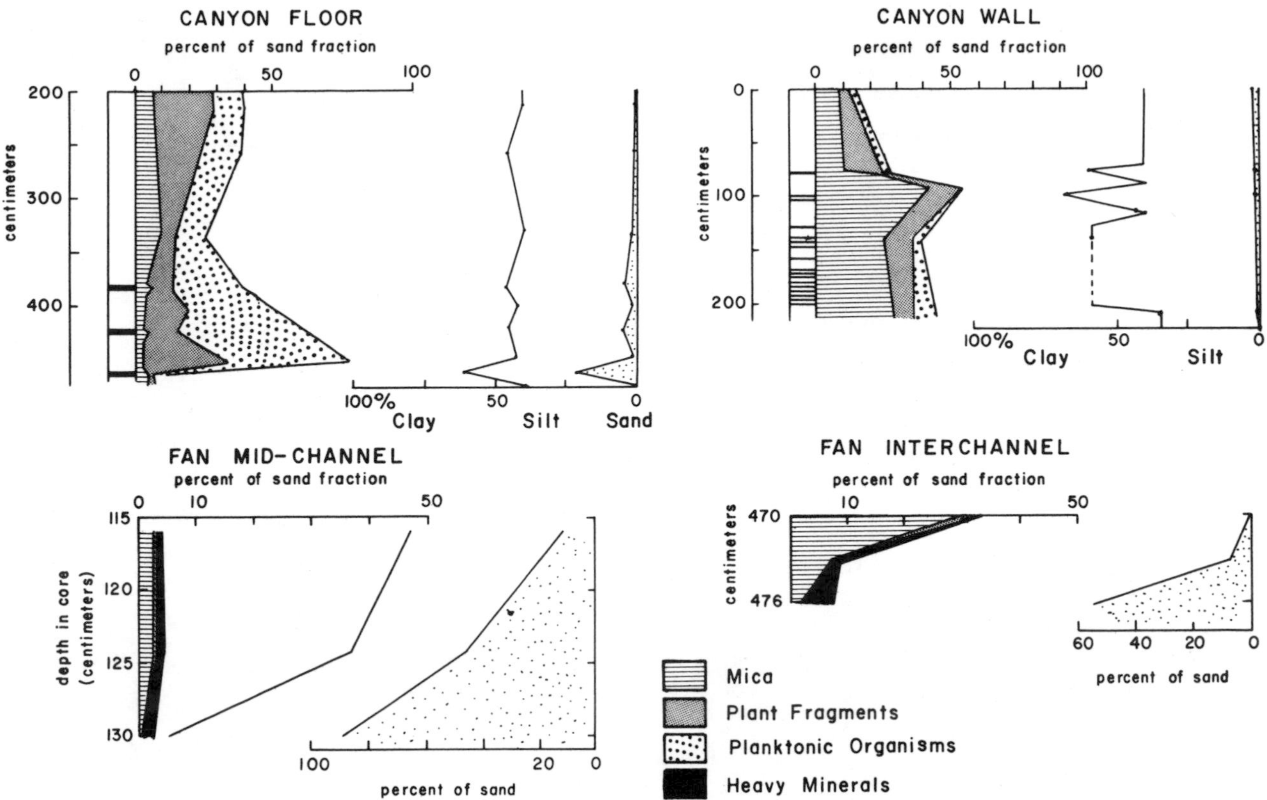

Figure 13-4. Textural and compositional variation in late Pleistocene core segments from Astoria Canyon and fan environments.

Major acoustic properties of a thin-bedded turbidite series on reflection records include the continuity and parallelism (or lack thereof) of reflectors, the depth of acoustic penetration, and the distinctness of individual reflectors. The depth of acoustic penetration and the continuity and parallelism of reflectors is related in part to the amount of sand in the near-surface sediments (Normark 1970; Normark and Piper 1972; Damuth and Kumar 1975).

Poor acoustic penetration characterizes the thalwegs of fan valley floors and suprafan areas where sand layers are thick; the valley walls and levees of some smaller fans commonly produce poor reflections, probably because of the dominance of sand deposition (A-A' and B-B' in Figure 13-5). Penetration typically is intermediate in upper fan valley walls and levees of larger fan systems (Bouma 1973; Hess and Normark 1976).

The depth of penetration in interchannel and interlobe areas increases progressively as the distance from levees and suprafans increases (Figure 13-6). Penetration also is good on basin slopes adjacent to fans. Good penetration and continuity of reflectors (see D-D' in Figure 13-5) normally occurs in ponded basin plain or abyssal plain areas (Normark and Piper 1972; Bouma 1973; Damuth and Kumar 1975).

A 3.8 kHz reflection profile taken by the deep-tow instrument shows that acoustic penetration increases rather uniformly across the levee with increasing distance from Ascension Fan Valley (Figure 13-7). However, in other areas (see Figure 13-5) when the sediments are sandy, bedding continuity does not increase significantly from the smooth levee crest to the backslope of the levee. An undulating or rolling topography is common on the backslope of higher levees on the right side of deepsea channels in the northeast Pacific (Hamilton 1967; Normark 1974) and reflects a depositional process presumed to be related to overflowing turbidity currents. The internal structure of the rolling topography in the Ascension channel region on Monterey Fan (Figure 13-7) suggests that this dune-like type of relief appears to form as the bedforms migrate upslope, toward the levee crest, during growth (Hess and Normark 1976). This process results in the onlap of lenticular beds as much as 1 km long; similar features may be common in smaller leveed valleys where narrow levee widths would make them less conspicuous (Hess and Normark 1976).

Deep-tow profiles over the western part of the suprafan on San Lucas Fan at the tip of Baja California, Mexico, show many disconnected depressions that resemble erosional channels (see Figure 13-6). Track lines over this suprafan are closely spaced, and analysis of narrow-beam echo-sounding and side-scanning sonar records show conclusively that these depressions do not form a simple

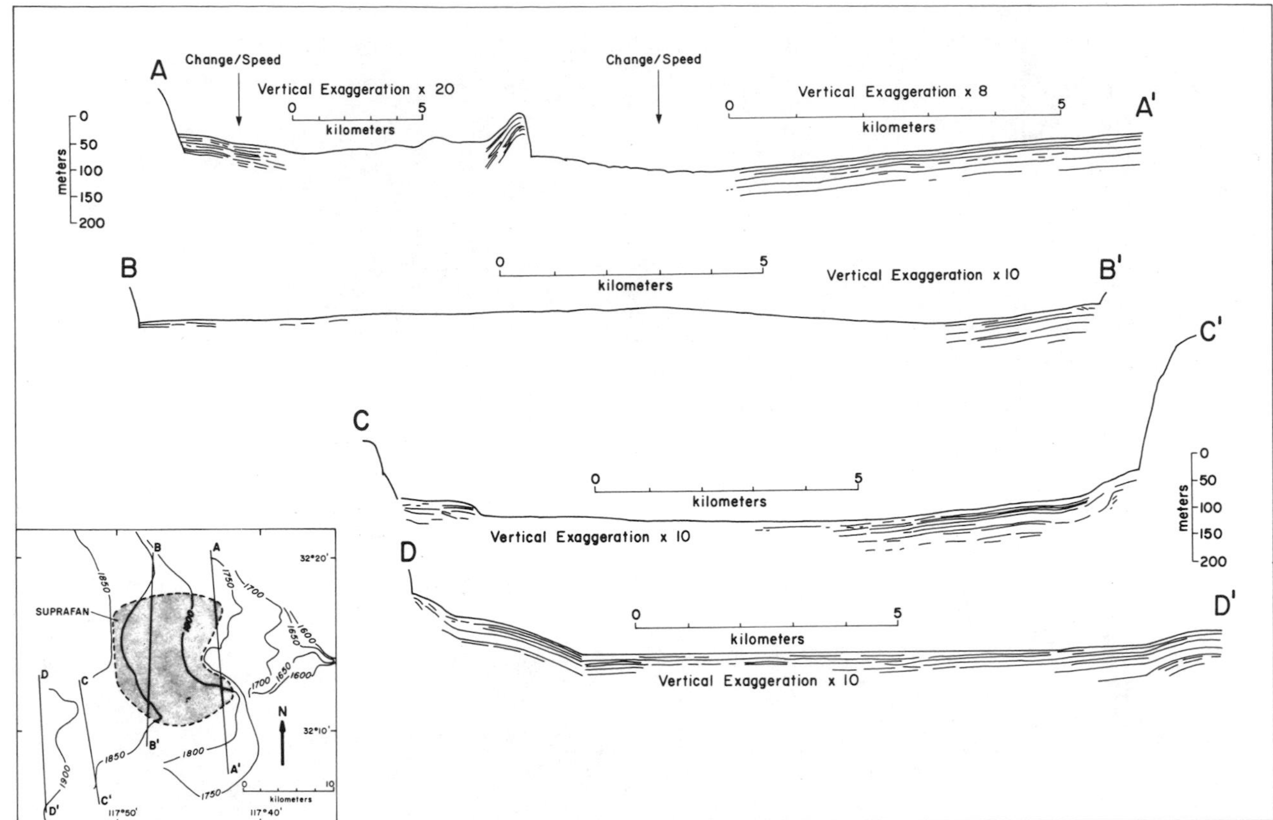

Figure 13-5. Line drawings of 3.5 kHz high-resolution acoustic profiles from Navy Fan (Southern California Boderland). Note: In inset map, contours are in meters; dashed line delineates suprafan. In the profile drawings, no acoustic penetration appears over upper fan and suprafan area of Navy Fan (A-A', B-B'); penetration improves toward lower fan, which is a basin plain in this case (C-C', D-D'). One or both ends of all profiles show numerous, gently dipping, parallel-bedded, deeper reflecting horizons within hemipelagic muds and older basin sediments predating Navy Fan (Source: Normark and Piper 1972).

continuous channel network (Normark 1970). The distributary channel remnants reach 2 km in length and 50 m in relief, are arcuate and asymmetric, and appear to have a remnant "thalweg" and terraced walls. Sand to pebble grades are found within these depressions and over the central part of the suprafan (Normark 1970). The depth of acoustic penetration, using the deep-tow reflection system, suggests that sands become less common toward the margins of the suprafan (Figure 13-6). The continuity of individual reflectors increases both downfan and across the lower suprafan, thereby suggesting increasing extent of thin-bedded turbidites.

DEPOSITIONAL PROCESSES

Sediment gravity flows (slump, debris flow, grain flow, fluidized sediment flow, and turbidity current) generated in canyons transport sediments to submarine fans (Middleton and Hampton 1973; also see Figure 13-1 and Chapter 8 of this volume). Detailed seismic and sediment core data from canyon-fan systems indicate that depositional processes in the canyon and on the upper fan result primarily from the channelized flow down the canyon thalweg and the main fan valleys. The thin-bedded turbidites among canyon wall sediments may have been deposited from the dilute suspension flow of higher parts of gravity flows lapping up on the canyon walls. Slumping onto the canyon and channel floors from the surrounding steep and commonly undercut canyon and fan valley walls disrupts thin-bedded turbidites locally along valleys (Moore 1965; Carlson and Nelson 1969; Normark 1970; J. Yount, U. S. Geological Survey, personal communication, 1975). Because the maximum traction and bedload flow is localized in the axis of thalweg(s) on large fan valley floors, overbank deposition may be a significant process within fan valleys as well as over the fan surface. Correlation of beds between fan valley walls, levees, and adjacent channels indicates that suspension sheet flows (see Figure 13-1) of turbidity currents may overtop the deepest upper fan valleys and deposit thin-bedded turbidites over widespread interchannel areas (Nelson et al. 1968; Griggs and Kulm 1970; Piper 1970). Gradients over levees laterally away from the channel axis may be steeper than downvalley

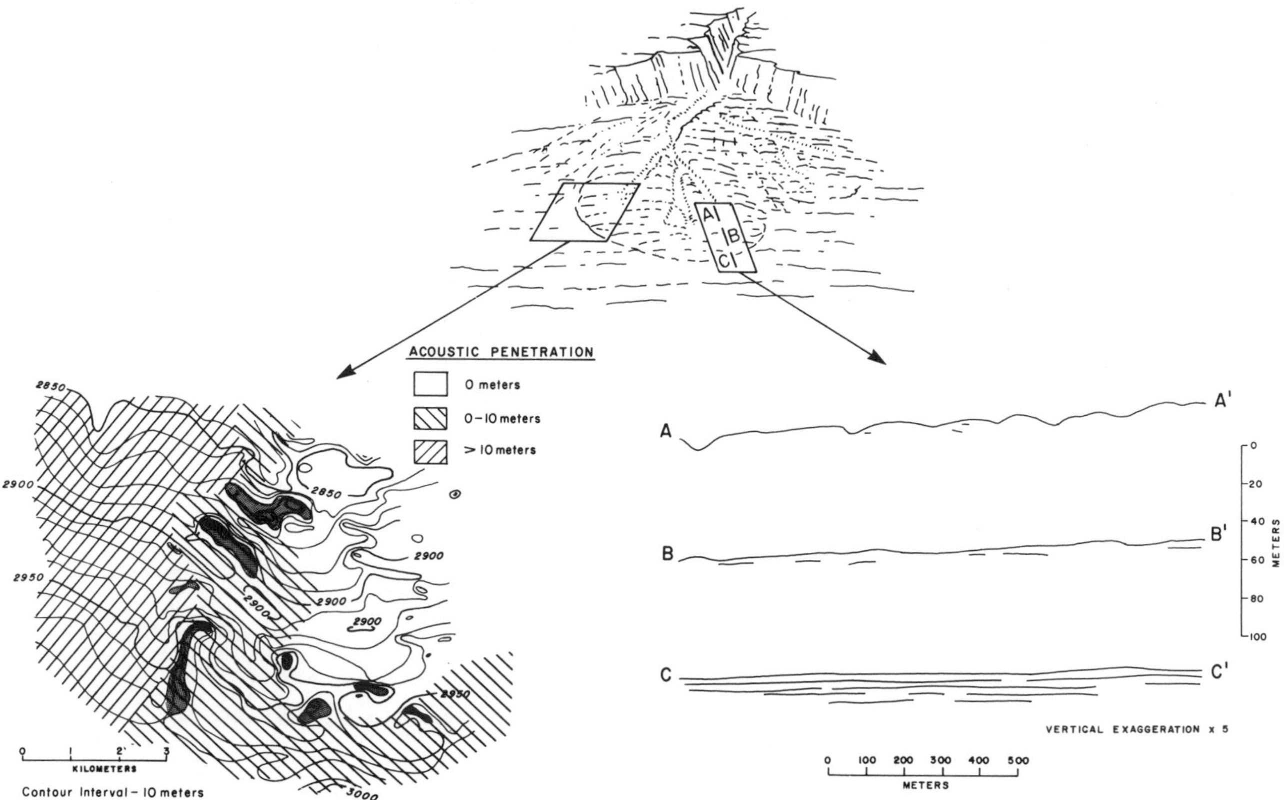

Figure 13-6. Physiographic sketch of submarine fan by Tau Rho Alpha with insets showing change in depth of acoustic penetration away from the suprafan, which appears as a low bulge on fan surface. Note: Contour map (lower left) based on San Lucas Fan, Baja, California (Normark 1970) shows isolated depressions (shaded areas) characteristic of suprafans. Line drawings of 4.0 kHz reflection profiles (lower right) are based on deep-tow profiles from Navy Fan (Normark et al. 1976). Both map and line drawings show a downfan decrease in local relief and an increase in acoustic penetration.

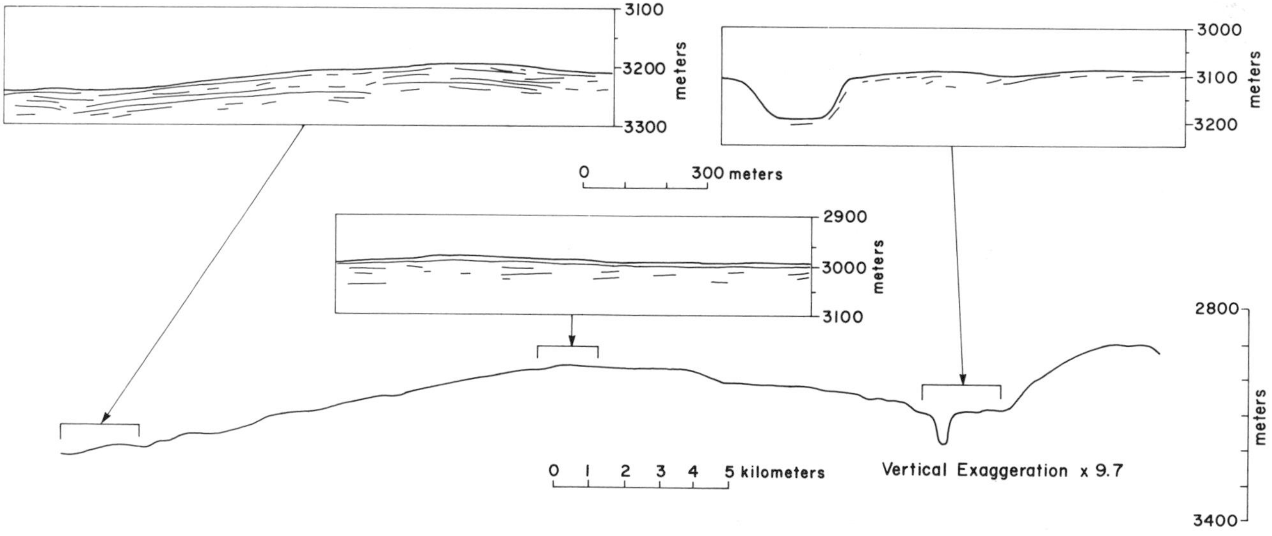

Figure 13-7. Enlargements of line drawings from deep-tow 3.8 kHz reflection profiles laterally across Ascension fan valley, Monterey Fan (central California). Note: Vertical exaggeration of insets is X3. Depth of penetration and bedding continuity increase from valley floor toward backslope of the levee.

Table 13-1. *Seismic and Sedimentologic Characteristics of Thin-Bedded Turbidites*

	Depositional Environment	*Morphology*	*Acoustic Character*	*Sedimentary Structures*	*Sediment Composition*	*Sediment Size*	*Literature Sources*[a]
Canyon	Walls	Steep and irregular.	Poor reflections because of steep slopes and side echoes with numerous hyperbolics.	Flat and cross-lamination in very thin lenses.	Mica dominant, plant fragments, organic debris.	Clayey and sandy silts.	2, 3, 8, 19, 26
	Floor	Straight to sinuous path, flat floor, some slump blocks.	Poor acoustic penetration. Reflectors discontinuous or absent, because of thick thalweg sands.	Cross-bedded, graded-bedded, and featureless thin to thick beds.	Detrital minerals and rock fragments in thicker beds; mica, plant fragments in thinner beds; displaced benthic foraminifers in both.	Fine sands, silts.	3, 8, 11, 13, 21, 23
Upper Fan	Valley thalweg	Meandering channels; (100 m to 500 m wide, 20 m to 50 m relief) valley floors to 20 km wide.	(as above)	(as above)	(as above)	(as above)	3, 8, 9, 11, 13, 17, 24
	Valley walls	Depositional slopes to 10°, undulating to irregular, erosionally formed walls with slopes near vertical.	Reflectors horizontal to subparallel along valley walls and may pinch out. Penetration slightly better than adjacent thalweg; poor reflector continuity.	Flat and cross-laminated in lenticular beds; thinner beds near top of wall, thicker, featureless beds at base.	Detrital grains, mica concentrations and displaced foraminifers.	(as above)	1, 5, 9, 14, 15, 24 Figure 7
	Levees	Smooth, slightly convex upward crests, undulating, some bed forms common on backslope.	Limited acoustic penetration and reflector continuity; both increase across backside of levee. Some large-scale bedforms common.	Thin lenses dominated by cross- and climbing ripple lamination, flat laminated beds.	Detritals in thicker beds; mica and plant fragment concentrations in thinner beds.	Silts and fine sands	3, 4, 5, 6, 8, 9, 11, 12, 17, 24
	Inter-channel	Gently undulating to smooth.	Good penetration and continuity of reflectors away from levee. Regular, flat-lying bedding.	Very thin, flat laminated.	Mica and plant fragments dominate.	Silts and very fine sands.	1, 4, 5, 9, 11, 12 15, 16, 19

Table 13-1. *continued.*

	Depositional Environment	*Morphology*	*Acoustic Character*	*Sedimentary Structures*	*Sediment Composition*	*Sediment Size*	*Literature Sources*
Mid-fan	Suprafan lobe	Irregular, channelled, hummocky.	Discontinuous reflectors and numerous hyperbolics; poor penetration because of sands.	Thin beds; Bouma $T_{b\text{-}e}$ sequences.	Detrital minerals.	Sands.	4, 15, 19
	Inter-lobe	Slight depressions and flat regions surrounding lobes.	Good penetration and continuity of reflectors away from lobes. Regular, flat-lying bedding.	Thin beds; Bouma $T_{b\text{-}e}$ grading into flat lamination.	Detrital with mica and plant fragment concentrations dominant away from lobes.	Silts and fine sands.	1, 4, 7, 15, 18, 19
Lower Fan	Basin plain and/or abyssal plain	Flat	Good penetration; flat, very continuous reflectors.	Thin even beds, flat lamination and some graded beds.	Mica and plant fragment concentrations common.	Clayey silts.	1, 4, 10, 19, 20, 22, 24, 25, 27

[a]1 = Bouma (1973); 2 = Carlson (1967); 3 = Carlson and Nelson (1969); 4 = Damuth and Kumar (1975); 5 = Griggs and Kulm (1970); 6 = Hamilton (1967); 7 = Haner (1971); 8 = Heezen et al. (1964); 9 = Hess and Normark (1976); 10 = Horn et al. (1971); 11 = Kelling and Stanley (1976); 12 = Maldonado and Stanley (1976); 13 = Middleton and Hampton (1973); 14 = Moore (1965); 15 = Nelson (1968); 16 = Nelson and Kulm (1973); 17 = Normark (1974); 18 = Normark (1976); 19 = Normark and Piper (1972); 20 = Piper (1970); 21 = Royse (1964); 22 = Ryan et al. (1965); 23 = Shepard and Dill (1966); 24 = Shepard et al. (1969); 25 = Stanley (1974); 26 = Stanley et al. (1974); 27 = Wilde (1965).

gradients (see Figure 13-1) and could provide a driving mechanism for extensive lateral flows that deposit thin beds tens of kilometers across interchannel areas (Nelson 1968).

The mid-fan is characterized by a depositional bulge (suprafan) that forms at the lower end of the active fan valley. Overbank flows are substantially more common in this area because wall relief has decreased significantly and fan valleys are in general broader and flatter and have split into distributaries. Bouma sequences are more complete (T_{b-e}) in interchannel regions on the middle fan (Haner 1971; Nelson and Kulm 1973; also see Figure 13-1), thereby suggesting that turbidity currents transport fine sand and silt by both traction and suspension into these areas.

Grain size distribution and rate of sediment supply may determine whether a suprafan is formed. The large Bengal Fan (Curray and Moore 1971) exhibits none of the characteristic features of the mid-fan division and has an extensive system of large leveed valleys. Many of the valleys cross almost the entire fan (2,500+ km), but none appear to terminate in a suprafan, and little coarse sediment is present within the main valley, even on the upper fan. The relative absence of internal levee structure (Curray and Moore 1971) may also reflect a rather uniform grain size in levee sediments. If little coarse sediment is present in the turbidity current, rapid deposition and formation of a suprafan may not occur. Instead, enhanced levee development near the upper fan valley terminus may form relatively straight channels that prograde rapidly downfan (Normark 1974, 1976).

Suspension sheet flows of turbidity currents reach the lower fan and basin plain to deposit thin-bedded turbidites consisting mainly of T_{d-e} sequences (Normark 1970; Nelson and Kulm 1973; Bouma 1973; also see Figure 13-1). These flows pond in restricted basins such as those of the Mediterranean Sea (Ryan et al. 1965), the Gulf of Mexico (Davies 1972), and the California Continental Borderland (Normark and Piper 1972) or build abyssal plains such as those of Hatteras and Sohm Abyssal Plains (Horn et al. 1971).

CONCLUSIONS

Thin-bedded turbidites occur in most recognized subenvironments of late Pleistocene submarine canyon-fan systems. The methods employed to study these deposits in modern depositional settings differ from those used to investigate equivalent ancient deposits. Piston and gravity cores from modern canyons and fans do not provide the lateral continuity possible in outcrops of ancient sediments and do not reach thickness visible in most outcrops. However, the combination of sediment cores and acoustic reflection data does provide a three-dimensional view of the modern system. Significant seismic and sedimentologic characteristics of these deposits are compiled in Table 13-1.

Distribution, concentration, and abundance of thin-bedded turbidites are influenced by factors such as sediment availability, basin geometry, sea level changes, tectonism, and age of the canyon-fan system. Mechanisms of deposition of thin-bedded turbidites are varied, depending on location within the canyon-fan system. Thin-bedded turbidites found in the interchannel areas largely represent overbank deposition, whereas those on the lower fan are deposits laid down by broad sheet flow of waning turbidity currents. Although the deposits in both cases represent a deceleration of the transporting density current, their depositional site on the fan surface, and consequently their paleogeographic significance, may differ greatly.

ACKNOWLEDGMENTS

The majority of the geophysical data was obtained with the deep-tow instrument package of the Marine Physical Laboratory under the direction of F. N. Speiss at Scripps Institution of Oceanography, University of California at San Diego. Tau Rho Alpha provided three-dimensional drawings and Gordon Hess and Bradley Larsen assisted with compilation of figures. The authors are indebted to Sam Clarke and Monty Hampton for beneficial review of the manuscript.

REFERENCES

Bouma, A. H., 1962. *Sedimentology of Some Flysch Deposits.* Elsevier, Amsterdam, 168 p.

——, 1973. Leveed-channel deposits, turbidites, and contourites in deeper part of Gulf of Mexico. *Proc. Trans. Gulf Coast Assoc. Geol. Soc.*, 23rd Annual Convention, Houston, pp. 368-76.

Carlson, P. R., 1967. *Marine geology of Astoria submarine canyon.* Unpublished Ph.D. thesis, Oregon State University, Corvallis, 259 pp.

——, and Nelson, C. H., 1969, Sediments and sedimentary structures of the Astoria submarine canyon-fan system, Northeast Pacific. *J. Sed. Petrol.*, 39: 1269-2182.

Curray, J. R., and Moore, D. G., 1971. Growth of the Bengal deep-sea fan and denudation in the Himalayas. *Geol. Soc. Amer. Bull.*, 82: 563-72.

Damuth, J. E., and Kumar, N., 1975. Amazon Cone: Morphology, sediments, age, and growth pattern. *Geol. Soc. Amer. Bull.*, 86: 863-78.

Davies, D. K., 1972. Deep-sea sediments and their sedimentation, Gulf of Mexico. *Amer. Assoc. Petrol. Geol. Bull.*, 56: 2212-39.

Griggs, G. B., and Kulm, L. D., 1970. Physiography of Cascadia deep-sea channel. *Northwest Science*, 44: 82-94.

Hamilton, E. L., 1967. Marine geology of abyssal plains in the Gulf of Alaska. *J. Geophys. Res.*, 72: 4189-213.

Haner, B. E., 1971. Morphology and sediments of Redondo submarine fan, Southern California. *Geol. Soc. Amer. Bull.,* 82: 2413-32.

Heezen, B. C., Menzies, R. J., Scheider, E. D., Ewing, W. M. and Granelli, N. C. L., 1964. Congo submarine canyon. *Amer. Assoc. Petrol. Geol. Bull.,* 48: 1126-49.

Hess, G. R., and Normark, W. R., 1976. Holocene sedimentation history of the major fan valleys of Monterey Fan. *Mar. Geol.,* 22: 233-51.

Horn, D. R., Ewing, M., Delach, M. N., and Horn, B. M., 1971. Turbidites of the Hatteras and Sohm Abyssal Plains, western Atlantic. *Mar. Geol.,* 11: 287-320.

Kelling, G., and Stanley, D. J., 1976. Sedimentation in canyon, slope, and base-of-slope environment. In: D. J. Stanley and D. J. P. Swift (eds.), *Marine Sediment Transport and Environmental Management.* Wiley, New York, pp. 379-435.

Maldonado, A., and Stanley, D. J., 1976. The Nile Cone: Submarine fan development by cyclic sedimentation. *Mar. Geol.,* 20: 27-40.

McKee, E. D., and Weir, G. W., 1953. Terminology for stratification and cross-stratification in sedimentary rocks. *Geol. Soc. Amer. Bull.,* 64: 381-90.

Menard, H. W., 1955. Deep-sea channels, topography, and sedimentation. *Amer. Assoc. Petrol. Geol. Bull.,* 39: 236-55.

Middleton, G. V., and Hampton, M. A., 1973. Mechanics of flow and deposition. In: G. V. Middleton and A. H. Bouma (eds.), *Turbidites and Deep-Sea Sedimentation.,* Soc. Econ. Paleont. Mineral., Pacific Section, Short Course, Anaheim, pp. 1-38.

Moore, D. G., 1965. Erosional channel wall in La Jolla Sea-Fan Valley seen from the bathyscaph "Trieste II." *Geol. Soc. Amer. Bull.,* 76: 385-92.

Nelson, C. H., 1968. *Marine geology of Astoria deep-sea fan.* Unpublished Ph.D. thesis, Oregon State University, Corvallis, 287 pp.

———, 1976. Late Pleistocene and Holocene depositional trends, processes, and history of Astoria deep-sea fan, Northeast Pacific. *Mar. Geol.,* 20: 129-73.

Nelson, C. H. and Kulm, L. D., 1973, Submarine fans and deep-sea channels. In: G. V. Middleton and A. H. Bouma (eds.), *Turbidites and Deep Water Sedimentation.* Soc. Econ. Paleont. Mineral., Pacific Section, Short Course, Anaheim, pp. 39-78.

———, and Nilsen, T. H., 1974. Depositional trends of modern and ancient deep-sea fans. In: R. H. Dott Jr., and R. H. Shaver (eds.), *Modern and Ancient Geosynclinal Sedimentation.* Soc. Econ. Paleont. Mineral., Sp. Publ. 19, pp. 54-76.

———, Kulm, L. D., Carlson, P. R., and Duncan, J. R., 1968. Mazama Ash in the northeastern Pacific. *Science,* 161: 47-49.

———, Carlson, P. R., Byrne, J. V., and Alpha, T. R., 1970. Development of the Astoria Canyon-Fan physiography and comparison with similar systems. *Mar. Geol.,* 8: 259-91.

Normark, W. R., 1970. Growth patterns of deep-sea Fans. *Amer. Petrol. Geol. Bull.,* 54: 2170-95.

———, 1974. Submarine canyons and fan valleys: Factors affecting growth patterns of deep-sea fans. In: R. H. Dott, and R. H. Shaver (eds.), *Modern and Ancient Geosynclinal Sedimentation.* Soc. Econ. Paleont. Mineral., Sp. Publ. 19, pp. 56-68.

———, 1976. Channels, suprafans and depositional lobes: Facies characters for turbidite sand environments on modern fans. *Sedimentary Environments and Hydrocarbons.* Amer. Assoc. Petrol. Geol., Short Course Notes, New Orleans, pp. 163-86.

———, and Piper, D. J. W., 1972. Sediments and growth pattern of Navy Deep-Sea Fan, San Clemente Basin, California Borderland. *J. Geol.,* 80: 198-223.

———, and Dickson, F. H., 1976. Sublacustrine fan morphology in Lake Superior. *Amer. Assoc. Petrol. Geol. Bull.,* 60: 1021-36.

———, Piper, D. J. W., and Hess, G. R., 1976. Distributary mesochannels, megaflutes, and microtopography of Navy Submarine Fan, California. (Abst.) *Geol. Soc. Amer. Prog.,* 8: 1032.

Piper, D. J. W., 1970. Transport and Deposition of Holocene Sediment on La Jolla Deep Sea Fan, California. *Mar. Geol.,* 8: 211-27.

Royse, C. F., 1964. Sediments of Willapa Submarine Canyon. Washington Univ., *Dept. Oceanog., Tech. Rept. III,* Seattle, 62 pp.

Ryan, W. B. F., Workum, F., Jr., and Hersey, J. B., 1965. Sediments on the Tyrrhenian Abyssal Plain. *Geol. Soc. Amer. Bull.,* 76: 1261-82.

Shepard, F. P., and Dill, R. F., 1966. *Submarine Canyons and Other Sea Valleys.,* Rand McNally, Chicago, 381 pp.

———, Dill, R. F., and von Rad, U., 1969. Physiography and sedimentary processes of La Jolla submarine fan and fan-valley, *Amer. Assoc. Petrol. Geol. Bull.,* 53: 390-420.

Stanley, D. J., 1974, Modern flysch sedimentation in a Mediterranean island arc setting. In: R. H. Dott, Jr., and R. H. Shaver (eds.), *Modern and Ancient Geosynclinal Sedimentation.* Soc. Econ. Paleont. Mineral., Sp. Publ. 19, pp. 240-59.

———, McCoy, F. W., and Diester-Haas, L., 1974. Balearic Abyssal Plain: An example of modern basin plain deformation by salt tectonism. *Mar. Geol.,* 17: 183-200.

Walker, R. G., 1967. Turbidite sedimentary structures and their relationship to proximal and distal depositional environments. *J. Sed. Petrol.,* 37: 25-43.

Wilde, P., 1965. Recent sediments of the Monterey Deep-Sea Fan. *Univ. of California, Hydraulic Eng. Lab., Rech. Rept. HEL-2-13,* Berkeley, 155 pp.

Chapter 14

Early Tertiary Deepwater Fans of Guipuzcoa, Northern Spain

ARTHUR VAN VLIET

Koninklijke/Shell Exploratie en Produktie Laboratorium Rijswijk, The Netherlands

ABSTRACT

The Lower Tertiary coastal belt of Guipuzcoa, northern Spain, consists of sediments deposited exclusively in a deep marine, east-west- oriented basin. Deepwater conditions prevailed from the late Albian until at least the early Eocene.

The basal part of this Lower Tertiary succession is characterized by pelagic deposits. Paleocene–early Eocene pelagic basin plain deposits are gradually succeeded by basin plain sediments of mixed pelagic and turbiditic origin. Turbidites composed of carbonate bioclasts and lithoclasts were emplaced by predominantly westward-flowing turbidity currents. During the early Eocene a major deepwater fan prograded from the northern margin of the basin. Fan fringe, outer fan, and middle fan environments can be recognized. Turbidites are quartz arenitic in composition with varying amounts of carbonate bioclasts. Locally, this fan is overlain by a fan that prograded from the southern margin of the basin. The turbidites of this second fan are lithic arenites with an admixture of carbonate bioclasts and lithoclasts. The Lower Tertiary succession is capped by further quartz arenitic fan deposits derived again from the northern margin of the basin.

INTRODUCTION

Since 1950, turbidites in ancient flysch formations have been the subject of considerable study. In addition to studies concerning the paleontology of flysch and its bathymetric implications, much attention has been paid to the characteristics of individual beds. This attention is hardly surprising, since these deposits, with their abundance of well-preserved internal structures and sole markings, represent one of the most spectacular sedimentary facies.

Only in the last few years have serious attempts been made to analyze ancient flysch deposits in terms of deepwater fan, base-of-slope, and basin plain sedimentation. This analysis was stimulated by the descriptions of the physiography and sedimentology of modern deepsea fans, particularly those off the Pacific coast of North America.

This chapter considers part of the fill of a deepwater basin that formed during the initial stages of the Pyrenean orogeny. This description and interpretation in terms of depositional processes and environments is preceded by a discussion of a deepwater fan model.

Regional Geologic Setting

The area studied forms a belt of Lower Tertiary sediments, 40 km long and 2 km to 5 km wide, along the Atlantic coast of northern Spain in the Basque province of Guipuzcoa whose capital is San Sebastian (map A, Figure 14-1). The Post-Hercynian sequence of Guipuzcoa rests on a plutonic and metamorphic basement. The oldest rocks of the overlying sequence consist of Permo-Triassic continental red beds that contain evaporitic facies of Keuper age. Conditions favoring deposition of shallow marine carbonates predominated through the Jurassic and early Cretaceous.

From the late Albian-early Eocene, deposition occurred in bathyal and shallow abyssal environments which resulted in thick sequences of turbidites, marls, pelagic carbonates, and to a lesser extent, olistostromes and slump deposits. In the deepwater deposits of the Upper Cretaceous, pillow lavas and volcaniclastics are found locally. Lower Eocene sediments are the youngest pre-Quaternary strata preserved in the area. The Upper Cretaceous-Lower Tertiary belt of deepwater deposits is located north of the axial zone of the Pyrenees and may be regarded as a remnant of the deep southern margin of the former Aquitaine basin.

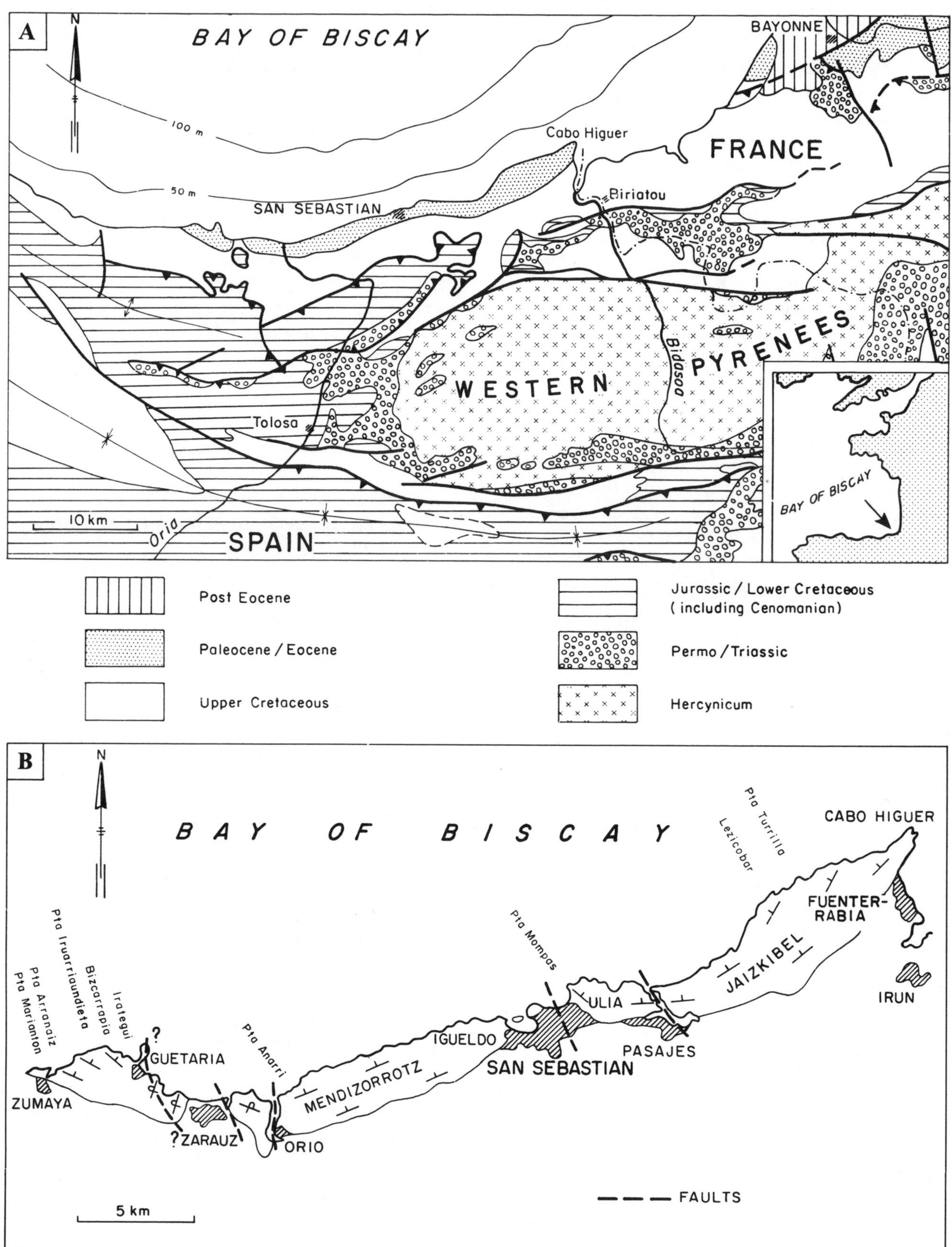

Figure 14-1. Lower Tertiary coastal belt of Guipuzcoa, northern Spain. Note: Map A is regional geological sketch map modified after Kruit et al. (1975). Map B shows locations cited in this study.

Structurally, the Lower Tertiary belt of Guipuzcoa represents a northward-dipping (20° to 50°) monocline (map B, Figure 14-1) complicated by disturbances near Zarauz and San Sebastian. At these disturbances the regional strike is diverted to the north, and dips are increased or even become overturned as in the case near Zarauz. The latter disturbance was studied by Hanisch (1974). Detailed stratigraphic and structural work led him to conclude that this disturbance results from Maastrichtian – Paleocene diapirism of Keuper evaporites. According to Hanisch the diapiric structure was subsequently modified during the Pyrenean phase (late Eocene-Oligocene) by the drag of a northward moving nappe. The remnants of this nappe occur a few kilometers south of the coastal belt (see map A, Figure 14-1).

Previous Work

The excellent coastal exposures of the Lower Tertiary of Guipuzcoa have not escaped previous sedimentological and paleontological scrutiny. Sedimentological studies have been made by Gomez de Llarena (1954), Crimes (1973, 1976), and Kruit et al. (1972, 1975), while Seilacher (1967) carried out a palichnological investigation. Von Hillebrandt (1965) studied the foraminiferal biostratigraphy of the Zumaya – Guetaria section, and Kapellos (1974) completed a calcareous nannoplankton zonation of the same sequence.

The Lower Tertiary deepwater fan deposits of Guipuzcoa were first recognized as such by Kruit et al. (1972). In subsequent years several reconnaissance trips were made by staff members of Koninklijke/Shell Exploratie en Produktie Laboratorium, Rijswijk (KSEPL), under the supervision of C. Kruit. Since 1974 the present author has been engaged on a sedimentological study aimed at further detailing KSEPL's work in the area. In 1975, on the occasion of the IX International Congress of Sedimentology at Nice, an excursion to the area was organized by KSEPL. Much of the information contained in the excursion field guide (Kruit et al. 1975) has been used in this chapter.

Chronostratigraphic Correlations

Regional correlations within the Lower Tertiary were made using planktonic foraminifera and calcareous nannoplankton, which are usually abundant within the hemipelagic interbeds. Age determinations in this study are in terms of nannoplankton zones. The study of the calcareous nannoplankton was focussed on the Lower Eocene interval. According to the standard Tertiary-calcareous nannoplankton zonation (Martini 1971), five zones have been distinguished within the Lower Eocene. These are, from top to bottom:

Discoaster sublodoensis zone
Discoaster lodoensis zone
Marthasterites tribrachiatus zone
Discoaster binodosus zone
Marthasterites contortus zone.

A zonation based on planktonic foraminifera was used by Kruit et al. (1975).

Paleobathymetry

Sedimentological criteria do not allow an accurate assessment of the depth of deposition during the early Tertiary other than "below-wave-base." To obtain a more precise depth determination, J. Brouwer (in Kruit et al. 1975) carried out an extensive study of the benthonic foraminifera from intercalated marls. His study indicates that throughout the early Tertiary deposition took place at bathyal to shallow abyssal depths (1,000 m to 4,000 m). According to Berger and Von Rad (1972), the Eocene carbonate compensation level in the North Atlantic was located between 3,600 m to 3,800 m.

MAJOR FEATURES OF DEEPWATER FANS AND THEIR DEPOSITS

General

In their ideal form, deepwater fans are cone-shaped sediment accumulations at the base of a submarine slope. Canyons cutting into the slope supply fans with sediment derived from shallower environments. Deepwater fans are a common feature along many present-day continental margins. Modern deepwater fans show a considerable variation in size. The world's largest fan is the Bengal Fan, which is 2,600 km long and 1,100 km wide. Smaller fans occur on the continental borderland of California, where several fans with lengths and widths of the order of 10 km have been observed.

Many modern deepwater fans were constructed almost entirely prior to the Holocene sea level rise. During the low sea levels of the Pleistocene many rivers dumped their sediment load close to the modern shelf edge, which thus lead to intensive resedimentation. Nelson (1976) emphasized this point in his comparison of Pleistocene and Holocene sedimentation rates on Astoria Fan in the northeast Pacific.

During the last decade a number of detailed studies of the physiography and sedimentology of modern deepwater fans have been made (e.g., Shepard et al. 1969; Carlson and Nelson 1969; Nelson et al. 1970; Normark 1970; Piper 1970; Haner 1971; Normark and Piper 1972; Cleary and Conolly 1974; Damuth and Kumar 1975). Nelson and Kulm (1973) provide an exhaustive survey of modern deepwater fans and channels. These studies indicate that the overall physiographies of modern fans are basically similar. Shallow cores have shown that fans consist predominantly of sediments deposited from sediment gravity flows (see Middleton and Hampton 1973). Classical turbidites have frequently been recognized in these cores.

Extensive studies by Horn et al. (1970, 1971, 1972) reveal that many modern abyssal plain sediments are similar to ancient flysch accumulations and that turbidites in these deposits can be related to deepwater fans on continental rises.

Mutti and Ghibaudo (1972) and Mutti and Ricci Lucchi (1972) revolutionized the study of turbidites in flysch formations with their presentation of a facies model for ancient deepwater fans. This facies model was based on observations in Tertiary flysch formations of the northern Apennines and the southern Pyrenees, in combination with data derived from the studies of modern deepwater fans. Jacka et al. (1968) had previously presented a deepwater fan model based on a study of Permian rocks from the Delaware Basin (Texas). This model, however, refers to a fan built by density currents (originating from salinity differences), rather than by sediment gravity flows.

In later papers, Ricci Lucchi and Parea (1973), Mutti and Ricci Lucchi (1974), Mutti and Ricci Lucchi (1975), and Ricci Lucchi (1975), the facies model underwent some subtle refinements. Mutti and Ricci Lucchi (1974) claim to have been able to apply their model in practically all the ancient turbidite basins of the circum - Mediterranean region and North America visited by them. This model again proved its value in this study of the Lower Tertiary of Guipuzcoa.

The fan model utilized in this study is a modified version of the original Italian one. The classification of Walker and Mutti (1973) used in Kruit et al. (1975) has not been applied here.

Physiographic Aspects of the Deepwater Fan Model

With respect to deepwater fan systems, the following proximal-distal environments can be distinguished: inner fan, middle fan, outer fan, fan fringe, and basin plain (Figure 14-2).

The *inner fan* is the part of a fan showing "large-scale" channeling features (depths of tens to hundreds of meters and widths of hundreds to thousands of meters). A leveed fan valley splits into a number of main distributaries. The fan valley is the basinward continuation of the feeder canyon. In a downfan direction, channel gradients and levee heights decrease, while channels become increasingly sinuous.

The *middle fan* is a transitional area between the channeled and the nonchanneled parts of a fan. It is characterized by the presence of numerous "shallow"-braided and meandering channels (in this context *shallow* refers to depths of several meters). On modern fans, the middle fan is indicated by a distinct convex bulge in the longitudinal fan profile, termed *suprafan* by Normark (1970) and considered to be the apex of an outer fan depositional lobe. In Normark's model the fan valley does not divide into main distributaries, and only a single suprafan is present. In cases where the fan valley does divide, each distributary will develop its lobe apex or "suprafan."

The *outer fan* is devoid of channeling and consists of depositional lobes and interlobe depressions. Depositional lobes are convex, flow aligned, sediment bodies, fed by distributary channels, and analogous to the mouthbars of river-dominated deltas. A modern example of what might represent lobe-interlobe topography in the Magdalena outer fan was described by Bouma and Treadwell (1975).

The *fan fringe* is a transitional area occurring between the outer fan and the basin plain. Here depositional lobes gradually lose their "relief" in a basinward direction.

The *basin plain* is the flat, deepest part of the basin.

DESCRIPTION OF DEPOSITIONAL SEQUENCES IN THE LOWER TERTIARY FAN DEPOSITS OF GUIPUZCOA

This section is concerned with the lithologic features observed in the Lower Tertiary successions of Guipuzcoa and endeavors to relate these to different physiographic elements of the deepwater fan model described above. For the sake of completeness, the general features of deposits commonly assigned to the inner fan region are also summarized, although no deposits unequivocally ascribed to this subenvironment are known from the Lower Tertiary of Guipuzcoa.

Inner Fan

Coarse clastics are concentrated in channelized bodies surrounded by fine-grained slope and interchannel deposits. The products of sediment gravity flows other than turbidity currents may be well represented in the channelized bodies (see Middleton and Hampton 1973). Mutti and Ricci Lucchi (1972), Walker and Mutti (1973), and Mutti and Ricci Lucchi (1975) have developed classifications for such deposits.

According to Mutti and Ricci Lucchi (1972), channel deposits commonly consist of arrangements of beds in fining- and thinning-upward megasequences that are considered to be the result of gradual channel abandonment. In the opinion of the present author this explanation is not entirely satisfactory. Gradual channel abandonment, is certainly not indicated by the normal strong asymmetry of the coarsening- and thickening-upward megasequences of the outer fan lobes (see below). Moreover, examples are known of channel infill by coarsening- and thickening-upward megasequences (Ricci Lucchi 1975, Figure 24). A different explanation for fining- and thinning-upward

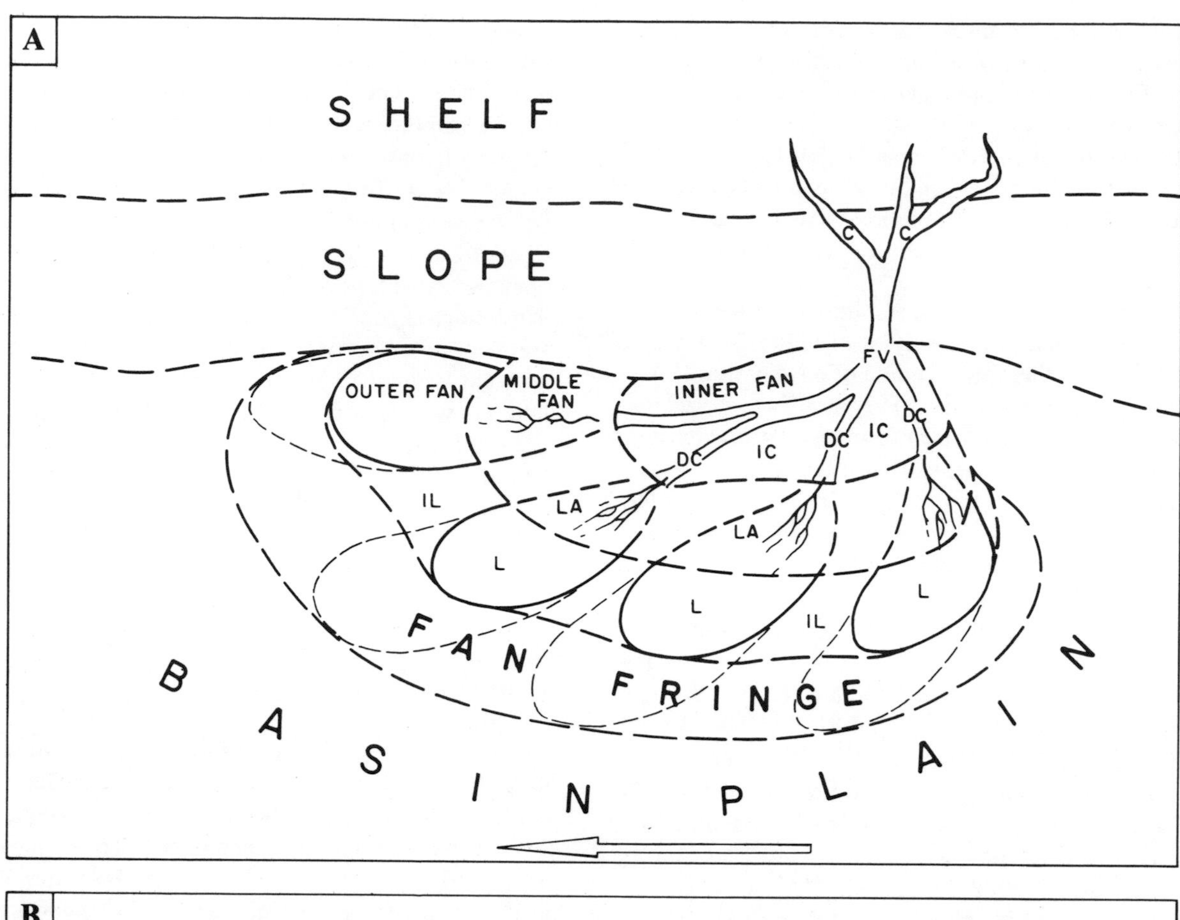

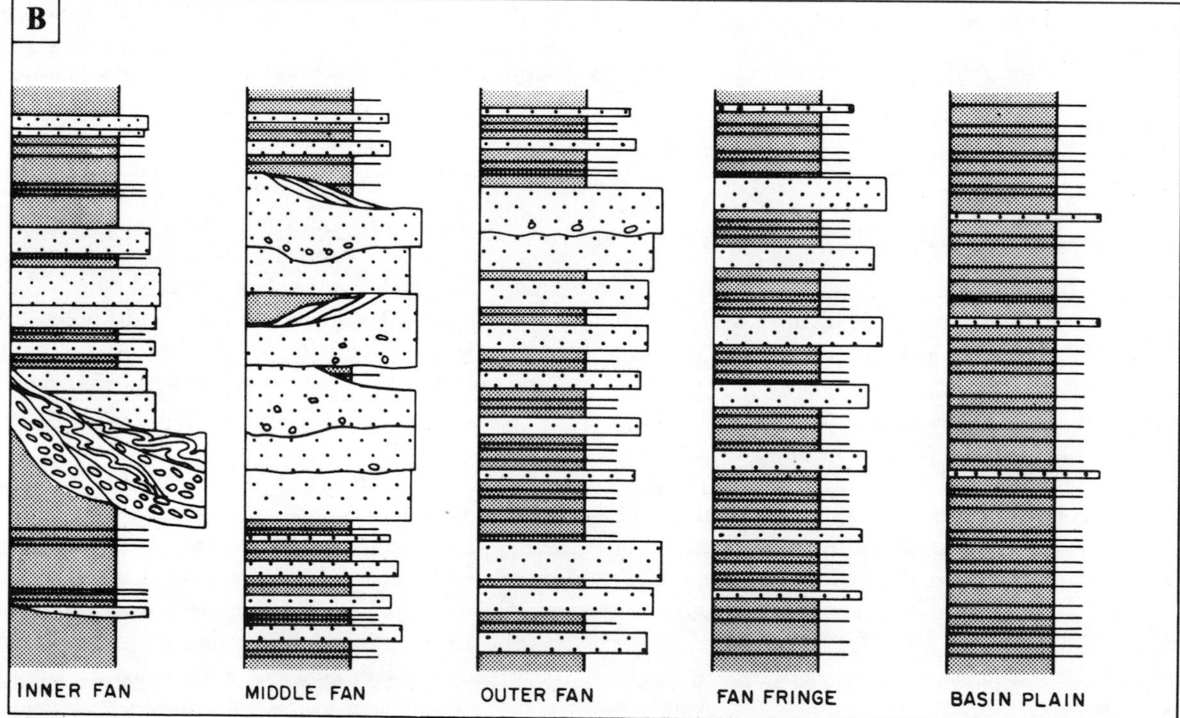

Figure 14-2. Deepwater fan model. Note: Diagram A shows the physiographic model of a deepwater fan and its related environments. C = feeder canyon of fan; FV = fan valley; DC = distributary channel; IC = interchannel area; LA = apex of depositional lobe ("suprafan"); L = depositional lobe; and IL = interlobe depression. Arrow indicates plunge of basin axis.
Chart B is a schematic representation of depositional sequences in various deepwater fan subenvironments.

megasequences could be channel migration, with the channel remaining active in a manner analogous to the growth of a fluvial pointbar. In this case, very low-angle lateral accretion units could produce a fining- and thinning-upward megasequence in a vertical section. Apart from their channelized geometry and possibly sequential arrangement, the active channel deposits are often characterized by slumping and by the presence of extremely coarse material (up to boulder size).

Interchannel deposits result mainly from overbank flooding and consist of shaley thin-bedded sequences of "distal" turbidites. Ancient and modern examples of interchannel deposits are described by Nelson and Nilsen (1974).

Examples of ancient inner fan deposits are provided by Stanley and Unrug (1972), Ghibaudo and Mutti (1973), and Ricci Lucchi (1975, here partially included with middle fan deposits).

Middle Fan

The coarser clastics (in Guipuzcoa, up to granule conglomerate) of middle fan deposits are concentrated in nonchannelized bodies interpreted as the apices of depositional lobes ("suprafans"). Such bodies have well-defined bases and consist of, in vertical section, a coarsening- and thickening-upward megasequence (Figure 14-3; photographs A and D, Figure 14-4). The well-defined base represents an abrupt increase in bed thickness. Megasequence thickness ranges from 20 m to 70 m. Individual sandstone beds in the upper parts of the sequences may attain thicknesses of the order of 15 m. Sandstones are usually massive or show consolidation laminations with broadly spaced pillars (photograph C, Figure 14-4; also Lowe 1975) and/or "ordinary" dish structures. The tops of the beds, if not eroded, are crudely horizontally laminated and often deformed into large undulations. Grading, if present, is of the coarse-tail type. These deposits are interpreted to be the product of high-density turbidity currents that passed through a stage of fluidized sediment flow during deposition (see Middleton and Hampton 1973).

Small-scale channeling is common in lobe apex deposits and can be divided into cut-and-fill structures at the base of individual beds (scouring depths up to 4 m were observed) and "true" channels, a few meters deep, eroded in the tops of thick sandstone beds (Figure 14-3; photograph E, Figure 14-4). Such channels commonly contain lateral accretion deposits overlying a lag of intraformational clasts (photograph F, Figure 14-4; also Kruit et al. 1975, p. 71). Accretion units may have dips of up to 15° and consist of parallel-laminated and/or cross-laminated sandstone. Channels normally have a postabandonment fill of marl containing some very thin turbidite sandstones (photograph E, Figure 14-4). Such fills are similar to the "clay plugs" of meandering river deposits. Their stratigraphic setting (Figure 14-3) makes clear that such channels develop near the lobe apex between major events.

In vertical section the thickly bedded lobe apex deposits alternate with relatively thin-bedded facies (photographs A and D, Figure 14-4). Most beds in this latter facies are crudely horizontally laminated. Subsequent deformation of the lamination by liquefaction is common. A characteristic feature of these deposits is the occurrence of erosively-based starved megaripples capping many beds (photograph B, Figure 14-4). In this respect they resemble Facies B2 of Mutti and Ricci Lucchi (1975). Occasionally, scoured top surfaces are overlain only by hemipelagic marl.

Apparently, the flanks of the lobe apices are areas of erosion and intensive tractional reworking during the waning phase of high-density turbidity currents. This phenomenon is perhaps related to the dechannelization of such currents. As a result of decreasing inertia, the width of the current will increase during waning conditions. While the convex lobe apex is emerging from the waning flow, the paths of maximum flow velocity will shift towards the flanks and there erode or rework the top of recently deposited turbidites.

Outer Fan

Ancient outer fan deposits were first described by Mutti and Ghibaudo (1972). These authors suggest that they are characterized by concentrations of sandstone beds in nonchannelized coarsening- and thickening-upward megasequences, produced by the basinward progradation of depositional lobes.

Coarsening- and thickening-upward megasequences are common in the Lower Eocene of Guipuzcoa and range in thickness from 10 m to 50 m. The uppermost beds of such sequences may attain thicknesses of 8 m. Outer fan lobe deposits are distinguishable from lobe apex deposits by the absence of channeling. Amalgamation is common near the top of the lobe megasequences. As with the lobe apex deposits, the sandstone beds are interpreted as the products of high-density turbidity currents that passed through final flow stages of fluidized sediment flow (thereby resulting in the ubiquitous dish structures and consolidation laminations).

The distal parts of depositional lobes prograding over the fan fringe produce gradually coarsening- and thickening-upward megasequences (chart A, Figure 14-5; charts A, B, and C, Figure 14-6; photograph B, Figure 14-7) that show a substantial, upward-declining, interbedding of thin-bedded distal turbidites. In the more proximal parts of the lobes, coarsening- and thickening-upward megasequences abruptly overlie interlobe deposits (chart D, Figure 14-6; photograph A, Figure 14-7).

The superposition of several lobe megasequences is frequently observed in outer fan deposits and is attributed

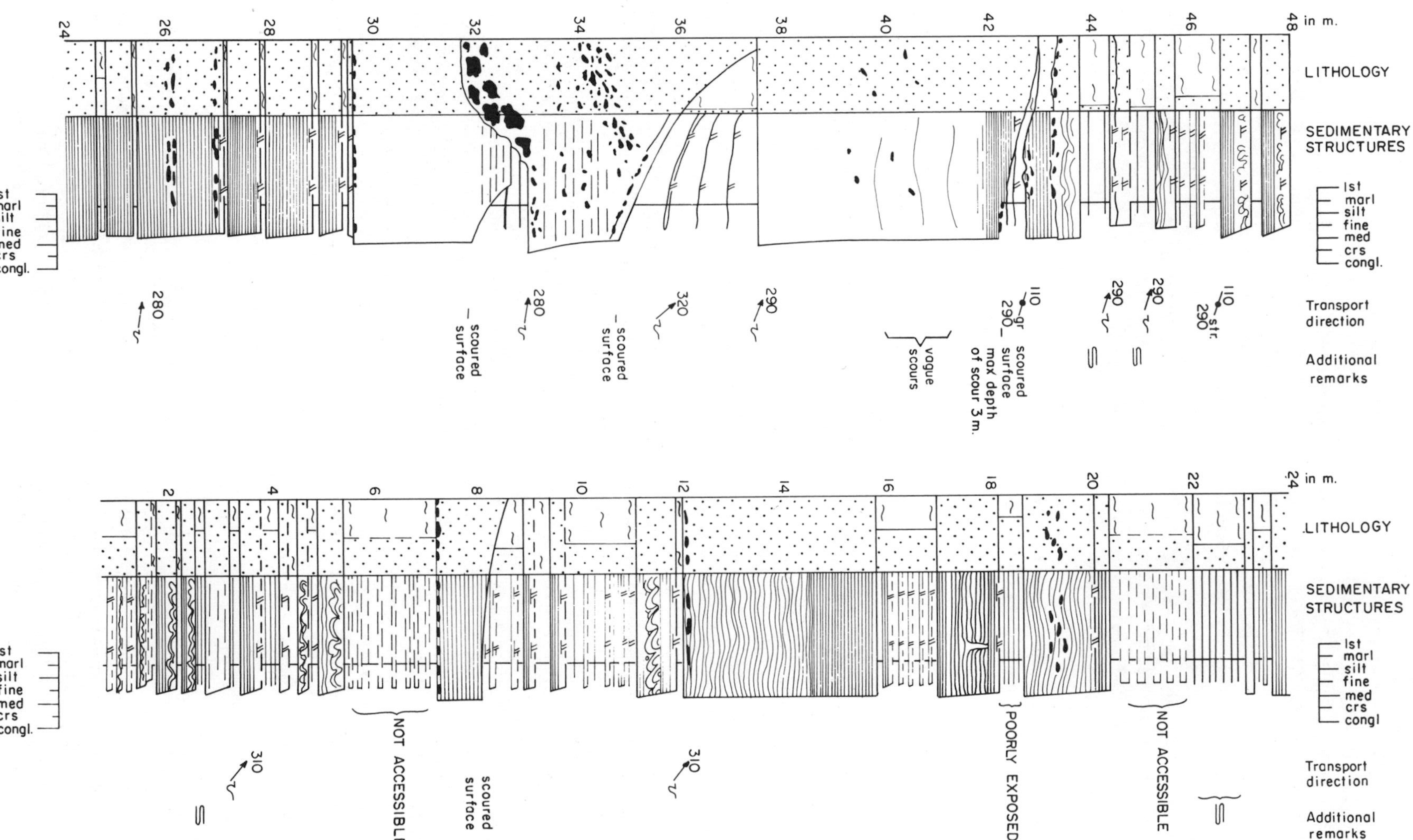

Figure 14-3. Detailed section of middle fan deposits at Bizcarrapia (Zumaya-Guetaria section). Note: The section shows the apex of a depositional lobe with small marl-filled channels cut into the topmost, very thick sandstone beds. (See also photographs D, E, and F, Figure 14-4.)

Figure 14-4. Middle fan deposits. Note: Photograph A shows massive body of very thick (up to about 15 m) turbidites interpreted as the apex of a depositional lobe. Basal scouring of individual beds and overall thickening-upward character are noteworthy. (Lezicobar, *M. tribrachiatus* zone.) Photograph B shows turbidite topped by a train of starved megaripples. Compass for scale is on lee-side of ripple. (Lezicobar, from "thin-bedded" interval in photograph A.) Photograph C shows dish structures (consolidation lamination, and broadly spaced "pillars") in a very coarse sandstone unit (top of Pasajes section, *M. tribrachiatus* zone). Photograph D shows apex of depositional lobe in middle fan deposits. Abrupt base of thickening-upward megasequence is noteworthy. (Bizcarrapia, *D. lodoensis* zone.) Photograph E shows cut bank and postabandonment fill (clay plug) of small erosive channel in upper part of megasequence in photograph D. Photograph F shows depositional bank of same channel as in photograph D. Arrow A denotes inclined accretion units resting on a lag (L) of marl clasts; CP is clay plug of another channel.

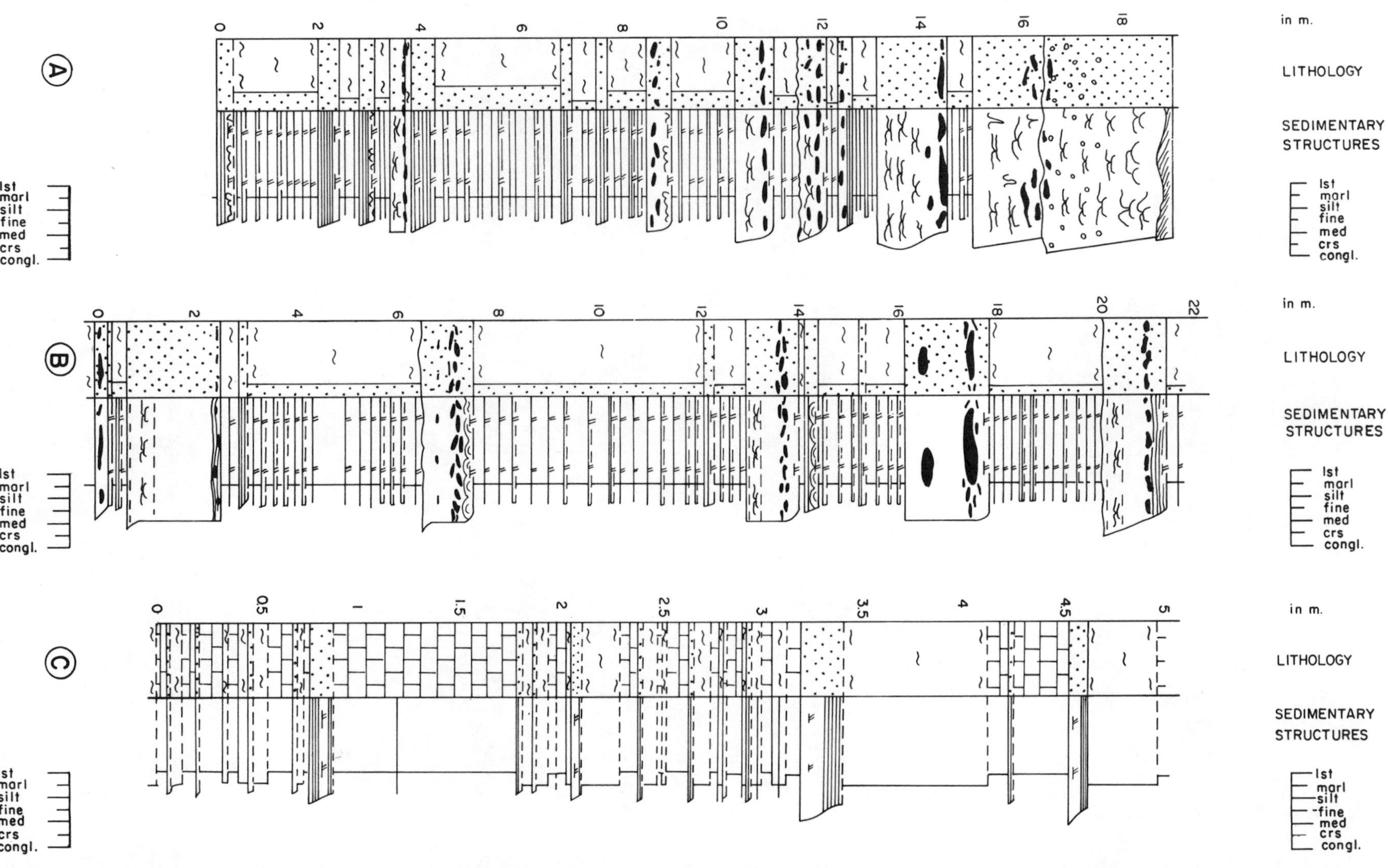

Figure 14-5. Detailed sections of outer fan, fan fringe, and basin plain deposits. Note: Chart A of outer fan deposits shows coarsening- and thickening-upward megasequence of turbidites (top of Orio section near Punta Anarri; see also charts B and C, Figure 14-6). Chart B of fan fringe deposits shows megasequence with a bimodal bed-thickness distribution. "Isolated" proximal turbidites in a matrix of distal turbidites are noteworthy. (Zumaya-Guetaria section near Pta Arranaiz; see also chart C, Figure 14-6; photographs C and D, Figure 14-7.) Chart C of basin plain deposits shows predominantly distal turbidites that alternate with pelagic limestones (highway cut near Orio; see also photographs E and F, Figure 14-7).

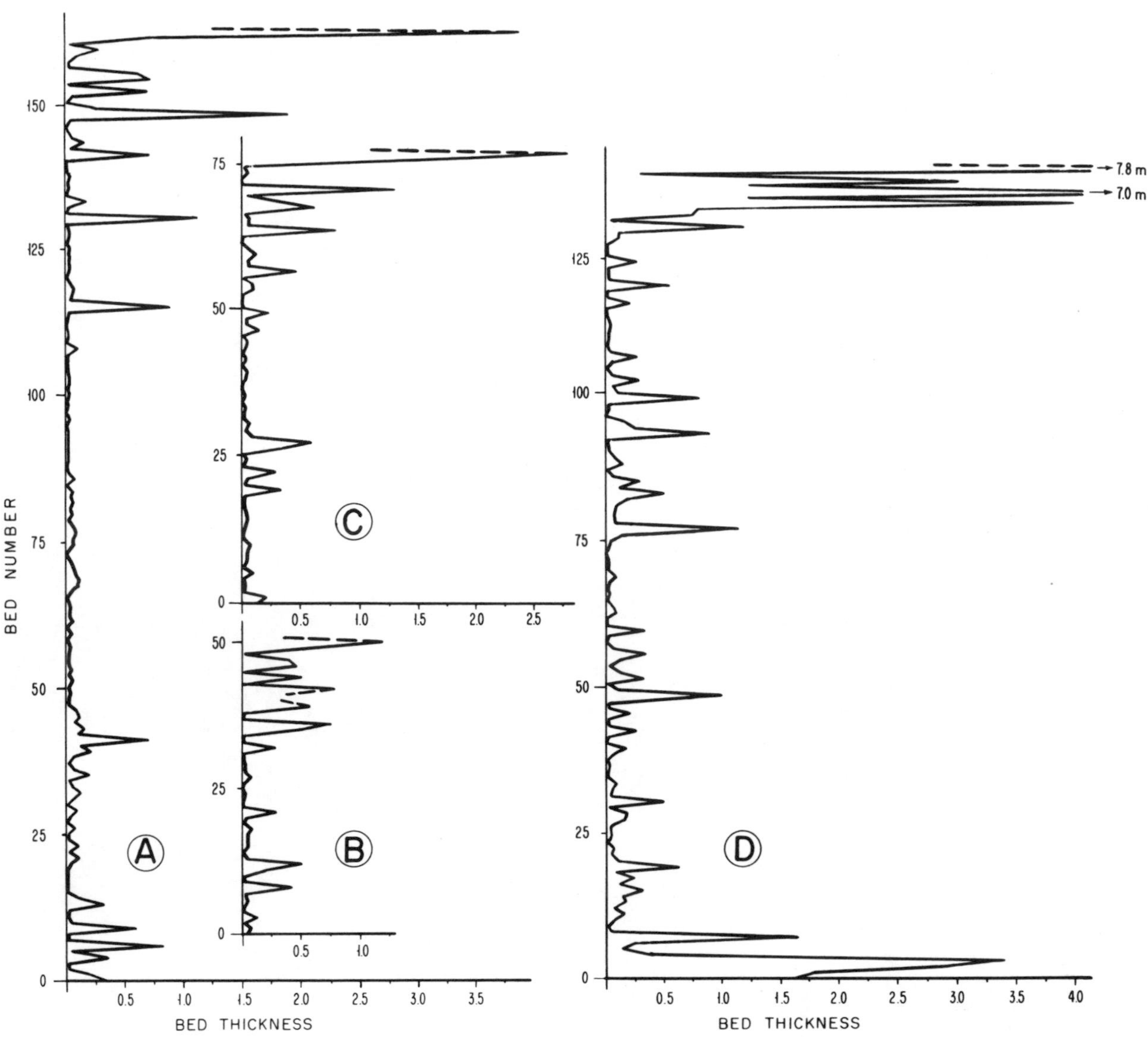

Figure 14-6. Bed-thickness logs of outer fan deposits. Note: "Bed thickness" refers to thickness of sandstone part of turbidites and in all charts is given in meters. Charts A, B, and C are of thickening-upward megasequences showing a considerable, upward-declining proportion of interbedded thin beds. Such "gradual" thickening-upward megasequences result from progradation of distal part of depositional lobe over the fan fringe (Chart A from Igueldo, *M. tribrachiatus* zone; charts B and C from Pta Anarri, *M. tribrachiatus* zone.) Chart D is a log of lobe (base and top of log) and interlobe (middle part of log) deposits. The interlobe deposits lack an obvious megasequential arrangement. Extreme bed thickness, absence of interbedding of thin beds, and abrupt increase in bed thickness at the base are typical of the more proximal parts of a depositional lobe. (Top of Pasajes section, *M. tribrachiatus* zone.)

to an autocyclic process involving lobe progradation and abandonment and lobe migration into interlobe depressions. Lobe megasequences are frequently strongly asymmetric—that is, the tops are marked by a sudden decrease in bed-thickness—which results in the typical cuesta landscape (photograph B, Figure 14-7). The strong asymmetry implies sudden abandonment of distributary channels feeding the lobes. Channel abandonment is supposedly caused by the increased topography of the lobes. A single depositional event can increase the height of the lobe apex by more than 10 m ("single-event beds" of such thicknesses are regularly observed!). Avulsion of the feeding distributary results in lobe abandonment and leads to the deposition of a new lobe in the topographically lower adjacent interlobe depression.

Fan Fringe

In fan fringe deposits, the megasequences produced by prograding depositional lobes are "diluted" by thin-bedded distal turbidites. This process can result in megasequences with bimodal bed-thickness distributions—that is, sequences

Figure 14-7. Examples of outer fan, fan fringe, and basin plain deposits. Note: Photograph A shows coarsening and thickening-upward megasequence of outer fan deposits. Uppermost bed is almost 8 m thick. This megasequence shows abrupt increase in bed thickness at base (see chart D, Figure 14-6) that is typical of more proximal parts of depositional lobe. (Top of Pasajes section, *M. tribrachiatus* zone.) Photograph B shows thickening-upward megasequence of outer fan deposits produced by progradation of distal part of depositional lobe over fan fringe. "Gradual" thickening of beds is noteworthy. Uppermost bed is approximately 1.50 m thick, (Iruarriaundieta, *D. lodoensis* zone.) Photograph C of fan fringe deposits shows "isolated" proximal turbidites in matrix of distal turbidites. Stratigraphic top is on the left. (Pta Arranaiz, *M. tribrachiatus* zone.) Photograph D shows detail of fan fringe deposits in photograph A. Concentration of rip-up clasts in upper part of proximal turbidite are noteworthy. Photograph E is an example of basin plain deposits. Distal turbidites alternate with pelagic limestones, and pronounced continuity of bedding is noteworthy. (Highway cut near Orio, *M. contortus* zone). Photograph F shows detail of the basin plain deposits in photograph C. Turbidite overlies pelagic lime mudstone.

of "isolated" proximal turbidites, up to 2 m thick, interbedded with groups of thin-bedded, cm-dm scale, distal turbidites, while beds of intermediate thickness are lacking (chart B, Figure 14-5; chart C, Figure 14-8; photograph C, Figure 14-7). This bimodality indicates different sources for thick and thin beds. Such megasequences are assumed to develop in situations where one fan lobe supplies the thick beds, while the thin beds are derived from more distant lobes of the same fan or from other fan systems. Megasequences, as described above, indicate an important physiographic feature of the fan fringe; namely, that the relief of depositional lobes is reduced to such an extent that they no longer form an obstruction to turbidity currents originating elsewhere. In megasequences produced by the progradation of distal portions of outer fan lobes, interbedded distal turbidites become less common upward, thereby reflecting the build-up of lobe relief (charts A, B, and C, Figure 14-6). In fan fringe megasequences such interbedding remains consistently present (charts B and C, Figure 14-8).

When several lobes occur in close proximity, a complex megasequence results that generally coarsens and thickens upward and contains substantial numbers of interbedded distal turbidites. An example of such a megasequence is illustrated in chart D, Figure 14-8.

A characteristic feature of fan fringe and distal lobe deposits is the abundance of intraformational clasts, normally concentrated above the massive or dish-structured divisions of proximal turbidites (photograph D, Figure 14-7; see also Kruit et al. 1975, p. 59). Intraformational clasts consist mainly of angular pieces of marl, siltstone, laminated fine sandstone or even (when they are sufficiently large) whole sequences of thin-bedded turbidites. Clasts up to 1 m were observed. Such intraformational "conglomerates" are common also within similar deposits in the Tertiary flysch of the northern Apennines (Mutti, personal communication). Ricci Lucchi (1965) and Marschalko (1970) have described in detail beds containing similar concentrations of clasts. Ricci Lucchi (1965) suggests they result from processes that are "intermediate between turbidity currents and slidings." Marschalko (1970) considers the clasts to have been ripped up from a soft substratum by high-density turbidity currents. The present author prefers the latter interpretation, as it directly explains the abundance of such clasts in fan fringe and distal lobe, deposits. Only in the fan fringe and distal lobes does the soft substratum of thin-bedded distal turbidites occur on which high-density turbidity currents can have their impact. In more proximal lobe areas, the thin-bedded distal turbidites were not deposited because of lobe relief. Where fan fringe deposits are interbedded with thin pelagic lime mudstones, rip-up clasts were not observed. The substratum in such cases is apparently less erodable. Rip-up clasts were also not observed in fan fringe deposits where thick, relatively proximal, turbidites are entirely laminated.

The fan fringe deposits described above are probably typical of fans in "narrow," elongated orogenic basins where lateral input from fan lobes swings around to parallel the basin axis.

Basin Plain

Mutti and Ricchi Lucchi (1972) suggest that basin plain deposits are characteristically monotonous successions of distal turbidites lacking obvious arrangements of beds in megasequences. Examples of such deposits from the Lower Tertiary of Guipuzcoa are illustrated in chart C, Figure 14-5; chart A, Figure 14-8; photograph E, Figure 14-7. The bedding continuity of basin plain deposits is rather pronounced (photograph E, Figure 14-8). These deposits range from predominantly pelagic to almost entirely turbiditic. Pelagic deposits consist predominantly of flat-bedded (on a dm scale) foraminiferal lime mudstones alternating with hemipelagic marls. Basin plain turbidites are characterized by base-absent Bouma sequences.

Pelagic Interbedding

No quantitative study was made of the amount of interbedded pelagic lime mudstones in the various fan subenvironments. Interbedded pelagic sediments are not a diagnostic feature of any of the environments discussed. During the Paleocene and earliest Eocene (*M. contortus* zone and *D. binodosus* zone) clastic input into the Guipuzcoan basin was broadly spaced in time, which thus resulted in substantial interbeds of pelagic lime mudstones, even in middle fan environments. In the Lower Eocene (*D. lodoensis* zone) pelagic interbeds are absent, even in basin plain deposits. The overall Paleocene-Lower Eocene trend of decreasing pelagic interbeds is probably related to increased clastic sedimentation rates resulting from the onset of the main phase of the Pyrenean orogeny.

REGIONAL SEDIMENTOLOGICAL DESCRIPTION

This part of the chapter consists of a general sedimentological description and interpretation of sections and relevant exposures in the Lower Tertiary of Guipuzcoa. The description is in geographical order from west to east. The facies interpretations and paleocurrent data are summarized in diagrams A and B, Figure 14-9.

Zumaya - Guetaria Section

The Zumaya - Guetaria section, together with the nearby Zarauz - Guetaria section, provides the most complete

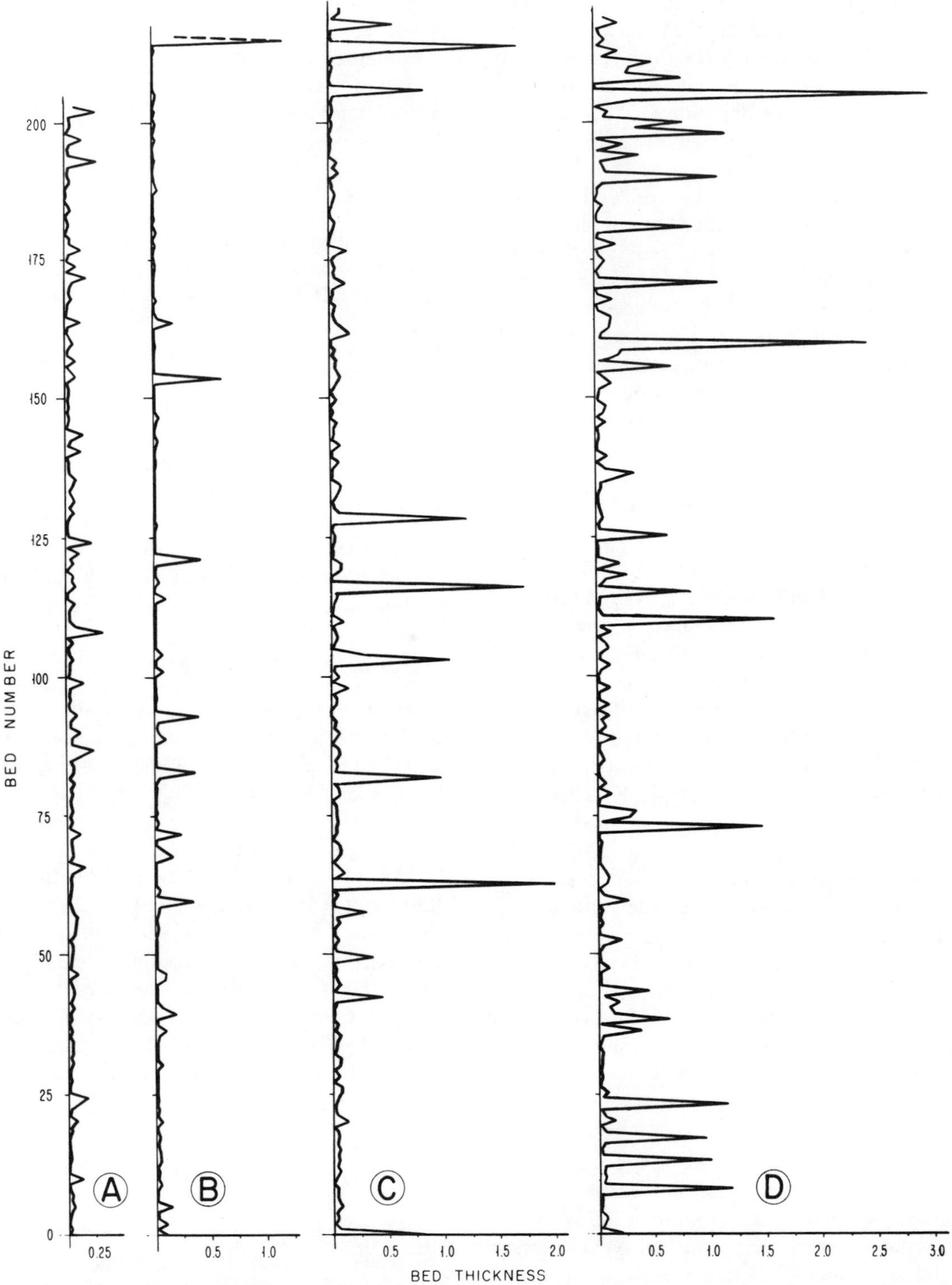

Figure 14-8. Bed-thickness logs of basin plain and fan fringe deposits. Note: "Bed thickness" refers to thickness of sandstone part of turbidites and in all charts is given in meters. Chart A shows basin plain deposits lacking an obvious megasequential arrangement of turbidites (Zumaya, *D. binodosus* zone). Chart B of basin plain to fan deposits shows gradual development of a diluted, thickening-upward megasequence (Zumaya, *M. tribrachiatus* zone). Chart C of fan fringe deposits is an excellent example of a megasequence with a bimodal bed-thickness distribution. "Isolated" thick beds in a matrix of thin beds are noteworthy; beds of intermediate thicknesses are lacking. (Pta Arranaiz, *M. tribrachiatus* zone.) Chart D of fan fringe deposits shows complex megasequence of overall thickening-upward character. Considerable interbedding of thin beds throughout the log is noteworthy. (Base of fan deposits near Igueldo, *D. binodosus* zone.)

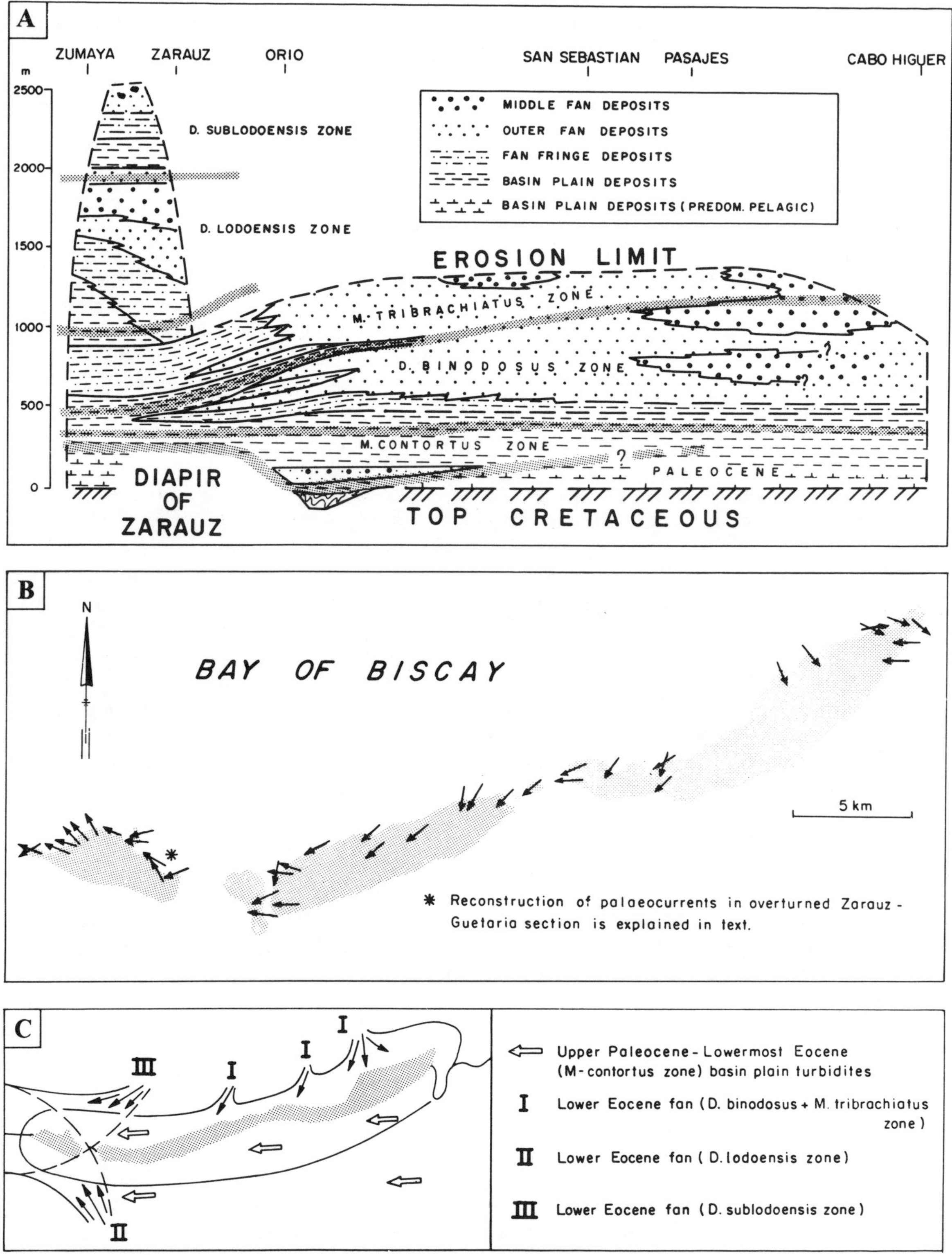

Figure 14–9. Facies interpretations and paleocurrent data. Note: Diagram A is a schematic stratigraphic cross-section through Lower Tertiary of Guipuzcoa. Diagram B is a paleocurrent map (based on a total of about 400 measurements). Diagram C is a schematic representation of early Tertiary basin plain and fan sedimentation in Guipuzcoa. (Deposits time-equivalent to fans II and III are not preserved in eastern part of area.)

stratigraphic record of the early Tertiary. The section includes the upper part of the Lower Eocene (*D. sublodoensis* zone), while other sections are truncated by erosion in the middle part of the Lower Eocene (*M. tribrachiatus* zone). The approximately 2,500 m thick section is almost continuously exposed along the San Telmo beach at Zumaya and the coastal road from Zumaya to Guetaria.

The Paleocene and the lowermost Eocene (*M. contortus* zone) consist predominantly of flat-bedded (cm-dm scale) lime mudstones, marls, and subordinately, thin turbidites. Bedding continuity of this facies association is excellent. In the Lower Paleocene the lime mudstones and marls are reddish in color. Lime mudstone interbeds decrease in abundance and thickness throughout the Paleocene and the lowermost Eocene.

The structureless lime mudstones contain abundant nonfragmented tests of planktonic foraminifera (photograph A, Figure 14-10). Scanning Electron Microscope studies showed the lime mud to consist largely of somewhat recrystallized calcareous nannofossils (coccoliths, and soforth). The limestones, hence, represent lithified pelagic oozes.

Based on the common presence of *Zoophycos,* Crimes (1973, 1976) interpreted the limestones as "shallow marine though below wave-base." Reineck (1973), however, reports *Zoophycos* from abyssal depths in the Indian Ocean, where it is exclusively and abundantly present in the pelagic carbonate oozes. The gradually decreasing occurrence of *Zoophycos* in the San Telmo section (Paleocene - Lower Eocene) is probably related to the diminishing amounts of lime mudstone rather than to a deepening of the depositional environment as suggested by Crimes (1973).

The Paleocene and lowermost Eocene sediments are interpreted to be predominantly pelagic and hemipelagic (the marls) basin plain deposits. Subordinate thin turbidites consist almost entirely of carbonate bioclasts and lithoclasts.

In the *D. binodosus* zone, pelagic basin plain deposits are gradually succeeded by turbiditic basin plain deposits containing about 30 percent pelagic interbeds. The turbidites are carbonate-cemented quartz arenites containing bioclastic material. All beds show base-absent Bouma sequences, and megasequences are absent (see chart A, Figure 14-6).

In the lower part of the *M. tribrachiatus* zone a gradual transition to fan fringe deposits is observed (see example in chart B, Figure 14-6). These deposits form a succession of approximately 350 m up to Punta Arranaiz. Pelagic interbeds disappear in the basal part of these fan fringe deposits. Examples of fan fringe deposits are illustrated in chart B, Figure 14-5; chart C, Figure 14-6; and photographs C and D, Figure 14-7. Only at Punta Marianton do the fan fringe deposits develop into a single outer fan lobe.

At Punta Arranaiz the quartz arenitic (photograph C, Figure 14-10) fan fringe deposits are onlapped by lithic arenitic (cf. photograph D, Figure 14-10) basin plain turbidites. The latter form a succession about 1,100 m thick, which over some 950 m changes from basin plain deposits, via fan fringe and outer fan deposits (see photograph B, Figure 14-7) into middle fan deposits at Bizcarrapia (see Figure 14-3; photographs D, E, and F, Figure 14-4) and Irategui. A characteristic feature of the lithic arenite units is their restricted grain size range (no material coarser than medium sand) and the sporadic occurrence of structureless and dish-structured divisions. In the top 150 m of the lithic arenite succession, the middle fan deposits are succeeded by outer fan deposits, which in turn are followed by basin plain deposits. In these basin plain sediments the lithic arenitic turbidites are gradually replaced by quartz arenitic turbidites. The latter develop upward via fan fringe deposits into the impressive outer fan lobes on the peninsula of Guetaria. Possibly middle fan deposits occur above, but the limited accessibility of the section does not permit this to be established with certainty.

In summary, the Zumaya-Guetaria section consists of a succession of predominantly pelagic basin plain deposits (Paleocene and the lowermost Eocene) overlain by three megacycles of Lower Eocene fan deposits. The megacycles begin with basin plain deposits and a change in petrographic composition of the turbidites and are progradational in character. Megacycles I and III are composed of quartz arenites; Megacycle II consists of lithic arenites. Megacycle I develops into fan fringe deposits; Megacycle II (and possibly III), into middle fan deposits.

Paleocurrent measurements in the section provide interesting additional information. Regional observations (see Kruit et al. 1975) make clear that Megacycle I represents a fan fed from the northern margin of the basin whose paleocurrents gradually swung around to parallel the basin axis until they point to west-northwest (N 295 E mean). Only in the lower part of megacycle I (*D. binodosus* zone) are some southward directed paleocurrents still observed.

Megacycle II represents a fan that was fed from the southern margin of the basin. The basin plain deposits in the lower part of this megacycle were deposited by currents flowing parallel to the basin axis, towards west-northwest (N 295 E mean). Fan fringe and outer fan paleocurrents change to a north-northwest orientation (N 330 E mean). The middle fan deposits in the upper part of megacycle II have paleocurrents that change to a west-northwest direction.

Megacycle III was most likely fed again from the northern margin of the basin. Its fan fringe deposits resulted from westward (N 270 E mean) flowing turbidity currents. If we

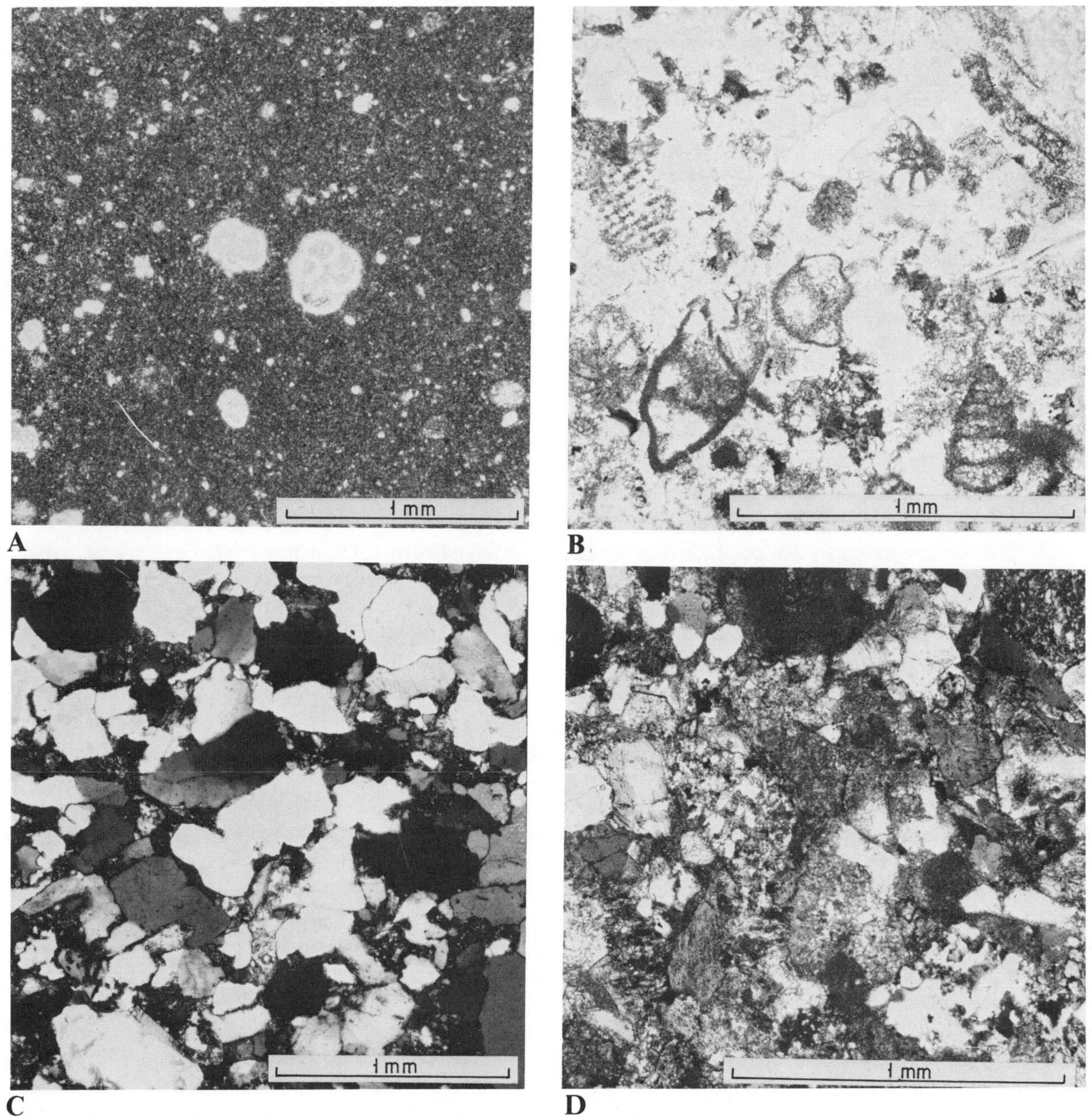

Figure 14-10. Petrographic examples. Note: Photomicrograph A shows lime mudstone with disseminated nonfragmented tests of planktonic foraminifera, which represents lithified pelagic ooze (Paleocene, Zumaya). Photomicrograph B shows bioclast grainstone with disseminated angular quartz grains. Bioclasts include fragments of echinoderms, bryozoans, dasycladaceans, and mainly benthonic foraminifera. (Lowermost Eocene axial basin plain turbidite, Fuenterrabia.) Photomicrograph C shows calcite-cemented quartz arenite. Sutured grain contacts are noteworthy. This petrography is characteristic of fan deposits derived from northern basin margin. (Pta Arranaiz, *M. tribrachiatus* zone). Photomicrograph D shows calcite-cemented lithic arenite. Lithics predominantly consist of chert, siltstone, and phyllite grains. This petrography is characteristic of fan deposits derived from southern basin margin. (Zarauz-Guetaria section, *D. lodoensis* zone).

assume that the orientation of the basin axis remained more or less constant throughout the early Eocene, a northern source for Megacycle III is implied. This assumption is supported by its petrographic composition, which is similar to that of Megacycle I.

Crimes (1976), referring to the San Telmo beach section, reports "additional" supply from the southern margin of the basin. In that part of the sequence no evidence of a substantial supply from the south was observed by the present author.

Zarauz - Guetaria Section

Along the coastal road from Zarauz to Guetaria only the middle part of the Lower Tertiary succession is exposed. The strata here are overturned and dip about 55° to the east. The exposed part of the section includes the *M. tribrachiatus* and *D. lodoensis* zone and shows the upper part of Megacycle I and a large part of Megacycle II.

The absence of basin plain deposits at the base of lithic arenitic Megacycle II represents a distinct difference with respect to the Zumaya-Guetaria section. Fan fringe deposits of Megacycle I are directly overlain by the fan fringe deposits of Megacycle II. The main problem, however, concerns the reconstruction of its paleocurrents. In this section, with its completely aberrant strike and dip (see map B, Figure 14-1), correction for dip alone is clearly not sufficient.

Due to the poor exposure, the structural relationship of the overturned Zarauz-Guetaria section to the normal-dipping Zumaya-Guetaria section is not clear. A faulted contact between normal and overturned dipping strata is exposed only on the peninsula of Guetaria. Hanisch (1975), basing his work on a lead-plate model, presented a reconstruction in which both sections form the flanks of a fold with a steeply plunging axis. In the present author's opinion, however, this model clearly suffers from scaling problems. It is hardly conceivable that the 2,500 m Lower Tertiary succession, containing competent fan sediments in its upper part, should be folded into such a tight synclinal fold without important faulting also taking place. The present author suggests that a major fault separates the normal dipping and the overturned sections. Such a fault makes it impossible to reconstruct the palaeocurrents of the overturned Zarauz-Guetaria section. Nevertheless, an interesting observation can be made with respect to paleocurrent trends in the section. The transition from Megacycle I to Megacycle II coincides with an abrupt change in paleocurrents (about 100° clockwise). This change confirms sediment supply from opposite basin margins for Megacycles I (N) and II (S). In the "downstream" Zumaya-Guetaria section, megacycle I and the basal part of Megacycle II were both deposited from axially flowing (towards west-northwest) turbidity currents. In the "upstream" Zarauz-Guetaria section, these turbidity currents were not yet reoriented parallel to the basin axis (see diagram C, Figure 14-9).

While accurate paleocurrent reconstruction in the Zarauz-Guetaria section is impossible, paleocurrent trends correspond with the regional picture when 150° are added to the dip corrected measurements.

Orio Section

Near Orio, the Oria River breaks through the coastal mountain chain of Lower Tertiary sediments along a north-south-trending fault. On the river banks the Lower Eocene succession is reasonably exposed. The Paleocene and the lowermost Eocene can be studied in exposures east of the village of Orio. The Upper Maastrichtian - Lower Paleocene, which in surrounding areas are basin plain deposits, consist of other facies near Orio. A detailed study of the complex stratigraphy near Orio was carried out by Hanisch (1974). Local occurrences of slump masses, olistostromes, pebbly mudstones, graded polymict conglomerates, breccias, and proximal limestone turbidites were recorded.

In an eastward direction these anomalous deposits gradually pass into "normal" flysch, while westward they appear to onlap abruptly against the diapir of Zarauz (cf. Hanisch 1974). Hanisch suggests these anomalous facies represent the fill of a deep rim syncline by material derived from the diapir. Feuillée and Mathey (1976), however, question the concept of Hanisch and Pflug (1974) that well-rounded pebbles of surrounding rocks could form in a salt mass during diapiric ascent.

A nearby shallow marine source for the well-rounded pebbles is an almost inevitable requirement. Hence, some uncertainty exists as to the nature of Hanisch's intrabasinal deepwater diapir, since it has more the character of a protrusion of the southern basin margin.

In the lowermost Eocene (basal *M. contortus* zone) a limited development of deepwater fan deposits is observed east of Orio. Poor exposure precludes their detailed study. The sandstone petrography (pure quartz arenites) suggests that they were derived from the northern basin margin.

These localized fan deposits are abruptly overlain by a thick (ca. 350 m) basin plain succession (see chart C, Figure 14-5; photographs E and F, Figure 14-7), which corresponds to the greater part of the *M. contortus* zone and the lower part of the *D. binodosus* zone. These deposits are of mixed turbiditic and hemipelagic origin. $T_{c\text{-}e}$ and T_{de} Bouma sequences predominate. Few sole markings are exposed due to the cementation of the turbidites to the underlying pelagic lime mudstones (see photograph F, Figure 14-7). The few soles exposed indicate axial (westward) transport. The turbidites are largely bioclastic with an admixture of terrigenous grains (mainly quartz).

In the upper part of the *D. binodosus* zone, the appearance of quartz arenitic turbidites in the section marks the base of Megacycle I. The upper part of the *D. binodosus* zone and the lower part of the *M. tribrachiatus* zone (ca. 300 m of section) comprise three asymmetric progradational cycles occurring within Megacycle I. The cycles begin with basin plain deposits and continue through fan fringe deposits into outer fan deposits. The depositional lobe megasequences, in the tops of the cycles, are 30 m to 40 m thick and contain individual beds up to 5 m thick. Flutes and parting lineations indicate transport towards the southwest and west-southwest.

The uppermost 250 m of the Orio section (from Pta Anarri onwards) are part of the same megacycle and were deposited in an outer fan environment. They consist of a number of stacked distal depositional lobe megasequences, 10 m to 20 m thick. Characteristic features of these megasequences are:

Beds that gradually coarsen and thicken upward,
Substantial interbedding of distal turbidites in the lower parts of megasequences, and
Concentrations of rip-up clasts overlying the massive divisions of the proximal turbidites.

Examples of these megasequences are shown in chart A, Figure 14-5; charts B and C, Figure 14-6.

From San Sebastian to Cabo Higuer

In the eastern part of the study area, continuous exposures in the Paleocene and the lowermost Eocene are rare. Lower Paleocene sediments can be studied in isolated exposures at San Sebastian, Pasajes, and near Fuenterrabia, where they consist of pelagic lime mudstones interbedded with reddish marls, similar to those of Zumaya.

The Upper Paleocene and the lowermost Eocene (*M. contortus zone* - lower *D. binodosus* zone) are well exposed only in cliffs near Fuenterrabia. Here, mixed turbiditic and pelagic basin plain deposits occur. The predominantly bioclastic turbidites (see photograph B, Figure 14-10) were deposited by westward flowing currents. The westward continuation of this facies from Fuenterrabia is indicated by the marked topographical depression that occurs between the coastal chain of Lower Eocene fan deposits and the Lower Paleocene limestones.

The deposition of quartz arenitic turbidites commences during the *D. binodosus* zone and continues into the *M. tribrachiatus* zone. At this level, the sections are truncated by the present coastline. The quartz arenitic turbidites form an impressive sequence (ca. 900 m) of outer and middle fan deposits. Fan fringe deposits are restricted to the basal part of the fan sequence and are particularly well exposed near Igueldo (see chart D, Figure 14-8) and San Sebastian (Pta Mompas).

The outer fan deposits consist of alternating proximal lobe and interlobe sediments. Characteristic features of the proximal lobe megasequences are:

Coarsening and thickening-upward megasequences 25 m to 50 m thick,
Individual turbidites as much as 8 m thick,
Very limited interbedding of distal turbidites, and
Bases that are marked by an abrupt increase in bed thickness (see chart D, Figure 14-6).

The best examples of such outer fan deposits are observed in the Pasajes section (see photograph A, Figure 14-7) and along the coastline of Mte Ulia.

Middle fan lobe apices and associated facies occur in coastal exposures at Mte Jaizkibel and Mte Igueldo. In particular at Lezicobar (see photographs A and B, Figure 14-4) and Punta Turrilla they form impressive amalgamated sand bodies up to 60 m thick. Shale interbeds are practically absent from such lobe apex bodies and basal scouring (cut and fill) of individual beds is common. Although the bases of such bodies are sharp, there is no evidence of overall channel geometries. Small channels (average depths of a few meters) with lateral accretion deposits are found near Cabo Higuer, in the Pasajes section, and on the North coast of Mte Igueldo.

A radiating paleocurrent pattern for the eastern part of the area was first recognized by Kruit et al. (1972). Paleocurrents diverge from a source north of Mte Jaizkibel. An additional source was probably located north of Mte Igueldo.

REGIONAL SYNTHESIS AND CONCLUSIONS

The Lower Tertiary belt of Guipuzcoa represents the remnants of a sedimentary basin formed during the initial stages of the Pyrenean orogeny. As time-equivalent slope, shallow marine, and continental deposits are not preserved in the area, the early Tertiary sedimentary history of this basin is recorded only by a narrow (a few km wide) strip of deepwater deposits. In spite of these severe limitations, some interesting conclusions can be reached with respect to the main trends of early Tertiary basin infill.

Deepwater environments of deposition prevailed from the late Albian until at least the late early Eocene. At the Cretaceous-Tertiary boundary a marked regional change occurred, thereby resulting in a predominance of pelagic sedimentation. Except for local anomalies, the Lower Tertiary succession commences with flat-bedded, lime mudstones consisting of calcareous nannofossils and planktonic foraminifera. The reddish color of these Lower Paleocene sediments and their intercalated marls may indicate conditions of slow sedimentation and oxidation. Turbidites occur sparsely in the Lower Paleocene and are almost entirely composed of shelf-derived carbonate fragments. In the Upper Paleocene a slight increase in turbidite deposition occurred. These turbidites were deposited from mainly westward (axially) flowing currents and still consisted predominantly of calcareous debris.

Terrigenous input into the basin was slight during the Paleocene, possibly as a result of the major transgression that occurred at the Cretaceous-Tertiary boundary (Plaziat 1975). Modern deepwater fans show similar prevalence of hemipelagic sedimentation in response to the Holocene transgression (see Nelson 1976).

The anomalous Upper Maastrichtian-Lower Paleocene facies near Zarauz, which may be associated with diapirism and a protrusion of the southern basin margin, have been considered previously.

On a regional scale, the lowermost Eocene is characterized by basin plain deposits of mixed pelagic and turbidite origin, in which the calcareous turbidites were deposited by westward flowing currents. Locally, between Orio and San Sebastian, a single progradational fan cycle occurs. Poor exposure precludes a detailed analysis of these deposits, which, on the basis of their quartz arenitic composition, were probably derived from the northern basin margin. In the lowermost Eocene, abrupt westward thinning of the succession is observed near Orio, which may be attributed to onlap against the relief of the Zarauz diapir (cf. Hanisch 1975a).

During the *D. binodosus* zone a deepwater fan began to prograde into the basin from the northern margin. The main source of this fan system was located north of Mte Jaizkibel, with an additional source to the north of Mte Igueldo. Paleocurrents in the westward portion of this fan system were strongly reoriented parallel to the basin axis. The fan deposits are quartz arenites containing varying amounts of shelf-derived bioclasts. Basin plain, fan fringe, outer fan, and middle fan environments are recognizable. The most proximal facies occur in the middle fan lobe apices of Mte Jaizkibel. The westernmost extension of this fan was probably not far beyond Zumaya, where it is represented by basin plain and fan fringe deposits. In the central and eastern parts of the area studied, no complete record of the history of this fan can be determined due to truncation by erosion.

In the western part (Zumaya-Guetaria and Zarauz-Guetaria sections) the more complete stratigraphic record indicates the gradual abandonment of this fan system, followed by onlap of lithic arenitic basin plain turbidites derived from another fan system. The source of this second fan was located on the southern basin margin, probably south of Zarauz. The outbuilding of this fan began during the upper part of the *M. tribrachiatus* zone and continued through the *D. lodoensis* zone into the lower part of the *D. sublodoensis* zone. This fan megacycle indicates the gradual progradation from basin plain to middle fan deposits. This progradational phase is succeeded by retreat and onlap of later basin plain deposits.

Finally, on the peninsula of Guetaria, quartz arenitic fan fringe and outer fan deposits, derived from the northern basin margin, complete the Lower Tertiary succession.

This superposition of fan deposits derived from opposite basin margins suggests that the early Eocene deepwater basin of Guipuzcoa was a relatively narrow one.

ACKNOWLEDGMENTS

The author would like to express sincere thanks to Dr. C. Kruit for his suggestion to undertake this study and for his continuous help and constructive criticism. The author is also grateful to Dr. A. P. Heward for his critical review of the manuscript. Drs. J. Brouwer, M. C. Budding, and W. Schöllnberger were of great support in discussing Guipuzcoa flysch problems in the laboratory. In the field Prof. J. D. de Jong, Prof. E. Mutti, and Dr. T. H. Nilsen assisted by providing many useful suggestions and comments. Finally, the author is indebted to Shell Internationale Petroleum Maatschappij B. V. for permission to publish this study.

REFERENCES

Berger, W. H., and von Rad, U., 1972. Cretaceous and Cenozoic sediments from the Atlantic Ocean. In: D. E. Hayes, A.C., Pimm, et al. *Initial Reports of the Deep Sea Drilling Project,* 14: 787-954.

Bouma, A. H., and Treadwell, T. K., 1975. Deep-sea dune-like features. *Mar. Geol.,* 19: M53-M59.

Carlson, P. R., and Nelson, C. H., 1969. Sediments and sedimentary structures of the Astoria submarine canyon-fan system, NE Pacific. *J. Sed. Petrol.,* 37: 1269-82.

Choukroune, P., and Seguret, M., 1973. Carte structurale des Pyrénées. *Elf-Erap, Boussens,* (Haute Garonne). Map.

Cleary, W. J., and Conolly, J. R., 1974. Hatteras deep-sea fan. *J. Sed. Petrol.,* 44: 1140-54.

Crimes, T. P., 1973. From limestones to distal turbidites: A facies and trace fossil analysis in the Zumaya flysch (Paleocene-Eocene), North Spain. *Sediment.,* 20: 105-31.

——, 1976. Sand fans, turbidites, slumps and the origin of the Bay of Biscay: A facies analysis of the Guipuzcoa flysch. *Palaeogeogr., Palaeoclim., Palaeoecol.,* 19: 1-15.

Damuth, J. E., and Kumar, N., 1975. Amazon Cone: Morphology, sediments, age, and growth pattern. *Geol. Soc. Amer. Bull.,* 86: 863-78.

Feuillée, P., and Mathey, B., 1976. The interstratified breccias and conglomerates in the Cretaceous flysch of the northern Basque Pyrenees: Submarine out-flow of diapiric mass – some comments. *Sed. Geol.,* 16: 85-87.

Ghibaudo, G., and Mutti, E., 1973. Facies ed interpretazione paleoambientale delle Arenarie di Ranzano nei dintorni di Specchio (Val Pessola, Appennino Parmense). *Soc. Geol. Ital. Mem.,* 12: 251-65.

Gomez de Llarena, J., 1954. Observaciones geologicas en el flysch cretacico – numulitico de Guipuzcoa, I. *Monogr. Inst. "Lucas Mallada" Invest. Geol.,* 13: 98 pp.

Haner, B. E., 1971. Morphology and Sediments of Redondo Submarine Fan, Southern California. *Geol. Soc. Amer. Bull.*, 82: 2413-32.

Hanisch, J., 1974. Der Tiefsee-Diapir von Zarauz (N-Spanien) im Spiegel von Sedimentation und Tektonik des Kreide/Tertiär Flyschs. *Geol. Jahrb.*, 11: 101-42.

———, 1975. Grabeneinbruch durch schichtparalleles Klaffen in Flyschfolgen. *N. Jahrb. Geol. Paläont. Mh.*, 11: 641-48.

———, and Pflug, R., 1974. The interstratified breccias and conglomerates in the Cretaceous flysch of the northern Basque Pyrenees: Submarine outflow of diapiric mass. *Sed. Geol.*, 12: 287-96.

Horn, D. R., Horn, B. M., and Delach, M. N., 1970. Sedimentary provinces of the North Pacific. *Geol. Soc. Amer. Mem.*, 126: 1-21.

———, Ewing, M., Horn, B. M., and Delach, M. N., 1971. Turbidites of the Northeast Pacific. *Sediment.*, 16: 55-69.

———, Ewing, J. I., and Ewing, M., 1972. Graded-bed sequences emplaced by turbidity currents North of 20° N in the Pacific, Atlantic and Mediterranean. *Sediment.*, 18: 247-275.

Jacka, A. D., Beck, R. H., St. Germain, L. C., and Harrison, S. C., 1968. Permian deep-sea fans of the Delaware Mountain Group (Guadalupian), Delaware Basin. In: *Guadalupian Facies, Apache Mountain Area, West Texas.*, Soc. Econ. Paleont. Mineral., Permian Basin Section., Publ. 68-11, pp. 49-90.

Kapellos, C., 1974. Uber das Nannoplankton im Alttertiär des Profils von Zumaya-Guetaria (Provinz Guipuzcoa, Nord Spanien). *Ecl. Geol. Helv.*, 67: 435-44.

Kruit, C., Brouwer, J., and Ealey, P., 1972. A deepwater sand fan in the Eocene Bay of Biscay. *Nature Phys. Sc.*, 240: 59-61.

———, Brouwer, J., Knox, G., Schöllnberger, W., and van Vliet, A., 1975. Une excursion aux cones d'alluvions en eau profonde d'age tertiaire près de San Sebastian (Province de Guipuzcoa, Espagne): Excursion 23. *Proc. IX Intern. Cong. of Sediment.*, Nice 75 pp.

Lowe, D. R., 1975. Water escape structures in coarse-grained sediments. *Sediment.*, 22: 157-204.

Marschalko, R., 1970. The origin of disturbed structures in Carpathian turbidites. *Sed. Geol.*, 4: 5-18.

Martini, E., 1971. Standard Tertiary and Quaternary calcareous nannoplankton zonation. In: A. Farinacci (ed.), *Proc. II Planktonic Conference,* Rome, pp. 739-86.

Middleton, G. V., and Hampton, M. A., 1973. Sediment gravity flows: Mechanics of flow and deposition. In: G. V. Middleton and A. H. Bouma (eds.), *Turbidites and Deep-Water Sedimentation,* Soc. Econ. Paleont. Mineral., Pacific Section, Short Course, Anaheim, pp. 1-38.

Mutti, E., and Ghibaudo, G., 1972. Un esempio di torbiditi di conoide sottomarina esterna: Le Arenarie di San Salvatore (Formazione di Bobbio, Miocene) nell' Appennino di Piacenza. *Mem. Acc. Sci. Torino, Classe Sci. Fis. Mat.,* Nat., Serie 4, 16, 40 pp.

———, and Ricci Lucchi, F., 1972. Le Torbiditi dell' Appennino settentrionale: introduzione all'analisi di facies. *Soc. Geol. Ital. Mem.*, 11: 161-99.

———, and Ricci Lucchi, F., 1974. La signification de certaines unités séquentielles dans les séries à turbidites. *Bull. Soc. Géol. Fr.*, XVI, 6, pp. 577-82.

———, and Ricci Lucchi, F., 1975. Turbidite facies and facies associations. In: E. Mutti, G. C. Parea, F. Ricci-Lucchi, M. Sagri, G. Zanzucchi, G. Ghibaudo, and S. Iaccarino (eds.), *Examples of Turbidite Facies and Facies Associations from Selected Formations of the Northern Appenines. Excursion 11.* IX Intern. Cong. Sediment., Nice, pp. 21-36.

Nelson, H., 1976. Late Pleistocene and Holocene depositional trends, processes, and history of Astoria deep-sea fan, Northeast Pacific. *Mar. Geol.*, 20: 129-73.

Nelson, C. H., Carlson, P. R., Byrne, J. V., and Alpha, T. R., 1970. Development of the Astoria Canyon-Fan physiography and comparison with similar systems. *Mar. Geol.*, 8: 259-91.

Nelson, C. H., and Kulm, L. D., 1973. Submarine fans and channels. In: G. V. Middleton and A. H. Bouma (eds.), *Turbidites and Deep-water Sedimentation.* Soc. Econ. Paleont. Mineral., Pacific Section, Short Course, Anaheim, pp. 39-78.

———, and Nilsen, T. H., 1974. Depositional trends of modern and ancient deep-sea fans. In: R. H. Dott and R. H. Shaver (eds.), *Modern and Ancient Geosynclinal Sedimentation.* Soc. Econ. Paleont. Mineral., Sp. Publ. 19, pp. 54-76.

Normark, W. R., 1970. Growth patterns of deep-sea fans. *Amer. Assoc. Petrol. Geol. Bull.*, 54: 2170-95.

———, and Piper, D. J. W., 1972. Sediments and growth pattern of Navy deep-sea fan, San Clemente Basin, California Borderland. *J. Geol.*, 80: 198-223.

Piper, D. J. W., 1970. Transport and deposition of Holocene sediment on La Jolla Deep-Sea Fan, California. *Mar. Geol.*, 8: 211-27.

Plaziat, J. C., 1975. Signification paléogéographique des "calcaires conglomères", des brèches et des niveaux à Rhodophycées dans la sédimentation carbonatée du Bassin Basco-Béarnais à la base du Tertiaire (Espagne-France). *Rev. Géogr. Phys. Géol. Dyn.*, 17: 239-58.

Reineck, H. E., 1973. Schichtung und Wühlgefüge in Grundproben vor der ostafrikanischen Küste. *"Meteor" Forsch. Ergebn.* C. 16: 67-81.

Ricci Lucchi, F., 1965. Alcune strutture di risedimentazione nella formazione Marnoso-arenacea Romagnola. *Giorn. di Geol.*, 33: 265-83.

———, 1975. Depositional cycles in two turbidite formations of Northern Apennines (Italy). *J. Sed. Petrol.*, 45: 3-43.

———, and Parea, G. C., 1973. Cicli deposizionali (megasequenze) nelle torbidite di conoide sottomarina: formazione della laga (appennino marchigiano – abruzzese). *Atti Soc. Nat. Mat. di Modena,* 104: 247-83.

Seilacher, A., 1967. Paleontological studies on turbidite sedimentation and erosion. *J. Geol.*, 70: 227-34.

Shepard, F. P., Dill, R. F., and U., 1969. Physiography and Sedimentary processes of La Jolla Submarine Fan and Fan-valley, California. *Bull. Amer. Assoc. Petrol. Geol.*, 53: 390-420.

Stanley, D. J., and Unrug, R., 1972. Submarine channel deposits, fluxoturbidites and other indicators of slope and base-of-slope environments in modern and ancient marine basins. In: J. K. Rigby and W. K. Hamblin (eds.), *Recognition of Ancient Sedimentary Environments.*, Soc. Econ. Paleont. Mineral., Sp. Publ. 16, pp. 287-340.

Von Hillebrandt, A., 1965. Foraminiferen-Stratigraphie im Alttertiär von Zumaya (Provinz Guipuzcoa, NW Spaien) und ein Vergleich mit anderen Tethys-Gebieten. *Bayr. Akad. Wiss., Math. Naturw. Kl., Abh., N.F.*, 123, 62 pp.

Walker, R. G., and Mutti, E., 1973. Turbidite facies and facies associations. In: G. V. Middleton and A. H. Bouma (eds.), *Turbidites and Deep-water Sedimentation.* Soc. Econ. Paleont. Mineral., Pacific Section, Short Course, Anaheim, pp. 119-57.

Chapter 15

Outer Fan Depositional Lobes of the Laga Formation (Upper Miocene and Lower Pliocene), East-Central Italy

EMILIANO MUTTI

Istituto di Geologia
Università di Torino
Torino, Italy

TOR H. NILSEN

U.S. Geological Survey
Menlo Park, California

FRANCO RICCI LUCCHI

Istituto di Geologia
Università di Bologna
Bologna, Italy

ABSTRACT

Outer fan sandstone bodies or lobes in the turbiditic Upper Miocene and Lower Pliocene Laga Formation in the Monte Bilanciere area, west of Teramo in east-central Italy, are geometrically sheet-like and sedimentologically diverse. The sandstone bodies form part of the upper sequence of a 3,000 m thick turbidite fill in the restricted and complex Laga basin. Active faults apparently formed the western and southern margins of the basin. Lateral and vertical variations in the thickness, facies, paleocurrents, and bedding sequence of five sandstone bodies have been studied in detail through bed-by-bed measurement of nineteen columnar sections in an area of about 16 km². Excellent exposures in a structurally uncomplicated area of the basin permitted the tracing of each body throughout the area and the accurate correlation of the bodies and individual sandstone beds within each body. The sandstone bodies, which contain more than 60 percent sandstone, range in thickness from 3 m to 15 m and typically consist of thick, individual sandstone beds definable by Bouma (1962) sequences that begin with the *a* division (Facies C turbidites). The bodies are separated by more pelitic intervals that typically contain thin scattered Facies D sandstone or siltstone turbidites that begin with the *b, c, d,* or *e* divisions of Bouma (1962).

The outer fan sandstone bodies form continuous, laterally extensive, sheet-like bodies throughout the Monte Bilanciere area. They record deposition downfan from channels and are constructional features that do not fill channels. Paleocurrents are laterally and vertically very consistent in orientation and indicate sediment transport toward the southeast. Individual sandstone beds are also laterally extensive and typically can be traced throughout the area. Thickening-upward, vertical bedding cycles are developed within these bodies at different scales. Thick cycles that comprise the entire sandstone body can typically be traced throughout the area, whereas thin cycles that comprise only parts of the sandstone bodies characteristically cannot be traced throughout the area. The thinner cycles are apparently affected primarily by local constructional topographic lows and highs with as much as 3 m of relief that develop within the bodies. Some low areas represent erosional scours, others, nondeposition or very slow deposition. Topographic highs or bulges result from more rapid local deposition of thicker beds. Differential compaction within individual sandstone bodies as well as within underlying parts of the Laga Formation provided control over local variations in geometry, thickness, and cyclicity within the sandstone bodies. Local erosion of topographic highs yielded deposition of slurried beds in adjacent topographic lows.

The geometry and sedimentology of outer fan sandstone bodies are more complex than previous studies indicated. Variations in amount and type of sediment supply, penecontemporaneous tectonic activity, differential compaction of sediments, changes in slope gradients, and rate of progradation of fans are some of the factors that affect the growth and development of outer fan lobes.

INTRODUCTION

Sandstone bodies that are laterally extensive, nonchannelized, thick bedded, and situated downfan from related channelized distributary systems are a major feature of ancient deepsea fan turbidite deposits. These turbidite sandstone bodies were first described and interpreted as outer fan depositional lobes by Mutti and Ghibaudo (1972) from the Miocene Bobbio Formation in the northern Apennines (Figure 15-1). Subsequent investigations have led to the recognition of similar bodies in other ancient sedimentary basins (Mutti and Ricci Lucchi 1972; Mutti et al. 1972; Ricci Lucchi and Parea 1974; Ricci Lucchi and Pialli 1973; Mutti 1974; Kruit et al. 1975; Ricci Lucchi 1975a, b; Nilsen and Abbate 1976; and chapter 14 of this volume.)

The outer fan sandstone bodies are depositional features comparable in many respects with channel mouth bars of constructive deltaic systems that are controlled mainly by fluvial processes. Each sandstone body, or lobe, forms downfan from an active fan channel located in the higher, or more proximal, distributary system of the fan. The thick-bedded and commonly coarse-grained sediments of each lobe grade radially downcurrent or downfan into finer-grained and thinner-bedded lobe fringe deposits (Mutti and Ghibaudo 1972). Gradual basinward progradation or lateral shifting of the sandstone lobe facies on top of the marginal lobe fringe facies yields distinct coarsening- and thickening-upward cycles that are particularly well developed in the distal parts of each lobe (Mutti and Ricci Lucchi 1974). The temporal shifting of active fan channels over the distributary system leads to the shifting of lobe sedimentation and formation of thick sequences of alternating lobe and lobe fringe deposits in the outer fan.

The abundance of sandstone bodies in ancient outer fan depositional sequences is puzzling because virtually no important sand accumulations have yet been reported from outer fan depositional areas of modern deepsea fans. Lack of deep penetration by coring devices and the fact that most modern deepsea fans are not presently receiving much sand deposition (Nelson and Nilsen 1974; Nelson 1976) may account for the apparently contradictory situation between present and past.

Careful studies of well-exposed ancient outer fan sequences are critically important for a better understanding of the depositional setting of both modern and ancient deepsea fans. Information is needed from different basins in order to determine the principal factors that control the deposition, size, morphology, and geometry of outer fan sandstone bodies and their associated finer-grained and thinner-bedded deposits.

In the Monte Bilanciere west of Teramo in east-central Italy, the Upper Miocene and Lower Pliocene Laga Formation provides excellent exposures of nonchannelized sandstone bodies (see Figure 15-1). These bodies were interpreted by Ricci Lucchi and Parea (1974) and Ricci Lucchi (1975a) as outer fan deposits on the basis of reconnaissance studies. Because these bodies are readily accessible and can be physically traced in the field for distances of at least 5 km, they were selected for detailed study. We illustrate and discuss in preliminary fashion herein the geometry and detailed correlation patterns of individual beds and groups of beds from five sandstone bodies that are particularly well exposed in the Monte Bilanciere area. We define the sandstone bodies as units with a sandstone content that exceeds 60 percent consistently throughout the study area. Field studies included the measurement of numerous vertical sections, precise stratigraphic correlations of individual beds and sandstone bodies, and paleocurrent analyses.

GEOLOGIC SETTING

The Laga Formation is the most voluminous sandstone sequence in the northern Apennines foredeep; it crops out over an area of approximately 2,000 km^2 in the Marche and Abruzzi regions (see Figure 15-1). The Laga is bounded by Mesozoic rocks to the west, south, and southeast; by other Miocene units to the north; and by middle Pliocene to Pleistocene sediments to the east. The basin has been deformed by Pliocene and Quaternary tectonic events and is elongate parallel to the northwest-southeast trending Apennine mountain chain.

The foredeep is one of the major paleogeographic features of the Periadriatic Apennines. In the early and medial Miocene it formed part of a submerged foreland on which hemipelagic muds were slowly deposited; to the south, biogenous and bioclastic shelf facies were deposited. Coeval turbidites were deposited in the Inner Basin of Ricci Lucchi (1975b), a deep subsiding basin southwest of the Laga Basin. The foredeep became an actively subsiding area in the late Miocene (Tortonian and Messinian) during and immediately after the major tectonic activity that uplifted the Apennines (Abbate et al. 1970). Differential subsidence led to topographic differentiation of the foredeep into subbasins of varying sizes that were separated by submarine highs and sills.

Turbidite bodies in the northern Padan Basin of the foredeep merge with the Marnoso-arenacea Formation of the Inner Basin because subsidence shifted eastward toward the foreland. The Laga Basin to the south, however, was separated from the Marnoso-arenacea Basin by the Umbria-Marche Ridge, a wide positive structure (see Figure 15-1).

The lateral contacts of the Laga Formation with adjacent units are tectonic except to the east, where middle Pliocene strata that dip gently northeastward toward the Adriatic

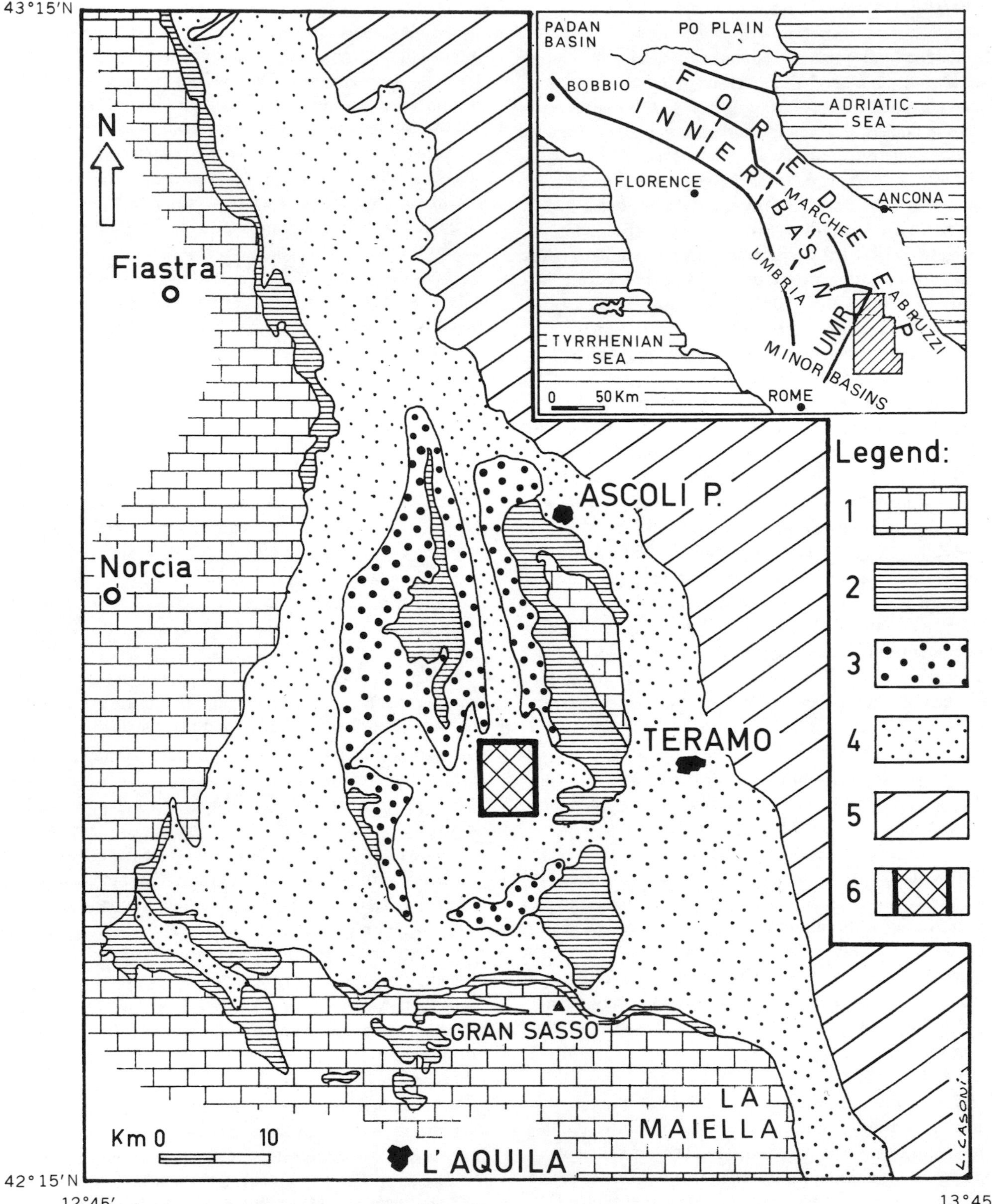

Figure 15-1. Index map (inset) and simplified geologic map showing location of Laga basin and Monte Bilanciere area. Note: Northeast-trending dashed lines on index map indicate syn- and postsedimentary tectonically active linear features; UMR = Umbria-Marche Ridge; 1 = Mesozoic and Lower Tertiary formations; 2 = Oligocene and Miocene preturbidite sequence; 3 = Laga Formation, channelized sandstone bodies; 4 = Laga Formation, outer fan sandstone bodies and basin plain turbidites; 5 = transgressive Pliocene deposits; and 6 = Monte Bilanciere area.

Sea overlap the Laga Formation. These younger strata include fan-delta to shallow marine deposits and indicate that the Laga was uplifted and eroded in the early Pliocene. Faulting subsequently affected this contact locally.

The western margin of the basin was severely deformed along a major tectonic lineament that has been interpreted as either a transcurrent fault (Ogniben 1969) or an overthrust (Dallan Nardi et al. 1971). In the southern part of the foredeep, the Laga Formation has been overthrust by the Mesozoic Gran Sasso and La Maiella carbonate platforms, which were derived from the west-southwest; the platforms probably constituted the original southern and southwestern margins of the foredeep (Ricci Lucchi 1975b).

To the north, the Laga grades abruptly into coeval thinner clastics and evaporites; however, the lack of precise correlation of stratigraphic units due to the absence of fossils, the presence of some faults with large vertical offsets, and the masking effect of vegetation and alluvial deposits make the interpretation of this boundary difficult. A major transverse fault was probably active in this area during deposition of the Laga Formation (Ricci Lucchi 1975a, Figure 13; 1975b, Figure 33d).

STRATIGRAPHY AND ENVIRONMENTS OF DEPOSITION

Pre-Laga Sequence

The pre-Laga deposits form a relatively thin sequence that was deposited in the time interval from Oligocene to late Miocene (Aquitanian to Tortonian) at an average sedimentation rate of 1 to 2 cm/1,000 years. These sediments conformably overlie similar Lower Tertiary sediments and can be subdivided into two units. The older unit consists of calcarenites (biosparites), calcisiltites (biomicrites), and marls thought to have been deposited on a shelf with a predominance of biogenous, inner shelf facies. The younger unit consists of dark shales, calcareous turbidites, and chaotic shelf deposits thought to have been deposited on a lower slope in a starved basin.

The older unit consists mainly of thick-bedded to massive bioclastic shelf carbonates in which primary sedimentary structures are completely obscured by bioturbation. Beds generally decrease in thickness and wedge out northward. The carbonates contain a mixture of benthonic and planktonic debris whose proportions vary with textural characteristics —that is, the finer grain sizes contain more planktonic remains and comminuted organic detritus as carbonate matrix. Interbedded marls thicken northward, where outer shelf facies are most common beneath the Laga Formation. A major tongue of hemipelagic inner shelf marl was deposited to the south during a middle Miocene transgressive phase.

The younger slope and basin deposits record an abrupt, almost catastrophic sinking of part of the shelf that resulted in the formation of a deep basin in the late Miocene (Tortonian). Collapse of the shelf and overlying sediments formed steep slopes, thereby causing slumping of the basin margins to form contorted and chaotic beds and huge olistoliths. Skeletal carbonate turbidites and dark shales rich in pteropods formed the initial deposits of the newly formed trough. Paleocurrent features within these turbidites indicate northward transport of sediments. The turbidites are laterally discontinuous, indicating that they accumulated in local, irregular topographic depressions and were derived from the edge of the remaining shelf. Apart from the minor irregularities created by slumping and sliding, one or more major furrows were produced by differential subsidence; they can be recognized by the geometry of the subsequent turbidite fill of the Laga Formation (Figure 15-2).

Laga Formation Turbidites

Preliminary data regarding the stratigraphy and depositional setting of the Laga Formation near Ascoli Piceno have been published by Colacicchi (1959), Girotti and Parotto (1969), Parea and Ricci Lucchi (1972), and Ricci Lucchi (1973, 1975a, b). The stratigraphic section most commonly thought to be representative of the Laga Formation (see Parea and Ricci Lucchi 1972, Figure 1; Ricci Lucchi 1973, Figure 2) is located in the northern part of the basin, west of Ascoli Piceno. It consists of an overall fining-upward sequence of sandy and pelitic turbidites as thick as 3,000 m. This fining-upward section was thought to reflect a receding or retrogradational suite, which is a common first-order depositional cycle in turbidite basins (Ricci Lucchi 1975a). The vertical arrangement of facies associations in the sequence was inferred to be in ascending order: (1) channelized mid-fan turbidites; (2) nonchannelized outer fan turbidites (thick-bedded sandstone bodies); and (3) interbedded nonchannelized basin plain turbidites (thin-bedded pelitic bodies) and outer fan turbidites. Paleocurrent directions indicate sediment transport from north-northwest to south-southeast.

The main aspect of this sequence is abrupt deposition of large volumes of massive, sandy turbidites in a previously deep but almost stagnant and starved marine basin. The channelized lower contact of the Laga extends for many kilometers, thereby suggesting the existence of either very wide submarine valleys or frequent lateral shifting of narrow valleys; in only a few sections does the lower Laga consist of basinal or fan fringe turbidites overlain by progradational outer fan lobes (Ricci Lucchi 1975a). The deepsea fans of the Laga probably were locally "choked" by overfeeding of sand and became asymmetric as a re-

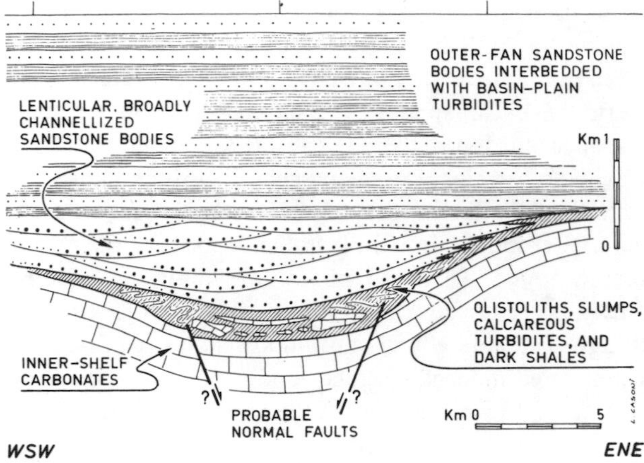

Figure 15-2. Diagrammatic cross-section of sedimentary fill of Laga Basin.

sult of lateral confinement caused by tectonically active structures.

The lower massive sandstones in the southern part of the basin fill a structural rather than erosional valley (see Figure 15-2). Beds of sandstone with little interbedded shale abruptly wedge out and lap against a marginal high for a distance of about 9 km. Many individual sandstone beds are as thick as 15 m to 20 m. Sets of beds are channelized and probably fill unstable systems of shallow minor channels that are broadly comparable with low-angle erosional surfaces of braided streams. This channelized sedimentation resulted in deposition of repeated and complicated sandstone lenses containing rare, thin pelitic intercalations that are representative of channel abandonment or overbank flow.

Within this composite sandstone body, channelization decreases upward and eventually disappears, thereby reflecting the smoothing of the submarine topography and a decrease in slope. If the original slope of the basin was 1°, which is comparable with slopes of modern deepsea basins, then the length of the distributary system of 60 km or more implies a difference of 1,000 m in depth from north to south.

Deposition of evenly bedded, thin sandstone turbidites characterized by low sandstone:shale ratios marks the complete filling of the structural valley and formation of a basin plain. Subsequently, a series of outer fan sandstone lobes prograded southward over this basin plain; the thick series of lobes in the upper part of the Laga Formation forms the object of this study (see Figure 15-2).

The average sedimentation rate for the Laga Formation exceeded 1 m/1,000 years. Lack of stratigraphic markers prohibits calculation of separate sedimentation rates for various parts of the basin or particular stratigraphic intervals. However, the rate was undoubtedly higher during deposition of the lower channelized facies than during formation of the upper progradational outer fan lobe facies, when the influx of coarse clastics was reduced.

Thus, in the Monte Bilanciere area, the vertical sequence of facies associations consists of lower amalgamated and channelized middle fan sandstone bodies that fill and smooth a deep (800 m) structural valley and upper alternating basin plain and outer fan sandstone bodies (see Figure 15-2).

Petrographic studies of Laga Formation sandstones are not complete. Based on stratigraphic and structural relations, ten Haaf (1964) suggested that the sand was recycled from older turbidites in the Apennines. Active transverse faults at the northwestern end of the basin may have served as transport routes and triggering mechanisms for sediment gravity flows that were deflected southward along the basin axis. The lack of pebbles in even the most proximal turbidite beds is problematical, since conglomerate is common in older turbidites of the northern Apennines, and gravel can be transported for long distances via submarine canyons and valleys (Nelson and Kulm 1973). The lack of pebbles may indicate a secondary rather than primary source for the Laga detritus.

Fossils are rare, and trace fossils are completely absent in the Laga Formation, although they are abundant in older and younger units, thereby suggesting that bottom waters were stagnant. The Laga was deposited partly during the late Miocene (Messinian) Mediterranean "salinity crisis," so that stagnant bottom conditions and highly saline waters were probably common. Discussion of the implications of this important paleoclimatic and paleogeographic event are beyond the scope of this chapter. Planktonic foraminifera collected from mudstone are presently being studied to ascertain whether they were derived from surface waters or exposed fossil-bearing beds and whether they were deposited by slow settling through the water column or by sediment gravity flows.

METHOD OF STUDY

Distinct sandstone bodies separated by pelitic intervals crop out on the east limb of a north-trending syncline in the Monte Bilanciere area (Figure 15-3). Five sandstone bodies, ranging in thickness from 3 m to 15 m, were selected for detailed study. They form prominent morphologic and stratigraphic units that can easily be recognized and traced in the field (photograph A, Figure 15-4). Each sandstone body consists of a group of individual sandstone beds separated by thin pelitic intervals (photograph B, Figure 15-4). Slurried beds are common in the sandstone bodies (photograph C, Figure 15-4).

The total thickness of the interval containing the sandstone bodies is about 150 m. Nineteen columnar sections that included, where possible, all five sandstone bodies,

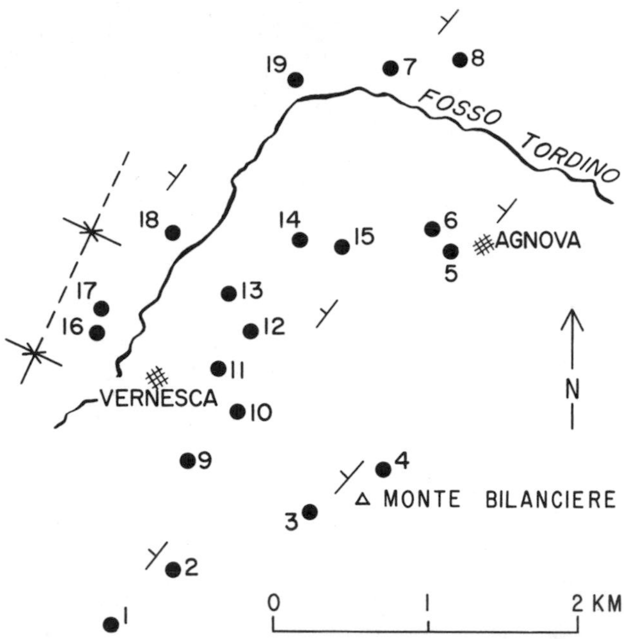

Figure 15-3. Sketch map of the Monte Bilanciere area showing location of measured sections (numbered) in outer fan sandstone bodies. Note: Approximate axis of north-trending syncline shown to west at boundary of area. Location of area is shown in Figure 15-1.

were measured on both sides of Fosso Tordino, a small stream cutting across the area (see Figure 15-3). For each bed, the thickness, lithology, Bouma sequence, and orientation of paleocurrent indicators were noted. Gross correlations of the sandstone bodies, and in many cases of individual beds, were established directly either by walking out the bodies or beds in the field or by tracing them on photographs taken in the field.

On the basis of the measured sections and subsequent correlations, a number of detailed stratigraphic cross-sections were prepared across the study area. Horizontal variations in the geometry of the five sandstone bodies were demonstrated by this method (Figure 15-5). We also attempted to establish more detailed correlations within individual sandstone bodies that show horizontal and vertical variations in the geometry of individual beds throughout the area.

Most of the cross-sections have a similar southwest-northeast orientation that is approximately perpendicular to the southeast-directed, average paleocurrent trends. This pattern permits convenient presentation of our preliminary results and enables discussion and comparison of similarly oriented correlation patterns of sandstone bodies and individual sandstone beds. Many different cross-sections have been prepared from the measured sections, but only a small number that illustrate some important geometric relations are presented herein.

PALEOCURRENT PATTERN

Paleocurrents in the nineteen sections shown in Figure 15-3 were systematically measured from sole markings; namely, flute casts, groove casts, and frondescent marks. The restored paleocurrent directions are remarkably similar in orientation throughout the Monte Bilanciere area and within both sandstone bodies and intervening pelitic units (Figure 15-6). Paleocurrents within sandstone bodies A, C, E, and H average 145°, and those within pelitic units B, D, F, and G average 155°; a few paleocurrent directions in sandstone body I indicate similar directions. The vertical and horizontal distribution of paleocurrent directions in sandstone and pelitic units is shown in Figure 15-5.

The similarity in orientations may be attributable to several important factors: (1) the narrow Laga Basin was confined by upland and shelf areas to the north, west, and south, thus limiting the ability of turbidity currents to change direction of flow during sediment transport and deposition; (2) a relatively steep south-southeast-trending basin slope may have controlled flow direction of turbidity currents; and (3) sediments were apparently derived only from provenance areas to the northwest, so that sediment was transported into the basin from only one direction (Ricci Lucchi 1975b).

LITHOLOGY, FACIES, AND GENERAL CORRELATION PATTERN

The overall correlation pattern, geometry, and variations in thickness for the sandstone bodies are shown on a generalized cross section in Figure 15-5. The stratigraphic subunits include both sandstone bodies (from base to top: A, C, E, H, and I) and associated more pelitic intervals (from base to top: B, D, F, and G.)

The turbidite facies of Mutti and Ricci Lucchi (1972 1975) are applied to sandstone bodies and beds examined in this study. These workers subdivided turbidites into seven major facies, A through G; various combinations of these facies form three major associations of facies characteristic of slope, deepsea fan, and basin plain environments. The outer fan facies association consists primarily of Facies C and D turbidites. These two facies resemble those graded sandstone turbidite beds described by Bouma (1962) that conform to the Bouma sequence ($T_{a\text{-}e}$), which consists in upward order of a massive *a* division, a lower parallel-laminated *b* division, a current-laminated *c* division, and upper parallel-laminated *d* division, and a massive pelitic *e* division. Facies C turbidites commence at the the base with the *a* division, whereas Facies D turbidites commence with the *b, c, d,* or *e* divisions. Beds of both facies are generally plane parallel, laterally extensive, and do not have lower contacts that are significantly chan-

Figure 15-4. Outer fan sandstone bodies and individual beds. Note: Photograph A shows outcrops of outer fan sandstone bodies in the Monte Bilanciere area; view from section 14 west-southwestward toward sections 16, 17, and 18 (see Figure 15-3 for location). Photograph B shows vertical sequence of beds in sandstone lobe A, section 9 (see Figure 15-3 for location of section and Figure 15-8 for measured section). Photograph C shows thick, slurried bed with a massive lower unit containing scattered shale clasts, slurried interval, and upper laminated unit. Staff is 1.5 m in length.

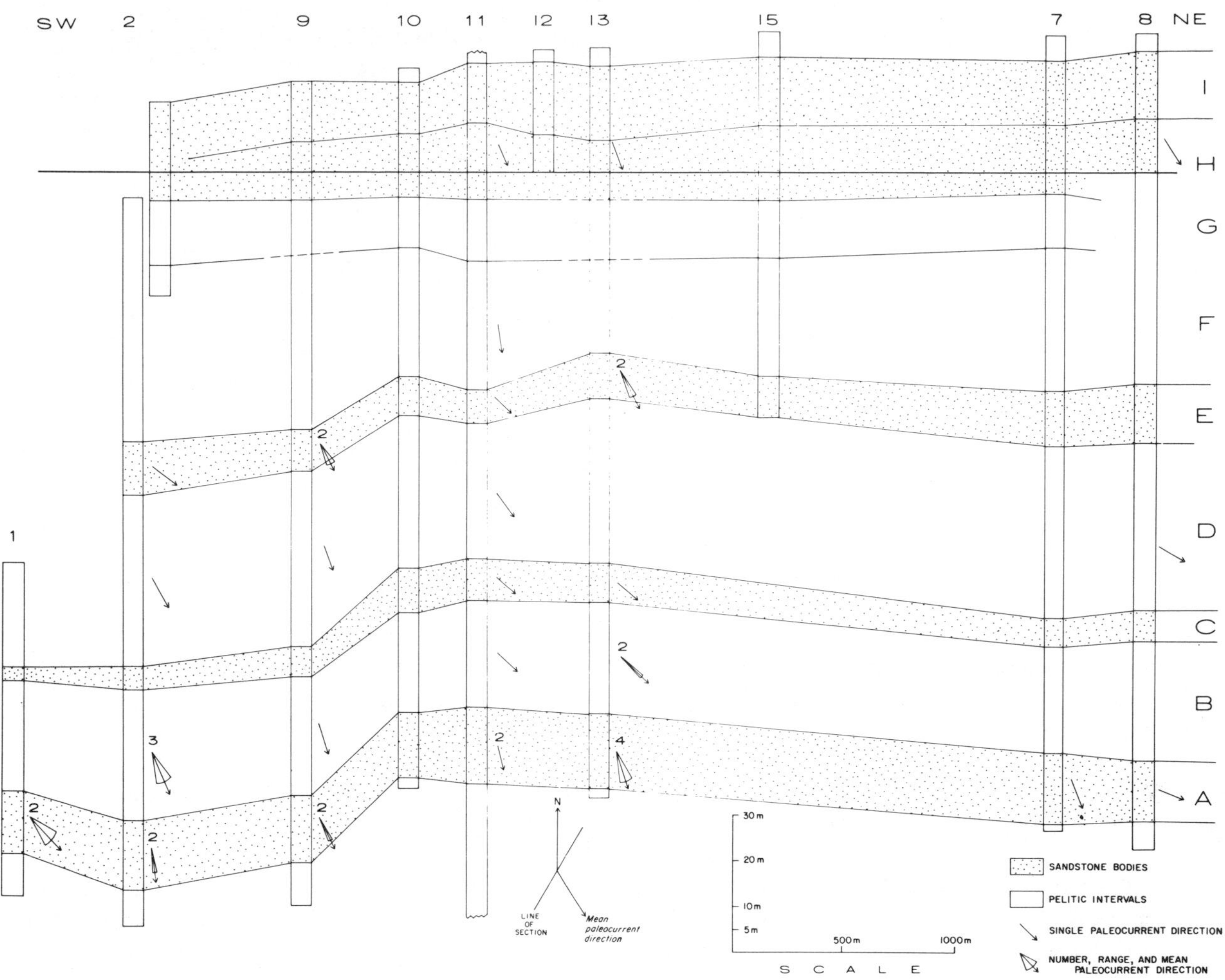

Figure 15-5. Generalized cross-section showing lateral continuity of outer fan sandstone bodies and intervening pelitic units. Note: Orientation of section relative to north and total range and mean of paleocurrent directions are shown below columns. Location of columnar sections is shown in Figure 15-3.

nelized. Facies C beds are generally thicker and coarser grained than Facies D beds.

The sandstone bodies typically consist of thick to massive sandstone beds with subordinate amounts of thin-bedded sandstone, siltstone, and shale. Most sandstone beds can be described in terms of complete Bouma (1962) sequences ($T_{a\text{-}e}$) and classified as Facies C turbidites. The coarse-tail graded, thick basal *a* division consists of fine- to medium-grained sandstone containing scattered, graded medium to coarse grains. This division is either structureless or contains faint parallel laminae that may be disrupted by fluid-escape structures. Above this lower division are thinner *b, c,* and *d* divisions deposited by traction plus fallout and characterized by parallel, current-rippled, and most commonly, convoluted laminae. These current-laminated divisions range from very fine sandstone to siltstone. Plant fragments are locally abundant on lamination surfaces. Highly concentrated sediment gravity flows that are transitional between liquefied flows and turbidity currents (in the sense of Lowe 1976) are thought to be responsible for transport and deposition of these beds.

The pelitic intervals between sandstone bodies contain 10 to 60 percent sandstone, with an average of 35 to 45 percent. The sandstone beds in the pelitic units are generally thinner than those within sandstone bodies and typically contain Bouma sequences that lack the basal divisions ($T_{b\text{-}e}$ through $T_{d\text{-}e}$). These beds are Facies D turbidites and are inferred to have been transported and deposited by relatively dilute turbidity currents.

The sandstone bodies do not have large-scale erosional basal contacts with underlying pelitic intervals. Correlation lines separating sandstone bodies and pelitic intervals were established and traced from more detailed cross sections.

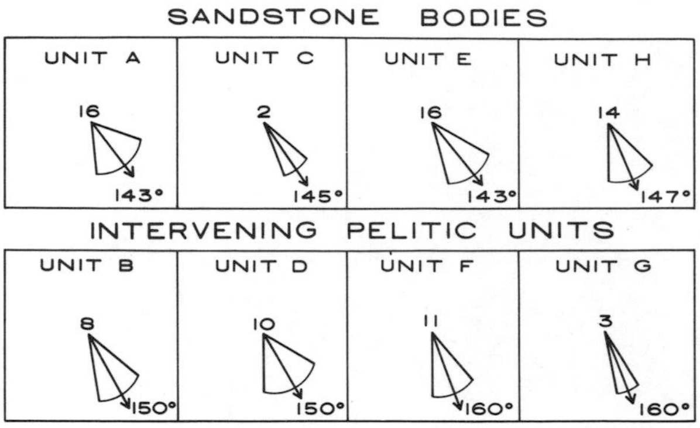

Figure 15-6. Range, mean, and number of paleocurrent directions from outer fan sandstone bodies and intervening pelitic units.

The correlation lines of Figure 15-5 represent tops or bases of distinct individual beds and are thus *time lines*. The datum for Figure 15-5 is a bed within the sandstone body H that can be traced easily from one section to another throughout the study area and may represent an original, nearly horizontal depositional surface. The same bed is also the datum for the detailed cross-sections of sandstone body H that follows.

The stratigraphic sequence comprising these sandstone bodies and pelitic units can be subdivided into two parts. The lower part deviates more from horizontality, probably because of greater bottom topography (Figure 15-5), which thus indicates that some deformation may have occurred during or immediately after deposition. The amount of this deformation decreases from base to top as indicated by smaller lateral changes in thickness of the sandstone bodies. The greater lateral changes in thickness in the lower part of the sequence probably reflect compaction or growth of penecontemporaneous structural features. The reconstructed lateral thickness changes exceed those estimated for basinal turbidite environments. However, presently available data do not permit a better understanding of these compactional or deformational events. The sandstone bodies are underlain by more than 1,000 m of Laga turbidites that include sizable sandstone bodies; the effects of geometry, lateral variations, and compaction within these underlying units on sandstone bodies A to I are not clear.

Except for body C and possibly body H, which thin southwestward (Figure 15-5), individual sandstone bodies can be traced at least 4.4 km perpendicular to the paleocurrent direction without substantial changes in thickness. Unit A is remarkably tabular or sheet-like. Our reconnaissance investigations indicate that sandstone bodies A and C can probably be traced an additional 3.5 km to the north-northeast without major changes in thickness. The precise correlation of bodies E, H, and I in the same direction is presently being studied.

INTERNAL GEOMETRY OF OUTER FAN SANDSTONE BODIES

The variations in bedding sequence for sandstone bodies A, C, H, and I as well as those for pelitic intervals B, D, and G, are shown in the detailed cross sections of Figures 15-7 to 15-10. Most individual sandstone beds match the overall sheet-like geometry of the sandstone bodies and can be traced throughout the study area, both within the sandstone bodies and intervening pelitic intervals. However, the thicker beds are geometrically more regular in some bodies than others, as demonstrated by a comparison of thicker beds in body A with those of body H (see Figures 15-7 and 15-9).

The thickness and geometry of the laterally continuous sandstone beds were affected by topographic irregularities on the surface of sedimentation, as demonstrated by the complex variations within body A (Figure 15-7). These lateral changes in thickness result from the filling of topographic depressions formed by either depositional or erosional processes. Most depressions related to depositional processes resulted from differential areal accumulation of sand by high-volume flows followed by differential compaction of underlying pelitic sediments. Depressions related to erosional processes were scoured by a few large turbidity currents before deposition of their sand load. These depressions are never deeper than 3 m. Erosion was more common on topographically higher areas but also affected and accentuated preexisting lows, as seen in the upper part of pelitic unit G (Figure 15-8). More extensive regional erosional surfaces and shallow scours can be seen at the top of sandstone body H (Figure 15-9). Similar observations of erosional depressions were made by Enos (1969) in the Cloridorme Formation of the Gaspé Peninsula, Canada.

Depressions formed by depositional processes are filled mostly with thin- or medium-bedded sandstone (see Figure 15-7, lower part of section 2, sandstone body A), whereas depressions formed by erosional processes are commonly filled with massive sandstone (see Figure 15-7, lower part of section 7, sandstone body A). The fillings in both types of depressions are organized into thin thickening-upward sequences.

The local deposition of sand as topographic bulges must have induced variations in pressure, known as the Bernoulli effect, in subsequent large turbidity currents (Enos 1969). However, increased flow velocities over the bulges did not always produce erosion because in some sandstone bodies the convex top of the bulge is preserved by nondeposition or thin pelitic-rich sediment drapes (see Figure 15-9, sandstone body H). The onlap of sandstone beds that fill adjacent depressions reveals the topographic bulges; these onlapping stratigraphic relations are best observed in pelitic unit D (Figure 15-8) and in

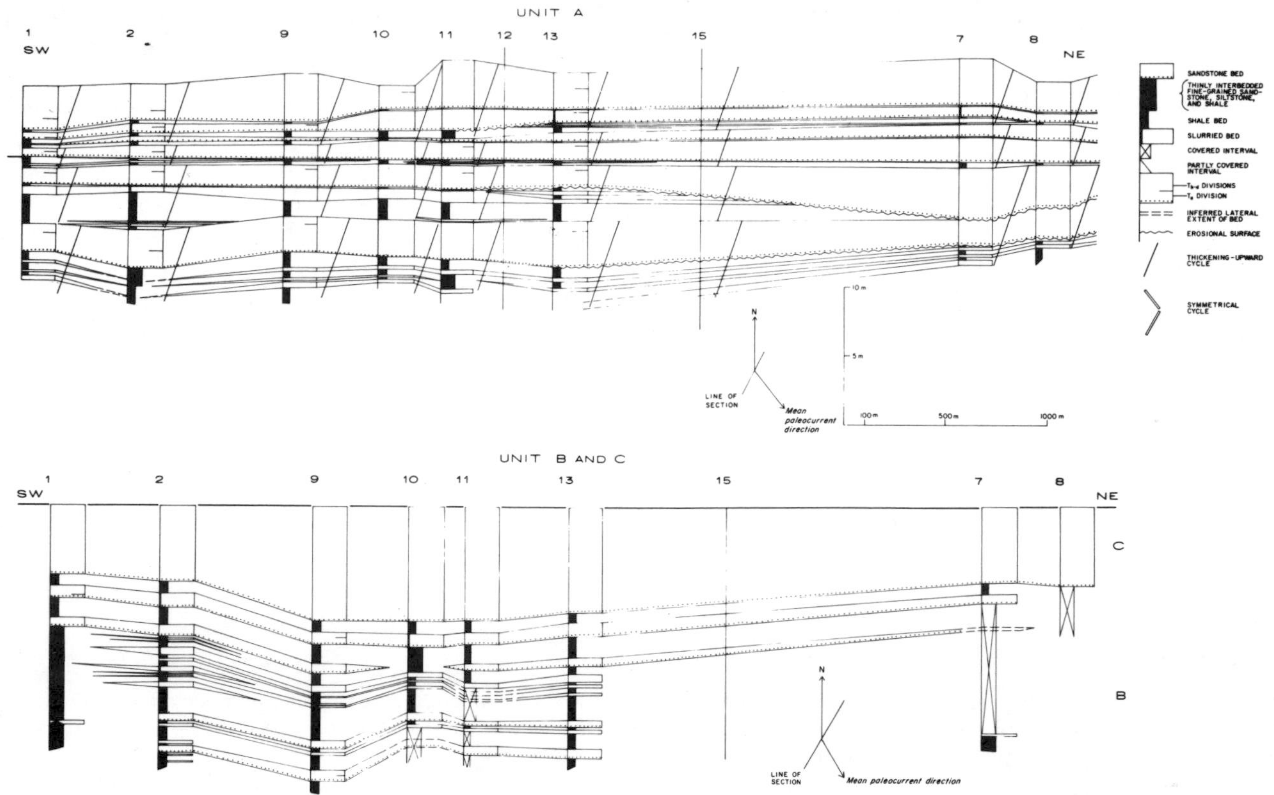

Figure 15-7. Detailed stratigraphic cross-section of sandstone body A, pelitic unit B, and sandstone body C. Note: Small diagram indicates orientation of section relative to north and paleocurrent directions within sandstone bodies and pelitic interval. Other symbols are as in Figure 15-5; location of columnar sections is shown on Figure 15-3.

pelitic sandstone body I (Figure 15-10), but can also be seen at a smaller scale from section 9 to 11 in pelitic unit B (Figure 15-7).

The thicker sandstone beds are less affected by bottom topography than are thinner beds, as is indicated by their relatively smaller lateral changes in thickness. Relatively greater lateral changes in thickness of thinner beds clearly result from more localized deposition. The thicker beds reflect both erosional and depositional smoothing action related to large, high-momentum turbidity flows, whereas the smaller, less-energetic flows were trapped within topographic lows.

Slurried beds, which are sandstone beds that include chaotic divisions containing outsize mudstone clasts between the massive *a* and the current-laminated *c* divisions of Bouma (see photograph C, Figure 15-4), are typically situated on both sides of depositional bulges (see Figure 15-7, sandstone body A, section 2; Figure 15-8, sandstone body D, sections 4 to 6). Because larger flows tend to smooth underlying topography by accelerating and eroding the tops of depositional highs, they incorporate rip-up clasts from the soft substratum and redeposit them as sand-clay slurries in adjacent depressions. The slurried beds indicate highly viscous flow conditions and may represent local or subsidiary debris flows that become incorporated in the trapped tail of a large turbidity current. They may also develop as smaller turbidity currents that have become detached from the margins of a larger current and flow down or along the flanks of bulges.

Massive sandstone beds thicker than 100 cm in sandstone bodies A, C, H, and I can be traced across the entire outcrop area, although locally with major changes in thickness. However, in sandstone body I (Figure 15-10) and pelitic interval D (Figure 15-8), massive sandstone beds lap onto and pinch out against features considered to be preexisting depositional highs having a relief on the order of 2.5 m to 3.0 m. The disappearance or pinch-out of sandstone beds thinner than 100 cm, particularly beds 10 to 30 cm thick, commonly takes place where these beds are associated with massive beds (see Figures 15-7 and 15-10, sandstone bodies A and I). Pelitic unit G is composed mainly of laterally continuous beds that are 30 cm to 100 cm thick and bounded by even and parallel bedding surfaces (Figure 15-8); the most persistent beds with the least change in thickness are between sections 2 and 15 on Figure 15-8. However, even in this cross-section three thin beds are discontinuous and pinch out where depositional bulges with relief of about 30 to 50 cm are present.

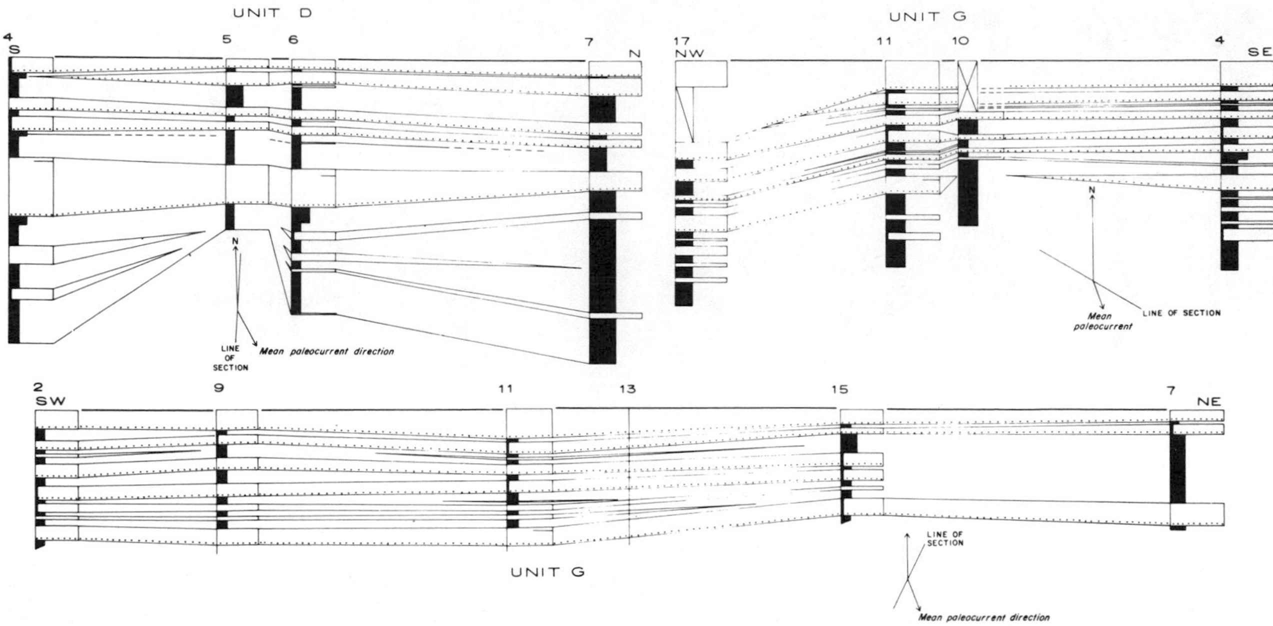

Figure 15-8. Detailed stratigraphic cross-section of pelitic intervals D and G. Note: Small diagram indicates orientation of section relative to north and paleocurrent directions within sandstone body. Other symbols are as in Figures 15-5 and 15-7. Horizontal and vertical scales are as in Figure 15-7; location of columnar sections is shown in Figure 15-3.

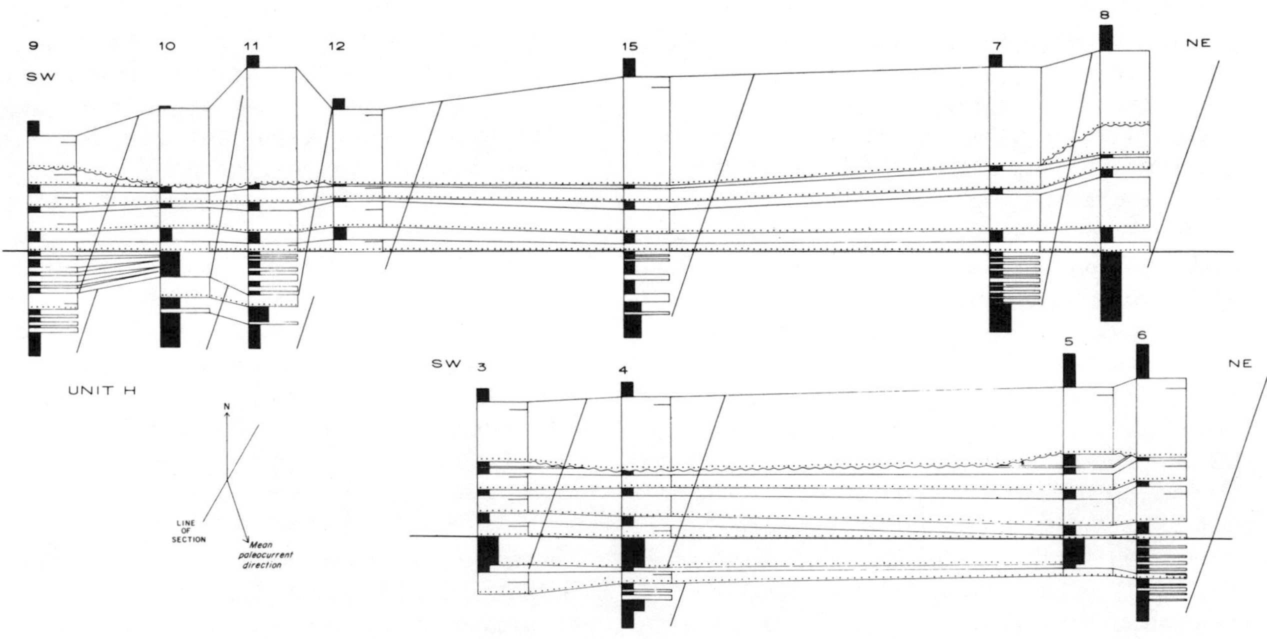

Figure 15-9. Detailed stratigraphic cross-sections of sandstone body H. Note: Small diagram indicates orientation of section relative to north and paleocurrent directions within sandstone body. Other symbols are as in Figures 15-5 and 15-7. Horizontal and vertical scales are as in Figure 15-7; location of columnar section is shown in Figure 15-3.

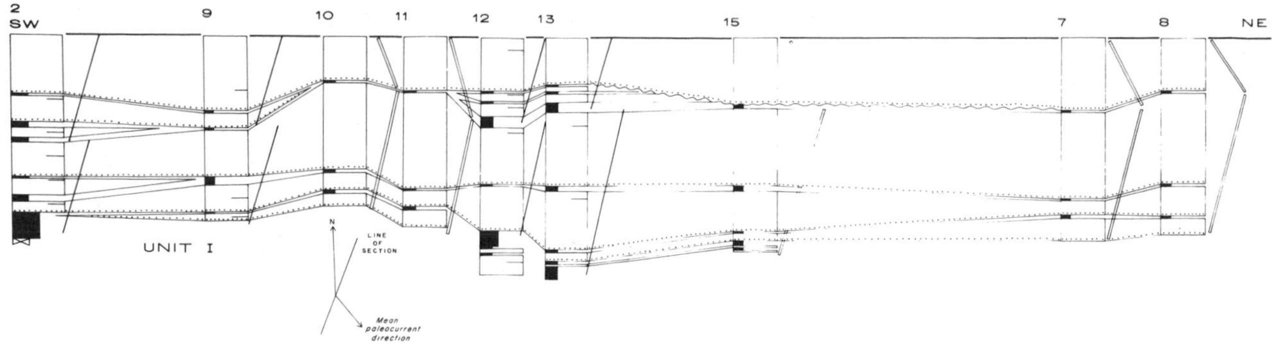

Figure 15-10. Detailed stratigraphic cross-section of sandstone body I. Note: Small diagram indicates orientation of section relative to north and paleocurrent directions within sandstone body. Other symbols are as in Figures 15-5 and 15-7. Horizontal and vertical scales are as in Figure 15-7; location of columnar sections is shown in Figure 15-3.

Compactional features are clearly observed where thick sandstone accumulations are found on top of thinner-bedded, more pelitic accumulations. The underlying beds are gently bent downward as a result of compaction caused by the weight of the overlying sandstone body. This effect is seen clearly in sandstone body C, where the compaction is associated with channeling, and in the lowermost part of sandstone body A, where channeling is not apparent (Figure 15-7, section 2).

The foregoing discussion makes it clear that, depending upon which features one wishes to emphasize, the choice of a datum for a cross-section is very critical. For example, the channeling and compactional features exhibited by pelitic unit B and sandstone body C (Figure 15-7) are conspicuous if the top of body C is chosen as a datum and arbitrarily flattened. However, the cross-section of pelitic unit D (Figure 15-8) clearly indicates that the top of unit C was not actually flat and that its convex-up shape affected deposition and geometry of the lowermost beds of pelitic unit D.

With the exception of sandstone body C, which probably represents a single depositional event (Figure 15-7), all of the sandstone bodies examined in this study display more or less distinct, vertically ordered variations or cycles in the thickness of individual sandstone beds. Sandstone body C consists of a single massive bed of possible composite character —that is, formed by two or more flows that either followed each other very quickly within the same depositional event or a succession of separate flows that eroded the pelitic tops of underlying flows and thus yielded a single thick, amalgamated bed (Figure 15-7).

Sandstone bodies H and I display well-developed thickening-upward sequences that terminate with massive, coarse-grained sandstone beds (see Figures 15-9 and 15-10). Sandstone body H shows an almost perfect development of a single thickening-upward cycle through the entire unit and across the entire study area (Figure 15-9). Only near the base of body H is there local development of a similar minor cycle within several thin sandstone beds.

Sandstone body I contains a distinct thickening-upward cycle in its lower half that terminates upward with a sandstone bed locally thicker than 5 m (Figure 15-10). Another thickening-upward cycle is present above the convex-upward depositional top of this bed and can be seen in sections 2, 9, 12, and 13, but not in sections 10 and 11 because of lack of deposition or in sections 7, 8, and 15 because of the combined effects of nondeposition and erosion. In sections 7, 8, 10, 11, and 15, where nondeposition or erosion occurred, the lower thickening-upward cycle has a tendency to develop upward into a symmetric one that is characterized by a thickening-upward lower section and a thinning-upward section.

Sandstone body A contains four distinct thickening-upward cycles, as thick as 5 m, west of section 7 (see Figure 15-7). The lower two cycles are partially truncated by erosion eastward (see sections 7 and 8). The uppermost cycle can be traced across the entire measured width of sandstone body A, although it contains irregularities caused by both erosion (sections 11 to 13) and nondeposition (section 8).

Within pelitic interval G, a distinct thickening-upward trend is present only in the uppermost part of section 11, where a topographic depression is filled (Figure 15-8). Otherwise, no distinct thickening or thinning vertical trends in bedding are observed in unit G.

GENERAL CONCLUSIONS

Outer fan sandstone bodies, as seen in the Laga Formation in the Monte Bilanciere area, typically consist of tabular, laterally extensive, sheet-like sandstone units. The bodies are at least 4.5 km wide in a direction perpendicular to the paleocurrent direction. Progressive changes in thickness towards the southwest are apparent in sandstone bodies C and H. However, whether such changes reflect an overall tendency for these sandstone bodies to thin or wedge out toward the southwest or whether they reflect random thickness changes resulting from pre- and syn-depositional topographic irregularities is impossible to determine with certainty.

The extensive lateral development of outer fan sandstone bodies may be caused by the large volume of the turbidity currents relative to basin size, and the high content of silt- and clay-size material. The first possibility is supported by the lateral continuity of most individual beds. The second possibility is reinforced by the low sandstone:shale ratio of the interlobe facies or pelitic intervals, in which most sandstone beds resemble thin and fine-grained basin plain deposits (Facies D_2 of Mutti and Ricci Lucchi 1975). The interlobe facies of the Laga Formation in the Monte Bilanciere area does not exhibit cyclic patterns in bed thickness and relatively high sandstone:shale ratios, which are characteristic of other ancient turbidite basins (Mutti, in press).

Although the outer fan sandstone bodies of the Laga generally are sheet-like in geometry and can be traced across the entire study area, in detail the individual beds and groups of beds exhibit many small-scale variations in thickness that reflect a tendency to smooth out irregular depositional and erosional topographic surfaces. Depositional lows or highs are smoothed through an increase or decrease, respectively, in the thickness of subsequently deposited sandstone beds. Distinct thickening-upward cycles within parts of the sandstone bodies are minor features related to this topographic control of deposition of sandstone bodies. These minor cycles are related to simple vertical accretion and do not necessarily reflect progradational events. Clearly ascertaining whether these cycles also reflect a more general mechanism by which larger-scale thickening-upward trends are developed within the entire thickness of a sandstone body is impossible.

Three general conclusions can be drawn from this study: (1) outer fan sandstone bodies as well as individual sandstone beds within the bodies form laterally continuous and horizontal sheet-like deposits; (2) the bodies are non-channelized and do not fill large-scale erosional depressions; (3) thickening-upward trends may develop at different scales within these bodies. Major thickening-upward cycles that affect the entire thickness of a sandstone body appear to extend for much greater distances than minor cycles within parts of sandstone bodies, which are apparently affected by local topographic irregularities. Much measuring and correlating of detailed stratigraphic sections is needed in order to gain a complete understanding of turbidite depositional settings in deepsea fan deposits. In our opinion, only detailed field work will resolve geometric problems and related correlation patterns.

ACKNOWLEDGMENTS

The authors thank M. Sonnino and H. Giradello for assistance in the preparation of the stratigraphic cross-sections, and A. Quintili, R. Riccioni, S. Santi, and L. Carillo for assistance in the field. Funding for Mutti and Ricci Lucchi was provided in part by Consiglio Nazionale delle Ricerche in Rome. Discussions in the field with A. van Vliet clarified many concepts, particularly the significance of slurried beds in outer fan sequences. G. R. Winkler and J. H. Stewart reviewed the manuscript and provided many helpful suggestions to improve it.

REFERENCES

Abbate, E., and Sagri, M., 1970. The eugeosynclinal sequence. Introduction to the Geology of Northern Apennines. *Sediment. Geol.*, 4: 207-49.

Bouma, A. H., 1962. *Sedimentology of Some Flysch Deposits.* Elsevier, Amsterdam, 168 pp.

Colacicchi, R., 1959. Osservazioni stratigrafiche sul Miocene del confine marchigiano-abruzzese. *Boll. Soc. Geol. Ital.*, 77: 59-69.

Dallan Nardi, L., Elter, P., and Nardi, R., 1971. Considerazioni sull' arco dell'Appennino settentrionale e sulla "linea" Ancona-Anzio. *Boll. Soc. Geol. Ital.*, 90: 203-12.

Enos, P., 1969. Chloridorme Formation, Middle Ordovician flysch, northern Gaspé Peninsula, Quebec. *Geol. Soc. Amer. Spec. Paper* 117, 66 pp.

Girotti, O., and Parotto, M., 1969. Mio-Pliocene di Ascoli Piceno. *Atti. Accad. Gioenia Sci. Nat. Catania,* (Suppl. Sci. Geol.), 7: pp. 127-74.

Kruit, C., Brouwer, J., Knox, G., Schöllnberger, W., and van Vliet, A., 1975. Une excursion aux cones d'alluvions en eau profonde d'age tertiaire près de San Sebastian (Province de Guipuzcoa Espagne).*Field Trip Guidebook Z23.* IX Intern. Sediment. Cong., Nice, 75 pp.

Lowe, D. R., 1976. Subaqueous liquefied and fluidized sediment flows and their deposits. *Sediment.*, 23: 285-308.

Mutti, E., 1974. Examples of ancient deep-sea fan deposits from circum-Mediterranean geosynclines. In: R. H. Dott, Jr. and R. H. Shaver (eds.), *Modern and Ancient Geosynclinal Sedimentation.* Soc. Econ. Paleon. Mineral., Sp. Publ. 19, pp. 92-105.

——, 1975. Distinctive thin-bedded turbidite facies and related depositional environments in the Eocene Hecho Group (South-central Pyrenees, Spain). *Sediment.*, 24: 107-131.

——, and Ghibaudo, G., 1972. Un esempio di torbiditi di conoide sottomarina esterna: le Arenarie di S. Salvatore (Formazione di Bobbio, Miocene) nell' Appennino di Piacenza. *Mem. Acc. Sci. Torino. Cl. Sci. Fis. Mat. Nat.* 4, 16, 40 pp.

——, and Ricci Lucchi, F., 1972. Le torbiditi dell'Appenino settentrionale: introduzione all'analisi di facies. *Mem. Soc. Geol. Ital.*, 11: 161-99.

——, and Ricci Lucchi, F., 1974. La signification de certaines unités séquentielles dans les séries à turbidites. *Bull. Soc. Géol. France,* 16: 577-82.

——, and Ricci Lucchi, F., 1975. Turbidite facies and facies associations. In: *Examples of Turbidite Facies and Facies Associations from Selected Formations of the Northern*

Apennines, Field Trip Guidebook A-11. IX Intern. Sediment. Cong., Nice, pp. 21-36.

———, Luterbacher, H. P., Ferrer, J., and Rosell, J., 1972. Schema stratigrafico e lineamenti di facies del Paleogene marino della zona centrale sudpirenacia tra Tremp (Catalogna) e Pamplona (Navarra). *Mem. Soc. Geol. Ital.,* 11: 391-416.

Nelson, C. H., 1976. Late Pleistocene and Holocene depositional trends, processes, and history of Astoria deep-sea fan, northeast Pacific, *Mar. Geol,* 20: 129-73.

———, and Kulm, L. D., 1973. Submarine fans and channels. In: G. V. Middleton and A. H. Bouma (eds.), *Turbidites and Deep Water Sedimentation.* Soc. Econ. Paleont. Mineral., Pacific Section, Short Course, Anaheim, pp. 39-78.

———, and Nilsen, T. H., 1974. Depositional trends of modern and ancient deep-sea fans. In: R. H. Dott, Jr., and R. H. Shaver (eds.), *Modern and Ancient Geosynclinal Sedimentation.* Soc. Econ. Paleont. Mineral., Sp. Publ. 19, pp. 54-76.

Nilsen, T. H., and Abbate, E., 1976. The Gottero Sandstone, a Late Cretaceous and Paleocene deep-sea fan complex in the Ligurian Apennines, northern Italy. (Abst.) *Geol. Soc. Amer. Prog.,* 8: 1028-29.

Ogniben, L., 1969. Schema introduttivo alla geologia del confine calabrolucano. *Mem. Soc. Geol. Ital.,* 8: 453-764.

Parea, G. C., and Ricci Lucchi, F., 1972. Resedimented evaporites in the Periadriatic Trough (Upper Miocene, Italy). *Israel J. Earth-Sci.,* 21: 125-41.

Ricci Lucchi, F., 1973. Resedimented evaporites: Indicators of slope instability and deep-basin conditions in periadriatic Messinian (Apennines Foredeep, Italy). In: C. W. Drooger (ed.), *Messinian Events in the Mediterranean.* North-Holland, Amsterdam, pp. 142-49.

———, 1975a. Depositional cycles in two turbidite formations of northern Apennines. *J. Sed. Petrol.,* 45: 3-45.

———, 1975b. Miocene paleogeography and basin analysis in the periadriatic Apennines. In: C. Sguyres (ed.), *Geology of Italy.* Petrol. Explor. Soc. Libya, Tripoli, pp. 5-111.

———, and Parea, G. C., 1974. Cicli deposizionali (megasequenze) nelle torbiditi di conoide sottomarina: Formazione della Laga (Appennino marchigiano-abruzzese). *Atti Soc. Nat. Mat. Modena,* 104: 247-83.

———, and Pialli, G., 1973. Apporti secondari nella Marnoso-arenacea: 1. Torbiditi di conoide e di pianura sottomarina a Est-Nordest di Perugia. *Boll. Soc. Geol. Ital.,* 92: 669-712.

ten Haaf, E., 1964. Flysch formations of the northern Apennines. In: A. H. Bouma and A. Brouwer (eds.), *Turbidites.* Elsevier, Amsterdam, pp. 126-31.

Chapter 16

Prodelta Turbidite Fan Apron in Borden Formation (Mississippian), Kentucky and Indiana

ROY C. KEPFERLE

U. S. Geological Survey
Cincinnati, Ohio

ABSTRACT

The Kenwood Siltstone Member of the Borden Formation (Mississippian) is a wedge of sedimentary rock that thins from 33.5 m to 0 m across a width of 16 km and that extends southeastward for 80 km along depositional strike from southern Indiana into north-central Kentucky.

The Kenwood is made up of two interbedded rock types: illitic silty shale and illitic subarkosic siltstone. The shale is similar to shale of the New Providence Shale Member of the Borden Formation that underlies and locally intertongues with the Kenwood. Siltstone of the Kenwood is poorly sorted, immature, and medium grained. Cumulative grain size curves are very fine skewed and leptokurtic. Textural data plotted on a C-M diagram fall near the end of the turbidite field and in the field for pelagic suspension.

Internal bedding sequences are mostly planar laminae, but locally some are complete Bouma sequences. Sole marks consist mainly of trace fossils and groove casts; the latter indicate that paleocurrents moved generally west-southwest. Trace fossils and geomorphic reconstruction suggest water depths during deposition were circalittoral (16 m to 180 m). Sedimentary structures and an upward-coarsening sequence through the overlying terrigenous clastic members of the Borden Formation suggest that the siltstone beds of the Kenwood were deposited as turbidites that fanned out from two centers along the front of a prograding platform of sediment.

The cessation of progradation is recorded in the rocks by a glauconite-rich layer. The glauconitic layer is the Floyds Knob Bed, which this author regards as both a rock-stratigraphic and time-stratigraphic unit. The surface on which the glauconite is strewn preserves the morphology of the subsea part of the delta in which a delta platform, slope, and base of slope can be delineated. This fossil delta front, as outlined by the glauconitic layer, is the template for reconstructing relief, orientation, and slope of the delta front at an earlier stage during deposition of the turbidite fan apron of the Kenwood Siltstone Member.

This front marks an edge of the Catskill-Pocono-Price delta system, which began building westward in Devonian time. The glauconite probably formed during a hiatus in deposition of terrigenous clastic sediment in Kentucky that coincided with the end of the Acadian orogeny.

INTRODUCTION

Examples of modern and ancient deepsea fan deposits are the subject of recent studies, many of which are relevant to plate tectonics and continental margin sedimentation. This study provides an example of inferred small fan deposition in a fairly well defined base-of-slope environment on a stable cratonic plate.

The areal distribution and stratigraphic relations of the rocks containing this Early Mississippian example of a subsea fan apron were mapped in more than a dozen 7.5-minute quadrangles in north-central Kentucky (Peterson 1966a, 1966b, 1967, 1968, 1972; Kepferle 1968, 1969, 1972a, 1972b, 1974, 1977) as a part of the geologic mapping program of the U. S. Geological Survey in cooperation with the Kentucky Geological Survey.

The Borden Formation, where it contains the Kenwood Siltstone, was chosen for this study for several reasons: Surface mapping completed in the area shows that the Kenwood is restricted laterally and vertically, is well defined, is easily recognized in that it appears to contain turbidites, has undergone little structural deformation, and through use of a glauconite marker bed higher in the section, can be related to the paleogeomorphology of the site of deposition. This relationship is shared by at least two other turbidite sequences in the Borden Formation in eastern and south-central Kentucky, but is more difficult to demonstrate for them except through comparison with the Kenwood. Supplemental detailed petrologic, trace fossil, and paleocurrent studies of the Kenwood have been used in reconstructing the depositional environments of the rocks in the study area.

A detailed study of the Kenwood relative to the remainder of the Borden will give us a better understanding of sedimentary environments on the stable craton during Early Mississippian time and will enable us to develop a model of cratonic fan sedimentation.

SETTING

The Borden Formation (Mississippian) in Kentucky and southern Indiana is mainly an upward-coarsening sequence of terrigenous clastic rocks (Kepferle 1971) for which a marine deltaic origin has been proposed (Rich 1951b; Swann et al. 1965; Peterson and Kepferle 1970). The Borden paraconformably overlies the Late Devonian New Albany Shale or, locally, the Early Mississippian Rockford Limestone (diagram A, Figure 16-1). At its base is an illitic silty shale, the New Providence Shale Member, that intergrades vertically and laterally with the Nancy Member, a silty shale, which, in turn, is overlain by a more resistant shaly siltstone unit, the Holtsclaw Siltstone Member. The upper contact of the Holtsclaw with the base of the overlying dolomitic Muldraugh Member is marked nearly everywhere by a glauconite-rich unit, the Floyds Knob Bed.

Regionally, three rock units are among several exceptions to the general upward-coarsening grain-size profile of the Borden. They are, from oldest to youngest, the Farmers Siltstone Member in eastern Kentucky, the Kenwood Siltstone Member in north-central Kentucky and southern Indiana, and the "Rockcastle Freestone" in the Wildie Member in south-central Kentucky (Figure 16-2). Turbidites commonly contribute to such exceptions to a general upward-coarsening profile (Visher 1965, pp. 56-57), and turbidites are common to all three of these members. Siltstone turbidite beds are essential to the Farmers and the Kenwood.

The Kenwood, at or in the upper part of the New Providence Shale in the lower part of the Borden, is bounded by sharp contacts. It stands out lithologically and topographically as a discrete, easily recognized sequence of siltstone and shale.

That part of the Borden Formation containing the Kenwood Siltstone Member extends southeastward for 80 km in a zone about 16 km wide from southern Indiana to north-central Kentucky. This zone will be referred to as the study area.

Ten to 15 km southwest of the western limit of the Kenwood, the unit composed mainly of carbonate rocks at the top of the Borden thickens, and the units of terrigenous clastic rocks thin. Recognition of the thinning is facilitated by the thin, glauconite-rich Floyds Knob Bed, which is both a rock-stratigraphic and a time-stratigraphic marker between the underlying terrigenous clastic rocks and the overlying carbonate rocks. This line has been mapped in the outcrop in Kentucky (Peterson and Kepferle 1970) and extends in the subsurface northwest into Indiana and Illinois (Swann et al. 1965) and southeast to Tennessee (Sedimentation Seminar 1972) as the Borden delta front (Figures 16-2 and 16-3).

Relief on the front near the study area is about 60 m for distances ranging from 6.5 km to 13 km. The slope ranges from about 5 to 10 m/km over the entire distance. Locally the slope is as steep as 23 m/km, which is slightly more than 1° (Peterson and Kepferle 1970). Similar slopes are common on modern deltas (Fisk 1961; Mathews and Shepard 1962; Allen 1970).

Compaction may have reduced the original relief on the depositional front. Asquith (1970) considered the compaction in shale and marlstone on the Cretaceous shelf margin in Wyoming to be as much as 50 percent. Mathews and Shepard (1962) also indicated that muds tend to undergo compaction with a loss of 50 percent of their original volume. From their studies of the Fraser River delta, however, they concluded that most of this process takes place within 2,000 years of deposition. Accumulation on major deltaic cones ranges from 10 cm to 30 cm/1,000 years (Ewing et al. 1958; Huang and Goodell 1970; Stanley and Huang 1971). Thirty cm/1,000 years is estimated as an average rate of accumulation for the Borden clastics. At this rate only the top meter of sediment in the Borden would be relatively uncompacted. The remaining postdepositional compaction would be small and would have little effect on the relief of the front. In other words, the slope preserved in the geologic record is virtually the slope that was present during the deposition of the Kenwood.

The Borden delta front, as delineated by the Floyds Knob Bed, is regarded as a template for much of the adjacent Borden deposition east of the front and for several units in the Fort Payne Formation west of the front (Sedimentation Seminar 1972; Pryor and Sable 1974). The configuration of the front is in accord with the concept of a deltaic complex prograding westward into a sediment-starved basin during Late Devonian and Early Mississippian time (Lineback 1966a, 1966b). Such a configuration has been divided into the fondo, clino, and unda environments by Rich (1951a) or the basin, slope, and platform environments of more recent usage (modified from Wilson 1969, p. 18).

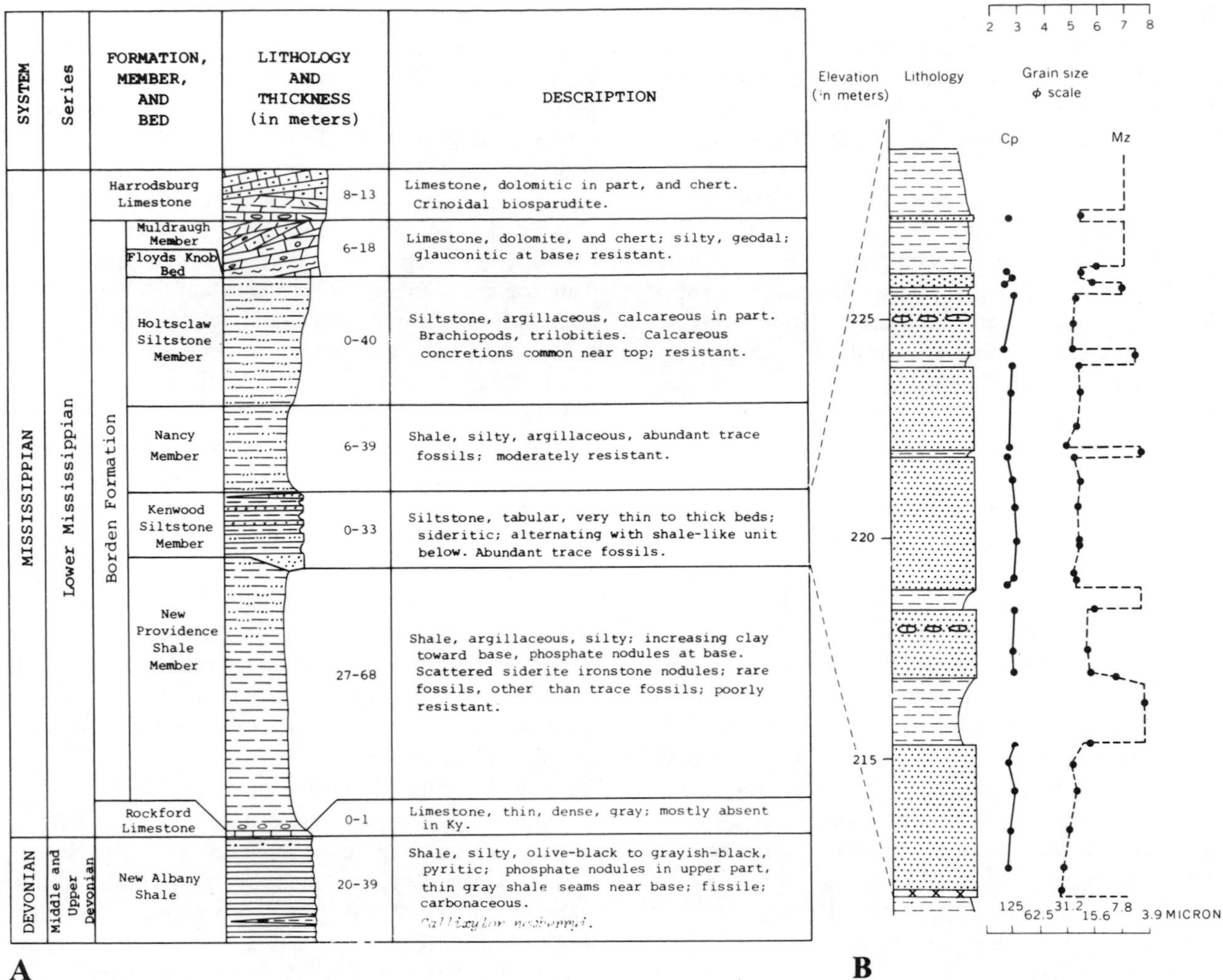

Figure 16-1. Generalized columnar section of the Borden Formation in north-central Kentucky (diagram A) and columnar section of Kenwood Siltstone Member at its type locality in Louisville, Kentucky (diagram B). Note: Diagram B shows vertical variation in coarsest grain (C_P), estimated from thin section, and graphic mean grain size (M_z).

Rich (1951b) also recognized that the New Providence Shale typifies the basinal fondo deposit. The overlying siltstone and silty shale units, including the turbidites, he considered to be deposits of the clino or slope environment. The Kenwood is here considered as a base-of-slope deposit, however, because of its stratigraphic position between the New Providence and the Nancy and because of local intertonguing with the New Providence.

KENWOOD SILTSTONE MEMBER OF THE BORDEN FORMATION

Butts (1915) named the Kenwood for interbedded siltstone and shale exposed on Kenwood Hill in Louisville, Kentucky. He assumed that the unit was at the top of a uniform thickness of New Providence Shale and that it was about 12 m thick everywhere. Subsequent work has modified these earlier generalizations. A graphic section of the type exposure is shown in diagram B, Figure 16-1.

Stratigraphy and Facies Distribution

The wedge of New Providence Shale Member on which the Kenwood rests thins southwestward from as much as 75 m to less than 10 m (diagram A, Figure 16-4). The Kenwood, too, is grossly wedge shaped and thins irregularly southwestward from 33.5 m to 0 m (diagram B, Figure 16-4). The Kenwood is mainly siltstone only in the easternmost area of outcrop; the siltstone:shale ratio decreases in the direction of thinning (diagram C, Figure 16-4). The thickness of the thickest siltstone bed also decreases in the direction of thinning of the Kenwood (diagram D, Figure 16-4).

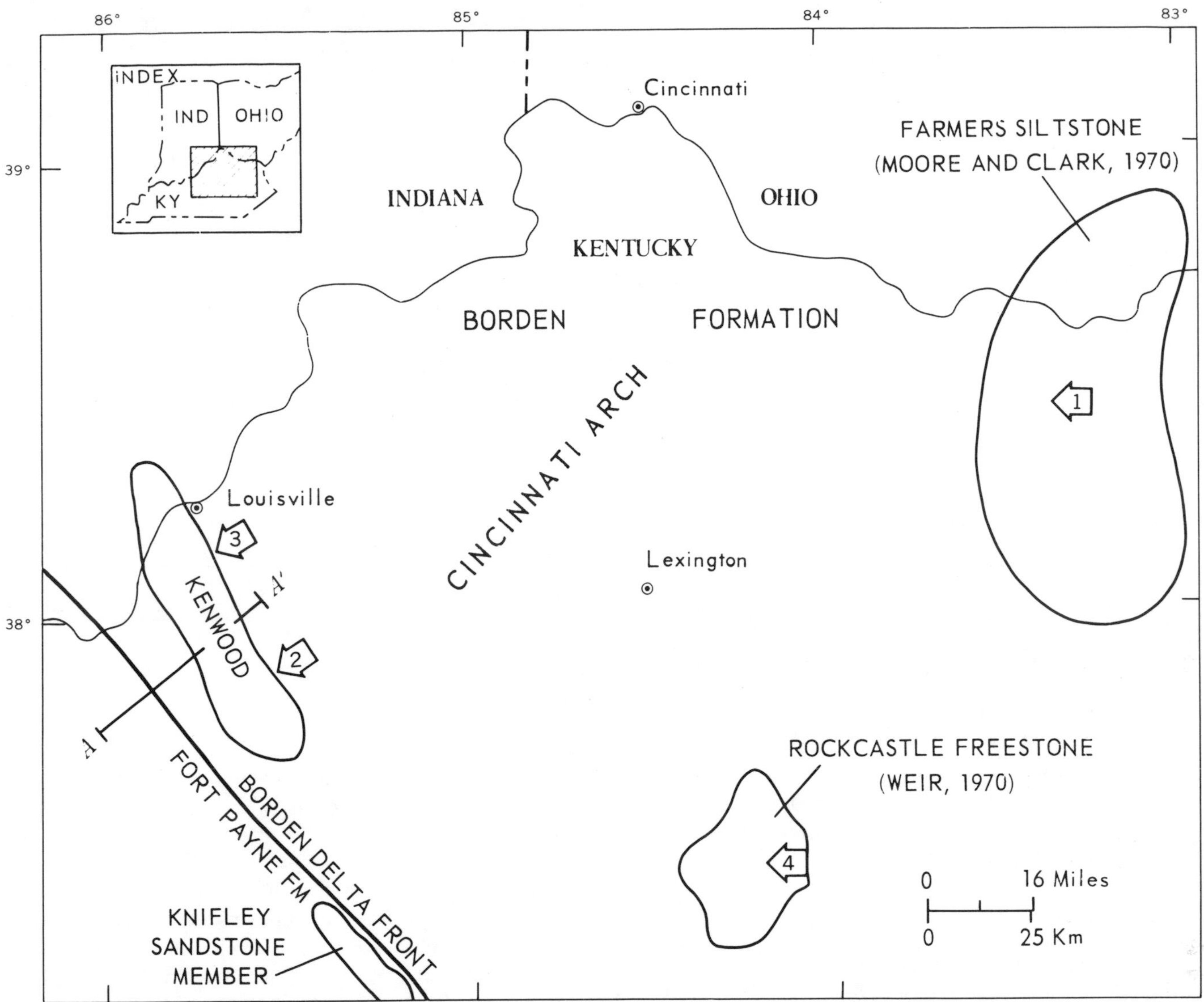

Figure 16-2. Areal distribution of the turbidite fan apron of the Kenwood Siltstone Member of the Borden Formation and other depositional features discussed in text. Note: Line of section A-A' is shown in Figure 16-3. Arrows indicate paleocurrent direction; numbers indicate relative depositional sequence.

The thickness of the average individual siltstone bed in the Kenwood is 20 cm from 892 measurements; one bed is 6 m thick.

Individual beds and bundles of beds show somewhat obscure westward imbrication. The depositional dip of most of the beds relative to the base of the Borden (diagram A, Figure 16-4) is from 1 to 4 m/km. Anomalous dips within the Kenwood range from 3° to 10° at four localities in the east-central part of the study area.

Texture

The siltstone beds of the Kenwood show subtle grading and are locally inversely graded (diagram B, Figure 16-1; diagram A, Figure 16-5). The average mean grain size (M_Z) is that of medium-grained silt (5.36 phi or 0.024 mm) on the Wentworth scale. The siltstone is poorly to moderately sorted. The frequency curves are strongly fine skewed and are leptokurtic to extremely leptokurtic (diagram A, Figure 16-5).

The shale beds of the Kenwood in samples from Kenwood Hill (diagram B, Figure 16-1; diagram A, Figure 16-5) range from mud-to silt-shale. Mean grain size is 7.23 phi (0.0067 mm). The shale is poorly to very poorly sorted and, like the siltstone, is strongly fine skewed. The frequency curves are platykurtic to mesokurtic.

A plot of the coarsest percentile (C_P) with the mean grain size (M_Z) on a C-M diagram (after Passega 1957) shows two clusters of samples, one near the fine end of

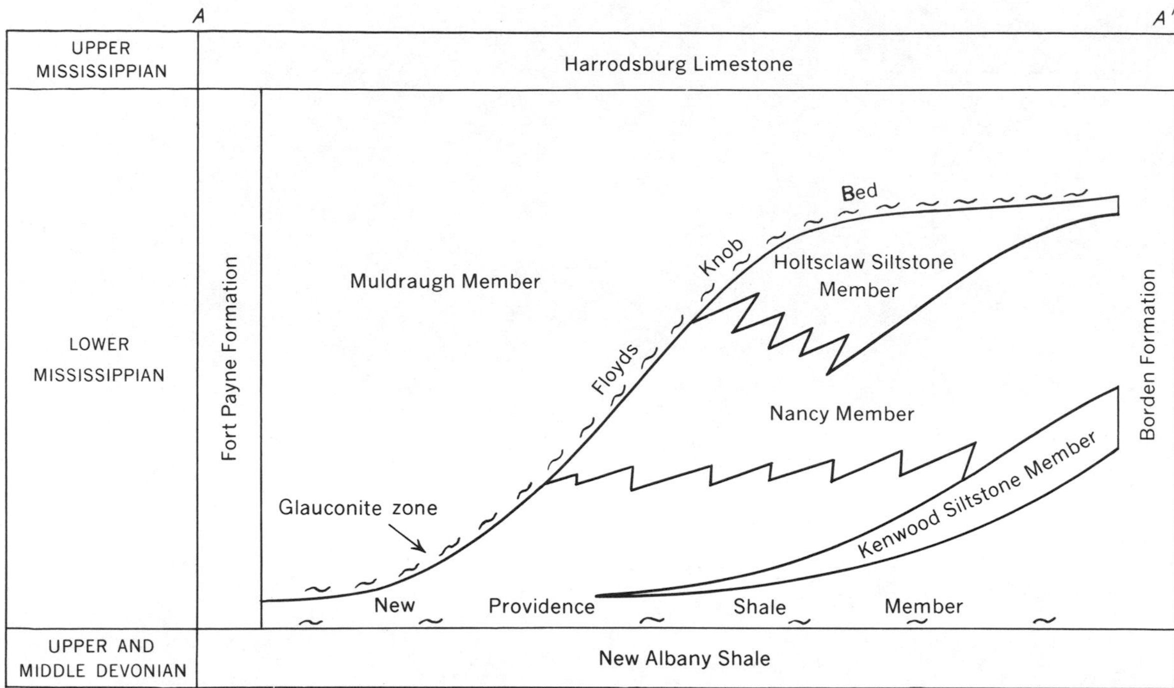

Figure 16-3. Diagrammatic cross-section of the Borden Formation in north-central Kentucky. Note: Line of section is shown in Figure 16-2.

the turbidite field and the other in and near the field for quiet water (diagram B, Figure 16-5). Otherwise, plots of various grain-size-curve statistical parameters show no strong discriminating tendencies.

Composition

The siltstone of the Kenwood was studied in forty six thin sections of samples from fifteen beds. Most beds were sampled at the top and base, and additional samples were taken from within some of the thicker beds. Mineral composition as percentage of framework grains (>30 microns with distinct boundary) and matrix (<30 microns) was determined from 300 point counts per thin section. Qualitative mineralogy was corroborated by X-ray diffraction of slabs from which thin sections had been made. Weathering effects precluded study of cement.

The major framework constituents, determined on the basis of the thin section study of the siltstone, average 68 percent and consist chiefly of quartz, feldspar, and rock fragments. The matrix consists mainly of sericitic illitic clay, limonite, microcrystalline quartz, calcite, and pyrite. Mica is conspicuous in most beds but, following Folk (1968 p. 123), was not included with framework grains when plotting compositional data. Data are summarized on a triangular diagram (Figure 16-6) that includes samples from other units of the Borden and from the Knifley Sandstone Member of the Fort Payne Formation, which is a slightly younger unit lying seaward of the Borden delta front (Sedimentation Seminar 1972; Kepferle and Lewis 1974).

Quartz makes up 54 to 93 percent of the framework grains; feldspar makes up 5 to 36 percent (in which potassium feldspar is 10 to 20 times more abundant than plagioclase); and rock fragments (which include mudstone, phyllitic quartzite, and chert) range from 6 to 10 percent. Glauconite, zircon, tourmaline, rutile, and hornblende were found in trace amounts and were excluded in calculations of the framework constituents.

The siltstone is submature and subarkosic and originated from reworked sediment as well as from a tectonically active granitic terrane. This interpretation is based on the greater abundance of potassium feldspar than plagioclase and the mixing of fresh and weathered feldspar.

The approximate composition of the clay fraction of the five shale interbeds of the Kenwood was determined by X-ray diffraction using the technique proposed by Schultz (1964). The clay is 87 percent illite, 10 percent kaolinite, and 3 percent chlorite.

Bedding Structures and Trace Fossils

Internal sedimentary structures and sole marks, including trace fossils, furnish some evidence for interpreting process and environment of deposition. Internal bedding

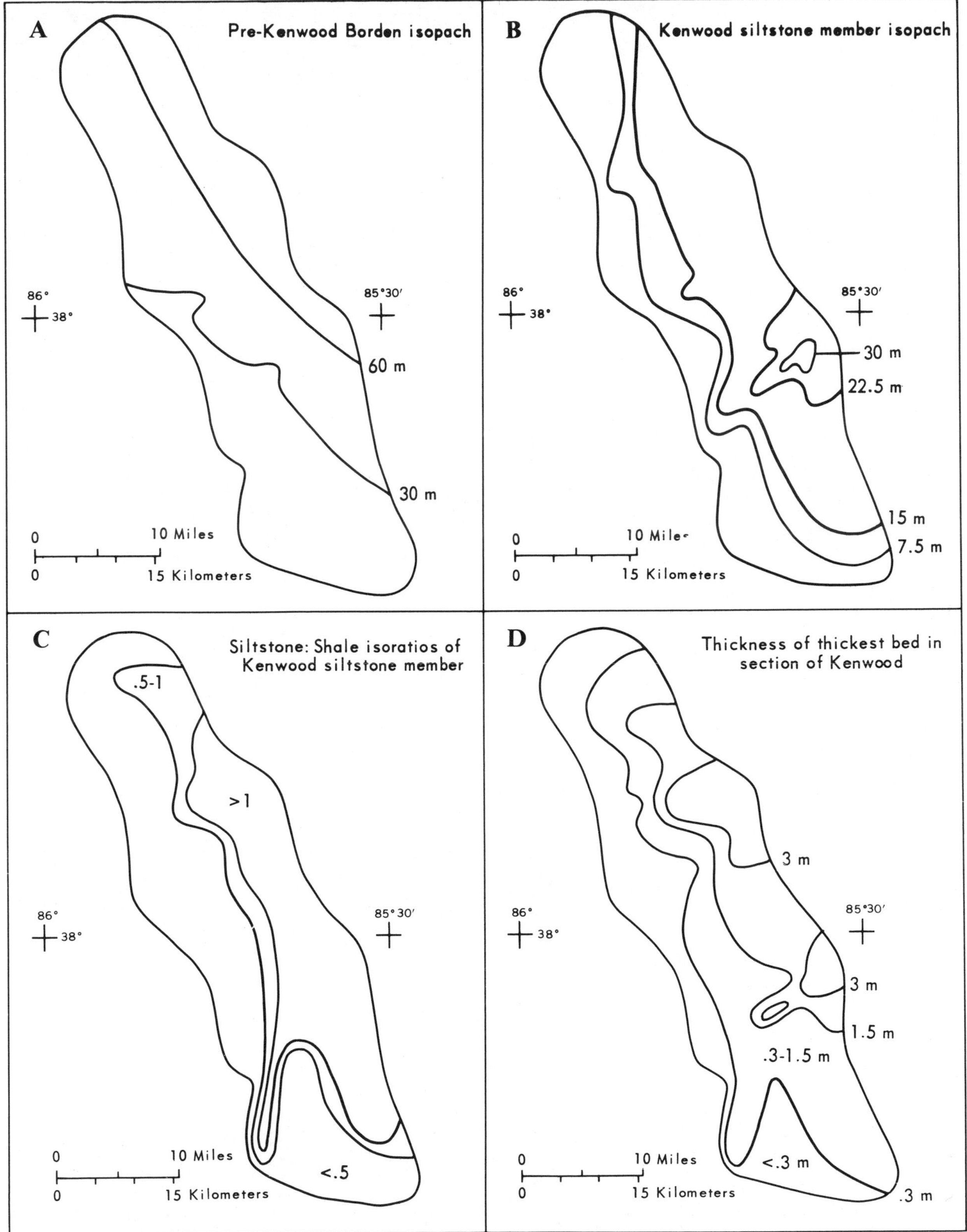

Figure 16-4. Stratigraphy and facies distribution. Note: Diagram A is an isopach map of the interval between the base of the Kenwood Siltstone Member and the base of the Borden Formation (see Figure 16-2 for location of study area.) Diagram B is an isopach map of the Kenwood Siltstone Member. Diagram C shows siltstone:shale isoratios of the Kenwood Siltstone Member. Diagram D shows thickness of thickest siltstone bed in the Kenwood Siltstone Member (not everywhere the same bed).

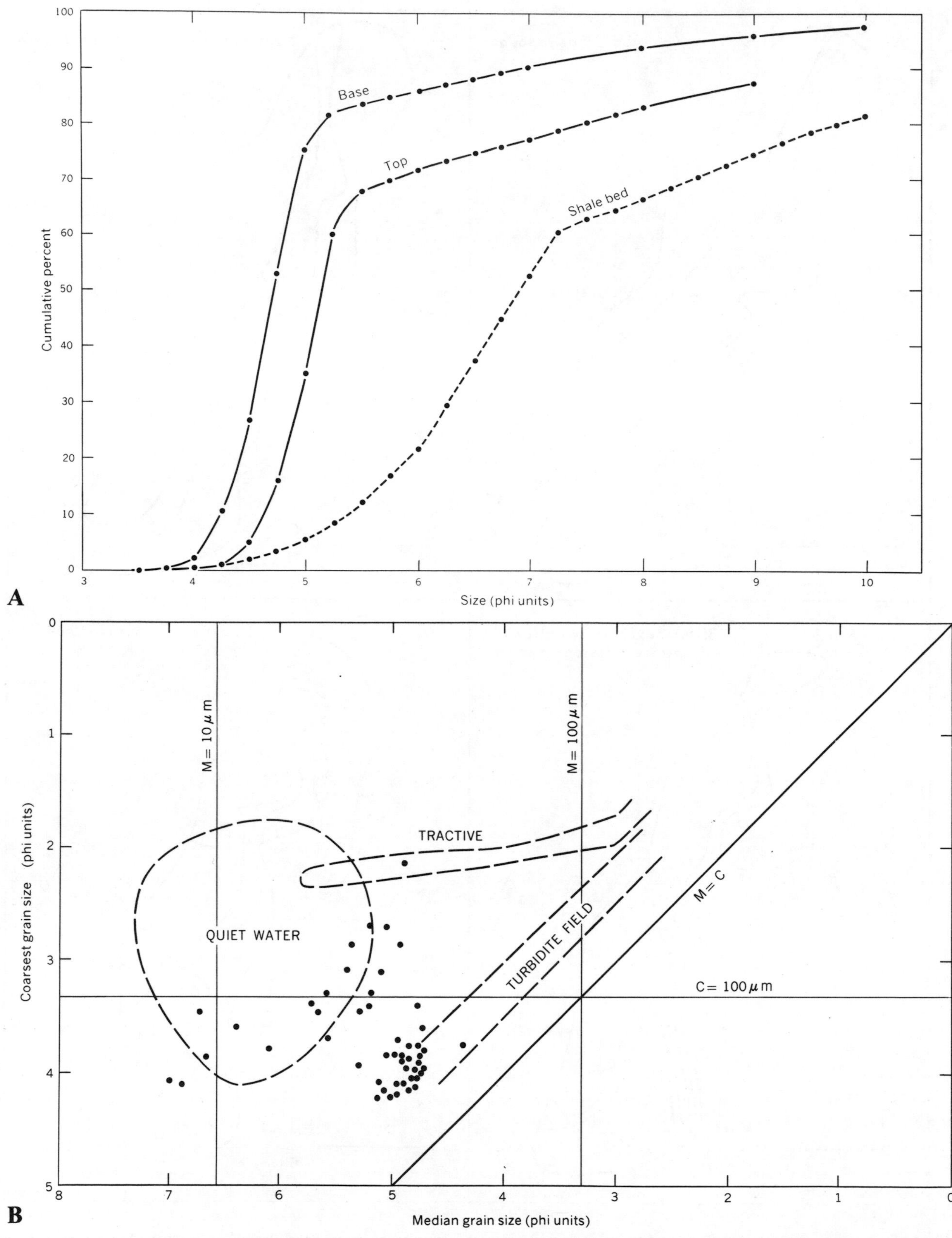

Figure 16-5. Textural data for the Kenwood Siltstone Member of the Borden Formation. Note: Diagram A shows cumulative grain-size curves of one sample from base and top of separate siltstone beds and of a typical silty shale interbed. Diagram B shows grain-size image on the C-M diagram. (Source: Adapted from Passega 1957.)

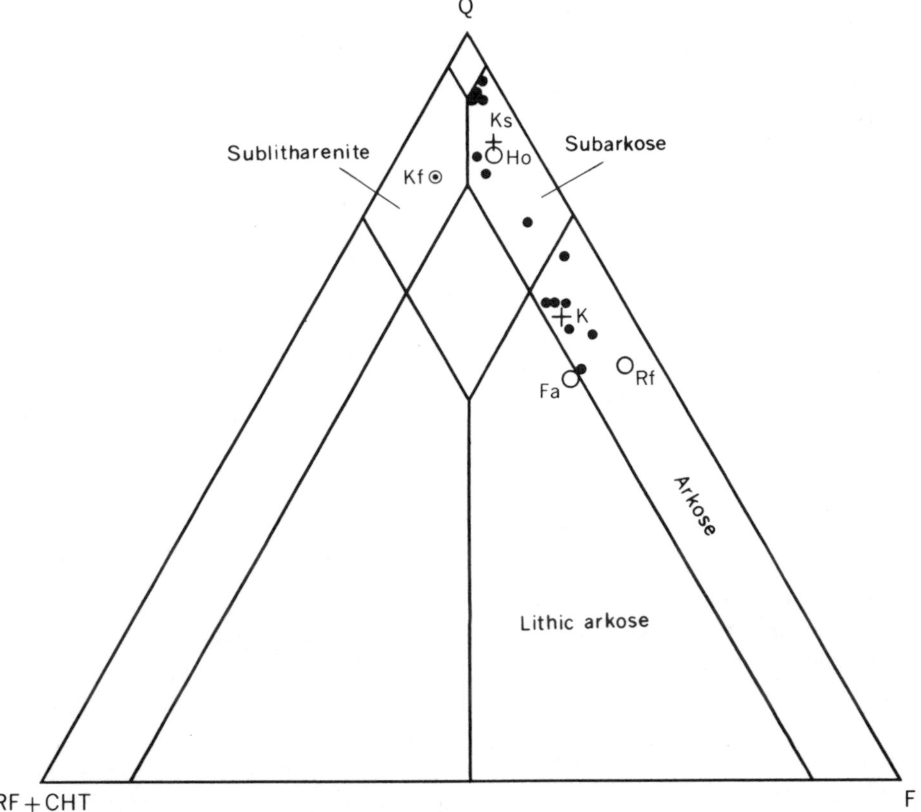

Figure 16-6. Composition of siltstones of Kenwood (indicated by dots) compared with that from Holtsclaw (Ho), Farmers (Fa), and Wildie (Rf = "Rockcastle Freestone") members of the Borden Formation and the Knifley Sandstone Member (Kf) of the Fort Payne Formation. Note: Q = quartz plus polycrystalline quartz; RF + CHT = rock fragments plus chert; F = feldspar; +Ks = average Kenwood composition in south area; +K = average Kenwood composition on Kenwood Hill. (Source: Classification after Folk 1968.)

laminae in the Kenwood, however, are commonly obscured by the fine grain size of the silt, by the homogeneity of the rock, and by *Liesegang* rings. Some beds contain rare, basal, clay clasts. Planar laminae of Bouma division *b* (Bouma 1962) appear to be most common. Less common are convolute or ripple-laminated beds (Bouma division *c*). Some of the thicker beds appear massive in field exposures, but X-radiographs of thin slabs of these commonly show that they are laminated (Bouma division *b*). The complete Bouma sequence $T_{a\text{-}e}$ is rare. Siderite is common as a secondary feature in the middle of some beds (diagram B, Figure 16-1) that are possibly in an otherwise obscure Bouma division *c*.

The upper surface of some siltstone beds have low-amplitude compound or cuspate types of ripples. Internally, most of the ripples show dipping lamination, from which a current direction can be inferred. Because the upper surface of the siltstone beds appears mostly gradational with the overlying shale, external current-produced structures on the upper surfaces of beds are sparse.

The basal surface of siltstone beds in the Kenwood is abrupt and planar, with the exception of sole marks. In order of abundance these are casts of grooves and brush, prod, bounce, and roll marks made by tools as well as casts of crescents, flutes, ruffled groove marks, and furrow flutes made by currents. All of these features are current controlled. Some give an orientation; others, a vector sense to the paleocurrents that formed them. A few beds, notably the basal bed in the Kenwood, show a complete lack of sole marks; their contacts with the underlying shale are planar and smooth.

A variety of trace fossils appear on the upper and lower surfaces of the siltstone beds, within the beds, and in the shale between the beds. Trace fossils commonly noted in the Kenwood Siltstone Member of the Borden Formation include those listed by Häntzschel (1962) as: *Scalarituba missouriensis* Weller, 1899; *Cosmorhaphe?* Fuchs, 1895; *Zoophycus* Massalongo, 1855; *Chondrites* Sternberg, 1833; *Nereites* MacLeay, 1839; *Lophoctenium* Reinhold Richter, 1850; *Teichichnus* Seilacher, 1955; *Paleophycus?* Hall, 1847; and *Planolites* Nicholson, 1873. In addition, several types of unidentified burrows were found. These include vertical burrows, most of which are fairly straight and nonbranching. Common straight trails on the upper surface of the siltstone beds are tentatively attributed to orthoconic cephalopods. Using criteria outlined by Seilacher (1967), this author concludes that the trace fossils indicate water depths in the circalittoral range (45 m to 180 m).

Turbidites

The siltstone beds of the Kenwood are believed to have been deposited from pulses of sediment in a series of turbidity currents. The turbidite nature of these beds is indicated by a comparison of the geometry and petrology of the beds, their internal and external structures, associated fauna, and nature of the enclosing sediments, with a list of key features that characterize turbidites in ancient rocks (Cline 1970, pp. 89-91), as well as modern sediments (Kuenen 1964, pp. 10-13). Only two of the listed criteria have not been recognized in the Kenwood: One is a correlation between bed thickness and maximum grain size; the other is the presence of plant remains. Both of these criteria are controlled by the nature of the sediment entrained in the current, not in the nature of the sedimentation process.

The Kenwood shows characteristics that are in accord with the following listed criteria compiled from Cline (1970) and Kuenen (1964): (1) beds appear in alternating fine- and coarse-grained strata; (2) the finer-grained beds are mainly normal pelagic clay or shale and lack all evidence of shallow-water deposition; (3) the finer-grained interbeds tend to be uniform; (4) the original sediments in the coarser-grained beds tend to be sand, but may be as fine as silt or as coarse as conglomerate; (5) the coarser beds are generally graded, but grading varies and may be obscure; (6) lower surfaces of the coarser beds commonly show sole marks; (7) the bases of these beds are abrupt, with the tops somewhat less abrupt to gradational into the overlying shale; (8) the sole marks have directional properties in which vectoral properties (flute, prod, and bounce casts) are less common than are linear (grooves); (9) internal bedding shows the entire Bouma sequence (Bouma 1962) locally, with truncated base or missing parts more common than a complete sequence; (10) the laminated divisions (*b* or *d*) are common in fine sand and silt beds; (11) trace fossils may occur on the upper bed surfaces where gradation occurs and may be missing where beds are truncated or are immediately overlain by another turbidite; (12) sorting is poor; (13) subaerial and shallow-water features such as mud cracks and algal colonies are absent; (14) shallow-water fauna are absent except where displaced and incorporated into the coarser-grained beds; (15) large-scale cross-bedding is absent; (16) bed thickness may vary from a few millimeters to 6 meters; (17) nearly all subsea sands are found on slopes of less than one degree; (18) mainly laminated structures, obscure grading, and gradational tops characterize turbidites on basin floors; (19) mud may occur as lumps in beds; (20) the character of the turbidite depends on the character of the source; (21) the sand or silt may be rich in feldspar, angular quartz, glauconite, mica and pyrite; (22) thinner beds are farther from the dispersal center; (23) proximal beds show more top-truncated Bouma sequences and fewer fines than do distal beds.

Paleocurrents

Sole marks, ripple foresets, and flame structures were used to determine that the paleocurrents from which the siltstone beds were deposited flowed mainly to the southwest (Figure 16-7), normal to the strike of the Borden delta front (see Figure 16-2). If the lineation means shown on Figure 16-7 are projected upslope, at least two local points of convergence are indicated in an area where the Lower Mississippian rocks are no longer preserved (compare with Sullwold's [1960] study of the Tarzana Fan). These points lie on the upcurrent direction, east of the concentrations of the thickest beds in the Kenwood (diagram D, Figure 16-4). They are inferred to have been dispersal centers for the siltstone of the Kenwood and mark the mouths of two submarine canyons, as in the model presented by Normark (1974, p. 58). Overlap and coalescence of the two fans resulted in the elongate body of interbedded turbidites termed here the *fan apron.*

TURBIDITE FAN APRON

The apron is 80 km long and more than 16 km wide. The proximal edge is lost to erosion. The thickness ranges from 33.5 m to 0 and appears thickest where sediments from separate dispersal centers merge. Individual siltstone beds, ranging in abundance from 1 to 37 in a single section, average 20 cm in thickness, and proximally attain a thickness of as much as 6 m. The depositional dip with reference to the base of the underlying hemipelagic prodelta shale is 1 to 4 m/km. The silt:shale ratio decreases away from dispersal centers as does the thickness of individual beds and the apron as a whole (Figure 16-4). The initial sequence of turbidites thins upward in the section (diagram B, Figure 16-1), as in the inner and middle parts of the fan model of Mutti (1974).

The depth of the basin into which a delta has prograded can be found from the total thickness of the prograding sequence (Klein 1974). This thickness, from the base of the Borden to the top of the Holtsclaw Siltstone Member in the central part of the study area (Kepferle 1972a) is 114 m; the thickness of the sequence from the base of the Kenwood fan apron to the top of the Holtsclaw is 81 m. These depths are within the circalittoral range indicated by trace fossils and are in tair accord with the relief of

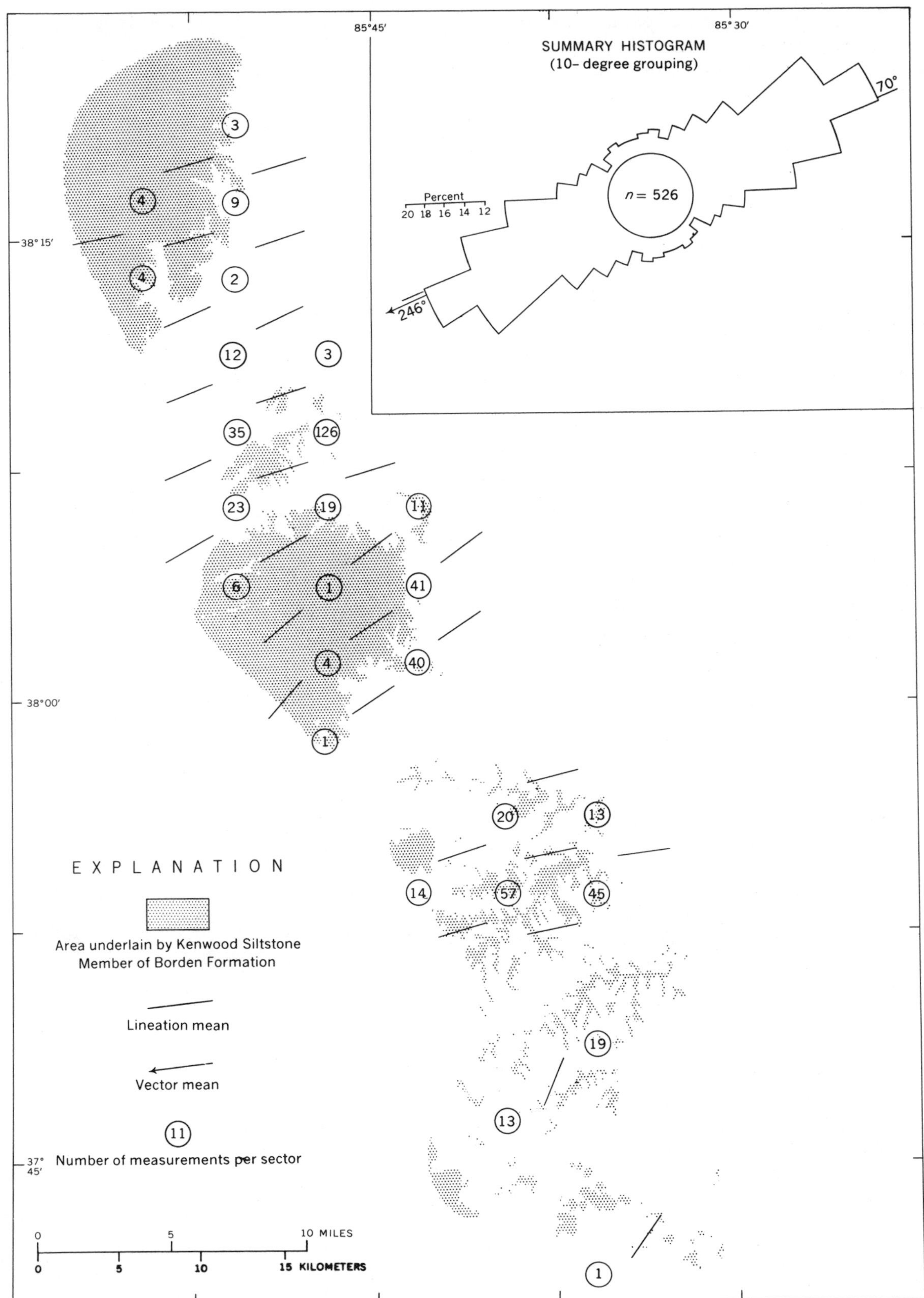

Figure 16-7. Map and summary histogram of paleocurrent lineations and vectors, Kenwood Siltstone Member of the Borden Formation. Note: Data were grouped according to subareas consisting of one-sixth of a 7.5-minute quadrangle; averages of contiguous subareas are plotted at their common corners.

60 m found on the "fossil" delta front in outcrop. Thus, the turbidite fan apron represented by the Kenwood was deposited in relatively shallow water.

Channels are recognized in the Kenwood mainly from the nature of the beds filling the channel. Three types of channel fill in the Kenwood are characterized either by local thickening along a single bed (Figure 16-8, Type I), by local distribution of thick beds (Figure 16-8, Type II), or by local inclined beds (Figure 16-8, Type III). All are found in the eastern, proximal part of the fan apron.

The types of channel fill indicate an increase from Type I to Type III in the magnitude and variability of the depositing currents as well as the magnitude of the current eroding the original channel. Type I fill occupies a channel previously cut by a current having minor erosive power; the lack of persistence of this directed current is shown by absence of any similar influence in the overlying beds. Channels occupied by fills of Types II and III, on the other hand, by their depth reflect considerably more initial erosive scour. Type II fills appear to have been deposited from currents confined to the channels, inasmuch as the tops of the beds are horizontal. Type III fills, however, appear to have a far more variable sequence of deposition than either of the other two types. Type III fills appear to be the result of alternating scour and deposition. Some of the deposition was from currents not entirely confined to the channel, as shown by Reineck and Singh (1975, p. 63) for other rocks. Should a current reoccupy a partly filled channel, it might scour anew some or all of the silt deposits and repeat the depositional process, or abandon the channel to pelitic accumulation (Figure 16-8, Type III rescoured).

The presence of channels in the proximal part of the Kenwood is in accord with the fan models of both Walker (1967, 1970) and Haner (1971). No channel could be traced extensively in the field, although the elongate body delineated in the southwestern part of the Kenwood by thickness maps (see Figure 16-4) may be a channel. If so, it is an exception to the cited models.

Outside the channels described above, the turbidity flows from which the siltstone beds were deposited were relatively weak. Supporting evidence includes a near-absence of flutes and other current-produced flow marks, sparsity of easily recognized Bouma *a* and *c* divisions, low range of grain size within the fan apron, low relief of the slope from platform to base (perhaps as low as 30 m), and small lateral extent of the turbidite sequence across the basin floor.

Paleocurrents, facies distribution, bed thickness, and stratigraphic relations all indicate an eastern source for the sediment from two dispersal centers along the prograding delta front. The northward increase in thickness of the hemipelagic sediment beneath the turbidite sequence suggests either that the sediment was thicker to the north

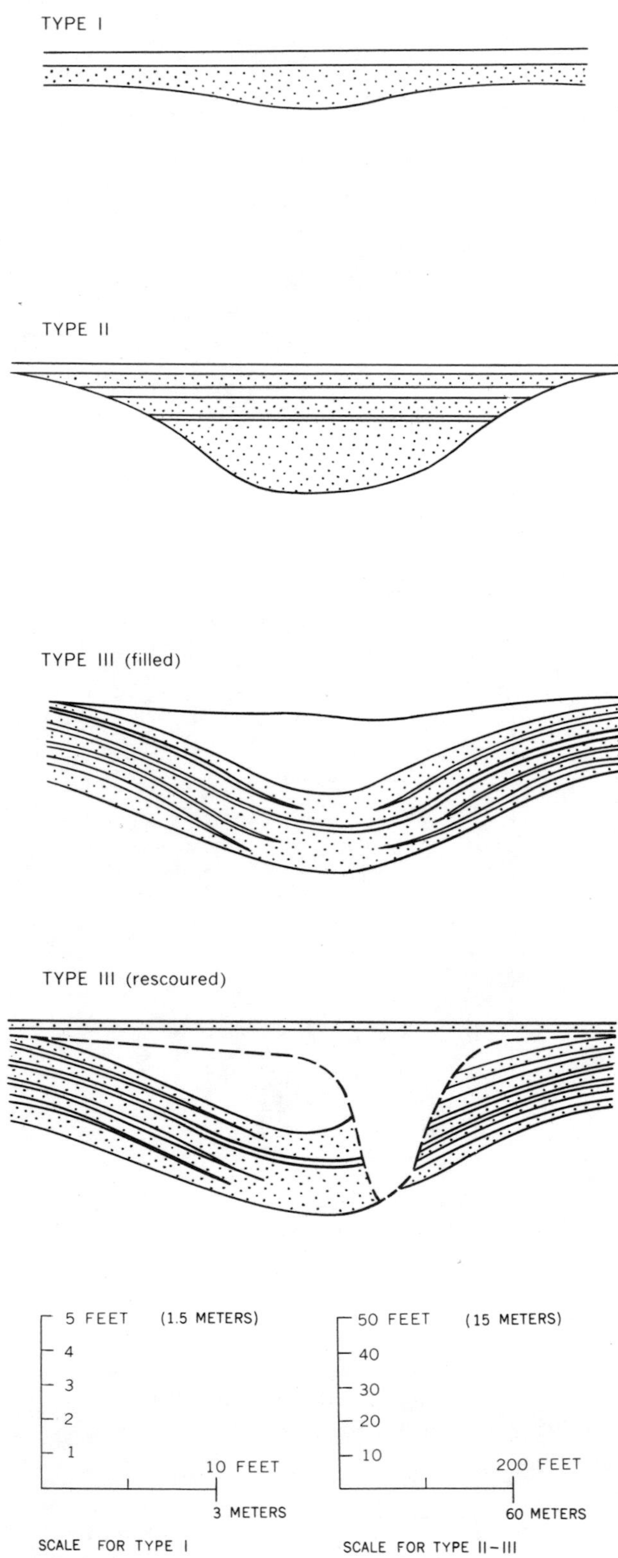

Figure 16-8. Channel fill types in the Kenwood Siltstone Member of the Borden Formation.

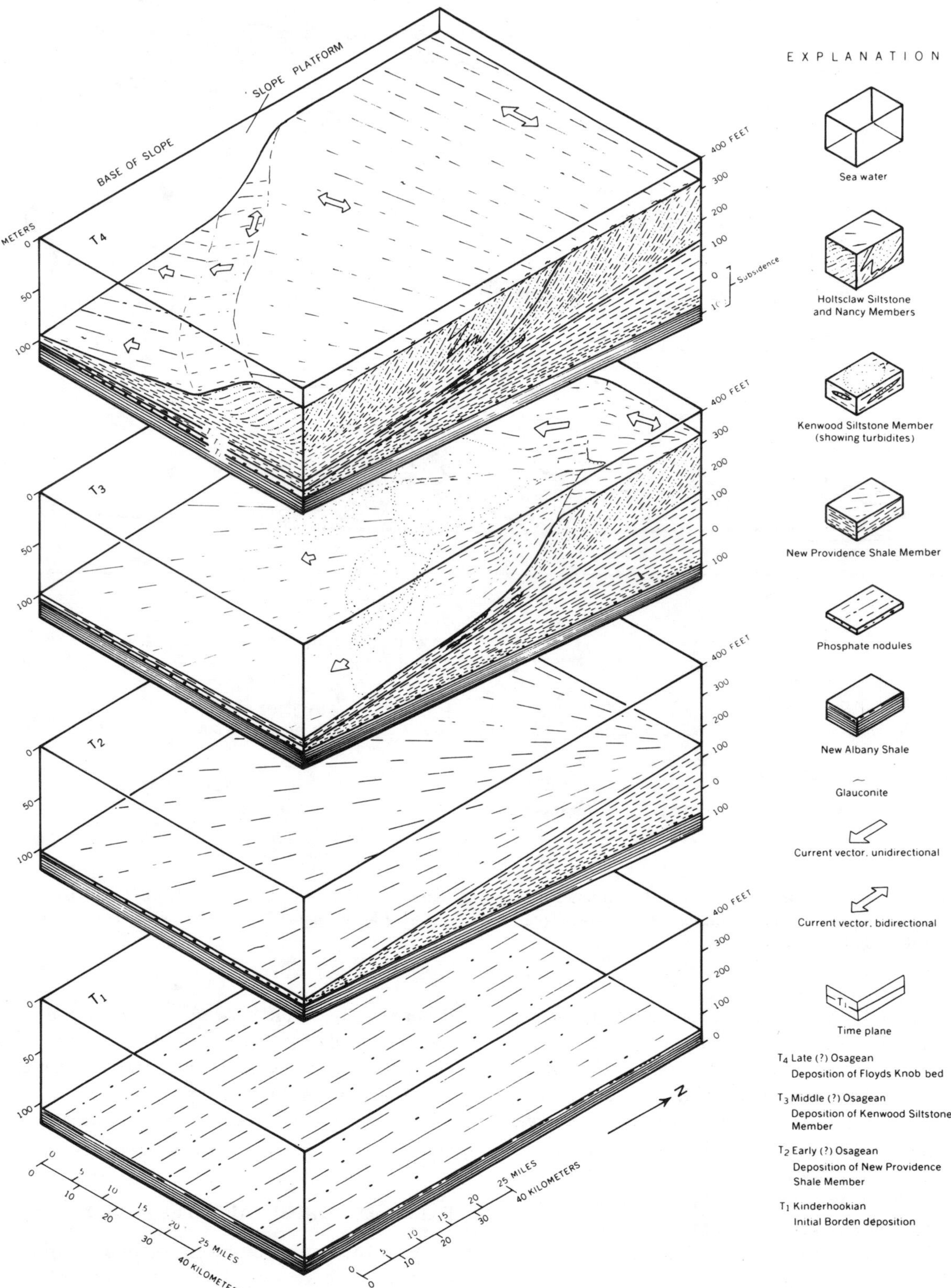

Figure 16-9. Sequential block diagrams of study area showing depositional environments of the Borden Formation. Note: T_1 = basin; T_2 = base of slope; T_3 = Prodelta turbidite fan apron; T_4 = Borden delta front in its westernmost position.

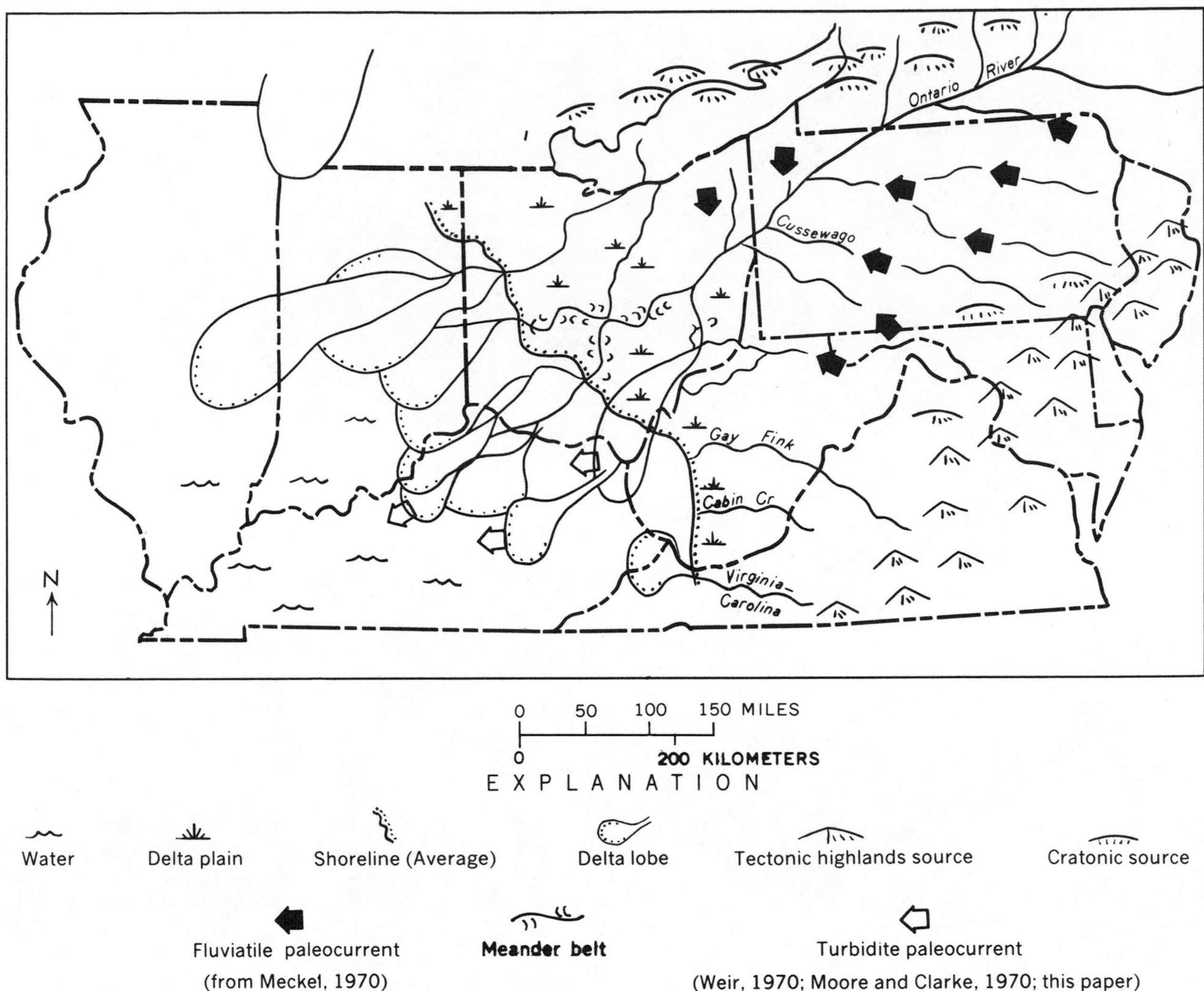

Figure 16-10. Paleogeographic diagram of Eastern Interior Basin during deposition of Kenwood Siltstone Member and related Lower Mississippian rocks prior to final establishment of Borden delta front. (Source: Channel names from Pepper et al. 1954; see also Pryor and Sable 1974, p. 311.)

before Kenwood time, or that the southern source was active earlier than the northern source – that is, the Kenwood is time transgressive to the north. This author holds to the latter interpretation, although the sequential block diagram (Figure 16-9) shows the former interpretation. A combination of the two presents a third possible interpretation. The thickening of the apron and a slight increase in the number of beds in the central part of the apron suggest interfingering from the two sources and persistence of the southern source after the northern source became dominant.

The grain size and composition of the Kenwood and adjacent rock units (Figures 16-1, 16-5, and 16-6) indicate that sediment coarser than silt was not available on the delta front. The size and composition of the framework grains of the Holtsclaw Siltstone Member in the one sample studied resemble the Kenwood. The Kenwood could have been derived from cannibalism of the upper slope facies of the delta front, similar to the later Holtsclaw, or the two may have shared a common source. The absence of coarse grains in the section indicates that the ultimate source was distant.

PALEOSLOPE AND PALEOTECTONICS

A persistent west-southwest paleoslope during the deposition of the Kenwood Siltstone Member of the Borden Formation is indicated by the stratigraphic descent of the Kenwood Siltstone Member southwestward within the Borden, the southwest imbrication of the individual siltstone beds in the Kenwood, the west-southwest direction of sole marks and other paleocurrent vestiges within the

Kenwood, and analogy with the west-southwest slope of the Borden delta front as established by mapping (see sequential block diagram, Figure 16-9).

A comparison of the paleocurrents for the Kenwood with those of the older Farmers Siltstone member of the Borden in northeastern Kentucky (see Figure 16-2) indicates that the paleocurrent system was virtually constant in Kentucky during the Early Mississippian.

The implication of this westward persistence of paleoslope is that the tectonic setting of the Eastern Interior and Appalachian basins in the Late Devonian continued into Early Mississippian time. The final stages of the Acadian orogeny in and east of New England fostered the maximum westward advance of the Pocono alluvial wedge (Meckel 1970, p. 54). A comparison with studies of Early Mississippian paleocurrents and paleotectonics of the Appalachian basin indicates that the terrigenous clastics of the Borden constitute the distal marine edge of this westward-prograding terrigenous clastic wedge. Seismic shock, centered perhaps far to the east, initiated slope failure and turbidity currents along the front of the active prograding delta complex. Two incipient canyons on the delta front in central Kentucky were the dispersal centers for the turbidites of the Kenwood Siltstone member (Figure 16-10).

CONCLUSIONS

The Kenwood Siltstone Member of the Borden Formation (Mississippian) in north-central Kentucky and southern Indiana embodies a base-of-slope apron of turbidite fans. The fans in this apron are smaller by one or two orders of magnitude than most deepsea fans described in the literature (summarized by Nelson and Nilsen 1974), whether in terms of dimensions, bed thickness, number of beds, maximum grain size, or depth of water. The Lower Mississippian deltaic sequence in Ohio, Kentucky, and Indiana contains at least three lithic units similar to and separate from the Kenwood. They are the Farmers Siltstone Member (Moore and Clarke 1970), the "Rockcastle Freestone" beds of the Wildie Member (Weir 1970), and beds within the Carwood and Locust Point Formations of the Borden Group of Indiana usage (Suttner and Hattin 1973). The Kenwood typifies fan deposition in this shallow cratonic Mississippian sea and is proposed as an example of small fans in this environment.

ACKNOWLEDGMENTS

This study was prepared in cooperation with the Kentucky Geological Survey. The author gratefully acknowledges the encouragement and probing discussion of Wayne A. Pryor, under whose direction this study was completed. Paul Edwin Potter furnished impetus and critically read an early version of this manuscript. The author also benefitted from field discussions with Francis J. Pettijohn, Alan F. Thomson, and students from the Johns Hopkins University and the University of Cincinnati. Warren L. Peterson shares the credit for discovering field exposures and delineating the Borden front in outcrops in north-central Kentucky. Financial support in part was furnished by a Penrose Grant from the Geological Society of America. This report was prepared in cooperation with the Kentucky Geological Survey.

REFERENCES

Allen, J. R. L., 1970. Sediments of the modern Niger delta: A summary and review. In: J. P. Morgan and R. H. Shaver (eds.), *Deltaic Sedimentation, Modern and Ancient.* Soc. Econ. Paleont. Mineral., Sp. Publ. 15, pp. 138-51.

Asquith, D. O., 1970. Depositional topography and major marine environments, Late Cretaceous, Wyoming. *Amer. Assoc. Petrol. Geol. Bull.,* 54: 1184-224.

Bouma, A. H., 1962. *Sedimentology of Some Flysch Deposits.* Elsevier, Amsterdam, 168 pp.

Butts, Charles, 1915. Geology and mineral resources of Jefferson County, Kentucky. *Kentucky Geol. Surv.,* ser. 4, 3, 270 pp.

Cline, L. M., 1970. Sedimentary features of Late Paleozoic flysch, Ouachita Mountains, Oklahoma. In: J. Lajoie (ed.), *Flysch Sedimentology in North America.* Geol. Assoc. Canada, Spec. Paper 7, pp. 85-101.

Ewing, Maurice, Ericson, D. B., and Heezen, B. C., 1958. Sediments and topography of the Gulf of Mexico. In: Lewis Weeks (ed.), *Habitat of Oil.* Amer. Assoc. Petrol. Geol., Tulsa, pp. 995-1058.

Fisk, H. N., 1961. Bar-finger sands of Mississippi Delta. In: J. A. Peterson and J. C. Osmond (eds.), *Geometry of Sand Bodies.* Amer. Assoc. Petrol. Geol., Tulsa, pp. 29-52.

Folk, R. L., 1968. *Petrology of Sedimentary Rocks.* Hemphill, Austin, 170 pp.

Haner, B. E. 1971. Morphology and sediments of Redondo Submarine Fan, southern California. *Geol. Soc. Amer. Bull.,* 82: 2413-432.

Häntzschel, Walter, 1962. Trace fossils and Problematica. In: R. C. Moore (ed.), *Treatise on Invertebrate Paleontology–Part W., Miscellanea.* Geol. Soc. America (and Univ. Kansas Press), New York, pp. W177–W245.

Huang, Ter-Chien, and Goodell, H. G., 1970. Sediments and sedimentary processes of eastern Mississippi Cone, Gulf of Mexico. *Amer. Assoc. Petrol. Geol. Bull.,* 54: 2070–100.

Kepferle, R. C., 1968. Geologic map of the Shepherdsville quadrangle, Bullitt County, Kentucky. *U. S. Geol. Surv. Geol. Quad. Map* GQ–740.

_____, 1969. Geologic map of the Samuels quadrangle, north-central Kentucky. *U. S. Geol. Surv. Geol. Quad. Map* GQ–824.

_____, 1971. Members of the Borden Formation (Mississippian) in north-central Kentucky. *U. S. Geol. Surv. Bull.,* 1354–B, 18 pp.

_____, 1972a. Geologic map of the Brooks quadrangle, Bullitt and Jefferson Counties, Kentucky. *U. S. Geol. Surv. Geol. Quad. Map* GQ–961.

_____, 1972b. Geologic map of the Valley Station quadrangle and part of the Kosmosdale quadrangle, north-central Kentucky. *U. S. Geol. Surv. Geol. Quad. Map* GP–962.

_____, 1972c. Stratigraphy, petrology and depositional environment of the Kenwood Siltstone Member, Borden Formation

(Mississippian), Kentucky and Indiana. *U. S. Geol. Surv. Open-file Rept.*, 233 pp.

——, 1974. Geologic map of parts of the Louisville West and Lanesville quadrangles, Jefferson County, Kentucky. *U. S. Geol. Surv. Geol. Quad. Map* GQ–1202.

——, 1977. Geologic map of the Pitts Point quadrangle, Bullitt and Hardin Counties, Kentucky. *U. S. Geol. Surv. Geol. Quad. Map* GQ–1376 (in press).

——, and Lewis, R. Q., Sr., 1974. Knifley Sandstone and Cane Valley Limestone: Two new members of the Fort Payne Formation (Lower Mississippian) in south-central Kentucky. In: G. V. Cohee and W. B. Wright (eds.), *Changes in stratigraphic nomenclature by the U. S. Geological Survey, 1972.* U. S. Geol. Surv. Bull., 1394–A: 63–70.

Klein, G. de V., 1974. Estimating water depths from analysis of barrier island and deltaic sedimentary sequences. *Geology,* 2: 409–12.

Kuenen, Ph. H., 1964. Deep sea sands and ancient turbidites. In: A. H. Bouma, and A. Brouwer (eds.), *Turbidites.* Elsevier, Amsterdam, pp. 3–33.

Lineback, J. A., 1966a. Deep-water sediments adjacent to the Borden Siltstone (Mississippian) delta in southern Illinois. *Illinois Geol. Surv. Circ.* 401, 48 pp.

——, 1966b. Illinois basin–sediment-starved during Mississippian. *Amer. Assoc. Petrol. Geol. Bull.,* 53: 112–26.

Mathews, W. H., and Shepard, F. P., 1962. Sedimentation of Fraser River delta, British Columbia *Amer. Assoc. Petrol. Geol. Bull.,* 46: 1416–43.

Meckel, L. D., 1970. Paleozoic alluvial deposition in the central Appalachians: a summary. In: G. W. Fisher, F. J. Pettijohn, J. C. Reed, Jr., and K. N. Weaver (eds.), *Studies of Appalachian Geology: Central and Southern.* Interscience-Wiley, New York pp. 49–67.

Moore, B. R., and Clarke, M. K., 1970. The significance of a turbidite sequence in the Borden Formation (Mississippian) of eastern Kentucky and southern Ohio. In: J. Lajoie (ed.), *Flysch Sedimentology in North America.* Geol. Assoc. Canada Spec. Paper 7, pp. 211–18.

Mutti, Emiliano, 1974. Examples of ancient deep-sea fan deposits from circum-Mediterranean geosynclines. In: R. H. Dott, Jr., and R. H. Shaver (eds.), *Modern and Ancient Geosynclinal Sedimentation.* Soc. Econ. Paleont. Mineral. Sp. Publ. 19, pp. 92–105.

Nelson, C. H., and Nilsen, T. H., 1974. Depositional trends of modern and ancient deep-sea fans. In: R. H. Dott, Jr., and R. H. Shaver (eds.), *Modern and Ancient Geosynclinal Sedimentation.* Soc. Econ. Paleont. Mineral., Sp. Publ. 19, pp. 69–91.

Normark, W. R., 1974. Submarine canyons and fan valleys: Factors affecting growth patterns of deep-sea fans. In: R. H. Dott, Jr., and R. H. Shaver (eds.), *Modern and Ancient Geosynclinal Sedimentation.* Soc. Econ. Paleont. Mineral., Sp. Publ. 19, pp. 56–68.

Passega, R., 1957. Texture as characteristic of clastic deposition. *Amer. Assoc. Petrol. Geol. Bull.,* 41: 1952–84.

Pepper, J. F., de Witt, Wallace, Jr., and Demarest, D. F., 1954. Geology of the Bedford shale and Berea sandstone in the Appalachian basin. *U. S. Geol. Surv. Prof. Paper* 259, 111 pp.

Peterson, W. L., 1966a. Geologic map of the New Haven quadrangle, Nelson and Larue Counties, Kentucky. *U. S. Geol. Surv. Geol. Quad. Map* GQ–506.

——, 1966b. Geologic map of the Nelsonville quadrange, central Kentucky. *U. S. Geol. Surv. Geol. Quad. Map* GQ–564.

——, 1967. Geologic map of the Lebanon Junction quadrangle, central Kentucky. *U. S. Geol. Surv. Geol. Quad. Map* GQ–603.

——, 1968. Geologic map of the Cravens quadrangle, Bullitt and Nelson Counties, Kentucky. *U. S. Geol. Surv. Geol. Quad. Map* GQ–737.

——, 1972. Geologic map of the Loretto quadrangle, central Kentucky. *U. S. Geol. Surv. Geol. Quad. Map* GQ–1034.

Peterson, W. L., and Kepferle, R. C., 1970. Deltaic deposits of the Borden Formation in central Kentucky. *Geological Survey Research, 1970.* U. S. Geol. Surv., Prof. Paper 700–D: D49–D54.

Pryor, W. A., and Sable, E. G., 1974. Carboniferous of the Eastern Interior basin. *Geol. Soc. Amer. Spec. Paper* 148, pp. 281–355.

Reineck, H. E., and Singh, I. B., 1975. *Depositional Sedimentary Environments.* Springer-Verlag, New York, 439 pp.

Rich, J. L., 1951a. Three critical environments of deposition, and criteria for recognition of rocks deposited in each of them. *Geol. Soc. Amer. Bull.,* 62: 1–19.

——1951b. Probable fondo origin of Marcellus-Ohio-New Albany-Chattanooga bituminous shales. *Amer. Assoc. Petrol. Geol. Bull.,* 35: 2017–40.

Schultz, L. G., 1964. Quantitative interpretation of mineralogical composition from X-ray and chemical data for the Pierre Shale. *U. S. Geol. Surv. Prof. Paper* 391–C: C1–C31.

Sedimentation Seminar, 1972. Sedimentology of the Mississippian Knifley Sandstone and Cane Valley Limestone in south-central Kentucky. *Kentucky Geol. Surv. Ser. 10, Rept. Inv.* 13, 30 pp.

Seilacher, Adolph, 1967. Bathymetry of trace fossils. *Mar. Geol.,* 5: 413–28.

Stanley, D. J., and Huang, Ter-Chien, 1971. Multiple origin of hemipelagic mud fill in Mediterranean Basin. (Abst.) *Am. Assoc. Petrol. Geol. Bull.,* 55: 365.

Sullwold, H. H., Jr., 1960. Tarzana Fan, deep submarine fan of Late Miocene age, Los Angeles County, California. *Amer. Assoc. Petrol. Geol. Bull.,* 44: 433–57.

Suttner, L. J., and Hattin, D. E. (eds.), 1973. *Field Conference on Borden Group and Overlying Limestone Units, South-central Indiana.* Soc. Econ. Paleont. Mineral., Great Lakes Section, Ann. Meeting Field Conference, 3, 113 pp.

Swann, D. H. Lineback, J. A., and Frund, E., 1965. The Borden Siltstone (Mississippian) delta in southwestern Illinois. *Illinois Geol. Surv. Circ.* 386, 20 pp.

Visher, G. S., 1965. Use of vertical profile in environmental reconstruction. *Amer. Assoc. Petrol. Geol. Bull.,* 49: 41–61.

Walker, R. G., 1967. Turbidite sedimentary structures and their relationship to proximal and distal depositional environments. *J. Sed. Petrol.,* 37: 25–43.

——, 1970. Review of the geometry and facies organization of turbidites and turbidite-bearing basins. In: J. Lajoie (ed.), *Flysch Sedimentology in North America.* Geol. Assoc. Canada, Spec. Paper 7, 219–51.

Weir, G. W., 1970. Borden Formation (Mississippian) in southeast-central Kentucky. *Guidebook for Field Trips.* Geol. Soc. Amer. Southeastern Sec., 18th Ann. Mtg., Lexington, Kentucky Geol. Surv. pp. 29–48.

Wilson, J. L., 1969. Microfacies and sedimentary structures in "deeper water" lime mudstone. In: G. M. Friedman (ed.), *Depositional Environments in Carbonate Rocks.* Soc. Econ. Paleont. Minderal., Sp. Publ. 14, pp. 4–19.

Chapter 17

Nile Cone Depositional Processes and Patterns in the Late Quaternary

ANDRES MALDONADO

Instituto "Jaime Almera", C. S. I. C.
University of Barcelona
Barcelona, Spain

DANIEL JEAN STANLEY

Division of Sedimentology
Smithsonian Institution
Washington, D. C.

ABSTRACT

Late Quaternary Nile Cone facies include both gravity-induced and suspensite (hemipelagic-sapropel) deposits that record the effects of major climatic cycles and eustatic oscillations. During the past 58,000 years, submarine fan deposition has been most pronounced on the Rosetta Fan in the western sector of the Cone as a result of enhanced input from the Rosetta Branch of the Nile.

The Rosetta Fan exemplifies a specific type of submarine fan model that is characterized by rapid sedimentation in a tranquil tectonic setting, low surface relief, and the absence of well-developed lobes, but includes thick pods of sand associated with a network of shallow channels restricted to the lower fan. Sediments forming the fan were conveyed across broad sectors of the shelf edge and were not strictly channelized downslope as is the case in many fans studied to date.

Two contrasting, regionally important lithofacies patterns are defined on the Nile Cone: a large, essentially downslope-trending belt consisting mainly of gravity-controlled deposits covers much of the Rosetta Fan, while a broad east-west zone comprising a high proportion of suspensites occupies much of the remaining cone surface. Intermittent phases of water mass stratification appear to have controlled to a considerable extent the concentration of thick deposits on the slope seaward of the major Nile sediment input. There is no clear-cut lithofacies distinction between prodeltaic deposits on the upper slope, developed mainly during low sea level stands, and the more characteristic Rosetta Fan facies, with which these are gradational. Distal-type turbidites are found throughout most of the cone. Salt tectonics in the Levant Platform sector may, in part, explain the reduced sediment supply to the central and eastern parts of the cone during the Late Quaternary.

INTRODUCTION

Considerable information on the origin of submarine fans has been amassed in recent years as a result of increased investigation of the base-of-slope province in modern oceans (Kelling and Stanley 1976). Marine geological studies of recent fans, or *cones* as larger bodies are sometimes termed, show that the physiographic and structural framework and physical oceanographic conditions under which these features form are highly variable and that depositional patterns result from a plexus of transport processes (Menard et al. 1965; Shepard et al. 1969; Huang and Goodell 1970; Normark 1970, 1974; Haner 1971; Curray and Moore 1971, 1974; Bouma 1972, 1973, 1975; Nelson and Kulm 1973; Damuth and Kumar 1975; Nelson 1976).

This chapter focuses on the Late Quaternary depositional history of the Nile Cone, largest of the Mediterranean submarine fans, which has accumulated in a relatively small enclosed sea, the Levantine Basin. To date there has been no regional sedimentological study of the recent evolution of this feature, and the purpose of this investigation is to summarize Late Quaternary lithofacies patterns of the Nile Cone based on a systematic core analysis.

As in the case of most submarine fans studied to date, an investigation of core sections seaward of such an important fluvial source might be expected to record fluctuations of sediment input related to marked Pleistocene eustatic sea level oscillations (cf. Moore and Curray 1974; Nelson and Nilsen 1974; Nelson and Kulm 1973; Normark and Piper 1972). The present study specifically considers the depositional variations in time and space that record modifications of sediment input and Nile River headwater migration and flow, the large-scale Würm to recent climatic events, including eustatic oscillations, and changes in physical oceanography, such as current patterns and stratification, in the southeastern Mediterranean (Olausson 1961; Herman 1972; Ryan 1972; Milliman and Müller 1973; McCoy 1974; Stanley et al. 1975). Our preliminary investigations revealed both the remarkable lithological variety of Nile Cone sediment types and their cyclic nature (Maldonado and Stanley 1975, 1976a) and that certain core transects on the cone show a gradual downslope transition from characteristic prodeltaic units to more typical distal facies (Stanley and Maldonado 1977). Definition of the regional lithofacies distribution during six different periods, from about 58,000 yrs B. P. to the present, calls attention to significant changes in the transport mechanisms and depositional patterns during the recent formation of this large submarine fan.

STUDY AREA AND METHODOLOGY

The study area, about 600 km wide, lies between the Egyptian Shelf and the Mediterranean-Cyprus Ridge (a distance of about 220 km) in the eastern Mediterranean (Figure 17-1). The Nile River has formed an extensive subaerial delta covering an area of about 22,000 km^2 (Wright and Coleman 1973; Orlova and Zenkovich 1974; Summerhayes and Marks 1975). It drains a region of about 3 million km^2 and, prior to the construction of the Aswan High Dam in 1964, had a discharge in excess of 12,000 m^3/sec and carried about 120 to 140 million tons of sediment per year that was mostly fine-grained (Aleem 1972). The Egyptian Shelf refers to the broad (about 40 km to 65 km wide) platform between the Nile Delta and the slope; it is one of the widest shelves bordering the Levantine Basin.

The Nile Cone presently occupies a seismically tranquil margin seaward of the Egyptian Shelf, although the post-Miocene tectonic history of this region is complex (Finetti and Morelli 1973; Kenyon et al. 1975; Neev et al. 1976; Smith 1976; Ross and Uchupi 1977; and Chapter 18 of this volume). The Plio-Quaternary development of the cone is closely related to the evolution of the Nile River, which has served as its major terrigenous sediment source. Seismic surveys of the cone (Ryan et al. 1970; Biju-Duval 1974; and Chapter 18 of this volume) have shown that the unconsolidated Plio-Quaternary sedimentary section of the cone has an average thickness of 2,000 m and in places exceeds 3,000 m; it is locally absent above some diapirs in the Levant Platform.

We apply the term "Nile Cone" to the entire immense arcuate submarine bulge off the Egyptian Shelf (cf. Emery et al. 1966; Ryan et al. 1970; Maldonado and Stanley 1975, 1976a). The term "Rosetta Fan" (referred to as the fan) is applied here to the region seaward of the Rosetta Branch of the Nile (RF in Figure 17-1); this name has been applied by Emery et al. (1966) to an area that extends further to the east than considered here. The area we designate as Rosetta Fan closely corresponds with the "Nile Cone" zone as defined by Ross et al. in Chapter 18 (see Figure 18-4). The immediately adjoining area to the east is termed the "Levant Platform" by Ross et al. in Chapter 18; it includes a zone of ridge-and-valley systems off the Damietta Branch that extend toward the northwest diagonally across the cone (Carter et al. 1972; Kenyon et al. 1975). The sector east of the Levant Platform off Sinai, Israel, and Lebanon is sometimes referred to as the Damietta Fan, but is herein termed the *distal continental margin.* Other important features on or contiguous with the cone and discussed in this study are the Herodotus Basin plain and the Eratosthenes Seamount and Basin plain (see Figure 17-1).

The Rosetta Fan has a gentle to hummocky relief except on the distal, lower fan area where salt tectonics have deformed the Plio-Quaternary cover (Smith 1976; Ross and Uchupi 1977). Downslope transects (Emery et al. 1966) show concave-up profiles extending from the shelf edge to the lower fan where it merges with the Herodotus Basin plain, whose depth exceeds 3,000 m. The fan covers an area of about 70,000 km^2 and is broader than it is long. A narrow, relatively low gradient slope, extending from the shelf edge to depths of 300 m to 600 m, merges with the uppermost fan. The Nile Cone surface is not intensely dissected except in the broad ridge-and-valley sector of the Levant Platform and in the region comprising a network of low relief ($<$20 m) channels on the lower Rosetta Fan sector. The concave-up fan surface does not display a marked suprafan or fan lobe physiography (see Figure 17-1). The slope above the fan is relatively smooth, and only one large valley, the Alexandria Canyon, appears to cross the slope and upper fan seaward of the Rosetta Branch; it cannot be traced to the base of the fan.

This investigation includes detailed petrologic and chronostratigraphic analyses of 47 piston and gravity cores (1 m to 15 m in length) and a more general lithofacies study of an additional 27 piston cores (average length, 7 m) collected in this region on R/V *Chain* 1975 cruise 119 (D. A. Ross and C. Summerhayes, personal communications, 1976) as well as of drill cores from Site 131 of the 1970 D.S.D.P. Leg 13 (see Figure 17-1). Core sample techniques utilized X-radiography, size analysis, total carbonate and organic matter, SEM, compositional analysis of the clay

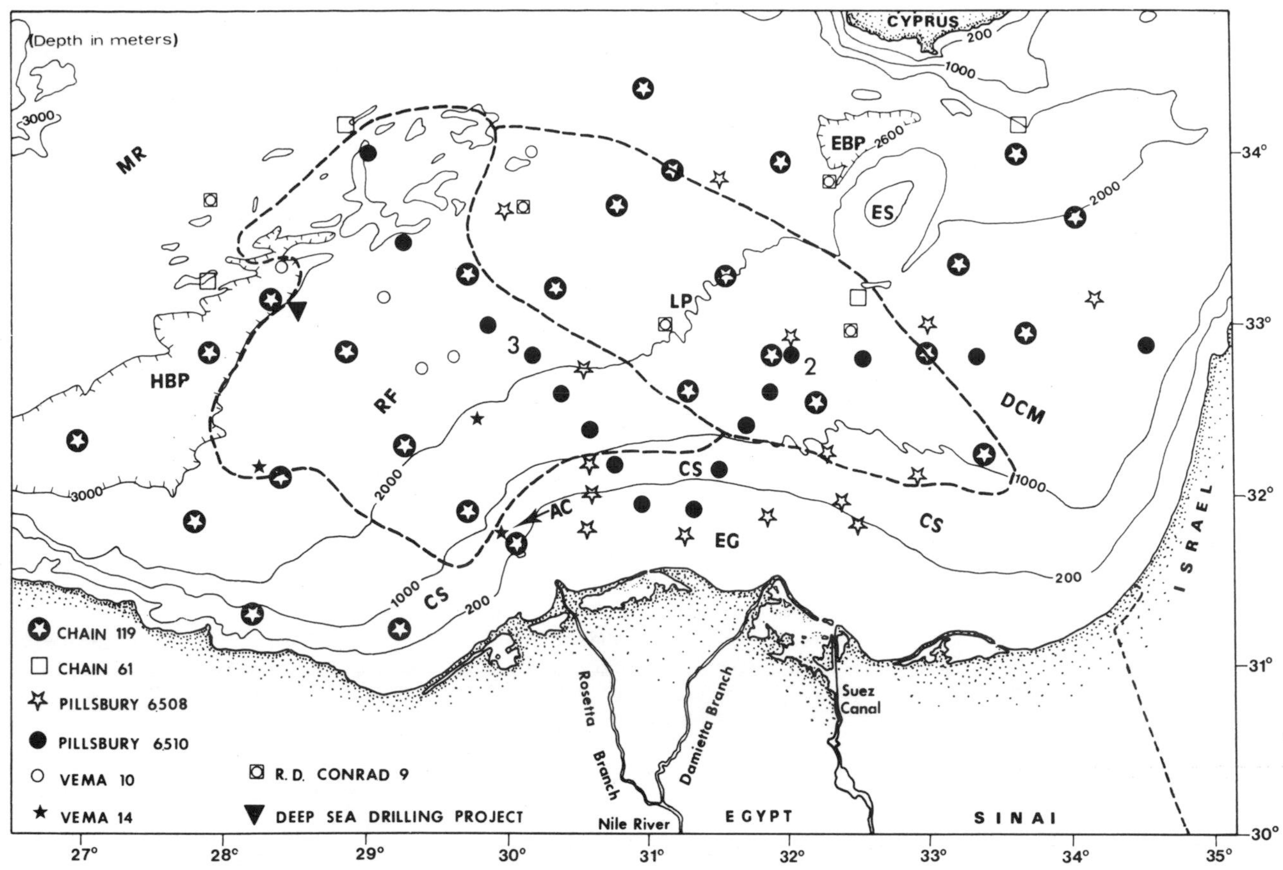

Figure 17-1. Chart showing bathymetric configuration of the southeastern Mediterranean (depth in meters) and position of Nile Cone cores examined in this study. Note: RF = Rosetta Fan; LP = Levant Platform; DCM = distal continental margin; CS = continental slope; EG = Egyptian Shelf; AC = Alexandria Canyon; HBP = Herodotus Basin plain; ES = Eratosthenes Seamount; EBP = Eratosthenes Basin plain; MR = Mediterranean Ridge. Numerals (2 and 3) provide location of stratigraphic cores logged in Figure 17-2. (Chart base: Defense Mapping Agency Hydrographic Center Chart N. O. 310.)

and sand fraction of more than 170 samples, and radiocarbon dating (data in Maldonado and Stanley 1975, 1976a; Stanley and Maldonado 1977). Data for echo sounding and subbottom records such as 3.5 kHz, not available to us, are presented in Chapter 18 of this volume.

SEDIMENT TYPES AND CHRONOSTRATIGRAPHY

Cores on the shelf and on the slope to depths of about 450 m include three major sediment types in order of decreasing importance: gray homogeneous mud (usually bioturbated) that is characterized by pelecypods and other benthonic and planktonic components; bioclastic, neritic calcareous sand and gravel (*maerl*) that display lithological transitions with the gray mud type; and terrigenous shelly sand and silt, including neritic and prodeltaic deposits. Two other neritic sediment types, well represented on the shelf (Summerhayes and Marks 1975) but not usually recovered in cores, are coarse, well-sorted, quartz-rich sands (reworked relict coastal sediments) and calcareous deposits of encrusting algal (reef) origin. A distinct neritic calcareous sand, radiocarbon dated between about 2,700 to 2,000 years B. P., is commonly found near the top of cores from the outer shelf and upper slope; on the inner shelf, cores display this horizon and in addition may be capped by a similar calcareous layer of subrecent to modern age. In mid-shelf cores, a lower layer of calcareous sand, separated from the upper calcareous sand by a section of gray mud, is dated at about 16,000 years B. P. The rapid rate of sedimentation and short length of available shelf cores preclude an analysis of the late Pleistocene and older Quaternary neritic facies off the Nile Delta.

Cores collected at depths beyond 550 m on the slope, the outer margin and in basin plains consist primarily of six sediment types: turbiditic sand and silt, turbiditic mud, hemipelagic mud, calcareous ooze, organic ooze with protosapropel, and sapropel. Also distinguished is a yellowish-

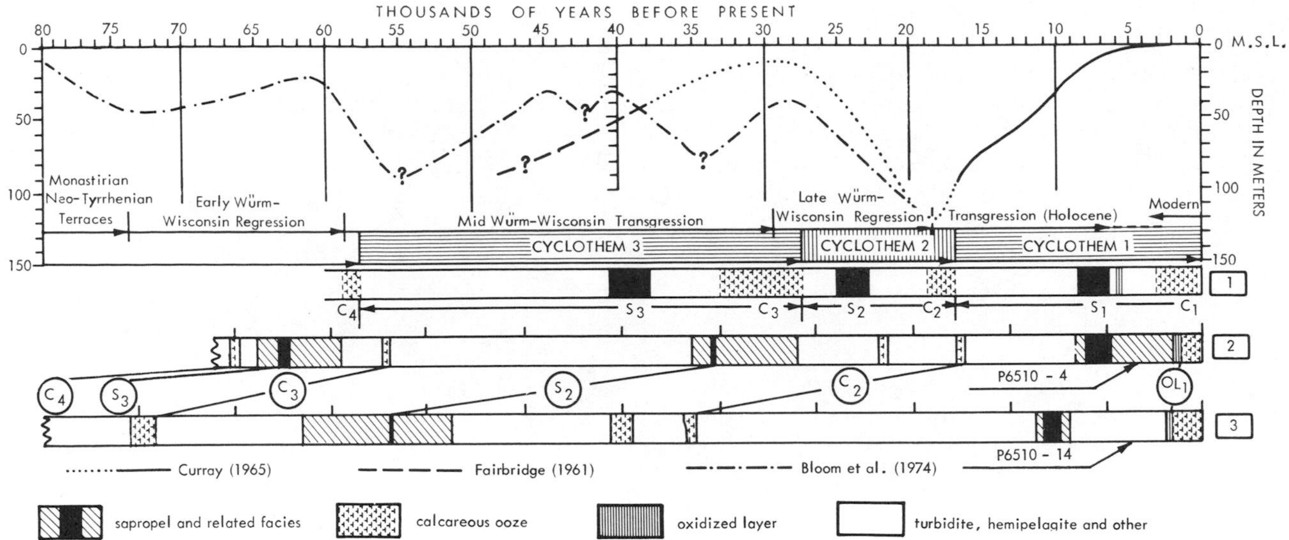

Figure 17-2. Correlation between two Nile Cone stratigraphic type cores (logs 2 and 3) and a generalized eustatic sea level curve modified from several authors. Note: Log 1 shows the age (based on radiocarbon dating) of stratigraphic horizons and the upper three cyclothems in thousands of years before present; also shown is the position of the upper three cyclothems relative to sea level. OL_1 = oxidized layer above uppermost sapropel sequence; C_1, C_2, C_3 = calcareous ooze layers; S_1, S_2, S_3 = sapropel layers. (Source: Stanley and Maldonado 1977.) The location of cores 2 and 3 is shown in Figure 17-1.

orange to light brown mud termed *oxidized layer* (OL); where present, it is commonly found capping a sapropel sequence. Clean, well-sorted quartz sands also are present in cores on the lower sector of the Rosetta Fan.

Sediment types found in cores commonly form repetitive sequences as follows (bottom to top): gray hemipelagic mud → sapropel and organic ooze → light brown hemipelagic mud and calcareous ooze. Each complete sequence is defined as a cyclothem (Maldonado and Stanley 1976a). Generally, incomplete silt-to-mud turbidite sequences occur irregularly throughout each cyclothem.

A distinctive suite of lithologic attributes enables each of the upper three cyclothems to be identified in cores throughout the study area (Stanley and Maldonado 1977). As shown in Figure 17-2, radiocarbon dating of regionally important horizons provides the following ages (in years before present): an upper calcareous ooze (C_1) ranges in age from 2,700 years B. P. to present; an oxidized layer (OL_1) is about 5,700 years B. P.; an upper sapropel (S_1) ranges from about 6,600 to more than 8,200 years B. P.; the calcareous ooze (C_2) capping the second cyclothem is dated about 17,000 to 19,000 years B. P.; the second sapropel (S_2) is about 23,000 to 25,000 years B. P.; the calcareous ooze (C_3) forming the top of the third cyclothem is dated between about 28,000 and 33,000 years B. P.; the sapropel (S_3) in this cyclothem is roughly dated at 38,000 to 40,000 years B. P. (sapropel terminology from McCoy 1974). The top of the fourth cyclothem (C_4) is considerably older than the resolution of the carbon-14 method; a maximum age of 55,000 to 58,000(?) years B. P. is estimated, based on its position relative to the overlying dated horizons (Figure 17-2). For the calculation of sedimentation rates of core sequences, the ages of the top of the six regionally prevalent horizons are used: OL_1 = 5,700; C_2 = 17,000; S_2 = 23,000; C_3 = 28,000; S_3 = 38,000 and C_4 = 55,000 to 58,000.

Comparing the combined litho- and chronostratigraphic sequence with a generalized eustatic sea level curve (modified after Fairbridge, 1961; Curray 1965; Mörner 1971; Bloom et al. 1974) is useful. Calcareous ooze layers, consisting of abundant calcareous nannofossils and Mg-calcite in the form of poorly defined crystals, appear to develop at times of maximum and minimum sea level stands, while the dark, organic-rich sapropel layers accumulate during phases of rising as well as falling sea level. Although calcareous ooze layer C_4 remains undated, we speculate that it accumulated during an early Würm low sea level stand (Stanley and Maldonado 1977). The younger, more reliably dated cyclothems 1 and 2 appear to correlate directly with major climatic oscillations—that is, they are initiated, with slight offset, at times of maximum low and high sea levels.

NILE CONE FACIES

Sequences

Sediment types of the Nile Cone generally occur in four natural groupings termed sediment sequences (Maldonado and Stanley 1975, 1976a): *channel, turbidite, hemipelagic,* and *sapropel.*

The channel-type sequence includes relatively thick layers of moderately to well-sorted, medium- to coarse-grade sands usually ranging from more than 1 m to 9 m thick (cf. D. S. D. P. Site 131, Ryan, Hsü et al. 1973; Bartolini et al. 1975). In general, these relatively matrix-free sands do not display well-developed graded bedding, while cut-and-fill structures are commonly observed. Two subtypes are recovered locally: the first, a coarsening-upward sequence (possibly formed by diversion and crevassing of a new channel), records an increase in energy level with time; the second, a fining-upward sequence (possibly representing the infilling of a channel after diversion and occupation of a new depression). The structures and textures reveal the composite nature of the sand layers and suggest a multiplicity of transport mechanisms. Two main types of transport are postulated: (a) gravity-induced flows, mainly sand flows (fluidized and/or grain flow; cf. Middleton and Hampton 1973) and high-density turbidity currents, and (b) bottom traction (cf. Kelling et al. 1978).

Turbidite sequences, which account for a significant proportion of cored Rosetta Fan sections, are deposited by both high and low concentration flows. The most common types are turbidites that consist of silt and clay and usually are micaceous in nature and that display the *c* or *d* divisions of the Bouma (1962) turbidite sequence at their base. Core analysis shows distinct areal distribution patterns for the different turbidite sequence types; these patterns, however, do not correlate strictly with distance from the source area. For example, thin and finer-grade sandy and silty turbidites occur on the fan proper as well as on the slope and in deep basin plains (Herodotus, Eratosthenes, and so forth). Thus, the terms *proximal* and *distal*, as herein applied to turbidites, refer to the lithologic nature of the deposit, and do not necessarily imply distance from source (cf. Bouma and Hollister 1972; and Chapter 13 of this volume).

Thin-bedded, fine-grade sand-to-mud units, also characteristic of the Nile prodelta facies, are concentrated on the outer shelf and slope. Lithofacies differences between distal turbidites and thin-bedded prodeltaic deposits are subtle; both types may occur together in a core, and gradational transitions from one type to the other appear to develop downslope. However, the mechanisms involved in these two types of deposit are probably different: silty prodeltaic layers can result from (a) direct supply by short-term, high-density flows initiated at or near the river mouth, or (b) by reworking and concentration of sediments by wave action and bottom currents, and subsequent seaward transport by shelf-edge spillover processes, or (c) both.

Hemipelagic mud is the most ubiquitous Nile Cone lithofacies; calcareous ooze and foraminiferal sands also are associated with this facies. A gradual lithologic transition from one specific hemipelagic sediment type to another is observed (Maldonado and Stanley 1976a). Transport, primarily from suspension, may be influenced by bottom currents as suggested by cross-lamination, moderate to good sorting, and orientation of grains in biogenic sand layers (cf. Bartolini and Gehin 1970; Huang and Stanley 1972).

Distinctive sapropel sequences are recovered in most cores, although they generally represent only a small proportion, in terms of thickness, of the late Quaternary section. As many workers have shown, these units record large-scale climatic and oceanographic fluctuations in the eastern Mediterranean. The prevalent olive-gray sediment types below sapropels accumulated under conditions of partial mixing and oxygenation of water masses that sustained a moderate degree of benthic activity. The dark gray to black sapropel layers, on the other hand, record major anaerobic phases related to regional stratification of water masses (van Straaten 1972; Ryan 1972; McCoy 1974; Stanley et al. 1975). The progression from anaerobic to more normal open ocean conditions is indicated by the subsequent accumulation of organic ooze. This layer is followed by deposition of a light brown to orange oxidized layer above the sapropel-organic ooze horizons; this unit records a return to oxygen-rich, vertically mixed bottom water conditions.

Hemipelagic and sapropel sequences, whose components include both biogenic and fine-grained terrigenous material, are sensitive indicators of major regional and, quite probably, basinwide events. The main process responsible for their deposition is settling from suspension that has been influenced by fine-grained sediment input, currents, biogenic productivity and the stratification and movement of water masses. The general term *suspensite* herein refers to both hemipelagic and sapropel sequences, while *gravitite* essentially refers to gravity-induced deposits (i.e., turbidites and sand flow units) that are indicators of important terrigenous input to the basin.

Facies Assemblages

Late Quaternary lithofacies ratios, facies assemblages, and sedimentation rates on the Nile Cone are contoured on four charts using geometric scales to emphasize regional variations (Figure 17-3). The facies distribution patterns on western and eastern Nile Cone sectors are markedly different. The Rosetta Fan area has been most active in terms of submarine fan deposition and of volume of sediment transported downslope during the past 55,000 to 58,000 years. The distribution of these fan deposits appears on the lithofacies charts as a broad southeast-northwest-trending lobe on the Rosetta Fan that extends seaward from the slope toward the eastern part of the Herodotus Basin plain. The broad, highly dissected southeast-northwest-trending Levant Platform east of the

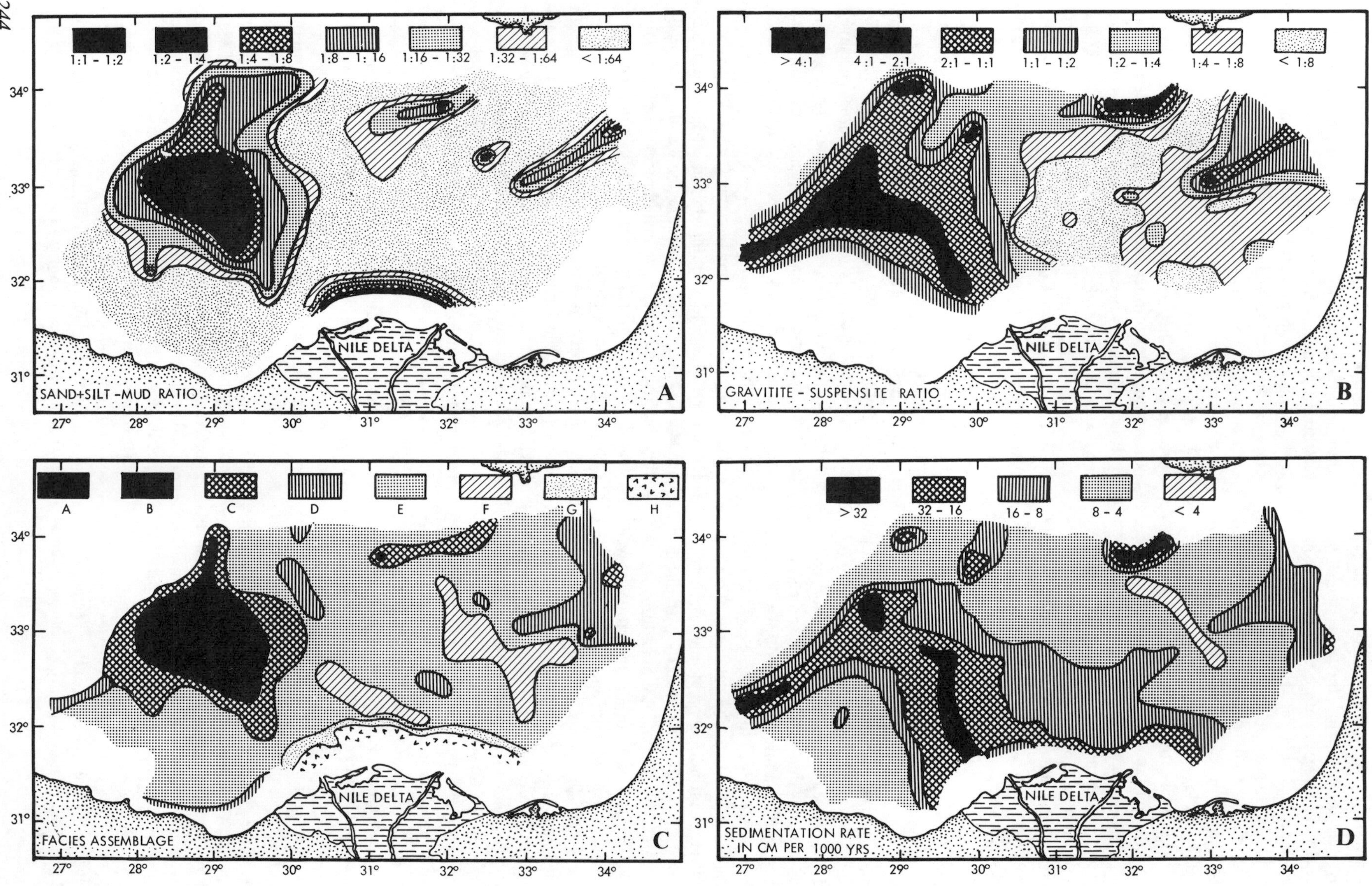

Figure 17-3. Late Quaternary lithofacies distribution off the Nile based on analysis of 75 cores. Note: The four charts summarize the sand and silt to mud ratio (chart A); gravitite to suspensite ratio (chart B); facies assemblages (chart C); and sedimentation rates for the period from about 58,000 years B. P. to present (chart D).

fan and north of the Damietta Branch does not appear to have played a significant role in downslope channelization of sediment during this period. Cores on the distal continental margin further to the east do not record active fan deposition.

Eight distinct facies assemblages, defined primarily on sediment types and sequences recovered at the 75 core sites, are distinguished. The data in Figure 17-3 take into account sediment thickness and proportion of sequence types per total core length and show the regional lithofacies distribution from stratigraphic horizon C_4 to present. Many cores fail to recover stratigraphic sections as old as horizon C_4, either because they are too short (gravity cores) or because of high sedimentation rates. In compiling these charts, the rates of sedimentation and facies in the stratigraphic section at short core stations have been extrapolated back in time—that is, the assumption is that the sedimentary trends cored are generally representative of the time span from present to horizon C_4.

Facies A. Cores attributed to assemblage A include thick sand strata alternating with complete turbidite sequences ($T_{a\text{-}e}$) and hemipelagic muds. In some instances, this facies also includes distal turbidite sequences ($T_{c\text{-}e}$, $T_{d\text{-}e}$) and, less commonly, thin-bedded sand and silt turbidites. The sand plus silt to mud ratio is high (1:2 to 1:1, or higher), and the gravitite to suspensite ratio exceeds 1 (usually ranges from 1 to 2). Sedimentation rates are generally high (>16 cm/10^3 year), although lower values (8 to 16 cm/10^3 year) also are recorded. Facies A is concentrated in an oval-shaped area on the middle to lower Rosetta Fan (Chart C, Figure 17-3). Examination of available cores suggests that this facies does not extend upslope as a canyon (Alexandria) or valley infill, as is commonly observed in modern and ancient submarine fans (Shepard et al. 1969; Stanley 1969; Nelson and Kulm 1973; Damuth and Kumar 1975; Kelling and Stanley 1976; Kelling et al. 1978). The coarse sand bodies appear concentrated on the lower Rosetta Fan sectors as in the case of the La Jolla Fan (Piper 1970).

Correlating individual sand units between relatively close-spaced cores does not appear to be possible, which thus indicates that the distribution of this facies may be related to a network of narrow, shallow channels: Each sand unit represents the infilling stage of a superimposed network of meandering channels, and deposition has resulted from continuous diversion during active periods of fan growth.

X-radiographic and textural analyses suggest an interplay of various transport mechanisms, including gravity-induced and bottom current processes. The bypassing of clean, coarse sands from the shelf to the lower fan may have occurred by a glacier-like creep or other types of gravity-induced process (cf. Shepard et al. 1969) through canyons or low-relief valleys. Long cores (to 9.5 m) on the upper slope have not recovered a single layer of clean coarse sand; either coarse sands, originating on the outermost shelf, have completely bypassed the slope or, if present, are buried at depth below the recovered Late Pleistocene and Holocene sections.

Facies B. Cores displaying Facies B comprise a proximal to distal turbidite and hemipelagic mud association. Sand turbidites 10 to 50 cm thick (some showing the complete $T_{a\text{-}e}$ sequence) are abundant and alternate with silt and mud turbidites and suspensites. The sand plus silt to mud ratio varies from 1:8 to 1:4, and the gravitite to suspensite ratio varies between 1:1 to 4:1. Sedimentation rates range from 8 to more than 32 cm/10^3 year. This association is confined to a narrow belt around the channel sand association; it also occurs in small basins at the base of the cone. It has not been cored on the slope.

Two related depositional processes are suggested: thicker, coarser turbidites are transported in the main body of density flows within channels, by crevasse-splay processes along channel margins, and by nonchannelized flows from the upper slope; thinner, fine-grained turbidites are deposited from less dense currents (cf. Moore 1969) or from the fine tail of density currents. In the latter case, material was presumably transported beyond channels by overbank flow and accumulated in alternating layers of mud and fine sandy units. The latter display small flaser structures, wavy and lenticular bedding or climbing ripples, and starved ripples. Mud rip-up clasts and deformed layers are occasionally noted in coarser sand turbidites, thereby indicating erosion and disruption by the high-density flows. Structures in both fine- and coarse-grade turbidites record a rapid diminution of turbulence, possibly associated with crevasse-splay and overbank flows.

Facies C. This facies includes primarily distal turbidites and suspensites, and can be distinguished from Facies B by the absence of complete $T_{a\text{-}e}$ sand turbidite sequences. Cores recover composite sequences of suspended sediment types interrupted by distal turbidite incursions. Sand plus silt to mud ratios generally range from 1:32 to 1:8, although some cores show a broader range. The gravitite to suspensite ratio is on the order of 1:2 to 2:1, but may exceed 4:1; some cores recovered from this facies usually display the highest gravitite to suspensite ratios. Sedimentation rates range from 4 to 32 cm/10^3 year. The regional facies distribution is quite irregular, and covers different sectors of the Nile Cone as well as isolated basin plains.

Transport mechanisms include low-density turbidite incursions, channel overbank flow, and deposition from the tail of turbidity current flows; deposits from the latter are concentrated in deep basin plains. Some lithologic sections that comprise these sediment types are similar to those consisting of prodeltaic facies cored on the slope.

Facies D, E, and F. Three facies assemblages, characterized by different proportions of turbidite mud and suspensite deposits, form a continuous spectrum. Facies D comprises predominantly mud turbidites with a variable proportion of hemipelagic-sapropel units; thin silt turbidites are subordinate in this facies. Sand plus silt to mud ratios are generally lower than 1:64 but a few core sections display ratios between 1:4 and 1:2. The gravitite to suspensite ratio is generally between >1:8 and 1:2 although values range between 1:8 to 4:1. Sedimentation rates are usually between 8 to 16 cm/10^3 year and less commonly are 4 to 8 cm/10^3 year; a few sections cored in the Herodotus Basin plain accumulated at rates in excess of 32 cm/10^3 year. Facies D is recovered in isolated areas of the slope and cone and more extensively on the distal margin off Lebanon and in the Herodotus Basin plain.

Facies E and F are distinguished by lower proportions of gravitites relative to suspensites; the low proportions characterize Facies F. Sand plus silt to mud ratios are generally lower than 1:64, and are rarely as high as 1:32 in Facies E. Gravitite to suspensite ratios in both facies commonly range from 1:8 to 1:4, but may be somewhat higher (1:4 to 2:1) in the case of Facies E. Sedimentation rates of 4 to 8 cm/10^3 year prevail, but a few exceptions are noted: <4 cm/10^3 year in the vicinity of Eratosthenes Seamount, and >32 cm/10^3 year in cores on the slope off the Nile Delta. *Facies E and F form much of the late Quaternary cover of the Nile Cone.* The only area where these facies do not prevail is the broad Rosetta Fan.

Settling from suspension is the major process involved in the deposition of these facies, and the hemipelagic and sapropel units record fluctuations of lutite and silt input, current patterns, water mass stratification, and biogenic productivity. Sporadic incursions of turbidites, more commonly noted in basin plain cores, modified the slower but more continuous accumulation of fine-grained suspensates. Facies E and F show a general decrease in sedimentation rate downslope from the shelf edge (except off the Rosetta Branch); this records a decrease in the amount of suspensate transferred seaward away from major sediment sources.

Facies G. Facies G consists largely of prodeltaic units but includes lithologically gradational units characteristic of the slope proper. Two subtypes are differentiated: The first is characterized by thin-bedded, sandy and silty laminae in mud (prodeltaic subtype), and the second consists largely of gray hemipelagic homogeneous mud. The sand plus silt to mud ratio usually ranges between 1:64 to 1:32, but may be as high as 1:4 in the case of the prodeltaic subtype. Gravitite to suspensite ratios are variable—that is, from 1:1 to more than 2:1 in the case of prodeltaic deposits and less than 1:8 for the homogeneous mud subtype. Sedimentation rates usually range from 8 to 32 cm/10^3 year and are much higher in the case of prodeltaic deposits. The deposition of this latter facies was accelerated during sea level low stands and records the seaward displacement of Nile depocenters toward the shelf edge.

Facies H. Cores recovering this facies assemblage include several neritic deposits typical of the Egyptian Shelf, including bioclastic sand, fine terrigenous sand, and shallow-water mud. These sediment types display both upward-fining and upward-coarsening sequences. The vertical transition between the different divisions of these sequences generally appears gradual in cores and is accompanied by an upward increase or decrease in sand and coarser fractions (largely bioclastic), including shells and calcareous algae. Similar core sequences on the shallow platform environments of the Strait of Sicily have been described (Maldonado and Stanley 1976b). More detailed descriptions of surficial shelf sediment types are presented elsewhere (Summerhayes and Marks 1975; and Chapter 18 of this volume).

The prodeltaic influence is recorded in shelf deposits by thin-bedded silt and sandy laminae and by starved ripples in a predominantly muddy matrix. The sand plus silt to mud ratio is extremely variable, from <1:64 to 1:2, and sedimentation rates are relatively high, generally between 8 and 32 cm/10^3 year, and are even higher locally (Einsele and Werner 1968). The nature and distribution of these sequences are closely related to the Quaternary eustatic oscillations that have periodically induced the displacement of the Nile depocenters and associated lithotypes and have altered bottom currents on the shelf. These oscillations, which are a response to climatic changes, have modified fluvial input and organic productivity, and thus have been a prime depositional control factor both on the shelf and in deeper contiguous Nile Cone sectors.

FACIES VARIATION IN TIME AND SPACE

A series of charts depict the facies distribution on the Nile Cone and adjacent regions during six time periods between about 58,000 years B.P. and the present (Figures 17-4 through 17-7). The following data are delineated: sand plus silt to mud ratio, gravitite to suspensite ratio, facies assemblages, and sedimentation rates. Aspects of the data shown in these and the more simplified charts in Figure 17-3 are discussed below.

Facies A, comprising thick layers of coarse, moderately to well-sorted sand, is best developed between stratigraphic horizons C_3 and C_2 (28,000 to 17,000 years B.P.)—that is, cyclothem 2 deposition during the last Würm regression (Figure 17-2; charts C and D, Figures 17-4 and 17-6). This facies accumulated on middle and lower sectors of the Rosetta Fan. Development of Facies A during the time span S_2 to C_2 was restricted to the lower fan (chart C, Figure 17-6). Available cores on the upper fan and slope are too short to recover underlying horizons C_3 to S_2 (chart D, Figure 17-6). This facies is recovered in a single core on the lower fan during time span C_2 - OL_1 (chart B, Figure 17-6). We can suggest that at times of accelerated

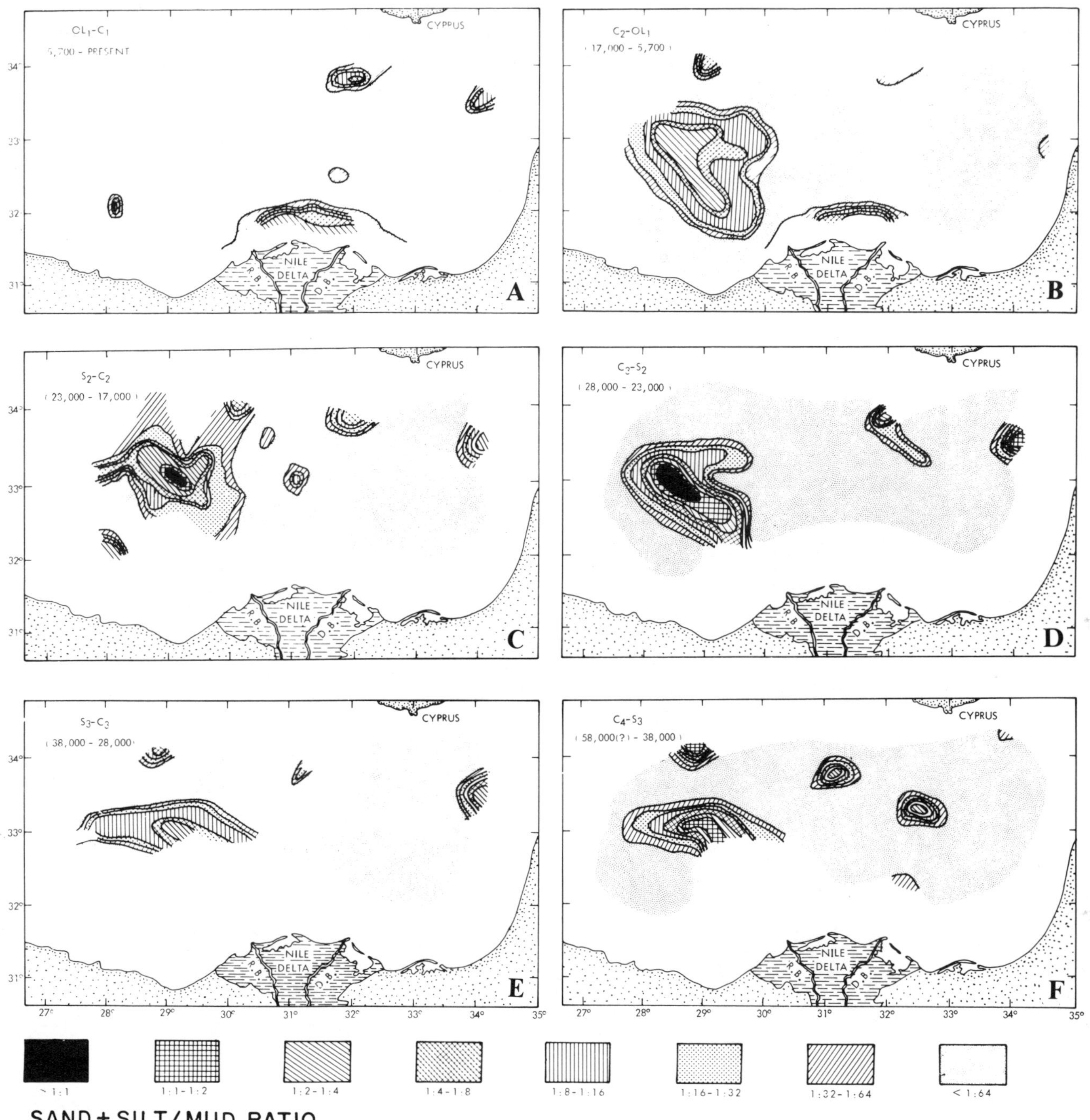

Figure 17-4. Sand plus silt to mud ratio for six periods in the Late Quaternary. Note: R. B. = Rosetta Branch; D. B. = Damietta Branch.

sediment input during lower sea level stands, the slope and canyon were characterized by transient accumulations of sandy sediments. Furthermore, increased numbers of sandy layers were deposited on the fan, and these were probably transported by poorly segregated gravity-induced flows, by immature high-density turbidity currents, and possibly by bottom currents. In the same areas, enhanced proportions of hemipelagic and complete ($T_{a\text{-}e}$) turbidites, and deposits showing the effect of increased biologic activity (reworking by benthic organisms as well as concentrations of planktonic tests), accumulated during periods of intermediate or high sea level stands. Cores show a periodic increase of sand and silt in Facies A sequences transported from the outer shelf and upper slope; these units generally provide evidence of traction, perhaps as a result of indigenous bottom currents (cf. discussion on deep bottom currents by Shepard and Marshall 1969, 1973, and Chapter 1 of this volume; Keller et al. 1973; and Chapter 2 of this volume).

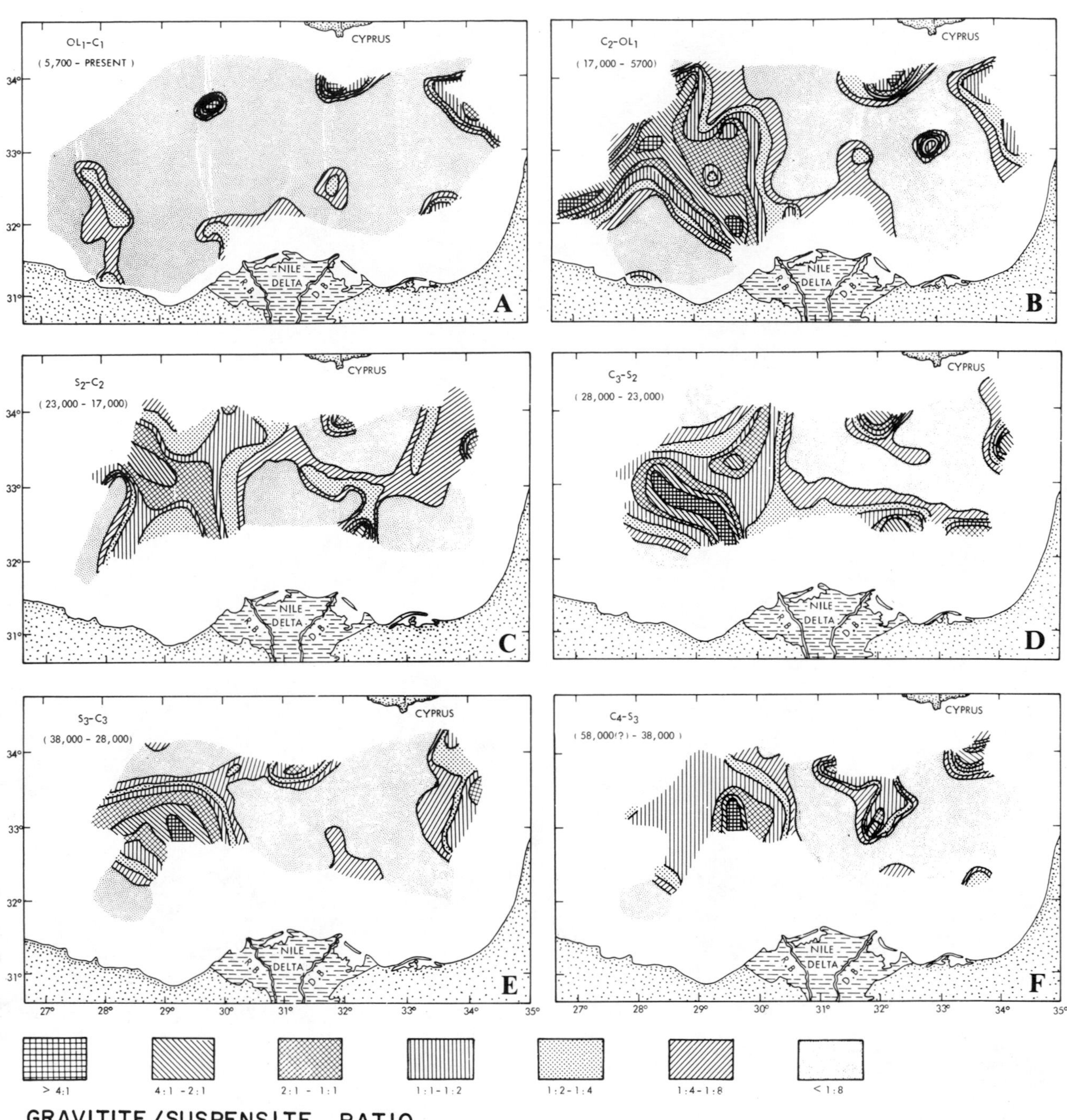

Figure 17-5. Gravitite to suspensite ratio for six periods in the Late Quaternary. Note: R. B. = Rosetta Branch; D. B. = Damietta Branch.

Several considerations may shed some light on the particular concentration of thick sand layers on the mid and lower fan. As has been postulated, for instance, as a turbidity current flow passes from a steep to a reduced gradient, there is a relative increase in turbidity current body thickness, and as a result, the head of the flow where coarse material is concentrated is likely to release this fraction on the more gentle slope (Komar 1972). Furthermore, turbidity flows may display hydraulic jump phenomena (passage from a supercritical state, with a Froude number greater than unity, to a subcritical condition) at the break in slope between the slope and fan. A possibility in the case of the Rosetta Fan is that the distribution of coarse sediment records preferential deposition just seaward of a hydraulic jump point (cf. Middleton 1970; Komar 1971). Another possible factor is related to water mass stratification. Only the densest flows emanating from the Nile, or triggered at the shelf edge, may have moved to deep distal

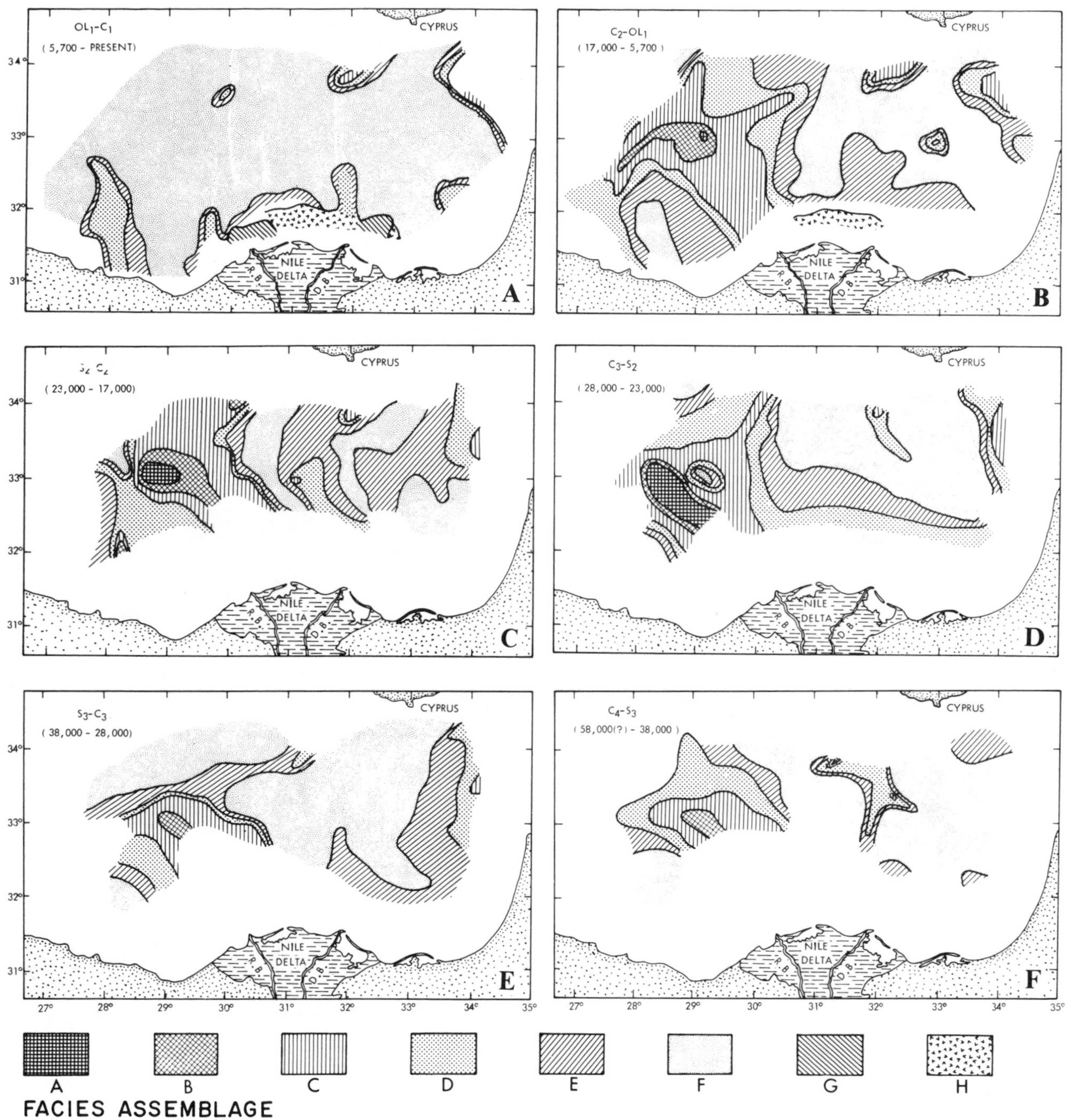

Figure 17-6. Facies assemblages for six periods in the Late Quaternary. Note: R. B. = Rosetta Branch; D. B. = Damietta Branch.

environments. Lower sediment concentrations would have been restricted, at least to some degree, along discrete density interfaces that intersect the slope (cf. McCave 1972; Pierce 1976). In summary, the distribution pattern of Facies A records coarse sediment input at the outermost shelf and upper slope on a periodic basis, most likely during low eustatic stands when the Nile Delta front advanced toward the shelf edge.

Facies B, restricted in its distribution to the same general area as Facies A on the Rosetta Fan (Figure 17-6), was deposited during most of the time span considered (but not from about 5,700 years B.P.,horizon OL_1, to present). It accumulated primarily during phases of lower fan growth (cyclothems 3 and 1, Figure 17-2; charts B, E, and F, Figure 17-6) or at times of enhanced sedimentation in areas more distal from sectors of high sediment input

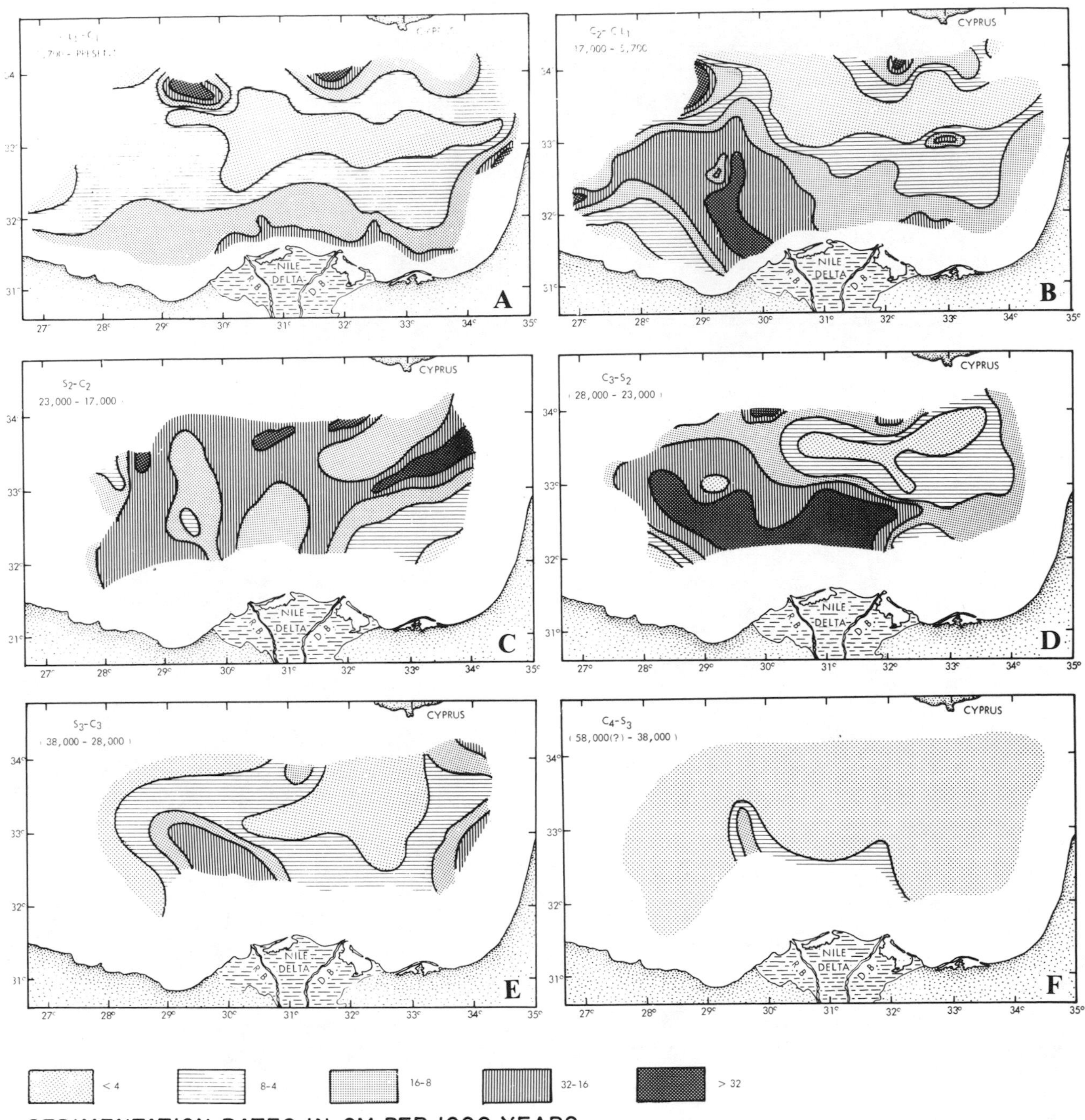

Figure 17-7. Sedimentation rates (in cm/1000 yrs) for six periods in the Late Quaternary.

(cyclothem 2, Figure 17-2; charts C and D, Figures 17-5 and 17-6). The relative thickness of the proximal turbidite assemblage may reflect, in part, proximity to an area of enhanced input. The typical Facies B distribution pattern is elongate, or elliptical, and generally extends upslope; this pattern suggests that sediment was transported seaward over broad sectors of shelf edge and that high-density flows were not well channelized downslope. The thick Facies A sand layers of the middle and lower fan merge laterally with the proximal turbidite sequences of Facies B. These proximal turbidites, in turn, are gradually transitional with "base cut-out" Bouma (1962) sequences and are recovered as thinner, finer-grade units in more distal areas.

Facies C and D, closely associated with the above two lithofacies in the sector of active Rosetta Fan development, also occur in areas close to major sediment input. These facies prevailed during the last major Würm regression—that is, from about 30,000 years B.P. (cf. interstadial suggested by Curray 1965; Bloom et al. 1974) to horizon OL_1 (about 5,700 years B.P.). This period corresponds to a lowering → rising eustatic oscillation (includes cyclothem 2 and the base of cyclothem 1; see Figure 17-2). The thin-graded sandy turbidites on the fan tend to display distal characteristics (cf. Walker 1967) and together with much of the associated lutite are interpreted as deposits from turbid overbank sheet flows that bypassed submarine levees bordering channels (cf. Ryan, Hsü et al. 1973, Site 131, p. 385). A further proposal is that similar turbidites on the distal continental margin on the eastern Nile Cone originated from turbid flows of low-density initiated by shelf-edge spillover processes (Moore 1969; Southard and Stanley 1976) and by settling from the tail of nonchannelized downslope-moving turbid flows. Dilute turbidity flows may be generated from slumps and slides triggered on the upper slope off the Nile as a result of failure of thick, rapidly accumulated prodeltaic sections. The resulting suspensions of predominantly fine deposits resulting from such failures should consist, a priori, of a limited range of particle sizes and these would be concentrated by flows likely to deposit distal-type graded units (cf. Hampton 1972). Mass failure in the form of slides and slumps, a major process on some submarine fans and on large cones off deltas (cf. Walker and Massingill 1970), is locally important on the Nile Cone. A study of physical properties of Nile Cone sediments coupled with radiocarbon dating (Einsele and Werner 1968) suggested that factors such as organism-induced porosity and increased water content in zones of rapid deposition might lead to failure. However, displaced units, more than a few tens of centimeters thick, were not detected in our core survey.

Facies E and F were deposited extensively on the Nile Cone during the Late Quaternary period considered, and as much as two-thirds of the cone surface displays this predominantly hemipelagic-sapropel sedimentation type. Facies E occurs as a narrow belt around areas of high sediment input, —that is, on the Rosetta Fan, upper slope, and in some basin plains. Facies E and D record two periods when sediment input from the Damietta Branch was of some importance: S_2 - C_2, which is 23,000 to 17,000 years B.P., and OL_1 - C_1, which is 5,700 years B.P. to present (Figure 17-2; charts C and A, Figures 17-5 and 17-6). Facies E sedimentation off the Damietta Branch also was important during time intervals S_3 - C_3, which is 38,000 to 28,000 years B.P. and C_2 - OL_1, which is 17,000 to 5,700 years B.P. (charts B and E, Figure 17-6).

Facies E and F were transported primarily by suspended processes. Hemipelagic deposits record fluctuations of fine-grained input, bottom currents, biogenic productivity, and changes in water mass stratification including the periodic formation of anaerobic conditions. The prevailing regime of suspensite sedimentation was interrupted by distal turbidity current incursions in sectors bounding zones of high sediment input and by turbid flows of low density (cf. Moore 1969) originating from shelf edge spillover onto the upper slope.

Facies D and E also occur in some isolated Cone sectors (see charts A, B, D, and F, Figures 17-5 and 17-6) that did not have direct access to major sediment input. Possibly in certain of these isolated sections, the deposits resulted from mass slope failure that at times may have been important (see Chapter 18).

Facies G, found primarily on the slope proper, includes sediment types transitional with deepwater facies—that is, deposits displaying incipient cyclothem development and mass gravity units, together with thin-bedded prodeltaic series. The latter are gradational into neritic facies consisting of homogeneous (bioturbated) muds, biogenic calcareous layers, and typical shelf sand units. Selected downslope core transects record a gradation from prodeltaic deposits on the slope to thin-bedded ("distal") turbidites and eventually sandy ("proximal") turbidites on the lower Rosetta Fan (chart C, Figure 17-3; chart B, Figures 17-4, 17-5, and 17-6). As yet, there is insufficient information to interpret the transport mechanism that resulted in the transition between the different types of thin-bedded deposits. However, prodeltaic units appear to have accumulated on the upper slope primarily during low sea level stands as a result of the steady increase of sediment input to this sector, either directly by increased outflow from the river mouths, by wave action and shelf edge spillover processes, or by both (Southard and Stanley 1976).

The preferential accumulation of sediment at depths of 450 m to 600 m off the Rosetta Branch, as recorded by the extremely high sedimentation rates (to >80 cm/1000 years; chart B, Figure 17-5 and 17-7) and the development of sapropels at and beyond these depths are evidence that deposition in this sector has been influenced by stratification of upper water masses. We believe that the coincidence of the two observations cited above is best explained by a concentration of turbid or suspensate-rich layers along density interfaces. We can envisage that such layers flowed, perhaps as detached turbid plumes, along the pycnocline-slope contact (cf. discussions by Bouma et al. 1969; Drake et al. 1972; McCave 1972; Nelson et al. 1973; Normark 1974; Pierce, 1976).

Interpretations of the shallow-water deposits included in Facies H, restricted to various Egyptian Shelf subenvironments and associated to varying degrees with prodeltaic deposits, are presented elsewhere (Summerhayes and Marks 1975).

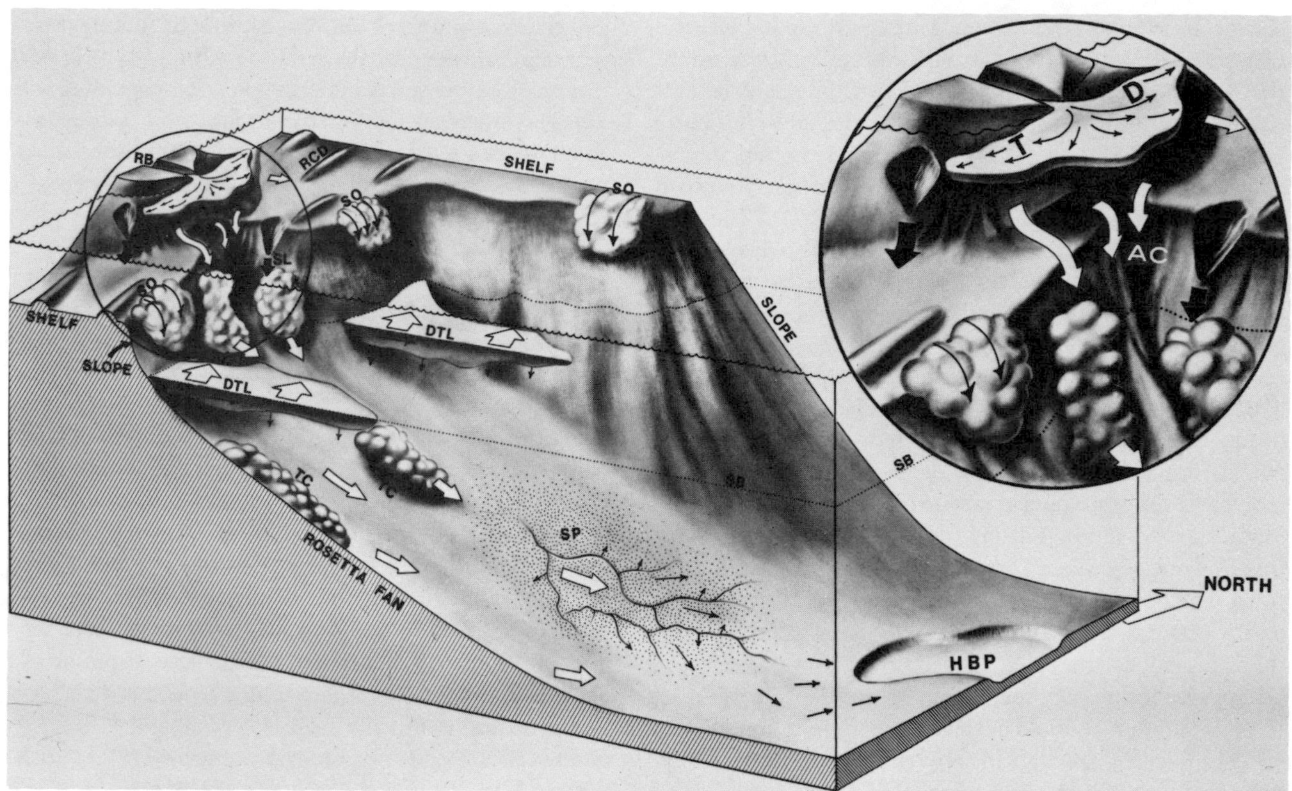

Figure 17-8. Schematic diagram showing predominant sedimentation patterns off the Rosetta Branch of the Nile during a eustatic sea level rise and phase of water mass stratification. Note: Processes include gravity-induced transport (slump = SL and heavy black arrow; high-density turbidity current = TC and open arrow; low-density turbid flow = thin arrow), shelf edge spillover (SO), and formation of detached turbid layers (DTL) along the pycnocline (SB = stratification boundary). RCD = relict coastal and neritic deposits. The inset shows outflow and movement of sediment turbid discharge (TD) off the Rosetta Branch (RB) and into the Alexandria Canyon (AC). The gradient of the continental margin to the west is greater than that of the more gentle fan surface. Sand pods (SP) and a low relief channel network are depicted on the mid- to lower fan sectors. HBP = Herodotus Basin plain. Note the general absence of marked channelization of the various downslope-moving turbid flows.

DISCUSSION

The origin of large submarine cones such as those deposited off major deltas including the Amazon, Congo, Ganges, Indus, Mississippi, and Rhône, and of smaller deepsea aprons and fans has been discussed by, among others, Ewing et al. (1958), Heezen and Menard (1963), Menard et al. (1965), Huang and Goodell (1970), Bouma (1973), Jipa and Kidd (1974), and Damuth and Kumar (1975). As a result of these and other studies in both the modern oceans and the ancient rock record, considerable attention has been paid to the formulation of fan models. However, the applicability of a generalized depositional model to all fans may be questioned.

In order to conceptualize Late Quaternary Nile Cone deposition, we must consider the orderly, but repetitive – or cyclothemic – nature of the sediment distribution that primarily reflects the interplay of sedimentation and highly variable climatic conditions in a quasi-enclosed silled basin (Mars 1963; Ryan 1972; McCoy 1974; Vergnaud-Grazzini 1975; Maldonado and Stanley 1975, 1976a). The regional climatic overprint considerably modified those physical oceanographic parameters affecting sedimentation offshore (Wüst 1961; Stanley et al. 1975) as well as the biogenic versus terrigenous input factors (Stanley and Maldonado 1977). The tectonic influence on sedimentation during this time span, although not negligible, has been of lesser importance in controlling depositional patterns than in the more mobile settings where some fans have developed.

The principal sedimentary processes on the Rosetta Fan include (a) gravity-controlled flows, differentiated into mass gravity (sliding, slumping, creep), high-density flows (sand flow, turbidity currents), and low-density flows (which result from shelf-edge spillover, overbank channel sheet flow, near-bottom turbid flow, and turbid discharge from the river mouth); (b) suspended sediment processes, influenced by fluctuations of fine-grained input, bottom currents, biogenic productivity and water mass stratification; and (c) bottom traction processes that also play a role in the deposition of clean, coarse channel sands and some finer units.

Deposits on the Rosetta Fan exemplify one type of submarine fan evolution (Figure 17-8). Core analysis shows that gravity flow processes prevailed in this sector during the period from about 38,000 to 5,700 years B.P. (charts B and F, Figure 17-5). Sediments forming the Rosetta Fan are distributed as an elongate to oval-shaped area that trends to the northwest—that is, away from the slope toward the Herodotus Basin plain (Figure 17-8). The overall concentric facies patterns within this lobe-shaped area displays thick sand channel units restricted to the middle and lower fan (charts B and F, Figure 17-4); these merge successively into sand turbidites, thin-bedded silt turbidites, and finally mud turbidites toward the margins of the fan. The most common turbidite facies are usually of the Bouma (1962) type $T_{c\text{-}e}$ and Rupke and Stanley (1974) type T_et sequences. Relatively enhanced proportions of suspensite sequences accumulated primarily during periods of less active sedimentation such as at present (chart A, Figures 17-4 and 17-5), and these are interbedded with gravity-induced deposits; fine-grade hemipelagic deposits include well-developed sapropel layers.

The fan is characterized by a gentle, smooth surface dissected by a broad, shallow valley on the slope (Alexandria Canyon) and a network of smaller channels on its lower sector. This Late Quaternary fan does not show well-developed fan lobes. The muted physiography is a function of high depositional rates over much of the fan surface, the fine-grained nature of the Quaternary cover, and the relatively tranquil tectonic setting.

The evolution of the Rosetta Fan is best explained in terms of Quaternary climatic-eustatic oscillations and the resulting displacement of the Nile Delta depocenters across the shelf. The accumulation of thick deposits on the upper slope seaward of areas of major sediment input is associated with the position of the pycnocline during periods of water mass stratification (see Figure 17-8).

Changes in deposition of style in time and space both on the fan and elsewhere on the cone, depicted by charts showing depositional rates (see Figure 17-7), reflect the overprint of Late Quaternary climatic-eustatic oscillations on offshore sedimentation. The period of overall highest sedimentation extended from about 28,000 to 17,000 years B.P. (charts C and D, Figure 17-7). Sedimentation rates over large areas of the cone during this time interval, which includes the last major Würm regression, are roughly equivalent to the high, long-term rates (approximately 30 cm/1000 years) previously postulated for this region (Ryan, Hsü et al. 1972; and Chapter 18 of this volume) and some other major submarine fans (Huang and Goodell 1970; Damuth and Kumar 1975; Nelson 1976) during the Pleistocene. Lowest depositional rates are recorded at times of high sea level stands, such as at present (Figure 17-2; chart A, Figure 17-7), as a result of landward regression of the prodelta lobe away from the shelf edge and reduced fresh water discharge (McCoy 1974; Stanley and Maldonado 1977).

The most active depositional zone on the Nile Cone during the time span considered here is the Rosetta Fan, where high sedimentation rates have persisted from about 58,000 to 5,700 years B.P. (charts B through F, Figure 17-7). Diminished deposition since about 6,000 years resulted from a marked reduction of sediment supplied to this area (chart A, Figure 17-7). Core sections recovered seaward of the Damietta Branch also recorded several short periods during which the sediment supply increased steadily. This sector, however, never received quantities comparable to those on the Rosetta Fan during the Late Quaternary, and does not display typical fan deposition. Furthermore, the broad, highly dissected Levant Platform east of the fan does not appear to have played a significant role in downslope channelization of sediments during the time period considered in this study; it appears that this zone may have served as a barrier to terrigenous material transported toward the north and northeast away from the Nile Delta. The reduced sediment accumulation on the eastern part of the Cone in part may have resulted from active salt tectonics affecting this area (see Chapter 18); sediments may have been directed preferentially toward the Rosetta Fan.

A broad east-west-trending belt of slow accretion has developed on the Cone during the last 6,000 years in response to a reduction of turbiditic flow incursions, the relatively enhanced role of hemipelagic sedimentation, and the lateral transport of suspensites and smoothing by predominantly east-west-trending surface water flow above the margin (Venkatarathnam and Ryan 1971; Lacombe and Tchernia 1972). All the above have accompanied the stabilization of sea level near its present stand (Mörner 1971).

Thus, two major contrasting sedimentary trends are observed on the Nile Cone. One involves a broad belt of rapid sediment accumulation on the Rosetta Fan seaward of the Rosetta Branch that extends downslope across the entire margin; this records the predominant seaward dispersal brought about primarily by gravity-induced turbid flows (charts B and D, Figure 17-3). A second trend is observed over much of the remaining cone surface. Here sedimentation rates show a decrease downslope as a result of progressively reduced deposition in a direction away from major source areas (see Figures 17-5 and 17-7). However, some sectors of the basin plains show anomalously high sedimentation rates that may indicate access to enhanced proportions of turbiditic flows. The presence of contour-following bottom geostrophic currents over the Nile Cone at some time during the Late Quaternary is not excluded, but this activity does not appear to be a major factor in the distribution of facies on this continental margin as previously suggested by some authors (Ryan, Hsü et al. 1973; Bartolini et al. 1975).

The occurrence on the Rosetta Fan of relatively thick, clean sand bodies isolated in fine-grained sediment masses containing organic-rich sapropel muds presents some ideal source rock-reservoir conditions for hydrocarbon accumulation and is thus of interest to petroleum exploration geologists.

SUMMARY

1. The Late Quaternary deposits on the Nile Cone reveal the degree to which slope and base-of-slope lithofacies patterns represent direct response to the overprint of regionally important climatic cycles and eustatic oscillations. Gravity-controlled deposits alternate with suspensite (hemipelagic-sapropel) deposits as a result of deposition in a quasi-enclosed silled basin, a setting where the interplay of climate, physical oceanography, and biogenic versus terrigenous input factors is particularly critical.

2. Submarine fan deposition during the past 60,000 years is best developed on the Rosetta Fan sector of the cone reflecting enhanced sediment input from the Rosetta Branch of the Nile, particularly during lowered sea level stands.

3. The fluctuations of fine-grade sediment input are related to the shifting of Nile prodeltaic depocenters across the shelf and changes in Nile headwater migration. Concomitantly, marked alteration of physical oceanographic parameters, including stratification and possible reversal of currents, has resulted in the distinct depositional patterns of the Rosetta Fan. The Late Quaternary fan, which exemplifies one fan model, is characterized by (a) deposition on a thick (2,000 m to 3,000 m) underlying wedge of Plio-Quaternary sediment in a region of relatively low structural mobility; (b) the absence of well-developed fan lobes, the presence of a shallow channel network on the lower fan surface, and a generally smooth surface configuration; and (c) accumulation of sediment successions consisting of thick sand pods (restricted to lower sectors of the fan), T_{c-e} and T_et turbidite sequences, well-developed sapropel layers, and relatively important proportions of suspensite as well as gravitative deposits.

4. There is no clear-cut lithofacies distinction between prodeltaic sections on the outer shelf to slope and those comprising thin-bedded turbidites further seaward on the fan; although transitions between these two facies are gradational, transport processes involved may be different.

5. Most of the extensive Rosetta Fan surface has been influenced by fan deposition during the Late Quaternary, thereby suggesting that sediments, transported across a broad sector seaward of the Rosetta Branch, were moved downslope over wide areas of the slope—that is, materials were not strictly channelized as they are in most of the fans studied to date.

6. Two contrasting depositional patterns are apparent on the Nile Cone: a broad, essentially downslope-trending belt consisting largely of gravity-induced deposits that cover much of the Rosetta Fan; a wide east-west zone, comprising largely suspensate facies, that occupies much of the remaining cone.

7. The cyclic repetition of sapropel with well-oxidized, bioturbated deepwater hemipelagic mud of the type recovered on much of the Nile Cone surface records phases of basinwide stratified water mass and anaerobic conditions that periodically alternated with phases of vertical mixing, as at present.

8. Water mass stratification appears to have modified the movement of suspensites and may have been of importance in concentrating thick, rapidly deposited units on the slope seaward of important areas of Nile River input and of shelf-edge erosion, primarily off the Rosetta Branch.

9. Distal turbidite types recovered throughout much of the Nile Cone may have accumulated as a result of several factors: low-density turbid flows initiated at a time of reduced sediment supply, deposition triggered by shelf-edge spillover and by overbank sheet flow, as well as deposition at increased distances away from the point of major sediment entry. Thus, the terms *proximal* and *distal* as applied to turbiditic sequences on the cone describe specific lithofacies types rather than distance from input sources or paleobathymetric conditions.

ACKNOWLEDGMENTS

This study, part of an ongoing Mediterranean Basin (MEDIBA) Project, was funded by Smithsonian Research Foundation grants 450137, 460132, and 71500532 (DJS), by National Geographic Society grant 5678 (DJS), and by the Consejo Superior de Investigaciones Científicas of Spain (AM). The following institutions generously provided core materials: University of Miami, Lamont-Doherty Geological Observatory (supported by ONR grant N00014-73-0210 and NSF grant DES72-01568), and Woods Hole Oceanographic Institution (NSF grant GS-13212 and OIP75-02516); the Deep-Sea Drilling Project provided selected Leg 13 samples. Particular appreciation is expressed to Dr. D. A. Ross (W.H.O.I.) for enabling A. Maldonado to participate in R/V *Chain* (1975) cruise 119, and to Dr. C. P. Summerhayes for furnishing unpublished core data from this cruise. The manuscript was reviewed by Drs. F. W. McCoy, W. R. Normark, and G. Kelling.

REFERENCES

Aleem, A. A., 1972. Effect of river outflow management on marine life. *Mar. Biol.*, 15: 200-08.

Bartolini, C., and Gehin, C. E., 1970. Evidence of sedimentation by gravity-assisted bottom currents in the Mediterranean Sea. *Mar. Geol.*, 9: M1-M5.

Bartolini, C., Malesani, P. G., Manetti, P., and Wezel, F. C., 1975. Sedimentology and petrology of Quaternary sediments from the Hellenic Trench, Mediterranean Ridge and the Nile Cone, DSDP, Leg-13 cores. *Sediment.*, 22: 205-36.

Biju-Duval, B., Letouzey, J., et al. 1974. Geology of the Mediterranean Sea basins. In: C. A. Burk and C. L. Drake (eds.), *The Geology of Continental Margins.* Springer-Verlag, New York, pp. 695-721.

Bloom, A. L., Broecker, W. S., Chappell, J. M. A., Mathews, R. K., and Mesolella, K. J., 1974. Quaternary sea level fluctuations on a tectonic coast: New ^{230}TH/^{234}Li dates from the Huon Peninsula, New Guinea. *Quater. Res.*, 4: 185-205.

Bouma, A. H., 1962. *Sedimentology of Some Flysch Deposits, a Graphic Approach to Facies Interpretation.* Elsevier, Amsterdam, 168 pp.

———, 1972. Recent and ancient turbidites and contourites. *Trans. Gulf Coast Assoc. Geol. Soc.*, 22: 205-21.

———, 1973. Leveed-channel deposits, turbidites, and contourites in deeper parts of Gulf of Mexico. *Trans. Gulf Coast Assoc. Geol. Soc.*, 23: 368-76.

———, 1975. Deep-sea fan deposits from Toyama trough, Sea of Japan. In: D. E. Karig, J. C. Ingle, Jr., et al. (eds.), *Initial Reports of the Deep Sea Drilling Project, 31.* U. S. Govt. Print. Off., Washington, D. C., pp. 489-95.

———, Rezak, R., and Chmelik, R. F., 1969. Sediment transport along oceanic density interfaces. (Abst.) *Geol. Soc. Amer. Prog.*, 7: 259-60.

———, and Hollister, C. D., 1973. Deep ocean basin sedimentation. In: A. H. Bouma and C. D. Hollister (eds.), *Turbidites and Deep Water Sedimentation.* Soc. Econ. Paleont. Mineral., Pacific Section, Short Course, Anaheim, pp. 79-118.

Carter, T. G., Flanagan, J. P., Jones, C. R., et al., 1972. A new bathymetric chart and physiography of the Mediterranean Sea. In: D. J. Stanley (ed.), *The Mediterranean Sea – A Natural Sedimentation Laboratory.* Dowden, Hutchinson & Ross, Stroudsburg, Pa., pp. 1-23.

Curray, J. R., 1965. Late Quaternary history, continental shelves of the United States. In: H. E. Wright and D. G. Frey (eds.), *The Quaternary of the United States.* Princeton University Press, Princeton, N. J., pp. 723-35.

———, and Moore, D. G., 1971. Growth of the Bengal Deep-Sea Fan and denudation in the Himalayas. *Geol. Soc. Amer. Bull.*, 82: 563-72.

———, and Moore, D. G., 1974. Sedimentary and tectonic processes in the Bengal Deep-Sea Fan and geosyncline. In: C. A. Burk and C. L. Drake (eds.), *The Geology of Continental Margins.* Springer-Verlag, New York, pp. 617-27.

Damuth, J. E., and Kumar, N., 1975. Amazon Cone: Morphology, sediments, age, and growth pattern. *Geol. Soc. Amer. Bull.*, 85: 863-78.

Drake, D. E., Kolpak, R. L., and Fischer, P. J., 1972. Sediment transport on the Santa Barbara-Oxnard Shelf, Santa Barbara Channel, California. In: D. J. P. Swift, D. B. Duane, and O. H. Pilkey (eds.), *Shelf Sediment Transport.* Dowden, Hutchinson & Ross, Stroudsburg, Pa., pp. 307-31.

Einsele, G., and Werner, F., 1968. Zusammensetzung, Gefüge und mechanische Eigenschaf-en rezenter Sedimente vom Nildelta, Roten Meer und Golf von Aden. *Meteor Forschung.*, C, 1: 21-42.

Emery, K. O., Heezen, B. C., and Allan, T. D., 1966. Bathymetry of the eastern Mediterranean Sea. *Deep-Sea Res.*, 13: 173-92.

Ewing, M., Ericson, D. B., and Heezen, B. C., 1958. Sediments and topography of the Gulf of Mexico. In: L. Weeks (ed.), *Habitat of Oil.* Amer. Assoc. Petrol. Geol., Tulsa, pp. 995-1053.

Fairbridge, R. W., 1961. Eustatic changes in sea level. In: *Physics and Chemistry of the Earth, 4.* Pergamon Press, New York, pp. 99-185.

Finetti, I., and Morelli, C., 1973. Wide scale digital seismic exploration of the Mediterranean Sea. *Boll. Geof. Teor. Appl.*, 56, 14: 291-342.

Hampton, M. A., 1972. The role of subaqueous debris flow in generating turbidity currents. *J. Sed. Petrol.*, 42: 775-93.

Haner, B. E., 1971. Morphology and sediments of Redondo Submarine Fan, Southern California. *Geol. Soc. Amer. Bull.*, 82: 2413-32.

Heezen, B. C., and Menard, H. W., 1963. Topography of the deep-sea floor. In: M. N. Hill (ed.), *The Sea, 3.* Wiley-Interscience, New York, pp. 233-80.

Herman, Y., 1972. Quaternary eastern Mediterranean sediments: micropaleontology and climatic record. In: D. J. Stanley (ed.), *The Mediterranean Sea – A Natural Sedimentation Laboratory.* Dowden, Hutchinson & Ross, Stroudsburg, Pa., pp. 129-47.

Huang, T.-C., and Goodell, H. H., 1970. Sediments and sedimentary processes of eastern Mississippi Cone, Gulf of Mexico. *Amer. Assoc. Petrol. Geol. Bull.*, 54: 2070-100.

———, and Stanley, D. J., 1972. Western Alboran Sea: Sediment dispersal, ponding and reversal of currents. In: D. J. Stanley (ed.), *The Mediterranean Sea – A Natural Sedimentation Laboratory.* Dowden, Hutchinson & Ross, Stroudsburg, Pa., pp. 512-59.

Jipa, D., and Kidd, R. B., 1974. Sedimentation of coarser grained interbeds in the Arabian Sea and sedimentation processes of the Indus Cone. In: R. B. Whitmarsh, O. E. Weser, D. A. Ross, et al. (eds.), *Initial Reports of the Deep Sea Drilling Project, 23.* U. S. Govt. Print. Off., Washington, D. C., pp. 471-95.

Keller, G. H., Lambert, D., Rowe, G., and Staresinic, N., 1973. Bottom currents in the Hudson Canyon. *Science*, 180: 181-83.

Kelling, G., and Stanley, D. J., 1976. Sedimentation in canyon, slope, and base-of-slope environments. In: D. J. Stanley and D. J. P. Swift (eds.), *Marine Sediment Transport and Environmental Management.* Wiley-Interscience, New York, pp. 379-435.

———, Maldonado, A., and Stanley, D. J., 1978. Salt tectonics and basement fractures: Key controls of recent sediment distribution on the Balearic Rise, Western Mediterranean. *Smithsonian Contr. Mar. Sci.* (in press).

Kenyon, N. H., Stride, A. H., and Belderson, R. H., 1975. Plan views of active faults and other features on the lower Nile Cone. *Geol. Soc. Amer. Bull.*, 85: 1733-39.

Komar, P. D., 1971. Hydraulic jumps in turbidity currents. *Geol. Soc. Amer. Bull.*, 82: 1477-88.

———, 1972. Relative significance of head and body spill from a channelized turbidity current. *Geol. Soc. Amer. Bull.*, 83: 1151-56.

Lacombe, H., and Tchernia, P., 1972. Caractères hydrologiques et circulation des eaux en Méditerranée. In: D. J. Stanley (ed.), *The Mediterranean Sea – A Natural Sedimentation Laboratory.* Dowden, Hutchinson & Ross, Stroudsburg, Pa., pp. 25-36.

Maldonado, A., and Stanley, D. J., 1975. Nile Cone lithofacies and definition of sediment sequences. *Proc. IX Intern. Congr. Sediment.*, 6, Nice, pp. 185-91.

_____, and Stanley, D. J., 1976a. The Nile Cone: Submarine fan development by cyclic sedimentation. *Mar. Geol.*, 20: 27-40.

_____, and Stanley, D. J., 1976b. Late Quaternary sedimentation and stratigraphy in the Strait of Sicily. *Smithsonian Contr. Earth Sci.*, 16, 73 pp.

Mars, P., 1963. Les faunes et la stratigraphie du Quaternaire méditerranéen. *Rec. Trav. Station Mar. d'Endoume,* 28: 61-97.

McCave, I. N., 1972. Transport and escape of fine-grained sediment from shelf areas. In: D. J. P. Swift, D. B. Duane, and O. H. Pilkey, (eds.), *Shelf Sediment Transport – Process and Pattern.* Dowden, Hutchinson & Ross, Stroudsburg, Pa., pp. 225-48.

McCoy, F. W., Jr., 1974. *Late Quaternary sedimentation in the eastern Mediterranean Sea.* Unpublished Ph.D. thesis, Harvard University, Cambridge, Mass., 132 pp.

Menard, H. W., Smith, S. M., and Pratt, R. M., 1965. The Rhône Deep-Sea Fan. In: W. F. Whittard and R. Bradshaw (eds.), *Submarine Geology and Geophysics.* Proc. 17th Symp. Colston Res. Soc., pp. 271-84.

Middleton, G. V., 1970. Experimental studies related to problems of flysch sedimentation. In: J. Lajoie (ed.), *Flysch Sedimentology in North America.* Geol. Assoc. Can., Spec. Paper 7, pp. 253-72.

_____, and Hampton, M. A., 1973. Mechanics of flow and deposition. In: G. V. Middleton and A. H. Bouma (eds.), *Turbidites and Deep Water Sedimentation.* Soc. Econ. Paleont. Mineral., Pacific Section, Short Course, Anaheim, pp. 1-38.

Milliman, J. D., and Müller, J., 1973. Precipitation and lithification of magnesian calcite in the deep-sea sediments of the eastern Mediterranean Sea. *Sediment.*, 20: 29-45.

Moore, D. G., 1969. Reflection profiling studies of the California Continental Borderland: Structure and Quaternary turbidite basins. *Geol. Soc. Amer. Spec. Paper* 107, 142 pp.

_____, and Curray, J. R., 1974. Midplate continental margin geosynclines: Growth processes and Quaternary modifications. In: R. H. Dott, Jr., and R. H. Shaver (eds.), *Modern and Ancient Geosynclinal Sedimentation.* Soc. Econ. Paleont. Mineral., Sp. Publ. 19, pp. 26-35.

Mörner, N. A., 1971. Eustatic and climatic changes during the last 20,000 years and a method of separating the isostatic and eustatic factors in a uplifted area. *Paleogr., Paleoclimat., Paleoecol.,* 9: 153-81.

Müller, C., 1973. Calcareous nannoplankton assemblages of Pleistocene to recent sediments of the Mediterranean Sea. *Bull. Geol. Soc. Greece,* 10: 133-44.

Neev, D., Almagor, G., Arad, A., Ginzburg, A., and Hall, J. K., 1976. The geology of southeastern Mediterranean. *Geol. Surv. Israel Bull.*, 68: 1-51.

Nelson, D. D., Pierce, J. W., and Colquhoun, D. D., 1973. Sediment dispersal by cascading coastal water. (Abst.) *Geol. Soc. Amer. Progr.*, 8: 423-24.

Nelson, H., 1976. Late Pleistocene and Holocene depositional trends, processes, and history of Astoria Deep-Sea Fan, Northeast Pacific. *Mar. Geol.*, 20: 129-73.

_____, and Kulm, L. D., 1973. Submarine fans and channels. In: G. V. Middleton and A. H. Bouma (eds.), *Turbidites and Deep Water Sedimentation.* Soc. Econ. Paleontol. Mineral., Pacific Section, Short Course, Anaheim, pp. 39-78.

_____, and Nilsen, T., 1974. Depositional trends of modern and ancient deep-sea fans. In: R. H. Dott, Jr., and R. H. Shaver, (eds.), *Modern and Ancient Geosynclinal Sedimentation.* Soc. Econ. Paleontol. Mineral., Sp. Publ. 19, pp. 54-76.

Normark, W. R., 1970. Growth patterns of deep-sea fans. *Amer. Assoc. Petrol. Geol. Bull.*, 54: 2170-2195.

_____, 1974. Submarine canyons and fan valleys: Factors affecting growth patterns of deep-sea fans. In: R. H. Dott, Jr., and R. H. Shaver (eds.), *Modern and Ancient Geosynclinal Sedimentation.* Soc. Econ. Paleontol. Mineral., Sp. Publ. 19: 56-68.

_____, and Piper, D. J. W., 1972. Sediments and growth pattern of Navy Deep-Sea Fan, San Clemente Basin, California Borderland. *J. Geol.*, 80: 198-223.

Olausson, E., 1961. Sediment cores from the Mediterranean Sea and the Red Sea. *Reports of the Swedish Deep-Sea Expedition, 1947,* 8: 335-91.

Orlova, G. and Zenkovich, V., 1974. Reports: Erosion of the shores of the Nile Delta. *Geoforum,* Pergamon Press, Oxford, pp. 68-72.

Pierce, J. W., 1976. Suspended sediment transport at the shelf break and over the outer margin. In: D. J. Stanley and D. J. P. Swift (eds.), *Marine Sediment Transport and Environmental Management.* Wiley-Interscience, New York, pp. 437-58.

Piper, D. J. W., 1970. Holocene sedimentation on La Jolla Fan. *Mar. Geol.*, 8: 211-27.

Ross, D. A., and Uchupi, E., 1977. The structure and sedimentary history of the southeastern Mediterranean Sea – Nile Cone area. *Amer. Assoc. Petrol. Geol. Bull,* 61: 872-902.

Rupke, N. A., and Stanley, D. J., 1974. Distinctive properties of turbiditic and hemipelagic mud layers in the Algéro-Balearic Basin, western Mediterranean Sea. *Smithsonian Contr. Earth Sci.*, 13, 40 pp.

Ryan, W. B. F., 1972. Stratigraphy of Late Quaternary sediments in the eastern Mediterranean. In: D. J. Stanley (ed.), *The Mediterranean Sea – A Natural Sedimentation Laboratory.* Dowden, Hutchinson & Ross, Stroudsburg, Pa., pp. 149-69.

_____, Stanley, D. J., Hersey, J. B., Fahlquist, D. A., and Allan, T. D., 1970. The tectonics and geology of the Mediterranean Sea. In: A. Maxwell (ed.), *The Sea 4.* Wiley, New York, pp. 387-492.

_____, Hsü, K. J., et al. (eds.), 1973. Western Nile Cone – Site 131. In: *Initial Reports of the Deep Sea Drilling Project, 13.* U. S. Govt. Print. Off., Washington, D. C., pp. 383-401.

Shepard, F. P., Dill. R. F., and von Rad, U., 1969. Physiography and sedimentary processes of La Jolla Submarine Fan and Fan-valley. *Amer. Assoc. Petrol. Geol. Bull.*, 53: 390-420.

_____, and Marshall, N. F., 1969. Currents in La Jolla and Scripps submarine canyons. *Science,* 165: 177-78.

_____, and Marshall, N. F., 1973. Currents along floors of submarine canyons. *Amer. Assoc. Petrol. Geol. Bull.*, 57: 244-64.

Smith, S. G., 1976. Diapiric structures in the eastern Mediterranean Herodotus Basin. *Earth Planet. Sci. Letters,* 32: 62-68.

Southard, J. B., and Stanley, D. J., 1976. Shelf-break processes and sedimentation. In: D. J. Stanley and D. J. P. Swift (eds.), *Marine Sediment Transport and Environmental Management.* Wiley-Interscience, New York, pp. 351-77.

Stanley, D. J., 1969. Sedimentation in slope and base-of-slope environments. In: D. J. Stanley (ed.), *The New Concepts of Continental Margin Sedimentation.* Amer. Geol. Inst., Short Course, Washington, D.C., pp. DJS8.1-DJS8.25.

_____, and Maldonado, A., 1977. Nile Cone: Late Quaternary stratigraphy and sediment dispersal. *Nature,* 266: 129-35.

_____Maldonado, A., and Stuckenrath, R., 1975. Strait of Sicily depositional rates and patterns, and possible reversal of currents in the late Quaternary. *Paleogr., Paleoclimat., Paleoecol.,* 18: 279-91.

Summerhayes, C. P., and Marks, N., 1975. Nile Delta: Nature, evolution and collapse of continental shelf sediment system, a preliminary report. *Proc. UNESCO Seminar on Nile Delta Sedimentology,* Alexandria, October 1975, 42 pp.

van Straaten, L. M. J. U., 1972. Holocene stages of oxygen depletion in deep waters of the Adriatic Sea. In: D. J. Stanley (ed.), *The Mediterranean Sea – A Natural Sedimentation Laboratory.* Dowden, Hutchinson & Ross, Stroudsburg, Pa., 631-43.

Venkatarathnam, K., and Ryan, W. B. F., 1971. Dispersal patterns of clay minerals in the sediments of the eastern Mediterranean Sea. *Mar. Geol.,* 11: 261-92.

Vergnaud Grazzini, C., 1975. ^{18}O changes in foraminifera carbonates during the last 10^5 years in the Mediterranean Sea. *Science,* 190: 272-74.

Walker, R. G., 1967. Turbidite sedimentary structures and their relationship to proximal and distal depositional environments. *J. Sed. Petrol.,* 37: 25-43.

Walker, J. R., and Massingill, J. V., 1970. Slump features on the Mississippi Fan, northeastern Gulf of Mexico. *Geol. Soc. Amer. Bull.,* 81: 3101-08.

Wright, L. D., and Coleman, J. M., 1973. Variations in morphology of major river deltas as functions of ocean wave and river discharge regimes. *Amer. Assoc. Petrol. Geol. Bull.,* 57: 370-98.

Wüst, G., 1961. On the vertical circulation of the Mediterranean Sea. *J. Geophys. Res.,* 66: 3261-71.

Part IV

The Tectonic Setting of Some Submarine Canyons and Fans

Chapter 18

Sedimentation and Structure of the Nile Cone and Levant Platform Area*

DAVID A. ROSS
ELAZAR UCHUPI
COLIN P. SUMMERHAYES
DONALD E. KOELSCH

Woods Hole Oceanographic Institution
Woods Hole, Massachusetts

E. M. EL SHAZLY

Egyptian Academy of Science and Technology and
Egyptian Atomic Energy Commission
Cairo, Egypt

ABSTRACT

A detailed seismic and sedimentologic study of the continental margin off Egypt has emphasized the sedimentary and structural framework of the area and, in particular, the origin and development of the large submarine fan off the Nile River. This fan can be subdivided into two principal subprovinces: the Nile Cone and the Levant Platform. A third feature, the Mediterranean Ridge, also contains significant amounts of Nile-derived sediments.

The present configuration of the Nile Cone, Levant Platform, and Mediterranean Ridge is a direct result of vertical and horizontal movements of the underlying Messinian evaporites, due in part to the considerable movements due to subduction of the African lithospheric plate and small-scale folding and faulting.

*Woods Hole Oceanographic Institution Contribution Number 3848.

INTRODUCTION

The Nile, with a length of 6,600 km, is the world's longest river and presently drains an area of about 3 million square kilometers, or about one-tenth of Africa. Nile distributaries have constructed one of the world's largest deltas (22,000 km^2), and it was to this triangular-shaped alluvial plain that the term delta was first applied by Herodotus in the fifth century B. C. (Figure 18-1).

Prior to the building of the Aswan High Dam in 1964, the Nile had a discharge of over 420,000 ft^3/sec. (11,894 m^3/sec) and carried as much as 140 x 10^6 tons of sediment per year; its delta front was prograding seaward at a rate of about 15 m per year (Aleem 1972). Since the completion of the dam, river discharge has generally been prevented by dams a few kilometers from the sea, and erosion and retreat of the shoreline is now common along much of the delta front. River discharge in the late Pleistocene was probably higher than during the Holocene, as marshes, lakes, and generally humid conditions were common to much of the Nile's drainage basin in the late Pleistocene; at present much of the drainage basin is desert.

On the basis mainly of bathymetric data, previous authors (Emery et al. 1966; Ryan et al. 1970) applied the term "Nile Cone" to the large submarine fan between the Egyptian and Israeli coasts extending to the Mediterranean Ridge. The seismic profiles from Ross and Uchupi (1977) and in this chapter suggest that the fan can be divided into two distinct physiographic features: the Nile Cone (in the west) and the Levant Platform (in the east). The combined area of these two features is about 70,000 km^2.

Prior to the study reported here, little was known about the structural-stratigraphic configuration of the area,

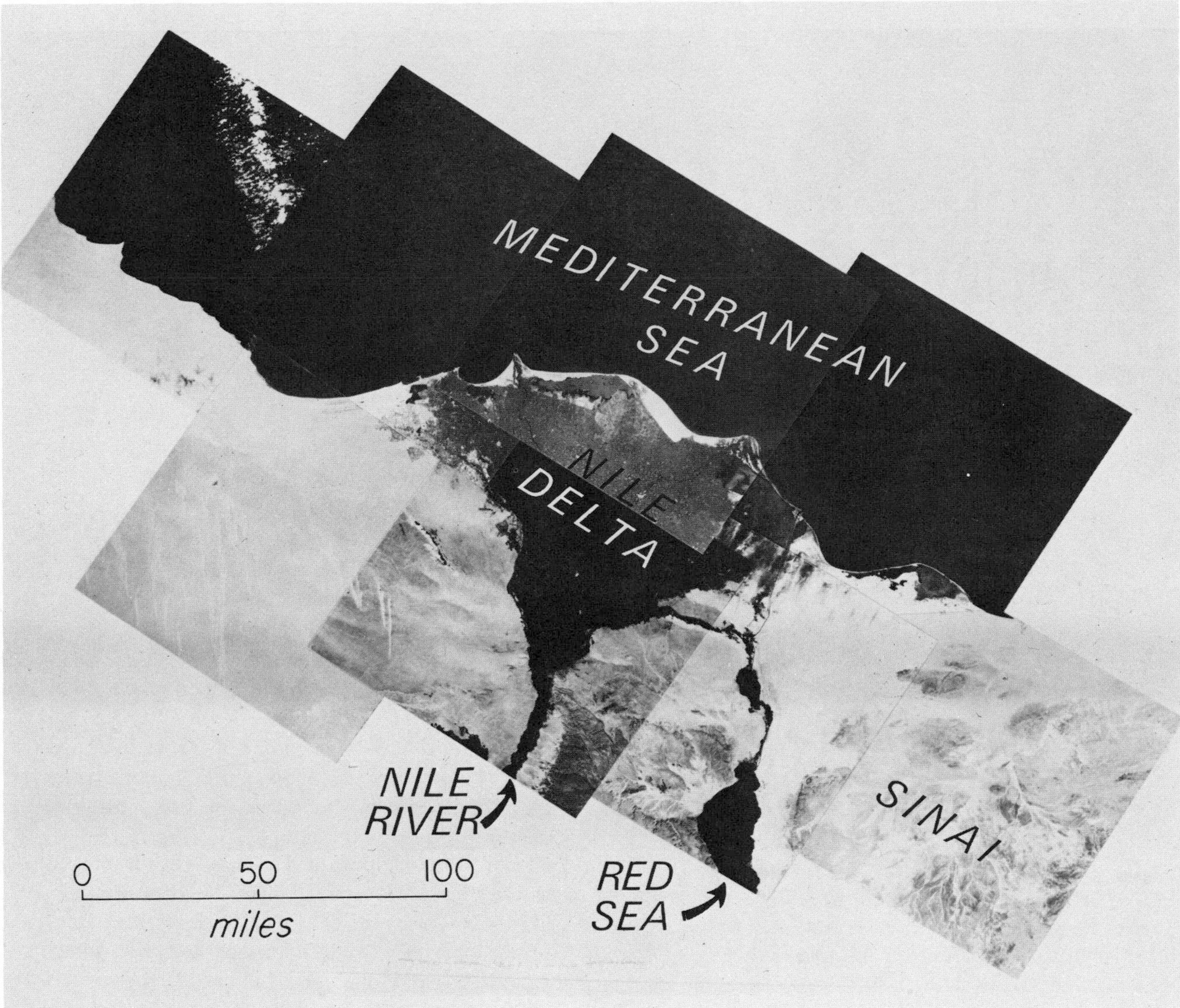

Figure 18-1. Satellite composite of the Nile Delta region.

although there are some excellent studies of the recent offshore sedimentary regime (for example, McCoy 1974; Maldonado and Stanley 1976; Stanley and Maldonado 1977 and Chapter 17 of this volume) and structure (see Finetti and Morelli 1973; Neev et al. 1973; Morelli 1975). In this chapter we present new data on the near-shore sedimentation, physiography, and shallow structure of the Levant Platform and Nile Cone – the largest submarine fan in the Mediterranean Sea. The important aspect of our study concerns the interrelationship of tectonics and sedimentation within a relatively restricted basin.

METHODS

The data were collected during cruise 119 of the R/V *Chain* (February-March 1975). The expedition and subsequent analysis were funded by the National Science Foundation (Grant No. DES 74-13212) and by the Office of International Programs of the National Science Foundation (Grant No. OIP 75-02516).

Two legs were made during cruise 119; the first obtained 32 seismic profiles, 27 oblique reflection-refraction sonobuoy profiles, magnetic data, and 3.5 kHz bathymetric data. The second leg was mainly for coring, water and suspended matter studies, and 12 kHz echo sounding.

The echo soundings (both 3.5 kHz and 12 kHz) were recorded on a Precision Graphic Recorder generally at a 100 or 200 fathom sweep. Depths were recorded at five-minute intervals and at changes in slope. The data were digitized and corrected for variations in sound velocity using Matthews (1939) tables and plotted along the navigation track by computer. A bathymetric chart was

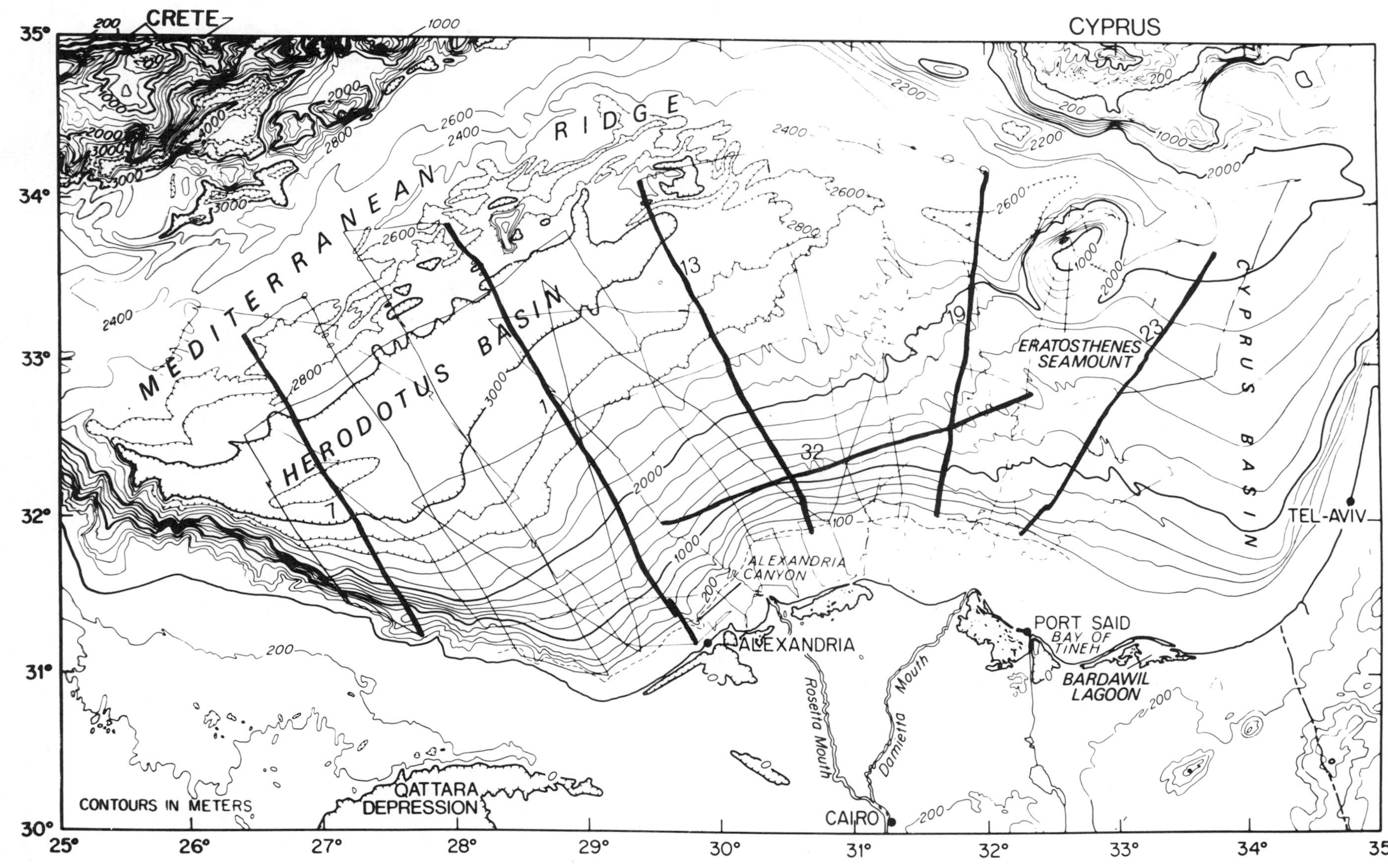

Figure 18-2. Bathymetry of the continental margin and adjacent deep sea in the eastern Mediterranean. (Source: Ross and Uchupi 1977.) Note: Map was compiled using data from R/V *Chain* Cruise 119 (cruise tracks indicated) supplemented by data from charts of Carter et al. (1972). Contours in meters corrected for sound velocity using Matthews (1939) tables. Location of seismic profiles shown in Figures 18-7 and 18-8 are indicated.

prepared using these data (Figure 18-2) and those of previous cruises. Navigation was principally by satellite, supplemented by star sights and radar.

Seismic profiles were obtained by means of a 300 cubic inch air gun (4.9 liters) charged to between 1300 to 1500 psi and fired every 10 or 12 seconds. The returning signals were summed, filtered at between 15 to 160 cycles and recorded on an X-Y recorder. The profiles were analyzed by accentuating the various reflectors on copies of the originals and reducing them, after correcting for variations in ship speed (usually between 11 to 14 km/hour).

About one hundred bottom samples were collected from the continental shelf by grab and underway samplers, as well as ten gravity cores. From the surrounding deep-sea floor we collected thirty four piston cores and occupied thirty hydrographic stations where near-bottom suspended matter was collected. Over the shelf we collected surface suspensates at sixty stations and near-bottom suspensates at fifteen stations.

GENERAL PHYSIOGRAPHY

Coastal Region

Within historical time the Nile appears to have had up to seven main distributaries, only two of which, the Damietta and Rosetta branches, are still active. Even these two do not always reach the sea, due to the construction of coastal barrages that are rarely opened. The other five branches have silted up, in part due to human activities. The main sedimentary provinces of the present delta are shown in Figure 18-3.

The coastal region of Egypt can be divided from west to east into four main features: the Western or Libyan Desert, the Nile Valley and Delta, the Eastern or Arabian Desert, and the Sinai Plateau or Peninsula (Figure 18-4). The western end of the delta is near Alexandria, whereas the eastern limit lies just east of the Suez Canal at the Bay of Tineh (see Figure 18-2).

Offshore Region

The principal physiographic features seaward of the Nile Delta are: the continental shelf, continental slope, continental rise, and farther seaward, the Eratosthenes and Herodotus Abyssal Plains, Mediterranean Ridge, and the Hellenic Arc. The continental rise can be divided into two major features: the Nile Cone (Rosetta Fan in Chapter 17) and the Levant Platform.

Continental Shelf. The continental shelves bordering the eastern Mediterranean are generally narrow, about 20 km or less in width. The exception is off the Nile, where between the Rosetta mouth and the Bardawil Lagoon, the shelf ranges in width from 48 to 64 km; farther east off Sinai it narrows to about 40 km (see Figure 18-2).

The Egyptian shelf can be divided into an inner and outer segment (Sestini 1976). The inner part extends from the shoreline out to a depth of about 36 m and consists of (1) a coastal slope, 2.5 to 7 km wide, that terminates on (2) a flat or gently sloping surface (upper terrace) 5 to 20 km wide that reaches a depth of 18 to 27 m. Seaward of the upper terrace, the seabed slopes gently down to another terrace at a depth of 45 m to 50 m, west of 31°E longitude and a series of terraces at depths ranging from 75 m to 100 m, east of this longitude. There is also a series of small linear ridges or highs off Alexandria (Sestini 1976) at depths of 9 m, 18 m, and 37 m. All these topographic features on the shelf are apparently relict, having been formed during or prior to the Holocene transgression. Morphologic features related to the present sedimentological regime are restricted to areas immediately off the Rosetta and Damietta branches and to the inner slope in the Bay of Tineh region.

The depth of the shelf edge ranges from 75 m to 90 m west of Alexandria, but deepens to 150 m to 265 m between the Rosetta and Damietta branches. This deepening is clearly due to down-warping by the weight of the thick section of recent sediments. In this area the post-Miocene Nile sediments reach their greatest thickness (Ross and Uchupi 1977). To the east of this area the shelf edge shoals to depths between 130 m and 90 m.

Continental Slope. The continental slope has its maximum relief off the Western Desert region of Egypt where it ranges in width from 34 km to 56 km and in depth from 75 m to 2,400 m. In this region the topography is fairly irregular and numerous gullies are found. Acoustic returns from 3.5 kHz echo-sounding records show a strong surface reflector here with little subbottom penetration (Figure 18-5).

In the area immediately seaward of the delta the slope is relatively smooth and about 20 km wide. The slope here has little relief and only extends to depths of 300 m to 600 m. Only one significant submarine canyon, Alexandria Canyon, appears to cross the slope and upper rise. The 3.5 kHz records from this portion of the slope show a fairly well-developed stratification with many closely spaced normal faults. East of the delta, depths and morphologic characteristics of the slope are intermediate to those of the Nile Delta and the western area.

Continental Rise. The continental rise off the Nile is a broad sedimentary feature, basically a submarine fan, that can be divided into two principal subprovinces: the Nile Cone and the Levant Platform (a new term applied to this portion of the continental rise). To the east and west of these features (see Figure 18-4), a more typical continental rise occurs.

The Nile Cone has gentle to hummocky relief except where salt intrusions have deformed sedimentary strata at the toe of the cone. The Levant Platform is distin-

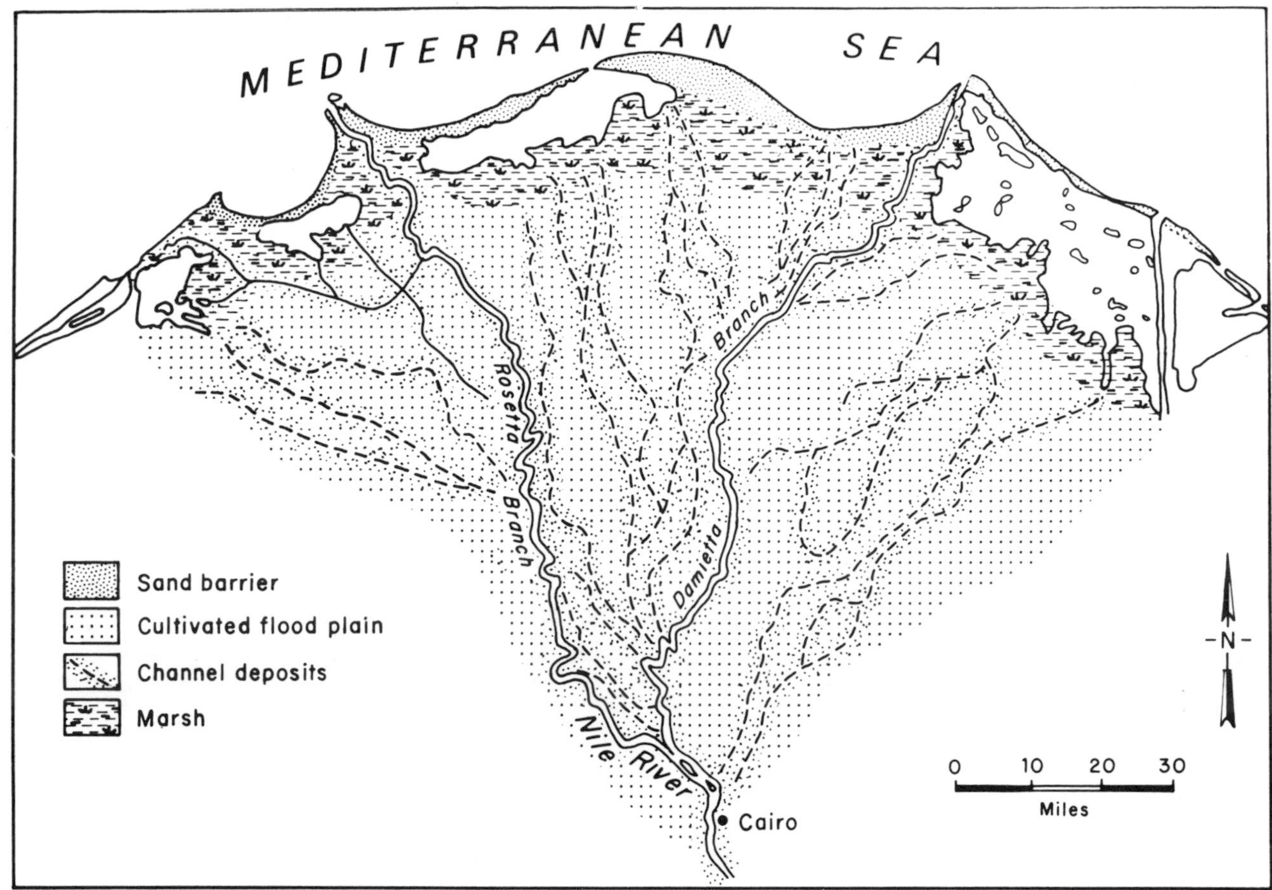

Figure 18-3. Main geomorphic features of the Nile River Delta. (Source: Adapted from Wright and Coleman 1973.)

guished by its more deformed and uplifted character. As shown in a later section, these differences are strongly controlled by salt tectonics affecting the underlying Upper Miocene (Messinian) evaporites.

The continental rise is well developed with gentle relief to the east of the Levant Platform in the Cyprus Basin. To the west of the Nile Cone the continental rise is relatively narrow and poorly developed.

Previous workers have noted the topographic difference between the Nile Cone and Levant Platform and the similarity of the cone to the continental rise in the Cyprus Basin (see Figures 18-2 and 18-4). These findings led Emery et al. (1966) to suggest that there were two fans, the Rosetta and Damietta Fans, separated by the irregular topography of what we call the Levant Platform. Our seismic profiles (see Figures 18-7 and 18-9) clearly show that both the Nile Cone and Levant Platform consist of Nile-derived sediments and can be distinguished by their underlying structure.

To the northeast of the Levant Platform is the Eratosthenes Seamount, a broad topographic high rising about 1 km above the continental rise. It is separated from the platform by a 50 km wide channel. Seismic profiles from the seamount show that it has a blocky, probably fault-controlled aspect.

Abyssal Plains. There are two abyssal plains in the area: the Herodotus in the west and the Eratosthenes in the east. They are separated by the Nile Cone, where it extends out to the Mediterranean Ridge. The Eratosthenes Abyssal Plain, the smaller of the two, has a maximum depth somewhat in excess of 2,600 m. The Herodotus Abyssal Plain parallels the Mediterranean Ridge and slopes gently down to the west from a depth of 3,070 m at its eastern end to a depth of 3,100 m at its western end.

SEDIMENTS

Shelf Sediments

As can be seen in Figure 18-6, in the area off Alexandria and to the west, the continental shelf is mantled by typical North African carbonate sands and muds (El Wakeel et al. 1974; El Sayed 1974; Summerhayes et al. 1976). Off

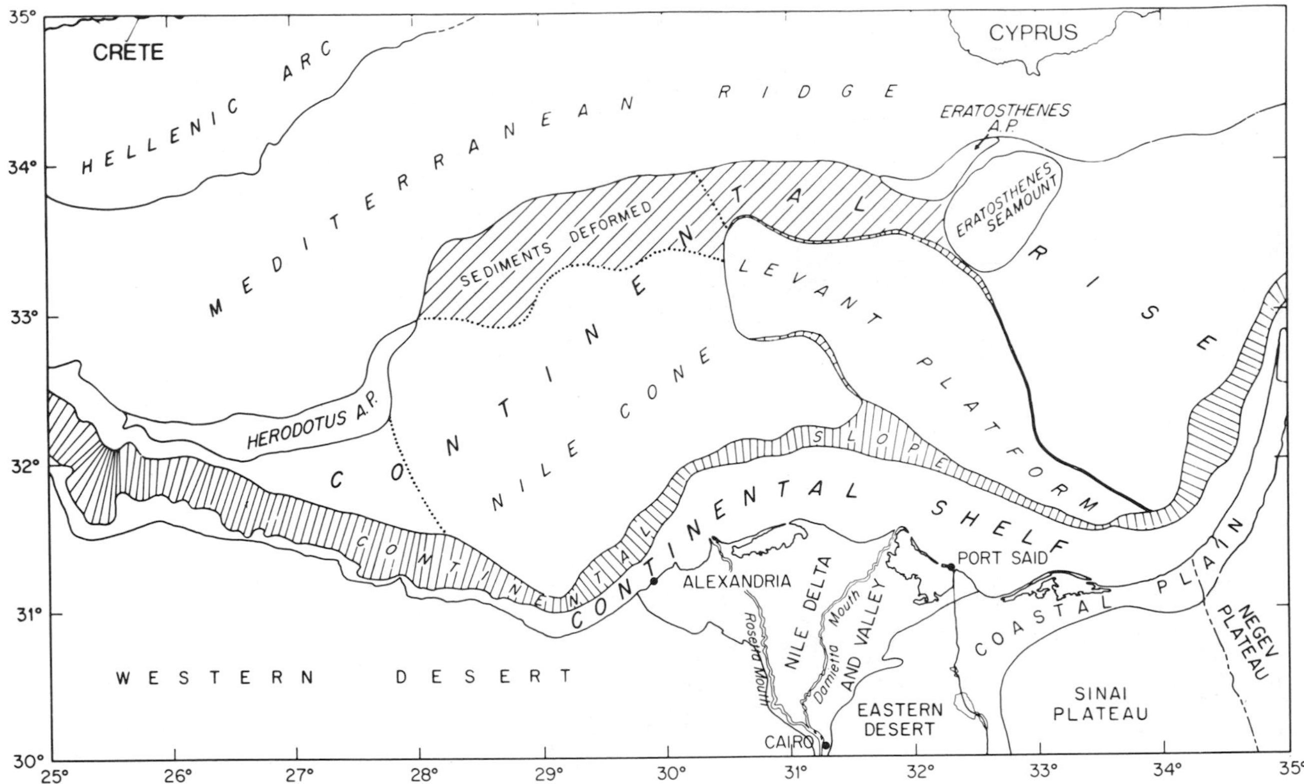

Figure 18-4. Physiographic provinces of coastal region and adjacent sea floor. (Source: Ross and Uchupi 1977.)

Egypt, these sediments are biodetrital mixtures of coralline algae, mollusks, benthonic foraminifera, echinoid remains, and fecal pellets, with the addition of planktonic foraminifera, pteropods, and bryozoa near the shelf edge. The sand fraction of middle shelf sediments is dominated by carbonate pelletoids.

As also shown in Figure 18-6, in the area off Abu Quir Bay (just east of Alexandria), there is a dramatic facies change to terrigenous sands near the coast and terrigenous silty muds further offshore (El Wakeel et al. 1974; Summerhayes et al. 1976). Off the Rosetta branch of the Nile, terrigenous muds cover the entire shelf; further east they are reduced in extent and form a middle shelf mud belt that appears to be more or less continuous. Surveys by El Wakeel et al. (1974) and Misdorp and Sestini (1976) show that terrigenous muds also occur close inshore off Damietta and on the middle shelf east of Damietta. Terrigenous hemipelagic muds cover the outermost shelf and uppermost continental slope off the delta area. These deeper deposits are separated from the muds of the middle shelf by slightly muddy carbonate sands that consist mainly of encrusting and/or branching coralline algae, with subordinate amounts of mollusks and bryozoa. Branching coralline algae tend to predominate west of Lake Burullus (east of Rosetta); encrusting coralline algae predominate to the east and form not only a coarse blanket of sediment, but also numerous algal reefs (Summerhayes et al. 1976).

There are two main types of terrigenous sand. The dominant one is a fine, dark, arkosic to subarkosic subangular sand, colored by abundant heavy minerals and flakes of mica. The other sand, which is most common on the beaches near Lake Burullus and seaward, is a coarse, usually iron-stained, rounded to subrounded and highly quartzose sand with no mica and less than 1 percent of heavy minerals. In the muds, the clay minerals are dominated by montmorillonite (ca. 75 percent) with subordinate illite and kaolinite (Stoffers et al. in preparation).

Offshore and Submarine Fan Sediments

Together, the Levant Platform and Nile Cone form one of the world's largest submarine fans (see Chapter 17 in this volume). Drilling through the distal part of the fan, near the edge of the Herodotus Abyssal Plain (DSDP Site 131), and on the Mediterranean Ridge, which consists of uplifted fan sediments (DSDP Site 130), shows that the bulk of the fan probably consists of terrigenous sediments carried by the Nile (Ryan, Hsü et al. 1973). According to Bartolini et al. (1975) most of these distal sediments are low-carbonate turbidites, with typical Nile mineralogy,

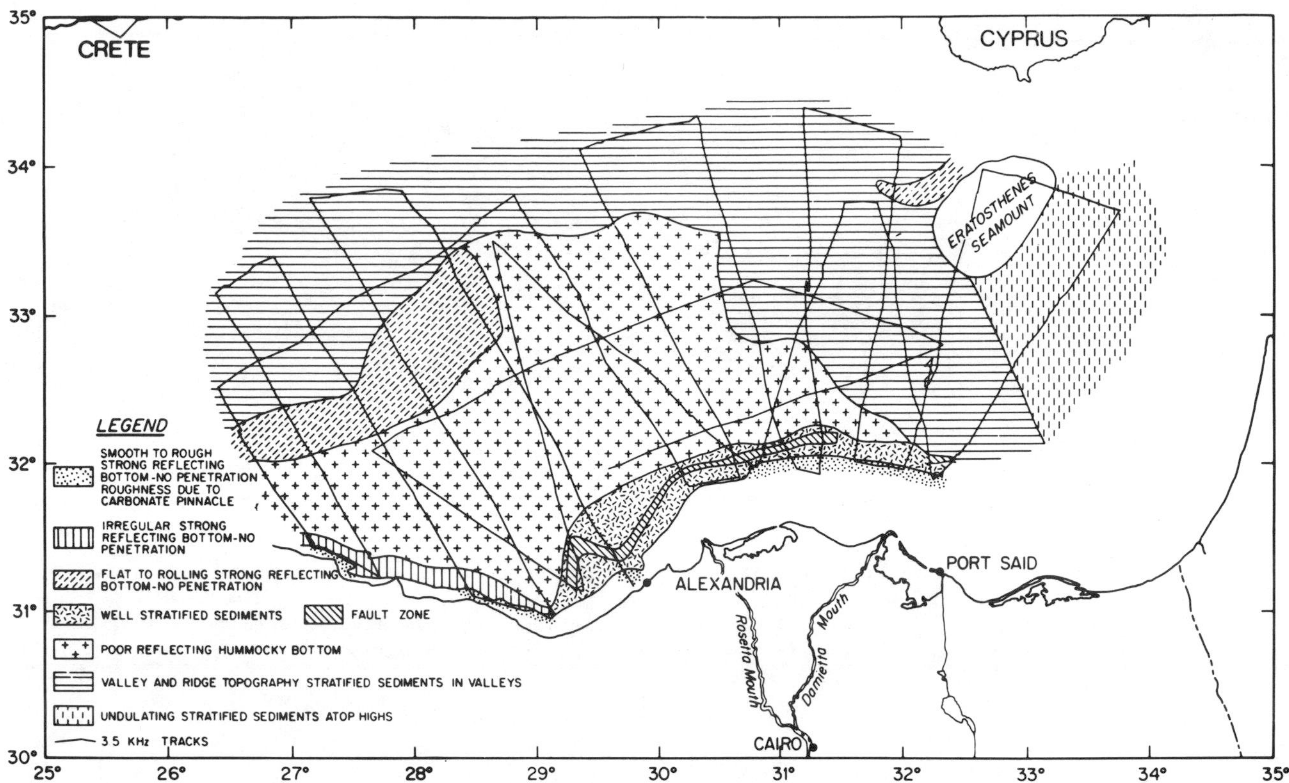

Figure 18-5. Morphological features of the sea floor as determined from 3.5 kHz records. (Source: Ross and Uchupi 1977.)

interbedded with pelagic nanno-oozes. Pelagic foraminiferal marl ooze containing some sapropels and volcanic ash covers the uplifted Nile sediment on the Mediterranean Ridge (Site 130). Perhaps more important in a historical sense is the fact that Site 130 bottomed in lower Pleistocene pelagic nanno-oozes, thereby signifying that fan development is probably a Quaternary phenomenon (Bartolini et al. 1975). Site 131 also terminated in a Quaternary foraminiferal limestone. The main difference between Nile sediments from the two sites is the abundance of sand in the hole nearest the delta (131), which included several thick, coarse-grained, massive sand beds that apparently are fan channel facies.

The terrigenous offshore sediments all contain substantial amounts of microcline and metamorphic rock fragments and thus reflect derivation from the Precambrian high-grade metamorphics of the Sudan-southern Egypt area. Total feldspar content in the sand fraction is rarely more than 15 percent, which led Bartolini et al. (1975) to classify most of these sediments as sublabile lithic feldspathic arenites and feldspathic subgreywackes. Heavy minerals are dominated by diopside, augite, hornblende, and epidote, and they fall into two distinct suites indicating both volcanic and metamorphic sources. Montmorillonite dominates the clay fraction and together with mixed-layer illite/montmorillonite makes up about 75 percent of the clay assemblage. Bartolini et al. (1975) found minor amounts of illite and chlorite, but no kaolinite in their samples. The predominance of montmorillonite in Nile sediments is well-known; however, piston core samples from the surface of the fan contain 10 to 30 percent of kaolinite (Venkatarathnam et al. 1972). Bartolini et al. (1975) explain this discrepancy by invoking eolian supply for the kaolinite.

The most recent sedimentological observations on the fan have been made by Maldonado and Stanley (1975, 1976, and Chapter 17 of this volume; Stanley and Maldonado 1977) who used fifty four assorted piston and gravity cores collected by a number of different research organizations. Most of their cores were less than 2 m long. The bulk of their stratigraphic data comes from thirteen selected cores that average 6 m long. These cores are located along several transects across the face of the fan and provide a regional picture of recent fan development. Our data are based on thirty four more piston cores with an average length of 7 m a maximum length of 12 m.

Maldonado and Stanley (1975; 1976) identify four major facies units (i) turbidite, (ii) channel, (iii) sapropel, and (iv) hemipelagic. These units are usually grouped in the following repetitive sequence or cyclothem: (i) basal, olive-grey hemipelagic muds and turbidites, (ii) sapropel sequence, (iii) pale yellowish-orange hemipelagic sequence (including calcareous oozes) and turbidites. These latter turbidites are fewer and finer than in the

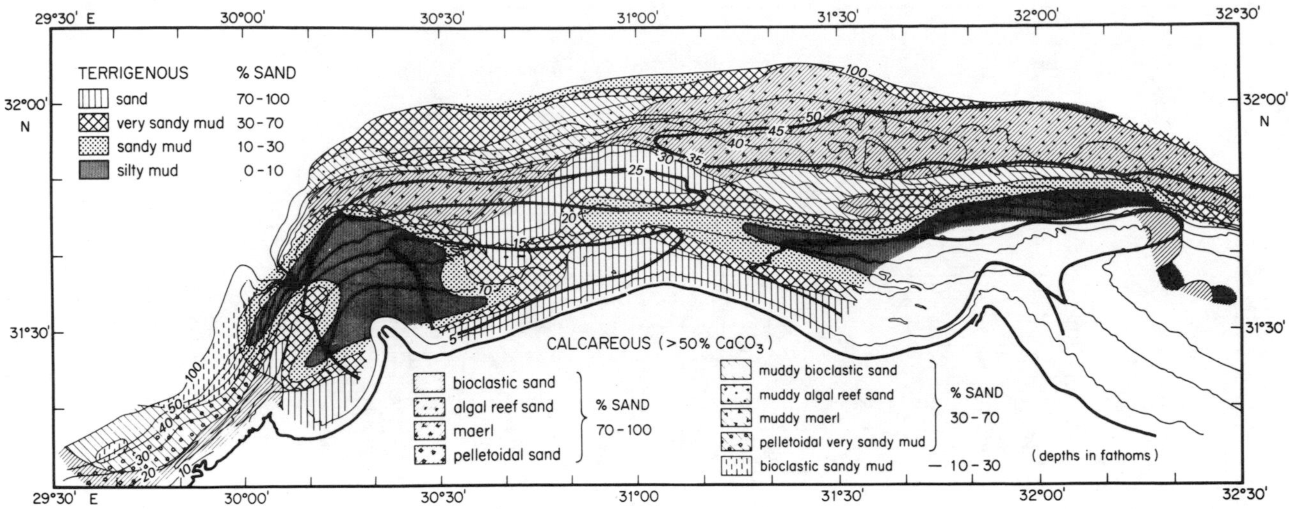

Figure 18-6. Distribution of major sedimentary facies on the Egyptian continental shelf. Note: Data are based mainly on analyses of 100 samples from *Chain* Cruise 119 (Summerhayes et al. 1977). Textural divisions are based on sand content: 0-10% = silty mud; 10-30% = sandy mud; 30-70% = very sandy mud; and 70-100% = sand. Carbonate sediments contain more than 50% $CaCO_3$. Dark lines represent outlines enclosing major terraces on the outer and inner shelves and small sedimentary cones off the two mouths of the Nile.

basal unit. Channel units consist of coarse clean terrigenous sands about 1 m thick that occur in only two adjacent cores near a submarine channel downslope from the Rosetta Canyon. The sapropels consist of thin laminae of coccolith-rich calcareous mud alternating with thicker layers of terrigenous mud low in carbonate; the organic content in the sapropels averages 15 percent organic matter. The modern surface sediment belongs in unit (iii), is pale yellow-orange in color and contains about 30 to 60 percent carbonate except on the upper part of the fan where carbonate content may drop to 25 percent (Venkatarathnam et al. 1972; *Chain* cruise 119 data, Peter Stoffers, personal communication, 1976).

Available mineralogical data from surface samples show that the prevailing circulation transports montmorillonitic Nile clay east and northeast towards Cyprus and Lebanon. Surface sediments from the western parts of the fan, particularly the lower fan, contain substantial kaolinite, probably blown into the area from the western desert by the Khamasin, which is a hot, dry, strong wind that blows for short periods from the south during the spring and summer and causes widespread haze due to its content of airborne dust (Venkatarathnam et al. 1972; *Chain* 119 data, Peter Stoffers, personal communication, 1976).

Another interesting feature of the *Chain* 119 cores is that several of them show increases in the salinity of interstitial waters with depth in the core (Elderfield 1975). Upward diffusion of salt from buried Messinian salt deposits appears to be the cause for the observed increase, although no obvious areal pattern was observed.

STRUCTURE

Basement

Acoustic basement on the Egyptian continental margin (Figures 18-7 and 18-8) is made up of two reflectors that Ross and Uchupi (1977) designated reflectors P and M (the term "M" was first applied by Ryan et al. 1966). Most of the inner margin is underlain by reflector P, and the rest of the margin, adjacent abyssal plains, and the Mediterranean Ridge is underlain by reflector M. In places these two reflectors are in faulted contact, but in other locations the lateral transition from one horizon to another is marked only by a gradual increase in the relief of the acoustic basement. Based on data from wells drilled on the shelf and measured compressive velocities of 4.2 to 5.0 km/sec for the strata beneath reflector P, Ross and Uchupi (1977) suggested that this reflector may be the top of a carbonate sequence of middle to late Miocene age that prograded eastward from the Western Desert during the Messinian regression (Salem 1976; Said 1973). Ross and Uchupi (1977) also suggested the possibility that reflector P may represent an erosional surface carved in pre-Messinian marls and limestones that were uplifted during the late Miocene (Mulder et al. 1975; Mulder 1973). Whereas reflector P is a relatively smooth to gently undulating reflector, M displays considerable relief, which Ross and Uchupi (1977) interpreted as being due to the plastic flow of evaporites and the collapse of the crests of some piercement structures due to solution of the evaporites. DSDP drilling in the Mediterranean

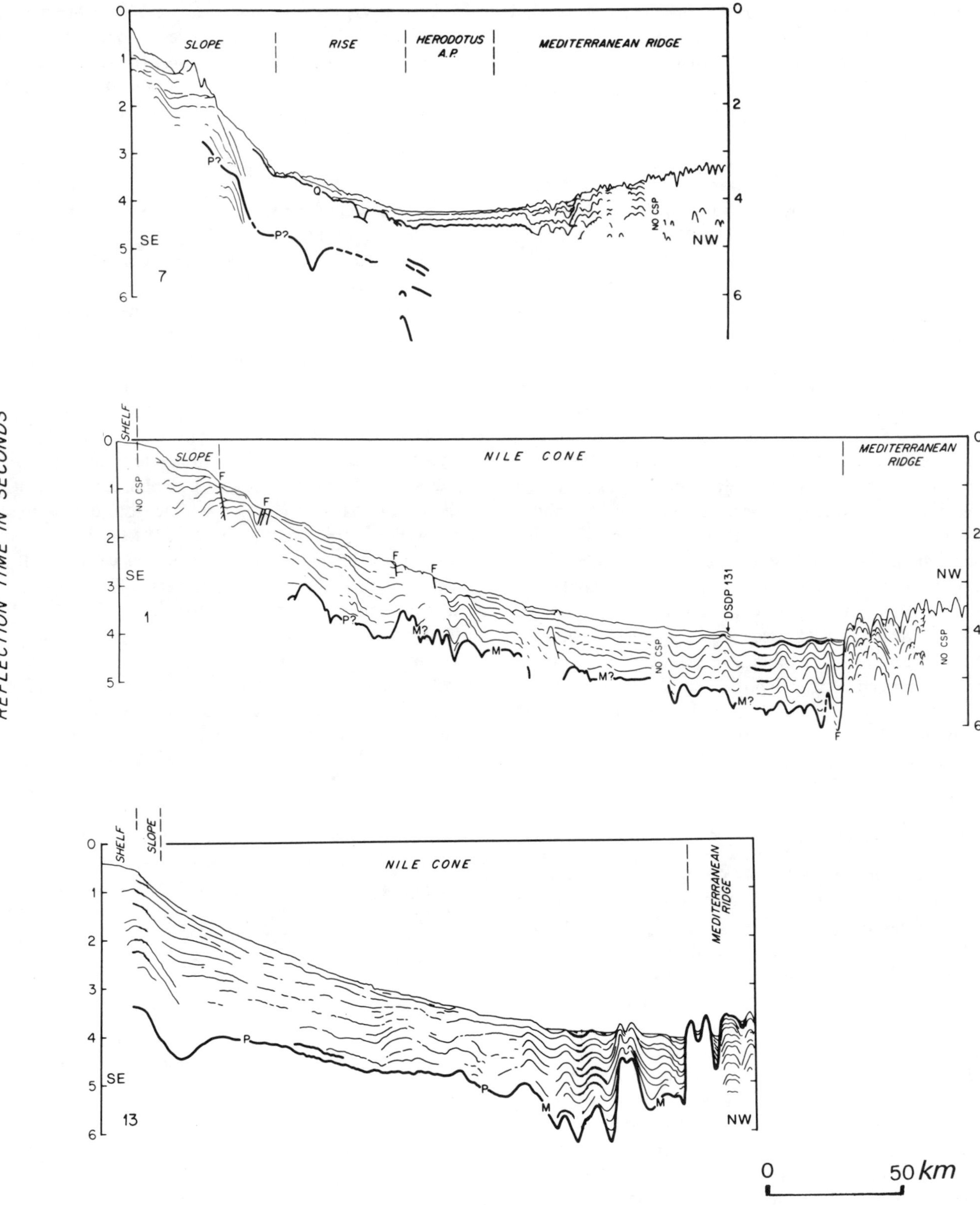

Figure 18-7. Seismic reflection profiles of the continental margin and adjacent deep sea off western Egypt. (Source: Adapted from Ross and Uchupi 1977, Figures 6, 7, and 8.) Note: For location of these profiles see Figure 18-2.

indicates that reflector M, which is common to most of the Mediterranean, delineates the top of a Messinian (late Miocene) evaporitic sequence (Ryan, Hsü et al. 1973). If the interpretation postulated by Ross and Uchupi (1977) is correct, then the sediments above these reflectors provide information on the Pliocene-Quaternary depositional history of the Egyptian margin.

The surfaces delineated by reflectors M and P slope gently seaward from an elevation at sea level near Cairo to 3.0 seconds (two-way reflection time) near the coast (Figure 18-9). Near the shelf edge the acoustic basement gradually rises slightly to form a ridge paralleling the shelf edge. Seaward of this ridge, the basement drops to a depth of about 5.0 seconds (profile 13, Figure 18-7). Whereas the segment of the acoustic basement defined by reflector P is generally smooth, the segment defined by reflector M displays considerable relief, with basement shoaling to depths of less than 2.0 seconds below sea level along the crests of some of the diapiric structures (profile 32, Figure 18-8). Sediment cover on the acoustic basement is thickest on the outer shelf north of the Nile Delta where it exceeds 3.5 seconds. On the Nile Cone itself, the sediment accumulations exceeding 3.0 seconds occur along the boundary between the cone and the Levant Platform. Ross and Uchupi (1977) estimated the volume of post-Messinian sediments, assuming a compressive velocity of 2 km/sec., to be about 387,000 km^3, which calculates to an average thickness of 1.98 km over the area and an average depositional rate of about 40 cm/1,000 years (for the last 5 million years). This rate is comparable to values determined from other sediment studies (Parker 1958; Olausson 1961; Emery et al. 1966; McCoy 1974; Stanley and Maldonado 1977).

Sediment Blanket

The sedimentary framework of the Egyptian continental margin varies considerably in an east-west direction. West of the Nile Cone (profile 7, Figure 18-7), a few poorly reflecting horizons are truncated by the continental slope. Seaward of the slope, the western rise sequence is masked by a near-surface undulating reflector, called Q, which because of its shallowness probably is within the Quaternary section. The Nile Cone (indicated by profiles 1 and 13, Figure 18-7) is better stratified, and individual reflectors can be traced laterally. Along profile 1 the upper reflectors are disrupted by small faults dipping both landward and seaward. No evidence of such faulting can be seen on profile 13. On both of these profiles, strata beneath the outer shelf and upper slope have been arched. On profile 13 this folding appears to be due either to uplift of the basement or to differential compaction above a preexisting basement high. On all these profiles, folding of the rise strata slowly increases in a seaward direction until it becomes so intense on the Mediterranean Ridge that the subbottom structure is completely obscured. Since the evaporite sequence below reflector M is involved, folding of the Mediterranean Ridge could be due to either plastic flow of the salt, crustal deformation, or both. The incorporation of Nile sediments into the Mediterranean Ridge has reduced the Nile Cone to its present dimensions. As the uppermost sediments on the ridge are involved in this deformation, folding must have taken place rather recently or is still continuing.

Deformation of the sediments, exclusive of the Mediterranean Ridge, is most intense within the Levant Platform (profiles 19, 23 and 32, Figure 18-8). The seismic profiles clearly demonstrate that the rough topography of this region, first noted by Emery et al. (1966) and later by Carter et al. (1972) and Kenyon et al. (1975), is due in part to the vertical migration of Messinian salt. Further topographic relief comes from the collapse features caused by solution of the salt beneath the crests of some diapiric structures (Figure 18-10). Additional structural complexities have arisen due to the lateral migration of the evaporites, with the northern edge of the Levant Platform representing the front of this migrating salt wedge. Structurally, this uplifted segment of the continental rise resembles the margin of the western Gulf of Mexico with a salt front along its seaward edge, complex structures due to plastic flow atop the platform, and a faulted continental slope along its southern boundary. The nature of the contact between the platform and the rest of the undisturbed Nile Cone, displayed on profile 32 (Figure 18-8) shows that the Levant Platform was formed by the intrusion of evaporites into the Nile Cone sediments and their consequent elevation. A similar uplift has also occurred to the east, between the Levant Platform and the continental rise in the Cyprus Basin. Much of the sedimentary section of the rise in the Cyprus Basin northeast of the Levant Platform appears to be folded. These undulations do not appear to be due to bottom currents, but are probably a result of basinward sliding of the strata along reflector M, concurrent with the formation of the Levant Platform. In general, the seismic profiles are indicative of a formerly much larger Nile Cone that has been reduced to its present dimensions by incorporation of the cone sediments into the Mediterranean Ridge to the north and deformation of the eastern part of the cone by salt tectonics.

DISCUSSION

Origin of Nearshore Sediments

At present, the Nile is a completely self-contained system. Barrages at the Rosetta and Damietta mouths of the river have prevented much discharge of sediment-laden fresh water into the Mediterranean since 1964, when the

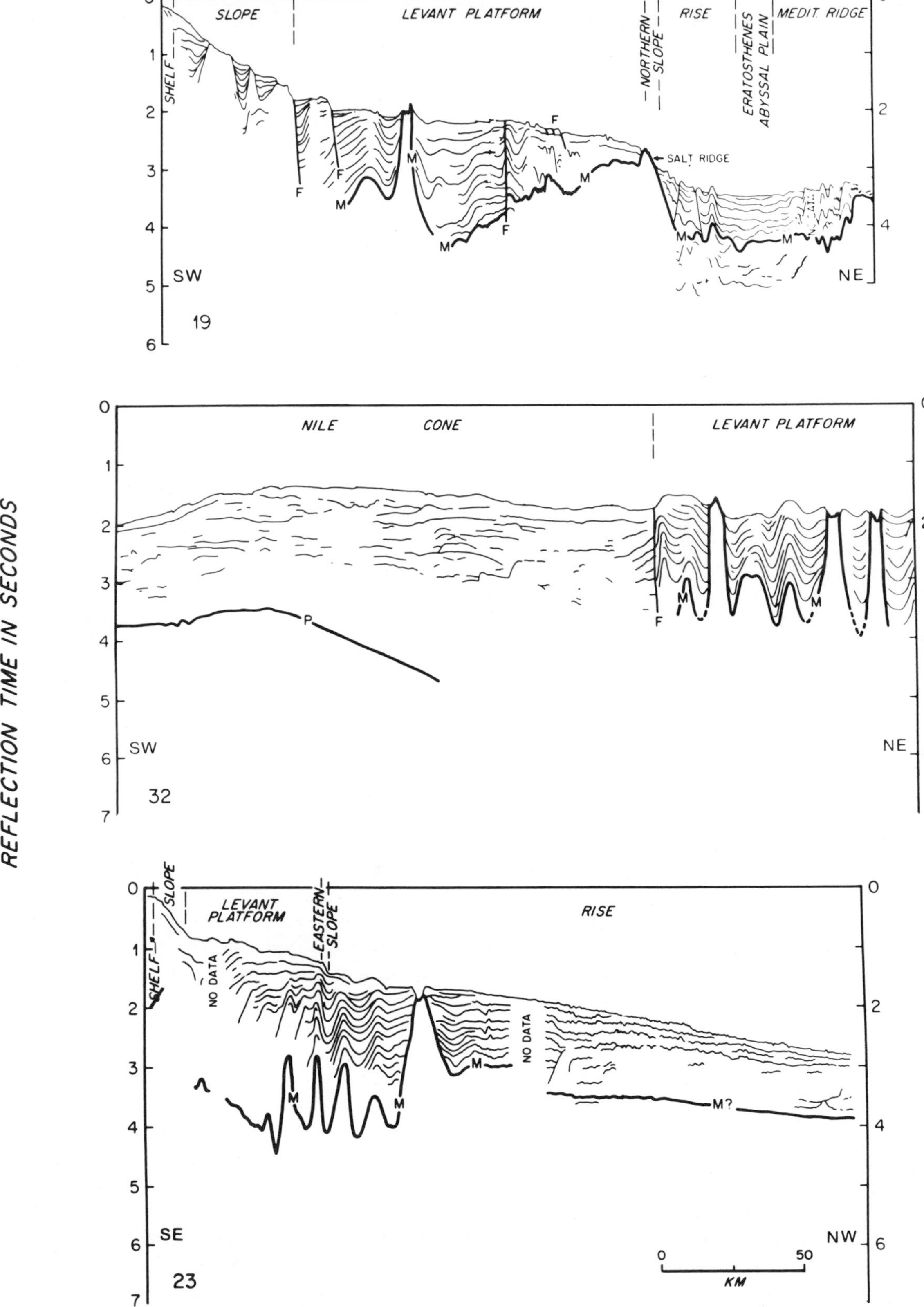

Figure 18-8. Seismic reflection profiles of the continental margin and adjacent deep sea off eastern Egypt. (Source: Adapted from Ross and Uchupi 1977, Figures 10, 11, and 12.) Note: For location of these profiles see Figure 18-2.

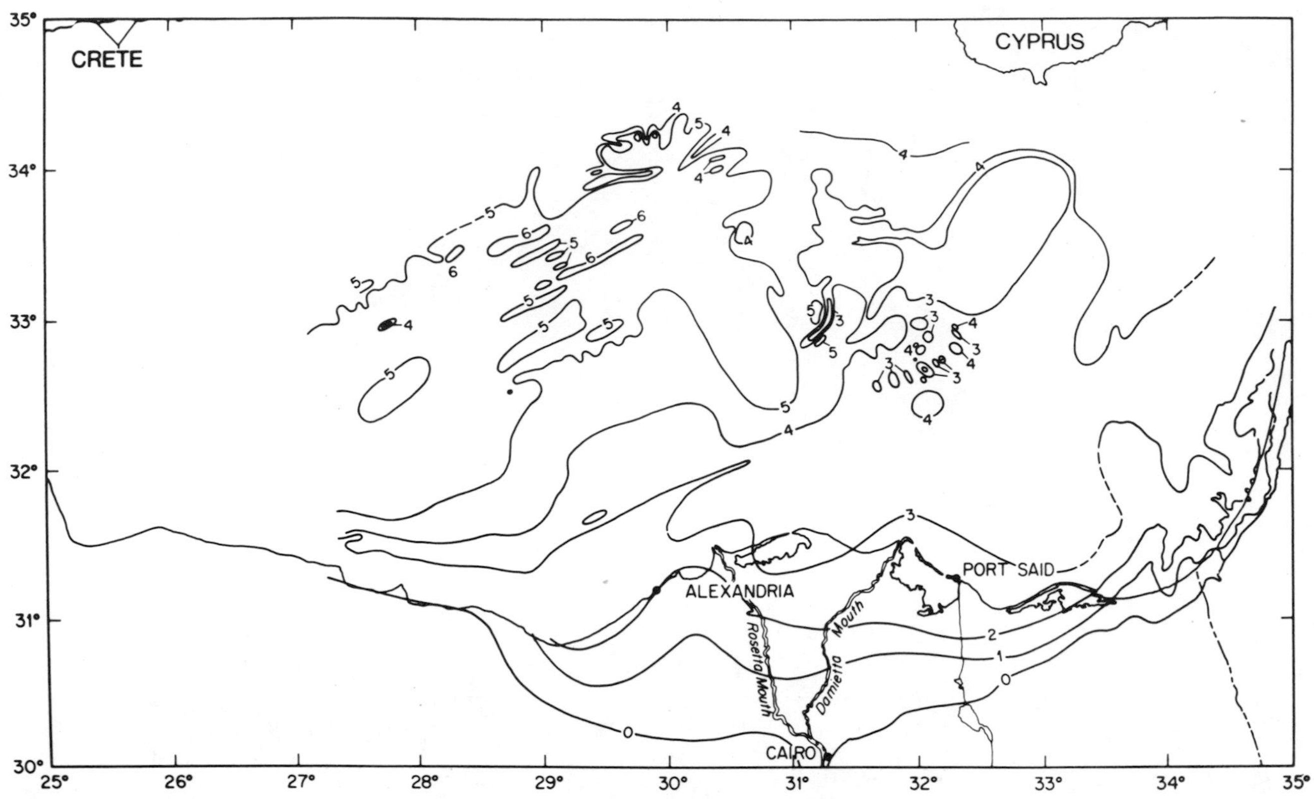

Figure 18-9. Structure map of acoustic basement (reflectors P, M and B). (Source: Ross and Uchupi 1977.) Note: Reflector B is restricted to the Eratosthenes Seamount area and probably is older than Miocene in age. Contours show depth in two-way travel time (secs.) to acoustic basement (based on data obtained during *Chain* 119 Cruise and that from Neev et al. 1973 and from Ginsburg et al. 1975.)

Aswan High Dam was completed (Aleem 1972). As a result erosion now prevails over deposition, thereby causing locally spectacular coastal retreat (UNESCO 1973, 1976) and resuspending nearshore muds in substantial quantities (Summerhayes and Marks 1976). Past hydrographic observations showed that prior to construction of the Aswan High Dam, the flood waters of the Nile were deflected east along the Egyptian coast by the prevailing surficial east-flowing currents and northwesterly winds and eventually reached the coasts of Israel and Lebanon (Aleem 1972; El Din 1974, 1977). Unfortunately no suspended load measurements were made until recently, although available data suggest that Nile sediment was carried eastward in a broad plume located more or less over the middle shelf (Emelianov and Shimkus 1972; Summerhayes et al. 1976). Studies of bottom sediments from the Cyprus Basin confirm that clays are carried northeast away from the Nile (Venkatarathnam et al. 1972; McCoy 1974). An easterly longshore current has carried Nile sand a considerable distance to reach the coast of Israel (Emery and Neev 1960).

The evolution of the sedimentary blanket on the shelf in front of the Nile Delta can be interpreted in terms of several stages (Summerhayes and Marks 1976; Summerhayes et al. 1977). When sea level was at its lowest, some 17,000 years ago, the Nile discharged its sediment load through channels and canyons directly onto the continental slope. As sea level rose (17,000 to 5,000 years ago), the coastal system became submerged, and fluvial muds were trapped in coastal lagoons and in drowned river mouths that migrated landward with the transgressing sea. At this time outer shelf waters were clear, thereby allowing extensive algal growth. As sea level stabilized at more or less its present level, continued river discharge caused the shoreline to begin to prograde again. Sediments were discharged simultaneously through as many as six or seven different distributaries, according to chroniclers like Herodotus. Offshore, these sediments spread out and moved eastwards forming the prodeltaic mud facies of the middle shelf, landward of the still growing algal deposits of the outer shelf. Repetition of this pattern during the Pleistocene is indicated by the mixture on the outer shelf of fresh (live) and relict (iron-stained and reworked) coralline algal remains.

Increasing control of the river by man resulted in reduction of the number of active distributaries to two (Rosetta and Damietta branches) and the localization of coastal progradation at these two mouths. Between the two

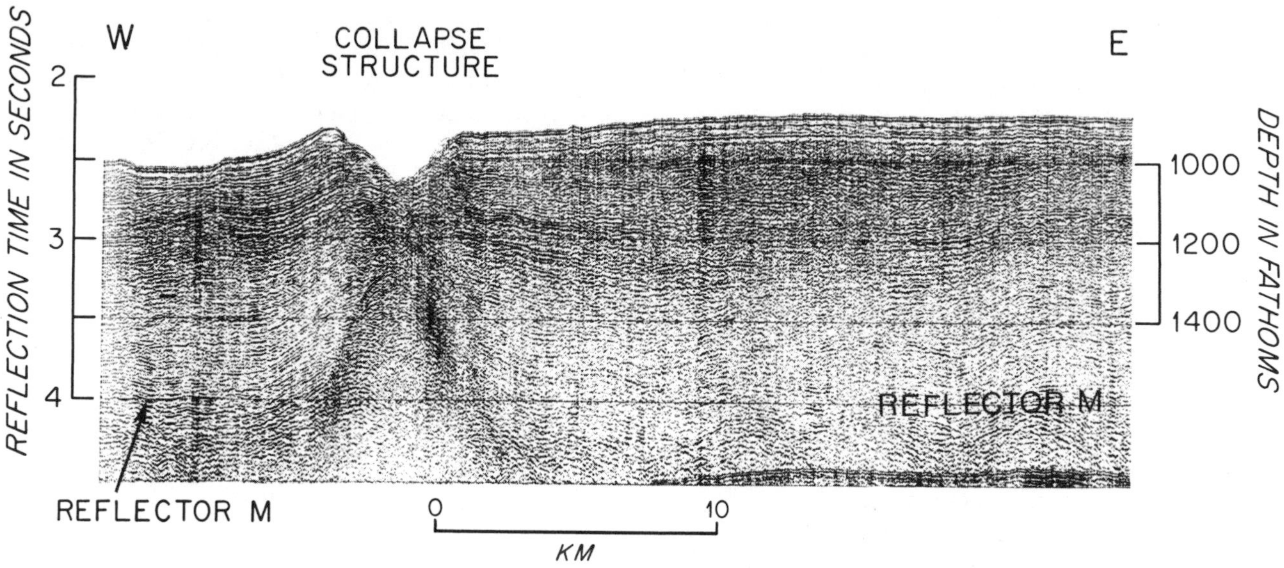

Figure 18-10. Collapse or solution structure, a common feature on the Levant Plateau. Note: This structure is probably due to upward motion of salt (see tilted layers around structure) followed by solution resulting in the depression.

mouths and in Abu Quir Bay, there was corresponding coastal retreat and destruction of old settlements at the mouths of former distributaries. Cessation of discharge in 1964 resulted in erosion that focused on the Rosetta and Damietta promontories (Orlova and Zenkovitch 1974; UNESCO 1973, 1976). This erosion should continue until a wave-straightened shore is approached, after which there may be gradual coastal retreat accompanying gradual subsidence.

Origin of Offshore Sediments

The cyclothemic stratigraphic pattern of Nile Fan cores can be interpreted in terms of the climatic controls that were imposed on the eastern Mediterranean, and the Nile Delta, during the late Pleistocene (cf. Maldonado and Stanley 1976, 1977; Stanley and Maldonado 1977; McCoy 1974). During glacial maxima, sea level was low, and the Nile River discharged directly onto the upper fan through canyons and shelf-edge channels, thereby creating widespread deposition of interchannel turbidites and hemipelagic muds. As the climate warmed and sea level rose, terrigenous sedimentation on the fan became much reduced because the transgressing sea drowned the coastline and lower courses of river valleys, thereby trapping terrigenous material near shore. Between 9,000 and 7,500 years B. P. the terrigenous input declined but the input of organic matter did not, and since the Mediterranean bottom water was anaerobic at this time (at depths below about 500 m), the organic-rich black muds termed *sapropels* accumulated over the fan and adjacent features. The upper sequence, consisting mainly of hemipelagic mud and calcareous ooze, is postglacial and represents the period during which most Nile mud has been trapped in the terrestrial delta or on the continental shelf. Much the same sequence of events is recognized on other fans, for example, the Amazon Cone (Milliman et al. 1975; Damuth and Kumar 1975), except that the sapropels are lacking because the western equatorial Atlantic was not a closed basin like the Mediterranean during glacial maxima.

Development of the Nile Cone and Levant Platform

The evolution of the Nile Cone and Levant Platform as reconstructed from our seismic reflection profiles, other geophysical measurements, and sediment cores, appears to be as follows. In the Messinian, the Mediterranean basin was fragmented and at least partially isolated from the Atlantic probably by the uplift of the Strait of Gibraltar. From our data we cannot determine whether this Messinian Mediterranean had depressions as deep as those at present or was a more shallow sink that has subsequently subsided to its present depths. This isolation and fragmentation led to dessication of the Mediterranean and the deposition of the Messinian evaporites. During this regression the Nile carved a deep trough (Said 1973), but the present location of the sediments deposited by this ancient Nile cannot be recognized from our seismic profiles. During this time the tectonic setting off Egypt consisted of a positive element at the present site of the Nile Cone, the surface of which is now represented by reflector P. The present site of the Levant Platform on the other hand, was an evaporitic sink – that is, a basin that extended around the northern periphery of the high, along much of the present Mediterranean Ridge and the remaining continental rise area.

During the early Pliocene, the sea transgressed through the Nile Valley, well into the interior of Egypt as far south as Aswan. Following the transgression, but still in the early Pliocene, the sea slowly regressed and sediment progradation kept pace with the regression. By the early Pleistocene the shoreline had reached its present position and deltaic and cone deposition began. Sediments filled the sink occupied by the present Levant Platform and prograded westward, burying the platform. Continued sediment deposition ultimately resulted in the formation of a broad fan extending from the Cyprus Basin in the east to the present western margin of the Nile Cone and northward beyond the present limit of the fan to the region partially occupied by the Mediterranean Ridge. As the sediment sequence thickened, vertical and lateral plastic flow of the Messinian evaporites uplifted the eastern part of the fan to form the Levant Platform. Lateral migration of the Messinian evaporites from under the platform was due to overburden pressure exerted by the thick sediment accumulation on the outer shelf. The present northern margin of the Levant Platform represents the front of this advancing evaporitic wedge. Along the seaward (northern) edge of the fan a combination of the vertical migration of salt and the horizontal compression associated with the subduction of the African Plate by the Eurasian Plate folded the fan sediments and incorporated some of them into the Mediterranean Ridge. Formation of these two highs composed of Nile-derived sediments, the Mediterranean Ridge to the north and the Levant Platform to the east, has reduced the fan to its present dimensions to form the Nile Cone.

ACKNOWLEDGMENTS

The opportunity to work off the Egyptian Coast was greatly aided by the Egyptian Academy of Science and Technology. Dr. E. M. El Shazly, besides being an author of this chapter, was the principal coordinator of the Egyptian phase of the program and merits special acknowledgment from his four co-authors.

The authors also wish to thank the following participants in the expedition: Ed Boyle and Dave Drummond of M.I.T., Harry Elderfield of the University of Leeds, Andrés Maldonado of the Instituto "Jaime Almera" C.S.I.C., Barcelona, and Peter Stoffers of the University of Heidelberg. Participants from the Woods Hole Oceanographic Institution included: Syed Ali, Alan Driscoll, Jeff Ellis, Mark Flora, Francy Forrestel, John Forrestel, C. W. Grant, Robert Groman, Kurt Holmes, Robert McGirr, Lois Toner, and Cliff Winget. Mr. Abraham Maghib merits special thanks for aiding the land-based operations. The authors are especially grateful to the captain (M. Palmeiri), officers, and crew of the R/V *Chain* for their help and patience during the expedition. Eight Egyptian scientists also participated in the expedition: Dr. A. A. Amar of the Egyptian Atomic Energy Commission; Dr. N. M. H. Ali of Ain Shams University; Dr. M. G. Barakat of Cairo University; G. A. Dabbour of the Egyptian Atomic Energy Commission; Dr. O. El Badry of the Petroleum Research Institute; Dr. R. M. Kebeasy of Helwan Observatory; Dr. M. A. Mamdouh of UNESCO; and Dr. R. A. Schereef of Mansura University. The manuscript was reviewed by Drs. A. Maldonado, D. Milliman, A. J. Erickson, and D. J. Stanley.

REFERENCES

Aleem, A. A., 1972. Effect of river outflow management on marine life. *Mar. Biol.* 15: 200-08.

Bartolini, C., Malesani, P. G., Manetti, P., and Wezel, F. C., 1975. Sedimentology and petrology of Quaternary sediments from the Hellenic Trench, Mediterranean Ridge and the Nile Cone. DSDP, Leg 13 cores. *Sediment.* 22: 205-36.

Carter, T. G., Flanagan, J. P., Jones, C. R., Marchant, J. P., Murchison, R. R., Rebman, J. H., Sylvester, J. C., and Whitney, J. C., 1972. A new bathymetric chart and physiography of the Mediterranean Sea. In: D. J. Stanley (ed.), *The Mediterranean Sea – A Natural Sedimentation Laboratory.* Dowden, Hutchinson & Ross, Stroudsburg, Pa., pp. 1-23.

Damuth, J. E., and Kumar, N., 1975. Amazon Cone: Morphology, sediments, age, and growth pattern. *Geol. Soc. Amer. Bull.*, 86: 863-78.

Elderfield, H., 1975. The chemistry of pore fluids in sediments of the Nile Cone. *19th Ann. Rep. Res. Inst. Afr. Geol., Univ. Leeds,* pp. 3-5.

El Din, S. H. Sharaf, 1974. Longshore sand transport in the surf zone along the Mediterranean Egyptian coast. *Limnol. Oceanog.* 19: 182-89.

_____, 1977. Effect of the Aswan High Dam on the Nile flood and on the estuarine and coastal circulation pattern along the Mediterranean Egyptian coast. *Limnol. Oceanog.*, 22: 194-207.

El Sayed, M. K., 1974. *Littoral and shallow water deposits of the continental shelf area of Egypt, off Alexandria.* Unpublished M. Sc. Thesis, University of Alexandria, Egypt, 150 pp.

El Wakeel, S. K., Abdou, H. F., and Mohamed, M. A., 1974. Texture and distribution of recent marine sediments of the continental shelf off the Nile Delta. *Jour. Geol. Soc. Iraq.*, 7: 15-34.

Emery, K. O., and Neev, D., 1960. Mediterranean beaches of Israel. *Israel Min. Ag. Div. Fish., Sea Fish. Res. Stn. Bull.*, 28, 22 pp.

_____, Heezen, B. C., and Allan, T. D., 1966. Bathymetry of the eastern Mediterranean Sea. *Deep-Sea Res.*, 13: 173-92.

Emelianov, E. M., and Shimkus, K. M., 1972. Suspended matter in the Mediterranean Sea. In: D. J. Stanley (ed.), *The Mediterranean Sea – A Natural Sedimentation Laboratory.* Dowden, Hutchinson & Ross, Stroudsburg, Pa. pp. 417-40.

Finetti, I., and Morelli, C., 1973. Geophysical exploration of the Mediterranean Sea. *Boll. Geof. Teor. Appl.*, 15: 263-341.

Ginzburg, A., Cohen, S. S., Hay-Roe, H., and Rosenzweig, A., 1975. Geology of Mediterranean shelf of Israel. *Amer. Assoc. Petrol. Geol. Bull.*, 59: 2142-60.

Kenyon, N. H., Stride, A. H., and Belderson, R. H., 1975. Plan views of active faults and other features on the lower Nile Cone. *Geol. Soc. Amer. Bull.*, 86: 1733-39.

Maldonado, A., and Stanley, D. J., 1975. Nile Cone lithofacies and definition of sediment sequences. *Proc. IX Intern. Cong. Sediment.*, Nice, 10 pp.

——, and Stanley, D. J., 1976. The Nile Cone: Submarine fan development by cyclic sedimentation. *Mar. Geol.*, 20: 27-40.

Matthews, D. H., 1939. Tables of velocity of sound in pure water and sea water. *Admiralty Hydrog. Dept.*, London, 32 pp.

McCoy, F. W., Jr., 1974. *Late Quaternary sedimentation in the eastern Mediterranean Sea.* Unpublished Ph.D. Thesis, Harvard University, Cambridge, Mass., 132 pp.

Milliman, J. D., Summerhayes, C. P., and Barretto, H. T., 1975. Quaternary sedimentation on the Amazon continental margin: A model. *Geol. Soc. Amer. Bull.*, 86: 610-14.

Misdorp, R. and Sestini, G., 1976. The Nile Delta: Main features of the continental shelf topography. *Proc. UNESCO Seminar on Nile Delta Sedimentology.*, Alexandria, October 1975.

Morelli, C., 1975. Geophysics of the Mediterranean. *Newsletter of the Cooperative Investigations in the Mediterranean,* Sp. Issue 7, Monaco, pp. 27-111.

Mulder, C. J., 1973. Tectonic framework and distribution of Miocene evaporites in the Mediterranean. *Koninkl. Nederl. Akad. Wetensch.*, Amsterdam, pp. 44-59.

Mulder, C. J., Lehner, P., and Allen, D. C. K., 1975. Structural evolution of the Neogene salt basins in the eastern Mediterranean and the Red Sea. *Geol. Mijnb.*, 54: 208-21.

Neev, D., Almagor, G., Arad, A., Ginzburg, A., and Hall, J. K., 1973. The Geology of Southeastern Mediterranean Sea, *Israel Gol. Surv. Rept. MG/73/5,* 43 pp.

Olausson, E., 1961. Studies of sediment cores from the Mediterranean Sea and the Red Sea, *Swedish Deep-Sea Exped. Report, 1947-1948,* 8 (4): 335-91.

Orlova, G., and Zenkovitch, V., 1974. Erosion of the shores of the Nile Delta. *Geoforum,* 18: 68-72.

Parker, F. L., 1958. Eastern Mediterranean Foraminifera: *Swedish Deep-Sea Exped. Report, 1947-1948,* 8(2): 217-87.

Ross, D. A., and Uchupi, E., 1977. The structure and sedimentary history of the southeastern Mediterranean Sea – Nile Cone area. *Amer. Assoc. Petrol. Geol. Bull.*, 61: 872-902.

Ryan, W. B. F., Ewing, M., and Ewing, J. I., 1966. Diapirism in the sedimentary basins of the Mediterranean Sea. (Abst.) *Trans. Amer. Geophys. Un.*, 47: 120.

——, Stanley, D. J., Hersey, J. B., Fahlquist, D. A., and Allan, T. D., 1970. The tectonics and geology of the Mediterranean Sea. In: A. Maxwell (ed.), *The Sea,* 4. Wiley, New York, pp. 387-492.

Ryan, W. B. F., Hsü, K. J., et al., 1973. *Initial Reports of the Deep Sea Drilling Project,* 13. U. S. Govt. Print. Off. Washington, D. C., 514 pp.

Said, R., 1973. The geological evolution of the River Nile. Intern. Conf. Northeast African and Levantine Pleistocene Prehistory (unpublished report), Cairo, 128 pp.

Salem, R., 1976. Evolution of Eocene-Miocene sedimentation patterns in parts of northern Egypt. *Amer. Assoc. Petrol. Geol. Bull.*, 60: 34-64.

Sestini, G., 1976. *Proc. UNESCO Seminar on Nile Delta Sedimentology.* Alexandria, October 1975.

Stanley, D. J., and Maldonado, A., 1977. Nile Cone: Late Quaternary stratigraphy and sediment dispersal. *Nature,* 266: 129-35.

Stoffers, P., Summerhayes, C. P., and Fitzgerald, M., Geochemistry and mineralogy of Nile shelf sediments. (in preparation).

Summerhayes, C. P., Sestini, G., Misdorp, R., and Marks, N., 1977. Nile Delta: Nature and evolution of continental shelf sediment. *Mar. Geol.*, (in press).

——, and Marks, N., 1976. Nile Delta: Nature, evolution and collapse of continental shelf sediment system, a preliminary report. In: *Proc. UNESCO Seminar on Nile Delta Sedimentology.*, Alexandria, October 1975.

UNESCO 1973. *Arab Republic of Egypt, Project EGY/70/581, Coastal Erosion Studies.* Tech. Rept. i, United Nations Dev. Corp., Alexandria, 66 pp.

——, 1976. *Proc. of Seminar on Nile Delta Sedimentology.* Tech. Rept. United Nations Dev. Prog., Alexandria.

Venkatarathnam, K., Biscaye, P. E., and Ryan, W. B. F., 1972. Origin and dispersal of Holocene sediments in the Eastern Mediterranean Sea. In: D. J. Stanley (ed.), *The Mediterranean Sea – A Natural Sedimentation Laboratory.* Dowden, Hutchinson & Ross, Stroudsburg, Pa., pp. 455-69.

Wright, L. D., and Coleman, J. M., 1973. Variations in morphology of major river deltas as functions of ocean wave and river discharge regimes. *Amer. Assoc. Petrol. Geol. Bull.*, 57: 370-98.

Chapter 19

Timing of Tertiary Submarine Canyons and Marine Cycles of Deposition in the Southern Sacramento Valley, California

ALVIN A. ALMGREN

Union Oil Company of California
Ventura, California

ABSTRACT

The Tertiary sedimentary rocks of the southern Sacramento Valley, California, include three marine cycles of deposition that were similar in their development. Each cycle began with a marginal marine to marine sand transgression followed by subsidence to bathyal depths and then a shoaling of varying amounts. A fourth and last cycle began with shallow marine deposition, but this soon gave way to nonmarine deposition.

Each cycle of deposition was preceded by a period of tectonism and erosion, and three of the cycles were also preceded by the cutting of a canyon, which was subsequently filled by marine sediment. The three submarine canyons and their provincial ages are:

1. The Martinez Canyon, cut in late early to middle Paleocene and filled in early late Paleocene time.
2. The Meganos Canyon, cut and filled in latest Paleocene time.
3. The Markley Canyon, cut in early Oligocene (early Refugian) and filled in late Oligocene (late Refugian) to early Miocene time.

The filling of each submarine canyon preceded a new cycle of deposition, and the timing of the development of these canyons in relation to the cycles of deposition is apparently related to the tectonic activity of central California and its borderland.

INTRODUCTION

The Sacramento Valley constitutes the northern half of the Great Valley of California (Figure 19-1, inset). This major tectonic trough is a hydrocarbon-rich province in which several thousands of wells provide biostratigraphic data that make possible establishing the depositional history of the area. Marine sedimentary rocks present range in age from Jurassic to middle or late Miocene. Above this marine sequence, the latest Miocene and younger sediments are all nonmarine.

This chapter is concerned with the timing of the deposition of the Tertiary sediments and the development of three associated ancient submarine canyons of different ages in the southern part of the valley (Figure 19-1). The presence of these erosional features, the Martinez, Meganos, and Markley canyons, within this comparatively small area, together with the abundant subsurface data, make this area a classic one for the study of ancient submarine canyons and their relationships to tectonics and marine sedimentation. The western extent of these canyons is unknown, since all evidence of them has been removed by subsequent erosion west of the Sacramento Valley. These erosional features have been termed gorges, channels, or canyons, but the size of two of them qualifies them to be termed sea valleys. For simplicity, all three features are referred to herein as *canyons*. Discussing the origin, erosion, and filling of the canyons is beyond the scope of this chapter.

A fourth erosional feature, the Princeton Submarine Valley (Figure 19-1) located in the northern part of the Sacramento Valley, was cut into late Cretaceous (Campanian) rocks and was filled with marine sediments during the early Eocene. This feature has not been included in the present study because the time of cutting of the valley is uncertain. This uncertainty is the result of the large time gap that exists between the age of the rocks into which the valley was cut and the age of the fill. According to Redwine (1972), the time of origin was either subsequent to deposition of the Paleocene Martinez and prior to the Meganos Formation or sometime during deposition of the Martinez Formation, or both.

Frick et al. (1959) reported on the Princeton Submarine Valley, and Redwine (1972) documented this erosional valley in great detail. An index of wells used in the present study is listed in Table 19-1.

REGIONAL SETTING

Structurally, the Great Valley is an elongate asymmetric megasyncline (Dickinson 1971) or synclinal trough with the axis near and parallel to its western border (Figure 19-2). It lies between the Sierra Nevada to the east and the Coast Ranges to the west. The Sierra Nevada is essentially an immense block of granitic rock that has been faulted upward along its eastern edge, tilted slightly to the west, and extends under the eastern part of the valley. Some Paleozoic to Mesozoic sedimentary and volcanic rocks, older than the granitic intrusive rocks and in places strongly metamorphosed, crop out along the lower western slopes of the Sierra Nevada. Scattered, small exposures of Tertiary rocks are also present along the eastern border of the valley.

West of the valley, the Coast Ranges make up an anticlinorium in which Mesozoic and Cenozoic sedimentary rocks are complexly folded and faulted as a result of periodic tectonic activity. The eastern Coast Ranges border the west side of the valley and consist of easterly dipping, homoclinal strata of late Mesozoic (latest Jurassic and Cretaceous) age (Figure 19-2). These Mesozoic strata dip under the valley floor and onlap basement rocks to the east. The metamorphosed, eugeosynclinal facies of these rocks, the Franciscan Formation (Figure 19-2), form the dominant rock sequence in the western Coast Ranges (Bailey et al. 1964).

The Tertiary rocks in the southern Sacramento Valley are widely distributed throughout the subsurface and a nearly complete sequence of these strata crops out on the north flank of Mt. Diablo at the southwest end of the valley (Figure 19-2). Safonov (1962) pointed out that the main tectonic factors affecting the geology of the Sacramento Valley, including the Tertiary sedimentation, were continual southerly and westerly tilting of the valley and periodic uplift, particularly on the west side, including the Mt. Diablo area. As a result of this tilting, the southwest part of the valley was most deeply depressed structurally, and it contains the thickest and most complete stratigraphic section. Tertiary deposition extended west of the present Sacramento Valley, but subsequent uplift and erosion removed most evidence of it. All Tertiary units thin to the north, and only part of the Eocene section extends into the northern part of the valley. The Midland Fault, trending north-south in the south-central part of the valley (Figure 19-3), also affected the Tertiary stratigraphy. This is a normal fault downthrown to the west with approximately 610 m of vertical separation due to episodic movement since Paleocene time. Tectonic uplifts with faulting and subsequent periods of erosion in the Kirby Hills area also had an effect on the stratigraphy, as is apparent in east-west cross-sections. The Stockton Arch Fault, a reverse separation fault upthrown to the south about 610 m, marks the approximate southern boundary of the Sacramento Valley.

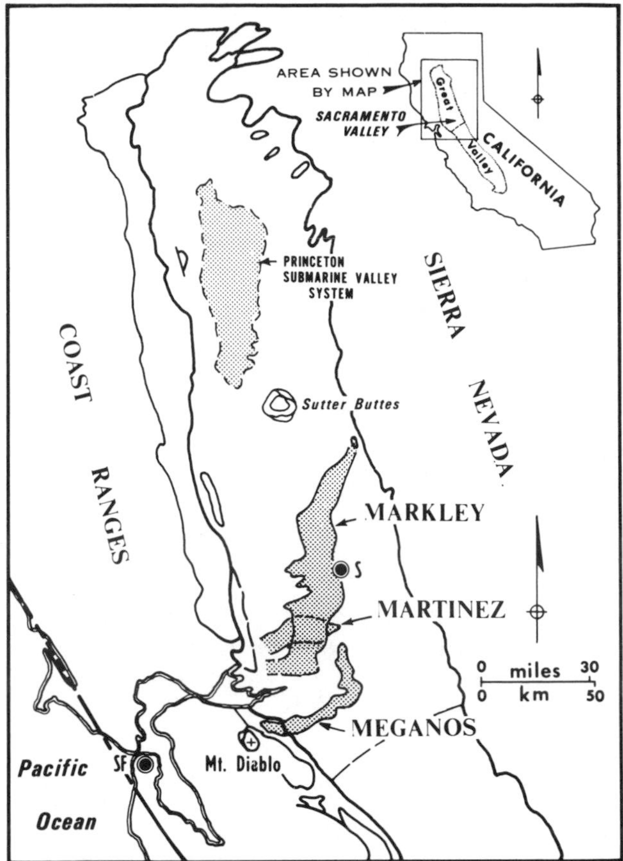

Figure 19-1. Setting of Tertiary canyons (Meganos, Martinez, Markley, and Princeton) in Sacramento Valley, northern California. Note: S = Sacramento; S. F. = San Francisco. Inset shows location of Sacramento Valley.

Table 19-1. *Index of Wells in Cross-Sections*

Cross-Section A-A' (Figure 19-4)

1. Kadane, Maine Praire Gas Unit A-1; Sec. 19, T 6N/R2E, MDBM.
2. Signal, Hastings Farms #1; Sec. 15, T5N/R2E, MDBM.
3. Amerada, Union (W) Unit #1; Sec. 4, T4N/R2E, MDBM.
4. Amerada, Drourin #8; Sec. 23, T4N/R2E, MDBM.
5. Amerada, McCormick Estate #4; Sec. 36, T4N/R2E, MDBM.
6. Occidental, Upham #1; Sec. 34, T3N/R2E, MDBM.
7. Shell, Heidorn #2; Sec. 4, T1N/R2E, MDBM.
8. Shell, Costello #2; Sec. 23, T1N/R2E, MDBM.

Cross Section B-B' (Chart A, Figure 19-6)

1. Kadane, Main Praire Unit A-2; Sec. 19, T6N/R2E, MDBM.
2. Amerada, Zimmerman Gas Unit #1; Sec. 29, T6N/R2E, MDBM.
3. Signal, Lilienthal #1; Sec. 9, T5N/R2E, MDBM.
4. Signal, Hastings Farms #4; Sec. 21, T5N/R2E, MDBM.
5. Signal, Peterson Estate #3; Sec. 28, T5N/R2E, MDBM.
6. Amerada, Union (W) Unit #1; Sec. 4, T4N/R2E, MDBM.
7. Dow Chemical, Phillips-Sumpf-Kroutch #1; Sec. 19, T4N/R2E, MDBM.

Cross-Section C-C' (Chart B, Figure 19-6)

1. Standard, Suisun Community #16; Sec. 4, T3N/R1W, MDBM.
2. Shell, Peterson #1; Sec. 32, T5N/R1E, MDBM.
3. Signal, Hastings Farms #4; Sec. 21, T4N/R2E, MDBM.
4. Signal, Hastings Farms #3; Sec. 22, T5N/R2E, MDBM.
5. Union, Rio Minerals #1; Sec. 20, T5N/R3E, MDBM.
6. Burmah, Ostman #1; Sec. 1, T5N/R3E, MDBM.
7. Burmah, Peck #1; Sec. 7, T5N/R4E, MDBM.

Cross-Section D-D' (Chart A, Figure 19-8)

1. Shell, Ginochio #4-4; Sec. 4, T1N/R1E, MDBM.
2. Shell, Sullinger #31-9; Sec. 9, T1N/R2E, MDBM.
3. Occidental, Tom Davis #2; Sec. 15, T1N/R2E, MDBM.
4. Occidental, Shell Enos #1; Sec. 8, T1N/R3E, MDBM.
5. Humble, Riverview Investment #1; Sec. 33, T2N/R4E, MDBM.
6. Westates, Zuckerman #1; Sec. 13, T2N/R4E, MDBM.

Cross-Section E-E' (Chart B, Figure 19-8)

1. Union, Delta Properties #6; Sec. 6, T2N/R3E, MDBM.
2. Signal, Signal-Burroughs #1; Sec. 20, T2N/R3E, MDBM.
3. Occidental, Machado #1; Sec. 30, T2N/R3E, MDBM.
4. Quintana, Knightsen #1; Sec. 5, T1N/R3E, MDBM.
5. Occidental, Shell Enos #1; Sec. 8, T1N/R3E, MDBM.
6. Occidental, Great Yellowstone-Bloomfield #1; Sec. 29, T1N/R3E, MDBM.

Cross-Section F-F' (Chart B, Figure 19-10)

1. Kadane, Maine Praire Gas Unit A-2; Sec. 19, T6N/R2E, MDBM.
2. Amerada, Zimmerman Gas Unit #1; Sec. 29, T6N/R2E, MDBM.
3. Signal Lilienthal #1; Sec. 9, T5N/R2E, MDBM.
4. Signal, Hastings Farms #4; Sec. 21, T5N/R2E, MDBM.
5. Signal, Peterson Estate #3; Sec. 28, T5N/R2E, MDBM.
6. Amerada, Union (W) Unit #1; Sec. 4, T4N/R2E, MDBM.
7. Dow Chemical, Sage-Phillips-Sumpf-Kroutch #1; Sec. 19, T4N/R2E, MDBM.

Cross-Section G-G' (Chart A, Figure 19-10)

1. Shell, Peterson #1; Sec. 32, T5N/R1E, MDBM.
2. Signal, Peterson Estate #3; Sec. 28, T5N/R2E, MDBM.
3. Signal, Hastings Farms #3; Sec. 22, T5N/R2E, MDBM.
4. S. M. Reynolds, Liberty Farms #1; Sec. 13, T5N/R2E, MDBM.
5. Reserve, Liberty Farms #7; Sec. 18, T5N/R3E, MDBM.
6. Arcady, Prospect Island #4; Sec. 9, T5N/R3E, MDBM.

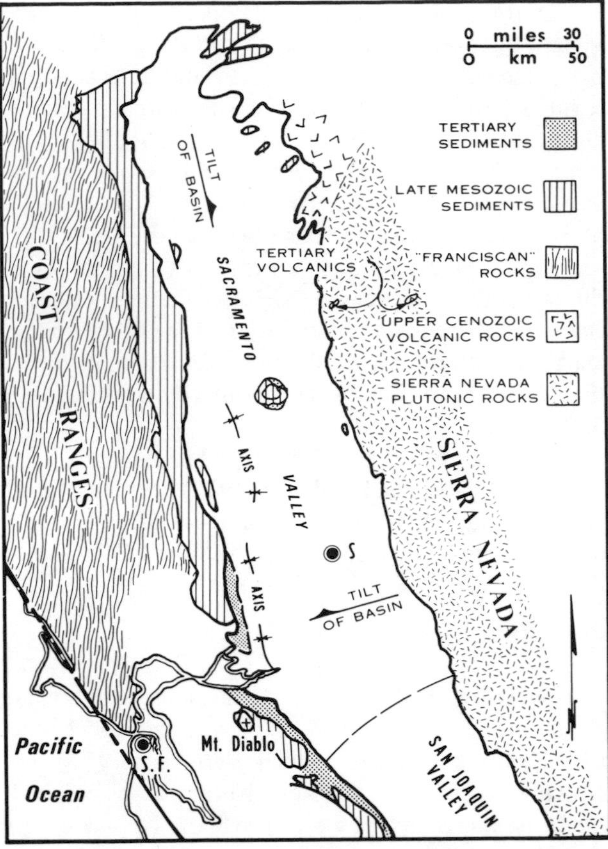

Figure 19-2. Geologic setting of Sacramento Valley showing generalized distribution of dominant rock types. Note: S = Sacramento; S. F. = San Francisco. (Source: Geologic map modified after Bailey 1966.)

TERTIARY SEDIMENTARY SEQUENCE

The Tertiary sequence of strata is represented by four cycles of deposition, shown in cross-section A-A' (Figure 19-4), each of which was initiated by the deposition of a transgressive shallow water sand. Each cycle was preceded by a period of tectonism, primarily in the vicinity of the western edge of the present valley, but also including a slight increase in the southerly tilt of the trough, as shown by the regional unconformities at the base of the transgressive sands (Figure 19-4). The periods of tectonic activity were followed by erosional periods, three of which were accompanied by the cutting of a large canyon. The submarine filling of these canyons preceded the next cycle of deposition. The relationship of these submarine canyons to their associated depositional cycle is apparent in cross-section A-A' (Figure 19-4). Note that in this and all other cross-sections, the necessary vertical exaggeration of the vertical scale seriously distorts the profiles of the canyons by reducing their width to depth ratios.

The timing of the Tertiary sedimentation, tectonism, and periods of erosion will be demonstrated by presenting

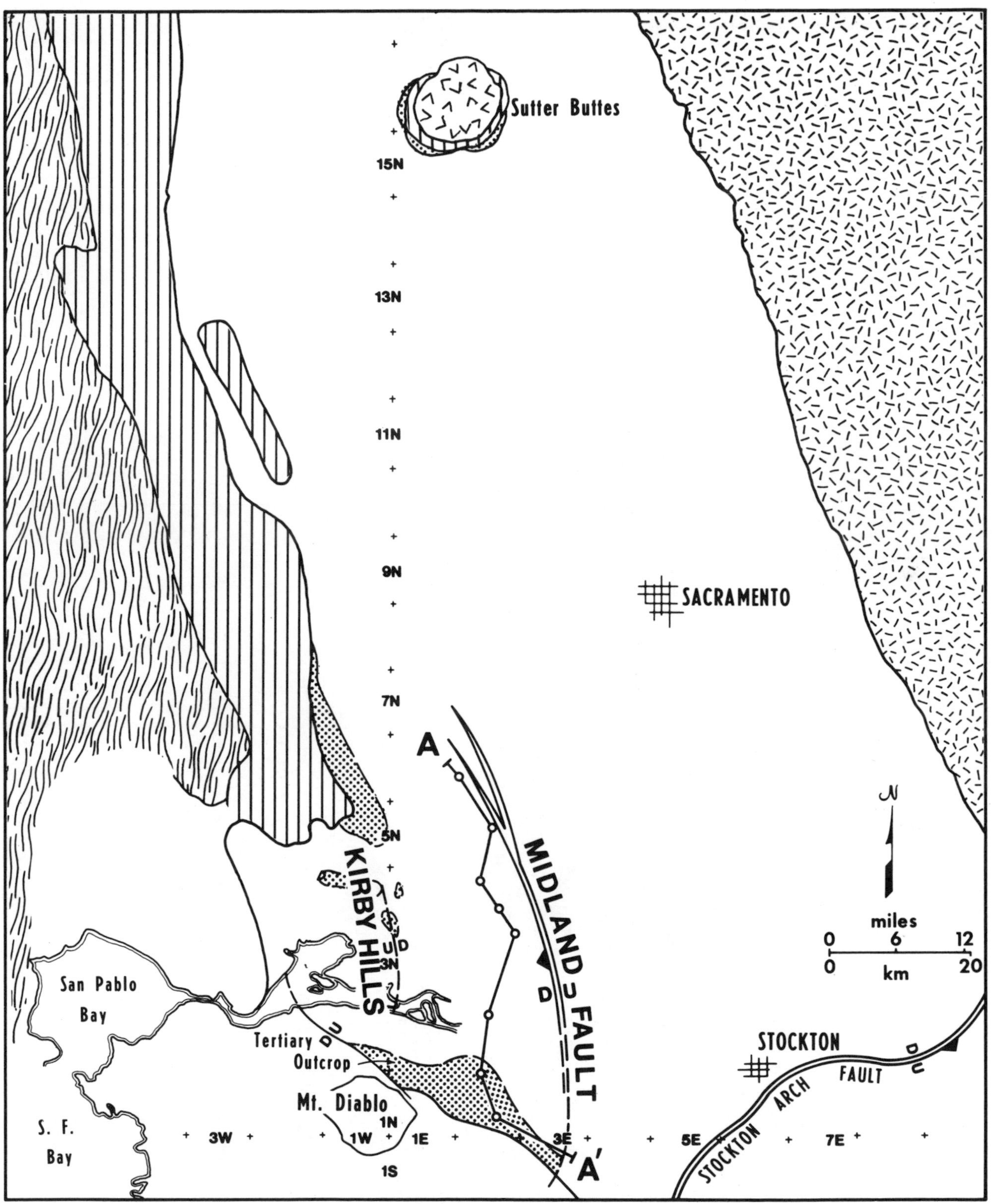

Figure 19-3. Map showing position of faults that affected sedimentation during Tertiary time. Note: Location of cross-section A-A' shown in Figure 19-4 is indicated; "D" and "U" on faults refer to Downthrow and Upthrow sides, respectively.

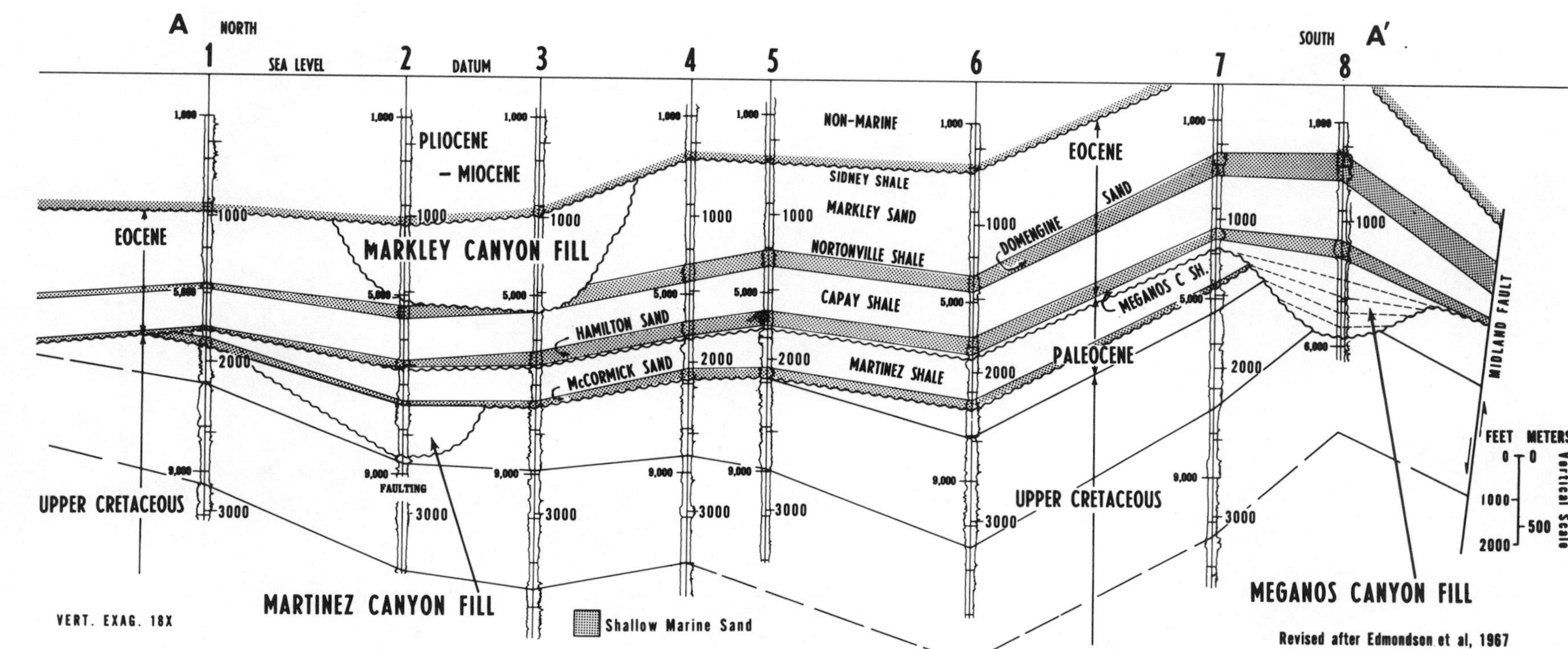

Figure 19-4. Stratigraphic cross-section A-A' showing relationship of Tertiary canyon fills to associated cycles of sedimentation. Note: See Figure 19-3 for location of section, and Table 19-1 for index of wells.

each cycle of deposition and its associated submarine canyon together chronologically. The series ages used are the provincial time-rock correlations used by Kleinpell (1938), Weaver et al. (1944), Mallory (1959), and Goudkoff (1945) and are based on benthic foraminiferal correlations. Recent studies of planktic microfossils indicate that the provincial series age designations used in California need revision. The relationship of these correlations and potential revisions are summarized in Figure 19-11. The main revision is the expansion of the Eocene. The Eocene top is raised to include the Refugian Stage (provincial Oligocene) as Upper Eocene (Warren and Newell 1976) and the base is lowered to include the Meganos Canyon fill as Lower Eocene. The datum in each cross-section is the base of the transgressive shallow-water sand that initiated each cycle of deposition.

SEDIMENTARY HISTORY

Martinez Canyon and the Late Paleocene Cycle of Sedimentation

The late Paleocene cycle of deposition was preceded by a phase of fully marine to marginal marine late Cretaceous to early Paleocene sedimentation (see A-A', Figure 19-4), which ended with shoaling of the entire area. Deltaic conditions existed to the east and a shallow sea to the west.

A period of tectonism in late early to middle Paleocene followed, with a strong uplift on the west side of the trough, as shown by the strong angular discordance to the west (see chart B, Figure 19-6). Subsequent erosion truncated more than 2,000 m of Upper Cretaceous strata in the area of strong uplift to the west (see chart B, Figure 19-6), and it gently beveled the early Paleocene and some of the underlying Upper Cretaceous strata in a northerly direction (see Figure 19-4; chart A, Figure 19-6). This erosion reduced the entire southern Sacramento Valley to an approximate peneplane. Evidence is the relatively uniform thickness of the superjacent, late Paleocene transgressive sand, the McCormick sand, which thins to the north and west, as shown in cross-section A-A' (Figure 19-4) and cross-sections B-B' and C-C' (charts A and B, Figure 19-6).

During this early Paleocene erosion a west-trending canyon, the Martinez Canyon (Edmondson 1967), was cut (Figure 19-5), probably under subaerial conditions at least in the eastern part of its course. This canyon, the smallest of the three, has a known length of about 33 km, a width up to 12 km, and a maximum depth of about 360 m.

The filling of this canyon took place in early late Paleocene time, as indicated by the sparse foraminiferal assemblage present and its relative stratigraphic position. In the eastern part of the canyon, the fill is dominantly marginal marine sandstone with some carbonaceous silt and shale interbeds. In the central and western part, the lower half of the fill contains a dominant shale section, with a few foraminifers indicating that at least part of the fill was deposited in marine waters at least as deep as outer neritic. Isochores of the fill are shown in Figure 19-5. The notch on the north side of the fill is the result of erosion during the cutting of the overlying Markley Canyon.

Following the infilling of the Martinez Canyon, late Paleocene sedimentation continued with the deposition of the shallow water, transgressive McCormick sand, which thins uniformly to the north and to the west (charts A and B, Figure 19-6). Rapid subsidence of the entire basin to bathyal depths, west of the Midland Fault (chart B, Figure 19-6) followed and a deep depositional area developed along the trend from Kirby Hills to Mt. Diablo (see Figure 19-3). Approximately 900 m of deepwater Martinez sand and shale were deposited in this deep western depositional area. Planktic foraminifers in the lower part of this unit are indicative of the *Globorotalia pseudomenardii* zone (P4). To the east, just west of the Midland Fault, about 300 m of section is preserved, but essentially none is present east of the fault. Shallow-water facies of these late Upper Paleocene sediments were undoubtedly deposited on the east side of the Midland Fault, but subsequent erosion removed all evidence of them prior to the cutting of the Meganos Canyon, except immediately adjacent to the east side of the fault along its northern trace.

Meganos Canyon and the Late Paleocene - Early Eocene Cycle of Sedimentation

Following the early late Paleocene cycle of deposition (chart B, Figure 19-6), tectonism deformed these sediments and the underlying Upper Cretaceous sediments (see chart A, Figure 19-8), particularly to the west in the vicinity of the eastern edge of the present Coast Ranges, including the Mt. Diablo area (see Figure 19-2). This tectonism was followed by a period of probable subaerial erosion that removed all of the Paleocene and the uppermost part of the Cretaceous sediments east of the Midland Fault, as mentioned above, nearly peneplaning the southern part of the valley (see chart A, Figure 19-8). During and/or following this period of subaerial erosion, the Meganos Canyon was cut, probably under subaerial conditions, at least to the east of the Midland Fault (Figure 19-7). However, Dickas and Payne (1967) have interpreted paleontologic and regional stratigraphic data to indicate that this canyon was cut in a marine environment.

Cross-section E-E' (chart B, Figure 19-8), a north-south section, shows the northward tilting of the Paleocene and underlying late Cretaceous strata that took place along

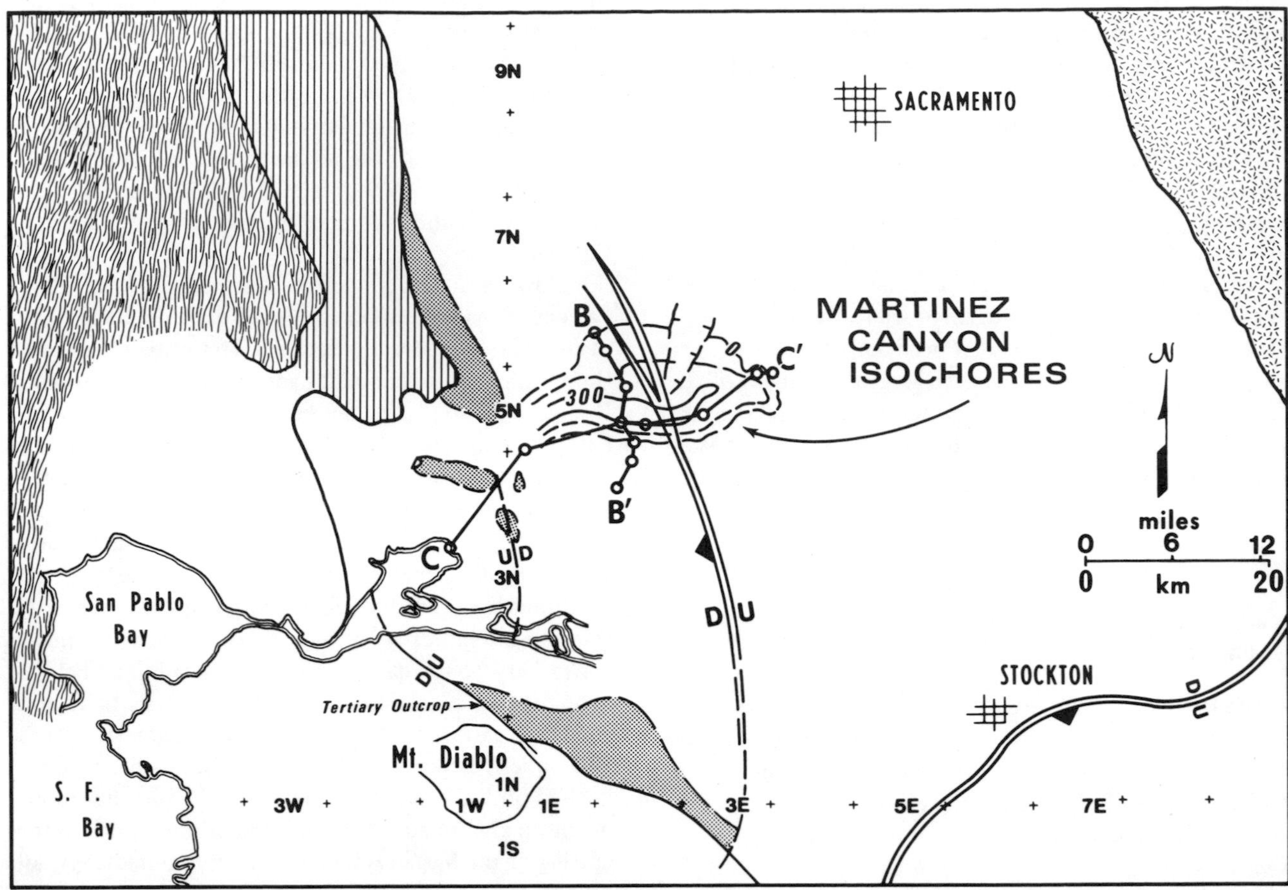

Figure 19-5. Map showing location of Martinez Canyon, with approximate isochores (in meters). Note: The notch on the north side of the fill is the result of erosion during the cutting of the overlying Markley Canyon; locations of stratigraphic cross-sections B-B' and C-C' shown in Figure 19-6 are also indicated.

the north flank of Mt. Diablo, prior to the erosional period and cutting of the Meganos Canyon. Edmondson (1965) and Dickas and Payne (1967) have described this canyon and given interpretations of its origin and filling. This canyon can be traced for over 80 km. It varies from about 2 km to 13 km in width, has a depth of over 760 m in its western course, and has an average gradient of 2° along the axis (Dickas and Payne 1967). Fischer (1972) estimated a minimum depth of 1,125 ± m in the westernmost exposures of the canyon where it is exposed on Mt. Diablo.

Sediments that fill the canyon were divided by Fischer (1972) into two facies: (1) an early submarine fan facies of coarser clastics, 460 m thick, present in the westernmost exposures of the canyon fill on Mt. Diablo; and (2) a late claystone fill sequence at least 670 m thick along its western course where it crops out on Mt. Diablo (Figure 19-7), which was named the C member of the Meganos Formation by Clark and Woodford (1927). The fill is dominantly claystone throughout most of the west-trending part of its course, and the fill in most of the south-trending eastern part of the canyon is sand. Edmondson (1965) and Dickas and Payne (1967) did not include the "submarine fan facies" of Fischer (1972) in the canyon fill. However, Johnson (1964) suggests that in the outcrop on the north flank of Mt. Diablo, the Meganos C and all or part of the Meganos B sand of Clark and Woodford (1927) represent surface exposures of the Meganos Canyon fill. Foraminifers present in the fill suggest that it was deposited in a marine environment ranging in depth from neritic to upper bathyal (Dickas and Payne, 1967). Planktic foraminifers present in a clay shale interval in the middle part of the fill in the Oil Canyon area of Mt. Diablo are representative of the lower part of the *Globorotalia subbotinae* zone (P6), which is early Eocene in age. Provincial age assignments place these strata in the late Paleocene.

The infilling of the canyon was accompanied by the deposition of the thin (15 m to 90 m) Meganos C shale section outside the canyon (chart B, Figure 19-8). This shale interval, present on the west side of the Midland Fault only, extends to the north for about 25 km where it is onlapped by the overlying early Eocene Hamilton sand,

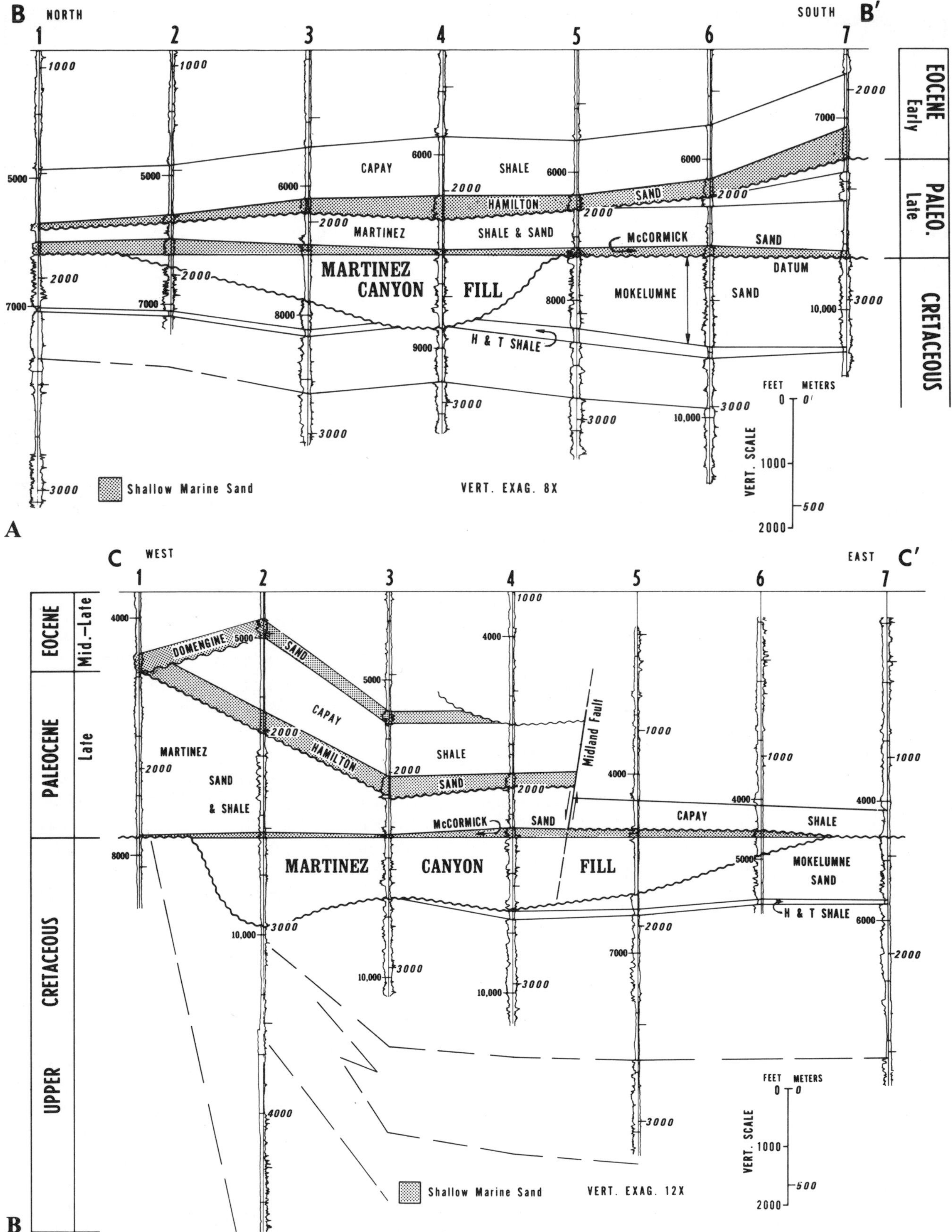

Figure 19-6. Martinez Canyon stratigraphic cross-sections. Note: Chart A shows cross-section B-B' (normal to canyon axis) and the relationship of fill to underlying and overlying cycles of sedimentation. Chart B shows cross-section C-C' (parallel to canyon axis) and the relationship of fill to underlying and overlying cycles of sedimentation. See Figure 19-5 for location of sections and Table 19-1 for index of wells.

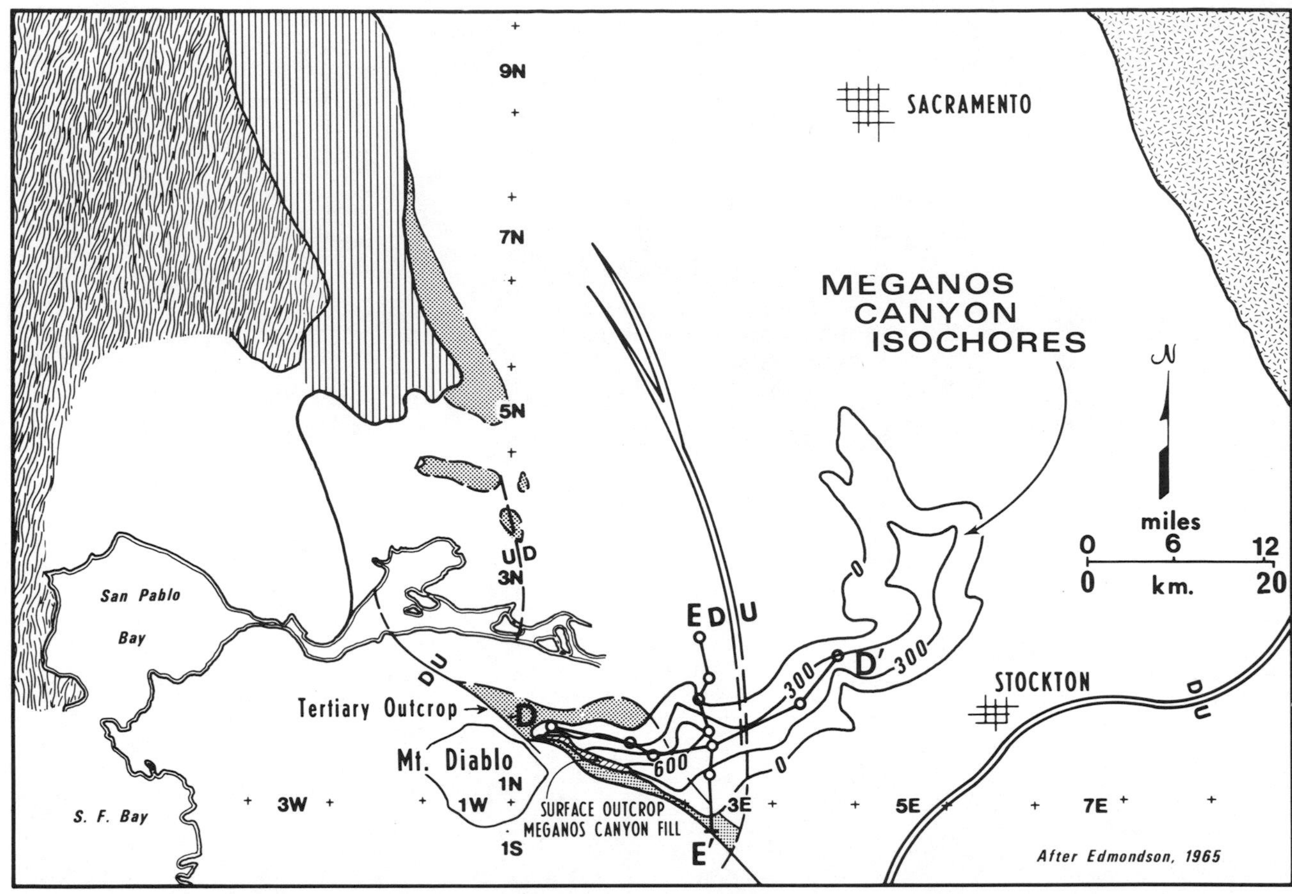

Figure 19-7. Map showing location of Meganos Canyon, with approximate isochores of the shale fill (in meters). Note: Locations of stratigraphic cross-sections D-D' and E-E' shown in Figure 19-8 are also indicated. (Source: Adapted from Edmondson 1965.)

as shown in cross-section A-A' (Figure 19-4). A thin veneer of this shale may have been deposited on the east side of the Midland Fault outside the canyon but, if so, all evidence of it was removed by erosion prior to the deposition of the Capay shale.

Isochores demonstrate that there was no measurable movement along the Midland Fault during the cutting and filling of the Meganos Canyon (see Figure 19-7), thereby indicating a period of relative tectonic stability. The isochores also indicate that there has been no apparent strike-slip separation on the Midland Fault since the filling of the canyon.

Early Eocene Cycle

After the filling of the Meganos Canyon sedimentation continued with the deposition of an early Eocene sequence of strata which began with the transgressive, shallow marine Hamilton sand, which is preserved only west of the Midland Fault (chart A, Figure 19-8). This early Eocene sand crops out on Mt. Diablo as the Meganos D sand (Clark and Woodford 1927), where it represents the start of a transgressive cycle with shallow-water deposits at the base (Johnson 1964). Here it contains several highly fossiliferous beds, and in some well sections to the north the upper part of the sand contains the benthic foraminifers *Amphistegina* and *Discocyclina,* which indicate a probable mid-neritic environment of deposition. This sand has a maximum thickness of about 200 m in the Mt. Diablo area, thins regionally, and wedges out about 65 km to the north. The absence of the Hamilton sand on the high (east) side of the Midland Fault indicates probable vertical separation on the Midland Fault just prior to the deposition of the sand, with the Hamilton sand being deposited against a bluff formed by a fault scarp. Movement on the Midland Fault during the early Eocene and its effect on the sedimentation is apparent in Section D-D' (chart A, Figure 19-8).

Rapid subsidence followed the Hamilton sand transgression, and sedimentation continued with the deposition of the Capay shale in upper bathyal to perhaps mid-bathyal water depths. The lower 30 m to 90 m of sediment west of the Midland Fault is dominantly clay shale with some glauconite and abundant planktic and benthic foraminifers. The planktic foraminifers represent the upper part of the *Globorotalia subbotinae* zone (P6). As deposition continued, the sea gradually shoaled to inner neritic depths, thereby completing this early Eocene cycle of deposition.

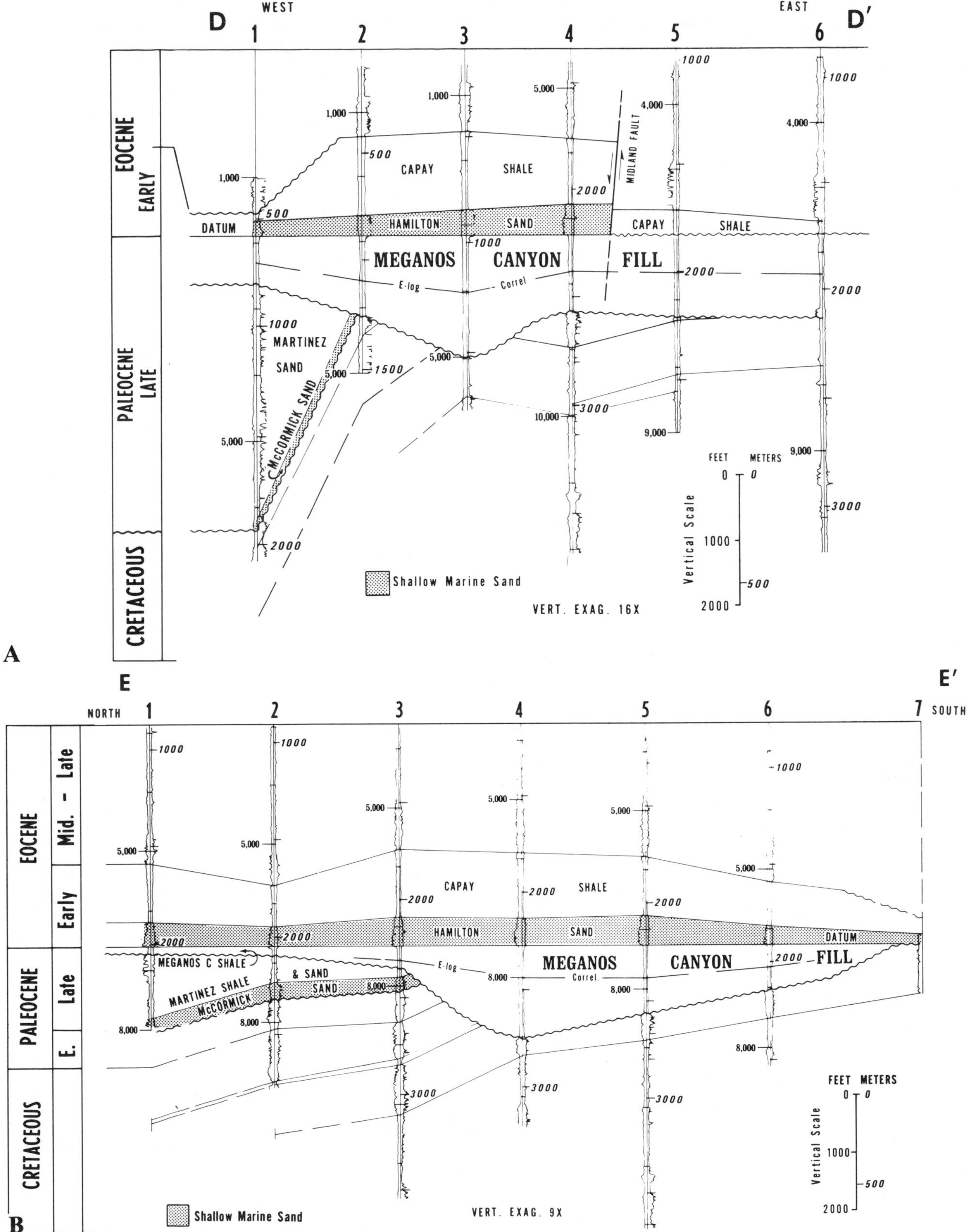

Figure 19-8. Meganos Canyon stratigraphic cross-sections. Note: Chart A shows cross-section D-D' (parallel to canyon axis) and the relationship of the fill to underlying and overlying cycles of sedimentation. Chart B shows cross-section E-E' (normal to canyon axis) and the relationship to underlying and overlying cycles of sedimentation. See Figure 19-7 for location of section, and Table 19-1 for index of wells.

A period of tectonism and erosion, particularly to the west in the Kirby Hills and Mt. Diablo area, followed the early Eocene deposition (see chart B, Figure 19-6). The truncation of the early Eocene Capay shale and Hamilton sand by the middle Eocene, Domengine sand in well #1 (chart B, Figure 19-6) is noteworthy. No evidence of a submarine canyon has been detected in the southern Sacramento Valley after the middle Eocene tectonism and erosion that preceded the middle to late Eocene cycle of deposition. In the northern Sacramento Valley, the filling of the lower part of the Princeton Submarine Valley preceded middle Eocene sedimentation.

Middle and Late Eocene Cycle of Deposition

Middle to late Eocene deposition (see chart B, Figure 19-10) began with marginal marine, lower Domengine Formation sands, which locally contain siderite nodules, lignite, and even thin coal beds in the Mt. Diablo area. Open marine sedimentation began with transgression of shallow-water sediments (Johnson 1964), which were deposited over all but the northern and northwestern part of the Sacramento Valley. Todd and Monroe (1968) recognized five lithologic subunits in the Domengine Formation. These subunits were all sublittoral and included prodelta, deltaic, tidal shelf, and coastal marsh deposits with open marine deposits at the top of the formation.

Following this shallow-water transgression, rapid subsidence, particularly to the southwest, resulted in the development of a deep depositional trough or plain in which 210 m of the radiolarian and foraminiferal-rich Nortonville shale was deposited in deep bathyal conditions. To the east, upper bathyal and neritic conditions probably existed during the deposition of a thinner shale section. Sedimentation continued under bathyal conditions to the west with the deposition of about 910 m of the arkosic, mica-rich Markley Formation sandstone. At the same time, the area east of the Midland Fault was an area of sediment bypass that received little sediment. The Markley sand was followed by deposition of the plankticrich, diatomaceous Sidney shale over most of the south end of the valley. It is 210 m thick in the Mt. Diablo area and wedges out to the north and to the east, where it onlaps the Markley Formation sandstone and rests on the Nortonville shale (Almgren and McDougall 1975). The Eocene cycle of deposition ended with the emplacement of over 100 m of upper Markley sandstone in the viciinty of Mt. Diablo where it crops out.

The Markley Canyon and Middle (?) - Late Miocene Cycle of Deposition

After completion of the middle to late Eocene cycle of deposition, a period of tectonism including some continued movement on the Midland Fault took place (see chart A, Figure 19-10). Volcanism accompanied this tectonic activity and emplaced basalts and andesites to the northeast on the western slopes of the Sierra Nevada. Durrell (1966) presented evidence indicating that the isolated exposures of the Tertiary volcanic basalt and andesite of the Lovejoy Formation, which are exposed on the surface (see Figure 19-2) about 130 to 150 km north and northeast of the City of Sacramento, are late Eocene and early Oligocene in age.

During and/or after this tectonism in post-Eocene (post-Markley Formation) time, erosion of some late Eocene sediment took place and the Markley Canyon (Figure 19-9) was cut (Almgren and Schlax 1957). This erosion continued throughout most of the Refugian Stage of the Oligocene (chart A, Figure 19-10). Travers (1972) suggested that the cutting of this canyon was caused by tectonic activity related to a nearby oceanic trench and states: "Development of the trench probably increased the regional submarine slope, and this may have initiated the erosion of the valley" (p. 179).

The Markley Canyon is the largest of the three canyons in the southern Sacramento Valley. It has a known length of over 130 km, averages 9 km to 12 km in width, and is over 760 m deep (Figure 19-9).

The filling of this submarine canyon began in the late Oligocene, late Refugian, as determined by benthic foraminifers present in the lower part of the fill (chart A, Figure 19-10; also see Almgren and Schlax 1957). These foraminifers suggest that this sedimentation took place under bathyal conditions. Similar conditions of deposition continued during the early Miocene, Zemorrian Stage, but with continued infilling of the canyon, the environment shoaled, perhaps during the Saucesian (?), to shallow marine conditions. There is no foraminiferal evidence to indicate a Saucesian (?) age, as the upper part of the fill contains only rare, shallow-water nondiagnostic forms or is barren.

Late Eocene foraminifers have been obtained from the fill, but these are considered to be from reworked shale pebbles that were derived from the Sidney shale exposed in the wall rock of the canyon (Almgren and Schlax 1957).

The sandstone intervals, particularly in the lower part of the fill, contain abundant andesitic and basaltic debris derived from the volcanic rocks emplaced during early Refugian time. The outcrop of the Wheatland Formation, about 55 km north-northeast of Sacramento (Figure 19-9) is probably a surface exposure of the Markley Canyon fill (Almgren and Schlax, 1957). Significantly, the sands of the Markley Formation, into which the canyon is cut west of the Midland Fault, are comprised dominantly of arkosic debris.

After the infilling of the Markley Canyon was completed, a middle (?) to late Miocene, restricted shallow

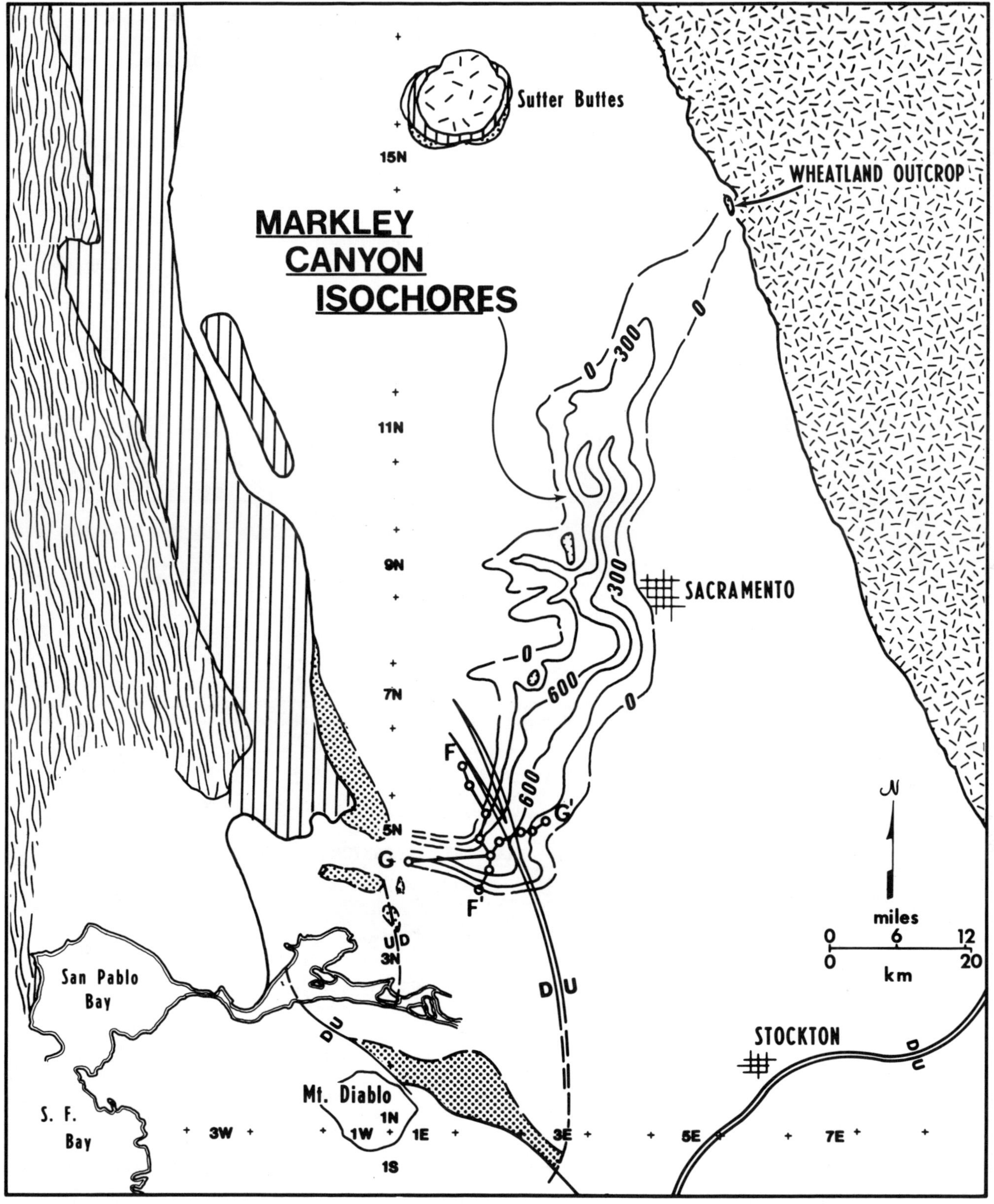

Figure 19-9. Map showing location of Markley Canyon, with approximate isochores (in meters). Locations of stratigraphic cross-sections F-F' and G-G' shown in Figure 19-10 are also indicated.

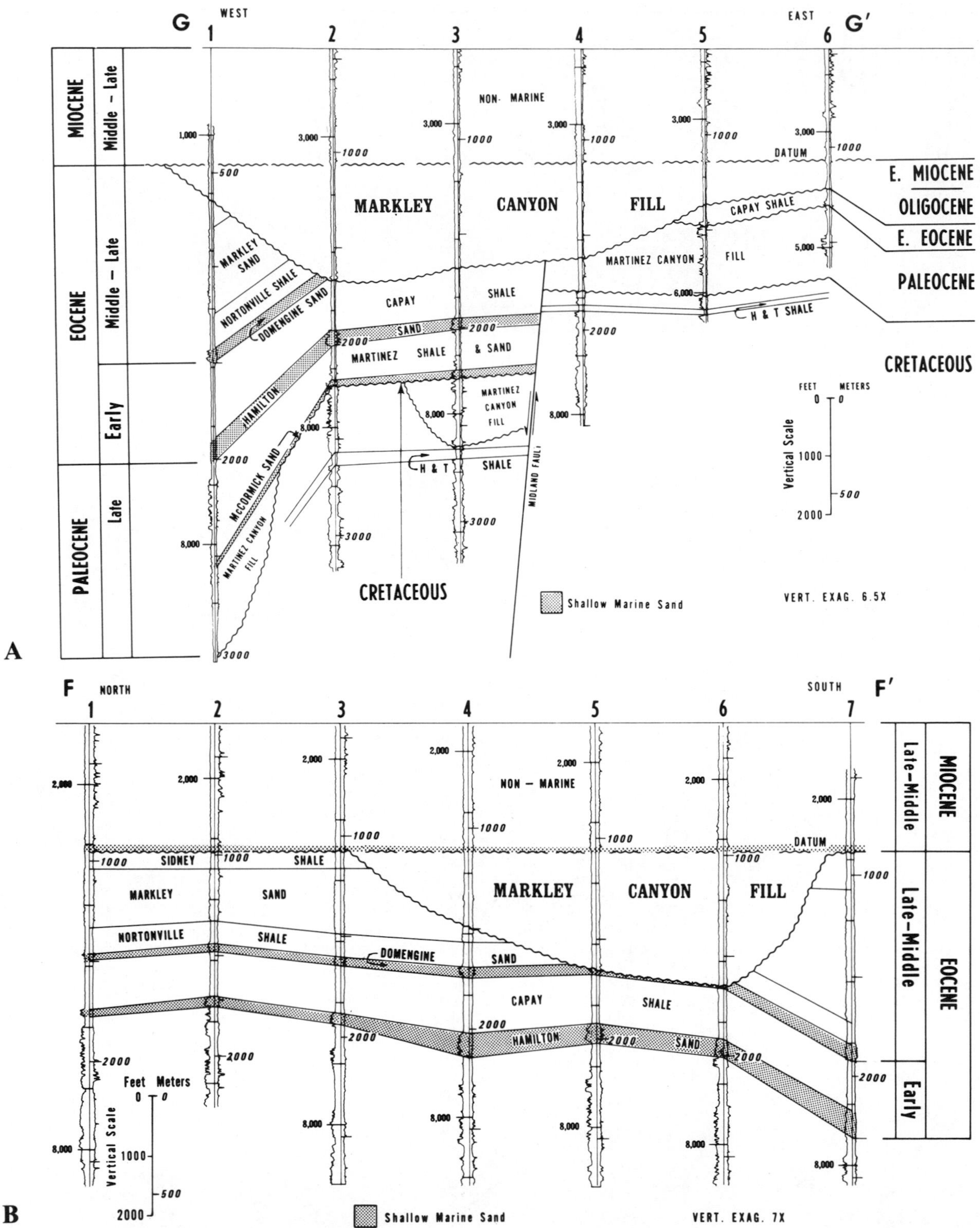

Figure 19-10. Markley Canyon stratigraphic cross-sections. Note: Chart A shows cross-section G-G' (parallel to canyon axis) and the relationship of canyon fill to underlying and overlying cycles of sedimentation. Chart B shows cross-section F-F' (normal to canyon axis) and the relationship of canyon fill to underlying and overlying cycles of sedimentation. See Figure 19–9 for location of section and Table 19–1 for index of wells.

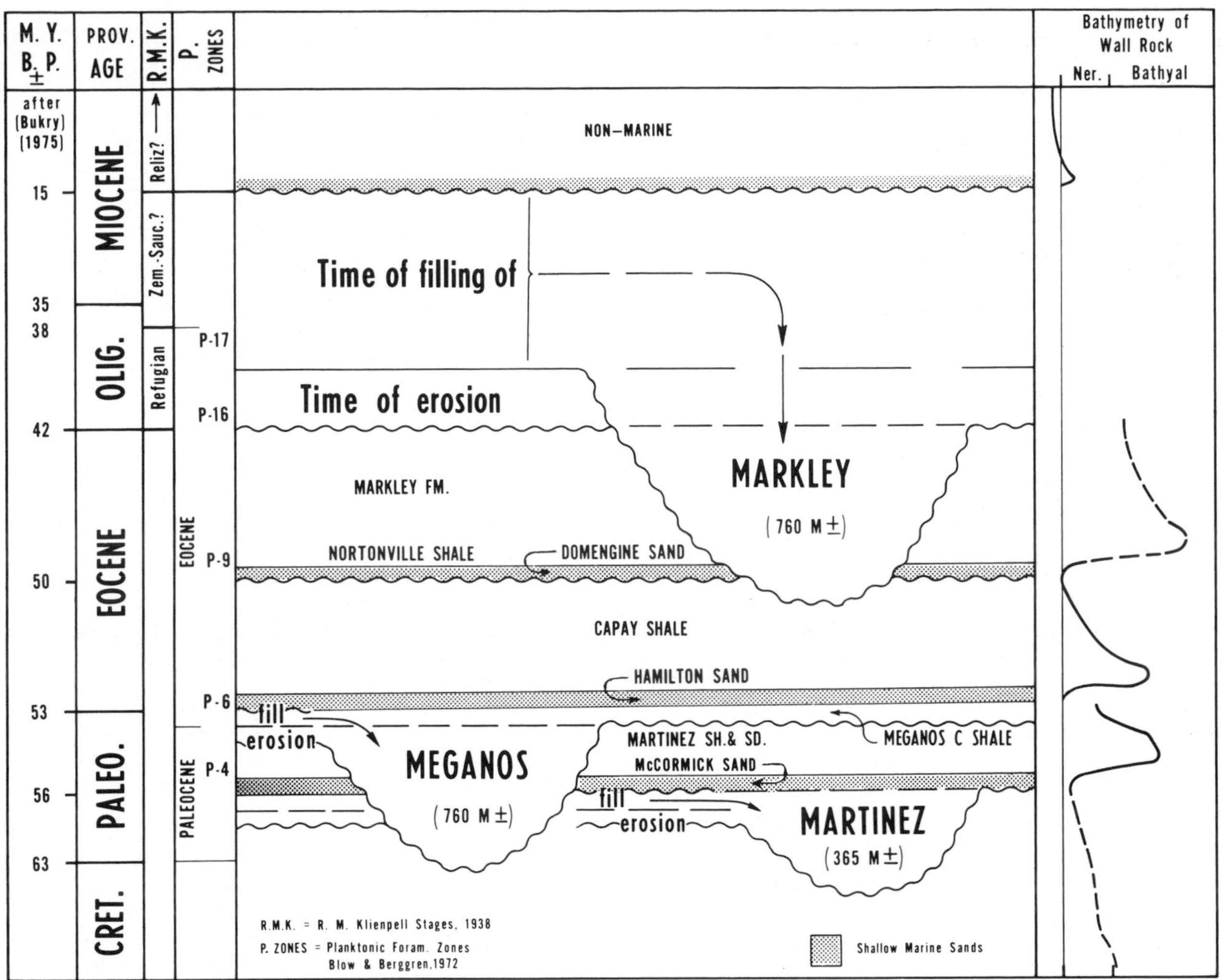

Figure 19-11. Schematic chart showing relationship of Tertiary canyons to periods of erosion and cycles of sedimentation. Note: The thickness of the fill in each canyon is indicated. Suggested absolute ages and position of planktic foraminiferal zones are tentative.

sea transgressed over part of the southern Sacramento Valley including the area of the filled canyon. Following the deposition of a thin sequence of sediment, this shallow sea receded to the west and south, giving way to the non-marine conditions which have since existed in the southern Sacramento Valley (charts A and B, Figure 19-10).

SUMMARY AND COMMENTS

Figure 19-11 diagrammatically summarizes the data presented, and shows the similarity of the timing of the three ancient submarine canyons in relation to their associated cycles of deposition. The same sequence of geologic events is associated with each submarine canyon. These events are summarized in Table 19-2.

In Figure 19-11, the planktonic zones (P zones) indicated are tentative assignments based on planktic foraminifers. The assignment of the Refugian of this area to late Eocene and planktonic zones ?P 16 and ?P 17 is uncertain, as no planktic foraminifers have been observed in the late Refugian in the Markley Canyon fill. It is based on the correlation of benthic foraminifers present in the lower part of the Markley Canyon fill to the late Refugian Stage, and the assignment of the type Refugian to the late Eocene is based on nannoplankton (Warren and Newell 1976). The suggested bathymetry shown in the right column of Figure 19-11 is a generalized interpretation based on benthic foraminifers present in the sediments of the wall rock of the canyons.

A significant fact may be that the period of time encompassing the development of the three canyons (late Paleocene to early Miocene) is the approximate period in which Nilsen and Link (1975) postulated that no major right lateral displacement occurred on the San Andreas Fault in central and northern California. This period appears to be one of convergence of the North American

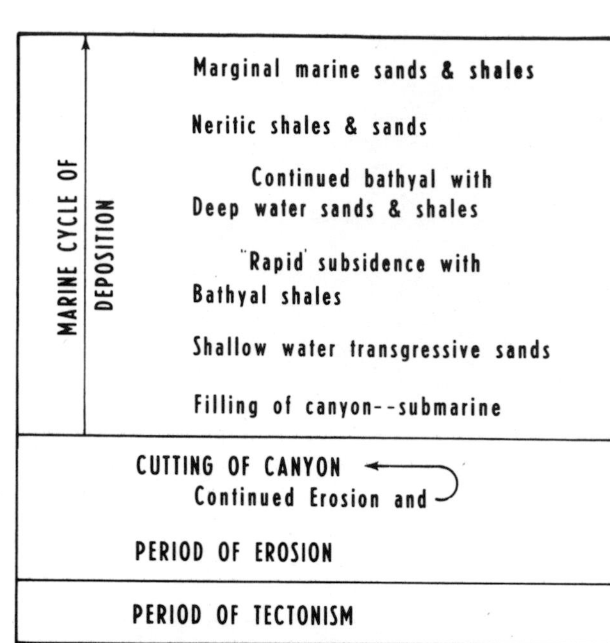

Table 19-2. *Relationship of Submarine Canyons to Cycles of Deposition and Tectonism*

continental margin and the Pacific plate and is also the approximate period during which Hamilton (1969), McKenzie and Morgan (1969), and Atwater (1970) postulated the existence of an oceanic trench off the central California coast.

Data presented here suggest that the episodic tectonic movements, including synclinal trough development, in the Sacramento Valley area were similar in nature and repetitive throughout this period. The tectonic forces active during this time apparently ceased by late Miocene, which is the approximate time that movement on the San Andreas Fault was resumed.

Travers (1972) concluded that the geologic phenomena that he discussed suggest the presence of a trench in late Eocene and early Oligocene time, because of the associated volcanic activity. However, pre-late Eocene trench activity is not supported by Travers' interpretation of the onshore data from central California, because of the lack of volcanic activity. In this writer's opinion, onshore data indicates that if a trench existed off central California in late Eocene to early Oligocene time, it probably also existed as early as late Paleocene time. This opinion derives from the writer's interpretation that the geologic events associated with development of the submarine canyons under discussion were affected by similar tectonic forces that were periodically active throughout this period of time.

ACKNOWLEDGMENTS

The author wishes to thank the Union Oil Company of California for permission to publish this study. Beneficial discussions were held with C. R. Carlson, William F. Edmondson, Lowell Redwine, Grosvernor C. Brown, Robert G. Douglas, Donn S. Gorsline, and Martin Link. The manuscript has benefited from review by Lowell Redwine, William F. Edmondson, William G. Reay, and Joan I. Winterer. A special acknowledgment is due John C. Crowell for his suggestions and critical editing of the manuscript.

REFERENCES

Almgren, A. A., and McDougall, K., 1975. Stratigraphic relationship of the middle Eocene Kellogg and Sidney Flat shales of northern California. *Future Energy horizons of the Pacific Coast.,* A. A. P. G., S. E. P. M., and S. E. G., Pacific Sections, Long Beach, pp. 367-99.

———, and Schlax, W. N., 1957. Post-Eocene age "Markley Gorge" fill, Sacramento Valley, California. *Amer. Assoc. Petrol. Bull.,* 41: 326-30.

Atwater, T., 1970. Implications of plate tectonics for the Cenozoic tectonics of western North America. *Geol. Soc. Amer. Bull.,* 81: 3513-36.

Bailey, E. H., 1966. Geology of northern California. *Calif. Div. Mines Geol. Bull.,* 190: 508.

———, Irwin, W. P., and Jones, D. L., 1964. Franciscan and related rocks, and their significance in the geology of western California. *Calif. Div. Mines Geol. Bull.,* 183: 177.

Blow, W. H., and Berggren, W. A., 1972. In: W. A. Berggren (ed.), A Cenozoic time scale – Some Implications for regional geology and paleobiogeography. *Lethaia,* 5: 195-215.

Bukry, David, 1975. Coccolith and silicoflagellate stratigraphy, northwestern Pacific Ocean, Deep Sea Drilling Project Leg 32. In: R. W. Larson, R. Moberly, et al., *Initial Reports of the Deep Sea Drilling Project,* 32, U.S. Govt. Print. Off., Washington, D. C., pp. 677-701.

Clark, B. L., and Woodford, A. O., 1927. The geology and paleontology of the type section of the Meganos Formation (lower middle Eocene) of California. *Calif. Univ. Bull., Dept. Geol.,* 17: 63-142.

Dickas, A. B., and Payne, J. L., 1967. Upper Paleocene buried channel in Sacramento Valley, California. *Amer. Assoc. Petrol. Geol. Bull.,* 51: 873-82.

Dickinson, W. R., 1971. Clastic sedimentary sequence deposited in shelf, slope and trough settings between magmatic arcs and associated trenches. *Pacific Geol.,* 3: 15-30.

Durrell, C., 1966. Tertiary and Quaternary geology of the northern Sierra Nevada, California. *Calif. Div. Mines and Geol. Bull.,* 190: 185-97.

Edmondson, W. F., 1965. The Meganos Gorge. *San Joaquin Geol. Soc., Selected Papers,* 3: 36-51.

Edmondson, W. F. (Chairman), 1967. Correlation section 16, Sacramento Valley, Winters to Modesto, California. *Amer. Assoc. Petrol. Geol.,* Pacific Section, Los Angeles, Calif. (chart).

Fischer, P. J., 1972. An ancient (late Paleocene) submarine fan: The Meganos channel, Sacramento Valley, California. *Pacific Petrol. Geol. Newsletter,* 26: 5.

Frick, J. D., Harding, T. P., and Marianos, A. W., 1959. Eocene gorge in northern Sacramento Valley (Abst.). *Amer. Assoc. Petrol. Geol. Bull.,* 43: 255.

Goudkoff, P. P., 1945. Stratigraphic relations of Upper Cretaceous in Great Valley, California. *Amer. Assoc. Petrol. Geol. Bull.,* 29: 956-1007.

Hamilton, W., 1969. Mesozoic California and the underflow of the Pacific mantle. *Geol. Soc. Amer. Bull.,* 80: 2409-30.

Johnson, W. S., 1964. Paleocene and Eocene geology of the north flank of Mt. Diablo. *Guidebook and Field Trip to the Mt. Diablo Area.,* Geol. Soc. Sacramento, pp. 23-32.

Kleinpell, R. M., 1938. *Miocene Stratigraphy of California.* Amer. Assoc. Petrol. Geol., Tulsa, 450 pp.

Mallory, V. S., 1959. *Lower Tertiary biostratigraphy of the California Coast Ranges.* Amer. Assoc. Petrol. Geol., Tulsa, 416 pp.

McKenzie, D. P., and Morgan, W. J., 1969. Evolution of triple junctions. *Nature,* 224: 125-33.

Nilsen, T. H., and Link, M. H., 1975. Stratigraphy, sedimentology and offset along the San Andreas Fault of Eocene to Lower Miocene strata of the northern Santa Lucia Range and the San Emigdio Mtns., Coast Ranges, central California. *Future Energy Horizons of the Pacific Coast.* A. A. P. G., S. E. P. M., and S. E. G., Pacific Sections, Long Beach, pp. 367-99.

Redwine, L., 1972. *The Tertiary Princeton Submarine Valley system beneath the Sacramento Valley, California.* Unpubl. Ph.D. Dissertation, University of California, Los Angeles, 480 pp.

Safonov, A., 1962. The challenge of the Sacramento Vallev, California. *Calif. Div. Mines and Geol. Bull.,* 181: 77-97.

Todd, T. W., and Monroe, W. A., 1968. Petrology of Domengine Formation (Eocene) at Potrero Hills and Rio Vista, California. *J. Sed. Petrol.,* 38: 1024-39.

Travers, W. B., 1972. A trench off central California in late Eocene/early Oligocene time. *Geol. Soc. Amer. Mem.,* 132: 173-82.

Warren, A. D., and Newell, J. H., 1976. Nannoplankton biostratigraphy of the upper Sacate and Gaviota Formations, Arroyo El Bulito, Santa Barbara County, California (Abst.). *Amer. Assoc. Petrol. Geol. Bull.,* 60: 2191.

Weaver, C. E. (Chairman), 1944. Correlation of the marine Cenozoic Formations of western North America. *Geol. Soc. Amer. Bull.,* 55: 569-98.

Chapter 20

The Red Oak Sandstone: A Hydrocarbon-Producing Submarine Fan Deposit

STEPHEN G. VEDROS
GLENN S. VISHER

Department of Earth Sciences
The University of Tulsa
Tulsa, Oklahoma

ABSTRACT

An integrated subsurface analysis of the geologic setting, electric log shapes, lithologic format, and sedimentary structures of the Red Oak Sandstone in southeastern Oklahoma reveals that this hydrocarbon-producing member represents the channeled portion of a submarine fan. This fan was formed in a subsiding foreland trough in which the marginal slope appears to have acted as a facies boundary throughout much of Paleozoic time. Lower and middle Atoka deposition is reflected by prograding deltaic deposits on the shelf, with gravity-induced units on the slope and in the basin.

The Red Oak sediments are characterized by fluidized and liquefied sedimentary structures. Sharp blocky bases on electric logs indicate channel patterns. The interbedded sandstones and shales indicate interchannel areas. Straight channels on the upper fan reflect confinement by levees while mid-fan deposition is characterized by a complex distribution of sandstone lenses resulting from channel migration and aggrading channel fill.

The consequences of this reinterpretation of the Red Oak Sandstone for hydrocarbon exploration and exploitation are discussed briefly and some criteria enabling recognition of similar types of sandstone reservoir are outlined.

Note: Stephen G. Vedros is currently with Northern Natural Gas Co., Exploration Division, Tulsa, Oklahoma.

INTRODUCTION

The study of depositional processes and sedimentation in modern submarine canyons and deepsea fans has been greatly expanded in recent years. Modern fan systems have been shown to differ in geometry and scale, but the character of the sediments and nature of the depositional processes appear to be comparable. Different tectonic and paleogeographic settings produce similar stratigraphic sequences, sedimentary structures, and sedimentary associations and it is thus possible to recognize ancient sediments deposited in these paleoenvironments.

Recently, sediment sequences of canyon-fan type have been recognized in deeper wells drilled in outer shelf areas. Since transitional zones linking the shelf to the deep basin are of limited areal extent, and since such slopes may be modified by subsidence, faulting, or thrusting, the recognition of clastic-filled channels on the continental slope is essential for correct interpretation of the paleogeographic and paleoenvironmental framework. Moreover, since many of these frontier exploration areas contain petroleum accumulations in channelized units of this type, their recognition and interpretation are of the utmost economic importance. To date, however, there have been few *subsurface* studies of hydrocarbon-yielding sequences of this type.

The present study provides a case history of a thoroughly explored, prolific gas producer, the Pennsylvanian middle Atoka Red Oak Sandstone of southeastern Oklahoma, which is here interpreted as a deepsea fan deposit on the basis of an integrated stratigraphic subsurface study involving consideration of the tectonic setting, electric logs, and sedimentary structures. The Red Oak thus provides an opportunity to determine the depositional pattern of a fan deposit within a producing field.

PREVIOUS WORK

Early studies in the area of the Red Oak focused on the description of lithologies and rock types within the Atoka. The term *hinge line*, introduced by Weirich (1953), referred to an area of more rapid thickening of the Atoka and was interpreted as the continental slope (Figure 20-1).

Atoka outcrops from the central Ouachitas to the flanks of the Ozark Uplift afford the opportunity to study rock types that show the transition of environments from shelf, through slope, to basin (Cline 1966). Shallow marine sediments are observed on the shelf, while slump and sedimentary flow features are seen in the slope sediments. Stark (1966) applied the term "Ouachita Graded Bedding" to Atokan slope and upper rise sediments. Growth faulting is common, with Atoka strata thickening basinward in a series of successively and progressively thicker overlapping sediment wedges. Detailed work by Koimn and Dickey (1967) on Atoka growth faults demonstrated thickening on the downthrown side of these faults. Faulting occurs on the mid- to outer shelf, and the Red Oak field is located south of the last major Atoka growth fault.

Paleocurrent data from Atoka outcrops (Briggs and Cline 1967) show an abrupt westward deflection near the southern margin of the subsiding Arkoma Basin, north of the Choctaw Fault (Figure 20-1). This deflection from south to west has been interpreted by Briggs and Cline to coincide with a marked increase in water depth. The Red Oak field is situated north of the Choctaw Fault, and north of this change in paleogeographic configuration.

Chamberlain (1971) identified four trace fossil assemblages from the Atoka, thereby defining the basin-to-shoal bathymetric profile. The bathyal-abyssal environment is characterized by the "Chondrites facies," which is found in the frontal Ouchitas, south of the Choctaw Fault.

GEOLOGIC SETTING

Structurally, the Arkoma Basin is bounded on the north by the Ozark Uplift, and on the west by the Tishomingo Uplift (see Figure 20-1). The most important structural element bounding the Arkoma Basin to the south is the Ouachita Mountains. The Choctaw Fault, a large thrust fault, separates the Ouachita system from the Arkoma Basin.

The Red Oak Sandstone member is middle Atokan in age, ranging in thickness from a feather edge on the Ozark Uplift to more than 3,000 m in the Ouachita Mountains.

In the inferred outer shelf and slope area, the Atoka unconformably overlies the fossiliferous Wapanucka Limestone of late Morrowan age. Regionally, the Atoka-Morrow boundary can be identified paleontologically. Morrowan faunal elements have been identified (Honess

Figure 20-1. Arkoma Basin index map.

1924; Cline 1960) in the Atoka sequence in the central Ouachitas. The Atoka overlies the Johns Valley Shale in the central Ouachitas (Shideler 1970). Exotic boulders, ranging in age from late Cambrian to early Pennsylvanian, can be found in the Johns Valley Shale. The stratigraphic age and structural history of the Johns Valley Shale are controversial and make the definition of the basinal Atoka difficult.

Older Pennsylvanian time-stratigraphic units, including the Springeran and Morrowan, are limited to basinal areas, whereas the Atokan and Desmoinesian units are present both in the basin and on the shelf. Desmoinesian and Atokan sediments filled the Ouachita geosyncline and Arkoma basin. This fill occurred as a series of cyclic regressions and transgressions of the sea across a continental margin. During Atokan time, there were periods of deltaic progradation from north and east. The Red Oak Sandstone was deposited beyond the distal margins of these prograding Atoka deltas on the slope and rise.

RED OAK FORMAT ANALYSIS, LOG MOTIFS, AND SANDSTONE GEOMETRY

General

A format (a marker-defined unit) or genetic sequence approach to study clastic intervals, such as the Atoka,

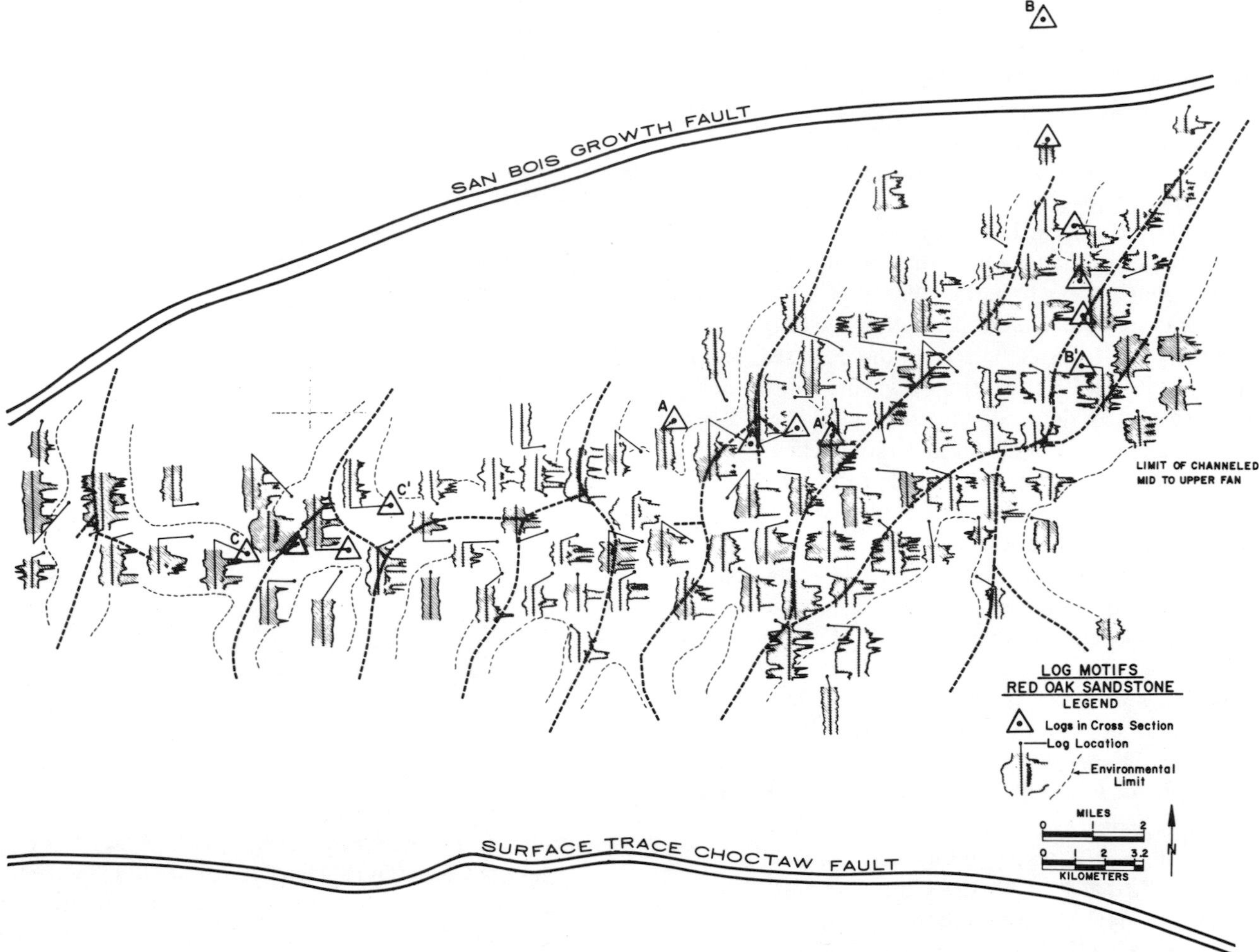

Figure 20-2. Red Oak Sandstone log motif map. Note: See Figure 20-9 for explanation of logs.

must rely on time-rock correlations. Transgressive or onlapping shale units can be distinguished on electric logs in the Atoka by low resistivity responses. These key beds are interpreted as parallel to time lines and are used to divide the Atoka into time-rock increments. The Red Oak format, or lithozone, is used to interpret the tectonic framework, history of sedimentation, and paleogeography.

In recent years, electric log motifs have proved to be useful in subsurface studies. An isolith map, coupled with log motifs, aids in determining the depositional pattern of the Red Oak Sandstone. The shapes or signatures of electric logs or gamma ray responses correspond to different sedimentary sequences (Figure 20-2). The Red Oak exhibits a great number of sharp-bottom, blocky, electric log signatures suggestive of channeling. Log motifs that indicate alternating sands and shales are interpreted as sediments deposited in interchannel areas.

An isopach map of the Red Oak format shows the depositional framework of the Red Oak Sandstone. Channel patterns can be distinguished by thick trends (Figures 20-3 and 20-4). Away from the channeled area to the south, the format thins, and to the north the format also thins in close proximity to the growth fault (Figure 20-3).

The Red Oak Sandstone, due to the numerous cores and abundant well control, provides a unique opportunity for study of the channelized portion of an ancient submarine fan, and consideration of the sandstone geometry suggests comparison with the tripartite division of modern fans (cf. Normark 1970).

Upper Fan

A few large and deep channels characterize the upper fan valleys (see Figures 20-3 and 20-4). These fan valleys range in width from hundreds of meters to two or three kilometers, and depths range from 10 m to a maximum of about 150 m. Fan valleys were incised by erosion and aggraded with deposition on the levees. In the Red Oak Sandstone, the channels within the upper

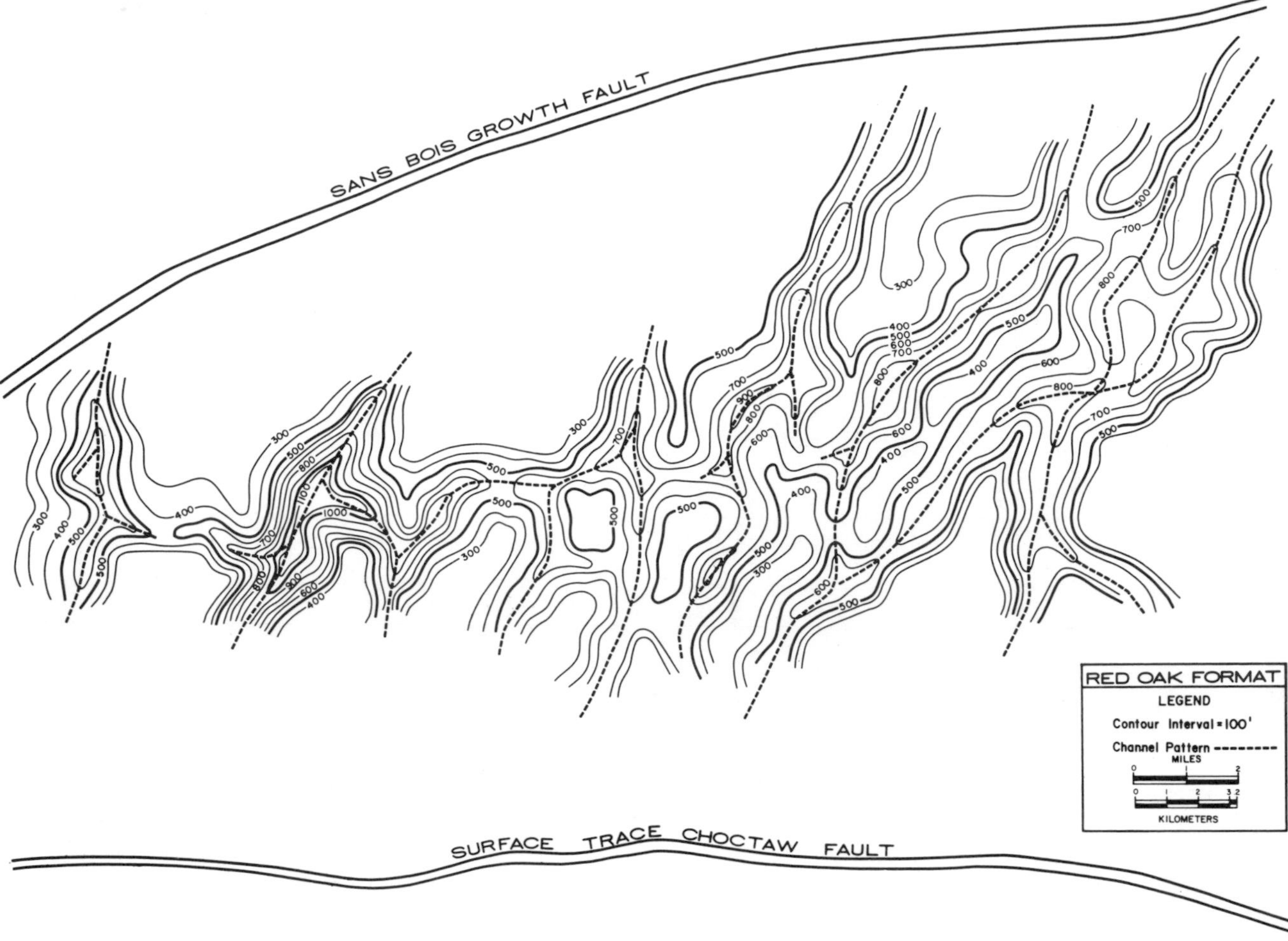

Figure 20-3. Red Oak Sandstone format map.

fan are found within mudstones or shale intervals only slightly older or contemporaneous with cutting and subsequent filling of the valley. The upper fan valleys have a relatively straight-line configuration from the shelf to the basin. The Red Oak Sandstone isolith and format thickness maps indicate four to five upper fan valley channels with relatively straight-line patterns trending approximately north to south (Figures 20-3 and 20-4). Lack of well control prevents tracing of the upper fan channels into canyons.

Middle Fan

In the Red Oak Sandstone the middle fan is characterized by a decrease in fan slope and aggradation resulting in lateral distribution of sediment (Figure 20-4). Channels of the middle portions of modern fans vary greatly in size, but dimensions are always less than for the upper fan valley channels. Active mid-fan channels have a branching pattern, which helps distribute sediment more evenly across the fan. The mid-region is the sandiest portion of the fan, and passes downfan into shaly thin-bedded sandstones separated by intervening thick shales. Lateral shifting of channels and spreading of depositional lobes result in overlap of later sedimentary wedges. These processes have led to a complex pattern of lateral and longitudinal sand trends on the Red Oak mid-fan (Figure 20-4).

Lower Fan

The lower portion of the Red Oak fan is characterized by thin sedimentary units. Channels, which dominate modern upper and mid-fans, are absent on the lower fan. Stratigraphic units consist of persistent parallel beds of mudstone and shale that alternate with fine-grained sandstones. Rock displaying this pattern can be seen in Atoka outcrops in the frontal Ouachitas. This facies primarily lies south of the Red Oak gas field and is consistent with the lithologic patterns established by surface work.

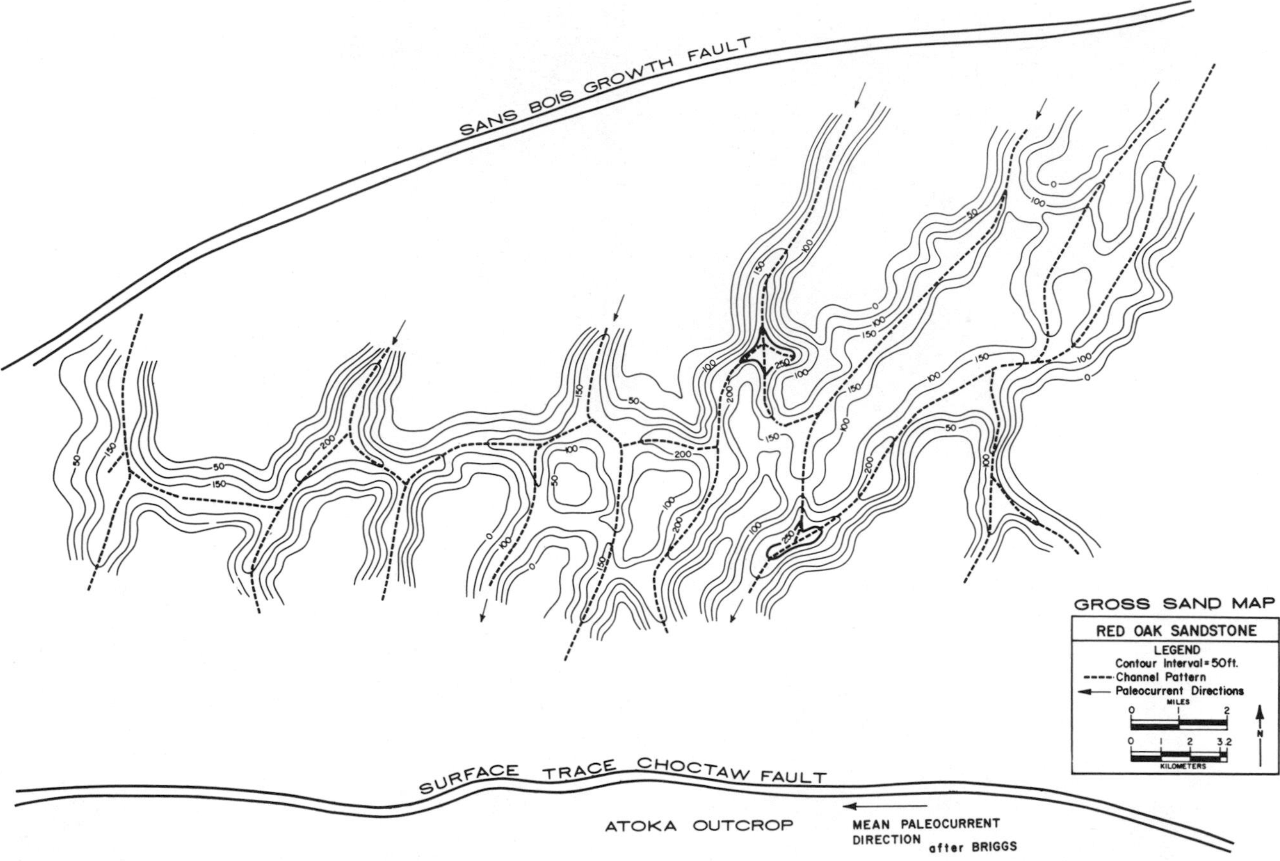

Figure 20-4. Red Oak Sandstone gross sand map.

RED OAK LITHOLOGY AND SEDIMENTARY STRUCTURES

Mineralogy and Lithology

The Red Oak unit is a very fine-grained quartz-rich (90 to 95 percent) sandstone. In the six intervals sampled in cores, the sandstone has a mean size of approximately 0.125 mm. It is moderately well sorted, with mica and carbonaceous material present in the matrix. Minor constituents include composite quartz grains, heavy minerals of zircon and tourmaline, and carbonate rock fragments. Shale clasts up to several millimeters in size also are present. The porosity of the Red Oak Sandstone measured in cored sections ranges from 13 to 20 percent, and permeability ranges from less than 0.1 to 20.0 md.

Sedimentary Structures

A characteristic association of sedimentary structures is present in the Red Oak Sandstone.

Dish structures are recognized by their pattern of successive arcs that are concave-up or dish-shaped (photograph A, Figure 20-5). Commonly, flat or parallel laminations grade upward into this pattern, then into extreme dish structures that are cup-shaped with vertical or even overturned sides. Upward, these structures pass into massive sandstone, or the structure is truncated by flat laminae. The base of the "dish" is more carbonaceous, and sometimes there is a slight increase in grain size just above the dark cup outline.

Dish structures have been interpreted to result from consolidation of a fluidized bed (Wentworth 1967) produced during the last stages of compaction. Disrupting forces have been suggested to be (1) homogeneous shearing just before the moving mass "gels," (2) internal load deformation, or (3) localized upward dewatering of clay-rich zones affecting the less cohesive, slightly larger grain sizes above the dark dish outlines. Dewatering of the sediment after semiconsolidation may be the primary factor in producing dish structures.

Shale inclusions and stringy lignite materials range in size from small flakes to large clasts. Fragments may be angular, rounded, or elongate and either isolated or strung out along bedding planes in thin planar groups (photographs B and C, Figure 20-5). These clasts are interpreted as related to the scouring by clastic mass fluid flows (Stauffer 1967; Dzulynski et al. 1959).

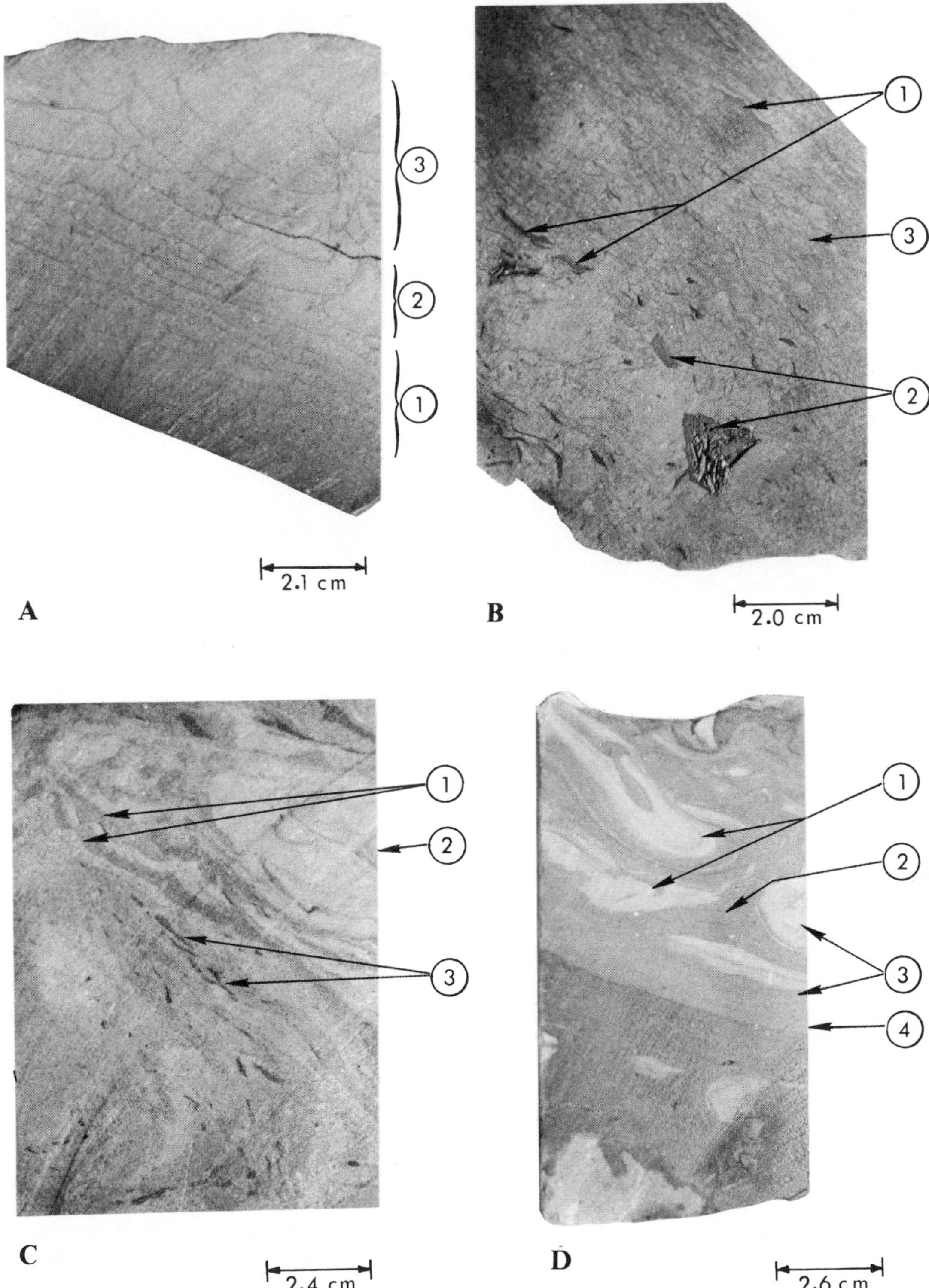

Figure 20-5. Sedimentary structures of Red Oak Sandstone. Note: Photograph A shows dish structures, Midwest #1 Ryder, 17-6N-22E (7013-7014). 1 = Slightly wavy parallel laminations; 2 = Transition zone; and 3 = Normal dish structures. Photograph B shows shale clasts and lignite detritus, Midwast #1 Martain 3-6N-22E (6982). 1 = Step structures in shale layers; 2 = Sand fold or tongue overlying flow structures; and 3 = Carbonaceous, lignite detritus or planar shale clasts, cutting across core. Photograph C shows flow structures, Midwest Sentry #1 Royalty, 17-6N-21E (7938). 1 = Step structures in shale layers; 2 = Sand fold or tongue overlying flow structures; and 3 = Carbonaceous, lignite detritus or planar shale clasts, cutting across core. Photograph D shows deformed laminations, Midwest Gradner, 13-6N-20E (8397-8398). 1 = Recumbent folds or tongues in finer grained sediment; 2 = Fine-grained sand matrix; 3 = Sand folds with horizontal fold axes; and 4 = Scour contact with slightly finer-grained sediment below.

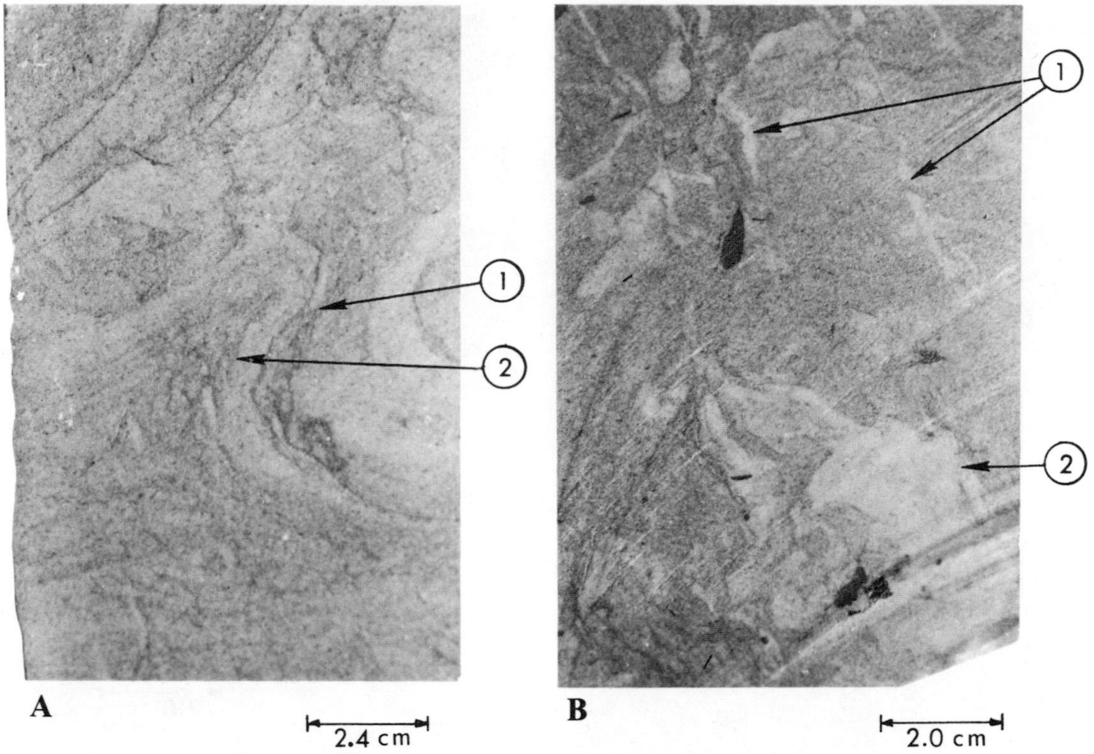

Figure 20-6. Sedimentary structures of Red Oak Sandstone. Note: Photograph A shows fluid heave, Midwest #1 McFerran, 2-6N-22E (7012). 1 = Heave structure, and 2 = Organic-rich material concentrated in center of structure. Photograph B shows mottled sediment, Midwest #1 Ryder, 17-6N-22E (6985). 1 = Isolated sand veins, avenue for escape of fluid, and 2 = Large sand mass in finer-grained matrix.

Lignitic plant detritus is a conspicuous accessory constituent. Plant fibers are layered or randomly oriented within the sandstone (point 3 on photograph B, Figure 20-5). The pattern of randomly oriented plant fibers gives the impression that they were emplaced by mass flow.

Distorted laminations and mottled sediments can be observed in Red Oak sandstones. They occur in sediments with an admixture of sand, silt, shale, and organic-rich carbonaceous material. Fine clays and organic detritus are concentrated in the central part of the structure (photograph A, Figure 20-6). Mottles and irregular layers of sand and shale also are present (photograph B, Figure 20-6).

Similar structures in sediments of the Gulf Coast (Coleman and Gagliano 1965) consist of slightly coarser grained sands in tongue-like masses, or folds, generally symmetrical and with near-horizontal axial planes (photographs C and D, Figure 20-5). These folds or tongue-like masses are completely surrounded by finer grained sediments (Figure 20-7A). Frequently, the coarser-grained sediments delicately reflect the microrelief (depressions) or the depositional surface (point 1 on photograph C, Figure 20-7). McKee et al. (1962) have reproduced these tongue-like sand masses or folds in tank studies and have referred to them as intraformational recumbent folds. The folds are attributed by McKee et al. (1962) to (1) overriding forces (fluidized clastic-laden currents), (2) differential overloading, and (3) mass movement of saturated, fluidized sediments on an inclined slope.

Climbing ripple laminae are characterized by ripple crests that appear to be offset from each other and advance upward and across the bedform (point 1 on photograph D, Figure 20-7). The bounding planes in climbing ripples form parallel curving surfaces (point 2, photograph D, Figure 20-7). Steeply dipping laminae between these bounding planes usually show tangential relationships to the top and bottom of the layer (point 3, photograph D, Figure 20-7). Changes in the form of ripple lamination (climbing, in-phase, or horizontal) occur both laterally and vertically. McKee (1965) has shown that horizontal laminae, with constant accumulation, commonly develop into in-phase ripple laminae. These in-phase ripples may then change laterally or vertically into climbing ripples where deposition occurs exclusively on the leeside of the ripple. This change from in-phase ripples to climbing ripples indicates a slight progressive increase in current velocity (points 3, 4, and 5, photograph D, Figure 20-7; also see Walker 1963).

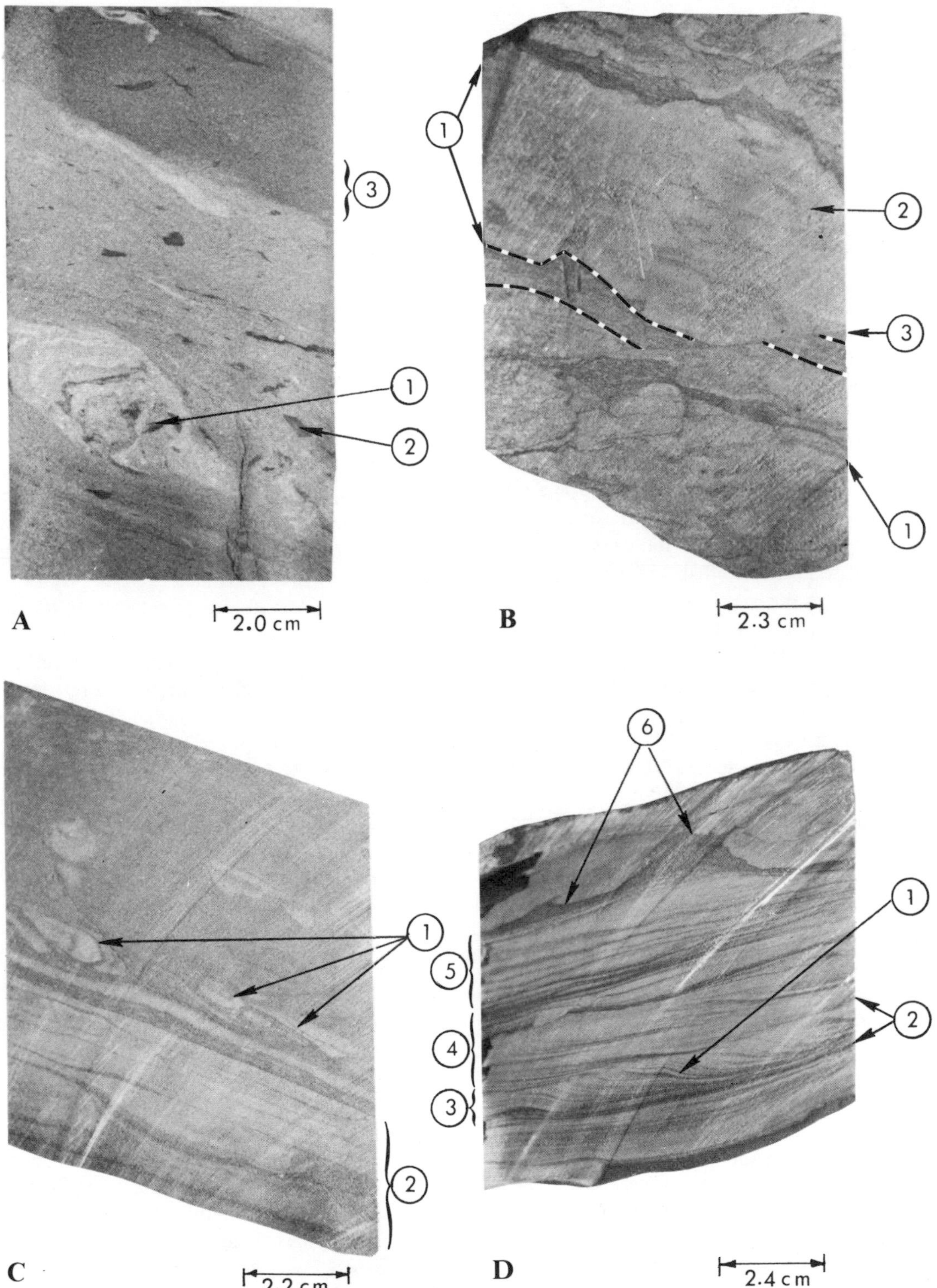

Figure 20-7. Sedimentary structures of Red Oak Sandstone. Note: Photograph A shows mottled sediment, Midwest Sentry #1 Royalty, 17-6N-21E (7935-7936). 1 = Large sand fold in finer grained sediment; 2 = Shale clasts; and 3 = Contact with slightly finer grained sediment. Photograph B shows flame structures and horizontal slip plane, Midwest Sentry #1 Royalty, 17-6N-21E (7791). 1 = Flame structures in shale layers; 2 = Fine-grained sandstone; and 3 = Horizontal slip plane (note offset in shale layer, dashed outline). Photograph C shows depression filled by fluidized flow, Midwest #1 Ryder, 17-6N-22E (6991-6992). 1 = Sand tongues, and 2 = Consistent flow of fluid grain layers. Photograph D shows climbing ripples and horizontal laminae, Midwest Sentry #1 Royalty, 17-6N-21E (7909). 1 = Climbing ripple (leeside); 2 = Parallel boundary planes; 3 = Set of climbing ripples (note tangential relationship to bounding planes); 4 = Transition zone → decreasing ripple amplitude; 5 = Horizontal laminae; and 6 = Flame structure in shale layer.

Abundant sand supply from upstream or from suspension is essential for the development of ripple laminations. This process implies that sand or silt is introduced in such quantities as to cause rapid burial with only minor modification of the ripple form. Environments of periodic rapid accumulation favor the development of climbing ripples.

Slump structures can be recognized (point 2, photograph A, Figure 20-8). Carlson and Nelson (1969) have described laminae of sand and shale from the walls of a distributary on the Astoria Fan, which are similar to the slumped block in photograph A, Figure 20-8.

Massive sandstone shows homogeneous sands with diffuse laminations or faint flow structures (photographs A, B, and C, Figure 20-8). Very finely disseminated mica and carbonaceous material are present throughout these sandstones. Only a slight change in color, from light to dark tan, marks the flow features. Very faint dish structures (point 1, photograph C, Figure 20-8) can be observed at the top of some massive homogeneous sandstones; other sandstones contain lighter "bands" or "veins" that cut across the core (point 2, photograph C, Figure 20-8; point 1, photograph B, Figure 20-6). These massive sands are interpreted as resulting from fluidized grain flow, and the veins of lighter sandstone are regarded as avenues of fluid escape.

Flame structures (photographs B and D, Figure 20-8) are plumes or waves of shale squeezed irregularly upward into an overlying layer. The deformation of these shale layers is caused in part by the sinking of coarser clastic sediments into the soft shale substratum. Flame structures in some cases are accompanied by horizontal slip or drag (point 3, photograph A, Figure 20-7).

DEPOSITIONAL PROCESSES

The sedimentary structures from the Red Oak Sandstone provide insight into the mode of deposition. Evidence indicates that a majority of these features are products of mass flow (liquefied and fluid) in conjunction with a gravity current process. Graded bedding characteristic of "classical" turbidity current deposits is not present. Slumping and submarine sliding are only locally developed. The internal structures can only be associated with mass flow. Mass flow processes have been divided into fluidized and liquefied grain flows (Middleton and Hampton 1975).

Recently, Lowe (1976) has made the distinction between liquefied and fluidized systems. Liquefied beds are characterized by solids settling downward through the fluid, while in fluidized beds, the fluid moves upward through the granular solid. Lowe (1976) cited examples of liquefied flow deposits from the Atoka and Jackfork sandstones of southeastern Oklahoma. He described the sandstone beds as follows:

> composites of individual sedimentation units up to 50 cm thick which contain structures suggesting hydroplastic shear in the late stages of deposition followed or accompanied by partial liquefaction, foundering of superposed layers due to density instabilities, and vertical water escape [Lowe 1976, p. 300].

The differentiation between fluidized and liquefied systems goes beyond the scope and purpose of this chapter, but the depositional processes described by Lowe are similar to those indicated by structures from the Red Oak Sandstone.

DEVELOPMENT OF THE RED OAK FAN

The data on both gross and detailed lithologic attributes of the Red Oak, given above, allow interpretation of depositional events and facies and facilitate comparison with modern and other ancient canyon–fan systems. For example, Hamilton (1967) in the Gulf of Alaska described depositional valleys bounded by levees with convex-upward bedding surfaces. Channels within depositional fan valleys display erosional contact with the valley floor or show contemporaneous lateral relationships. Mutti (1974) described similar stratigraphic relationships in the upper portion of a submarine fan of the Ranzano Sandstone of the northern Italian Apennines.

In the Red Oak Sandstone, coarse-grained and thick-bedded sandstones are confined to the channelized portion of the submarine valley and display erosional contacts only along their base. Thin-bedded sandstones flank the coarser-grained channel fill and dip toward the deepest part of the channel. This relationship suggests that the two facies are laterally equivalent and demonstrates a facies pattern in which the lower sandstone filled the local submarine valley through lateral accretion. A similar depositional sequence of events is suggested by cross-section A-A' (Figure 20-9), which is interpreted as a channel fill at the lower end of an upper fan valley. Well No. 3 contains 100 m of continuous sand that shows an erosional contact at its base, but exhibits lateral continuity with thinner-bedded levee sandstones and shales of Well No. 4. In a downfan direction, increased interfingering of sands with alternating sands and shales at channel margins indicates that the fan valley is considerably less incised.

The branching pattern of mid-fan channels, coupled with rapid aggradation, leaves isolated channel remnants. The higher concentration of sand on the mid-fan portion of the Red Oak complex results from the accumulation of many thin sand beds. Eroding currents of migrating channels removed interbedded shales of previous channel fills and deposited sand intervals. Individual sand beds reach a thickness of 6 to 9 m, but channel fill often is

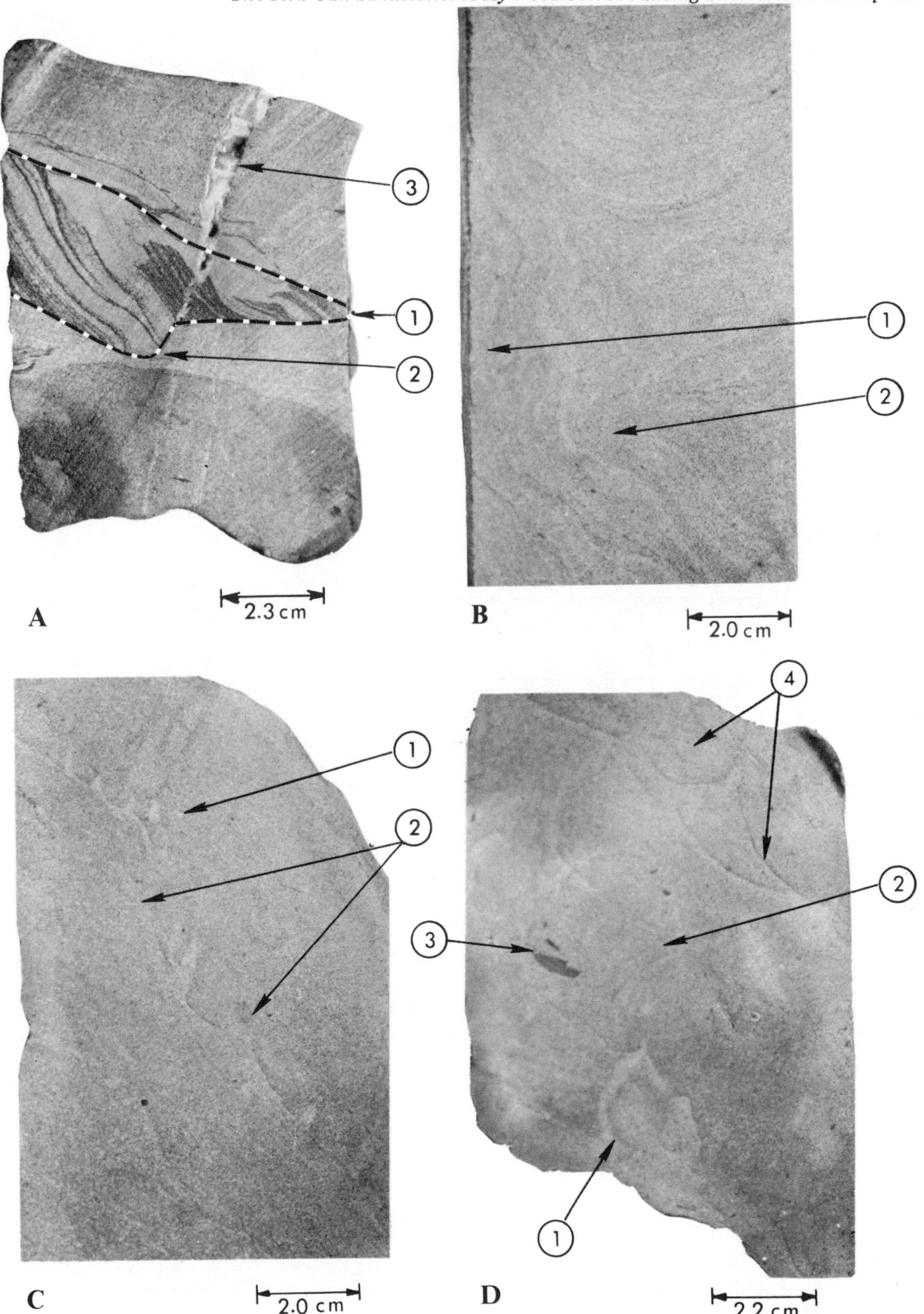

Figure 20-8. Sedimentary structures of Red Oak Sandstone. Note: Photograph A shows slump block, Midwest Sentry #1 Royalty, 17-6N-21E. 1 = Slump block (dashed outline); 2 = Depression in finer-grained material; and 3 = Late fracture, filled with kaolinite. Photograph B shows massive sandstone, Midwest Sentry #1 Royalty, 17-6N-21E (7793). 1 = Darker tan-colored sandstone laminae, and 2 = Flow feature. Photograph C shows massive sandstone, Midwest Ryder, 17-6N-22E (7010). 1 = Homogeneous fine-grained sandstone, and 2 = Continuous lighter sandstone vein, cutting across core = fluid escape "pipe." Photograph D shows massive sandstone, Midwest #1 Ryder, 17-6N-22E (7009). 1 = Lighter sandstone vein; 2 = Homogeneous tan sandstone; 3 = Shale clasts; and 4 = Faint dish structures with slightly darker cup outline.

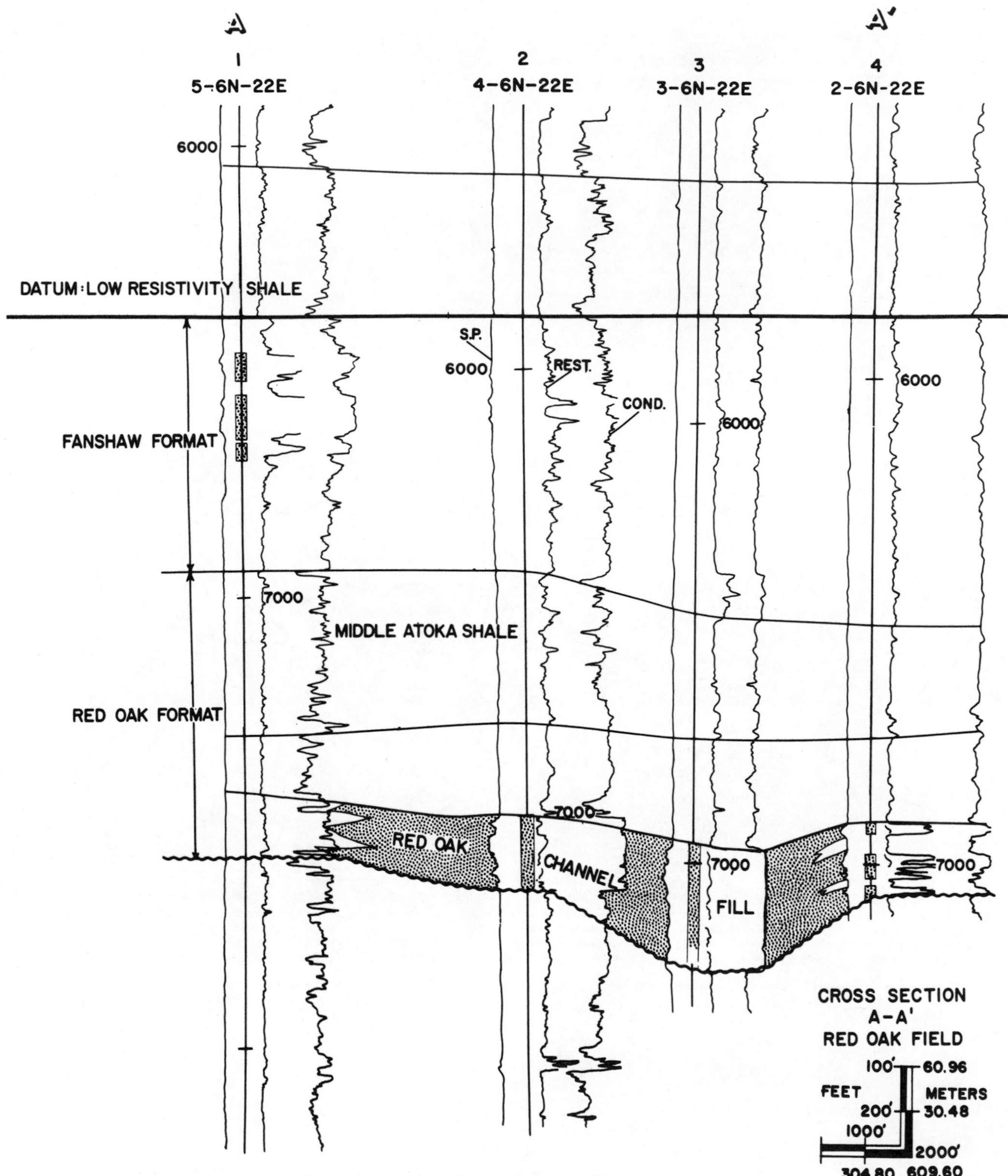

Figure 20-9. Cross-section A-A' – profile across lower end of upper fan valley. Note: S. P. = self potential; Rest. = resistivity; and Cond. = conductivity. See Figure 20-2 for location.

more than 50 m. "Stacking" of mid-fan channels might be related to greater subsidence in channel areas caused by higher rates of sedimentation. This stacking could be self-perpetuating until some upfan process curtailed the sand supply. Walker (1966) described similar stacking of sand units in Upper Carboniferous deposits of northern England. Cross-section C-C' (Figure 20-10) is a profile across a mid-fan channel complex in the western portion of the mapped area, which exhibits both east-west and north-south sand trends. The thick sandstone unit in Well No. 2 is about 55 m thick, while an adjacent well (No. 3) has three distinct sand units each about six to nine meters thick. These mid- and upper fan channels consist of massive fine-grained sandstones displaying dish structures, faint parallel stratification, and flow features.

Shifting of active channels within the fan (upper and middle) resulted in aggradation of the fan, similar to the convex-upward bulge of sediment (the suprafan) described from modern fans by Normark (1970). Aggradation of the Red Oak fan influenced the deposition of the overlying Fanshaw Sandstone. A compensatory depositional pattern resulted between the Red Oak and Fanshaw submarine fans. This typical relationship is illustrated by cross-section A-A' (Figure 20-9), where Well No. 1 has a thick Fanshaw Sandstone overlying a Red Oak interval with no sand. Wells No. 2, 3, and 4 have Red Oak Sandstone representing aggradational fill of an upper fan valley channel, while the Fanshaw interval exhibits thin sandstone units interbedded with thicker shale intervals.

RED OAK SEDIMENT INPUT

The Red Oak deposits discussed above are associated with a shelf of intermediate width. Clastics are fine grained, exhibit depositional processes associated with fluidized flow mechanisms, and contain terrigenous carbonaceous material from prograding Atoka deltas of the shelf.

Typically, the input of clastics into the canyons is believed to have resulted from (1) interception of the southwest littoral drift on the shelf and (2) progradation of deltaic clastics into canyon heads. Rising sea level and subsidence by growth faulting resulted in shifting of sediment source away from canyon heads. The Red Oak canyon-fan system was able to maintain the canyon head in the near-shore area by headward erosion, and for a short period of time during the middle Atoka, clastics were available for the development of a complex of fans on the slope and rise.

FANS AND PETROLEUM GEOLOGY

Few stratigraphic studies are available of hydrocarbon-producing fan depositional sequences. Deeper drilling in sedimentary basins and on the continental shelves indicates that the canyon-fan sequence is commonly developed, but published reports on stratigraphic and sedimentary patterns are lacking. Most studies of ancient submarine fans, for example the Tarzana Fan of the Los Angeles Basin (Sullwold 1960), Shale Grit and Grindslow shales of northern England (Walker 1966), Cambro-Ordovician strata in Quebec (Hubert et al. 1970), Eocene of the Bay of Biscay of Spain and France (Kruit et al. 1972), the Butano Sandstone of California (Nelson and Nilsen 1974), and chapters in Part III of this volume are based primarily on outcrop investigations. Data on detailed subsurface geometry and lithologic sequences within regions of hydrocarbon potential (cf. Bloomer 1977; also Chapter 19 of this volume) are generally less available. For a fuller review of such subsurface investigations of canyon-fan systems, see Whitaker (1976).

A submarine fan origin has been suggested for the Bell Canyon Formation of the Delaware Basin (Jacka et al. 1968), the Pennsylvanian of north-central Texas (Galloway and Brown 1973), and the Paleocene of the North Sea basin (Parker 1975). As indicated by the electric logs and cores interpreted by the writers, producing Lower Pennsylvanian (Springeran) strata of the Ardmore Basin (Goddard Sandstone), and the Anadarko Basin (Cunningham Sandstone) of southern Oklahoma also are of a fan origin. Developmental drilling in these producing areas will provide additional subsurface control to determine the depositional history of these fans.

RED OAK FIELD

The Red Oak field data illustrate the depositional history, the syn- and postsedimentary tectonic development, and the history of generation and entrapment of hydrocarbons. Whether this developmental history can be the basis for a model for petroleum occurrence in fans cannot be determined from the available data, but many observations from the Red Oak Sandstone are similar to those from other producing fan trends.

Reservoir Patterns

Submarine mass flow channel sandstones in the Red Oak field are excellent reservoirs. These sandstones are low in clay matrix (less than 10 percent), and even with burial to depths of more than four kilometers, core analysis data indicates porosity from 10 to 15 percent, and permeability up to 100 md. The lenticular geometry of the channels provides natural stratigraphic traps for hydrocarbons. The topographic pattern of channels and channel fill allows the updip migration of hydrocarbons into the massive channel sands (Figures 20-4 and 20-11). Overbank deposits and distal fan sand

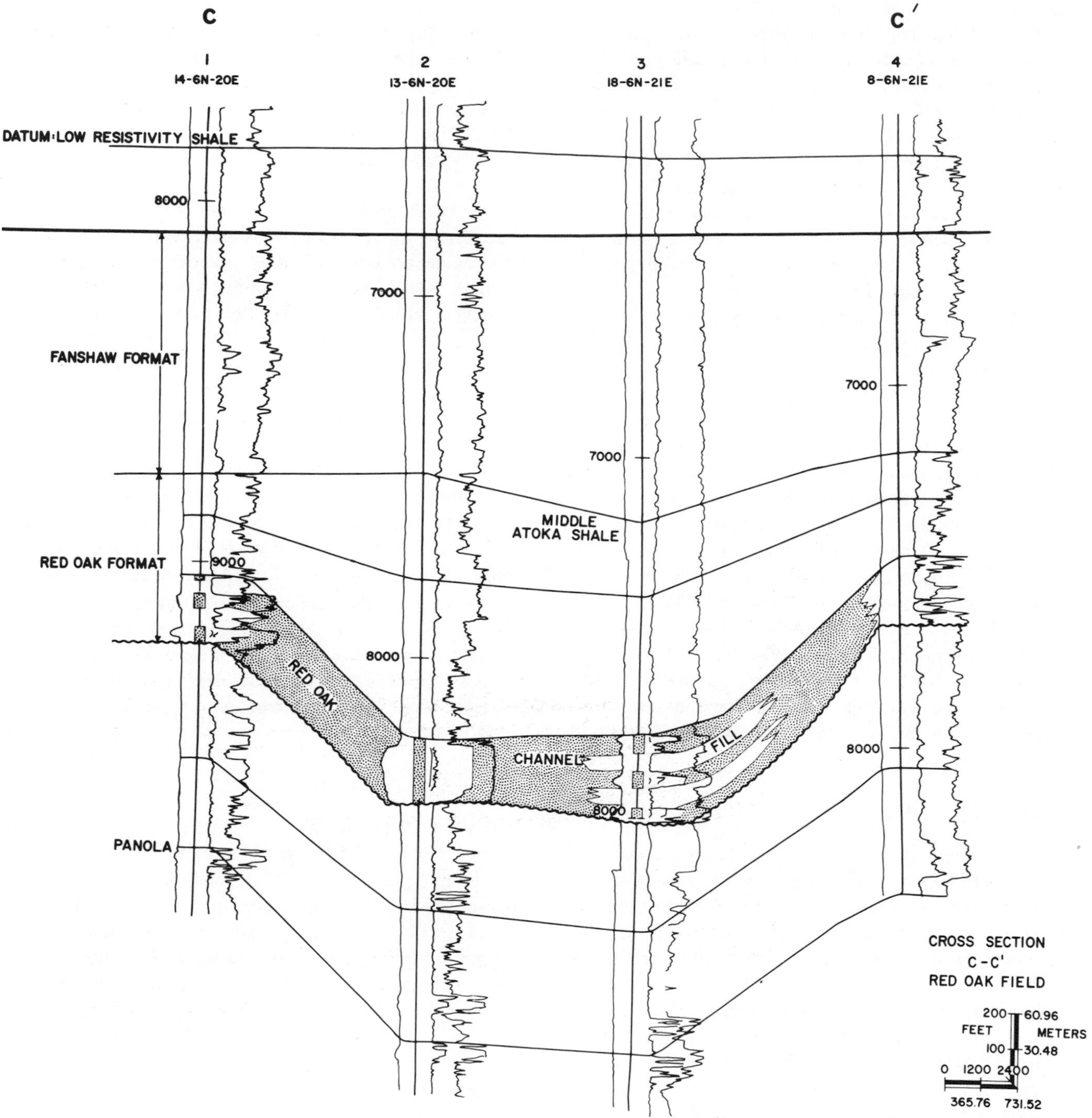

Figure 20-10. Cross-section C-C' – profile across mid-fan channel complex. Note: See Figure 20-2 for location.

and silt laminae are continuous in an updip direction, and provide a natural feeder system for the migration of generated hydrocarbons.

The pinchout of the Red Oak Sandstone updip against a growth fault is the main trapping mechanism (cross-section B-B', Figure 20-11). The fan is wedge shaped against this growth fault, and correlation of equivalent stratigraphic markers across the fault demonstrates a 1,300 m loss in section. Electric logs indicate that porosity is present in nearly all sandstone units south of the fault and commercial reservoir patterns reflect channel geometry.

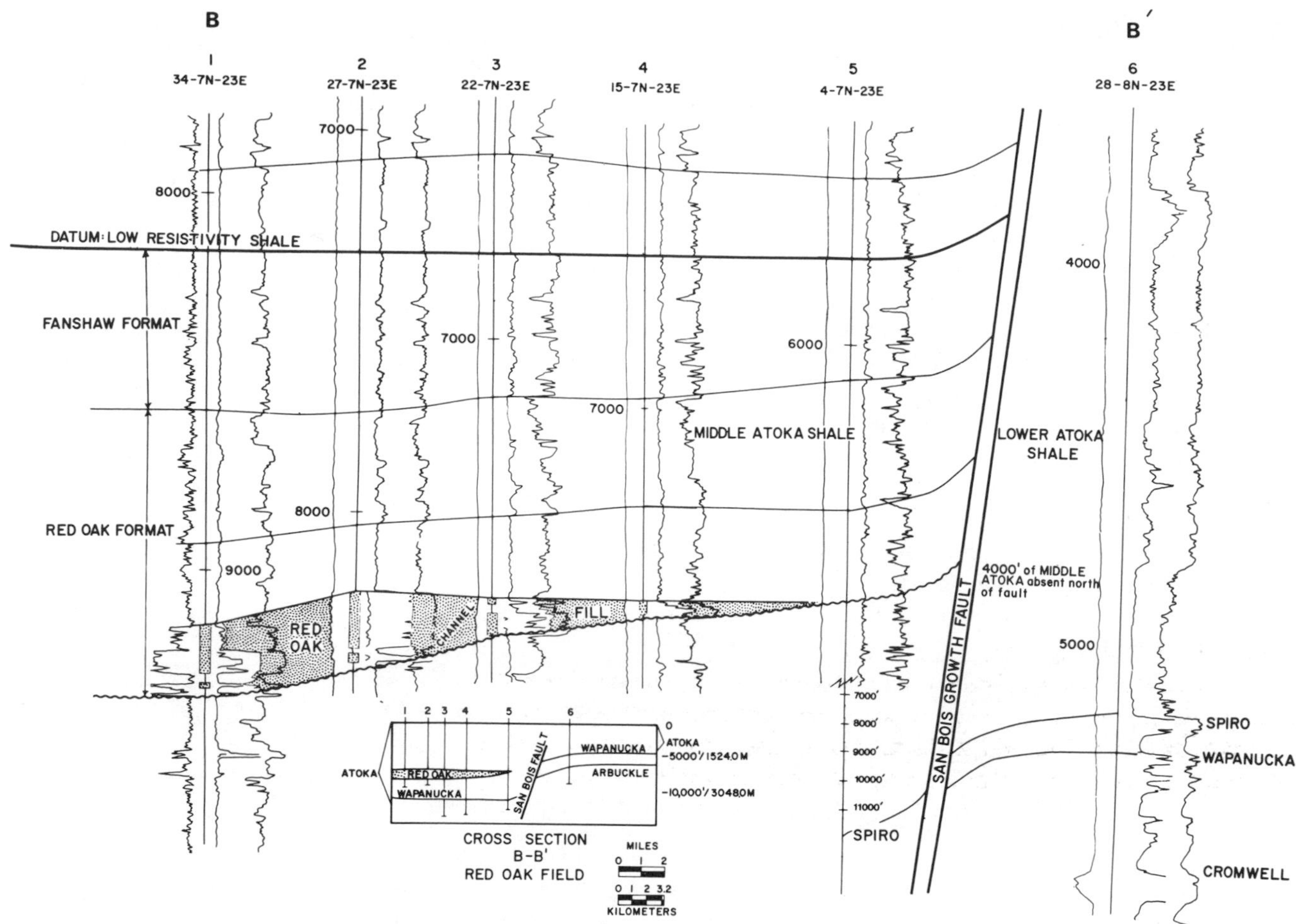

Figure 20-11. Cross-section B-B' – near updip limit of Red Oak Sandstone (upper fan). Note: See Figure 20-2 for location.

Structural Patterns

The Brazil Anticline, a surface structure, trends almost east-west in a line parallel to the surface trace of the Choctaw Fault, the frontal thrust of the Ouachita orogenic belt. The surface structure is nearly coincident with the Red Oak field. Stratigraphic sections suggest that the early history of the Brazil Anticline was related to a marginal growth fault and a mounded lens-shaped buildup resulted from the formation of the fan.

Similar mounded uplifts occur at the base of the continental slope marginal to the shelf in the Gulf of Mexico. They also have been reported by Menard (1955) in the Cascadia basin, and by Hamilton (1967) in the Gulf of Alaska. Such ridges, resulting from a combination of depositional and synsedimentary tectonic and diapiric processes, produce structural and topographic relief across the fan. Folds of this nature are the result of either compressional folding from gravity sliding of very plastic material or vertical diapiric movement of this material in response to loading.

The presence of a depositional fan bounded by an updip growth fault apparently influenced the later tectonic history of the area. Later thrust faulting resulted in repetition of the Red Oak section and renewed growth of the anticlinal structure. Thus, the Brazil Anticline is interpreted as a combination of depositional, diapiric, and post-depositional thrust patterns.

Seismic exploration in the area was concentrated on the subsurface definition of the Brazil Anticline, and dip reversal was thought to be the trapping mechanism. The association of production with the sandstone distribution and the absence of a gas-water contact suggest that the primary control for production was the presence of reservoir and not structure. Pinch-out of the sandstone updip onto the fan, or possibly against the basin slope, provided the trap. The presence of a depositional or tectonic ridge seaward of the fan could also have influenced the depositional pattern, and produced a ponding of sediment. The sandstone geometry pattern (Figure 20-4) indicates that multiple channels developed and that in some instances sand was deposited parallel to the slope rather than distributed only in a downslope direction.

Developmental History

Tectonic patterns in part controlled and in part were controlled by the sedimentation. The postulated ridge may have been responsible for accretion of the Red Oak fan. Channels show scour followed by an accretionary fill (Figure 20-10). The ridge also could have controlled the basinward flow of sediments and resulted in the accumulation of thick reservoir sandstones.

The formation of thick deltaic deposits at the shelf edge, or at the head of a canyon, controlled the development of growth faults. Growth faulting restricted the updip extension of the fan, thereby producing updip pinch-outs of reservoirs and a trap. Source rocks associated with organic-rich basinal shales generated hydrocarbons that could only migrate into the reservoirs higher on the fan. Later thrusting, controlled by the depositional history, produced the final distribution of the hydrocarbons.

Exploration Methods

The complexity of the developmental history makes exploration for hydrocarbon accumulations in canyon and fan systems difficult. The seismic tool can be usefully applied to the definition of the stratigraphic and structural framework. Seismic modeling, utilizing the differences in velocity between shale and sandstone, migration of reflection events, and separating topographic and structural patterns, can provide a basis for interpretation. Well logs can be used to discriminate lithologies, compaction history, paleocurrent patterns, and continuity of reservoir sandstones, thus forming the basis for interpreting the fan paleogeography. Mapping channel fill across the fan produces a pattern than can be used to determine the thickness, quality, and distribution of hydrocarbon reservoirs.

PALEOGEOGRAPHIC HISTORY AND CONCLUSIONS

The southern half of the Arkoma Basin and the frontal Ouachitas represent a foreland trough where Atoka sediments are the thickest. A marginal basin formed during the Devonian and migrated northward toward the craton during the later Paleozoic. Ouachita facies (post-Silurian - pre-Pennsylvanian) geosynclinal strata are distinct from shelf or Arbuckle facies rocks. Resistance of the craton to the northward migration of the geosyncline resulted in steepening of the slope. This steepened slope was associated with faulting of pre-Mississippian strata, which in turn resulted in slumping and the deposition of the Johns Valley boulder beds in the subsiding trough.

The southward regional dip of the Ozark uplift and associated pre-Atoka topography caused Atoka rivers, draining the eastern interior of the continent, to flow into Arkansas and Oklahoma. The Atoka fluvial-deltaic system on the shelf exhibits regressive-transgressive clastic deposition in a series of fluvial-deltaic lobes. Gravity processes moved Atoka sediment down the marginal slope and into the subsiding Arkoma Basin. Following the deposition of the shallow-water Morrowan Spiro sandstone, growth faulting and gravity tectonics enhanced subsidence and localized sedimentation into depocenters. Within this framework the Red Oak submarine fan was deposited.

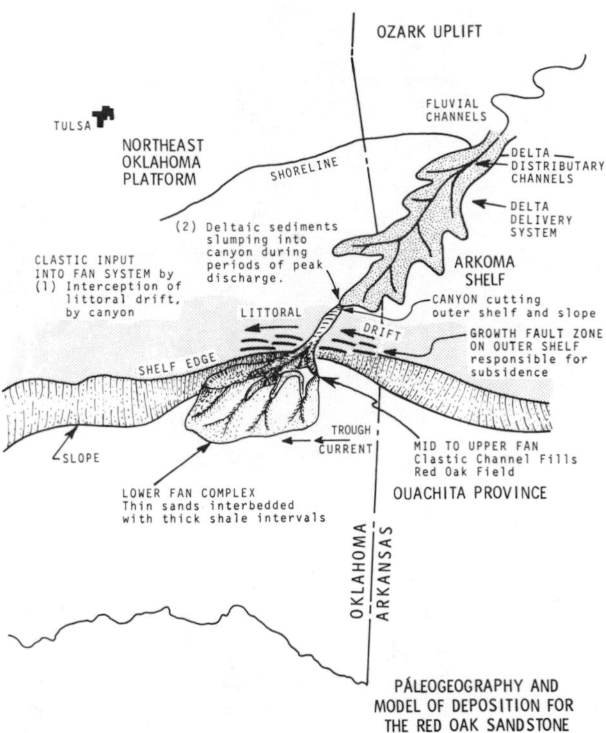

Figure 20-12. Paleogeography and model of deposition for the Red Oak Sandstone.

The Red Oak fan was formed by the interception of sands on the shelf by submarine canyons and downslope distribution through submarine fan channels resulted in deposition of coarse clastics in an environment dominated by basinal silts and shales (Figure 20-12). Closing of Atoka deposition saw the fill of the basin and the decreasing importance of the slope as an influence on deposition. Subsequent filling of the basin by deltaic processes during the upper Atoka and Desmoinesian is recorded in overlying stratigraphic units.

Evidence that the Red Oak Sandstone is a submarine fan is based on (1) deposition in an environment dominated by basin and slope shales, (2) abundant fluidized and liquefied flow structures, (3) development of channel and interchannel facies, (4) an upward bulge in upper and mid-fan portions that is indicated by aggradational fill of channels, and (5) proximal to distal stratigraphic facies including clastic-filled channels on the upper and mid-fan, and alternating sands and shales on the lower fan and basin floor.

Subsurface indicators of a canyon-fan system include (1) thick sandstone units surrounded by monotonous sequences of basinal shales and silts, (2) an abrupt increase in thickness of a mappable stratigraphic interval, (3) well log signatures that demonstrate thick channel fill with sharp contacts at the bottom, interbedded sands and shales, and isolated channel remnants, (4) straight-line channel geometry on the upper fan, a complex lateral and longitudinal branching pattern on the mid-fan, (5) sedimentary structures indicative of fluidized flows, including dish, dewatering, climbing ripple, and flame structures, shale clasts, and carbonaceous material, and (6) massive sandstones lacking either graded bedding or a clay matrix.

Petroleum exploration and field development within a fan system require an understanding of the interrelation of depositional and tectonic processes. The inferred developmental history of the Red Oak field indicates a genetic pattern that may be useful in developing a model applicable to other fields having more limited stratigraphic control.

ACKNOWLEDGMENTS

Appreciation is expressed to the Northern Natural Gas Company for assistance in the reproduction of maps and well logs, to the Oklahoma Well Core Library for use of their facilities and to the reviewers whose help is gratefully acknowledged.

REFERENCES

Bloomer, R. R., 1977. Depositional environments of a reservoir sandstone in west-central Texas. *Amer. Assoc. Petrol. Geol. Bull.*, 61: 344-59.

Briggs, G., and Cline, L. M., 1967. Paleocurrents and source of late Paleozoic sediments of the Ouachita Mountains, southeastern Oklahoma. *J. Sed. Petrol.*, 37: 985-1000.

Carlson, P. R., and Nelson, C. H., 1969. Sediments and sedimentary structures of the Astoria submarine canyon-fan system, northeast Pacific. *J. Sed. Petrol.*, 39: 1269-82.

Chamberlain, C. K., 1971. Bathymetry and paleoecology of Ouachita geosyncline of southeastern Oklahoma as determined from trace fossils. *Amer. Assoc. Petrol. Geol. Bull.*, 55: 34-50.

Cline, L. M., 1960. Stratigraphy of the late Paleozoic rocks of the Ouachita Mountains, Oklahoma. *Oklahoma Geol. Surv. Bull.*, 85, 113 pp.

———, 1966. Late Paleozoic rocks of Ouachita Mountains, a flysch facies. *Flysch Facies and Structures of the Ouachita Mountains Field Conference Guide Book.* 29th Kansas Geol. Soc., pp. 91-111.

Coleman, J. M., and Gagliano, S. M., 1965. Sedimentary structures; Mississippi River deltaic plain. In: G. V. Middleton (ed.), *Primary Sedimentary Structures and their Hydrodynamic Interpretation.* Soc. Econ. Paleont. Mineral., Sp. Publ. 12, pp. 133-48.

Dzulynski, S., Ksiazkiewicz, M., and Kuenen, P. H., 1959. Turbidites in flysch of the Polish Carpathian Mountains. *Geol. Soc. Amer. Bull.* 70: 1089-18.

Galloway, W. E., and Brown, L. F., Jr., 1973. Depositional systems and shelf-slope relations on cratonic basin margin, uppermost Pennsylvanian of north-central Texas. *Amer. Assoc. Petrol. Geol. Bull.* 57: 1185-218.

Hamilton, E. L., 1967. Marine geology of abyssal plains in the Gulf of Alaska. *J. Geophys. Res.* 72: 4189-213.

Honess, C. W., 1924. Geology of southern Leflore and northwestern McCurtain Counties, Oklahoma. *Oklahoma Geol. Surv. Circ.* 3.

Hubert, C., Lajoie, J., and Leonard, M. A., 1970. Deep-sea sediments in the lower Paleozoic Quebec Supergroup. In: J. Lajoie (ed.), *Flysch Sedimentology in North America.* Geol. Soc. Canada, Spec. Paper 7, pp. 103-25.

Jacka, A. D., et al., 1968, Permian deep-sea fans of the Delaware Mountains Group, Delaware basin. *Field Trip Guidebook for 1968 Symposium, Guadalupian Facies Apache Mountain Area, West Texas.* Soc. Econ. Paleont. Mineral., Permian Basin Section, 11 pp.

Koimn, D. N., and Dickey, P. A., 1967. Growth faulting in McAlester basin of Oklahoma. *Amer. Assoc. Petrol. Geol. Bull.*, 51: 710-18.

Kruit, D., Brouwer, J., and Ealey, P., 1972. A deep water sand fan in the Eocene Bay of Biscay. *Nature,* 240: 59-61.

Lowe, D. R., 1976. Subaqueous liquefied and fluidized sediment flows and their deposits. *Sediment.* 23: 285-308.

McKee, E. D., 1965. Experiments on ripple lamination. In: G. V. Middleton (ed.), *Primary Sedimentary Structures and Their Hydrodynamic Interpretation.* Soc. Econ. Paleontol. Mineral., Sp. Publ., 12, pp. 66-83.

———, Reynolds, M. A., and Baker, C. H., Jr., 1962. Experiments on intraformational recumbent folds, in cross-bedded sand. *U.S. Geol. Surv. Paper* 450-D: 155-60.

Menard, H. W., Jr., 1955. Deep-sea channels, topography, and sedimentation. *Amer. Assoc. Petrol. Geol. Bull.*, 39: 236-55.

Middleton, G. V., and Hampton, M. A., 1976. Subaqueous sediment transport and deposition by sediment gravity flows. In: D. J. Stanley and D. J. P. Swift (eds.), *Marine Sediment Transport and Environmental Management.* Wiley, New York, pp. 197-218.

Mutti, E., 1974. Examples of ancient deep-sea fan deposits from circum-Mediterranean geosynclines. In: R. H. Dott, Jr., and R. H. Shaver (eds.), *Modern and Ancient Geosynclinal Sedimentation.* Soc. Econ. Paleont. Mineral., Sp. Publ. 19, pp. 92-105.

Nelson, C. H., and Nilsen, T. H., 1974. Depositional trends of modern and ancient deep-sea fans. In: R. H. Dott, Jr. and R. H. Shaver (eds.), *Modern and Ancient Geosynclinal Sedimentation.* Soc. Econ. Paleont. Mineral., Sp. Publ. 19, pp. 69-91.

Normark, W. R., 1970. Growth patterns of deep-sea fans. *Amer. Assoc. Petrol. Geol. Bull.*, 54: 2170-2195.

Parker, J. R., 1975. Lower Tertiary sand development in the central North Sea. In: A. W. Woodland (ed.), *Petroleum and the Continental Shelf of Northwest Europe.* Wiley, New York, pp. 447-53.

Shideler, G. L., 1970. Provenance of Johns Valley boulders in late Paleozoic Ouachita facies, southeastern Oklahoma and southwestern Arkansas. *Amer. Assoc. Petrol. Geol. Bull.*, 54: 789-806.

Stark, P. H., 1966. Stratigraphy and environment of deposition of the Atoka in the Ouachita Mountains, Oklahoma. *Flysch Facies and Structures of the Ouachita Mountains Field Conference Guide Book.* 29th Kansas Geol. Soc., pp. 164-76.

Stauffer, P. H., 1967. Grain-flow deposits and their implications, Santa Ynez Mountains, California. *J. Sed. Petrol.* 37: 487-508.

Sullwold, H. H., 1960. Tarzana Fan, deep submarine fan of late Miocene age, Los Angeles, County, California. *Amer. Assoc. Petrol. Geol. Bull.,* 44: 433-57.

Walker, R. G., 1966. Shale Grit and Grindslow Shales: Transitions from turbidite to shallow water sediments in the upper Carboniferous of northern England. *J. Sed. Petrol.,* 36: 90-114.

_____1966, Deep channels in turbidite-bearing formations. *Amer. Assoc Petrol. Geol. Bull.,* 50: 1899-917.

Weirich, T. E., 1953. Shelf principles of oil origin, migration, and accumulation. *Amer. Assoc. Petrol. Geol. Bull.,* 37: 2027-45.

Wentworth, C. M., 1967. Dish structure, a primary sedimentary structure in turbidites (Abst.). *Amer. Assoc. Petrol. Geol. Bull.,* 51: 485.

Whitaker, J. H., McD. (ed.), 1974. *Submarine Canyons and Deep-Sea Fans, Modern and Ancient.* Benchmark Pap. Geol. 24. Dowden, Hutchinson & Ross, Stroudsburg, Pa., 460 pp.

Part V

Tectonics and Sedimentation in Arc and Trench Basins

Chapter 21

Depositional Patterns and Channelized Sedimentation in Active Eastern Pacific Trenches

WILLIAM J. SCHWELLER
LAVERNE D. KULM

School of Oceanography
Oregon State University,
Corvallis

ABSTRACT

Trenches associated with convergence zones can be modeled as elongate sediment basins that undergo simultaneous deposition and deformation. Data from seismic reflection and bathymetric profiles, sediment cores, and DSDP sites in the Peru-Chile and Aleutian Trenches and the Oregon-Washington filled trench have been utilized to analyze convergence zone deposits along continental margins.

The shape of the trench basin is determined by the downbending oceanic plate and the deformation front at the base of the continental slope. Four main sediment facies can be distinguished: pelagic plate, terrigenous plate, trench wedge, and fan. These can occur in a variety of facies associations. Large trench wedges and fans seem to develop mainly by deposition from channelized turbidity flows. Fans grow radially outward in low cone-shaped deposits, while trench wedges are restricted to the narrow, asymmetrical trench basins. Both fans and trench wedges lap seaward over the plate sediments, thereby creating slight angular unconformities.

The sedimentation system in the Chile Trench resembles that of a deepsea fan rather than a series of ponded turbidite basins as suggested by earlier studies. Axial channels in both the Chile and Aleutian Trenches appear to transport sediment hundreds of kilometers along trench axes without producing levees of high relief. The Chile Trench channel changes form in response to variations in the axial gradient along its 1,000 km length. The position of these channels in the axis affects the distribution of channel sands within the trench wedge. Sand bodies in fans are less continuous than those in trench wedges because the distributary fan channels migrate laterally at more frequent intervals.

A continual supply of turbidites to the trench axis is needed to sustain a constant volume of undeformed trench sediments. Interactions between sediment supply and convergence rates determine the type of trench deposits and sedimentary bodies present. In general, trench wedges tend to develop along rapidly converging margins, while fans form along margins with slower convergence rates and greater sediment supply.

INTRODUCTION

Submarine fan and fan channel sedimentation has been recognized as an important depositional system in a number of studies of passive continental margins and margins of translation (e.g., Nelson and Kulm 1973; Normark 1974; Nelson and Nilsen 1974; and references therein). Relatively few efforts have been directed toward understanding comparable processes occurring in trenches along active or convergent continental margins (Piper et al. 1973; von Huene 1974; Scholl and Marlow 1974; other chapters in this section of the volume). Fan sediments located along passive margins generally remain undisturbed following their deposition, while deposits in convergent zones are continually deforming and changing position even as new sediments are deposited over them. This dynamic interaction of sedimentation and convergent tectonism can create major deviations from the more thoroughly studied patterns of sediment transport and deposition along passive margins.

This study presents a conceptual model for sedimentation along convergent continental margins, followed by a more empirical analysis of sediment bodies produced by channelized sediment flows in ocean trenches. In order to fully gauge the variety of possible deposits and facies relationships, we use data from three tectonically and sedimentologically diverse settings in the eastern Pacific Ocean: the eastern Aleutian Trench, the Oregon-Washington Trench, and the Peru - Chile Trench (Figure 21-1). The eastern Aleutian Trench is an area of moderate, 5 to 6 cm/year, convergence (Minster et al. 1974) with high rates of sedimentation in the trench and on the incoming oceanic plate. The Oregon-Washington convergence zone is believed to be an ancient trench that has been completely filled due to a high rate of sedimentation and slow, 2 to 3 cm/year, convergence (Silver 1969; Atwater 1970). Sediment input to the Peru-Chile Trench is low to moderate along most of its 5,000 km length, while convergence is a rapid 9 to 11 cm/year (Minster et al. 1974). Only the southernmost portion of the Chile Trench has a sediment input rate comparable to the Aleutian and Oregon-Washington trenches.

Data used to evaluate these areas include four Deep Sea Drilling Project (DSDP) sites, seismic reflection profiles, piston cores, and detailed bathymetric surveys. Except for the DSDP sites, the bulk of the data from the Oregon-Washington and Peru-Chile areas was collected on various cruises by Oregon State University (OSU). Seismic reflection profiles in the Aleutian Trench are from published studies by von Huene (1972) and Piper et al. (1973).

CONCEPTUAL MODEL

Basement Structure

The form and distribution patterns of sediments deposited in trenches along convergent margins are dependent to a large extent upon the configuration of the oceanic basement. The downward curvature of most trenches can be modeled as a thin elastic slab with a vertical load on one end (Caldwell et al. 1976). A slight topographic rise (the outer swell) is often produced just seaward of the trench because of this flexure (Parsons and Molnar 1976). The dip of the oceanic plate steepens from nearly horizontal at the seaward edge of the trench to an average of 5° to 7° near the trench axis. As the dip increases, strata deposited in originally horizontal layers on the oceanic crust will be tilted progressively more landward with time.

Downbending of the oceanic plate creates superimposed structures that can disrupt existing oceanic plate sediments and control the supply of terrigenous material to various segments of the trench axis (Schweller 1976). For example, basaltic basement ridges up to 900 m high protrude through axial sediments in the Peru Trench, at times dividing the axis into an inner and an outer basin (Kulm et al. 1973; Prince and Kulm 1975). These ridges are thought to be caused by basement thrust faults resulting from plate collision. Large graben structures with vertical offsets of up to 1,000 m have been observed on the seaward trench slope of the Chile Trench (Schweller 1976). These grabens occur further seaward from the collision zone than the thrust ridges and are probably produced by extensional rupture of the oceanic plate as it begins to downbend into the trench.

The landward boundary of most trench deposits is the zone at the base of the continental slope where the sediments riding on the oceanic plate collide with the continental block. This collision zone is commonly thought of as a nearly vertical front of deformation that is approximately stationary relative to the overriding block but that apparently advances through the sediments carried on the oceanic plate at a speed equal to the convergence rate (Helwig and Hall 1974). Although the fate of sediments after they pass through this deformation front is important to the evolution of the continental margin, that topic is beyond the scope of this study. We will be concerned primarily with the basically undeformed trench sediments seaward of this deformation front.

Sediment Facies

Sediments deposited in the tectonically active deepsea trenches are highly diverse with respect to source and composition, but for our purposes they can be categorized into four main facies: (1) pelagic plate, (2) terrigenous plate, (3) trench wedge, and (4) fan. Each of these facies can be conceptually and lithologically defined as a major component within an idealized trench sediment deposit (diagram A, Figure 21-2). They can also occur in various combinations with or without the other three facies. We will first briefly define each of the four component facies, present some actual occurrences, then proceed to the various possible facies associations.

The pelagic plate facies is the open ocean sediment component, predominantly deposited by settling through the water column. Accumulation rates are very low, typically 2 to 5 mm/1,000 years, or 2 to 5 m/million years for regions below the carbonate compensation depth (CCD).

The terrigenous plate facies begins to dominate plate sediments as the oceanic plate approaches the continent. This facies includes hemipelagic sediments deposited by settling of silt- and clay-sized particles from a relatively turbid water column, fine- to coarse-grained turbidity current deposits, and ice-rafted glacial erratics (in higher latitudes). The transition from pelagic to terrigenous plate facies usually occurs gradually over a distance of several hundred kilometers and is accomplished by overprinting and dilution of the pelagic components rather than by their exclusion. Accumulation rates vary from the typical pelagic values given above to as high as 175 m/million years for the terrigenous deposits on the Aleutian Abyssal Plain (Kulm, von Huene et al. 1973).

The trench wedge facies consists predominantly of horizontally layered, graded sand and silt turbidites.

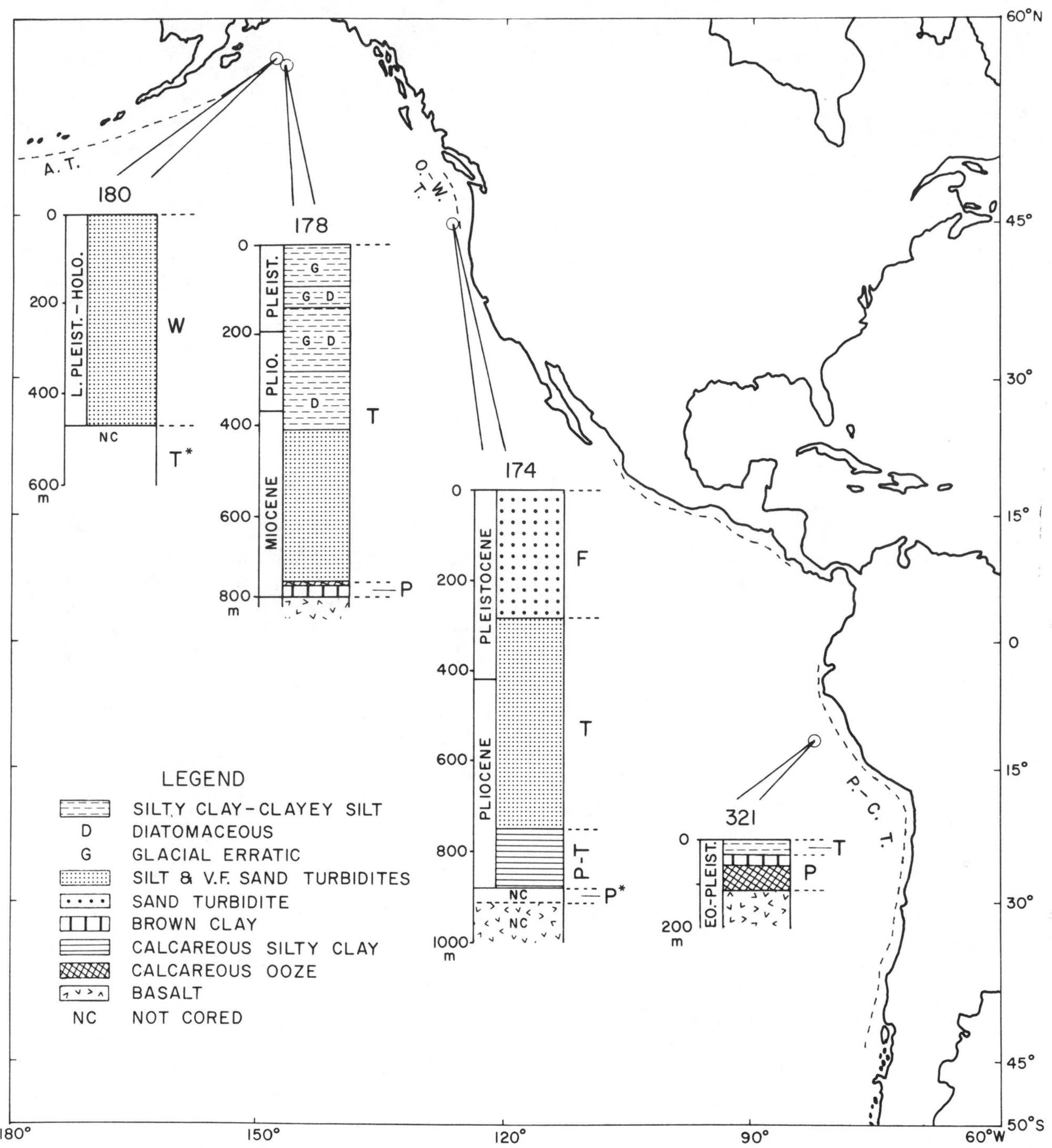

Figure 21-1. DSDP sites in eastern Pacific convergence zones, with lithologic columns and facies. Note: P = pelagic plate; T = terrigenous plate; W = trench wedge; and F = fan. Dashed lines are trench axes: A. T. = Aleutian; O. -W. T. = Oregon-Washington; and P. -C. T. = Peru-Chile.

Turbidity currents originate on the outer continental shelf and upper slope and travel down submarine canyons to form a wedge within the trench axis. The shape and extent of the wedge is dictated principally by the structure of the landward and seaward trench walls, the rate of sediment influx to the axis, and the convergence rate. A slight angular discordance between the wedge and the underlying pelagic or terrigenous plate facies is formed because of the downbending of the oceanic crust prior to encountering the area of trench wedge facies deposition. This discordance represents a seaward transgressive onlap of trench turbidites over the plate facies with time (Figure 21-3). Sedimentation rates can range from 300 to 3,000 m/million years (Kulm, von Huene et al. 1973) or

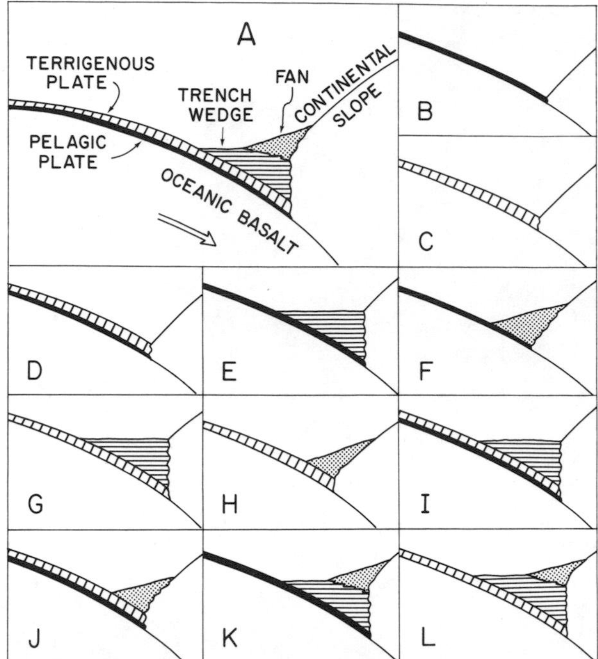

Figure 21-2. Conceptual models of sedimentary facies configurations along convergent margins. (See text for discussion.)

as much as an order of magnitude greater than for the terrigenous plate facies. As a result, wedges more than a kilometer thick have been deposited in both the Aleutian and southern Chile Trenches during the last few hundred thousand years or less (Scholl et al. 1970; von Huene 1972). Seismic reflection profiles across these wedges often show a large axial channel disrupting the otherwise horizontal reflectors (Figures 21-3 and 21-4). Sands are most abundant in and near the channels with increasing proportions of silt and clay laterally away from the channel banks (Nelson and Kulm 1973; Piper et al. 1973).

Deepsea fans along convergent margins are formed by similar processes from the same types of turbidity current deposits as trench wedges. The flattened cone-shaped fan builds outward away from the base of the continental slope either directly upon pelagic or terrigenous plate facies or over a trench wedge (Figure 21-5). An angular discordance would be expected between a fan and either the pelagic or terrigenous plate facies, but probably not between a fan and a trench wedge. The fan channel system usually consists of a main channel and a system of branching distributaries. Details of fan morphology and sedimentation have been summarized by Normark (1970, 1974) and Nelson and Kulm (1973). Accumulation rates are greatest from near the fan apex to the edge of the suprafan (Normark 1970) and decrease to much lower values at the fan edge. Maximum rates of 940 m/million years have been reported for the Astoria Fan (Kulm, von Huene et al. 1973).

Facies Patterns

The four main facies (pelagic plate, terrigenous plate, trench wedge, and fan) can combine in various ways to form many configurations or arrangements of convergence zone deposits in response to different conditions of tectonism and sedimentation. Some of the more common forms are shown in Figure 21-2. If no turbidity current deposits reach the trench axis, pelagic, and terrigenous plate facies enter the subduction zone as a simple blanket of sediment on the oceanic plate (diagrams B, C, and D, Figure 21-2). Turbidite deposits added to the trench axis usually form a horizontally layered trench wedge that onlaps the plate facies (diagrams E, G, and I, Figure 21-2). Fans can be superimposed on a trench wedge (diagrams A, K, and L, Figure 21-2) or deposited directly over pelagic or terrigenous plate sediments (diagrams F and J, Figure 21-2).

Other combinations of these facies are possible but are much less likely to occur in nature because they require rather unusual conditions. Faulting of the oceanic basement either as thrust ridges or horst-graben blocks can affect the distribution of trench sediments during and after their deposition. One example of trench wedge facies resting directly on basalt in the Peru Trench apparently resulted from thrust faulting within the axis (Prince and Kulm 1975). Large structural blocks can temporarily isolate sections of the axis from regions of active turbidite deposition by preventing axial transport. Uplift of small areas of crust above the level of turbidite deposition in the trench axis can result in localized hiatuses in an otherwise continuous section. However, the low frequency of such scattered occurrences renders them only of minor importance.

FACIES ASSOCIATIONS FROM EASTERN PACIFIC TRENCHES

Present-day examples of the different facies lithologies and distribution patterns can be seen in four DSDP sites in the eastern Pacific (see Figure 21-1). Sites 321 (Peru) and 178 (Aleutians) will be used to illustrate pelagic and terrigenous plate facies, while sites 180 (Aleutians) and 174 (Oregon) provide information on trench wedge and fan deposits. The locations, lithologic columns, and diagrammatic cross-sections for these four sites are shown in Figures 21-1, 21-5, and 21-6.

Site 321 is located about 250 km west of the axis of the Peru Trench at 12°S, well seaward of the edge of the downbending plate (see Figure 21-1; diagram A, Figure 21-6). Because of its position, the site has remained isolated from land-derived turbidity currents throughout its depositional history (Yeats, Hart et al. 1976). The lower 91 m of sediment represents a typical pelagic facies section,

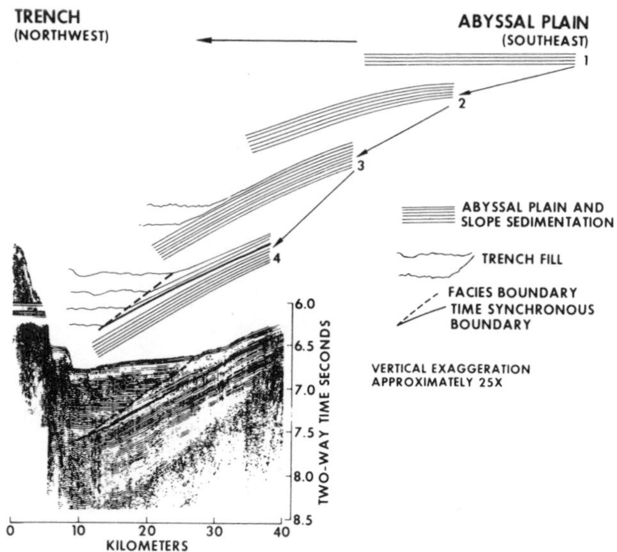

Figure 21-3. Seismic reflection record across the eastern Aleutian Trench showing time-transgressive onlap of trench wedge facies over abyssal plain deposits and axial channel at the base of the continental slope. (Source: Piper et al. 1973.)

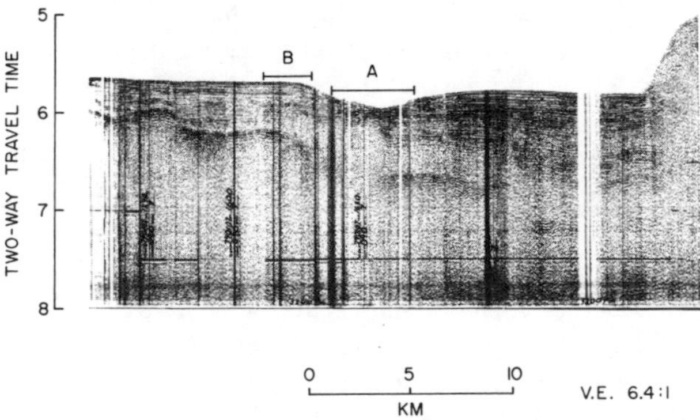

Figure 21-4. Seismic reflection record across the Chile Trench at 39.4° S showing axial channel (A) and buried channel reflectors (B). Note: Location is along profile 2 of Figure 21-9. Vertical light and dark lines are noise in the receiver system.

consisting of nannofossil ooze, brown zeolitic clay and siliceous glass-rich brown clay. The upper 35 m is a layer of greenish gray siliceous fossil-rich detrital clay characteristic of hemipelagic sedimentation, which we categorize as terrigenous plate facies. Sedimentation rates for the pelagic brown clays and oozes range from 1 to 5 m/million years, while the hemipelagic clays accumulated at 5 to 20 m/million years. The transition from pelagic to terrigenous plate facies occurred approximately 5 million years ago, when the site was about 500 km west of its present location and 750 to 800 km seaward of the trench axis (Kulm et al. 1976). The site 321 facies association is represented by diagram D, Figure 21-2.

Two sites in the Aleutian Trench region provide examples of terrigenous plate and trench wedge facies deposition along a margin of high sedimentation. Site 178 is about 200 km seaward of the axis of the Aleutian Trench on the gently landward dipping Aleutian Abyssal Plain (see Figure 21-1; diagram B, Figure 21-6). A thin 35 m layer of pelagic plate facies consisting of brown clay and chalk directly overlies the basalt basement. The remaining 742 m of the section is entirely terrigenous plate facies with various units of hemipelagic muds, fine sand to silt turbidites, and ice-rafted erratics. Sedimentation rates for these terrigenous sediments range from 170 to 180 m/million years during the Pleistocene and Holocene and 35 m/million year for the older Plio-Miocene sediments. Much of the terrigenous material is apparently transported to this area via the Surveyor Deep-Sea Channel that eventually flows into the Aleutian Trench further to the west (Ness and Kulm 1973).

Horizontally layered trench wedge turbidites onlap these abyssal plain deposits as the Pacific Plate bends downward to descend beneath the Aleutian Trench (see Figure 21-1; diagram B, Figure 21-6). DSDP Site 180 was drilled on the seaward side of this wedge, stopping just above the contact with the underlying abyssal plain facies. The trench wedge section consists of 470 m of graded silts and interbedded muds, all interpreted as turbidite sequences, with relatively few ice-rafted erratics. The estimated sedimentation rate is 1500 ± 300 m/million years, which is an order of magnitude greater than for the abyssal plain facies (Site 178) in this region. In terms of facies patterns, the eastern Aleutian Trench most closely resembles diagram I, Figure 21-2.

A distal fan facies overlying an abyssal plain sequence (terrigenous plate facies) was drilled at site 174, seaward of the slowly converging zone between the Juan de Fuca plate and the Oregon-Washington continental margin (see Figure 21-1; diagram C, Figure 21-6). Two major lithologic units were encountered: an upper section of late Pleistocene sand turbidites from 0 to 284 m, and a lower section of Pliocene to late Pleistocene thin-bedded silty clays with thin basal silt layers, extending down to 879 m. The fairly abrupt (less than 10 m) boundary between these two units marks the transition from terrigenous plate to fan facies as the Astoria Fan prograded out across the abyssal plain (Figure 21-7). Sedimentation rates were calculated at 370 to 940 m/million years for the distal fan sediments of the upper layer and 140 to 220 m/million years for the abyssal plain sequence below 284 m. Acoustic basement was estimated to be some 35 m below the bottom of the hole (Kulm, von Huene et al. 1973). If a pelagic plate facies section does exist, it would lie in this thin uncored section. Depending on the nature of these basal sediments, this site would resemble the facies pattern of either diagram H or J, Figure 21-2.

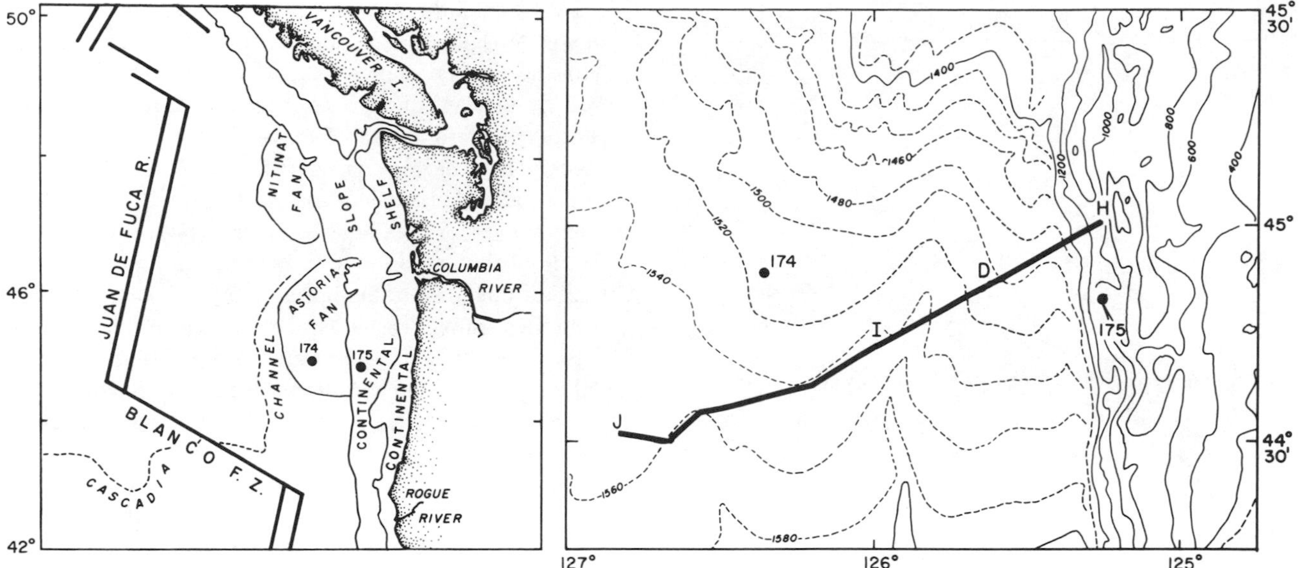

Figure 21-5. Regional setting of the Oregon-Washington convergence zone with locations of DSDP Sites 174 and 175 and seismic reflection profile J-H (Figure 21-7).

AXIAL TRANSPORT ALONG TRENCH WEDGES

Although pelagic and terrigenous plate facies are integral components of trench sediment deposits, they are formed on the oceanic plate away from the direct influence of depositional processes within the trench. Trench wedges and fans, on the other hand, form within the trench basin and are intimately tied to the local conditions of sediment supply and deposition. Both wedges and fans are subject to deposition from channelized turbidity currents and hemipelagic sedimentation. Turbidite deposition in deepsea fans has been extensively covered by several authors and summarized by Nelson and Kulm (1973), but turbidite deposition along trench wedges has been much less thoroughly studied.

Chile Trench Channel

Deepsea channels are integral parts of deepsea fans and many abyssal plains (Nelson and Kulm 1973); they also appear to be important in trench wedge development. Large axial channels have been recognized in both the Aleutian (von Huene 1972; Piper et al. 1973) and Chile Trenches (Scholl et al. 1970), but neither channel system has been studied in much detail. Von Huene (1974) was among the first to recognize the significance of lateral transport along the axial channels in these two trenches.

The Chile Trench channel can be seen in reflection records from at least 30°S to 40°S (Scholl et al. 1970; Lister 1971; Schweller 1976). Approximately 60 axial bathymetric profiles of the Chile Trench between 27°S and 42°S obtained from two recent Oregon State University cruises, and other sources provide data for a well-controlled plot of maximum depths along the trench axis (Figure 21-8). From this data set, twenty representative profiles were selected to portray the morphological changes in the channel and the trench axis (Figure 21-9).

The channel begins between 40°20' and 41°30'S and probably issues from one or more canyons coming off the high sedimentation region of the continental shelf and slope in these latitudes (Galli-Oliver 1969). No channels are seen in trench crossings south of 41°30'S. From 41°S to 34°S a single continuous deepsea channel occupies the center of a broad, sediment-filled axis (profiles 1 through 9, Figure 21-9). At three locales (profiles 3, 5, and 9, Figure 21-9), the channel abuts basement ridges protruding through the 1 to 2 km thick trench wedge. Both channel relief and width decrease north of 38°S from 200 m deep and 5 km wide to only 50 m deep and 1 km wide at 34°S.

A uniform gradient of 1:650 is maintained along the channel axis for almost 1000 km between 42°S and 33°S (Figure 21-8). Seismic reflection records show that there is no comparable smooth gradient of the oceanic basement beneath these sediments, although there is a slight regional increase in basement depth northward (Scholl et al. 1970). At 33°30'S (profile 10, Figure 21-9), the channel abruptly begins to deepen and cut a V-shaped notch in horizontally layered turbidites. Profile 11 (Figure 21-9) is similar to a high-resolution acoustic reflection profile near the same latitude discussed in detail by Lister (1971), who concluded that the channel was eroding through older trench fill. This erosional character is accompanied by a sharply steepened and uneven axial

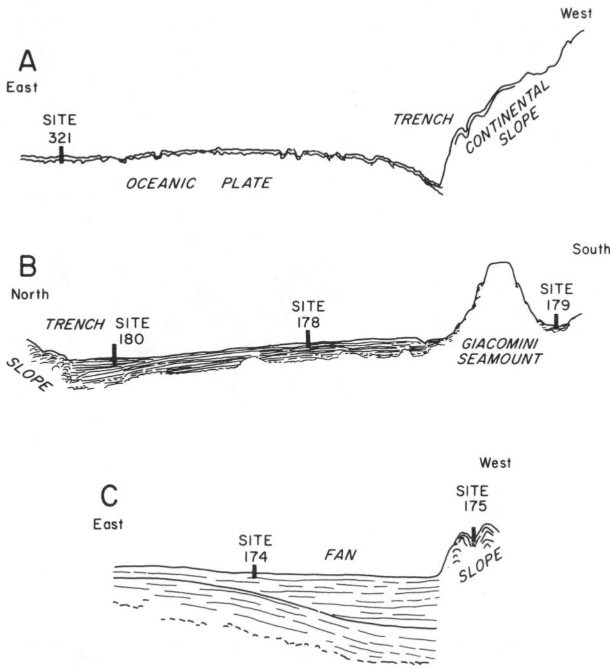

Figure 21-6. Diagrammatic sections (based on reflection records) across Peru Trench and Nazca Plate at 9° S, DSDP site 321 (diagram A); Aleutian Trench and adjacent abyssal plain, DSDP sites 178 through 180 (diagram B); and Astoria Fan and Oregon lower continental slope, DSDP Sites 174 and 175 (diagram C).

gradient of 1:200 or more (Figure 21-8). A partial structural blockage of the trench axis at 33°S by a basement high may be responsible for this steepened gradient and the 600 m jump in trench depth (Schweller 1976).

North of this constriction, the channel resumes a depositional character with a gradient of 1:650 along with a much smaller wedge of axial sediments than to the south (profiles 13 through 15, Figure 21-9). As the axial gradient flattens north of 30°S, the channel abruptly disappears from the axial wedge (profiles 16 through 18, Figure 21-9). The portion of the trench wedge between 30°S and 27° 30'S is the only section that has virtually no surface gradient and seems to be truly ponded. A basement ridge blocking the axis at 27°30'S (profile 19, Figure 21-9) is probably responsible for the ponding and abrupt termination of the axial sediment wedge. North of this point the trench axis is a deep V-shaped cleft with no horizontal axial sediments (profile 20, Figure 21-9), except for a few small basins further to the north that are now isolated from active axial turbidity current deposition (Schweller 1976).

The turbidites in the Chile Trench axis seem to be more closely related to channelized deposition systems than to a series of level turbidite ponds along the trench axis as envisioned in earlier concepts of wedge formation (Fisher and Hess 1963). Both the gradient and the continuity of the axial channel indicate sediment transport and deposition in a manner similar to that described for deepsea fans. There is a strong correlation between channel morphology and axial gradient that seems to reflect changing conditions of deposition or erosion down the length of the channel. Because an axial channel has been found in at least one other trench (Aleutian Trench), the channel system in the Chile Trench may be fairly typical of sedimentation in trenches with a moderate to high terrigenous sediment input.

Residence Time of Axial Deposits

The length of time sediments remain in an axial wedge along a convergent margin before they are completely deformed or removed from the wedge is a function of the convergence rate and the width of the wedge at the time of deposition. To illustrate these relationships better, we will examine a simplified case in which a trench with a convergence rate of 10 cm/year (100 km/million years) contains an axial wedge of sediments supplied from a single source at one end (Figure 21-10). The axial wedge is thickest and widest near the source and diminishes along the axis as less material is carried to the more distant portions. The end point of the wedge (point A, Figure 21-10) is where consumption of sediments exceeds the amount brought in by axial transport. If consumption of axial sediments along the entire wedge is equal to the sediment input to the axis, the wedge will be in equilibrium. It will neither grow nor shrink as long as both convergence and sediment supply remain constant.

Under these conditions, the maximum residence time for sediment of a single age or layer within the axial wedge is equal to the width of the wedge at the time of deposition divided by the convergence rate. In our example, sediment deposited on a 20 km wide section of the wedge would last a maximum of 200,000 years, while the same flow deposited on a 10 km wide section would be completely removed in only 100,000 years. At the end point of the wedge (point A, Figure 21-10), sediment is removed almost as soon as it is deposited, resulting in a residence time of essentially zero years. Reducing the convergence rate from 10 cm/year to 5 cm/year doubles the residence time for a given width.

If the sediment supply abruptly decreases, the trench wedge will begin to diminish in size because consumption will exceed supply. The wedge will decrease rapidly at first, then more slowly until it reaches a new equilibrium in balance with the smaller sediment supply. At a fixed location along the trench, the wedge cross-section will decrease through time (Figure 21-10). The end point of the wedge will migrate toward the source as the wedge shrinks, stopping at a new stable position as equilibrium is reached.

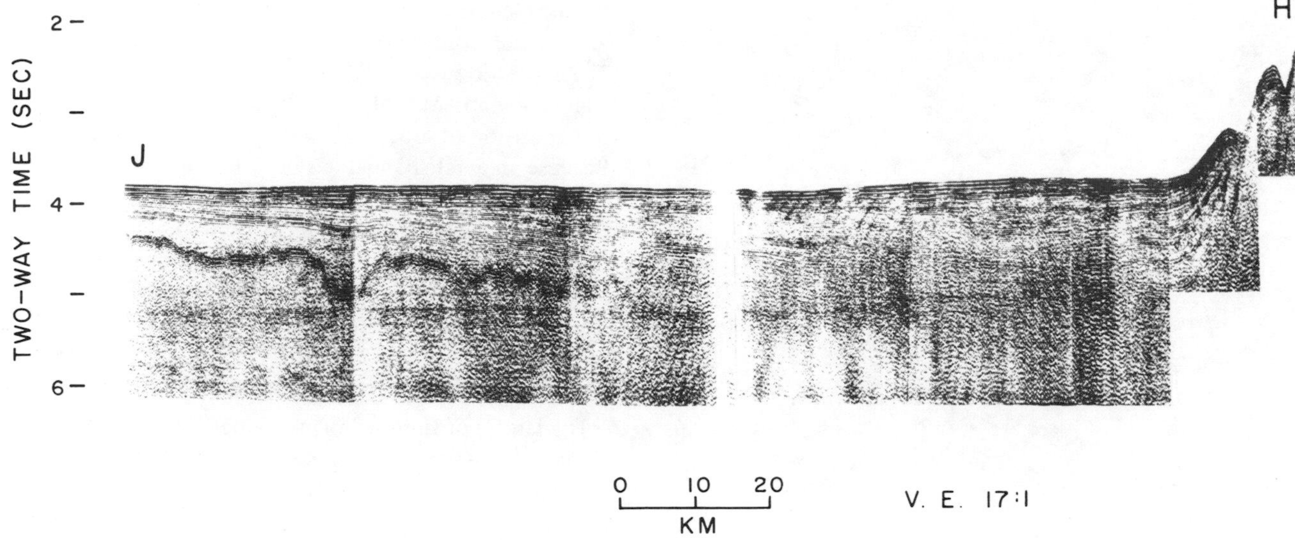

Figure 21-7. Seismic reflection record across Astoria Fan showing gently seaward dipping fan strata prograding over landward dipping abyssal plain deposits (terrigenous plate facies). (See Figure 21-5 for location.)

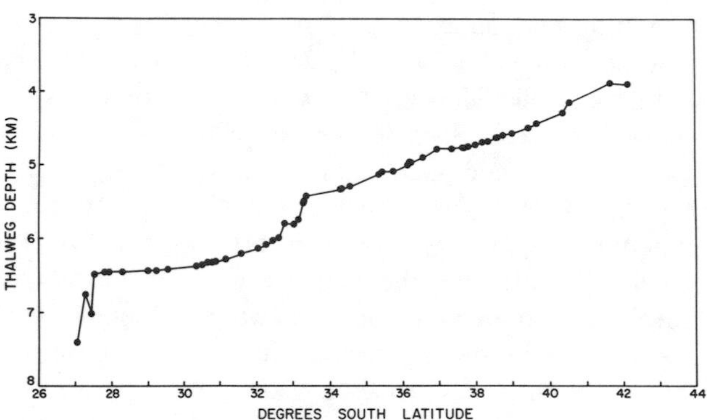

Figure 21-8. Thalweg depth along the axial channel in the Chile Trench.

FACTORS INFLUENCING FAN AND TRENCH WEDGE GROWTH

Geometry of Fan and Wedge Deposits

Evidence from drill hole and piston core lithologies, morphology, and structure suggests that channelized turbidity flows are the prime means of sediment transport and deposition along both the Chile and Aleutian Trench wedges (Piper et al. 1973; von Huene 1974). In terms of lithologies and presumed sedimentation processes, trench wedges appear to be very similar to deepsea fans. Many of the differences between wedge and fan deposits are in the fine structural elements and morphology associated with channel development (Table 21-1). The formation of wedges or fans may ultimately be tied to the interactions of convergence rate, sediment supply, and the downbending of oceanic crust as controlling factors for turbidity current flow patterns.

Most modern active oceanic trenches are nearly empty troughs containing 400 m or less of axial sediments (Scholl and Marlow 1974). Turbidity currents that flow into such trenches from the continental slope almost immediately encounter a reverse or landward gradient of the outer (seaward) trench slope. This reverse gradient diverts the flows and causes them to flow along the trench axis in one or both directions (Figure 21-11). A long, narrow wedge of turbidite sediment will be deposited along the axis and will decrease in thickness away from the source area or input point. Growth of the wedge will be predominantly linear along the axis because of the confining effects of the inner and outer trench slopes.

If sufficient sediment is supplied to the axis, the trench may eventually fill and disappear as a topographic feature even though the basement structure remains. The Oregon-Washington and southern Chile margins are two examples of actively converging regions with sediment-filled trenches. Since there is no reverse gradient to impede them, turbidity flows from the continental slope spread outward across the ocean floor and can form fan deposits similar to those along passive margins. Fan deposits thin away from the source both outward and along the base of the slope in a generally radial pattern.

As shown in Figure 21-11, comparative cross-sections of a trench wedge (X-X') and a fan (Z-Z') parallel and adjacent to the base of the continental slope are quite similar, with deposits gradually thinning away from the source area. Cross-sections perpendicular to the trend of the margin, however, show significant variations. A typical trench wedge section (W-W' in Figure 21-11) shows horizontal strata on the wedge surface, often with a leveed channel, underlain by layers with a slight landward dip. A cross-section of a fan (Y-Y' in Figure 21-11) shows similar

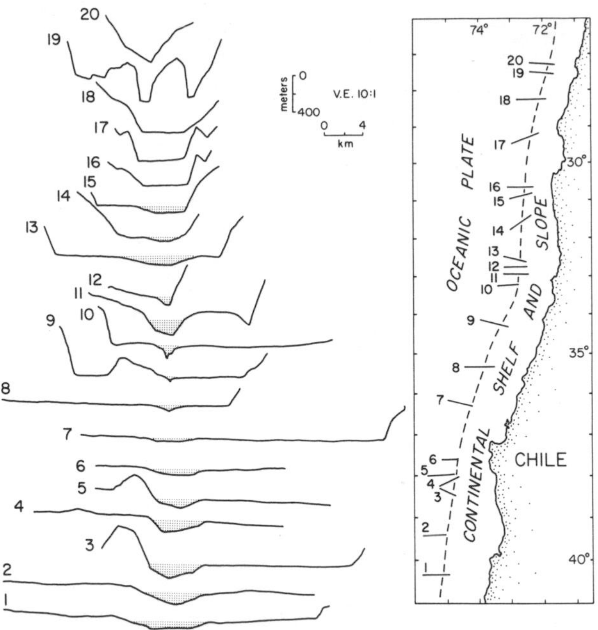

Figure 21-9. Bathymetric profiles of the axial channel in the Chile Trench with profile locations.

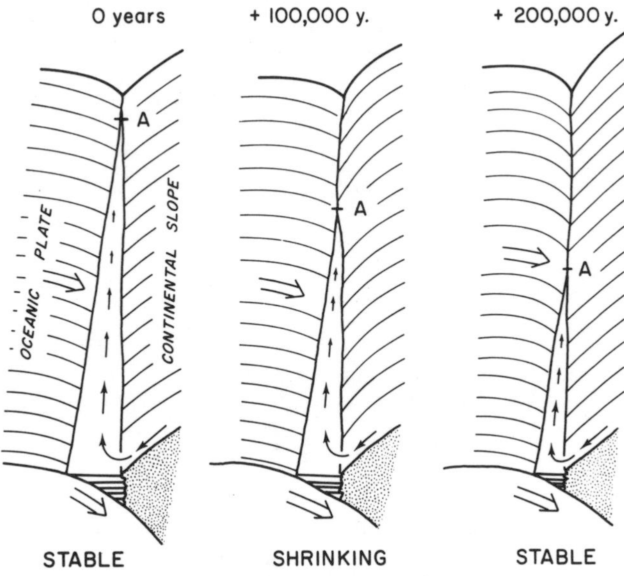

Figure 21-10. Schematic diagram showing the change in size of a trench wedge following an abrupt decrease in sediment supply. Note: Convergence rate is assumed constant.

strata at depth, but progressively steeper seaward dips approaching the surface of the fan. Trench wedges apparently have a single large channel, which is also characteristic of most fans, but fans have an additional system of distributary channels radiating away from the main channel, particularly near the fan apex (Nelson and Kulm 1973). The main fan channel and most of the distributary channels on fans appear to migrate frequently, thereby producing a series of chaotic discontinuous reflectors (see Figure 21-7).

Fan development along convergent margins may occur as a direct upward extension of trench wedge growth. A trench that receives more sediment than is consumed by convergence will eventually become completely filled and spill over the flat ocean floor. This process is apparently the case in the Chile Trench south of about 37°S, where level turbidites fill the trench basin and extend up to 100 km beyond the seaward edge of the downbending slab (Scholl and Marlow 1974). Additional input of sediment over these filled trenches may result in growth of deepsea fans over trench wedge sequences, as found off Oregon-Washington (Astoria and Nitinat Fans, Figure 21-5). The change from trench wedge to fan facies would probably be a gradual transition that would not produce a sharp facies boundary.

Continuity of Channel Sands

Other important differences between a trench wedge and a fan are found in details of wedge and fan channel structures controlled by sedimentation processes, some of which are revealed in the character of seismic reflectors (Table 21-1). In particular, channel levees have higher relief and are more sharply defined on the upper fan channels than on the channels observed along trench wedges, even though the sedimentation rates and the types of sediments are roughly the same. Seismic reflection records from the Chile Trench and Astoria Fan show that, in general, wedge channels migrate laterally less than mid- and upper fan channels over the short term (see Figures 21-4 and 21-7). Consequently, individual seismic reflectors in fans are short and discontinuous, while single horizons in trench wedges can be traced for tens of kilometers. From this we infer that channel sands in trench wedges should be more continuous, laterally and vertically, than those in fans because there appear to be fewer short-term fluctuations in the channel course along wedges. On the basis of these observations, channel sands in fans may be more subject to cut-and-fill processes than those in trench wedges.

Point Sources

One final factor that undoubtedly influences the growth of fans and wedges is the distribution of sediment input to the trench axis. Large deepsea fans seem to develop almost exclusively from point sources provided by submarine canyons connected to major rivers (e.g., Ganges Fan, Mississippi Delta Fan, and Astoria Fan). Both the southern Chile Trench and the eastern Aleutian Trench have high rates of sedimentation, but no major canyons to concentrate the input to the trench. Neither area has large fans,

Table 21-1. *Similarities and Differences between Fan and Trench Wedge Parameters*

Item	*Fan*	*Trench Wedge*
Shape	Flattened cone	Right triangular prism
Dimensions:		
length	10 to 2500 km	Few 100 km to 2000 km
width	10 to 2500 km	2 to 50 km
length:width	1:1 to 2:1	10:1 to 20:1
thickness	few 100 m to 12 km	Few 100 m to 2500 m
Age	10^5 to 10^7 years	10^4 to 10^5 years
Sedimentation rates (maximum)	1000 m/million years	3000 m/million years
Dominant sediment types	Silt and sand turbidites with interchannel and lower fan muds	Silt and sand turbidites with interbedded muds
Main sedimentation process	Channelized turbidity current flows	Channelized turbidity current flows
Main sediment source	Continental shelf and slope	Continental shelf and slope
Channels	One major channel with several distributaries	One major channel, no distributaries
Channel width	2 to 5 km	1 to 5 km
Channel relief	50 to 200 m	50 to 200 m
Channel gradient	1:250 to 1:1000	1:250 to 1:1000
Character of seismic reflectors	Short, discontinuous, chaotic	Continuous over tens of kilometers
Shape of sand bodies	Overlapping suprafan lobes	Long continuous lenses along channels

although small fans emanate from the mouths of several minor canyons along the Aleutian Trench (von Huene 1972; Piper et al. 1973). If either of these margins had a single large source comparable to the Columbia River, the possibility exists that they would also develop large fans comparable to Astoria Fan.

Channel Position and Trench Wedge Structures

Trenches can have significantly different types of trench wedge deposits depending on the position of the axial channel within the wedge. Piper et al. (1973) produced a facies model for the Aleutian Trench with a channel fixed against the base of the continental slope (diagram A, Figure 21-12). The wedge sediments become finer laterally away from the channel and grade from channel sands to silts and muds near the seaward edge of the trench. In cross-section, a coarsening of grain size should occur vertically upward through the complete trench wedge section at the base of the slope. Because the trench wedge is much wider than it is thick, the vertical gradation in grain size occurs over a shorter distance than the horizontal change.

The trench wedge facies laps seaward over the plate facies as the oceanic plate converges and descends into the trench at a shallow downward angle (see diagram A, Figure 21-2). There is an analogous time-transgressive onlap of sediment types within the trench wedge. The coarse-grained channel deposits override the fine-grained distal turbidites that are initially deposited at the seaward edge of the trench

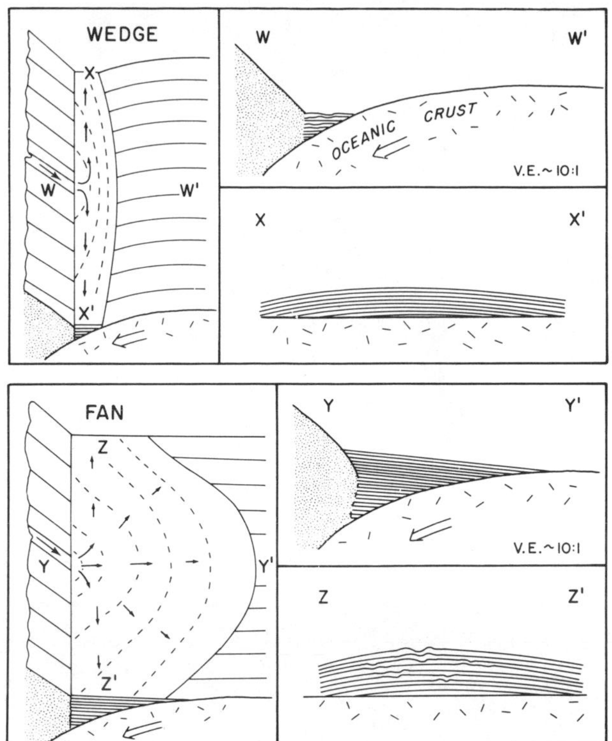

Figure 21-11. Schematic diagrams showing differences between trench wedge and fan structures along convergent margins. Note: Small arrows indicate sediment transport vectors; dashed lines are sediment isopachs. For simplicity, plate facies are not shown. See text for discussion of details.

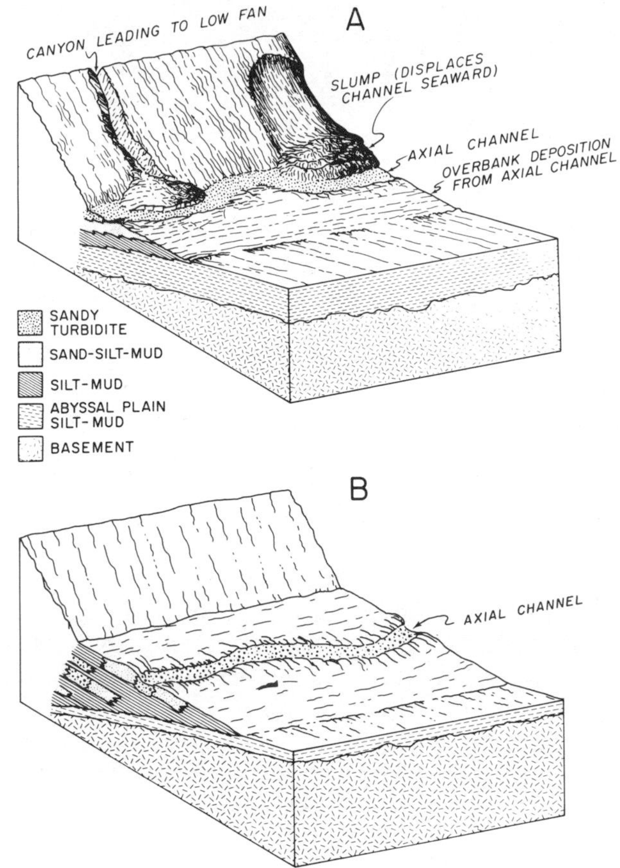

Figure 21-12. Models for trench wedge development. Note: Diagram A shows Aleutian Trench model with an axial channel at the base of the continental slope. (Source: Piper et al. 1973.) Diagram B shows Chile Trench model with a channel in the middle of the axial wedge and buried channels.

wedge (diagram A, Figure 21-12). An important aspect to recognize is that this succession of grain sizes within the wedge is due to landward movement of trench wedge and plate sediments beneath a stationary channel rather than to an absolute seaward migration of the channel away from the continental margin.

In contrast to the Aleutian Trench, the axial channel in the southern Chile Trench meanders along the center of a broad, 10 to 40 km wide trench wedge instead of against its landward edge. Seismic reflection records (see Figure 21-4) suggest that the channel remains stationary for a period of time, then abruptly shifts or jumps to a new location on the wedge surface. These periodic shifts to new channel courses leave abandoned channel sections to fill in with overbank deposits. The resulting trench wedge should show small, vertically continuous sections of channel sands distributed among layers of silt and mud from distal turbidites (diagram B, Figure 21-12). The overall lateral and vertical trends in grain size within the wedge are not as consistent as for the Aleutian Trench model given above. Channel sands may be slightly less continuous in a Chile-type trench wedge than in an Aleutian-type trench.

Alternative Hypotheses for Trench Wedge Deposition

The lack of well-defined levees along trench axis channels is the most critical problem left unresolved by the trench model outlined above. Deepsea channels and fan channels that transport large volumes of sediment generally have substantial levee relief because of rapid deposition from channel spillover (see Table 21-1). Trench axis channels more closely resemble the low-leveed abyssal plain channels than fan channels in morphology (e.g., Ness and Kulm 1973; Chough and Hesse 1976). However, trench wedges have sedimentation rates that are an order of magnitude greater than those on abyssal plains.

One possible explanation for the low relief of levees along the trench wedge channels is that these features do not necessarily reflect the system of deposition for the bulk of the turbidites within the trench wedge. If trench sedimentation rates during the last glacial maximum were up to an order of magnitude greater than at present (Kulm, von

Huene et al. 1973, p. 970), individual turbidity currents may also have been several times larger than those having formed the present channels. Such flows could have used the entire width of the trench axis (up to 30 or 40 km) as a channel. The flows would then funnel down the axial gradient, much like a sheet flow, and leave horizontal layers across the width of the trench wedge. As sedimentation rates declined after the glacial maximum, turbidity currents of greatly diminished volume cut erosional channels into these older layers, thereby creating the present configuration.

This model explains the rapid deposition of a trench wedge with an axial gradient without producing high-relief channel levees. However, it fails to account for the buried channel reflectors observed on some seismic reflection profiles (see Figure 21-4). Another possibility is that the percentage of sand in the trench wedge turbidites is much less than our few cores from the Chile Trench indicate. Turbidity currents, consisting mainly of silt and clay, would be capable of spreading across the wedge much further than sands and building broad, low levees. Unfortunately, we do not have the detailed stratigraphic control or deposition rates needed to prove or disprove either of these more radical hypotheses. A much more comprehensive set of data will be needed to resolve the sedimentation history of these significant trench channels.

RELATION OF CONVERGENCE RATE AND SEDIMENT SUPPLY TO TRENCH DEPOSITS

In general, the convergence rate and the supply of sediment to the trench are the two dominant factors that control the development of deposits along convergent margins. A predictable sequence of deposits, from a fan to a trench wedge to an empty axis, develops according to the interactions of these two variables. Figure 21-13 graphically summarizes the possible types of trench and fan deposits and their relationship to sediment supply and convergence rates. While the positions of the boundaries between deposits may vary somewhat, the types of deposits that accumulate should be consistent over almost any normal convergence conditions.

Continental margins with low to moderate convergence rates and moderate sediment inputs are characterized by fan development. Faster convergence rates require more sediment input to build and maintain fans of a given size; thus fans along faster converging margins tend to be smaller than those along margins of slow or zero convergence. Rapidly converging margins typically contain trench wedges or are devoid of turbidite deposits.

A rapid increase or decrease in sediment supply along a margin can radically transform the type of trench deposit. For example, a moderate (3 to 5 cm/year) convergence rate combined with a low rate of sediment input will usually result in a small trench wedge or a partially empty axis with discontinuous turbidite basins (Figure 21-13). Increasing the sediment supply without changing the convergence rate will cause some growth of the wedge with possible small fans radiating from canyon mouths. A further increase in sediment input may eventually build a large fan over a sediment-filled trench. Similarly, changes in convergence rates with a constant sediment supply can also cause alterations in the type of deposit. However, major variations in sedimentation rates are probably much more common and more rapid than significant changes in relative plate motions.

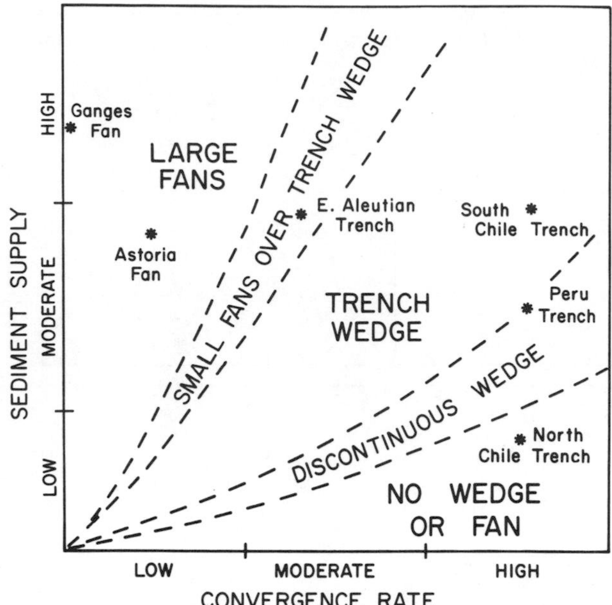

Figure 21-13. Empirical model for relationship of convergence rates and sediment supply to trench deposits. (See text for discussion.)

SUMMARY

Zones of convergence between oceanic and continental plates are commonly marked by trenches that contain variable types and amounts of sediments. Downbending of the oceanic plate forms an elongate asymmetrical trench basin that is deepest at the base of the continental slope. Sediment-filled trench basins along the eastern Pacific Ocean can be up to 50 km wide, 2 km deep, and a few hundred to more than one thousand kilometers long.

Sediments in trenches along continental margins can be separated into four main facies: pelagic plate, terrigenous plate, trench wedge, and fan. Each of these facies is a major conceptual unit that theoretically can occur with or without the other facies in various configurations (see Figure 21-2). Several examples of these facies combinations occur in eastern Pacific trenches (see Figure 21-1).

The pelagic plate facies includes brown clay and calcareous ooze and is deposited mainly by settling of suspended material out of the water column at low rates of accumulation. The pelagic facies usually grades upwards into the terrigenous plate facies as silt- and clay-sized particles settle on the converging plate near the continent. Continentally derived turbidites may be included in the terrigenous plate sediments if there are no topographic barriers to block turbidity flow transport to the oceanic plate. The trench wedge and fan facies consist predominantly of sand and silt turbidites with interbedded mud layers. Trench wedge and fan deposits typically accumulate rapidly at hundreds to thousands of meters per million years.

Despite their similar lithologies, trench wedge and fan deposits can have quite different structures and growth patterns (see Table 21-1 and Figure 21-11). The narrow, elongate trench wedge builds gradually seaward over the pelagic or terrigenous plate facies, thereby creating a time-transgressive angular discordance. The submarine fan deposits generally spread radially away from the base of the continental slope in a flattened cone-shaped feature. Strata in trench wedges have horizontal to slightly landward dips, while fan strata dip gently seaward.

That fan channels play a major role in the formation of deepsea fans is well known. Channels such as those along the axes of the Chile and Aleutian Trenches appear to be comparably important to the development of trench wedges. The channel in the Chile Trench runs one thousand kilometers along the trench wedge, and changes form in response to variations in the axial gradient. The shape and continuity of sand bodies in wedge deposits are dependent on the position of the channel in the trench axis (Figure 21-12). A channel at the base of the continental slope (Aleutian Trench) will be relatively stable in position, while a channel in the center of the axial wedge (Chile Trench) will shift periodically and have less continuous axial sands. Trench wedge channels are characterized by low-relief levees that are more typical of abyssal plain deepsea channels than fan channels. Very large turbidity flows during the last glacial maximum may partially account for this anomalous morphology.

Fan and trench wedge deposits adjacent to active margins are constantly being deformed, compressed, and subducted by plate convergence processes. Turbidites must be continually supplied to the trench axis in order to maintain a constant amount of undeformed trench sediments. The type and size of trench deposits (wedge and fan combinations) can be graphically related to balances between convergence rates and sediment supply (Figure 21-13). In general, trench wedges form along rapidly converging margins, while fans grow along margins with high sediment input and moderate to low convergence rates.

ACKNOWLEDGMENTS

The authors thank Paul D. Komar for helpful discussions and the staff and students of the School of Oceanography, Oregon State University and the captain and crew of the R/V *Yaquina* for help with data collections and processing. This research was supported under the Nazca Plate Project by the National Science Foundation, International Decade of Ocean Exploration (NSF Grants GX-28675, IDO 71-04208 and OCE 76-05903).

REFERENCES

Atwater, T., 1970. Implications of plate tectonics for the Cenozoic tectonic evolution of western North America. *Geol. Soc. Amer. Bull.,* 81: 3513-36.

Caldwell, J. G., Haxby, W. F., Karig, D. E., and Turcotte, D. L., 1976. On the applicability of a universal elastic trench profile. *Earth Planet. Sci. Lett.,* 31: 239-46.

Chough, S., and Hesse, R., 1976. Submarine meandering thalweg and turbidity currents flowing for 4000 km in the Northwest Atlantic Mid-Ocean Channel, Labrador Sea. *Geology,* 4: 529-33.

Fisher, R. L., and Hess, H. H., 1963, Trenches. In: M. N. Hill (ed.), *The Sea,* 3. Wiley-Interscience, New York, pp. 411-436.

Galli-Oliver, C., 1969. Climate—a primary control of sedimentation in the Peru-Chile Trench. *Geol. Soc. Amer. Bull.,* 80: 1849-52.

Helwig, J., and Hall, G. A., 1974. Steady-state trenches? *Geology,* 2: 309-16.

Kulm, L. D., Scheidegger, K. F., Prince, R. A., Dymond, J., Moore, T. C., Jr., and Hussong, D. M., 1973. Tholeiitic basalt ridge in the Peru Trench. *Geology,* 1: 11-14.

———, von Heune, R. E., et al., 1973. *Initial Reports of the Deep Sea Drilling Project,* 18. Washington, D. C., 930 pp.

———, Schweller, W. J., Molina-Cruz, A., and Rosato, V. J., 1976. Lithologic evidence for convergence of the Nazca Plate with the South American continent. In: R. S. Yeats, S. R. Hart, et al., *Initial Reports of the Deep Sea Drilling Project, 34,* U. S. Government Printing Office, Washington, D. C., pp. 285-301.

Lister, C. R. B., 1971. Tectonic movement in the Chile Trench. *Science,* 173: 719-22.

Minster, J. B., Jordan, T. H., Molnar, P., and Haines, E., 1974. Numerical modeling of instantaneous plate tectonics. *Royal Astron. Soc. Geophys. J.,* 36: 541-76.

Nelson, C. H., and Kulm, L. D., 1973. Submarine fans and channels. In: G. V. Middleton and A. H. Bouma (eds.), *Turbidites and Deep-Water Sedimentation.* Soc. Econ. Paleont. Mineral., Short Course, Anaheim, pp. 39-78.

———, and Nilsen, T. H., 1974. Depositional trends of modern and ancient deep-sea fans. In: R. H. Dott, Jr., and R. H. Shaver (eds.), *Modern and Ancient Geosynclinal Sedimentation.* Soc. Econ. Paleont. Mineral., Sp. Publ. 19, pp. 69-91.

Ness, G. E., and Kulm, L. D., 1973. Origin and development of Surveyor Deep-Sea Channel. *Geol. Soc. Amer. Bull.,* 84: 3339-54.

Normark, W. R., 1970. Growth patterns of deep-sea fans. *Amer. Assoc. Petrol. Geol. Bull.,* 54: 2170-95.

———, 1974. Submarine canyons and fan valleys: Factors affecting growth patterns of deep-sea fans. In: R. H. Dott, Jr., and R. H. Shaver (eds.), *Modern and Ancient Geosynclinal Sedimentation.* Soc. Econ. Paleont. Mineral., Sp. Publ. 19, pp. 56-68.

Parsons, B., and Molnar, P., 1976. The origin of outer topographic rises associated with trenches. *R. Astr. Soc. Geophys. J.,* 45: 707-12.

Piper, D. J. W., von Huene, R., and Duncan, J. R., 1973. Late Quaternary sedimentation in the active eastern Aleutian Trench. *Geology,* 1: 19-22.

Prince, R. A., and Kulm, L. D., 1975. Crustal rupture and the initiation of imbricate thrusting in the Peru-Chile Trench. *Geol. Soc. Amer. Bull.,* 86: 1639–53.

Scholl, D. W., Christensen, M. N., von Huene, R., and Marlow, M. S., 1970. Peru-Chile Trench sediments and sea-floor spreading. *Geol. Soc. Amer. Bull.,* 81: 1339-60.

——, and Marlow, M. S., 1974. Sedimentary sequence in modern Pacific trenches and the deformed circum-Pacific eugeosyncline. In: R. H. Dott, Jr., and R. H. Shaver (eds.), *Modern and Ancient Geosynclinal Sedimentation.* Soc. Econ. Paleontol. Mineral., Sp. Publ. 19, pp. 193–211.

Schweller, W. J., 1976. *Chile Trench: extensional rupture of oceanic crust and the influence of tectonics on sediment distribution.* Unpublished M. S. thesis, Oregon State University, Corvallis, 90 pp.

Silver, E. A., 1969. Late Cenozoic underthrusting of the continental margin off northernmost California. *Science,* 166: 1265-66.

von Huene, R., 1972. Structure of the continental margin and tectonism at the eastern Aleutian Trench. *Geol. Soc. Amer. Bull.,* 83: 3613-26.

——, 1974. Modern trench sediments. In: C. A. Burk and C. L. Drake (eds.), *The Geology of Continental Margins.* Springer, New York, pp. 207-11.

Yeats, R. S., Hart, S. R., et al., 1976. *Initial Reports of the Deep Sea Drilling Project, 34.* U. S. Govt. Print. Off., Washington, D. C., 814 pp.

Chapter 22

The Irpinids: A Model of Tectonically Controlled Fan and Base-of-Slope Sedimentation in Southern Italy

TULLIO PESCATORE

Istituto di Geologia e Geofisica
Università di Napoli
Naples, Italy

ABSTRACT

The marine Irpinian sequences of Miocene age in southern Italy provide examples of base-of-slope and submarine fan deposition in a highly compressive tectonic setting. Facies analysis serves (1) to differentiate the deposits that accumulated at neritic to basin depths, (2) to define paleocurrent dispersal, and (3) to assist in the interpretation of the regional paleogeography and paleotectonics. The Irpinids presently trend along two major belts, both oriented predominantly northwest-southeast. The southwestern one includes terrigenous units that rest unconformably upon older allochthonous thrust sheets. The northeastern belt includes calcareous units followed conformably by terrigenous flysch deposits. Deposition took place concurrently with the structural displacement of the trough from the southwest (i.e., more internal zone) toward the northeast; this movement occurred between early and middle Miocene time. The western margin of the trough was formed by superposed thrust sheets, while the eastern (more external) margin was a relatively undeformed carbonate shelf called the Abruzzi-Campania Platform. Large rockfall olistoliths accumulated at the base of the western margin. Seaward in the Irpinian Trough arenaceous-rudaceous and pelitic-arenaceous series were emplaced by gravity-induced and hemipelagic mechanisms; canyon mouth, fan valley, interchannel units, and depositional lobes are recognized. The geometry of fan sequences reflects structural control. The olistolith and fan facies distributions record the tectonic shift eastward of the slopes and trough axis with time as the internal margin was overthrusted above the external margin. The Irpinid sequences serve as a model to better understand present trench and basin deposits accumulating along mobile plate junctions.

INTRODUCTION

In recent years, studies in Italy have led to a better understanding of the origin and evolution of the Alpine and Apennine chains. As a result of this line of research, some paleogeographic models reveal the similarities that exist in the Alpine and Apennine geological development in spite of their apparent diversity (Accademia Nazionale dei Lincei 1973; Ogniben et al. 1975; Squyres 1975).

This chapter describes terrigenous sedimentary facies, the Irpinid Group, in southern Italy that accumulated in a deepwater setting during the early to middle Miocene. The name *Irpinid* is derived from *Irpinia,* the Latin name for southeastern Campania. The study area is located in the Campania and Lucania regions east and southeast of Naples (map A, Figure 22-1). Detailed mapping in this area reveals that base-of-slope and deepsea fan deposits developed in a highly mobile and compressive setting and that these facies were displaced eastward as a result of thrusting to the east during deposition. This type of sedimentation, directly controlled by compressive tectonics, probably has counterparts in present-day trenches and basins formed along some plate margins.

GEOLOGICAL FRAMEWORK

General

Several geological units, all oriented northwest-southeast, have been mapped in the Southern Apennines (Figure 22-1), and the paleogeographic evolution of this part of the chain has been delineated schematically (Ogniben 1969, 1973; Ogniben et al. 1975; Haccard et al. 1972;

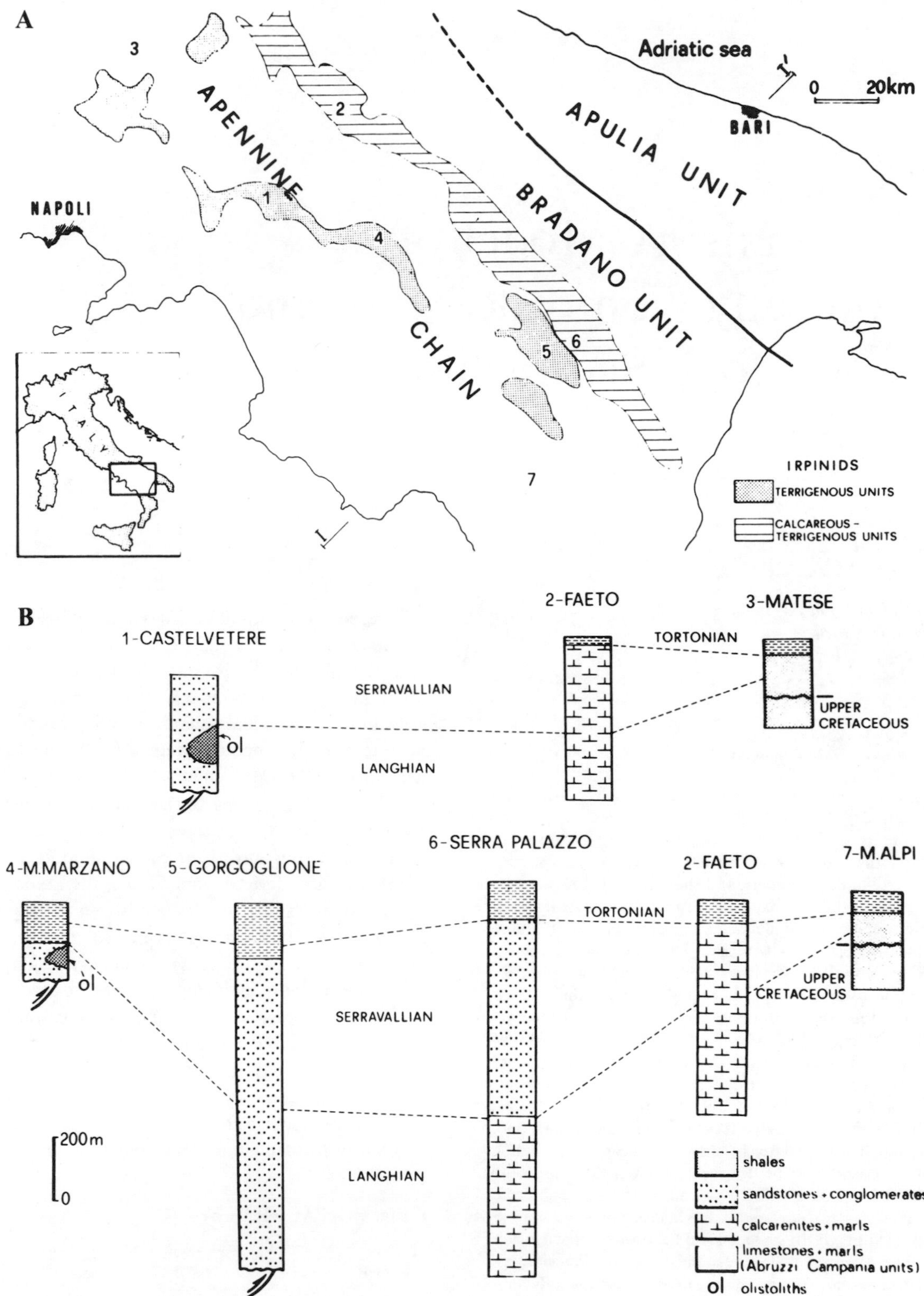

Figure 22-1. Location of the Irpinian units (map A) and stratigraphic columns (chart B), southern Apennines, Italy. Note: The numbers in map A correspond with columns in chart B. The line I-I' corresponds to the section in Figure 22-2. In chart B, sequences 1, 4, and 5 overlie the Langhian thrust sheets; sequences 2 and 6 overlie the Lagonegro basin deposits; sequences 3 and 7 overlie the Abruzzi-Campania Platform deposits, not yet deformed in Langhian times (see Figure 22-3 and text).

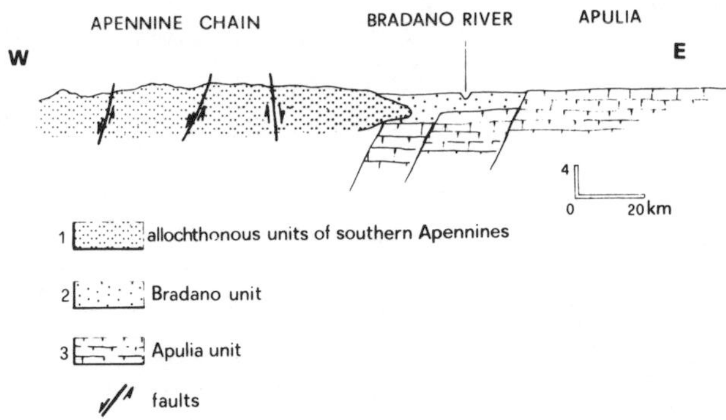

Figure 22-2. Schematic section across the southern Apennines. Note: The location of the section is shown in Figure 22-1. 1 = *Apennine Chain,* formed by several allochthonous geologic units (Mesozoic to Tertiary rocks) and deformed by compressive episodes from Cretaceous up to middle Pliocene. 2 = *Bradano Unit,* consisting of terrigenous sediments deposited in an upper Pliocene-Pleistocene foretrough between the frontal side of the Apenninic thrust sheets and the Apulia foreland. This unit has been only slightly deformed by vertical faults. 3 = *Apulia unit,* carbonate platform and scarp deposits (Upper Trias-Pleistocene), which from the middle Pliocene, constitutes the southern Apennine foreland. Mainly vertical faults deform this unit.

D'Argenio et al. 1973, 1975; Ippolito et al. 1975). Autochthonous successions (including foreland and foretrough units) and allochthonous series (thrust sheets) have been distinguished by Ippolito et al. (1975). The autochthonous units comprise the following:

The *Apulia Unit,* formed by a section (6,000 m thick) of Mesozoic and Cenozoic carbonate shelf deposits, was a foreland during the Pliocene; it has been a stable area subject primarily to vertical displacement.

The *Bradano Unit* consists of Pliocene and Quaternary terrigenous neritic to pelagic facies as much as 3,000 m thick. The Bradano Trough developed between the Apennine thrust sheet fronts to the west and the Apulia foreland after the middle Pliocene tectonic phase; its substratum was the foreland (Apulia) unit, depressed by a series of step faults towards the Apennines. This unit has been subject to vertical movement only.

The allochthonous units constitute the Apenninic chain, which presently forms the western part of the study area. It is formed by the superposition of thrust sheets, which have a total thickness of about 15,000 m. These thrusts were initially displaced westward during the Cretaceous to the early Miocene and deformed the internal part of the Apenninic chain (Haccard et al. 1972); this compression was then followed by a north and then northeast displacement (Ippolito et al. 1975). This more recent compressive phase occurred from the early Miocene to the Pliocene and deformed the external part of the Apenninic chain. Uplift of the entire Southern Apennine region has occurred since the late Pliocene (Figure 22-2).

Note that the chronostratigraphic units for the Miocene cited in this chapter are defined according to Pescatore et al. (1970) and are: (1) Aquitanian, (2) Langhian, (3) Serravallian, and (4) Tortonian; they correspond to the following biostratigraphic units: (1) *Catapsidrax dissimilis* zone, (2) *Globigerinoides trilobus* zone, (3) *Orbulina s.l.* zone, and (4) *Globigerina nepenthes* and *Globorotalia menardii* zones. The former zones in turn correspond to zones 4 (emended) to 17 of Blow (1969).

Irpinian Units

The marine Irpinian sequences (herein abbreviated Irpinids) include several tectonic units of the Apennine chain and were deposited in an elongate and narrow basin (Irpinian Trough) that was formed in the Langhian (early Miocene) as a result of important tectonic pulses (Cocco et al. 1972). Before this Langhian deformation phase took place, several paleogeographic units existed during the Mesozoic to the early Cenozoic; from west to east (diagram A, Figure 22-3), they have been termed as follows (Ippolito et al. 1975): Cilento Basin, Campania-Lucania Platform, Lagonegro Basin, Abruzzi-Campania Platform, Molise Basin, and Apulia Platform.

Sand and conglomerates, as well as shaly and calcareous deposits, accumulated in the Cilento Basin (Ogniben 1969; Cocco and Pescatore 1975). The sediments on the platforms and the intervening basins (Lagonegro and Molise) are similar to those on carbonate platforms and basins of the Bahama area (D'Argenio 1970; Bernoulli 1972; Bernoulli and Jenkyns 1974).

The tectonic pulses in the Langhian deformed the Cilento Basin, the Campania-Lucania Platform, and the western part of the Lagonegro Basin (diagram B, Figure 22-3); repeated thrusting displaced the respective sedimentary rock sequences toward the east and gave rise to a new elongate basin herein termed the Irpinian Trough. The eastern part of this new depression was emplaced in part on the preexisting Lagonegro Basin; thrust sheets formed the western part of the basin. Thus, an east-west profile (diagram B, Figure 22-3) shows the western margin formed by superposed thrust sheets and the eastern margin bounded by an almost undeformed carbonate sequence (Abruzzi-Campania Platform). In the Langhian, the axis of the Irpinian Trough lay eastward with respect to the older (Trias to Langhian) Lagonegro Basin.

The Irpinian successions presently crop out along two major belts, both trending northwest-southeast (map A, Figure 22-1). The southwestern belt includes terrigenous deposits unconformably resting on older allochthonous thrust sheets (Castelvetere, M. Marzano, and Gorgoglione sections in chart B, Figure 22-1), while the northeastern belt is formed by calcareous turbidites and terrigenous successions conformably following above the Lagonegro Basin deposits (Faeto, Serra Palazzo sections in chart B,

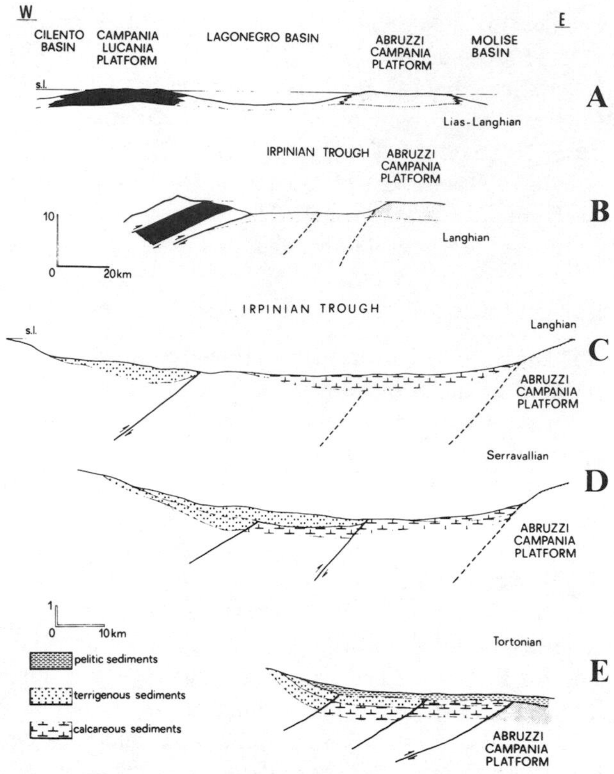

Figure 22-3. The Irpinian Trough evolution illustrated by some palinspastic sections. Note: *Upper* = diagram A – Jurassic to early Miocene configuration (before deformation); diagram B – early Miocene (after the Langhian tectonic phase). *Lower* = diagram C to E – distribution of the Irpinian sedimentary facies in the deformed trough, from Langhian to Tortonian time (diagram C is an expansion of diagram B). See text for explanation.

Figure 22-1). The terrigenous deposits of the southwestern belt were supplied by erosion of the thrust sheets and also were derived from a probable crystalline area farther to the west—that is, in the area of the present Tyrrhenian Sea, probably a northward prolongation of the Calabria thrust sheets. The calcareous sediments of the northeastern zone were fed from the Abruzzi-Campania carbonate platform.

During the Langhian to Tortonian (early to middle Miocene), an eastward displacement of the thrust sheets again occurred, thereby resulting in a renewed eastward migration of the trough axis, while the western margin of the Abruzzi-Campania platform collapsed along vertical faults (diagrams C, D, and E, Figure 22-3). Evidence for this complex tectonic evolution is provided by the eastward shift of Irpinian terrigenous units: Terrigenous facies first replaced the calcareous ones during the Serravallian (Serra Palazzo section in chart B, Figure 22-1; diagram D, Figure 22-3). Afterwards (during the Tortonian), the terrigenous facies extended to the east on the former margin—that is, the western margin of the Abruzzi-Campania carbonate platform (Matese and M. Alpi sections in chart B, Figure 22-1; diagram E, Figure 22-3). The stratigraphic columns in chart B, Figure 22-1, schematically depict the Irpinian successions geographically and in time and emphasize the diachronism of the Irpinid terrigenous facies.

Sedimentation in the Irpinian Trough terminated as a result of a new and important tectonic pulse in the early Tortonian, which again displaced the sedimentary basement and overlying basin deposits eastward. As a result, the Irpinian units overlap and also override the older sediments of the Abruzzi-Campania Platform. Subsequent tectonic displacements between Tortonian and Pliocene time have shifted the Irpinids to their present position—that is, almost to the western border of the Apulia unit (map A, Figure 22-1).

DESCRIPTION OF IRPINIAN FACIES

The following sections briefly describe the sedimentary characteristics of the terrigenous successions that accumulated between the neritic zone to the west and the trough to the east.

Sand and Conglomerate Facies (Neritic Environment)

Terrigenous deposits of shallow marine origin, which rest unconformably above the carbonate units of the Campania-Lucania Platform, constitute the westernmost outcrop belt of the Irpinian series. These units, described in detail elsewhere (Pescatore et al. 1970), have been interpreted as coastal to neritic platform deposits. Their thickness is generally a few tens of meters. In various localities (e.g., M. Marzano section in chart B, Figure 22-1), the succession begins with conglomerates; the clasts were derived from deposits of the Campania-Lucania Platform. Conglomerates are followed by sandy deposits with ripple marks and large-scale cross-lamination; these units yield neritic fossils (corals, lamellibranchs, *Pecten, Ostrea,* bryozoans, gastropods, and soforth). Shaly sediments are interbedded with sandy conglomeratic strata. Pelecypod borings occur locally on the base of the strata that overlie the unconformity. As has been postulated (Pescatore et al. 1970), these neritic deposits grade laterally to terrigenous units, which include the calcareous olistoliths described in the following section.

Sand, Conglomerate, and Olistolith Facies (Base-of-Slope)

One of the most characteristic Irpinian successions includes a formation with very large, exotic calcareous blocks, or olistoliths, interbedded in sandy and conglomeratic strata. This formation with olistoliths, the Castelvetere Flysch, has the same northwest-southeast trend as the Mesozoic carbonate deposits of the Campania-

A

B

C

D

Figure 22-4. Castelvetere Flysch, near Pesopagano, North of Monte Marzano (Figure 22-1). Note: Photograph A shows small erosional channel in conglomerate beds, with imbricated pebbles (transport direction from the west). Photograph B shows erosional base (white line) of a fining-upward sequence in arenaceous-rudaceous deposits (channel fill deposits). Photographs C and D show calcareous olistoliths in arenaceous-rudaceous deposits (in photograph D the variable dip of the calcareous blocks is noteworthy).

Lucania Platform that supplied the calcareous olistoliths. These large exotic blocks, whose volume may be a million cubic meters, are found at different stratigraphic levels in the Castelvetere Flysch; they are particularly large and common in the sediment of Serravallian age. The concentration of these blocks is believed to record a tectonic pulse.

This facies is well exposed in the Monte Marzano area (Figure 22-1), where it is dated as Langhian to Lower Tortonian. The olistolith-bearing deposits are unconformable upon thrust sheets emplaced during the Langhian tectonic pulse (Pescatore et al. 1970). The structural configuration of the units as presently observed is the consequence of various tectonic phases that have taken place since the Langhian. A particularly significant pulse during the Tortonian resulted in the development of several thrust faults and a major eastward shift of the Irpinian Trough deposits.

A recent study (Cocco et al. 1974) details the geological and sedimentological characteristics of the Monte Marzano area. The lithofacies of this region are described according to Mutti and Ricci Lucchi (1972) and Walker and Mutti (1973).

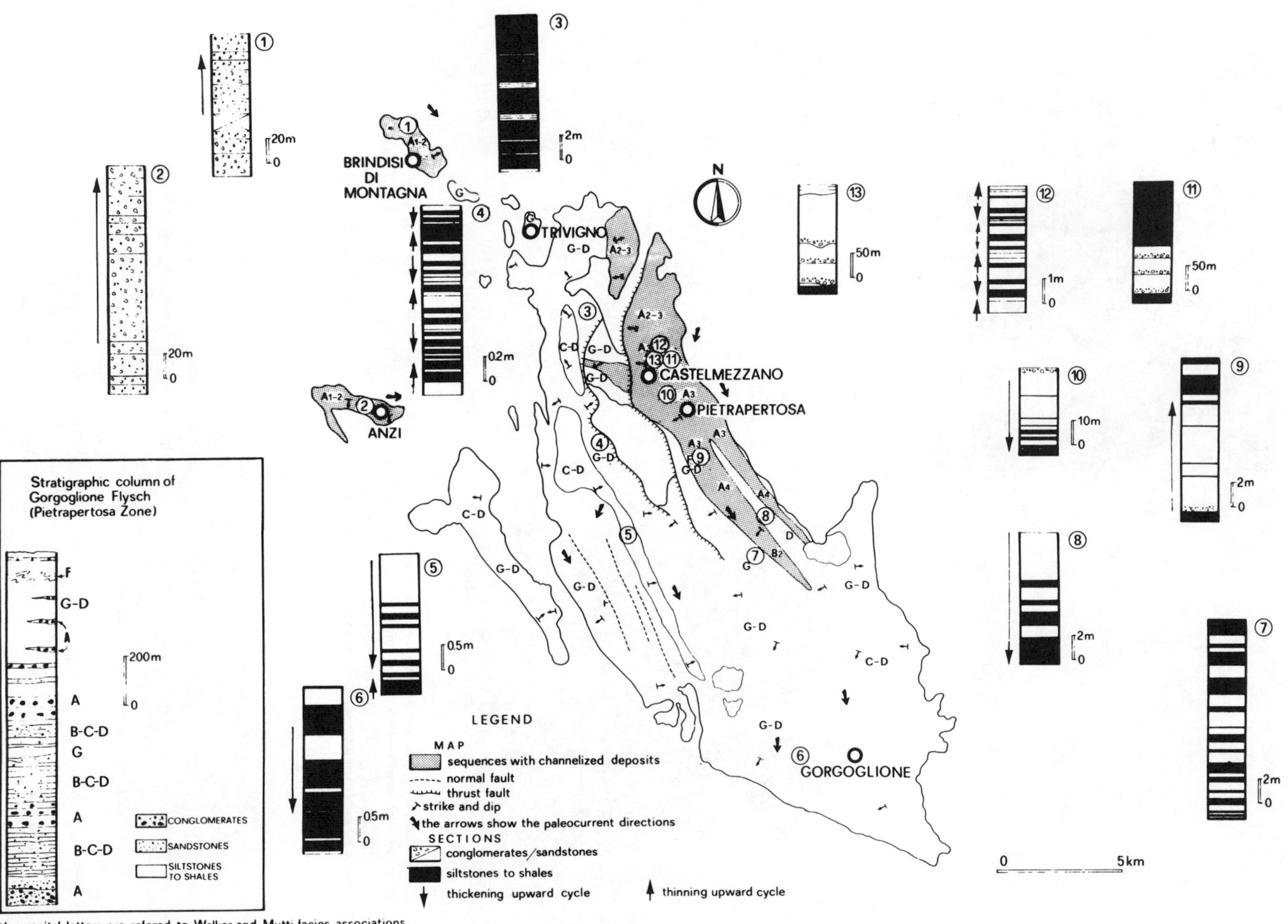

Figure 22-5. Gorgoglione Flysch facies distribution, paleocurrent directions, and lithologic sections. Note: The paleocurrent arrows indicate the mean of several measurements (5 to 10) except for the Brindisi di Montagna area (one measurement). Inset at lower left shows the stratigraphic column of the Gorgoglione Flysch in the Pietrapertosa zone.

Facies A: Arenaceous-rudaceous deposits. This facies includes both disorganized conglomerates and pebbly sandstones (Facies A1 and A3 of Walker and Mutti 1973) and organized conglomerates (photograph A, Figure 22-4) and pebbly sandstones (Facies A2 and A4 of Walker and Mutti 1973). Facies A1 corresponds to the disorganized conglomerate model of Walker (1975)—that is, a lack of stratified deposits and no gradation or imbrication of clasts. Facies A2 consists mainly of graded conglomeratic beds several meters thick locally with distinct imbrication and with large-scale cross-stratification in the upper part of the bed. This facies corresponds to the normally graded conglomerate model of Walker (1975). In Facies A both thinning-upward (to some tens of meters thick: photograph B, Figure 22-4) and thickening-upward sequences (up to ten meters thick) are present. The latter begins, at the base, with coarse laminated sandstones and terminates upward with conglomeratic lenses (clast diameter exceeds 10 cm). Facies A is sometimes intercalated in pelitic-arenaceous strata.

Facies E and G: Pelitic-arenaceous deposits. Facies E of Walker and Mutti is represented by siltstones and shales with thin interbeds of sandstone. The beds are more uniform than Facies A and display parallel or small-scale cross-lamination and rapid lateral pinch-out; their thickness ranges from 2 to 10 cm. They are probably overbank deposits (Mutti and Ricci Lucchi 1972).

Facies G is represented by shale, silty shale, and marl with few fine-grained and thin sandstone interbeds; pelagic foraminifera are sometimes abundant in the finer-grade units.

Facies F: Chaotic deposits. Facies F is characterized by large calcareous olistoliths (photographs C and D). Associated with these exotic blocks are deformed fragments of fine grade (Facies G) sediment and/or other blocks derived from thrust sheets eroded on the basin margins. These exotic carbonate blocks lie chaotically within arenaceous-rudaceous strata (Facies A), and one of their characteristics is their irregular dip (photograph D, Figure 22-4). In addition, Facies F also includes large slide and slump deposits.

Paleocurrent directions in these deposits show transport from west to east and from west-northwest to east-southeast. The succession in the Monte Marzano area is about 350 m thick (chart B, Figure 22-1). The lower deposits, primarily arenaceous and conglomeratic, are Langhian to Serravallian in age, while the upper units, mainly pelitic, are dated as Serravallian to Tortonian. The microfaunal assemblages show that the succession upward is a regressive one. These assemblages comprise mainly planktonic forams; benthonic faunas in the lower and middle part of the succession are indicative of deeper marine environments than those in the upper part (Cocco et al. 1974).

Arenaceous-Rudaceous and Pelitic-Arenaceous Facies (Deep Sea Fan Environment)

A formation, termed Gorgoglione Flysch (Selli 1962; Vezzani 1968, 1975; Ogniben 1969; Boenzi and Ciaranfi 1970; Ciaranfi 1972) accounts for nearly half of the terrigenous Irpinid facies exposed in the study area (map A, Figure 22-1). To date, this formation has received less attention as to its sedimentological origin than the sequences described in the two previous sections. Recent results show that such analysis is needed to interpret the paleogeography of the Irpinid Trough.

The Gorgoglione Flysch lies unconformably upon the allochthonous deposits of the Cilento Basin, and its thickness is over 1,200 m (chart B, Figure 22-1). The age ranges from Langhian to early Tortonian. The Gorgoglione Flysch is well exposed between Brindisi di Montagna and Gorgoglione (Figure 22-5). The Tortonian tectonic phase is well recorded in this zone, especially in the northern sector where thrust faults are present. The paleogeographic origin of this formation and its relations to other Apennine units of the same age in this area are not fully understood. According to Ogniben (1969), the Gorgoglione Flysch may have been deposited to the west of the Campanian-Lucania Platform. However, Pescatore (1971) and Cocco et al. (1972) suggest that the Gorgoglione Flysch was deposited in the Irpinian Trough on the basis of its specific lithologic and structural characteristics.

Coarse-grained deposits outcrop in the northern or northwestern area (Brindisi di Montagna and Anzi) and become progressively finer southward toward Pietrapertosa and Gorgoglione (Figure 22-5). Some of the terrigenous structures attributed to deepsea fans as described by, among others, Mutti and Ricci Lucchi (1972) and Walker and Mutti (1973) may be observed in the Gorgoglione Flysch. The following five sections describe the sedimentary characters of the formation from north to south.

1. Near Brindisi di Montagna and Anzi, the succession is formed almost completely by conglomeratic deposits. These deposits are comparable to Facies A1 and A2 of Walker and Mutti (1973). The three conglomerate models of Walker (1975) have been observed here, even if a clear distinction among the individual models is not always possible: graded-stratified, inverse to normally graded and disorganized. Generally the stratification is very irregular, the beds are always amalgamated, and large erosive channeling is commonly displayed at the base of beds (photograph A, Figure 22-6). Thickening- or fining-upward sequences are not easily distinguished in these successions. In the Brindisi di Montagna area there are some thinning-upward sequences (section 1, Figure 22-5), 40 to 50 m thick, which begin with disorganized or inversely graded conglomerates at their bases and pass upward to organized conglomerates, pebbly sandstones and sandstones with large-scale cross-stratification. In the Anzi area (section 2,

A

B

C

D

E

F

Figure 22-5), thinning-upward sequences have also been noted. The sequence begins with conglomerates, mostly disorganized, and with no distinctive stratification; lenses of coarse sandstone are interbedded in thick (50 m) units of conglomerate. Sequences terminate upward with arenaceous and conglomeratic strata. These upper arenaceous-conglomeratic units also show smaller scale (3 m to 10 m) thinning-upward sequences (section 2, Figure 22-5; photograph B, Figure 22-6), and both the thickness of the sequences and the mean diameter of the pebbles decrease upward.

2. About three kilometers to the south, in the zone of Trivigno and Pietrapertosa (Figure 22-5), the proportion of coarse, thick conglomeratic deposits (Facies A1 and A2) decreases in the succession, and the sediments are substituted by pebbly sandstone strata (Facies A3 and A4 passing to Facies B2). In this zone, Facies A1 and A2 are found only in the upper part of the series.

In this region, two distinct pebbly sandstone beds, each about 100 m thick, are well exposed. These two units merge northward and divide southward; the upper unit in particular splits into several layers interbedded with thin pelitic and arenaceous strata (photographs C and D, Figure 22-6). The massive appearance of the two units is the result of amalgamation of several pebbly sandstone deposits. No definite fining- or coarsening-upward sequences are noted in these deposits in the northern area (sections 11 and 13, Figure 22-5), while thickening-upward sequences (section 10, Figure 22-5; photograph D, Figure 22-6) are observed in the southern area. Locally, the base of these thick deposits is strongly channeled into the underlying sediment (photograph D, Figure 22-6); elsewhere it is flat and nonerosive. Rip-up clay clasts are commonly concentrated near the base of the beds where the pebbly sandstone rests on eroded underlying slumped pelitic units.

Still farther southward (to the south of Castelmezzano), pebbly sandstones are intercalated with pelitic deposits in the upper part of the sequence. Here the pebbly sandstone is about ten meters thick and shows fining- and thinning-upward sequences with well-developed erosion at the base of the beds (section 9, Figure 22-5).

3. Southward and westward of Pietrapertosa, the pebbly sandstone deposits pass into alternating arenaceous and pelitic units (Facies C and D of Mutti and Ricci Lucchi 1972). This type of facies is also present between the two above-cited pebbly sandstone units. In this area, thinning- and thickening-upward sequences alternate more or less regularly (section 12, Figure 22-5). The thickness of each sequence is about one meter. The beds are lenticular and wedge out abruptly. Erosive and cut-and-fill structures, typical of channelized deposits, are preserved at the base of a few thinning-upward sequences. Bouma sequences are developed in these sediments (Bouma 1962; Walker 1967); normally the turbidite sequence type T_{bcde} or T_{cde} is present. Some sections of the alternating arenaceous-pelitic sequences are mainly constituted by arenaceous sediment (sections 5 and 8, Figure 22-5). These show thickening- and coarsening-upward sequences and no erosive phenomena at the base of the beds; the Bouma sequences (T_{bcde} or T_{cde}) are also recognized. Moreover, large-scale cross-stratification (photograph E, Figure 22-6), which is festoon-like, is observed in the upper part of some sequences. The bed thickness ranges from a few decimeters to two meters. However, there are also thinning- and fining-upward sequences of pebbly sandstone units in this arenaceous-pelitic succession (Facies A4 and B); the thickness of the pebbly sandstone sequences is about 10 m.

The arenite-rich sections are associated with more pelitic deposits, the latter increasing to the south and southwest. These sediments include fine-grained distal turbidites, which normally begin with the *c* division of the Bouma (1962) sequence and are generally less than 10 cm thick.

The arenaceous and pelitic deposits described here also are found in the area to the north, adjacent to the conglomeratic and sandstone exposures.

4. The succession in the Gorgoglione area is constituted primarily by massive shales and shaly and silty marls with intercalations of alternating siltstone and sandstone. The bed thickness ranges from a few centimeters for the pelitic deposits to 50 cm for the sandstone strata. The beds are well stratified and flat based; the T_{cde} Bouma turbidite sequence is common. Sometimes the arenaceous beds form

---

Figure 22-6. Gorgoglione Flysch. Note: Photograph A (Brindisi di Montagna) shows large-scale erosion at the base of conglomeratic beds. Photograph B (Anzi) shows fining- and thinning-upward sequence in the arenaceous-rudaceous facies (channel fill deposits). The hammer provides a scale. Photograph C (Northern area of Castelmezzano) shows thick pebbly sandstone units (A, B, C), fan valley deposits splitting southward (downcurrent) as shown on lower left in the photograph (see arrows). Photograph D (Pietrapertosa) shows channel deposits (left, white line shows the base of the channel) and depositional lobe (thickening- and coarsening-upward sequence; see arrow). The section in this photo partly corresponds downcurrent to that photograph of C. The letters A, B, and C correspond to the same units in photographs C and D. Photograph E (southeast of Castelmezzano) shows large-scale cross-lamination in the upper part of a thickening-upward sequence (depositional lobe). Photograph F (southwest of Pietrapertosa) shows upper part of the Gorgoglione Flysch sequence (base-of-slope facies). Detail of a chaotic interval is pebbly mudstone with internally deformed hemipelagic deposits; the pebbles are both dark and light in color. The watch is provided for scale.

thickening-upward sequences (section 6, Figure 22-5), and locally the arenites are randomly interbedded in the pelitic sediments. The shale and marls commonly contain a rich planktonic foraminiferal fauna.

5. Shales and marls (Facies G) constitute the upper part of the Gorgoglione Flysch throughout the study area. These pelitic-rich deposits are associated with fine-grained arenaceous deposits (sections 3, 4, and 7, Figure 22-5). The sandstone strata are well sorted, from a few centimeters to one decimeter thick, and are characterized by small-scale cross-lamination. Flute casts are sometimes observed at the base of sandstone units, while asymmetrical ripple marks are noted on the top of beds. Additional work is needed to determine the origin of these sandstones, which may have been deposited by low-concentration turbidity currents (Middleton and Hampton 1973) and possibly modified by contour currents (Hollister and Heezen 1972; Bouma and Hollister 1973). In addition, chaotic intervals–slump folds, slump breccias, and pebbly mudstones (photograph F, Figure 22-6)–have been found associated with this pelitic facies (cf. upper part of stratigraphic log in lower left inset in Figure 22-5).

Paleocurrents in the Gorgoglione Flysch. In the earliest paleocurrent study of the Gorgoglione Flysch, Ten Haaf (1959) recorded a north to south orientation of flute casts and argued for a northern source. Boenzi and Ciaranfi (1970) also propose a source to the north and northwest on the basis of flute cast orientations and regional petrological considerations. More recently, Loiacono (1975) has completed an extensive study of linear sedimentary structures (flute casts, and so forth) as well as of grain orientation, and his results show that directions based on grain orientations are more consistent than those based on sole marks. The mean paleocurrent direction determined from sole markings is offset by 2° to 29° from the mean direction based on grain elongation. Grain elongation and imbrication both show transport primarily from northwest to southeast. Some sole markings, however, indicate transport from north and northeast. The north-northeast directions are observed primarily in the Castelmezzano and Pietrapertosa areas. West to east directions are measured in the conglomeratic deposits of the western zone (Anzi area, shown in Figure 22-5). The main transport directions from northwest to southeast shown by sedimentary structure and grain orientation directions vary irregularly up the stratigraphic column. There is a clock-wise rotation of direction from northwest to north-northeast as one proceeds up the section in the Pietrapertosa region. These north and north-northeast directions also appear to prevail in the upper part of the section elsewhere in the study area.

On the basis of all data available, Loiacono (1975) suggests that the main source area for the Gorgoglione Flysch was located to the west.

DEPOSITIONAL IMPLICATIONS

A structural-depositional model based on observations made in the study area is presented in Figure 22-7. The major aspects of this model include:

1. The western slope of the trough (tectonically active and undergoing thrusting) where sedimentation at the base of the slope is characterized by olistoliths;
2. The axial zone of the Irpinid Trough, where terrigenous deposits accumulated as a series of deep-sea fans;
3. The importance of lateral sediment supplied to the trough, but prevailing longitudinal transport within the narrow elongate trough.

Those arenaceous-conglomeratic facies interpreted as neritic on the basis of sedimentary and paleontological characters are not depicted on the model in Figure 22-7. These deposits, which accumulated west of the trough, probably were formed on a narrow, shallow platform at the edge of an emerging land mass that supplied the Irpinian Trough with terrigenous materials; a crystalline source area is believed to lie farther to the west. The western margin of the trough, including the platform and slope, was formed by easterly moving superposed thrust sheets. Sands and pebbles were transferred from the narrow platform onto the slope and farther seaward into the trough proper. The shallow origin of pebbles of the Irpinian units is demonstrated by boring structures caused by neritic organisms.

The terrigenous deposits at and beyond the base of the slope (probable bathyal depths based on microfauna; Pescatore et al. 1970) can be subdivided into two types: those with large olisoliths (Castelvetere Flysch) and those without olistoliths (Gorgoglione Flysch).

The Castelvetere Formation, which presents typical flysch attributes with large olistoliths, is comparable to the "base-of-slope deposits" described by Stanley and Unrug (1972) and, more specifically, with Facies F of Mutti and Ricci Lucchi (1972) and Walker and Mutti (1973). The spectacularly large blocks of Mesozoic neritic carbonates, which characterize the formation, are interpreted as rockfall deposits detached from the trough margin – that is, presumably from the thrust fronts forming the upper slope. This type of formation containing allochthonous blocks is also described in the Alps, Sicily, and elsewhere, where the allochthonous series are interpreted as materials sliding seaward on the moving fronts of thrust sheets (Lugeon 1916; Ogniben 1963; Pescatore et al. 1970). This interpretation (Pescatore et al. 1970) is supported by the normal stratigraphic relationships of the blocks (olistoliths) with the adjacent deposits and by diffused slump masses assoc-

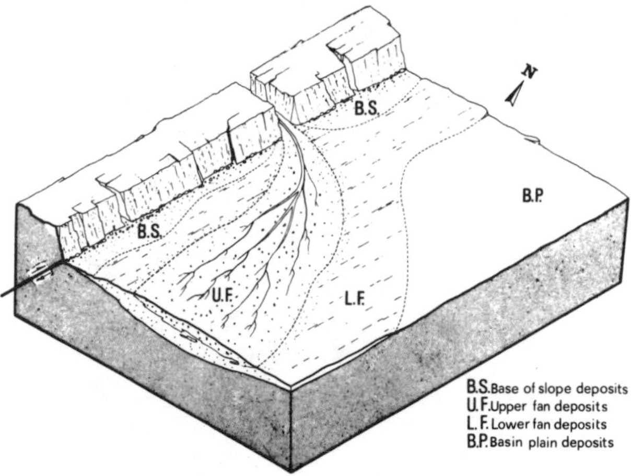

Figure 22-7. Simplified geotectonic-depositional model for the western margin of the Irpinian basin during the early Miocene.

iated with the olistoliths. Moreover submarine erosional channels can be observed below the blocks for a width of few tens of meters. The matrix of these deposits is pelitic or even sandy. The aforementioned characters and the absence of pervasive shearing indicate that the Castelvetere Flysch is an olistolith-bearing deposit and not a mélange (both as defined by Hsü 1974). These deposits may be analogous to slumps and related gravity mass deposits found at the base of the landward walls in modern trenches (Piper et al. 1973; Scholl 1974).

In addition, several other facies interpreted as channel (Facies A), interchannel (G), hemipelagic deposits (E), and finally chaotic deposits (F) (cf. Mutti and Ricci Lucchi 1972; Walker and Mutti 1973) are also observed. As should be noted, this flysch-with-olistoliths crops out along a northwest-southeast trend that approximately parallels the thrust sheet front. Thus we can conclude that the flysch-with-olistoliths roughly delineates the base-of-slope zone where rockfalls, slides, slumps, and several types of sand- to granule-size gravity flow deposits, including grain flows, debris flows, and so forth (Middleton and Hampton 1973), are concentrated.

In the proposed model of the Irpinian Trough, the olistolith-rich flysch is presumably localized at the base of a relatively steep slope that was migrating eastward during sedimentation. This movement, as stated earlier was the result of broad regional compression. However, olistoliths need not be restricted to basin margins where compressional tectonics prevail (cf. Stanley and Unrug 1972; Bernoulli and Jenkyns 1974; Cocco and Pescatore 1975).

The lateral transition between the Castelvetere Flysch-with-olistoliths and the Gorgoglione Flysch is not observed. The terrigenous facies forming the Gorgoglione Flysch can best be interpreted as deepsea fans; they show characteristics similar to ancient fan deposits, like the ones described by Mutti and Ricci Lucchi (1972), Walker and Mutti (1973), Ricci Lucchi (1975), and similar to modern fans, as described by Normark (1970) and others. The observations made in the Irpinid Basin suggest that there are some channels in the outer fan zone that die out on the distal fringe, such as described by Nelson and Kulm (1973) in the Astoria and Cascadia fans. In the Irpinid deepsea fans several sequences and facies associations are mapped. These include fining-upward sequences interpreted as channel fills in the middle fan, and thickening- and coarsening-upward sequences interpreted as progradational lobes (cf. Mutti and Ricci Lucchi 1972; Mutti and Ghibaudo 1972).

The transport mechanisms of the Irpinian units include primarily sediment gravity flows: debris and sand flows (grain or fluidized flows) that are largely responsible for the transport of coarser sand and pebbly sand deposits (Middleton and Hampton 1973; Stanley 1975) and turbidity currents that are responsible for the deposition of less coarse-graded sand sheets and some pelitic deposits. In addition, hemipelagic and contourite deposition may also be represented (Hollister and Heezen 1972).

The Gorgoglione Flysch extends from northwest to southeast with fining and thinning of deposits towards the southeast. This lateral change sheds light on the transport processes and dispersal. The conglomerates in the northern region display examples of the three resedimented conglomerate models of Walker (1975). That author interprets disorganized conglomerates as deposits resulting from rapid deposition on the base of relatively steep slopes; inverse to normally graded and graded stratified conglomerates may accumulate in somewhat more distal environments. In the Irpinid conglomeratic sections (Brindisi di Montagna and Anzi areas) the three types are interbedded. A significant volume of conglomeratic material is believed to have accumulated in the lower part of a submarine valley as it debouched into the Irpinid Trough, as shown in Figure 22-7, although some coarse units also may have been deposited further seaward on the fan itself (see Chapters 11 and 24 of this volume).

Downcurrent the conglomeratic deposits pass into disorganized pebbly sandstones. Two thick amalgamated pebbly sandstone units, mapped regionally as discussed earlier, are interpreted as channelized bodies, probably deepsea channel or fan valley deposits. Their bases are erosive upslope in the more proximal zone; downcurrent the deposits display depositional sequences. Some lateral variations may be explained by changes in flux of the flows or changes in the preexisting morphology of the channel (cf. Normark 1970). These channelized deposits develop for several kilometers along a northwest to southeast trend and are accompanied by reduced thicknesses downcurrent and eventual pinch-out. Facies

C and D, between the two pebbly sandstone units, represent interchannel deposition; they display both thinning- and thickening-upward sequences about one meter thick. Southward, the formation is not channelized, but includes Facies C and D that display coarsening- and thickening-upward sequences, from one to several meters thick. These sequences are interpreted as depositional lobes. Distal turbidites and hemipelagic deposits (Facies D and G) become progressively more important down-current (southward) and eventually are replaced by hemipelagic and thin turbiditic sandstones that are normally a few centimeters and rarely as much as one meter thick. These strata represent the nonchannelized, distal deposits of deepsea fans. Finally the upper Gorgoglione Flysch sequence, formed mainly by hemipelagic deposits with slumped and/or chaotic beds and pebbly mudstones, may be considered a slope facies association (cf. Mutti and Ricci Lucchi 1972).

The main transport directions in the Castelvetere Flysch-with-olistoliths are from the west and west-northwest toward the east. In the Gorgoglione Flysch, the sedimentary structures of the coarse deposits indicate a western source, while those of the finer deposits show flows derived primarily from northwestern and northern sectors (Loiacono 1975). All factors, including the regional facies patterns, tectonic trends, and paleocurrents, indicate that the bulk of sediment was laterally derived from the mobile, western margin of the basin. Materials were fed from eroded sedimentary terranes forming the thrust sheets and also possibly from an emergent crystalline zone located farther to the west.

Within the Irpinid Trough, longitudinal transport parallel to the basin axis is proposed during the development of the deepsea fan successions such as the Gorgoglione Flysch. Moreover, the linear channelized deposits and the southeast-trending paleocurrent directions substantiate predominant longitudinal transport along the trough as a direct result of the regional northwest-southeast tectonic trend.

PALEOGEOGRAPHIC CONSIDERATIONS AND SIGNIFICANCE

The base-of-slope and fan model here developed for the Irpinian Trough makes possible interpreting more accurately the paleogeographic evolution of this trough and the related southern Apennine region during the Miocene. The main point is that sedimentation occurred in an elongate basin while its axis migrated from west to east during deposition. The evidence for this tectono-sedimentary evolution is the diachronism of the deepwater terrigenous facies, which implies that such sediments, from the Langhian to Tortonian, become progressively younger eastward. Also the variation of the paleocurrents in the Gorgoglione Flysch may be associated with the above mentioned eastward shift of the trough axis, although such a variation also could be attributed to the migration of the fan axis.

Furthermore, the vertical evolution of the terrigenous deposits shows a regressive trend in the sedimentary sequence. Microfauna in the Castelvetere Flysch-with-olistoliths record deeper marine environments in the lower and middle parts than in the upper part of the succession. The Gorgoglione Flysch succession also is regressive in the upper part, as indicated by a vertical transition from deepsea fan to somewhat shallower deposition. The regressive trend of the entire Irpinian succession is the result of the contemporaneity of two major factors: the uplift and thrusting of the western side of the basin and the collapse of its eastern margin.

This evolution of the Irpinian Trough is depicted in diagrams C to E, Figure 22-3. The relationships between the Irpinian units shown in diagram E, Figure 22-3 are essentially those observed at present. The present configuration is the result of Tortonian and post-Tortonian tectonic pulses, which moved the Irpinian units and their sedimentary substratum *in toto* toward the east.

The Irpinian Trough developed during the last part of the geologic history of the Apennines and therefore followed collision of the European and African continental margins. This collision very probably caused a drastic change in the thrust direction: westward from Cretaceous to lower Tertiary, eastward from the Miocene. The Irpinian Trough behaved as a *peripheral basin* (Dickinson 1975) – that is, a sedimentary basin originated by crustal collision and emplaced at the crustal suture lines within the African continental margin depressed by partial subduction.

A comparison between the Irpinian and the other troughs developed in a compressional regime reveals analogy with continental margins of Pacific type. Models based upon some present basins of the Eastern Mediterranean, now in a compressional stage of their evolution, have been used previously for terrigenous deposits of the Alpine (Hsü and Schlanger 1971; Hsü 1972; Stanley 1973, 1974a, b) and Maghrebid (Wezel, 1973) chains. Nevertheless the Irpinian facies distribution and evolution do not seem comparable with the eastern Mediterranean basins.

A sedimentary model derived from the facies distribution of the Irpinian Trough suggests a terrigenous influx from the western margin and a carbonate influx from the eastern margin (Cocco and Pescatore 1975), as in the present day Gulf of Mexico. A tectono-sedimentary model, however, shows many similarities with Pacific-type continental margins notwithstanding the different dimensions of the processes and of the rock bodies and their different geotectonic conditions.

On the basis of the facies distribution in the Oregon continental margin, Nelson and Kulm (1973) propose a depositional model characterized by a base-of-slope style of sedimentation, which is distinct from the deepsea fan sedimentation style. In both the Irpinian and the Nelson and Kulm basin models, the base-of-slope deposits incorporate zones of olistoliths that mark the leading margin of the basin. The Irpinian deepsea fan deposits thus may be compared with the present Astoria fan – that is, formed by a thick wedge of sediments that develops parallel to the continental margin, rather than a simple cone-shaped body (Nelson and Nilsen 1974). The Astoria fan fills a trench-type depression at the eastern margin of the subducting Gorda-Juan de Fuca plate (Kulm and Fowler 1974a). The form of this depression in the basement partly filled by the Astoria fan is similar to the Aleutian and Peru-Chile Trenches (Scholl 1974; Kulm and Fowler 1974a).

The uplift of the trough plain deposits and the tectonic thrusting of the wedge-shaped rock bodies at the base of the slope, whose growth is thus tectonically produced, seem typical elements of the Pacific-type margins (cf. Karig and Sharmon 1975; Prince and Kulm 1975; Seely et al. 1974; Kulm and Fowler 1974a). Seely et al. (1974) propose a model that involves uplift of the continental margin due to an imbricated series of upthrusted sedimentary wedges at the base of the slope. This model explains (1) the uplift of the trough plain sediments and (2) the facies diachronism. Kulm and Fowler (1974b) have adopted this model for the Oregon continental margin, which from the Tertiary shows the structural and stratigraphic characters of the fore-arc as defined by Karig (1970). Such a structural configuration involving imbricated sedimentary thrust sheets is strongly supported, according to the foregoing authors, both by the stratigraphic position of some of the abyssal deposits of the Cascadia Basin on the continental slope and by the repetition of section.

Accordingly, the eastward migration of the terrigenous facies of the Irpinian trough is attributed to a compressional regime, which has produced thrusting of sedimentary wedges at the western margin of the trough. In fact, the structural model of the Irpinian Trough is characterized by a tectonically deformed *internal* margin and by an *external* margin affected only by downfaulting up to the Tortonian. Such a style of deformation, with eastward moving thrust sheets and contemporaneous subsidence of the eastern side of the basin, is invoked to account for the sedimentary facies migration and the uplift of the *internal* border.

Although our understanding of the relationships between the evolution of continental margins and that of mountain chains is still vague, the fact seems significant that there exist several modern analogs for the development of basins that have originated in a compressive regime but in different geotectonic settings.

SUMMARY

1. A study of the Miocene (Langhian to Tortonian) Irpinid terrigenous units in southern Italy provides a model for base-of-slope and deepsea fan sedimentation in an elongate trough formed in a compressive setting.

2. The internal (western) margin of the tectonically mobile Irpinid Trough is composed of several superposed thrust sheets moving eastward, while the eastern margin was formed by a subsiding carbonate platform.

3. The base of the western slope was characterized by the presence of large olistoliths, which slid on the thrust front, while the trough axial zone was filled by elongate, deepsea fans.

4. Sediment, supplied largely from the west, was transferred across the western margin. Once in the trough, materials were subsequently shifted toward the southeast, parallel to the major tectonic trend.

5. Irpinid sedimentation loci shifted in time toward the east as a direct result of the regional tectonic displacement in this direction; consequently, the terrigenous units are diachronous – that is, progressively younger toward the east.

6. The postulated sequence of Irpinian evolution presents significant analogies with geotectonic-depositional continental margin models of the Pacific type.

ACKNOWLEDGMENTS

This investigation has been funded by National Research Council of Italy grant 75.00131.05. Professors B. D'Argenio and E. Cocco are gratefully acknowledged for their critical reading of the manuscript. Special thanks are due to Professors D.J. Stanley, G. Kelling, and R.H. Dott, Jr., for their help in the review and revision of the text.

REFERENCES

Accademia Nazionale dei Lincei, 1973. Atti del Convegno sul tema: Moderne vedute sulla geologia dell'Appennino. *Acc. Nazion. Lincei, Quaderno* 183: 454 pp.

Bernoulli, D., 1972. North Atlantic and Mediterranean Mesozoic facies. In: C.D. Hollister, J.I. Ewing et al. (eds.), *Initial Report of the Deep-Sea Drilling Project, 11.* U.S. Govt. Print. Off., Washington, D.C., pp. 801-71.

——, and Jenkyns, H.C., 1974. Alpine, Mediterranean and Central Atlantic Mesozoic facies in relation to the early evolution of the Tethys. In: R.H. Dott, Jr., and R.H. Shaver (eds.), *Modern and Ancient Geosynclinal Sedimentation.* Soc. Econ. Paleont. Mineral., Sp. Publ. 19, pp. 129-60.

Blow, W.H., 1969. Late Middle Eocene to Recent Planktonic Foraminiferal Biostratigraphy. *Proc. 1st Intern. Conference on*

Planktonic Microfossils, Geneva, 1: 199–422.

Boenzi, F., and Ciaranfi, N., 1970. Stratigrafia di dettaglio del Flysch di Gorgoglione (Lucania). *Mem. Soc. Geol Italiana, 9:* 68-80.

Bouma, A.H., 1962. *Sedimentology of Some Flysch Deposits.* Elsevier, Amsterdam, 169 pp.

_____, and Hollister, C.D., 1973. Deep ocean basin sedimentation. In: G.V. Middleton, and A.H. Bouma (ed.), *Turbidites and Deep Water Sedimentation.* Soc. Econ. Paleont. Mineral., Pacific Section, Short Course, Anaheim, pp. 79-118.

Ciaranfi, N., 1972. Flysch di Gorgoglione. *Boll. Serv. Geol. Italiano*, 92: 101-14.

Cocco, E., and Pescatore, T., 1975. Facies pattern of the Southern Apennines flysch troughs. In: C. Squyres (ed.), *Geology of Italy.* Earth Science Soc., Libyan Arabic Republic, pp. 289-303.

_____, Cravero, E., Ortolani, F., Pescatore, T., Russo, M., Sgrosso, I., and Torre, M., 1972. Les facies sédimentaires du Bassin Irpinien (Italie méridionale). *Atti Acc. Pontaniana, Napoli*, 21: 13.

_____, Cravero, E., Ortolani, F., Pescatore, T., Russo, M., Torre, M., and Coppola, L., 1974. Le unità irpine nell'area a nord di Monte Marzano, Appennino meridionale. *Mem. Soc. Geol. Italiana*, 13: 607-54.

D'Argenio, B., 1970. Evoluzione geotettonica comparata tra alcune piattaforme carbonatiche dei Mediterranei Europeo ed Americano. *Atti Acc. Pontaniana, Napoli*, 20: 3-34.

_____, Pescatore, T. and Scandone, P., 1973. Schema geologico dell'Apprennino meridionale (Campania e Lucania). Atti del Convegno: Moderne vedute sulla geologia dell'Appennino. *Acc. Nazion. Lincei, Quaderno* 183: 49-72.

_____, Pescatore, T. and Scandone, P., 1975. Structural pattern of the Campania-Lucania Apennines. In: L. Ogniben, M. Parrotto, and A. Praturlon (eds.), *Structural Model of Italy.* Cons. Nazion. Ric., Quaderni de "La Ricerca Scientifica," 90: pp. 313-27.

Dickinson, W.R., 1975. Plate tectonics and sedimentation. In: W.R. Dickinson (ed.), *Tectonics and Sedimentation.* Soc. Econ. Paleont. Mineral., Sp. Publ. 22, pp. 1-27.

Haccard, D., Lorenz, C., and Grandjacquet, C., 1972. Essai sur l'évolution téctogénétique de la liason Alpes-Appennines (de la Ligurie à la Calabrie). *Mem. Soc. Geol. Italiana*, 11: 309-41.

Hollister, C.D., and Heezen, B.C., 1972. Geologic effects of ocean "bottom currents": Western North Atlantic. In: A.L. Gordon (ed.), *Studies in Physical Oceanography*, 2. Gordon and Breach, New York, pp. 37-66.

Hsü, K.J., 1972. Alpine flysch in a Mediterranean setting. *Proc. 24th Intern. Geol. Cong.* Section 6, pp. 67-74.

_____, 1974. Melanges and their distinction from olistostromes. In: R.H. Dott, Jr., and R.H. Shaver (eds.), *Modern and Ancient Geosynclinal Sedimentation.* Soc. Econ. Paleont. Mineral., Sp. Publ. 19, pp. 321-33.

_____, and Schlanger, S.O., 1971. Ultrahelvetic flysch sedimentation and deformation related to plate tectonics. *Geol. Soc. Amer. Bull.*, 82: 1207-18.

Ippolito, F., D'Argenio, B., Pescatore, T. and Scandone, P., 1975. Structural-stratigraphic units and tectonic framework of Southern Apennines. In: C. Squyres (ed.), *Geology of Italy.* Earth Science Soc., Libyan Arabic Republic, pp. 317-28.

Karig, D.E., 1970. Ridges and basins on the Tonga-Kermadec Island arc system. *J. Geophys. Res.*, 75: 239-54.

_____, and Sharman III, G.F., 1975. Subduction and accretion in trenches. *Geol. Soc. Amer. Bull.*, 86: 377-89.

Kulm, L.D., and Fowler, G.A., 1974a. Cenozoic sedimentary framework of the Gorda-Juan de Fuca plate and adjacent continental margin. A review. In: R.H. Dott, Jr., and R.H. Shaver (eds.), *Modern and Ancient Geosynclinal Sedimentation.* Soc. Econ. Paleont. Mineral., Sp. Publ. 19, pp. 212-29.

_____, and Fowler, G.A., 1974b. Oregon continental margin structure and stratigraphy: A test of imbricate thrust model. In: C.A. Burk and C.L. Drake (eds.), *The Geology of Continental Margins.* Springer Verlag, New York, pp. 261-84.

Lugeon, M., 1916. Sur l'origine des blocs exotiques du Flysch préalpin. *Eclog. Geol. Helv.*, 14: 217-21.

Lojacono, F., 1975. Osservazioni sulle direzioni delle paleocorrenti nel Flysch di Gorgoglione (Lucania). *Boll. Soc. Geol. Italiana*, 93: 1127-55.

Middleton, G.V., and Hampton, M.A., 1973. Sediment gravity flows: mechanics of flow and deposition. In: G. V. Middleton and A.H. Bouma (eds.), *Turbidites and Deep Water Sedimentation.* Soc. Econ. Paleont. Mineral., Pacific Section, Short Course, Anaheim, pp. 1-38.

Mutti, E., and Ghibaudo, G., 1972. Un esempio di torbiditi di conoide sottomarina esterna: le arenarie di S. Salvatore (Formazione di Bobbio, Miocene) nell'Appennino di Piacenza. *Mem. Acc. Scient. Torino, Classe Sc. Fis. Mat., Nat., serie* 4, n. 16: 40.

_____ and Ricci Lucchi, F., 1972. Le torbiditi dell'Appennino settentrionale: introduzione all'analisi delle facies. *Mem. Soc. Geol. Italiana*, 11: 161-99.

Nelson, C.H., and Kulm, L.D., 1973. Submarine fans and deep sea channels. In: G.V. Middleton and A.H. Bouma (eds.), *Turbidites and Deep Water Sedimentation.* Soc. Econ. Paleont. Mineral., Pacific Section, Short Course, Anaheim, pp. 39-70.

_____, and Nilsen, T. H., 1974. Depositional trend of modern and ancient deep sea fans. In: R.H. Dott, Jr., and R.H. Shaver (eds.), *Modern and Ancient Geosynclinal Sedimentation.* Soc. Econ. Paleont. Mineral., Sp. Publ. 19, pp. 69-91.

Normark, W.R., 1970. Growth pattern of deep-sea fans. *Amer. Assoc. Petrol. Geol. Bull.*, 54: 2170-95.

Ogniben, L., 1963. Le formazioni tipo Wildflysch delle Madonie (Sicilia centro-settentrionale). *Mem. Ist. Geologia, Univ. Padova*, 24: 58.

_____, 1969. Schema introduttivo alla geologia del confine calabro-lucano. *Mem. Soc. Geol. Italiana*, 8: 453-763.

_____, 1973. Conclusioni sullo stato attuale delle conoscenze sulla geologia dell'Appennino. Atti del Convegno: Moderne vedute sulla Geologia dell'Appennino. *Acc. Nazionale dei Lincei, Quaderno*, 183: 367-445.

_____, Parrotto, M., and Praturlon, A. (eds.), 1975. *Structural Model of Italy.* Cons. Nazion. Ric., Quaderni de "La Ricerca Scientifica," 90, 502 pp.

Pescatore, T., 1971. Considerazioni sulla sedimentazione miocenica nell'Appennino campano-lucano. *Atti Acc. Pontaniana, Napoli*, 20: 17.

_____, Sgrosso, I. and Torre, M., 1970. Lineamenti di tettonica e sedimentazione nel Miocene dell'Appennino campano-lucano. *Mem. Soc. Naturalisti in Napoli, suppl. Bull.* 78: 337-408.

Piper, D.J.W., Von Huene, R.E., and Duncan, J.R., 1973. Late Quaternary sedimentation in the active eastern Aleutian Trench. *Geology*, 1: 19-22.

Prince, R.A., and Kulm, L.D., 1975. Crustal rupture and the initiation of imbricate thrusting in the Peru-Chile trench. *Geol. Soc. Amer. Bull.* 86: 1639-53.

Ricci Lucchi, F., 1975. Miocene paleogeography and basin analysis in the Periadriatic Apennines. In: C. Squyres (ed.), *Geology of Italy.* Petrol. Explor. Soc. Libya, Tripoli, pp. 1-111.

Scholl, D.W., 1974. Sedimentary sequences in the North Pacific trench. In: C.A. Burk and C.L. Drake (eds.), *The Geology of Continental Margins.* Springer-Verlag, New York, pp. 493-504.

———, and Marlow, M.S., 1974. Sedimentary sequences in modern Pacific trenches and deformed circumpacific eugeosyncline. In: R.H. Dott, Jr., and R.H. Shaver (eds.), *Modern and Ancient Geosynclinal Sedimentation.* Soc. Econ. Paleont. Mineral., Sp. Publ. 19, pp. 193-211.

Seely, D.R., Vail, P.R., and Walton, G., 1974. Trench model slope. In: C.A. Burk and C.L. Drake (eds.), *The Geology of the Continental Margins.* Springer-Verlag, New York, pp. 249-60.

Selli, R., 1962. Il Paleogene nel quadro della geologia dell'Italia centro-meridionale. *Mem. Soc. Geol. Italiana,* 3: 737-89.

Squyres, C. (ed.), 1975. *The Geology of Italy.* Earth Science Soc., Libyan Arabic Republic, 799 pp.

Stanley, D.J., 1973. Basin plains in the eastern Mediterranean: significance in interpreting ancient marine deposits. I: Basin depth and configuration. *Mar. Geol.,* 15: 295-307.

———, 1974a. Basin plains in the eastern Mediterranean: significance in interpreting ancient marine deposits. II: Basin distribution. *Bull. Centre Rech. Pau-SNPA,* 8: 373-88.

———, 1974b. Modern flysch sedimentation in a Mediterranean island arc setting. In: R.H. Dott, Jr., and R.H. Shaver (eds.), *Modern and Ancient Geosynclinal Sedimentation.* Soc. Econ. Paleont. Mineral., Sp. Publ. 19, pp. 240-59.

———, 1975. Submarine canyon and slope sedimentation (Grès d'Annot) in the French Maritime Alps. *Proc. IX Intern. Cong. Sediment.,* Nice, 129 pp.

———, and Unrug, R., 1972. Submarine channel deposits, fluxoturbidites and other indicators of slope and base-of-slope deposit environments in modern and ancient basins. In: J.K. Rigby and W.K. Hamblin (eds.), *Recognition of Ancient Sedimentary Environments.* Soc. Econ. Paleont. Mineral., Sp. Publ. 16, pp. 287-340.

Ten Haaf, E., 1959. *Graded beds in Northern Apennines.* Published Doctoral Thesis, State University, Groningen, 102 pp.

Vezzani, L., 1968. Geologia della tavoletta Castronuovo di S. Andrea (Prov. di Potenza, F. 211-IV SE). *Atti Acc. Gioenia, Scienze Natur., Catania,* 19: 9-108.

———, 1975. Lithostratigraphic complexes and evidence for tectonic phases in the Molise-Puglia-Lucania Apennines. In: L. Ogniben, M. Parrotto, and A. Praturlon, *Structural Model of Italy.* Cons. Nazion. Ric., Quaderni de "La Ricerca Scientifica," 90, pp. 329-63.

Walker, R.G., 1967. Turbidite sedimentary structures and their relationship to proximal and distal despositional environments. *J. Sed. Petrol.,* 37: 25-43.

———, 1975. Generalized facies model for resedimented conglomerates of turbidite association. *Geol. Soc. Amer. Bull.,* 86: 737-48.

———, and Mutti, E., 1973. Turbidites facies and facies associations. In: G.V. Middleton and A.H. Bouma (eds.), *Turbidites and Deep Water Sedimentation.* Soc. Econ. Paleont. Mineral., Pacific Section, Short Course, Anaheim, pp. 119-57.

Wezel, F.C., 1973. Diacronismo deli eventi oligo-miocenici nelle Maghrebidi. *Riv. Miner. Siciliana,* 142-44: 219-32.

Chapter 23

Sedimentation Patterns in an Ancient Arc-Trench-Ocean Basin Complex: Carboniferous to Jurassic Rangitata Orogen, New Zealand

ROBERT M. CARTER
MURRAY D. HICKS
RICHARD J. NORRIS
IAN M. TURNBULL

Department of Geology
Otago University
Dunedin, New Zealand

ABSTRACT

The sedimentary facies developed in the belts of the Rangitata Orogen, New Zealand, are consistent with interpretation of the orogen as an arc-trench-ocean basin complex. The plate margin was the site of active subduction between at least the Permian (Sakmarian) and late Jurassic (Tithonian), and represents the contact between a southwesterly Gondwanaland plate and a northeasterly paleo-Pacific plate.

The three major belts of the orogen are the *Hokonui Assemblage* (southwesternmost), *Te Anau Assemblage* (medial) and *Alpine Assemblage* (northeasternmost). The southwesternmost part of the Hokonui Assemblage, the *Brook Street Supergroup*, is a pile of mainly Permian volcanics, volcanogenic sediments, and intrusives about 12 km thick; it represents at least part of the volcanic chain that bounded the Rangitata plate boundary to the southwest. On the northeastern side of the Hokonui Assemblage the late Permian to Triassic *Maitai Supergroup* rests with sedimentary contact on, or is faulted against, the Dun Mountain Ophiolite Belt; the Maitai Supergroup comprises mainly fine-grained, often laminated, sediments, with lesser redeposited coarse detritus, and is interpreted as the fill of a deepsea midslope basin. The Maitai and Brook Street Supergroups are overlain by *Murihiku Supergroup*, a northeasterly prograding wedge of nonmarine, shallow shelf, and deep marine slope and slope basin sediments that are described under five sedimentary facies associations. The major provenance for both the Hokonui and Te Anau Assemblages was a calc-alkaline volvanic terrain, probably the Brook Street terrain. The Hokonui Assemblage is relatively undeformed; it is in contact to the northeast with the *Te Anau Assemblage*, a strongly deformed zone of mainly deepsea pelagic, hemipelagic, hemiterrigenous, or redeposited sediment, with associated ultramafics and basic volcanics. Structural and metamorphic style and sediment facies are consistent with the Te Anau Assemblage being a subduction zone accretionary wedge. The northeasternmost *Alpine Assemblage* is also dominated by voluminous redeposited sediments, including flysch type sequences that correspond to continental margin fans; structural style and metamorphism are similar to those of the Te Anau Assemblage and the Alpine Assemblage is viewed as part of the same accretionary subduction wedge. The quartzofeldspathic detritus of the Alpine Assemblage must have been supplied from a different source to that of the other two assemblages. The source may have been a continental margin situated to the northeast of an ocean ridge, perhaps represented in the Haast Schist, itself consumed along the Rangitata subduction zone.

The major sedimentary belts of the Rangitata Orogen are now disposed according to a classic active plate margin model. A subduction zone wedge, comprising distant continental margin and deepsea deposits (Alpine Assemblage) and trench and lower trench slope basin fills (Te Anau Assemblage), has its upper (older) margin marked by the major ophiolite belt that demarcated the trench slope break (Dun Mountain Ophiolote Belt). Further southwest still follows the fill of the consequential midslope basin (Maitai Supergroup) and the regressive frontal arc wedge of upper slope and slope basin, shelf, and nonmarine sediments (Murihiku Supergroup).

Note: Ian M. Turnbull is currently with the New Zealand Geological Survey, Dunedin, New Zealand.

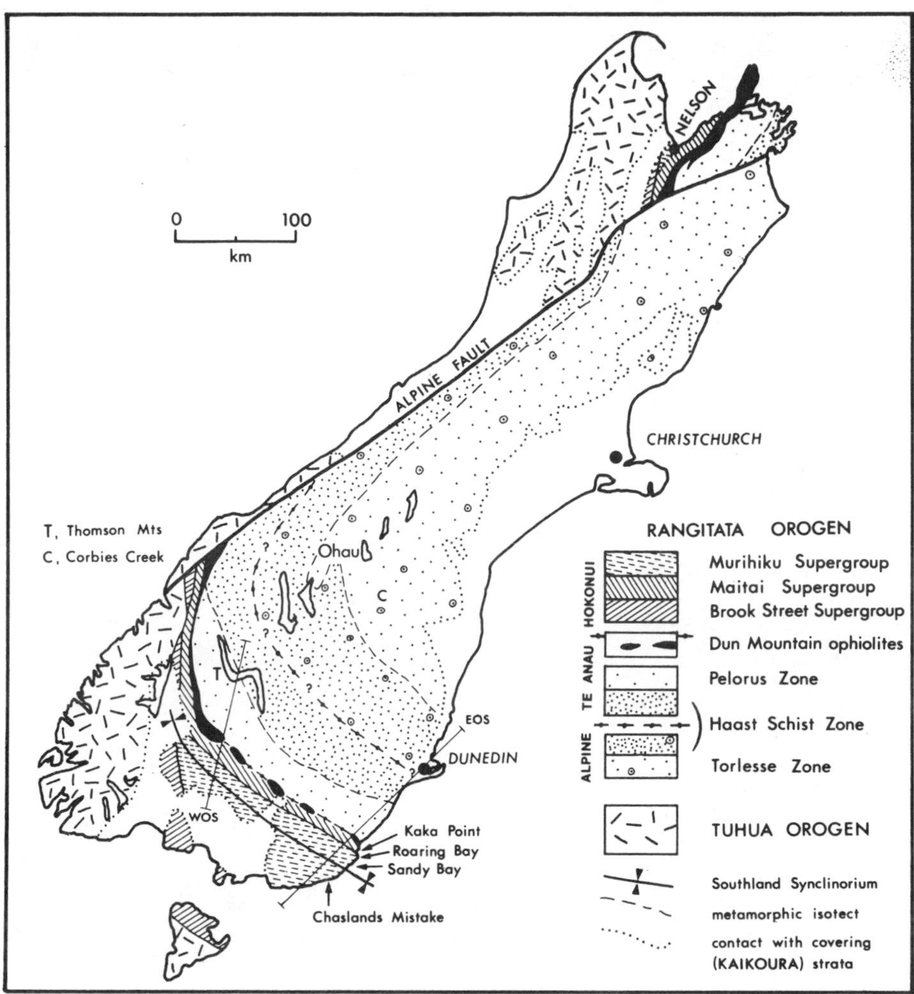

Figure 23-1. Outline geologic sketch map of South Island, New Zealand, showing the disposition of the major elements of the Rangitata Orogen, the main localities mentioned in the text, and approximate positions of the East (EOS) and West Otago (WOS) sections of Figure 23-2.

INTRODUCTION

Large areas of both main islands of New Zealand are underlain by basement rocks of the Carboniferous to Jurassic Rangitata Orogen (New Zealand Geosyncline of earlier writers; Figure 23-1). The orogen comprises the eastern end of a larger orogenic belt, shown on most continental drift reconstructions as fringing Gondwanaland to the south (e.g., Campbell 1975). This regional orogenic belt, first recognized by Wegener (1920) and later termed the Samfrau Geosyncline by Du Toit (1937), stretches from New Caledonia through New Zealand and West Antarctica into southern South America. Ties between New Zealand and Gondwanaland have been strengthened by the recent discovery of *Glossopteris* leaves in Permian rocks of the Productus Creek Group (Mildenhall 1976).

In this chapter we summarize the results of some recent sedimentological investigations of the three major depositional belts of the New Zealand South Island sector of the Samfrau Geosyncline. Our overall sedimentary interpretation is based on detailed lithofacies analyses of only small parts of the sequence, but reconnaissance studies, descriptions available in the literature, and plate tectonic considerations (cf. Coombs et al. 1976a), largely support our contention that the Rangitata Orogen represents a single major transect through the sedimentary and tectonosedimentary belts typical of modern arc-trench-ocean basin systems (cf. Figure 23-2).

Terminology

Many of the sedimentary rocks described in this chapter fall within the facies scheme for redeposited sediments of Mutti and Ricci Lucchi (1972) and Ricci Lucchi (1975). Other sedimentary terminology used is either conventional or follows Scholl and Marlow (1974) or Carter (1975).

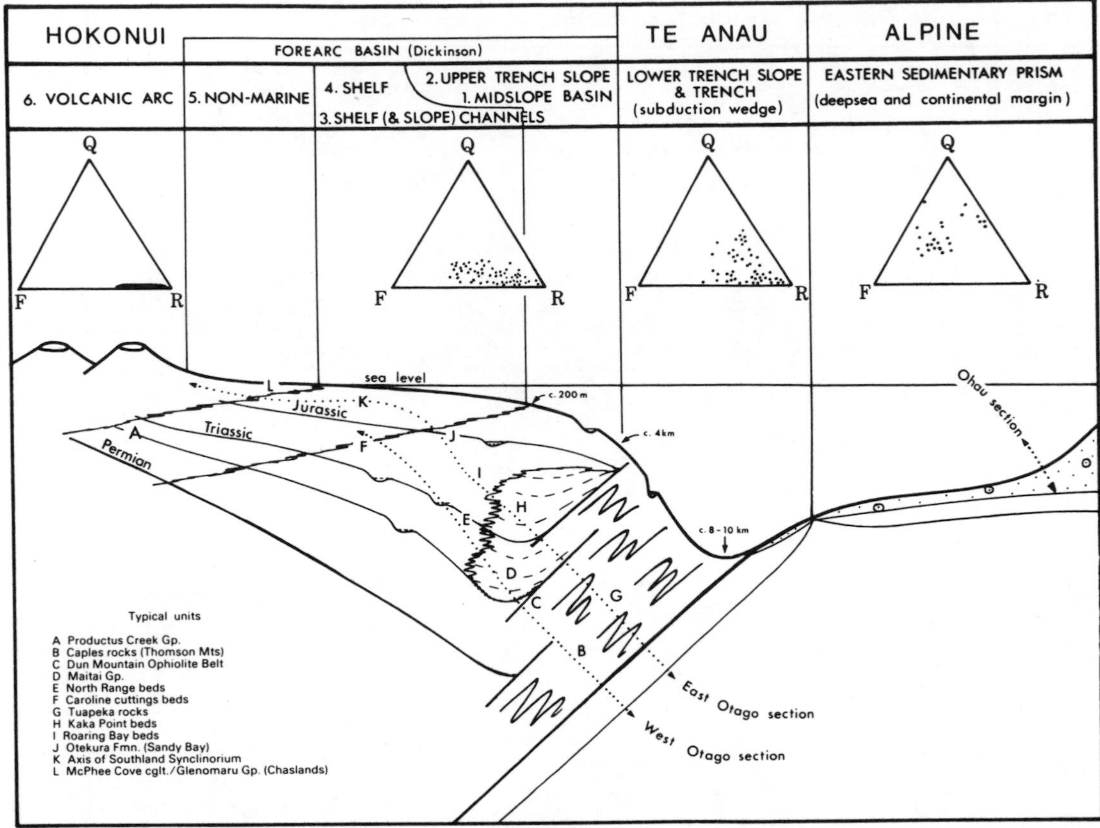

Figure 23-2. Pictographic cross-section through the major sedimentary units of the Rangitata Geosyncline. Note: Dotted lines approximate the positions of the main East and West Otago transects through the geosyncline; examples of major lithostratigraphic units, discussed further in text, are keyed by letters. Undeformed sediments of Te Anau and Alpine Assemblages stippled as in Figure 23-1.

Stratigraphic terminology mainly follows Carter et al. (1974), Coombs et al. (1976a), and Turnbull (in press). For consistency of nomenclature, we treat the Brook Street and Maitai "terranes" of Coombs et al. (1976a) as Supergroups, thereby following the recommendation in the International Stratigraphic Guide (Hedberg 1976) that a major lithostratigraphic unit "may be extended laterally beyond the area where it is divided into formations, if it seems probable that the equivalent interval will eventually be divided into formations". Where used herein, the term *terrain* has its conventional geographic meaning. Large parts of the Dun Mountain Ophiolite Belt are penetratively deformed and are therefore included in the Te Anau Assemblage (Figure 23-1); those parts of the ophiolite belt that are relatively undeformed, and followed unconformably by Maitai Supergroup sediments, are included in the Hokonui Assemblage.

Geologic Setting

The rocks of the Rangitata Orogen occur in a number of subparallel belts, each characterized by distinctive details of sedimentation, petrography, metamorphism and structural style (see Figures 23-1 and 23-2). We recognize three assemblages for the highest level grouping of rocks of the orogen. The characteristics of these assemblages are summarized in Table 23-1. We should note that most earlier synopses (e.g., Fleming 1970; Grindley 1974) follow the classic paper of Wellman (1956) and view the geosyncline as comprising two major sedimentary belts, a western "shelf" or Hokonui facies and an eastern "redeposited" or Alpine facies, with the latter including both the Alpine and Te Anau Assemblages of Table 23-1.

The western *Hokonui Assemblage* is predominantly volcaniclastic, relatively little deformed or metamorphosed, and contains nonmarine, shallow shelf, redeposited slope and deep marine sediments. At places in the northeast the basal sediments of the assemblage are in sedimentary contact with a relatively undeformed ophiolite suite (Blake and Landis 1973). The medial *Te Anau Assemblage* is similarly volcanogenic, but strongly deformed, and its sediments are entirely of deep marine origin. Ultramafics include the penetratively deformed parts of the Dun Mountain Ophiolite Belt and other smaller ultramafic pods or belts further northeast.

The eastern *Alpine Assemblage* has a predominantly quartzofeldspathic plutonic provenance, which contrasts

Table 23-1. *Summary of the Sedimentary Characteristics of the Major Assemblages of the Rangitata Geosyncline*

ASSEMBLAGE (SUPERGROUP)	HOKONUI (MURIHIKU)	HOKONUI (MAITAI)	TE ANAU	ALPINE
FACIES				
Chert	Not recorded	Occurs, but atypical	Occurs	Occurs
Laminated mdst/slst	Possible in NE	Typical	Typical	?
Flowite cglt/sst	Particularly in NE	Typical	Typical	Typical
Turbidites	Particularly in NE	Typical	Typical	Typical
Fan sequences	Probable in NE	Probable	Occur	Occur
Intraformational cglt	Particularly in NE	Typical	Typical	Typical
Shelf slst/sst	Typical in SW	Absent	Absent	Occurs, but atypical
Non-marine beds	Typical in SW	Absent	Absent	If present, atypical
SST PROVENANCE	Calc-alkaline	Calc-alkaline	Calc-alkaline	Quartzo-feldspathic
BENTHONIC FOSSILS	in situ in SW	Generally rare and redeposited, except for tubiculous worms		
IGNEOUS ROCKS	Tuff and ash	Mainly ash	Ophiolites Pillowed spilites	Pillowed spilites
STRUCTURE	Simple synclinorium		Complex; including extensive tectonic slides and melanges	

Note: For a more detailed but similar presentation, see Table 1 in Blake et al. 1974.

with that of the other two assemblages (see Figure 23-2); the rocks are strongly deformed and comprise mainly redeposited continental margin and deep marine sediments.

Early plate tectonic models of the Rangitata Orogen (Landis and Bishop 1972) viewed it as the record of an active island arc margin and provided many insights into the significance of the assemblage belts discussed above. However, difficulty was encountered in explaining the contrast of provenance between the volcanogenic Hokonui rocks and the voluminous quartzofeldspathic Alpine rocks (cf. Bradshaw and Andrews 1973). More recent plate tectonic models (Landis, in Blake et al. 1974; Coombs et al. 1976a) resolve the provenance problem by deriving the Alpine Assemblage from the northeastern (oceanward) side of the orogen. The model for the evolution of the Rangitata Orogen proposed by these authors forms a convenient setting within which to examine the sedimentary patterns of the orogen and is strongly supported by the studies described in this chapter.

HOKONUI ASSEMBLAGE

The volcanogenic and sometimes richly fossiliferous rocks of the Hokonui Assemblage (Table 23-1) have been the subject of detailed biostratigraphic and petrographic research for over 100 years (e.g., bibliography of Grant-Mackie et al. 1976). In southern South Island the rocks of the assemblage occur within a major synclinorium, the Southland Synclinorium (see Figure 23-1). For rocks of equivalent age, the shallower water facies occur on the southwestern side of the synclinorium (Mutch 1957; Speden 1971; Boles 1974). The Hokonui Assemblage, or at least its Triassic-Jurassic part, has often been viewed as a shallow-water "shelf" facies (Wellman 1956; Fleming 1970; Kear 1971), though recently some authors have suggested an offshelf origin for some sediments, particularly Permian, on the north side of the Southland Synclinorium (Blake and Landis 1973; Landis 1974a).

Sedimentary Facies

Recent and mostly unpublished sedimentary studies suggest that there are six major sedimentary facies associations developed in the Hokonui Assemblage (see Figure 23-2). From northeast to southwest across the Southland Synclinorium, they are (1) the *midslope basin* association, (2) the *upper trench slope* and *slope basin* association, (3) the shelf and slope *channel* association, (4) the *shelf* association, (5) the *nonmarine* association, and (6) the *volcanic arc* association. At the broadest level the sediments of the Assemblage represent a regressive wedge that prograded from southwest (arcwards) to northeast (trenchwards) during Permian-Jurassic time—that is, they

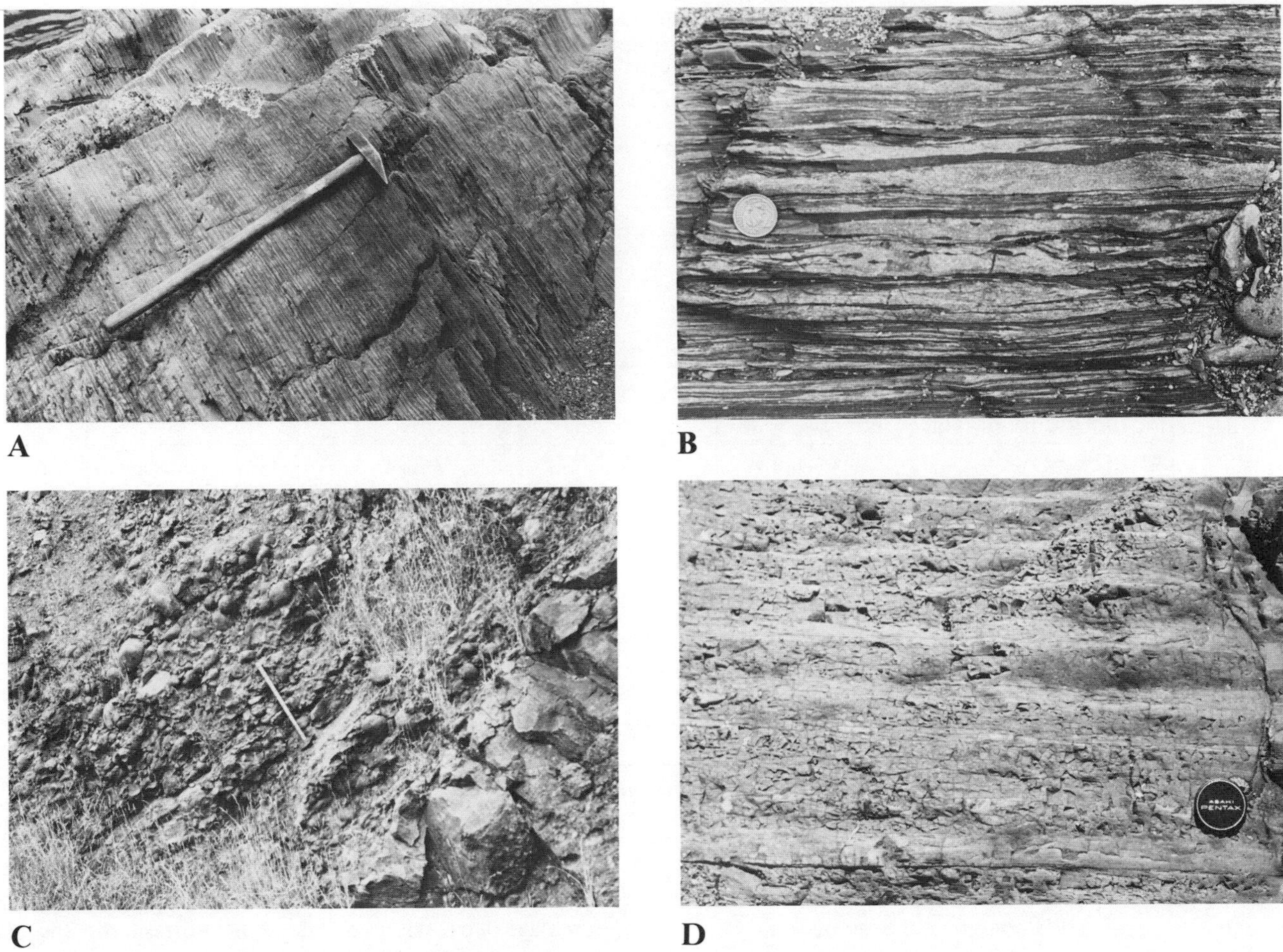

Figure 23-3. Photograph A shows thin-bedded sands and silts of the Waiua Formation, Lee River, Nelson (hammer = 85 cm long). B shows detail of photograph A. Note lenticular and wavy bedding (coin = 2.1 cm diameter). Photograph C shows redeposited roundstone conglomerates of Romohapa Conglomerate, Balclutha (hammer = 85 cm long). Photograph D shows overbank $T_{c\text{-}e}$ and $T_{d\text{-}e}$ turbidites adjacent to deepsea channel fill, Bates Siltstone, Kaka Point (lens cap = 5.4 cm diameter).

represent a regressive forearc basin sequence in the sense of Dickinson (1974). In the well-exposed coastal section, facies associations 1 to 5 follow one another in numerical and stratigraphic order, and reconnaissance studies suggest that similarly homotaxial sequences occur throughout the Hokonui terrain. Locally, the ages and positions of the facies associations may vary more spasmodically, presumably reflecting pulses of activity on the flanking subduction zone and island arc.

Facies Association 1. Association 1 broadly corresponds to the Maitai Supergroup, the oldest unit of the Hokonui Assemblage on the northeastern side of the Southland Synclinorium (see Figure 23-1). In inland regions of western Otago, and in Nelson, the 4.5 km thick Maitai Supergroup has been divided into seven distinctive formations (Grindley 1958; Waterhouse 1964; Landis 1974a). A thin volcaniclastic sequence at the base is followed by redeposited limestone and sandstone, and then by 2 km of fine-grained sediments, dominantly millimeter-scale parallel-laminated mudstone and siltstone interbedded with thin, graded, fine to very fine-grained sandstones and thin rippled siltstones to very fine sandstones (photographs A and B, Figure 23-3). The upper 1 km of the supergroup often includes coarser-grained sediment, both of sandstone and bioclastic limestone, variably thickly or thinly interbedded with finer sediments, and of flysch aspect.

The formations recognized in West Otago and Nelson are not traceable to eastern coastal districts, where limestones are rare and locally structure is more complex. Nonetheless, broad similarities persist along the Maitai terrain, the eastern region also having a basal igneous part, equated with the ophiolite belt by Coombs et al. (1976a), and an overlying sedimentary sequence including large amounts of dominantly argillaceous sediment associated with typical turbidites and with channelized packets of sandy or conglomeratic flysch (cf. Wood 1956; Bishop 1965; Boles 1974). The upper parts of the Maitai Super-

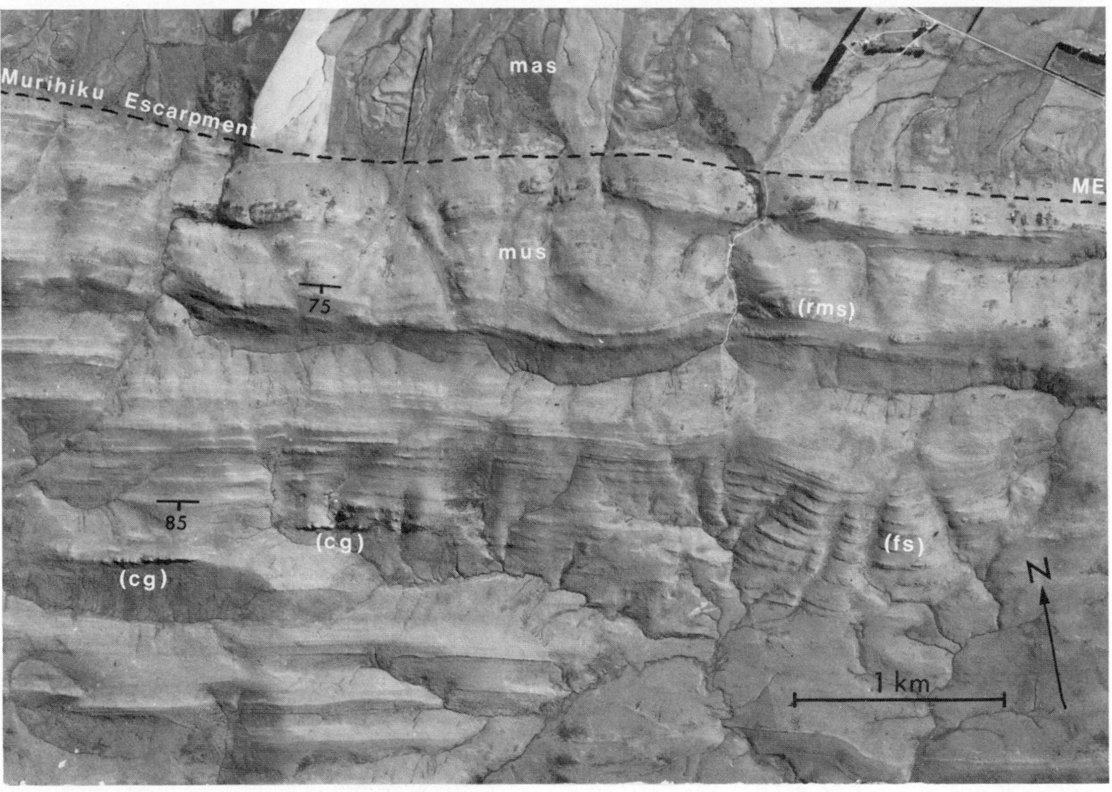

Figure 23-4. Aerial views of the upper Maitai and Murihiku Supergroups, north limb of the Southland Synclinorium between Gore and Lumsden. (Crown Copyright, reproduced with permission from the Surveyor General.) Note: Photograph A shows Ram Hill, an isolated lens of flowite and fluxoturbidite sandstones (rhs) (Mutti and Ricci Lucchi Facies A and B) occurring within silts and muds of the upper Maitai Supergroup (mas). Similar sandstones occur widely in the base of the overlying Murihiku Supergroup (mus). Photograph B shows Murihiku Escarpment about 20 km southeast of area in photograph A. The predominance of resistant massive sandstones (rms) in the base of the Murihiku Supergroup (mus), the presence of fan-shaped bodies of sand (fs), and isolated channels filled with hackly weathering redeposited conglomerate (cg) are noteworthy.

group in the east contain red, green, and black argillite, often with thin interbeds of rippled or convoluted siltstone to very fine sandstone. Lensoid masses of coarser-grained redeposited sandstone and conglomerate are prominent within the argillite-siltstone sequence (photograph C, Figure 23-3; photograph A, Figure 23-4); individual beds are poorly graded, often contain abundant intraformational detritus, and are up to several meters thick.

Siltstone and argillite again dominate in the coastal section at Kaka Point (1,800 m of a 1,906 m thick section; Campbell in Coombs et al. 1976b), where the early to middle Triassic fossils make this the youngest known part of the southern Maitai terrain. Deepsea channel and overbank facies, though minor, are well exposed near Kaka Point (photograph D, Figure 23-3; also see Bishop and Force 1969).

Interpretation (midslope basin). No detailed facies analyses have yet been published for any part of the Maitai Supergroup, though representative lithofacies are well figured in Waterhouse (1964). However available information suggests much of the finer grained sediment is hemipelagic or hemiterrigenous and hence points to a deep marine origin for the supergroup (cf. Bishop and Force 1969; Landis 1974a). The majority of the coarse-grained sediment is redeposited.

The basal volcanogenic facies of the Maitai Supergroup, up to 600 m thick, contains abundant detritus from the basement ophiolite suite and is interpreted by Coombs et al. (1976a) as an ocean floor accumulation at or near the site of formation of the underlying oceanic crust. The overall stratigraphic and sedimentary facies evidence suggests that most of the remainder of the Maitai sediments accumulated in a trench slope basin (Coombs et al. 1976a), probably a major midslope basin just above the trench slope break shown in Figure 23-2 (cf. Dickinson 1974; see also Te Anau Assemblage below).

Waterhouse (1975) has suggested that atypically richly fossiliferous Maitai sandstone and limestone that occurs near Arthurton (Wood 1956) may indicate shoaling water over an active submarine high such as a trench slope break. Although the suggestion is attractive and consistent with our general interpretation, the association of the fossil beds with lensoid exogenous conglomerates suggests as an alternative explanation that the fossil beds are channelized and redeposited. The Arthurton sediments are very poorly known, and more data are needed to resolve the question.

Facies Associations 2 to 5. Sedimentary Facies Associations 2 to 5 broadly correspond to the Murihiku Supergroup (Campbell and Coombs 1966), comprising at least 7 km of mainly Triassic and Jurassic sediment that occurs in the axial regions of the Southland Synclinorium (see Figure 23-1). The coastal sequence on the north limb of the synclinorium gives an almost continuously exposed transect across a magnificent example of a regressive upper trench slope to shelf to nonmarine sedimentary rhythm (see Figure 23-2).

Facies Association 2 (and 3). Association 2 contains large amounts of fine-grained sediment, particularly siltstone, but increased amounts of sand and conglomerate occur at various horizons, particularly near the base of the association. Though some of the massive siltstones may be hemipelagic or hemiterrigenous, by comparison with the underlying Maitai there is a general absence of fine parallel lamination. *Chondrites* is common throughout, and many apparently "massive" siltstones are pervasively bioturbated. In truly fresh exposures nonbioturbated siltstones can be seen to contain abundant silt to very fine sand layers deposited from gentle bottom currents (probably contourites) as well as graded silt and mud turbidites. The coarser-grained sediments are almost without exception redeposited.

A particularly well-exposed example of channel Association 3 occurs within argillites of the slope association at Sandy Bay (see Figure 23-1) in the Otekura Formation of Speden (1971) (Carter, in preparation). The sequence comprises two 20 m thick, channelized rhythms of medium-thick bedded sandy flysch (Mutti and Ricci Lucchi Facies B and C), each of which passes up into thin-bedded muddy flysch of overbank type (Mutti and Ricci Lucchi Facies D) (Figure 23-5). Directional indicators within the channel sequence suggest emplacement from the south-southwest; in contrast, rare turbidites that occur within the background argillites about 20 m below the lower packet of sandy flysch were emplaced from the east—that is, at right angles to the channelized flows. Reconnaissance studies indicate that the Otekura Formation is underlain by several kilometers of dominantly fine-grained, deepwater sediments containing occasional sand- and conglomerate-filled channels similar to those just described (e.g., the roundstone conglomerates near the base of the Roaring Bay section).

Exposure inland is poor, but facies similar to those of the coast section have been reported by Wood (1956) and Boles (1974), with the latter author figuring and describing a typical example of channelized redeposited conglomerate (Boles 1976, Figures 7 and 8). Aerial photographs of inland regions show that individual conglomerate or sandstone bodies occur as channelized lenses, generally a few tens of meters thick and one to several hundred meters long (photograph B, Figure 23-4; see also Boles 1974, Figures 2, 6, and 7; map in Coombs 1950).

Interpretation (upper trench slope). The position of Facies Association 2 (and 3) in a generally regressive sequence, above midslope basin Association 1 and below shelf Association 4, implies that the environments of deposition of Association 2 were those of an upper trench slope (see Figure 23-2). Such an interpretation is also consistent with the sedimentologic evidence, particularly

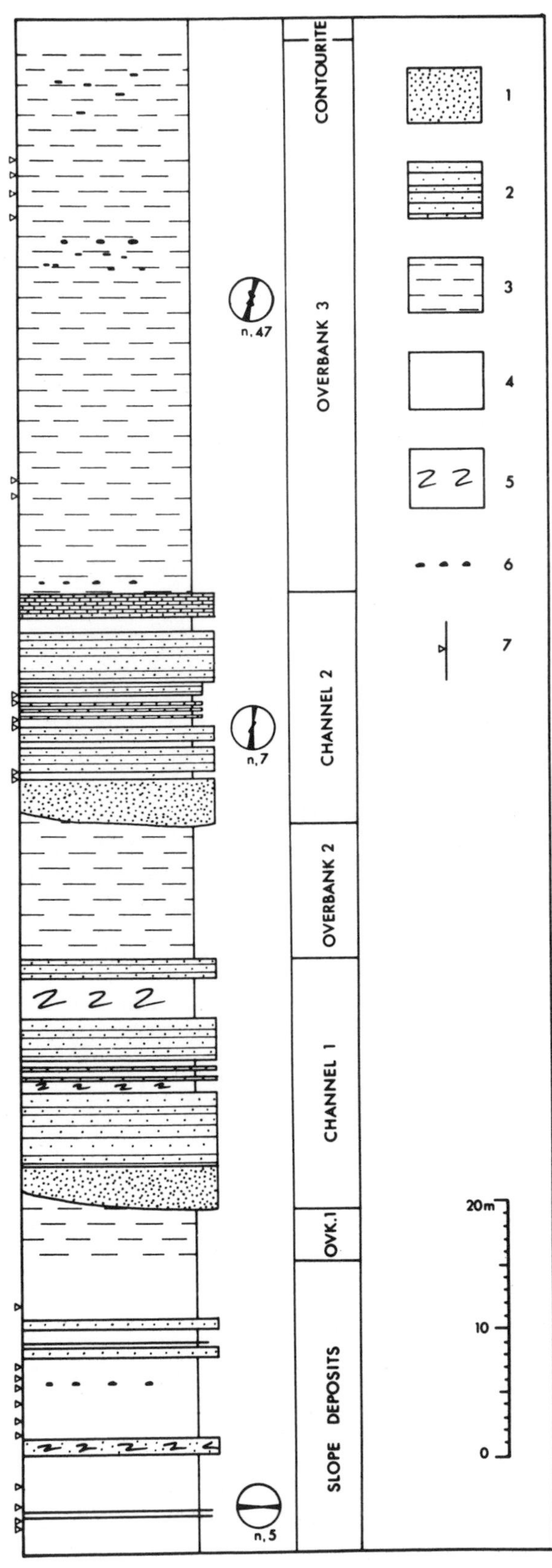

A

B

Figure 23-5. Upper slope channels occurring within Facies Association 2 of the Murihiku Supergroup, Sandy Bay. Note: Chart A shows measured section. 1 = Facies A (Mutti and Ricci Lucchi); 2 = Facies B, rare C; 3 = Facies D; 4 = muds and wispy rippled silts (contourites, hemiterrigenous beds, mud turbidites); 5 = Facies F; 6 = concretions; 7 = ash or tuff layers. Photograph B shows Facies A (left side of picture) and Facies B beds of channel 1. Photograph C shows Facies D turbidites of overbank packet 3 (hammer = 33 cm long). Photograph D shows detail of photograph C; note T_c climbing ripples (coin = 2.1 cm diameter).

C

D

Figure 23-5 Continued.

with the presence of redeposited sediments in both fine- and coarse-grained lithologies and with the abundance of contourites and volcanogenic hemipelagic or hemiterrigenous sediments amongst the dominant finer-grained lithologies.

The appreciable bodies of redeposited conglomerate and coarser sands towards the base of Murihiku (cf. Figure 23-4) are consistent with the presence of submarine fans that accumulated on the arcwards sides of the midslope basin represented by the underlying Maitai Supergroup. Higher in the Murihiku lenses of coarse-grained redeposited sediment are typically thinner and more isolated (cf. photograph B, Figure 23-4), such as those of the Otekura Formation at Sandy Bay, and they represent the infill of channels that crossed the upper trench slope and fed sediment further downslope (cf. Reimnitz et al. 1977). Comparison with modern examples suggests that the upper trench slope may have contained other small basins upslope from the main midslope basin and its infilling fans. The presence of the deposits of such basins is in fact suggested by the occurrence of some fan-shaped bodies of sand higher in the supergroup (cf. photograph B, Figure 23-4) and also by the presence of longslope turbidites at Sandy Bay.

Corresponding to the presence of appreciable coarse detritus at the base of the Murihiku Supergroup, the boundary between the Murihiku and Maitai Supergroups on the north side of the Southland Synclinorium occurs as a conspicuous escarpment, the Murihiku Escarpment, which can be traced for over 150 km from the coast (see Figure 23-1). Boles (1974) has argued that the contact between the two supergroups is, at least locally, conformable across the escarpment. The isolated sand and conglomerate lenses in the upper Maitai (cf. photograph A, Figure 23-4) therefore probably represent distal progradational lobes from lower Murihiku fans and/or the infills of deepsea channels that led from the fans onto the floor of the Maitai midslope basin.

Other workers, such as Wood (1956), have interpreted the Murihiku Escarpment as fault controlled, and a definite fault separating Maitai from Murihiku has been mapped in Nelson (Campbell 1974). Furthermore, at least in the east, the Maitai Supergroup is discordantly folded with respect to the superadjacent Murihiku (Wood 1956; Bishop 1965) and some of this folding may be syn - or early postdepositional. The presence of such structural disturbances is not unexpected, since the Murihiku Escarpment is located about 75 km to 100 km trenchwards from the Brook Street volcanic arc and thus corresponds in position to the upper trench slope tectonic discontinuity that Karig and Sharman (1975) have demonstrated for modern arc-trench systems.

Facies Associations 4 to 5 (and 3). Associations 4 to 5 approximate to the middle parts of the Murihiku Supergroup, comprising 4 km to 5 km of complexly interdigitated sedimentary rocks that pass up from bioturbated marine calcareous siltstones, often with rich *in situ* benthonic faunas, into sand-dominated tractionites that are generally faunally impoverished but may contain abundant plant material. The heterogeneous lithofacies that comprise the association represent shallow marine shelf, fluviomarine and non-marine environments of deposition. Associations 4 to 5 therefore correspond in fact with the original concept of Hokonui "facies" proposed by Wellman (1956) and followed by most later writers (e.g., Fleming 1970).

Associations 4 to 5 are also best exposed in the coastal section, particularly on the south limb of the Southland Synclinorium where they are of early and middle Jurassic age. A detailed stratigraphic and petrographic study by Speden (1971) has shown that the sediments are generally coarser grained, thicker, and more carbonaceous towards the south; paleocurrent data also support a southern source. A preliminary sedimentological interpretation of Speden's data is presented in diagram C, Figure 23-6.

A

B

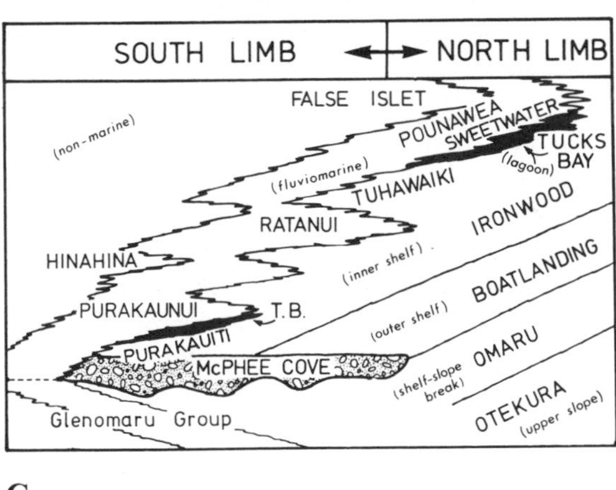

C

Figure 23-6. McPhee Cove conglomerate. Note: Photograph A shows channel Facies 3 within shelf Association 4 of the Murihiku Supergroup at Chaslands Mistake. Main outcrop in inaccessible cliff; height of face about 120 m. Base of channel is arrowed; fill of chaotically bedded sandy breccia-conglomerate (note large vertical block on right) and sharp-based, 1 m to 3 m thick conglomerate beds. Channel cut into mainly traction-emplaced fluviomarine sands of Glenomaru Group, which also contain occasional mass-emplaced conglomerates (mec). Photograph B shows large hydroplastically deformed sandstone clast. Diagram C shows summary stratigraphic nomenclature, with sedimentary interpretation for coastal exposures through the Jurassic beds of the Southland Synclinorium. (Source: Adapted from Speden 1971).

Locally, the inner shelf deposits of Facies Association 4 enclose bodies of conglomerate up to 130 meters thick, the McPhee Cove Conglomerate of Speden (1971), as shown in Figure 23-6. The conglomerate occurs in deep channels incised in the underlying shelf sediments. Channel walls are locally vertical or undercut, but the best exposed channel, at Chaslands Mistake (see Figure 23-1) has walls with an overall slope of about 30° to 40° (photograph A, Figure 23-6). The sedimentary fill of the channels is complex. In the Chaslands Mistake example, a circumferential zone up to several meters thick, and also filling undercuts in the walls, comprises disorganized roundstone conglomerate with only rare small penecontemporaneous clasts; the conglomerate may lap up against the wall of the channel and has a "poured-in" appearance. The main part of the channel fill comprises similar lensoid but semiorganized conglomerate in which some bed bases can be recognized. These semiorganized conglomerates are interspersed with zones of chaotic breccia-sands comprising blocks of penecontemporaneous sandstone and siltstone up to several meters in diameter set in a sandy to pebbly matrix and often with block-axes at high angles to the bedding (photograph A, Figure 23-6). Conglomerate beds may be separated by thin beds of siltstone, graded or rippled sandstone, or carbonaceous siltstone; these finer-grained lithologies sometimes occur draped over an undulating top to the underlying conglomerate, but are present in only small amounts. With the exception of the graded sands (probably turbidites), the fine-grained lithologies are traction and suspension emplaced, and are similar to Facies Association 4 shelf sediments that under-and overlie the McPhee Cove Conglomerate (cf. diagram C, Figure 23-6).

Closely similar lithologies to the disorganized parts of the fill of the McPhee Cove channels have been described from inferred continental margin submarine fan channels by Piper and Normark (1971) and Carter and Lindqvist (1975). More organized, redeposited roundstone conglomerates similar to those within the McPhee channels have also been described by these authors and by many others, nearly always from inferred deep marine fan sequences (see especially Scott 1966; Hendry 1973; Walker 1975). However, the stratigraphic and sedimentologic setting shows unequivocally that the McPhee Cove conglomerates occur in channels within a sequence of shallow marine traction and suspension emplaced sediments. Modern examples of such channels are not uncommon; they generally link with shelf-edge submarine canyons and continental margin fans and thus serve part of the major sediment transport system from continent to deep sea (Shepard and Dill 1966). Ancient examples of such shelf channels are apparently rare, though they have been described by Lewis (1976), Hyden (in preparation), and possibly Jeletsky (1975).

Overall interpretation of Association 3 to 5. The interpretation of these associations is clear-cut. They represent a broadly regressive wedge of nonmarine and shelf sediments that fronted onto a trench slope. The forearc shelf was cut by canyons and channels that fed sediment eastwards towards the trench slope and its basins. Fluctuations in the position of the regressive wedge (cf. diagram C, Figure 23-6), and the presence within it of minor unconformities, most likely relate to phases of tilting at the upper slope discontinuity and to volcanic activity in the adjacent arc.

Permian examples of Facies Association 4 have been described (Mutch 1957; Force 1975) from the Productus Creek Group on the western side of the Southland Synclinorium (see Figure 23-2). Other Jurassic examples of Facies Associations 4 to 5 occur in equivalent strata in the North Island (Fleming and Kear 1960).

Facies Association 6. The southwestern side of the Hokonui Synclinorium is demarcated by an early to middle Permian sequence of interlayered volcanogenic sedimentary, basaltic, and andesitic volcanic rocks, locally up to 12,000 m thick (Mutch 1957). Related intrusives in this, the Brook Street Supergroup, yield radiometric ages ranging from the Permian to the Jurassic (Aronson 1968; Devereux et al. 1968), and in all likelihood the Brook Street terrain represents at least part of the major volcanic arc system that fringed the Rangitata Orogen to the southwest (Grindley 1958; Landis and Bishop 1972; Waterhouse 1975).

The Brook Street Supergroup includes pillow lavas, hyaloclastites, and thick beds of pyroclastics, as well as massive flows and sills, and locally abundant volcanogenic epiclastic sediments (Mutch 1957). No detailed facies analyses have been published, but the sediments are apparently largely directly volcanic or redeposited, typical tractionites not yet being recorded.

Where a direct airfall-waterfall origin can be demonstrated, tuffs that occur interspersed amongst sediments of the other major facies associations are perhaps best viewed as corresponding to the "feather edges" of Facies Association 6.

Petrography

The petrography of the Hokonui Assemblage is dominated by volcanogenic detritus, largely of intermediate to acidic composition (see Coombs 1954; Speden 1971; Boles 1974). Associated plutonic detritus, particularly conspicuous in some occurrences of McPhee Cove Conglomerate, is mostly of cosanguineous granitoid types and there are only minor amounts of possibly more distantly derived metamorphic greywacke, hornfels, and schist (Speden 1971). The sandstones and conglomerates that infill the McPhee Cove channels have a similar volcanogenic provenance to the sediments that enclose them.

Table 23-2. *Summary of Characteristics of Selected Sedimentary Lithofacies of Te Anau Assemblage, Thomson Mountains, West Otago*

(1) Black mudstone. (2) Red & green mudstone. (3) Laminated slst-fs. (4) Thin-medium bedded flysch. (5) Sandy flysch. (6) Mass-emplaced cglt. (7) Thin rippled interbeds. (8) Chert. (9) Slump zones. (10) Pillow lava, hyaloclastite, breccia.

LITHOFACIES	1	2	3	4	5	6	7	8	9	10
Bed thickness	mm	mm	mm-cm	2cm-1m	1-5+m	1-10+m	.5-4cm	1-10cm	any	variable
Packet thickness	^300m	^30m	^200m	^1000m	^1000m	^800m	^2m	^20+m	^10m	^100+m
Grain size (1) typical	mdst	mdst	slst	f-ms	m-vcs	10cm	slst	-	any	-
(2) range	clay-silt	clay-vfs	^fs	^vcs	^gran	^1.8m	^fs	-	any	-
Sharp bases	-	-	yes	yes	yes	yes	grade	-	-	-
Graded (Normal/Reverse)	no	no	rarely	N common R rare	present	rarely	-	no	no	-
Bouma sequences	no	no	no	common	rare	no	no	no	no	-
Mdst/sst clasts	no	no	no	present	common	abundant	no	no	yes	-
Lateral wedging	no	no	no	no	yes	yes	yes	no	yes	-
Traction ripple x-beds	no	no	rare	T_c only	no	no	yes	no	no	-
Associated lithofacies	3-5	4-6,10	1-2,4	1,3,7	1,4,6	4-5	1,3-4	3,10	any	1-3
Mutti-RicciLucchi Facies	G	G	-	C,D,?E	A,B	A	-	-	F	-
Mode of deposition	pelagite/nephelite		overbank	turbidite	fluxotur. flowite	flowite	contourite	biopelagite	olisthanite	-

TE ANAU ASSEMBLAGE

In southern South Island the Te Anau Assemblage (Turnbull, in preparation) lies east of, and tectonostratigraphically below, the Hokonui Assemblage, the junction falling along the Dun Mountain Ophiolite Belt (see Figures 23-1 and 23-2; Table 23-1). The Assemblage comprises complexly deformed metagreywacke passing northwards into the Haast Schist; previously mapped units within the assemblage include the Tuapeka (Marshall 1918; Wood 1956; Bishop 1965), Caples (Grindley 1958; Kawachi 1974), Pelorus and Rai (Waterhouse 1964) and Morrinsville (Kear 1971) units ranging in age between Permian and Jurassic. Fossils are extremely rare, but the Assemblage is almost certainly marine throughout. The provenance is largely volcanogenic (see Figure 23-2).

Although Te Anau Assemblage rocks include extensive tectonic slides and mélanges and are always metamorphosed to at least prehnite-pumpellyite grade, sedimentary structures are locally well preserved. Turnbull (in preparation) has recently described distinctive lithofacies associations from the lower-grade rocks of the Thomson Mountains (see Figure 23-1), as briefly summarized in Table 23-2 (cf. Figure 23-7). Reconnaissance studies, and descriptions in the references cited above, suggest these and similar lithofacies and their metamorphosed equivalents are widespread throughout the Te Anau Assemblage.

Sedimentary Interpretation

Te Anau Assemblage rocks are largely if not entirely of deep marine origin. The background sedimentation type comprises hemipelagic or hemiterrigenous mudstone and siltstone and pelagic chert (lithofacies 1, 2, and 8). Associated sediments include mud turbidites (occurring within lithofacies 1), occasional more typical but generally "distal" turbidites (lithofacies 4), and the deposits of low-energy bottom currents (lithofacies 7). These fine-grained sediment types occur in sequences up to several hundred meters thick and may comprise as much as 25 percent of the sediment exposed in the Te Anau terrain of the Thomson Mountains. Interspersed amongst the finer sediments here are strongly channelized packets of thick-bedded sandy flysch (lithofacies 5) and mass-emplaced conglomerate (lithofacies 6), up to several hundred meters thick, associated with typical channel fill and overbank turbidites of lithofacies 4 and probable levee laminites (lithofacies 3; cf. Chough and Hesse 1976). Thinner packets of coarser-grained sediment (tens of meters) represent deepsea channels, whereas the thicker packets (hundreds of meters) probably accumulated as fans and longitudinal basin fills (Turnbull, in preparation). Thin zones of synsedimentary slumping occur mainly in fine-grained sediments (lithofacies 9). The volcanogenic association (lithofacies 10) is volumetrically minor in the Thomson Mountains, but includes at least three occurrences where pillow lavas are

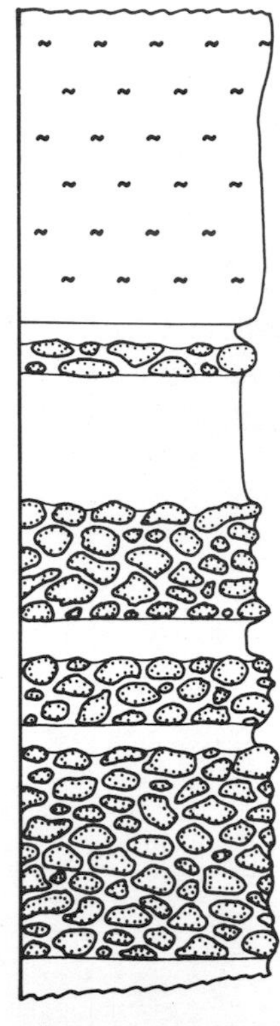

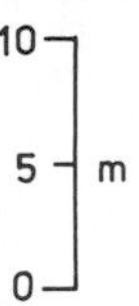

A

C

D

B

Figure 23-7. Selected sedimentary lithofacies of the Te Anau Assemblage, Thomson Mountains, West Otago. Note: Chart A shows section through pillow lava sequence and associated sediments, Rere Basin; tectonic contacts above and below. Photograph B shows massive sandstones of lithofacies 5 on West Tooth; obliquity of beds in lower left hand corner reflects presence of a tectonic slide (arrowed, chamois gives scale; beds upright). Photograph C shows laminated siltstone of lithofacies 3, with minor layers of wispy cross stratification (lithofacies 7). (Photo by K. D. Mason.) Photograph D shows alternating graded sandstone-mudstone of lithofacies 4; beds upright (hammer head = 17.5 cm long).

conformably overlain by sedimentary chert, laminated mudstone, claystone or siltstone (chart A, Figure 23-7). Volcanics (greenschist) and chert are locally more dominant rock types in the more metamorphosed Haast Schist Zone of the Te Anau Assemblage (cf. Henley et al. 1976).

Tectonostratigraphic Interpretation

The volcanogenic nature, sedimentary facies, and metamorphic and structural style of the Te Anau Assemblage is consistent with its representing a subduction zone prism (Turnbull 1974; Coombs et al. 1976a). The sedimentary facies preserved in the prism are closely similar to those described from modern trenches and lower trench slope basins (Piper et al. 1973; Connolly and Ewing 1967; von Huene 1974; and Chapter 12 of this volume). The tectonostratigraphic setting, below and east of the midslope Maitai basin and the trench slope break demarcated by the main ophiolite belt (cf. Figure 23-2), matches present models of arc-trench depositional systems (Dickinson 1974; Karig and Sharman 1975; Scholl and Marlow 1974).

Scholl and Marlow (1974) have criticized the interpretation of many circum-Pacific metagreywacke belts as deformed trench sediments. They assert that modern trench fills are pelagite dominated and that late Cenozoic trench fills only contain thick sandy flysch wedges as a response to Pleistocene sea level changes; supposed ancient examples, on the other hand, have abundant sandy flysch and a general lack of pelagic sediments. Neither of these criticisms apply to the Te Anau Assemblage. Calcareous plankton only evolved during the later stages of the history of the Rangitata Geosyncline, so that even without invoking deepwater dissolution, the general absence of biopelagic limestone is hardly surprising. Biopelagic chert is present (Benson and Chapman 1938), and there are also significant amounts of pelagic and hemipelagic claystone, mudstone, and siltstone. With respect to the abundant sandy flysch (Turnbull, in preparation) the possibility exists that Permian glacial events influenced its accumulation; however, another likelihood, Scholl and Marlow's comments notwithstanding, is that appreciable bodies of deepsea redeposited sand occur not uncommonly in lower trench slope canyons, channels, and basins and may thus become involved in subduction wedges of any age (cf. Karig and Sharman 1975; Moore and Karig 1976).

Although true stratigraphic thicknesses are not known, the width of the outcrop belt of the Te Anau Assemblage is at least several tens of kilometers (see Figure 23-1). Interpretation of the Assemblage as a subduction zone wedge implies the synsedimentary deformation and stacking of slices of sediment to form a northeastward (oceanward) younging tectonosedimentary belt. Fossils are too rare to indicate younging directions; other available evidence, particularly the widespread occurrence of tectonic slides associated with the earliest phase of folding (interpreted as having taken place whilst the sediments were still only partly lithified) is consistent with such an interpretation (Turnbull and Norris, in preparation).

ALPINE ASSEMBLAGE

Like the Te Anau Assemblage to the southwest, the Alpine Assemblage comprises tectonically complex sediments and metasediments that grade laterally into the Haast Schist Zone (see Figure 23-1). Although spilitic pillow lavas occur, the clastic sediments of the Alpine Assemblage are relatively nonvolcanogenic, and the terrigenous detritus of the Assemblage is quartzofeldspathic (see Figure 23-2). Fossils are rare, but indicate an age range of Carboniferous to Jurassic for the Assemblage (Campbell and Warren 1965; Landis and Bishop 1972; Speden 1976), which is thus broadly coeval with the Hokonui and Te Anau Assemblages.

The Alpine Assemblage covers large areas of both islands of New Zealand. Its cumulative thickness must be considerable; for example, 15,000 m as has been estimated at Dansey Pass (Bishop 1974). However, no basement is known, and accurate calculations are impossible. By far the majority of sediments within the unmetamorphosed Torlesse Zone are either of flysch type or are flysch-associated massive sandstone or mudstone-siltstone.

Few easily accessible areas within the Alpine Assemblage combine fresh and continuous exposure with low metamorphic grade and relative structural simplicity. We describe one such area (the Ohau Ski Basin) below and suggest that the sediments represent substantial deep marine fans. Although metamorphic grade at Ohau is largely prehnite-pumpellyite facies, sedimentary features are beautifully preserved and almost continuous sections up to 2 km thick can be measured (Hicks 1975).

The Ohau Ski Basin

Facies present. The majority of the Ohau sediments can be described by using Ricci Lucchi's (1975) sevenfold facies subdivision (Figures 23-8 through 23-10). Apart from suspension-emplaced argillites of Facies G, the sediments are largely redeposited. However, traction features (additional to those of the Bouma sequence) are not uncommonly present as reworked rippled tops to Facies C or D turbidites. The paleocurrent evidence (see below) indicates that the reworking resulted from slope-following bottom currents. In such contourites the Bouma *c* and *d* subdivisions are replaced by fine erosive-type cross-laminations that persist into the silt or even mud grain sizes, thereby indicating considerable reworking of the fine-grained fraction of the original turbidite.

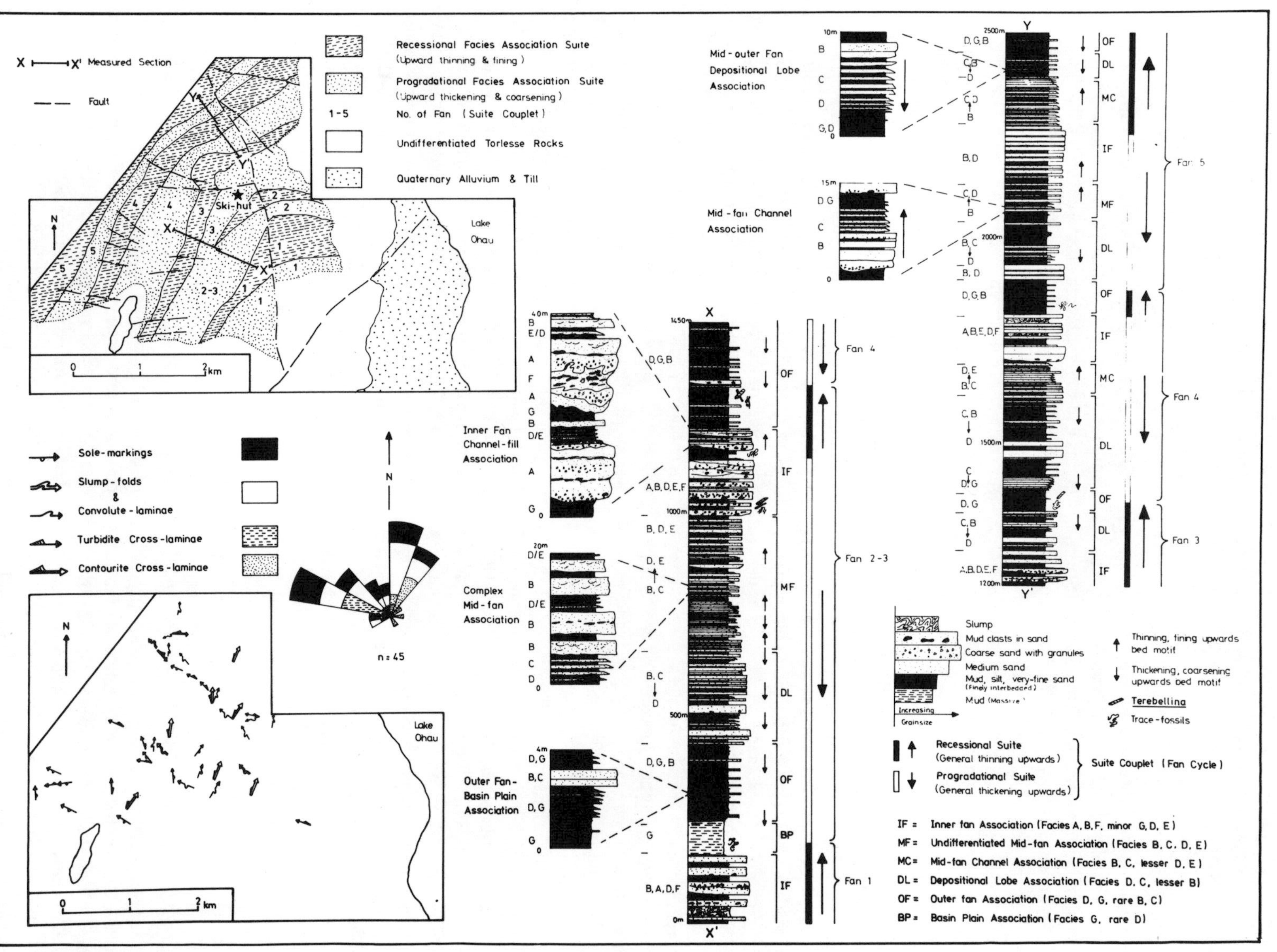

Figure 23-8. Measured sections spanning several fan cycles of the composite submarine fan sequence, Torlesse Zone, Ohau Ski Basin. Note: Insets show interpretative map of the distribution of facies association suites and paleocurrent and paleoslope indicators. Tilting and folding is too complex to allow unique restoration of paleodirectional indicators. The technique used assumed no complex folding or steeply plunging axes; as such folding may well have occurred, the data as plotted are probably misoriented, though all in the same sense and degree.

Facies associations. The beds at Ohau stand vertically, with Facies A, B, and C granule conglomerates and sandstones protruding as resistant buttresses, between which are the thinner-bedded and less resistant sediments of Facies D, E, and G (photographs A and B, Figure 23-9). As can be seen in Figure 23-9, the individual facies occur in packets up to several tens of meters thick. These packets are not randomly distributed; rather they are interbedded in rhythmic and recurring facies associations within which preferred sequences of facies can often be recognized. Based on studies of both ancient (e.g., Mutti and Ricci Lucchi 1972; Ricci Lucchi 1975) and modern (e.g., Normark 1970; Nelson and Kulm 1973) deepsea depositional systems, such progressive stratigraphic changes are thought to correspond to sedimentation on active submarine fans, particularly to changes in the loci of depositional lobes on the middle outer fan and in the inner fan channels that feed them.

Walker and Mutti (1973) distinguished six major sedimentary associations with relation to submarine fans. The slope-channel association has not been seen at Ohau. The other five associations are all recognizable and their distribution is shown in Figure 23-8. The facies associations of Walker and Mutti (1973) are mostly readily applied, though in parts of the Ohau section the characteristic thickening- or thinning-upward cycles of the middle fan are replaced by thick sandy packets of irregularly bedded Facies B and C with lesser D. Such variations are likely in any major sequence of active fans.

Suites of facies associations. Ricci Lucchi (1975) has shown that an individual fan-building or -migrating episode will result in the accumulation of a suite of facies associations. Such suites are commonly *progradational* and exhibit a general coarsening-upward trend (Facies G dominant at the base to Facies A and B dominant at the top), or they are *recessional* and exhibit a general fining-upward trend (A at the base to G at the top). In these terms, the Ohau section is consistent with the superposition of the sediments of several discrete submarine fans that collectively defined a large coalescing or migratory fan system. Suite couplets, with recessional overlying progradational suites within any one vertical section (Figure 23-8), can be related to the recurrent approach, presence, and departure of the main axis of sediment dispersal as it migrates across the fan.

At least five superposed progradational-recessional suite couplets can be distinguished within the Ohau Ski Basin area (see Figure 23-8). Individual couplets are generally between 500 m and 1,500 m thick.

Paleocurrents and paleoslope. Paleodirectional indicators include ripple cross-lamination (in both contourites and turbidites), convolute lamination, slump folds, flute casts, and tool marks. Restoration of the pattern by rotation of the bedding about strike indicates a regional paleoslope, and turbidity current transport, towards the northwest, with northeasterly flowing bottom currents parallel to the slope (see Figure 23-8).

Fossils. The only body fossil found is the tubiculous polychaete *Torlessia*, which may be abundant locally in Facies D or G argillites. The tubes are usually variably oriented within bedding planes, but occasionally occur in life positions that are oblique or even perpendicular to bedding; they also occur occasionally in the basal parts of redeposited sands (cf. Webby 1967). Trace fossils are also commonest in argillaceous beds and include *Helminthoides* type bedding plane trails.

Overall interpretation. The Alpine Assemblage at the Ohau Ski Basin represents a series of major, continental margin, deep marine fans. Facies analysis shows that the sediments closely match those described from such sequences elsewhere in the world. The *in situ* occurrence of *Torlessia* is reminiscent of the habitat of the recent deepsea polychaete *Hyalinoecia*, known from dredge hauls and bottom photographs to be a dominant, solitary macrofaunal element in continental slope and rise situations (cf. Menzies et al. 1973), and a deep marine origin for *Torlessia* has been argued independently by Webby (1967). The trace fossil assemblage is also consistent with a deep water origin.

Lithofacies Elsewhere in the Alpine Assemblage

Redeposited facies associations similar to those we have described from the Ohau Ski Basin are widespread throughout the Torlesse Zone of both islands of New Zealand (cf. Wellman 1956). Descriptions and illustrations have been given by many writers, particularly Andrews (1974), Andrews et al. (1974), Bishop (1974), Bradshaw (1972), Lillie and Gunn (1964), Speden (1972), Schofield (1974), Spörli (1975), Spörli and Barter (1973), Spörli and Lillie (1974), Spörli et al. (1974), Waterhouse (1966), and Webby (1959). Lithofacies that are not present at Ohau include red mudstone and chert (Bradshaw 1972; Schofield 1974), pillow lava (Bradshaw 1972), limestone (Bradshaw 1972), and marble (Bishop 1974). The majority of the known fossil localities (Campbell and Warren 1965; Speden 1976) occur in redeposited sediments.

Andrews (1964) and Bradshaw and Andrews (1963) have argued for a shallow marine and even deltaic origin for the Torlesse rocks described by them, and they cite the distribution of conglomerates and plant fossils as specific evidence for their interpretation. The absence of demonstrable shallow water sedimentary structures, depositional cycles and *in situ* fossils is an important objection to the shallow marine interpretation for the sequences these authors describe, or for any large amount of sediment elsewhere in the Alpine Assemblage. Also,

Figure 23-9. Selected characteristics of Alpine Assemblage rocks, Ohau Ski Basin. Note: Photograph A shows outer fan association with subordinate Facies B sandstones within Facies C and D turbidites (section X of Figure 23-8; youngs to right). Photograph B shows channelized inner fan association with dominantly Facies A sandstones and lesser facies B (section X of Figure 23-8; youngs to right; arrowed figure gives scale). Photograph C shows inner mid-fan association (youngs to right). Complexly interbedded Facies C and D turbidites (mostly overbank types), some reworked by bottom currents, are followed by slurry folded layer (adjacent figure for scale) and then by a massive Facies B sandstone. Photograph D shows inner mid-fan upward fining and thinning sequence; beds represent successive Facies B and C (sequence upright, hammer = 1 m).

in the absence of detailed facies analysis, the reported concentration of exogenous conglomerates towards the northeast is at least as likely to be related to the presence of active fan or deepsea channel sedimentation as it is to indicate shoaling water in that direction (cf. the distribution of conglomerates in the redeposited parts of the Hokonui Assemblage).

There do exist, however, a very few localities within the Alpine Assemblage at which rich *in situ* benthonic invertebrate faunas and shallow marine sedimentary facies have been described, notably towards the southeast at Corbies Creek (Ryburn, cited in Force 1974). The very rarity of such sequences in comparison with the typical redeposited facies and the fact that they occur "floating" in the midst of highly deformed metagreywackes of more typical Torlesse type are surely significant features. Although these shallow-water facies might indicate shelf conditions towards the southeast of the Torlesse terrain (cf. Bradshaw and Andrews 1973), concluding that such was the case would be premature. Such rare occurrences could be explained in several alternative ways; for example, as olistoliths. Kay (1974) has concluded, "Abyssal plains may have sediment that slumped from ocean shores. The deeps may contain anything that was free to slump from the beaches and deltas – the material has nothing to do with the site of deposition but with the source" (p. 379).

There is no particular a priori reason for a general lack of shallow marine or nonmarine sedimentation in the Alpine Assemblage, but the fact remains that the assemblage is dominated by flysch and associated sediment types. The presence of typical submarine fan sediments, probably derived from the east, adds weight to the arguments of Landis (in Blake et al. 1974) that the assemblage accumulated adjacent to a now-distant continental margin. Clearly, shallow marine or nonmarine sediments are not excluded by such an interpretation, but the probability exists that by far the greatest part of the 10^5 km^2 of the area of the Torlesse Zone metasediments comprises redeposited and associated deep marine sediment types.

Relationship between Hokonui - Te Anau and Alpine Assemblages

The complete contrast of provenance between the Alpine Assemblage and the Hokonui - Te Anau Assemblages requires separate source areas for each (cf. Landis and Bishop 1972; Bradshaw and Andrews 1973). Despite superficial similarities, due to their both containing large amounts of deep marine and redeposited sediment, there are also important contrasts in sediment facies between these two groupings. The Hokonui - Te Anau rocks correspond closely to an active arc-trench system; material derived from the volcanic arc and its fringing shelf was fed down through or past successive trench slope basins and finally into the trench, by way of submarine canyon and channel systems whose deposits are widely represented in the sedimentary record. The Alpine Assemblage, on the other hand, contains appreciable amounts of ocean basin and submarine fan sediment and corresponds more closely to a deepsea sequence adjacent to a distant continental margin (cf. Landis, in Blake et al. 1974).

A sedimentary rather than tectonic explanation for the petrographic contrasts was one of a number of models explored by Dickinson (1971), Landis and Bishop (1972), and Landis (1974b). These authors suggested that the quartzofeldspathic detritus in the Alpine belt was derived from the foreland to the southwest of the geosyncline and that it vaulted the Hokonui depositional realm by passing through submarine channels that led into the eastern Alpine Basin. We have presented evidence for the existence of just such sediment transport channels, particularly in Facies Association 3 of the Hokonui Assemblage; but the volcanogenic nature of the sediment fill of all channels so far investigated precludes this particular explanation for the contrast in provenance between the Hokonui - Te Anau and Alpine Assemblages.

The present juxtaposition of these dissimilar petrographic provinces may rather be tectonic, as suggested by several authors (Force 1974; Landis, in Blake 1974). Perhaps the most attractive explanation for the petrographic and sedimentologic differences between the Te Anau and Alpine Assemblages, which is at the same time consistent with their structural and metamorphic similarities, is that the Alpine Assemblage accumulated off a continental margin separated from the Rangitata plate margin by some form of ocean ridge or island chain. Such a ridge would have acted as a sediment barrier, and only after its subduction, or possible accretion, as the conspicuous metavolcanic belt within the Haast Schist (Henley et al., 1976; but see also Cooper 1976) would Alpine Assemblage sediment have been accreted at the subduction zone.

DISCUSSION

Actualistic Arc-Trench Models

We have attempted to show that the distribution of major sedimentary facies within the Rangitata Orogen matches that of modern actively subducting arc-trench systems. Karig and Sharman (1975) have developed a model for the evolution of such systems based on an analysis of the numerous recent examples of Indonesian and Pacific arc-trench couplets. The evidence is consistent with the Rangitata Orogen having followed a 130 my long evolutionary path closely similar to that implied by the Karig and Sharman model (cf. Figure 23-2).

Initially, during the early and middle Permian, the upper plate margin of the Rangitata arc system probably descended with a smooth, steep profile from the southwestern Brook Street volcanic arc to a northeastern trench, deposits of which have not yet been identified in New Zealand (Stage A of Karig and Sharman [1975]; example, the west Melanesian systems). Following the later Permian emplacement of the first major accretionary wedge along the inner trench wall, as the Dun Mountain Ophiolite Belt (Coombs et al. 1976a), a definitive trench slope break was established, upslope from which Maitai sediments accumulated in a midslope basin and downslope from which were emplaced successively younger accretionary wedges of the Te Anau Assemblage (Stage B of Karig and Sharman [1975]; example, the East Luzon system). During the latest Permian and Triassic, the Rangitata system passed through successively more mature steps of Stage B, with the continuing downslope emplacement of subduction wedges in the Te Anau and Alpine Assemblages, filling of the midslope basin by Maitai sediments of Facies Association 1, and the filling of upper slope basins and outbuilding of the forearc shelf by Facies Associations 2 to 5. With the final infilling of the upper and midslope basins during the later Triassic and Jurassic, the Rangitata system passed into Karig and Sharman's (1975) Stage C (example, the Japan system), where the arc-trench profile displays a trench slope break along the edge of a broad forearc shelf formed by arc-derived sediments finally filling and prograding out over the midslope basin of Stage B. Since shallow marine and nonmarine Associations 4 to 5 probably appear in the record at earlier times as one progresses along the Rangitata arc-trench system to the northwest (Fleming

A

B

Figure 23-10. Ohau Ski Basin facies. Note: Photograph A shows climbing ripples in T_c interval of turbidite bed in mid-fan overbank association. Photograph B shows finely laminated and wispy rippled silts and muds of distal fan to basin plain origin (coin = 2.1 cm diameter).

and Kear 1960; Campbell and McKellar 1960), a southeasterly directed component of along-slope regression was at least locally imposed upon the general arcwards to trenchwards sediment transport systems (cf. Boles 1974; Force 1974).

Facies Analysis of Redeposited Sediments

Mutti and Ricci Lucchi (1972) and Ricci Lucchi (1975) have developed a lithofacies scheme that greatly facilitates the shorthand description of redeposited sediments, particularly those in submarine fan settings. We have found this scheme valuable for the description of New Zealand sediments, particularly for the analysis of the fan sequences developed in the Alpine Assemblage and the slope channel sequences of Facies Association 3 of the Hokonui Assemblage. The descriptive aspects of the scheme are even more widely applicable. Some of Facies A to G of Mutti and Ricci Lucchi occur in at least the following depositional environments, as represented in the sediments of the Rangitata Orogen: (1) prodelta slope (Facies Association 5, Hokonui Assemblage); (2) shelf canyons and channels (Association 3, Hokonui Assemblage); (3) slope channels (Association 3, Hokonui Assemblage); (4) middle-upper slope basin fans, flysch wedge fills and basin floor channels (Associations 1 to 2, Hokonui Assemblage); (5) trench and lower trench slope basin fans, flysch wedge fills and basin floor channels (Te Anau Assemblage); and (6) continental margin fans (Alpine Assemblage). Similar facies are also known to occur in modern abyssal plain channel fills (Chough and Hesse 1976) and although not yet identified, such channels may be represented in either or both of the Alpine and Te Anau Assemblages. Although the facies scheme of Mutti and Ricci Lucchi may be applied in the description of redeposited sediments in at least these seven different settings, its use alone will seldom produce an unequivocal environmental interpretation. Careful analysis of facies groupings (Ricci Lucchi 1975) and the application of Waltherian principles are required for determinative results.

Faunal Patterns in Rangitata Geosyncline

By comparison with many late Paleozoic and Mesozoic rocks, those of the Rangitata Orogen are not richly fossiliferous. There is a particular lack of zonally important cephalopods, though isolated occurrences are known. If the reconstruction shown in Figure 23-2 is broadly correct, then large parts of the Alpine, Te Anau and northeastern Hokonui Assemblage probably accumulated below the contemporary carbonate compensation depths. Pelagic faunal elements, particularly aragonitic forms like cephalopods, would therefore have been dissolved as they settled through the water column, and calcareous-shelled benthos would have been largely absent.

Available faunal data falls into a pattern that is consistent with the sedimentary interpretation argued in this chapter. Three major faunal associations are known:

1. Pelagic elements, including especially cephalopods and epiplanktonic bivalves (*Halobia*, *Daonella*). Occurrences are scattered haphazardly throughout the stratigraphic column, but are always most common in the shallower water Facies Associations 2 and 4 of the Hokonui Assemblage. Where they occur in deeper water sediments, pelagic macrofaunas may often be redeposited.
2. Benthic brachiopod and molluscan communities of the shelf (? and upper trench slope). These communities form the basis for the local New Zealand zonation of the Permian, Triassic, and Jurassic (e.g., Campbell and McKellar 1960; Fleming and Kear 1960; Waterhouse 1964). They occur as life assemblages only in shelf

Facies Association 4 and, possibly, upper slope Association 2. They may occur as redeposited shellbeds or isolated valves in any of the other deeper water facies, with their scarcity increasing with the inferred water depth and distance from source.

3. Benthic deepwater communities. These communities are represented by largely unstudied trace fossil assemblages, by tubiculous terebellid worm assemblages (cf. Webby 1967), and by rare deepwater pleurotomariid gastropods. The commonest occurrences are in Facies Association 2 of the Hokonui Assemblage and in parts of the Te Anau and Alpine Assemblages. Facies Association 1 of the Hokonui Assemblage and parts of the Te Anau Assemblage include significant amounts of unfossiliferous finely laminated sediment that may represent truly abiotic conditions.

CONCLUSIONS

We conclude, with Waterhouse (1975), that "until the Rangitata Orogeny, the components of that orogeny were comparatively simple, and the volcanic arc and geosyncline developed along classic lines, and are so well preserved that they help serve as a model for other orogens . . . . The New Zealand pattern is superbly displayed, and, particularly as regards the volcanic arc and eugeosyncline, very much less tectonized along part of its length than most other closely studied examples."

Within this context, and within the now well-established plate tectonic framework of the Rangitata Orogen, the little-studied sedimentary patterns of this orogenic belt are of particular interest. We have attempted to show that at least the southwestern sedimentary belts of the orogen formed an interlinked transport and depositional complex that straddled a consuming plate margin, which we infer from a comparison between its sedimentary facies and those of modern arc-trench systems. Future detailed studies of the sediments of the Rangitata Orogen should provide valuable insights into arc-trench sedimentary mechanisms and patterns.

ACKNOWLEDGMENTS

The authors are grateful to the many colleagues who have discussed or otherwise helped shape this chapter. In particular, material improvements to the manuscript were effected by D.G. Bishop, D.S. Coombs, B.F. Houghton, C.A. Landis, J.C. Moore, and S. Nathan. Financial support for the work was provided by the New Zealand University Grants Committee and the University of Otago; in addition, M.D. Hicks acknowledges assistance towards fieldwork from the Benson Memorial Fund. Finally, the authors are most grateful to Professor Gilbert Kelling for his kind invitation to contribute to this symposium volume.

REFERENCES

Andrews, P.B., 1974. Deltaic Sediments, Upper Triassic Torlesse Supergroup, Broken River, North Canterbury. *N.Z.J. Geol. Geophys.*, 17: 881-905.

_____, Bishop, D.G., Bradshaw, J.D., and Warren, G., 1974. Geology of the Lord Range, Central Southern Alps, New Zealand. *N.Z.J. Geol. Geophys.* 17: 271-99.

Aronson, J.L., 1968. Regional geochronology of New Zealand. *Geochim. Cosmochim. Acta*, 32: 669-97.

Benson, W.N., and Chapman, F., 1938. Note on the occurrence of radiolarian limestone among the older rocks of south-east Otago. *Trans. Roy. Soc. N.Z.*, 67: 373-74.

Bishop, D.G., 1965. The geology of the Clinton district, South Otago. *Trans. Roy. Soc. N.Z.*, 2: 205-30.

_____, 1974. Stratigraphic, structural and metamorphic relationships in the Dansey Pass area, Otago, New Zealand. *N.Z.J. Geol. Geophys.*, 17: 301-35.

_____, and Force, E.R., 1969. The reliability of graded bedding as an indicator of the order of superposition. *J. Geol.*, 77: 346-52.

Blake, M.C., Jones, D.L., and Landis, C.A., 1974. Active continental margins: contrasts between California and New Zealand. In: C.A. Burk and C.L. Drake (eds.), *The Geology of Continental Margins.* Springer-Verlag, New York, pp. 853-72.

_____, and Landis, C.A., 1973. The Dun Mountain ultramafic belt – Permian oceanic crust and upper mantle in New Zealand. *J. Res. U.S. Geol. Surv.* 1: 529-34.

Boles, J., 1974. Structure, stratigraphy and petrology of mainly Triassic rocks, Hokonui Hills, Southland, New Zealand. *N.Z.J. Geol. Geophys.*, 17: 337-74.

Bradshaw, J.D., 1972. Stratigraphy and structure of the Torlesse Supergroup (Triassic-Jurassic) in the foothills of the Southern Alps near Hawarden. *N.Z.J. Geol. Geophys.*, 15: 71-87.

_____, and Andrews, P.B., 1973. Geotectonics and the New Zealand Geosyncline. *Nature*, 241: 14-16.

Campbell, J.D., 1974. Biostratigraphy and structure of Richmond Group rocks in the Wairoa River - Mount Heslington area, Nelson. *N.Z.J. Geol. Geophys.*, 17: 41-62.

_____, (ed.), 1975. *Gondwana Geology.* (Proc. 3rd Gondwana Symposium, Canberra, 1973.) ANU Press, Canberra.

_____, and McKellar, I.C., 1960. The Otamitan Stage (Triassic): Definition and type locality. *N.Z.J. Geol. Geophys.*, 3: 643-59.

_____, and Warren, G., 1965. Fossil localities of the Torlesse Group in the South Island. *Trans. Roy. Soc. N.Z.*, 3: 99-136.

_____, and Coombs, D.S., 1966. Murihiku Supergroup (Triassic-Jurassic) of Southland and South Otago. *N.Z.J. Geol. Geophys.*, 9: 393-98.

Carter, R.M., 1975. A discussion and classification of subaqueous mass-transport with particular application to grain-flow, slurry-flow and fluxoturbidites. *Earth Sci. Rev.*, 11: 145-77.

_____, Landis, C.A., Norris, R.J., and Bishop, D.G., 1974. Suggestions towards a high-level nomenclature for New Zealand rocks. *J. Roy. Soc. N.Z.*, 4: 5-18.

_____, and Lindqvist, J.K., 1975. Sealers Bay submarine fan

complex, Oligocene, southern New Zealand. *Sedimentol.* 22: 465-83.

Chough, S., and Hesse, R., 1976. Submarine meandering thalweg and turbidity currents flowing for 4000 km in the Northwest Atlantic mid-ocean channel, Labrador Sea. *Geology*, 4: 529-33.

Connolly, J. R., and Ewing, M., 1967. Sedimentation in the Puerto Rico Trench. *J. Sedim. Petrol.*, 37: 44-59.

Coombs, D.S., 1950. The geology of the northern Taringatura Hills, Southland. *Trans. Roy. Soc. N.Z.*, 78: 426-48.

———, 1954. The nature and alteration of some Triassic sediments from Southland New Zealand. *Trans. Roy. Soc. N.Z.*, 82: 65-109.

———, Landis, C.A., Norris, R.J., Sinton, J.M., Borns, D.J., and Craw, D., 1976a. The Dun Mountain Ophiolite Belt, New Zealand: Its tectonic setting, constitution, and origin, with special reference to the southern portion. *Amer. J. Sci.*, 276: 561-603.

———, Campbell, J.D., McKellar, I.C., Landis, C.A., and Bishop, D.G., 1976b. Regional geology of southern part of South Island, New Zealand Geosyncline. *Excursion Guide*, 59c. 25th Int. Geol. Cong., Sydney.

Cooper, A.F., 1976. Concentrically zoned ultramafic pods from the Haast Schist Zone, South Island, New Zealand. *N.Z.J. Geol. Geophys.*, 19: 603-23.

Devereux, I., McDougall, I., and Watters, W.A., 1968. Potassium – argon mineral dates on intrusive rocks from the Foveaux Strait area. *N.Z.J. Geol. Geophys.*, 11: 1230-35.

Dickinson, W.R., 1971. Detrital modes of New Zealand greywackes. *Sedimentol.*, 5: 37-56.

———, 1974. Plate tectonics and sedimentation. In: W.R. Dickinson (ed.), *Tectonics and Sedimentation.* Soc. Econ. Paleont. Mineral., Sp. Publ. 22, pp. 1-27.

Du Toit, A., 1937. *Our Wandering Continents.* Oliver & Boyd, Edinburgh, 366 pp.

Fleming, C.A., 1970. The Mesozoic of New Zealand: Chapters in the history of the Circum-Pacific Mobile Belt. *Quart. J. Geol. Soc. Lond.*, 125: 125-70.

———, and Kear, D., 1960. The Jurassic sequence at Kawhia Harbour, New Zealand. *N.Z. Geol. Surv. Bull.*, 67.

Force, E.R., 1974. A comparison of some Triassic rocks in the Hokonui and Alpine belts of the South Island, New Zealand. *J. Geol.*, 82: 37-49.

———, 1975. Stratigraphy and paleoecology of the Productus Creek Group, South Island, New Zealand. *N.Z.J. Geol. Geophys.*, 18: 373-99.

Grant-Mackie, J.A., Speden, I.G., Bradshaw, J.D., Campbell, J.D., and Stevens, G.R., 1976. Bibliography of New Zealand Mesozoic historic geology. *Int. Geol. Correl. Proj. 8 (Mesozoic chronostratigraphy, New Zealand - New Caledonia), Report 3.*

Grindley, G.W., 1958. The geology of the Eglinton Valley, Southland. *N.Z. Geol. Surv. Bull.*, 58, 68 pp.

———, 1974. New Zealand. In: A.M. Spencer (ed.), Mesozoic - Cenozoic orogenic belts: Data for orogenic studies. *Geol. Soc. Lond., Sp. Publ.* 4, pp. 387-416.

Hedberg, H.D. (ed.), 1976. *International Stratigraphic Guide.* Wiley, New York, 200 pp.

Hendry, H.E., 1973. Sedimentation of deep water conglomerates in Lower Ordovician rocks of Quebec – composite bedding produced by progressive liquefaction of sediment? *J. Sedim. Petrol.*, 43: 125-36.

Henley, R.W., Norris, R.J., and Paterson, C.J., 1976. Multistage ore genesis in the New Zealand Geosyncline. A history of post-metamorphic lode emplacement. *Mineral. Deposita*, 11: 180-196.

Hicks, M.D., 1975. *Geology of the Lake Ohau Ski Basin, South Canterbury, New Zealand.* Unpublished B.Sc. (Hons) thesis, Otago University, Dunedin, New Zealand, 168 pp.

Jeletzky, J.A., 1975. Hesquiat Formation (new): A neritic channel and interchannel deposit of Oligocene age, western Vancouver Island, British Columbia. *Geol. Surv. Canada Paper* 75-32, 54 pp.

Karig, D.E., and Sharman, G.F., 1975. Subduction and accretion in trenches. *Geol. Soc. Amer. Bull.*, 86: 377-389.

Kawachi, Y., 1974. Geology and petrochemistry of weakly metamorphosed rocks in the upper Wakatipu district, southern New Zealand. *N.Z.J. Geol. Geophys.*, 17: 169-208.

Kay, M., 1974. Geosynclines, flysch and melanges. In: R.H. Dott, Jr., and R.H. Shaver (eds.), *Modern and Ancient Geosynclinal Sedimentation.* Soc. Econ. Paleont. Mineral., Sp. Publ. 19, pp. 377-80.

Kear, D., 1971. Basement rock facies – northern North Island. *N.Z.J. Geol. Geophys.*, 14: 275-83.

Landis, C.A., 1974a. Stratigraphy, lithology, structure and metamorphism of Permian, Triassic and Tertiary rocks between the Mararoa River and Mt Snowdon, western Southland. *J. Roy. Soc. N.Z.*, 4: 229-51.

———, 1974b. Paleocurrents and composition of lower Bryneira strata (Permian) at Barrington Peak, Northwest Otago. *N.Z. J. Geol. Geophys.*, 17: 799-806.

———, and Bishop, D.G., 1972. Plate tectonics and regional stratigraphic-metamorphic relations in the southern part of the New Zealand Geosyncline. *Geol. Soc. Amer. Bull.*, 83: 2267-84.

Lewis, D.W., 1976. Subaqueous debris flows of early Pleistocene age at Motunau, North Canterbury. *N.Z.J. Geol. Geophys.*, 19: 535-67.

Lillie, A.R., and Gunn, B.M., 1964. Steeply plunging folds in the Sealy Range, Southern Alps. *N.Z.J. Geol. Geophys.*, 7: 403-23.

Marshall, P., 1918. Geology of the Tuapeka Subdivision. *N.Z. Geol. Surv. Bull.*, 19.

Menzies, R.J., George, R.W., and Rowe, G.T., 1973. *Abyssal Environment and Ecology of the World Oceans.* Wiley, New York, 488 pp.

Mildenhall, D.C., 1976. *Glossopteris ampla* Dana from New Zealand Permian Sediments. *N.Z.J. Geol. Geophys.*, 19: 130-32.

Moore, G.F., and Karig, D.E., 1976. Development of sedimentary basins on the lower trench slope. *Geology*, 4: 693-97.

Mutch, A.R., 1957. Facies and thickness of the upper Paleozoic and Triassic sediments of Southland. *Trans. Roy. Soc. N.Z.*, 84: 499-511.

Mutti, E., and Ricci Lucchi, F., 1972. Le torbiditi dell'Apennino Settentrionale: introduzione all'-analisi di facies. *Mem. Soc. Geol. Italia*, 11: 161-99.

Nelson, C.H., and Kulm, L.D., 1973. Submarine fans and deep-sea channels. In: G.V. Middleton and A.H. Bouma (eds.), *Turbidites and Deep Water Sedimentation.* Soc. Econ. Paleon. Mineral., Pacific Section, Short Course, Anaheim, pp. 39-78.

Normark, W.R., 1970. Growth patterns of deep-sea fans. *Amer. Assoc. Petrol. Geol. Bull.*, 54: 2170-95.

Piper, D.J.W., and Normark, W.R., 1971. Re-examination of a Miocene deepsea fan and fan-valley, southern California. *Geol. Soc. Amer. Bull.*, 82: 1823-30.

———, von Huene, R., and Duncan, J.R., 1973. Late Quaternary sedimentation in the active eastern Aleutian trench. *Geology*, 1: 19-22.

Reimnitz, E., Shepard, F.P., and Toimil, L.J., 1977. Slope valleys and sedimentary structures off the Rio Balsas of

Western Mexico. *Amer. Assoc. Petrol. Geol. Bull.*, 60: 712-13.

Ricci Lucchi, F., 1975. Depositional cycles in two turbidite formations of northern Apennines (Italy). *J. Sedim. Petrol.*, 45: 3-43.

Schofield, J.C., 1974. Stratigraphy, facies, structure and setting of the Waiheke and Manaia Hill Groups, east Auckland. *N.Z.J. Geol. Geophys.*, 17: 807-38.

Scholl, D.W., and Marlow, M.S., 1974. Sedimentary sequence in modern Pacific trenches and the deformed circum-Pacific eugeosyncline. In: R.H. Dott, Jr., and R.H. Shaver (eds.), *Modern and Ancient Geosynclinal Sedimentation.* Soc. Econ. Paleon. Mineral., Spec. Publ. 19 pp. 193-211.

Scott, K.M., 1966. Sedimentology and dispersal pattern of a Cretaceous flysch sequence, Patagonian Andes, Southern Chile. *Amer. Assn. Petrol. Geol. Bull.*, 50: 72-107.

Shepard, F.P., and Dill, R.F., 1966. *Submarine Canyons and Other Sea Valleys.* Rand McNally, Chicago, 381 pp.

Speden, I.G., 1971. Geology of the Papatowai Subdivision, southeast Otago. *N.Z. Geol. Surv. Bull.*, 81, 166 pp.

_____, 1972. New fossil localities in the Torlesse Supergroup, western Raukumara Peninsula, New Zealand. *N.Z.J. Geol. Geophys.*, 15: 433-45.

_____, 1976. Fossil localities in Torlesse rocks of the North Island, New Zealand. *J. Roy. Soc. N.Z.*, 6: 73-91.

Spörli, K.B., 1975. Waiheke and Manaia Hill Groups, east Auckland, Comment. *N.Z.J. Geol. Geophys.*, 18: 757-60.

_____, and Barter, T.P., 1973. Geological reconnaissance in the Torlesse Supergroup of the Kaimanawa Ranges along the lower reaches of the Waipakihi River, North Island, New Zealand. *J. Roy. Soc. N.Z.*, 3: 363-80.

_____, and Lillie, A.R., 1974. Geology of the Torlesse Supergroup in the northern Ben Ohau Range, Canterbury. *N.Z.J. Geol. Geophys.*, 17: 115-41.

_____, Stanaway, K.J., and Ramsay, W.R.H., 1974. Geology of the Torlesse Supergroup in the southern Liebeg and Burnett Ranges, Canterbury, New Zealand. *J. Roy. Soc. N.Z.*, 4: 177-92.

Turnbull, I.M., 1974. *Geology of the Thomson Mountains.* Unpublished PhD dissertation, Otago University, Dunedin, New Zealand, 462 pp.

_____, 1978. Nomenclature in the Rangitata Geosyncline (note). *N. Z. J. Geol. Geophys.*, 20 (in press).

von Huene, R., 1974. Modern trench sediments. In: C.A. Burk and C. L. Drake (eds.), *The Geology of Continental Margins.* Springer-Verlag, New York, pp. 207-11.

Walker, R.G., 1975. Generalized facies models for resedimented conglomerates of turbidite association. *Geol. Soc. Amer. Bull.*, 86: 737-48.

_____, and Mutti, E., 1973. Turbidite facies and facies associations. In: G.V. Middleton and A.H. Bouma (eds.), *Turbidites and Deep-Water Sedimentation*, Soc. Econ. Paleon. Mineral., Pacific Section, Short Course, Anaheim, pp. 119-57.

Waterhouse, J.B., 1963. Communal hierarchy and significance of environmental parameters for brachiopods: The New Zealand Permian model. *Roy. Ontario Museum, Life Sci. Contrib.*, 92, 49 pp.

_____, 1964. Permian stratigraphy and faunas of New Zealand. *N.Z. Geol. Surv. Bull.*, 72, 101 pp.

_____, 1966. The Häckel Syncline and neighbouring folds of the upper Tasman glacier. *Trans. Roy. Soc. N.Z.* (Geol.), 3: 183-95.

_____, 1975. The Rangitata Orogen. *Pacific Geol.*, 9: 35-73.

Webby, B.D., 1959. Sedimentation of the alternating greywacke and argillite strata in the Porirua district. *N.Z.J. Geol. Geophys.*, 2: 461-78.

_____, 1967. Tube fossils from the Triassic of south-west Wellington. *Trans. Roy. Soc. N.Z.* (Geol.), 5: 181-91.

Wegener, A., 1920. *Die Entstehung der Kontinente und Ozeane. (2nd ed.)* Vieweg & Sohn, Braunschweig, 135 pp.

Wellman, H.W., 1956. Structural outline of New Zealand. *N.Z. D.S.I.R. Bull.*, 121, 36 pp.

Wood, B.L., 1956. The geology of the Gore Subdivision. *N.Z. Geol. Surv. Bull.*, 53, 128 pp.

Chapter 24

Submarine-Fan Turbidites and Resedimented Conglomerates in a Mesozoic Arc-Rear Marginal Basin in Southern South America

ROBERT D. WINN, JR.
ROBERT H. DOTT, JR.

Department of Geology and Geophysics
University of Wisconsin
Madison, Wisconsin

ABSTRACT

Late Mesozoic flysch of Tierra del Fuego and South Georgia Island constitutes the infill of a small marginal ocean basin, which formed between the South American continent and an active island arc in the latest Jurassic (?) and Early Cretaceous. The former arc site is occupied now by the Patagonian batholith; ophiloites represent the former basin sea floor. Sediment gravity flow fabrics, structures, and bedding styles and deepwater trace fossils indicate deposition on deepsea fans. Lenses of pebbly sandstone, breccia, and conglomerate representing inner fan channel deposition are contained within the more typical flysch rocks. Sedimentary structures in the coarse clastics include grading, sole marks, parallel stratification, and medium-scale cross-stratification. Much of the coarsest debris was deposited from traction, probably at the base of turbulent flows. Paleocurrent and petrographic analyses indicates bilateral infilling of the basin. Debris shed southward from the continent came from older Jurassic silicic volcanic rocks and interbedded quartz-rich sedimentary material. Andesitic-dacitic debris was shed northward from a siliceous calc-alkaline parent suite. Continental basement west of the batholith and the siliceous nature of the arc-volcanic rocks suggest that the arc formed above a sliver of continental crust separated from the continent during basin formation. South Georgia was translated relatively eastward, probably as the result of a collission of the continent with the Drake Passage spreading zone during the Cenozoic.

Note: Robert D. Winn, Jr., is currently with the Denver Research Center, Marathon Oil Co., Littleton, Colorado 80120.

INTRODUCTION

Convergent plate boundaries are marked by the formation of magmatic arc-trench couplets. Magmatic arcs may form within oceanic regions, such as the Aleutian chain (Scholl et al. 1968) and the South Sandwich Islands (Barker 1972), as well as on or near the edges of continents, such as the Mesozoic to Recent central Andes (James 1971; Rutland 1971) and much of the Mesozoic of western North America (Kay 1951; Eardley 1962). Many present western Pacific arcs formed on or adjacent to continental areas, but later migrated away as new oceanic crust formed in the intervening zones (Karig 1971, 1974; Coleman 1973). The Kermadec-Tonga arc (Karig 1970; Shor et al. 1971), Japan (Minato et al. 1965), and the Philippines (Karig et al. 1973) are examples. Arc pieces or remnant arcs may be left behind during the outward migration. During some later stage, the same arcs may collide back into the remnant arcs or continents by the initiation of subduction within the marginal ocean basin behind the arc, which results in severe deformation in the intervening zones. Examples include the Cambro-Ordovician of Newfoundland (Dewey and Bird 1971) and deformation and thrusting on Taiwan caused by convergence of the Philippines and Ryukyu Islands microplates (Karig 1974). Repetitions

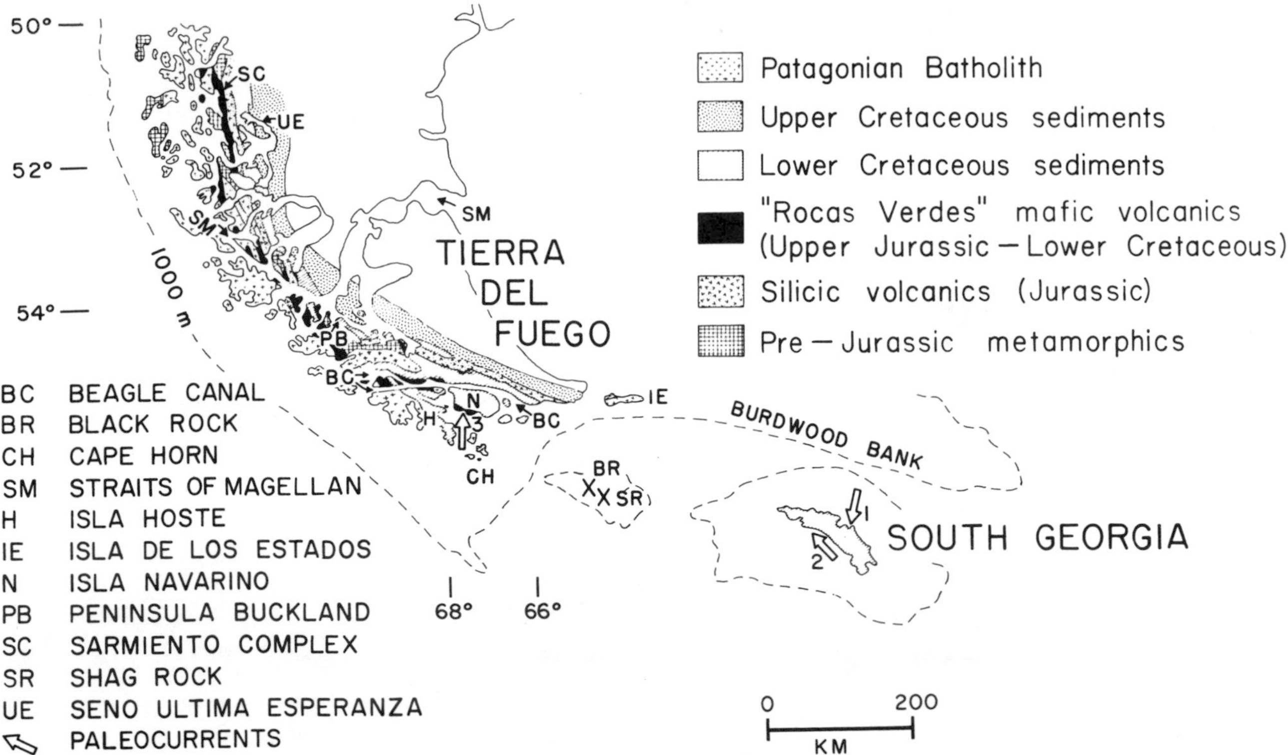

Figure 24-1. Geologic index map of the southern Andes of South America with the South Georgia micro-continent shown in its restored position. Note: Numbers by paleocurrent arrows indicate (1) Sandebugten; (2) Cumberland Bay; and (3) Yahgan formations. (Complete data can be found in Winn 1975; Winn 1978.)

of arc formation, divergence, and marginal basin formation and convergence result in complex tectonic, structural, and stratigraphic relations as exemplified, for example, in the Mesozoic of British Columbia (Monger et al. 1972; Anderson 1976). The relationships may be further obscured by lateral transform motions of portions of arcs.

Back-arc sedimentation differs significantly depending on the type and geometry of the basin formed behind the arc (see Dickinson 1974). While there are exceptions, magmatic arcs that form and remain on continental edges are typically characterized by shallow marine and/or molasse-like continental deposition; marginal ocean basins developed behind arcs, however, tend to be characterized instead by deepwater, flysch-like sedimentation.

Southern South America was the site of a magmatic arc with an arc-rear marginal ocean basin that originated in the mid-Mesozoic (Dalziel et al. 1974b). The basin was the site of deepsea fans that grew from both sides until basin closure in the mid-Cretaceous (Winn 1975; Winn and Dott 1976). South Georgia Island, now located about 2,000 km east of Tierra del Fuego, is thought to have once been adjacent and south of Burdwood Bank (Figure 24-1). The flysch-like section that underlies most of the island (Cumberland Bay and Sandebugten strata; Figure 24-2) is correlative with a Lower Cretaceous flysch-like section in Tierra del Fuego (Yahgan Formation).

In contrast to most ancient examples of back-arc marginal basins, the rocks of this South American example are only moderately deformed, which allows a fairly detailed reconstruction of depositional patterns and sedimentary environments. The area can thus serve as a model for *both* the ancient and modern for this variant of arc and basin. The basin fill has a range of bedding styles and structures; similar bedding types may be expected in deepsea fans in other settings. In particular the conglomerates and breccias are interesting because only recently have these been recognized as common in deepwater settings. Many workers may not be familiar with the possible ranges of structures in these very coarse deposits, and transport mechanisms are even less well understood. This chapter attempts to elucidate these matters by using the example of the Andean arc-rear basin fan deposits.

TECTONIC AND STRATIGRAPHIC SETTING

Northeast of the southern Andean cordillera, southern South America consists mostly of Paleozoic and older

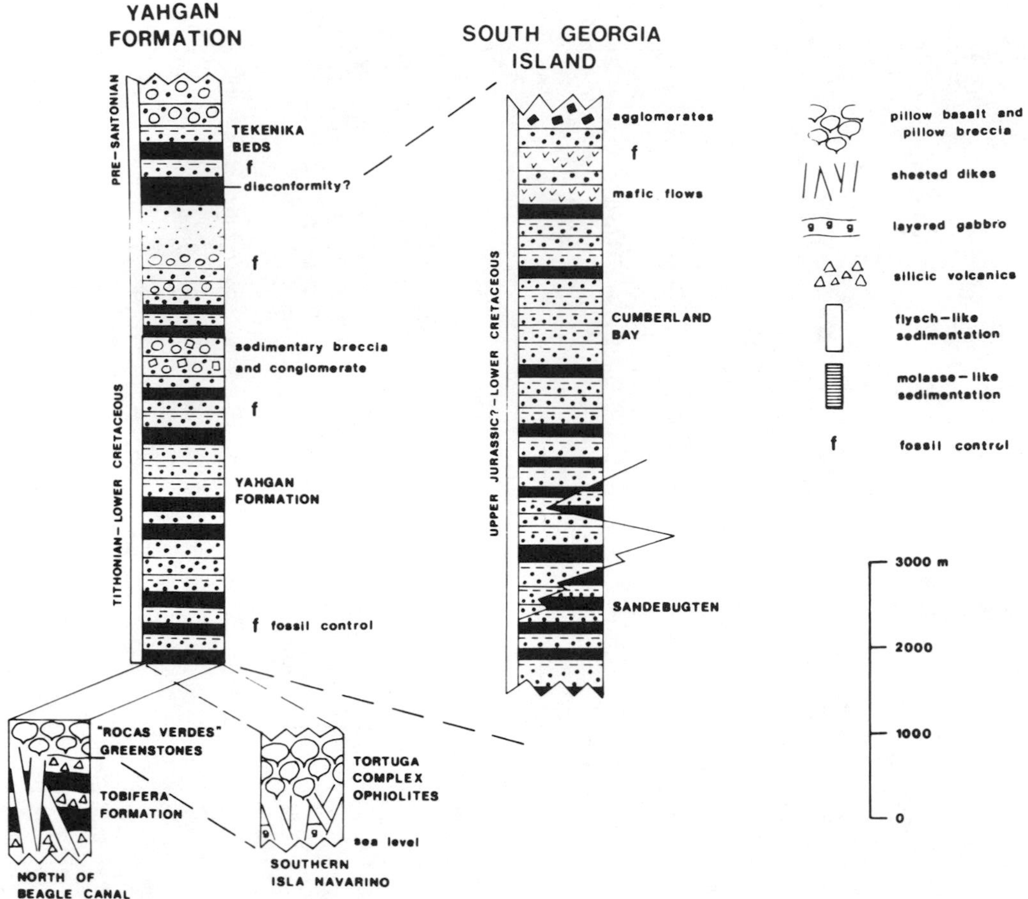

Figure 24-2. Correlation diagram for Mesozoic rocks of Tierra del Fuego (left) and South Georgia. Note: Two different pre-Yahgan sequences – silicic volcanics on cratonic margin (north) and ophiolites of the marginal basin (south) – are shown. (Compare with Figure 24-1). (Source: Winn 1978.)

rocks deformed during the Permo-Triassic Gondwanan Orogeny. These are overlain by extensive silicic volcanics and associated sedimentary rocks of Jurassic age (Tobifera Formation and equivalents) (Du Toit 1937; Dalziel 1974; Natland et al. 1974). Surface outcrops and drillhole and seismic-refraction data indicate that these volcanics extend over almost the entire southern 1,500 km of South America and are probably present in the shelf basins off the eastern coast and on Burdwood Bank (Ludwig et al. 1968). The volcanics consist of fragmental and crystal tuffs, agglomerates, lavas, and ignimbrites. With few exceptions, they are quartz latites to soda rhyolites in composition (Stepanicic and Reig 1955; Dalziel et al. 1974a). Typically they are porphyritic in quartz and sodic plagioclase, whereas biotite and potassic feldspar phenocrysts are rare and pyroxene and amphibole virtually absent. Trace element chemistry suggests that the volcanics were derived from melted continental crust (R. Bruhn, personal communication, 1976) and may be related to either the breakup of Gondwanaland or to the early stages of subduction beneath the western edge of South America.

A discontinuous belt of ophiolitic rocks, long called the "Rocas Verdes" by South American geologists, is present along the western edge of southern South America and is interpreted as representing ancient sea floor (Katz 1972; Dalziel et al. 1974b). Where best exposed, the "Rocas Verdes" consist of a layered gabbro zone at sea level that is cut by and passes upward into a sheeted dike complex, which is in turn overlain by basaltic pillows, breccias, and aquagene tuffs. Ultramafic components are not visible apparently only because of the level of exposure. The greenstones intrude and separate blocks and horsts of cratonic basement. Mafic volcanics on the southeastern edge of South Georgia Island (Trendall 1959) appear to be cogenetic with the "Rocas Verdes" (Suárez 1976). Subduction along the western edge of the continent, which commenced in the Jurassic, appears to have resulted in fragmentation of the continental margin and generation of new ocean crust in rift zones between continental blocks.

Overlying the ophiolites is a 7 km to 8 km thick, uppermost Jurassic to Lower Cretaceous, flysch-like section (Yahgan Formation and Tekenika Beds, Figure 24-2). The section is not well dated. The Yahgan Formation has fossils indicating a Cretaceous and possibly a Late Jurassic (Tithonian) age (Hoffstetter 1957). The Tekenika Beds concordantly overlie the Yahgan Formation, and paleontologic evidence suggests a Cretaceous age for them, too; timing of deformation sets a minimum age of 80 million years, the age of plutons intruded into nearby folded Yahgan rocks (Halpern and Rex 1972; Dott et al. 1977).

The correlative section on South Georgia Island (Cumberland Bay and Sandebugten strata) is even more poorly dated. The Cumberland Bay has Early Cretaceous fossils that may be as young as Albian in the upper part of the unit (Wilckens 1947, p. 56; Casey 1961; Pettigrew and Willey 1975). The Sandebugten has yet to be dated paleontologically, but the presence of only one deformation fabric and the virtual identity in sedimentary and deformational style to the Cumberland Bay strata seems to demand that it was deposited after the Gondwanan Orogeny (Dalziel et al. 1975). In addition, there is petrographic evidence indicating that the two units interfinger, and compositionally identical strata in South America contain Lower Cretaceous fossils (Winn 1975).

To the southwest of the flysch - greenstone belt are the intermediate-composition plutons of the Patagonian Batholith, which are mostly Cretaceous (Halpern 1973). The batholiths are thought to represent the roots of a calcalkaline volcanic arc, which formed on a sliver of pre-Mesozoic continental crust after rupture of the continent edge (note basement west of the batholith in Figure 24-1; also see Dalziel et al. 1974b). Volcanic and volcaniclastic rocks on southern Isla Hoste and on the Islas Wollaston may be part of the former surficial arc. The Antarctic Peninsula was probably a continuation of the late Mesozoic arc (Dalziel 1974; Suárez 1976; Suárez and Pettigrew 1976).

The flysch rocks of South Georgia and Tierra del Fuego are the infill of the basin; sediment was derived from cratonic South America to the east and north and from the andesitic - dacitic arc to the west and south (Winn 1975; Winn and Dott 1976). Cessation of basin deposition occurred when the arc collided with the continent during the Andean Orogeny. The early and major pulse occurred in mid-Cretaceous time before intrusion of the 80 million-year old plutons when the basin fill was deformed and metamorphosed up to prehnite-pumpellyite grade. The most intense deformation and metamorphism, such as along the Beagle Canal and along the northeast coast of South Georgia Island (see Winn 1975, Table 8), seem to be the result of faulting associated with aborted obduction of the marginal basin onto the craton (Bruhn and Dalziel 1977). South Georgia was translated relatively eastward during the mid and late Cenozoic (Winn 1975; Dott 1976).

BASIN FILL

Bedding Characteristics

The flysch-like units are made up dominantly of interbedded graywackes and mudstones; diamictites, small slumps, conglomerates, and sedimentary breccias are also present. Grading is common in the sandstones, as are sole markings, flames, convolute bedding, and rip-up mudstone intraclasts. Sandstones are all poorly sorted and most can be described using Bouma-interval terminology (Bouma 1962). The Yahgan Formation and Cumberland Bay strata contain quartz-prehnite beds that represent former tuffs and tuffaceous sediment.

The Sandebugten on South Georgia Island is about 60 percent mudstone, and the sandstones tend to be thin (mostly less than 50 cm). Estimating the percentages of Bouma interval types is difficult because of extreme cleavage, but a full range of lower (top missing) to complete (T_{a-e}) to upper (base missing) Bouma sequences is present. The Sandebugten superficially resembles the Cumberland Bay strata, although the latter are coarser. The Cumberland Bay comprises 50 to 60 percent sandstone and most of the beds display Bouma *a, ae, ab,* and *abe* sequences. Bed amalgamation of graded or ungraded *a* beds or *b* beds is uncommon. Channeling on a large scale was not observed, and most beds in both units tend to vary little for tens of meters across the outcrops. Evidence presented later suggests that the Sandebugten and Cumberland Bay interfinger. The Upper Cumberland Bay contains mafic flows, agglomerates, and sedimentary breccias interbedded with very coarse sandstones in the top of the sequence (Trendall 1959; Pettigrew and Willey 1975; Suárez and Pettigrew 1976).

Near and north of the Beagle Canal the Yahgan Formation of South America consists mostly of laminated mudstone to fine sandstone. The sandstones here are mostly upper Bouma-interval (*cde, de*) and ripple laminated beds, although very thin, graded *a* interval sandstones are present. The unit becomes sandier to the south with complete and base cut-out Bouma sequences dominant on central Isla Navarino and central Isla Hoste (see Figure 24-1). A prominent breccia - sandstone lens, interbedded with the more typical turbidite lithologies, is present in southwest Isla Navarino.

The Tekenika Beds concordantly overlie the Yahgan on Isla Hoste (Figure 24-1) and grade upward from a mudstone - sandstone sequence to a zone of sandstone and

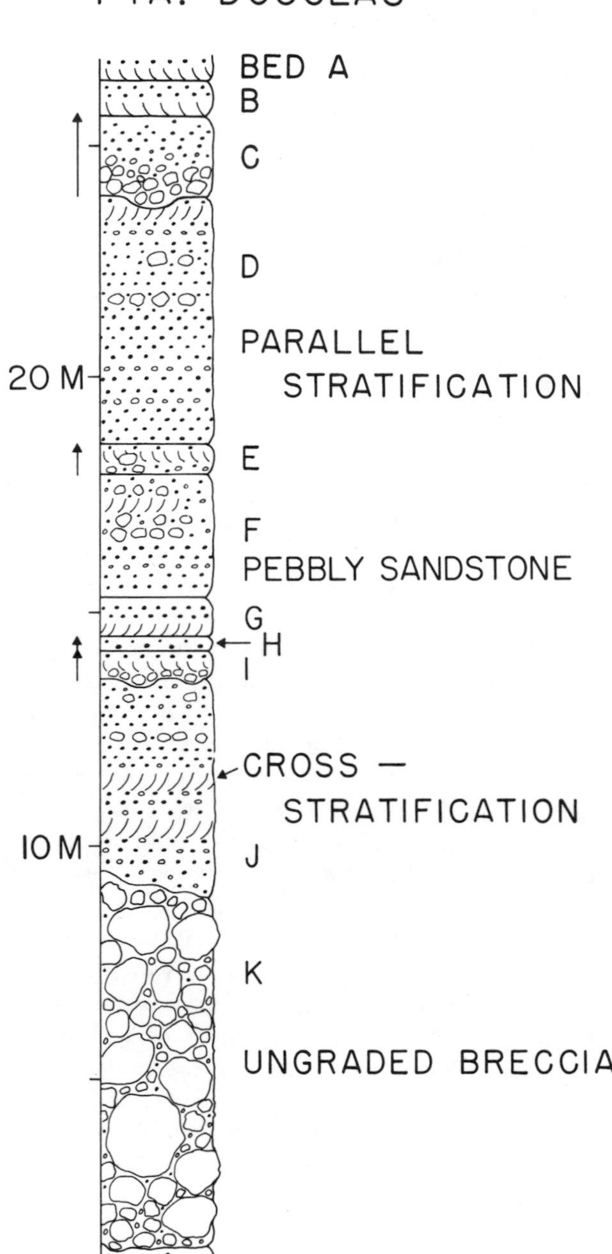

Figure 24-3. Partial columnar section of lens of coarse andesitic sandstones and breccias in Yahgan Formation around Punta Douglas, western Navarino Island, Tierra del Fuego, Chile. Note: Arrows indicate clear graded bedding; lettered units are referred to in text.

Figure 24-4. Graded andesitic conglomerate in Punta Douglas section (bed I, Figure 24-3). Note: Sharply scoured base, crudely graded lower part passes up into cross-stratified and parallel-stratified sandstone reminiscent of the Bouma sequence.

conglomerate (Figure 24-2). The unit, being mostly sandstone and conglomerate, is coarser than the Yahgan. The finer-grained, more thinly bedded basal zone is identical sedimentologically to the finer-grained sections of the Yahgan, Sandebugten, and Cumberland Bay. Laminated and ripple-laminated mudstones, siltstones, and very fine sandstones predominate. Graded and massive *a* interval and parallel- and cross-stratified sandstones interbedded with conglomerates are dominant in the upper portions of the Tekenika Beds. Resedimented plant debris is now preserved as thin lignitic coal seams and stringers in the tops of many graded sandstones. Many beds also contain displaced shallow marine shell fragments.

Occurrence of Conglomerates and Breccias

Interbedded with the rhythmically alternating graded graywackes and mudstones in the upper parts of the Yahgan Formation, Tekenika Beds, and Cumberland Bay strata are conspicuous coarse conglomerates and breccias (Winn and Dott 1975). The coarsest Cumberland Bay examples, exposed on Annenkov Island just southwest of South Georgia, were not observed by us in the field. The upper part of the succession there is reported to consist predominantly of massive cobble and boulder breccia beds, commonly tens of meters thick, alternating with lenses of coarse sandstone (Pettigrew and Willey 1975; Suárez and Pettigrew 1976).

Breccias, with clasts up to 2.5 m in diameter, interbedded with finer-grained clastics are present in the Yahgan (Figure 24-3). Lateral relations of the sequence could not be determined. The coarser clasts were all derived from intermediate volcanics. Individual beds are thick (up to 10 m) and are commonly channeled into underlying units. The conglomerates and breccias are massive (bed K, Figure 24-3) or graded (beds C, E, I, Figure 24-3; Figure 24-4). Imbrica-

Figure 24-5. Cross-stratification in coarse, andesitic sandstone of the Punta Douglas section (e.g., bed J, Figure 24-3). Note: Scale of cross-stratification is unusually large compared to that typical of flysch sequences (hammer = 27.5 cm long).

Figure 24-6. Well-developed imbrication in volcanic conglomerate of the Tekenika Beds on Burleigh Peninsula, Isla Hoste, Chilean Tierra del Fuego. Note: Top of photograph parallels bedding; height of view is about 25 cm.

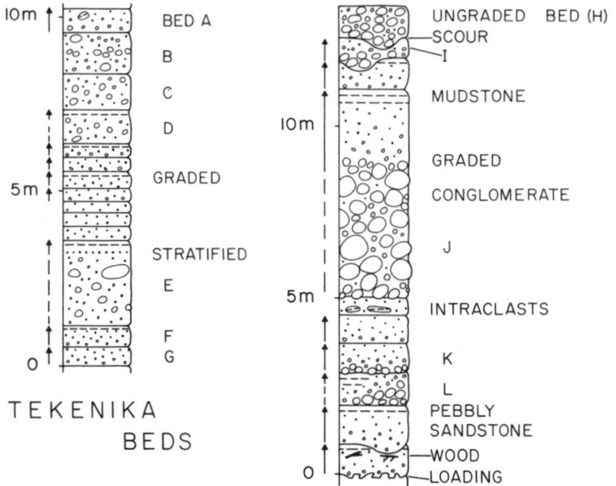

Figure 24-7. Representative partial columnar sections of conglomeratic Tekenika Beds. Note: South side Allen Gardner Bay is shown at left; south side Navidad Bay is shown at right. Symbols are the same as in Figure 24-3.

tion is not apparent and the great irregularity of clast shapes makes it doubtful that a significant fabric exists. Parallel stratification and/or medium-scale cross-stratification are present in the coarse sandstones to pebble conglomerates interbedded with the breccias (Figure 24-5); dish structures were not observed. The various structures may be present within the same bed and they commonly grade laterally and vertically (bed F, Figure 24-3). Cobbles are sometimes present within the sandstone beds (beds D, E, F, J, Figure 24-3).

The upper Tekenika Beds also are largely conglomerate and sandstone. These conglomerates tend to be well rounded, and imbrication is obvious in some beds (Figures 24-6 and 24-7). The clasts show a greater variety than the Yahgan Formation breccias (90 percent intermediate to mafic, 4 percent silicic volcanics, 5 percent gabbro clasts, and 1 percent sedimentary; Halpern and Rex 1972). In general, the Tekenika conglomerates also differ from the Yahgan breccias in displaying structures that are more readily comparable with those of the typical sandstones of flysch-like sequences. The conglomerates are graded or massive (beds A, E, I, J, and K, Figure 24-7), and some units are very poorly sorted (bed D). Both parallel- and cross-stratification are present and intraclasts are common. One large log, 4 m in length, was observed in a conglomerate bed.

Trace Fossils

The trace fossils found in all units were clearly made by more or less systematic deposit feeders (Winn 1975). *Phycosiphon, Scalarituba?, Lophoctenium?, Helminthopsis, Chondrites, Taenidium, Gyrochorte,* and *Palaeophycos* are present in the Cumberland Bay (also see Wilckens 1947); *Chondrites* and *Scalarituba* are present in the Sandebugten; and *Chondrites, Phycosiphon, Zoophycos* and *Helminthopsis* ichnogenera were identified in the Yahgan Formation. No forms were noted in the Tekenika Beds. The assemblage generally corresponds to the *Chondrites* facies of Chamberlain (1971a, 1971b), which he believed typical of bathyal depths. No characteristically shallow-water forms were found.

SEDIMENT GENESIS

Submarine Fan Deposition

The greatest volume of modern turbidite sequences is found on submarine fans or cones built at the bases of continental slopes (see Nelson and Kulm 1973; Normark

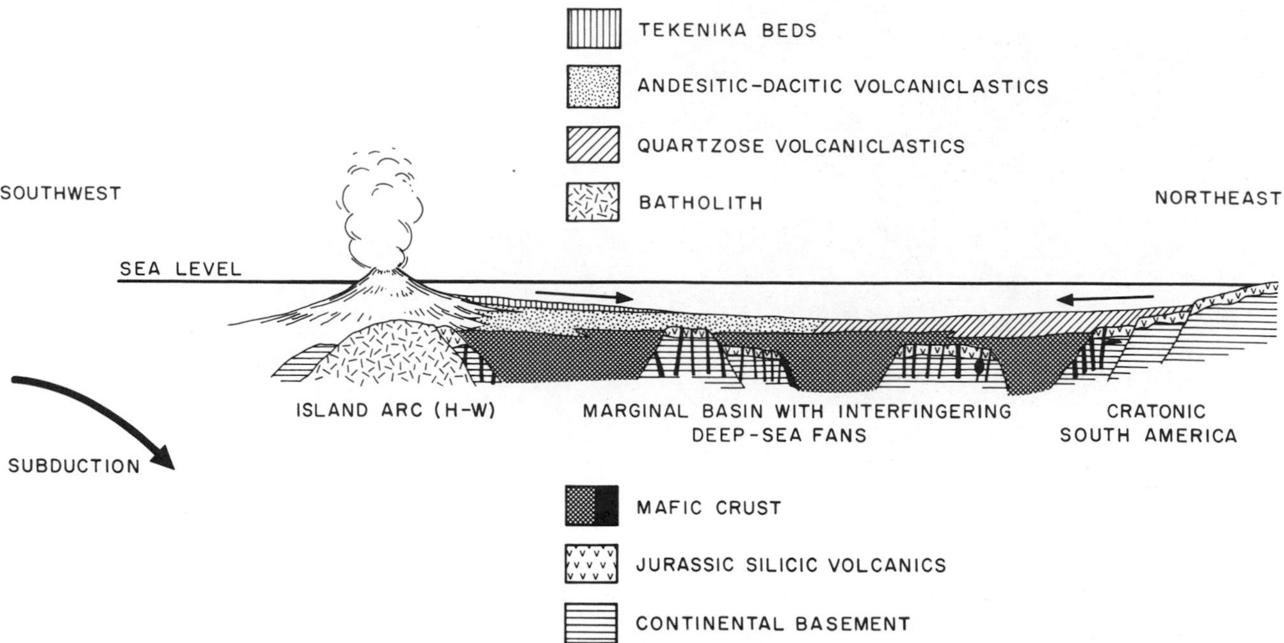

Figure 24-8. Restoration of the Early Cretaceous marginal ocean basin of Tierra del Fuego showing development of two deepsea fan systems over horsts of continental basement and graben of ophiolite basement. (Source: Adapted from Dalziel et al. 1975).

1974). The area and volume of modern deepsea fans supports the analogy for ancient deposits (Mutti 1974), and specifically for the Cumberland Bay, Sandebugten, Yahgan, and Tekenika strata. The southern Andean basin fill has deeper water trace fossils that are compatible with the ocean depths in which contemporary fans are being deposited, and units generally lack typical shallow-water bedding styles and structures. The bedding styles and structures of the greater volume of the Tierra del Fuego and South Georgia rocks are most typical of mid-fan deposition (as outlined by Walker 1970; Walker and Mutti 1973; Nelson and Nilsen 1974; Normark 1974). The coarsest deposits may correspond to inner fan channels. Petrographic evidence supported by paleocurrent data will show in a later section that the basin was being filled by fans growing inward from both sides.

Source Terranes

Petrographic analysis of the sandstones and clast composition of the conglomerates and breccias indicates that the clastics in the Cumberland Bay, Sandebugten, Yahgan, and Tekenika were derived from two distinct source areas (Figure 24-8). Most debris was derived from an active calc-alkaline arc, while lesser volumes originated from the silicic volcanics covering the craton (Tobifera Formation). The two modes have the following characteristics:

1. *Intermediate-volcaniclastic mode.* All of the Tekenika, over 95 percent of the Yahgan and Cumberland Bay, and a few Sandebugten sandstones fall in this category (Figures 24-9 and 24-10). This type consists dominantly of volcanic lithic fragments (Figure 24-11) with lesser amounts of plagioclase, small quantities of quartz, and uncommon mafic minerals (see Winn 1975). This mode has petrographic characteristics that indicate derivation from a moderately siliceous, but still intermediate-volcanic terrane (plagioclase/total feldspar ≈ 1, volcanic fragments/total lithics ≈ 1, mica content = 0, mafic mineral content < 1 percent, and the sandstones have appropriate QFL ratios; cf. Dickinson 1969). Almost all the rock fragments in these sandstones are andesitic and dacitic flow and tuff fragments. Basaltic and rhyolitic fragments are much less common.

2. *Quartz-rich mode.* This mode is represented by a few of the Yahgan and Cumberland Bay sandstones, interbedded within the intermediate volcaniclastics, together with more than 95 percent of the Sandebugten sandstones. The clast content of these sandstones is made up of subequal amounts of quartz, feldspar (chiefly plagioclase but some orthoclase and rare microcline), and lithic fragments (Figure 24-12). These sands were derived almost exclusively from a volcanic terrane ($P/F \approx 0.9$, $V/L \approx 1$, $M = 0$, mafic mineral content < 1 percent), but this terrane had a distinctly different composition from the type supplying the other sands. The quartz-rich mode covers a different area of the QFL diagram and has very different lithic fragment types. The rock fragments are almost all rhyolitic and silicic volcanics, mosaic quartz, and quartzose sedimentary grains. These clasts have identical counterparts in the silicic volcanic rocks of cratonic South America (Tobifera

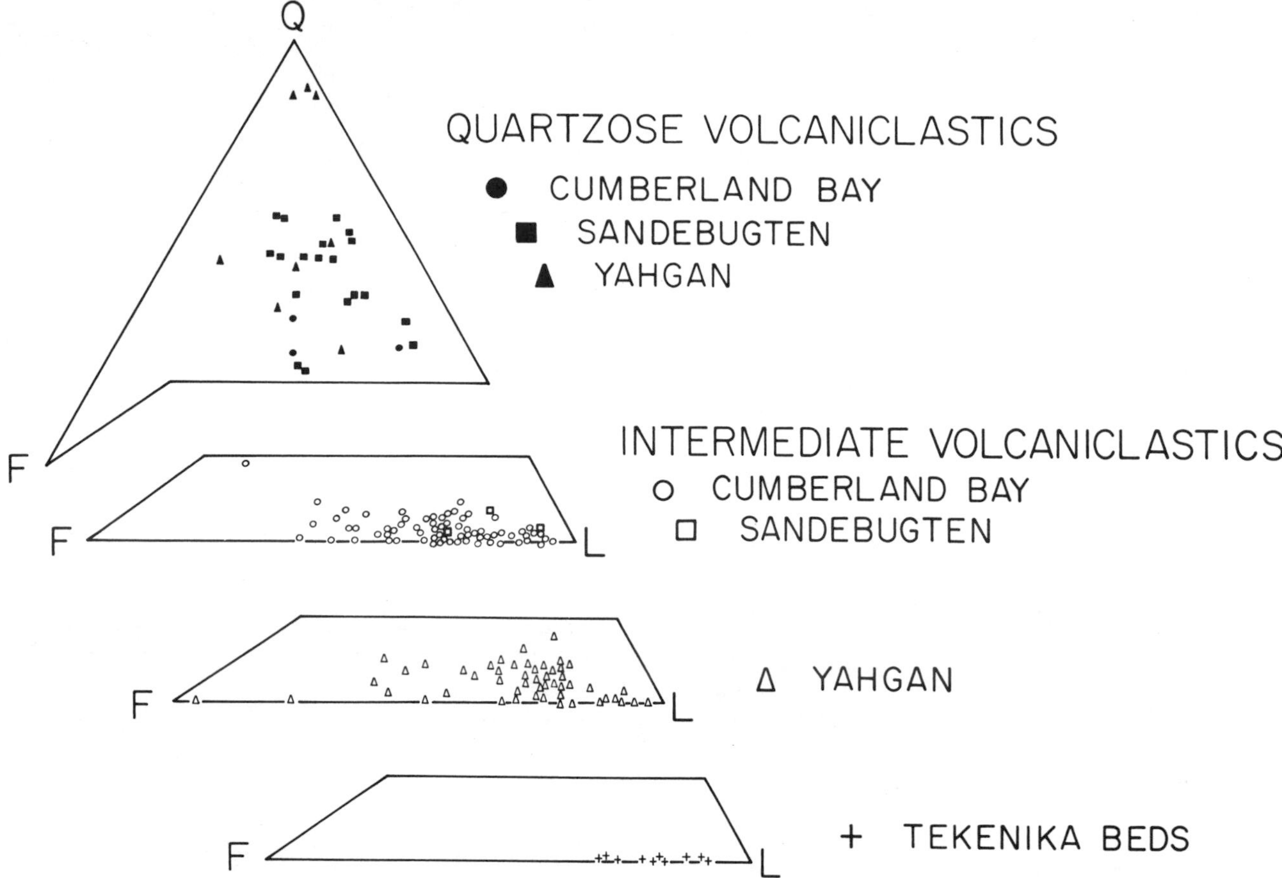

Figure 24-9. Triangular quartz-feldspar-lithic (chiefly volcanic) diagrams comparing sandstone compositions for Mesozoic formations in Tierra del Fuego and South Georgia. (Source: Winn 1978)

Formation), which are their obvious source. The uncommon mafic minerals include biotite, epidote, and amphibole. Trace heavy minerals include zircon, apatite, tourmaline, and sphene. The microcline and some of the trace heavies suggest a minor contribution from plutonic rocks, such as are known to be associated with the superficial Tobifera volcanics. Suárez and Pettigrew (1976, p. 317) also cite minor schist and gneiss fragments in the Sandebugten as reported by P. Stone (in press).

Very striking is the distinct difference between the two modes described above. In only a few sandstones was the source ambiguous. Some of the quartzose sandstones in the Yahgan and Cumberland Bay may have been derived from minor silicic volcanics within the andesitic - dacitic arc, and conversely, some of the intermediate-composition volcaniclastics in the Sandebugten and within the quartzose zones in the Yahgan Formation were probably derived from minor, slightly mafic extrusives in the Tobifera. The identity in composition of the Yahgan and Cumberland Bay quartzose sandstones with that of the Sandebugten and the presence of quartzose sedimentary fragments in many quartz-rich sands of the Yahgan and Cumberland Bay argues against derivation of all of these sands from the arc. These sedimentary grains are made up almost entirely of volcanic quartz, many of which are well rounded, with some plagioclase and silicic volcanic grains. Such grains were derived from preexisting sedimentary rocks and almost certainly from sediments interbedded in the Tobifera volcanics.

Paleocurrents

Paleocurrent analysis was based mostly on ripple cross-laminations and grooves from all units. Additional evidence was obtained from a few flutes (in the Yahgan and Cumberland Bay), from medium-scale cross-stratification and a rippled surface in the Yahgan, and from conglomerate imbrication in the Tekenika beds. Bipolar data was resolved by considering the cross-laminations and the few flutes.

The paleocurrent analysis confirms the source area directions determined petrographically (see Figure 24-1). In the reconstructed position of South Georgia Island southeast of Tierra del Fuego, the dominant southern paleo-

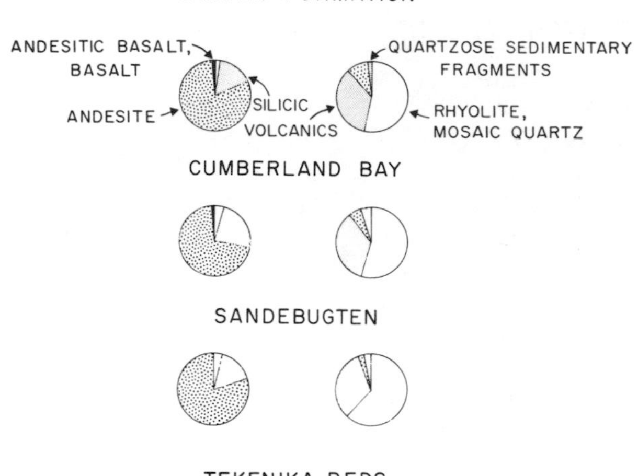

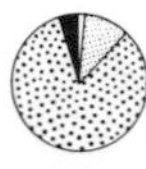

Figure 24-10. Comparison of lithic volcanic and quartz detritus in same formations as Figure 24-9. (Source: Adapted from Winn 1978)

current transport mode of the Sandebugten is consistent with a continental South American source to the north. Most of the Yahgan, Cumberland Bay, and Tekenika beds were deposited by currents generally flowing north or northwest. Quartzose beds in the lower part of the Yahgan Formation are in the zone of dominantly longitudinal fill (see Winn 1975, Figure 4).

DEPOSITIONAL MODELS FOR DEEPWATER CONGLOMERATES

The mechanics of transport and deposition of deepwater, very coarse clastics are less well understood than those responsible for the deposition of deepwater sands. The ideal turbidite bed, a sandstone unit displaying Bouma *a* through *e* divisions, is thought to be the result of deposition from a sediment - water mass moving downslope, driven by gravity, in which sediment appears to be supported mostly by water turbulence. As the current wanes, material drops from suspension and is subject to tractive movement by the still-active, though diminished, current overhead. The *a* interval is deposited too quickly for bedforms to develop, but the Bouma *b* through *d* divisions are the result of slower tractive deposition (Walton 1967). The final phase is the deposition of clay-sized material from the turbidity current tail, followed by pelagic fallout.

Figure 24-11. Microphotograph (approximately 25x) of andesitic sandstones typical of Cumberland Bay and most of Yahgan formations.

Sediment, even in dominantly sandy flows, is not kept in suspension by water turbulence alone. Other processes appear to be dominant, or at least very important, to explain the more unusual sedimentary structures and bedding styles within turbidite sequences (see Middleton and Hampton 1973). Very coarse clastics in particular have been thought difficult to ascribe to turbulent support alone (Davies and Walker 1974; Walker 1975, 1976). Support of material in sediment gravity flows can be conceptualized as due to five overlapping and interrelated mechanisms: (1) *fluid buoyancy* in a heterogeneous-sized sediment - water mixture in which larger clasts "see" smaller clasts as part of the fluid; (2) *fluid viscosity;* (3) *yield strength* in clay - water mixtures; (4) *fluid turbulence;* (5) direct *grain-to-grain stresses* among clasts. The internal structures of sediment gravity flow deposits appear to be determined chiefly by the relative importance of these support mechanisms at the final stages of transport, and by deposition rates.

Walker (1975, 1976) has argued that prior to speculating about flow mechanisms for the less-studied deepwater conglomerates, *descriptive* vertical models should be established. He suggested that four categories would encompass deep-water, clast-supported conglomerates (see Walker 1975, Figure 5): (1) *Inverse-to-normally* graded beds formed from cobbles and boulders consist of an inversely graded zone overlain by a normally graded zone. Imbrication is present in about half of such beds, and characteristically, A axes of elongate clasts dip upstream. Stratification is absent. This type grades into the (2) *simple graded* beds and (3) *graded-stratified* beds of pebbles and sand. Typically, the latter type consists of a normally graded basal zone overlain by "crude" oblique stratification, parallel stratification, and/or medium-scale cross-stratification. (4) The *disorganized bed* model lacks

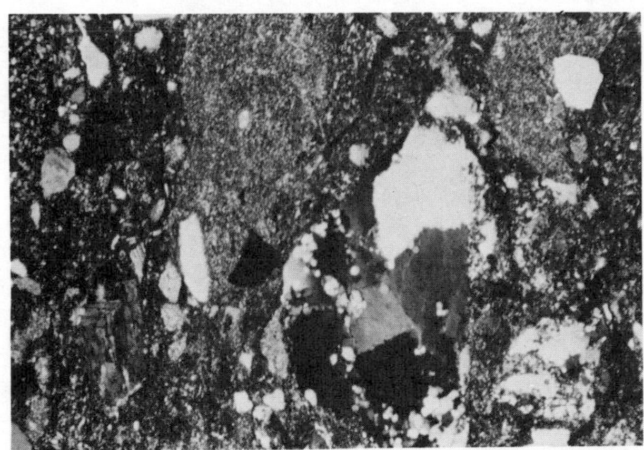

Figure 24-12. Microphotograph (approximately 25x) of quartzose sandstone typical of the Sandebugten and lowest Yahgan formations. Note: Porphyritic rhyolite and mosaic quartz clasts, both typical of Jurassic Tobifera silicic volcanics (Figure 24–2) are shown. (Source: Winn 1978)

grading and appears not to pass into the other two; imbrication is present only in a few examples. Coarse gravels are also found in deepwater settings within chaotic *diamictites* (very poorly sorted to unsorted mud-sand-gravel mixtures), and in at least one other class of clast-supported conglomerate – that is, the lensoid, well-sorted, imbricated, *cross-* and *parallel-stratified conglomerates* of up to cobble and boulder sizes recognized in the Silurian of Ireland (Piper 1970) and in the Upper Cretaceous in southern Chile (Winn and Dott 1977). The latter superficially resemble fluvial deposits.

Several of the above classes are represented in the Yahgan and Tekenika beds and presumably in upper Cumberland Bay breccias. Diamictities are present within coarse clastic zones and within finer-grained facies. Disorganized conglomerates (bed K, Figure 24-3; bed B, Figure 24-7) and inversely graded conglomerates are present, although uncommon. Graded-stratified beds (beds C, E, and I Figure 24-3; bed E, Figure 24-7) are well represented. Many conglomerates are graded but lack overlying tractive sandstone and conglomerate structures, although they may be overlain by siltstones and mudstones (beds A, I, J, and K, Figure 24-7). Conversely, a large number of the Tekenika Beds and Yahgan Formation conglomerates and pebbly sandstones have tractive structures without basal massive or graded conglomeratic zones. Such structures include well-developed, medium-scale, individual cross-sets up to 30 cm thick (Figure 24-5). Rare, medium-scale cross-sets also occur in coarse sandstones isolated within more typical, finer-grained sequences of the Yahgan.

The greater volume of conglomerates to pebbly sandstone in the coarse Tekenika Beds and Yahgan Formation almost certainly was deposited from traction, probably at the base of turbidity flows. The cross-stratification appears to be due to dune deposition rather than the result of the infilling of scour-hollows. Within the coarse Yahgan lens, apparent massive bedding, grading, and parallel- and cross-stratification commonly merge laterally (bed F, Figure 24-3), thereby suggesting deposition from the same flow. Most of the units in the Yahgan contain complex vertical alternations between structures. These beds lack pronounced bedding plane separations between structures and represent several surges of the same flow, or multiple flows closely related in time, or the amalgamation of deposits from several events. Turbulent flows are known to be fully capable of transporting cobbles and boulders (see Komar 1970; Winn and Dott 1977) and may be responsible for much of the graded or massive conglomerate (see Chapter 8).

The attempts by Walker and others (Walker and Mutti 1973; Davies and Walker 1974; Walker 1975, 1976) to differentiate the vertical sequence types in conglomerates is commendable, but we believe the classification may hinder and constrict the study and hydrodynamic interpretation of such deposits. An important fact to remember is that the structures in conglomerates are not distinct from those in sandstones. Sandy mudstones are common as are pebbly mudstones (diamictites). Sandstones and conglomerates with clast-supported fabrics generally display the same structures, including flames, grooves, flutes, graded and massive bedding, reverse grading, imbrication, and parallel- and cross-stratification. Many of the Tekenika conglomerates are closely analogous to graded or massive *a-* and *ae-*interval sandstone beds (see Figure 24-7). Medium-scale cross-stratification in sandstones and pebbly sandstones formed by deposition in dunes is common in the Tierra del Fuego examples, while large-scale cross-stratification in conglomerates has been reported elsewhere (up to 4 m high; Winn and Dott 1977). Some conglomerates have structure sequences identical to the Bouma interval progressions in sandstones. The graded basal conglomeratic zone in Figure 24-4 was probably deposited too rapidly from a turbulent flow for bedforms to develop, while the overlying parallel- and inclined-stratification in the pebbly sandstone resulted from slower (and/or thinner) current deposition. Similar structures suggest similar conditions of transport and deposition for sands and gravels.

Unfortunately, in a like manner, the almost universal acceptance and use of the Bouma sequence may have hindered our study of the processes occurring in the transport of deepwater sands. The Bouma sequence was meant as a descriptive model *only* (Bouma 1962), but the model has since become inseparable from genetic interpretation. In a subtle way it causes workers to think of deepwater transport as different in type from sediment transport in other environments. Not only are the sediment support processes identical (see Middleton and Hampton 1973), but flow parameters also may be almost identical in certain cases. The movement and deposition of sand or gravel are similar for dunes or flat beds regardless of whether deposition occurs at 2,000 m above or 2,000 m below sea level.

CONCLUSIONS

Calc-alkaline arcs tend not to be well preserved. Usually, only their debris and the intermediate plutons intruded along the locus of the former arc remain. The occurrence of huge volumes of andesitic and dacitic volcaniclastics is a reliable indicator of a former arc and thus of a former convergent plate margin and subduction zone. If the debris is flysch-like and located on the continent side of the magmatic arc (or arc roots), the area was almost certainly the site of an oceanic or quasi-oceanic basin. Basins floored by continental crust are less likely to accumulate thick sections composed exclusively of sediments deposited entirely below wave base. Back-arc basins are characterized by bilateral filling from two source areas – from the arc and from the craton. Interfingering of these deposits, as in the southern South American example, or later structural juxtaposition of the two types due to convergence of the arc and the continent may be diagnostic of a back-arc marginal basin even if the area later is strongly affected by lateral translation and tectonic relations are not clear.

Volcanic conglomerates tend to be common on the arc side of arc-rear basins. Careful attention to their bedding characteristics, together with the context of associated strata and fossils, can help distinguish deepwater fan deposition from shallow marine and nonmarine gravel deltas or alluvial fans.

ACKNOWLEDGMENTS

The research for this study was made possible by Grant No. DDP72-05799 A04 from the National Science Foundation. Field work was done around Ushuaia, Argentina, in April 1973 and in Chilean Tierra del Fuego from NSF's R/V *Hero* in May-June 1974. The authors thank D. J. Stanley for encouragement in preparing this chapter. Discussions with R. G. Walker, J. B. Southard, and J. C. Harms were helpful in assessing field observations. Columbia University colleagues in the Tierra del Fuego work, I. W. D. Dalziel, R. L. Bruhn, M. J. de Wit, and M. Winslow have all contributed valuable criticism.

REFERENCES

Anderson, P., 1976. Ocean crust and arc-trench gap tectonics in southwestern British Columbia. *Geology,* 4: 443-46.

Barker, P. F., 1972. A spreading center in the east Scotia Sea. *Earth Planet. Sci. Lett.,* 15: 123-32.

Bouma, A. H., 1962. *Sedimentology of Some Flysch Deposits.* Elsevier, Amsterdam, 168 pp.

Bruhn, R. L., and Dalziel, I. W. D., 1977. Destruction of the Early Cretaceous marginal basin in the Andes of Tierra del Fuego. In: M. Talwani and W. C. Pitman III (eds.) *Maurice Ewing Symposium, series I, Island Arcs, Deep Sea Trenches, and Back-Arc Basins.* Am. Geophys. Union, pp. 395–406.

Casey, R., 1961. A monograph of the Ammonoidea of the Lower Greensand, Part 2. *Palaeontogr. Soc.* London, pp. 45-118.

Chamberlain, C. K. 1971a. Bathymetry and paleoecology of Ouachita Geosyncline of southeastern Oklahoma as determined from trace fossils. *Bull. Amer. Assoc. Petrol. Geol.,* 55: 34–50.

———, 1971b. Morphology and ethology of trace fossils from the Ouachita Mountains, southeast Oklahoma. *J. Paleont.,* 45: 212–46.

Coleman, P. J. (ed.), 1973. *The Western Pacific Island Arcs, Marginal Seas, Geochemistry.* Crane, Russak, and Co. and University of Western Australia Press, New York, 675 pp.

Dalziel, I. W. D., 1974. Evolution of the margins of the Scotia Sea. In: C. A. Burk and C. L. Drake (eds.), *The Geology of Continental Margins.* Springer-Verlag, New York, pp. 567-77.

———, Caminos, R., Palmer, K. F., Nullo, F., and Casanova, R., 1974a. The southern extremity of the Andes: Geology of Isla de Los Estados, Argentine Tierra del Fuego. *Bull. Amer. Assoc. Petrol. Geol.,* 58: 2502-12.

———, de Wit, M. J., and Palmer, K. F., 1974b. A fossil marginal basin in the southern Andes. *Nature,* 250: 291-94.

———, Dott, R. H., Jr., Winn, R. D., Jr., and Bruhn, R. L., 1975. Tectonic relations of South Georgia Island to the southernmost Andes. *Geol. Soc. Amer. Bull.,* 86: 1034–40.

Davies, I. C. and Walker, R. G., 1974. Transport and deposition of resedimented conglomerates: The Cape Enragé Formation, Cambro-Ordovician Gaspé, Quebec. *J. Sed. Petrol.,* 44: 1200-16.

Dewey, J. F., and Bird, J. M., 1971. Origin and emplacement of the ophiolite suite: Appalachian ophiolites in Newfoundland. *J. Geophys. Res.,* 76: 3179-206.

Dickinson, W. R., 1969. Evolution of calc-alkaline rocks in the geosynclinal system of California and Oregon. *Oregon Dept. Geol. and Min. Ind. Bull.,* 65: 151-56.

———, 1974. Sedimentation within and beside ancient and modern magmatic arcs. In: R. H. Dott, Jr., and R. H. Shaver (eds.), *Modern and Ancient Geosynclinal Sedimentation.* Soc. Econ. Paleont. Mineral., Sp. Publ. 19, pp. 230-39.

Dott, R. H., Jr., 1976. Contrasts in tectonic history along the eastern Pacific rim. In: G. H. Sutton, M. H. Manghnani, and R. Moberly (eds.), *The Geophysics of the Pacific Ocean Basin and Its Margin.* Amer. Geophys. Union Mon., 19, pp. 299-308.

———, Winn, R. D., Jr., de Wit, M. J., and Bruhn, R. L., 1977. Tectonic and sedimentary significance of Cretaceous Tekenika Beds of Tierra del Fuego. *Nature,* 266: 620–22.

Du Toit, A. L., 1937. *Our Wandering Continents.* Oliver and Boyd, Edinburgh, 366 pp.

Eardley, A. J., 1962. *Structural Geology of North America.* Harper & Row, New York, 743 pp.

Halpern, M., 1973. Regional geochronology of Chile south of 50° latitude *Geol. Soc. Amer. Bull.,* 84: 2407-22.

———, and Rex, D. C., 1972. Time of folding of the Yahgan Formation and age of the Tekenika Beds, southern Chile, South America. *Geol. Soc. Amer. Bull.,* 83: 1881-86.

Hoffstetter, R., 1957. Clay-Slate Formation and Yahgan Formacion. *Lex. Strat. Int.,* (Chile) 5: 71-72; 375-77.

James, D. E., 1971. Plate tectonic model for the evolution of the central Andes. *Geol. Soc. Amer. Bull.*, 82: 3325-46.

Karig, D. E., 1970. Ridges and basins of the Tonga-Kermadec island arc system. *J. Geophys. Res.*, 75: 239-55.

——, 1971. Origin and development of marginal basins in the western Pacific. *J. Geophys. Res.*, 76: 2542-61.

——, 1974. Evolution of arc systems in the western Pacific. *Ann. Rev. Earth. Planet. Sci.*, 2: 51-75.

——, Ingle, J. C., Jr., Bouma, A. H., Ellis, H., Haile, N., Koizumi, I., MacGregor, I. D., Moore, C., Vjiie, H., Watanabe, T., White, S. M., Yasii, M., and Ling, H. Yi, 1973. Origin of the West Philippine Basin. *Nature*, 246: 458-61.

Katz, H. R., 1972. Plate tectonics – orogenic belts in the southeast Pacific. *Nature*, 237: 331.

Kay, M., 1951. North American Geosynclines. *Geol. Soc. Amer. Mem.*, 48, 143 pp.

Komar, P. D., 1970. The competence of turbidity current flow. *Geol. Soc. Amer. Bull.*, 81: 1555-62.

Ludwig, W. L., Ewing, J. I., and Ewing, M., 1968. Structure of Argentine continental margin. *Bull. Amer. Assoc. Petrol. Geol.*, 52: 2337-68.

Middleton, G. V., and Hampton, M. A., 1973. Sediment gravity flows: Mechanics of flow and deposition. In: G. V. Middleton and A. H. Bouma (eds.), *Turbidites and Deep-Water Sedimentation.* Soc. Econ. Paleont. Mineral., Pacific Section, Short Course, Anaheim, pp. 1-38.

Minato, M., Gorai, M., and Hunahashi, M. (eds.), 1965. *The Geologic Development of the Japanese Islands.* T. Shokan, Tokyo, 442 pp.

Monger, J. W. H., Souther, J. G., and Gabrielse, H., 1972. Evolution of the Canadian Cordillera: A plate-tectonic model. *Amer. Jour. Sci.*, 272: 577-602.

Mutti, E., 1974. Examples of ancient deep-sea fan deposits from circum-Mediterranean geosynclines. In: R. H. Dott, Jr., and R. H. Shaver (eds.), *Modern and Ancient Geosynclinal Sedimentation.* Soc. Econ. Paleont. Mineral., Sp. Publ., 19, pp. 92-105.

Natland, M. L., Gonzalez P., E., Cañon, A., and Ernest, M., 1974. A system of stages for correlation of Magallanes Basin sediments. *Geol. Soc. Amer. Mem.*, 139, 126 pp.

Nelson, C. H., and Kulm, L. D., 1973. Submarine fans and deep-sea channels. In: G. V. Middleton and A. H. Bouma (eds.), *Turbidites and Deep-Water Sedimentation.* Soc. Econ. Paleont. Mineral., Pacific Section, Short Course, Anaheim, pp. 39-78.

——, and Nilsen, T. H., 1974. Depositional trends of modern and ancient deep-sea fans. In: R. H. Dott, Jr., and R. H. Shaver (eds.), *Modern and Ancient Geosynclinal Sedimentation.* Soc. Econ. Paleont. Mineral., Sp. Publ. 19, pp. 69-91.

Normark, W. R., 1974. Submarine canyons and fan-valleys: factors affecting growth patterns of deep-sea fans. In: R. H. Dott, Jr., and R. H. Shaver (eds.), *Modern and Ancient Geosynclinal Sedimentation.* Soc. Econ. Paleont. Mineral., Sp. Publ. 19, pp. 56-68.

Pettigrew, T. H., and Willey, L. E., 1975. Belemnite fragments from Annenkov Island. *Brit. Ant. Sur. Bull.*, 40: 33-36.

Piper, D. J. W., 1970. A Silurian deep sea fan deposit in western Ireland and its bearing on the nature of turbidity currents. *J. Geol.*, 78: 509-22.

Rutland, R. W. R., 1971. Andean Orogeny and ocean-floor spreading. *Nature*, 233: 252-55.

Scholl, D. W., Buffington, E. C., and Hopkins, D. M., 1968. Geologic history of the continental margin of North America in the Bering Sea. *Mar. Geol.*, 6: 297-330.

Shor, G. G., Kirk, H. K., and Menard, H. W., 1971. Crustal structure of the Melanesian area. *J. Geophys. Res.*, 76: 2562-86.

Stepanicic, P. N. and Reig, O., 1955. Breve noticia sobre el hallazgo de anuros en el denominado Complejo Porfirica de la Patagonia extra-Andina con consideraciones acerca de la composicion geologica del mismo. *Rev. Assoc. Geol. Argent.*, 10: 215-33.

Stone, P. The geology of South Georgia: IV. Barff Peninsula and Royal Bay areas. *Scient. Rept. British Antarctic Survey* (in press).

Suarez, M., 1976. Plate-tectonic model for southern Antarctic Peninsula and its relation to southern Andes. *Geology*, 4: 211-14.

——, and Pettigrew, T. H., 1976. An upper Mesozoic island-arc-back-arc system in the southern Andes and South Georgia, *Geol. Mag.*, 113: 305-400.

Trendall, A. F., 1959. The geology of South Georgia-II. *Falkland Islands Depen. Surv. Sci. Rept.*, 19, 26 pp.

Walker, R. G., 1970. Review of the geometry and facies organization of turbidites and turbidite-bearing basins. In: J. Lajoie (ed.), *Flysch Sedimentation in North America.* Geol. Assoc. Canada, Spec. Paper, 7, pp. 219-51.

——, 1975. Generalized facies models for resedimented conglomerates of turbidite association. *Geol. Soc. Amer. Bull.*, 86: 737-48.

——, 1976. Facies models 2. Turbidites and associated coarse clastic deposits. *Geosci. Canada*, 3: 25-36.

——, and Mutti, E., 1973. Turbidite facies and facies associations. In: G. V. Middleton and A. H. Bouma (eds.), *Turbidites and Deep-Water Sedimentation.* Soc. Econ. Paleont. Mineral., Pacific Section, Short Course, Anaheim, pp. 119-57.

Walton, E. K., 1967. The sequence of internal structures in turbidites. *Scott. J. Geol.*, 3: 306-17.

Wilckens, O., 1947. Palaeontologische und geologische Ergebrisse der eise von Kohl-Larsen (1928-29) nach Süd-Georgien. *Abhand. Senc. Nat. Gesellschaft, Bd.*, 474: 1-66.

Winn, R. D., Jr., 1975. *Late Mesozoic Flysch of Tierra del Fuego and South Georgia Island: A Sedimentological Approach to Lithosphere Plate Restoration.* Ph. D. Thesis, University of Wisconsin, Madison.

——, 1978. Late Mesozoic flysch of Tierra del Fuego and South Georgia Island: A sedimentologic approach to lithosphere plate restoration. *Geol. Soc. Amer. Bull.*

——, and Dott, R. H., Jr., 1975. Deep-water conglomerates and breccias of late Mesozoic age in Chilean Tierra del Fuego. (Abst.) *Amer. Assoc. Petrol. Geol.*, Ann. Meeting Program, 2: 82.

——, and Dott, R. H., Jr., 1976. Submarine-fan sedimentation in Mesozoic arc-rear marginal basin in southern South America. (Abst.) *Amer. Assoc. Petrol. Geol.*, 60: 734.

——, and Dott, R. H., Jr., 1977. Large-scale tractive structures in deep-water fan-channel conglomerates in southern Chile. *Geology*, 5: 41-44.

Part VI

Synthesis and Prognosis

Chapter 25

Sedimentation in Submarine Canyons, Fans and Trenches: Appraisal and Augury

GILBERT KELLING
DANIEL JEAN STANLEY

Department of Geology
University of Keele
Staffordshire, England

Division of Sedimentology
Smithsonian Institution
Washington, D. C.

ABSTRACT

The study of sedimentation in ancient and modern submarine canyons, fans, and trenches is in a state of explosive growth. A review of the principal milestones in the development of this research and an evaluation of factors providing the impetus for the expansion of interest in this field suggest that there are a number of elements that have facilitated this evolution. These include (a) revived recognition of the intimate relationship between sedimentation and structure, brought about largely by the new global tectonics concept; (b) clarification and refinement of some existing terms and concepts that resulted from imprecise appreciation of deep marine environments and sequences; (c) development of conceptual models that systematize the description and interpretation of deep marine sediment sequences; and (d) recognition of the remarkable variety of those sediment processes that operate beyond the shelf edge. The uneven development in understanding aspects of this complex area of research results from disparities both in economic pressures and in the technological capacity to test prevailing hypotheses. Studies of ancient sequences have significantly contributed to our present understanding of outer margin sedimentation. A shift in emphasis to modern equivalents, including trench and arc margin basins, is anticipated and is likely to provide a more balanced and truly actualistic approach and to generate new concepts. These trends are exemplified by the current interest in developing a more sophisticated view of the transportation mechanisms for sands and muds, and in solving problems concerned with the occurrence of deepsea gravels.

INTRODUCTION

If, in retrospect, we can assign a date for the inception of concerted and organized research in the environments that are the concern of this volume, it probably should be placed somewhere between the early 1940s and late 1950s. That period witnessed the publication of an important series of detailed surveys of deeper marginal regions, most of which had a strongly morphological emphasis (Shepard and Emery 1941; Menard 1955; Heezen et al. 1959). The mid-1950s also was the period when, thanks primarily to the evangelistic efforts of Ph. H. Kuenen, the geological community began to realize that the thick piles of sand found in modern and ancient marine sequences were representative of deep sea environments (Heezen and Ewing 1952) equally as much as the dark muds and radiolarian oozes previously considered diagnostic of that sedimentary milieu.

During the intervening quarter-century, much has been learned concerning the processes and products that characterize this sector of the marine realm. Thanks to expanded research efforts, a more refined and flexible technology, and increased socioeconomic pressures, the past decade has seen a virtual quantum jump in our understanding of sedimentation in this area. Consequently a broad review of the state of our knowledge of sedimentation in submarine canyons, fans, and trenches – especially a review that includes both modern and ancient examples – appears both appropriate and opportune.

The collection of contributions gathered together in this volume is broadly representative, if not entirely comprehensive, of contemporary interests in this area of sedimentological research, and in this concluding chapter we attempt to assess the development, current status, and future growth of information and concepts relating to deeper marine margin environments. We trust that the reader will find this exercise to be a stimulant and not a soporific.

CANYONS, FANS AND TRENCHES: THE STATE OF PLAY

The study of this trinity—canyons, fans, and trenches—predates even the classic oceanographic cruise of the *Challenger* a century ago. During the intervening period much knowledge has been gained, but the advances have been achieved in rather uneven fashion and have favored logistically attainable zones.

The morphology and origin of submarine canyons early excited interest that culminated in a series of fierce debates prior to and shortly following World War II, led particularly by D. W. Johnson, R. A. Daly, F. P. Shepard, and Ph. H. Kuenen. The varying views on canyon genesis have been ably summarized by Francis P. Shepard, one of the pioneer proponents (Shepard 1963; Shepard and Dill 1966). Subsequently, the emphasis in the study of canyons has shifted to their role as sediment traps and funnels conveying sediment to the deep sea. Despite intensive (if rather localized) efforts directed to understanding processes operating within these features, our knowledge of these aspects remains limited both in the geographic sense (mostly heads of canyons; cf. the papers in Miller 1965) and in terms of the dynamics of downslope sediment transfer. Here, paradoxically, the ancient record still appears to be providing a key to the present.

The growth of concepts largely derived from the study of ancient analogs (cf. Whitaker 1976) has surpassed the technology required to test and quantify these sedimentological principles in the difficult environment of these submerged Grand Canyons. Parts I and II of the present volume provide an indication of the attempts now being made to throw light on this obscure area. While the advent of direct "ground verification" via submersibles and underwater television are obvious boons, they do not necessarily reveal the intricate plexus of physical, chemical, and biological processes that are presumably responsible, in the long term, for the downslope transfer of sediment (see Chapter 8).

In contrast to submarine canyons, the extensive features we now term deepsea fans eluded attention until the middle of this century. Thus, classic marine geological treatises such as Trask (1939), Shepard (1948), and Kuenen (1950) leave this topic untreated. The publication of large-scale physiographic maps of the ocean floor, so dramatically depicted in the work of Bruce C. Heezen and Marie Tharp, drew attention to these base-of-slope provinces. The generally muted topography of deepsea fans and their very magnitude may account for the failure of earlier workers to recognize the significance of these features. The quality, quantity, and density of bathymetric data required to highlight such features arose primarily from military needs during and following the Second World War and were supplemented by the intensive efforts of the U. S. Navy and the major oceanographic institutions (Nelson and Kulm 1973; cf. articles in Middleton and Bouma 1973; Dott and Shaver 1974). Simultaneously, limited knowledge of the internal structure and evolution of fans and rises has been gained through the Deep-Sea Drilling Project of the National Science Foundation.

At the same time, sedimentologists began to distinguish ancient canyon–fan systems (e.g., Stanley 1961; Walker 1966). This approach grew out of an attempt to refine the basinal studies of ancient deep marine sediments that had been stimulated by the turbidity current vogue (Dzulynski and Walton 1965) and paleocurrent reconstructions (Dzulynski et al. 1959) of the late 1950s and early 1960s. Subsequent studies, especially in mobile belts, have progressively detailed the criteria for recognizing fans and their subenvironments (Mutti and Ricci Lucchi 1972; Mutti 1974; Ricci Lucchi 1975). Since more is known of the depositional patterns in both modern and ancient deepsea fans than in the associated submarine canyons and slopes, we need to be aware that the examples most intensively studied to date are drawn from a relatively restricted geotectonic spectrum. Are these large fans of the Pacific and Indian Oceans truly representative of all such features? Are fans of this type likely to be preserved and ultimately exposed on land? Or are the smaller canyon-fan systems developed in more restricted marginal basins, such as those on the California Borderland (Gorsline and Emery 1959; Moore 1969), more likely to be encountered, or at least recognized, in the rock record?

Also important is directing greater attention to the search for fossil fans in less mobile settings, including cratonic situations (see Chapter 16). As in the case of canyons, the rock record provides directly observable details of fan organization that are based on time-dependent factors which are less readily assessed in modern fans with our existing technology. The above considerations underline the need for combined analyses of modern and ancient examples (cf. von Rad 1968).

Although recognized long before deepsea fans as major features of the ocean floor (Murray and Hjort 1912), trenches have attracted attention up to the present mainly in terms of their geophysical (cf. Vening Meinesz 1948; Hess 1938; Stille 1955) and geomorphological (Fisher 1974) attributes. In spite of an increasing number of submersible dives (Picard and Dietz 1961; Heezen and Rawson 1977) and more refined seismic surveys (see Chapter 21) and a few DSDP drill cores, our knowledge of the detailed processes of sedimentation has scarcely progressed beyond the statement by Revelle et al. (1955): "The flat bottoms of certain trenches . . . represent 'sediment lakes' filled with and forming catchment troughs for such (turbidity current) deposits" (p. 227). Few studies have adduced trench sedimentary processes in any detail (Conolly and Ewing 1967). This remarkable dearth of sedimentological data must be alleviated since trenches play such a vital role in the scheme of global tectonics, particularly in evaluating rates of displacement. A program of closely spaced stations is now required to gather the sedimentological samples and physical oceanographic data to be used in conjunction with high-resolution subbottom profiling. Trenches, as presently defined, occur in compressional settings and in terms of modern global tectonics are generally associated with subduction (Fisher 1974). How can we distinguish the sedimentary fill of these features from other deep linear troughs in noncompressional zones or other geotectonic settings (e.g., Red Sea, Gulf of California, and so forth)? To what degree is the internal anatomy of trench wedges comparable with other base-of-slope accumulations (see Chapter 21), and what are the criteria that will enable ancient trench deposits to be recognized?

Any attempt to assess the evolution of ideas on canyon, slope, fan, and trench sedimentation is fraught with difficulty, due to the rapidity of progress in this field: Writing history as it is being made is both rash and presumptuous! Nevertheless, we have attempted to evaluate the broad trends from the literature appearing during the last ten years (1966-1976) in some relevant journals (*Marine Geology, Journal of Sedimentary Petrology, Sedimentology, Bulletin of the American Association of Petroleum Geologists,* and *Geological Society of America Bulletin*). This analysis reveals a marked increase in the total number of marine sedimentological studies published. However, the relative proportions of papers dealing with continental, coastal and shelf, and more general deep marine basinal analyses have decreased, in contrast to a marked increase in the proportion of studies pertaining to deltas. There has been a parallel growth (from about 5 percent to 25 percent) in papers dealing with the outer margin and comprising the environments discussed here. Even this evaluation underestimates the real burgeoning of research in this area. As in the case of other rapidly expanding scientific fields, much of the literature appears in the form of special publications (Lajoie 1970; Dott and Shaver 1974; Dickinson 1974) or symposia (Stanley and Swift 1976) and short course handbooks (Middleton and Bouma 1973; Saxena 1976).

WHY PLAY AT ALL?

Terrigenous sediments attributable to deep outer margin environments constitute a major element of the modern oceans (Emery 1970; Heezen and Hollister 1971), and an increasing number of studies both at outcrop and in the subsurface clearly demonstrate that such sediments also represent a significant volumetric proportion of the geological record. Moreover, a rapidly expanding body of literature shows that ancient sediments of deep marine margin aspect are not confined to mobile belts but also occur in more stable, cratonic contexts (as discussed, for example, by Kepferle and McCabe in this volume). A brief perusal of recent geological publications strongly suggests that many such sequences carrying the hallmarks of modern canyon-fan deposits may have eluded identification (cf. Payne 1976).

We believe that careful evaluation of this trinity of environments is a matter of economic concern inasmuch as subsurface studies have demonstrated that some ancient canyon-fan complexes are important hydrocarbon producers (Chapter 20 of this volume; Parker 1975; Selley 1975; Bloomer 1977). Moreover the increasing thrust towards discovery of stratigraphic traps in the petroleum industry (Chapter 19) underlines the importance of accurately identifying source-rock/reservoir couples and trends. The combination of organic-rich hemipelagic, interchannel muds and silts abutting against coarser, potentially more porous channelized canyon and fan valley sand bodies represents an ideal hydrocarbon target (Wilde et al. 1976). Intensified land-based exploration of ancient traps of this sort, particularly in fields that are already under production, would clearly be economically more feasible than a logistically complicated search in modern deepwater equivalents.

Nevertheless, driven by economic necessity and the technological developments epitomized by the Deep-Sea Drilling Project, exploration for hydrocarbons in modern deep environments is inevitable and imminent. During the initial phases of such exploration, there must be considerable reliance on existing models of sedimentation that are based primarily on ancient analogs and on data derived mainly from Quaternary surficial sediments in canyons, fans, and trenches. At that point, the validity of such models in concepts of canyon-fan sedimentation becomes crucial.

Concomitantly, there is the threat of submergence under our own garbage. Such pressures are most evident off densely populated or highly industrialized coastal regions or in semiclosed bodies of water. Some of the latter group are of shallow, epi- or intracontinental nature (the Baltic and Caspian Seas, to name two areas where such problems are rampant!). Even the deeper enclosed basins, such as the Mediterranean, the Black Sea, the California Borderland and the Gulf of Mexico, are at risk. Here knowledge of processes operating in the canyon-fan systems known to be present is of vital concern to ensure permanent disposal of waste materials, perhaps by means of canyons, and to guarantee their burial in fans or more distal basin plains, without excessive disturbance of the existing ecosystems.

While canyon-related disposal systems are more obviously appropriate to pressure-areas adjoining narrow shelves (Mediterranean), the removal in this way of large volumes of materials, some toxic, soon may have to be effected off broad platforms. A case in point is New York City where the impact of ocean dumping in New York Bight (Gross 1976) has prompted the formation of an interdisciplinary task force (National Oceanic and Atmospheric Administration's Marine Ecosystems Analysis, or MESA, Program). Comparable governmentally funded programs are clearly needed to evaluate and monitor offshore dumping beyond the shelf-edge as well as to minimize the effects of possible blowouts in the course of hydrocarbon exploitation in deeper waters.

The recent proposal (Bostrom and Sherif 1970) to achieve permanent disposal of highly toxic materials, such as long-lived radioactive waste, along actively subducting plate margins appears far-fetched at present (Silver 1972). However, such a procedure could be viable soon in highly industrial or populated sectors. A case in point is Japan, which is bounded on its eastern and southern margins by canyons entering the Japan-Ryukyu trench systems.

This cursory outline has emphasized practical and economic aspects. It highlights the pressing need to document and understand the processes responsible for the transport and deposition of sediment beyond the shelf edge; succeeding sections consider more specific, albeit academic, facets.

THE TECTONIC PERSPECTIVE: GROUND RULES FOR THE GAME

In the scientific climate of the late 1950s, when both of us first became involved in research on ancient deep marine sediments, sedimentology appeared to constitute an end in itself. The field was almost entirely concerned with cataloging features observed at the outcrop and in the laboratory. Understanding of these features and their genetic significance appeared to be the goal of the sedimentologist. Turbidites, sole marks, paleocurrent analyses, and the like were of engrossing interest.

The revolution in the earth sciences brought about by the new global tectonics has had inevitable repercussions in the field of sedimentology, which perforce is now accommodating to this reality. The progression from cataloging to modeling of various sedimentary regimes has generated a broader, more genetic approach to virtually every class of sediment. The types of sequences discussed in this volume are especially susceptible to that most fundamental of genetic controls—that is, tectonics. Viewed in this context, sedimentology is basically a means to the larger end, and sedimentary process and structural factors become so intimately linked that neither can be adequately viewed in isolation. The importance of such links is apparent in several chapters by contributors to this volume (e.g., Carter et al.; Winn and Dott).

Marine geologists and geophysicists have been in the forefront in developing the new concepts of global tectonics. In doing so they have contributed much to the elucidation of the structure and evolution of continental margins. The recent compendium by Burk and Drake (1974) bears eloquent testimony to the fact that on a worldwide basis, subbottom surveys have greatly surpassed strictly sedimentological investigations in deepwater environments. In particular, subbottom profiling has significantly advanced our knowledge of the geometry and gross organization of deep marine sediment bodies.

How can marine sedimentologists contribute to this particular area of research? In terms of modern slope and base-of-slope environments, detailed interpretation of subbottom reflectors ultimately depends on sedimentological identification. An example where this might prove critical in resolving a major geotectonic problem is the currently controversial question of the genesis of the western Mediterranean. Determination of slope and canyon configuration in the late Miocene to Pliocene (cf. Drooger 1973) requires finer resolution through detailed study of sedimentological samples to amplify the presently available data from seismics and Deep-Sea Drilling Project cores. The recent suggestions concerning possible large-scale vertical tectonics in this region (Chapter 18 of this volume; Stanley et al. 1976); and elsewhere may prompt a reevaluation of large-scale crustal movements, now popularly envisaged as predominantly involving lateral displacement. An important related aspect incorporating sedimentological data concerns the rates of sediment accumulation in different geotectonic settings. Knowledge of such depositional rates would help resolve problems in zones of arc margin and trench type (see Chapter 21) and help determine the rate of deformation of the basin and trench fill at subducting margins (Scholl and Marlow 1974).

At present, the nature of the fill in modern marginal troughs such as fore-arc and back-arc basins remains enigmatic, both in terms of composition and detailed anatomy. As a result, ancient sedimentary sequences formed in these two environments are not readily differentiated. Comparable problems exist in other outer margin environments. How, then, can we identify fossil analogs of such geotectonically critical zones, particularly in mobile belts where later deformation has obscured and distorted original spatial relationships? In terms of this dynamic framework, how valid are the simplistic models of sediment provenance and dispersal proposed during the first flush of "turbiditomania"? Interpretation of ancient marginal basins that display longitudinal transport patterns remains a vexing question. Are such linear belts to be attributed exclusively to trench (i.e., compressional) settings (as discussed in studies of the Carpathians and Apennines), or were some of these linear terrigenous bodies generated on more stable margins by coalescence of asymmetrical fans?

The marine sedimentologist working with both modern and ancient deposits formed on mobile margins is in a unique position to resolve a variety of problems relating to the poorly understood area of early deformation of strata. Examples that involve distinction of features produced by penecontemporaneous deformation from those of tectonic origin include the separation of olistostromes (see Chapter 22) from mélanges (cf. Hsu 1974; Aalto 1976), and the development of fissility and early cleavage in soft slope sediments from brittle postlithification fractures (Davies and Cave 1976).

The new approach emphasizing the interrelationship between tectonics and sedimentation is exemplified by papers in the recent volume edited by Dickinson (1974). This trend appears to represent a revival (from a different viewpoint) of the concepts advanced by Krynine and others (see Pettijohn 1957, pp. 611-47) relating sediment composition and broad facies development to tectonic cycles. However, there is a disturbing lack of petrographic data from deposits formed in tectonically active zones such as modern marginal volcanic arcs. Compositional data from such zones is essential to a more detailed understanding of those ancient sequences formerly categorized as miogeosynclinal and eugeosynclinal (Kay 1951).

CALLING AND CLARIFYING THE PLAYS

In the early phases of scientific research, lack of data and imprecision of objectives leads to the use of broad, often equivocal terms that are initially useful but that ultimately may prove to be obstacles to progress. As knowledge increases, the need for such catch-all names declines. The demise of terms such as "eugeosyncline" and "miogeosyncline" has resulted not only from the development of plate tectonics but also from the impact of a more detailed insight into modern, deeper marine environments. Concurrently, long-established terms such as "flysch," valid in the context of earlier knowledge and thus useful as broad descriptors, become redundant as more precise environmental and geotectonic criteria are established from modern equivalents. The particular term *flysch*—the source of much confusion and the consumer of so much ink (see Hsü 1970)—has lost much of its initial utility, and its continued use must now be largely a confession of ignorance. The progressive abandonment of this term by sedimentologists in Western Europe and North America during the 1970s has followed the realization that flysch sequences are the heterogeneous products of a variety of now-separable marginal to distal marine environments (prodelta slopes, deepsea fans, basin plains, trenches, and so forth) which may be formed in diverse geotectonic settings. Some indication of the rapidity of this evolution can be gained by comparing studies reported in Lajoie (1970) with those contained in Dott and Shaver (1974).

The chapters in the present volume also reflect this change in attitude. Thus the ancient turbidite-rich sequences described in Part III make little or no reference to the term *flysch* in contrast to studies published a few years ago on the same or similar successions. As examples we may cite the sequences in the Apennines (Sestini 1970 versus Mutti et al., Chapter 15 of this volume) and the Borden Formation of the U.S. Eastern Interior Basin (Moore and Clarke 1970 versus Kepferle, Chapter 16 of this volume). However, the term tends to be retained for sequences where the geotectonic settings are most complex, such as trench-arc regions (see Part V of this volume). We submit that, if it is to be used at all, the term *flysch* now should be confined to descriptive, nongenetic use at the outcrop.

Catch-all terms applied to more specific processes or products, such as *wildflysch, fluxoturbidite,* or *pelagic* (as commonly used in the past), are also being relinquished, or refined, in the light of data acquired from modern and experimental investigations.

Even the study of modern sediments is not free from this tendency to overgeneralize. A rapidly increasing number of studies utilizing continuous reflection profiles designate as *turbidites* those subbottom sequences that display parallel reflectors; sequences that appear acoustically transparent frequently are termed *contourites* (cf. Ryan et al. 1973). As sedimentologists we counsel caution in the use of such specific genetic labels until more direct evidence is available of the sediment characteristics giving rise to these acoustic responses.

The terminology applied to some modern deep marine physiographic provinces also may obscure their real nature. A case in point is the term *rise,* as applied to extensive and volumetrically important base-of-slope wedges present in all the large ocean basins and also in smaller basins such as the Mediterranean Sea and the Black Sea. The use of this term may be regarded as a temporary expedient, required only until sufficient data is available to distinguish features formed by coalescence of megafans from linear sediment wedges conveyed and smoothed primarily by geotrophic contour currents or by other processes as yet unrecognized.

MODELS TO PLAY WITH

The interaction between research on modern and ancient sediments has enabled sedimentologists to keep pace with other advancing fields in geology, and depositional models are a measure of the progress achieved. These allow qualitative—and, in some instances, semiquantitative—

assessment of the processes and environments represented in sedimentary sequences. As indicated in the previous section, formulation of such models in the study of deeper marine sequences provides a means of systematizing and interpreting more precisely the observations. In turn, this more formal conceptualization stimulates research to probe, test, and refine presently speculative theories.

The 1970s has seen the proliferation of models depicting the temporal and spatial relations of sediments in deepsea fans. Seemingly random lithological variations—and aberrations—in ancient marine successions now have become explicable in terms of the subenvironments and the evolutionary history of fan and canyon systems. The growing number of studies of fans, both modern and ancient (see Chapters 12 through 16), demonstrates that, as in all natural systems, there is a spectrum of possible modes of organization. Accompanying the development of such models, the somewhat simplistic conception of depositional proximality in turbidite environments as advocated, for example, by Walker (1967) becomes more readily accommodated to observations. Thus, coarse channelized sand bodies occurring in distal (outer fan, trench, basin plain) settings may show "proximal" lithological characters (see Chapters 11, 17, and 21). Conversely, sediment facies displaying "distal" attributes may occur in the slope regions closer to source (see Chapter 8) or as overbank deposits of interchannel regions of deepsea fans (see Chapter 13).

The encouraging progress made in identifying and understanding fan sequences should not mislead us into a belief that we now possess a definitive knowledge of these phenomena. Information gaps become apparent, for example, in attempting to establish the precise nature of the process continuum involving canyon to fan valley sediment transfer. Another perplexing aspect concerns the dearth of canyon deposits reported in the geological record. Is this because they occur on margins that are prone to tectonic obliteration? Or is it because once canyons are inactive, the fill of such features is indistinguishable from adjacent slope sediments? Perhaps most ancient fans have originated from deltaic spillover and lack a permanently connected, incised canyon system (cf. Nile Cone, Chapters 17 and 18; Pennine Basin of England and U.S. Eastern Interior Basin, Chapter 9). At present, there is insufficient information from modern delta-fan couples to confirm this last speculation. In this respect the clear link between deltas formed off major rivers and their extension as cones or fans on the deeper adjacent margins is a topic ripe for investigation. Studies of rivers such as the Mississippi, Magdalena, Congo, and Ganges (Shepard 1960; Davies 1972; Shepard et al. 1968; Shepard and Emery 1973; Curray and Moore 1971) demonstrate the necessity of studying in conjunction the shallow and deep portions of these sediment input systems. There is a specific difficulty in establishing the genesis of the sequences termed prodelta turbidites (see Chapter 9). Do these deposits primarily represent river-borne sediment incursions accumulated on delta front slopes, or do they result from mobilization of sediment by failure on such metastable slopes, and the subsequent displacement of these materials onto relatively flat basin floors?

Depositional models relating to both modern and ancient noncarbonate sequences in shelf break and upper to mid-slope regions remain generally undefined. Similarly, there is uncertainty regarding the nature of the distal fan to basin plain juncture. Slope, fan, and basin plain sequences appear to merge in some cases, while in other instances the basin plain deposits may result from bypassing of the more proximal base-of-slope sectors (cf. Heezen and Hollister 1971). The ability to distinguish these two patterns is crucial in dealing with disconnected outcrops of ancient sediments that display evidence of both lateral and axial transport. An example is the early Silurian basin of central Wales (Kelling 1964; Kelling and Woollands 1969).

The problem of formulating different types of models for the various base-of-slope wedges termed *rises* has been exacerbated by a lack of modern data, which are essentially inadequate with respect to density, geographic extent and vertical sedimentological control. Applying the term *rise* to sediment bodies of the past may be rash until modern rises are better defined (cf. Bein and Weiler, 1976).

Another consideration in the study of modern deep marine deposits is the palimpsest character of surficial sediments—that is, the reworking of materials from earlier and possibly different environments (cf. Swift et al. 1971). As in the case of shelf sequences, the sediments in canyons, fans and trenches are not merely relict Pleistocene (or earlier) end-products, but record the combined effects of both former and ongoing hydrodynamic mechanisms. One example of this genetic complexity is provided by the Monterey Fan off California (Hess and Normark 1976).

Moreover, most of the sedimentary models advanced thus far for the deeper marine environments have relied upon a small, geographically restricted number of actualistic examples. In most of these, available techniques provide little information concerning detailed internal organization. For this type of information, we must still turn to ancient analogs. Although such an approach may be useful in understanding deep marine sequences, we need to be cognizant of the dangers of circular sophistry, and we need to ensure that models developed in this manner are truly actualistic and representative.

One final point concerns hitherto neglected deep marine environmental indicators, notably geochemical and biological (see Chapters 5 and 6), which with few exceptions, such as the organic markings and lebenspurren, have not been adequately catalogued. Geochemical parameters, in conjunction with compositional and textural indices (e.g., porosity, permeability), should provide valuable complementary information in formulating depositional models. Such parameters may be particularly useful in evaluating small samples derived from low-density sampling systems, such as boreholes, soft-sediment cores or dredge hauls. In areas of hydrocarbon exploration, the need to distinguish source rock from reservoir sediments and thus overbank from channelized deposits, renders this approach more than an esoteric exercise.

NEW TOYS AND GAME PLANS

Sedimentological process-response models require the ordered input of a body of quantitative data that is variable in nature. With respect to modern deep marine sediments, systematic aquisition of this data is routinely difficult. As a result hypotheses have outstripped the hard facts. However, that this gap is diminishing through the increasing use of new technologies is apparent. The expansion of "eyeball" or indirect observations of the sea floor by means of submersibles (see Chapter 4), underwater television and photography (see Chapter 6), side-scan systems (Belderson et al. 1972), and near-bottom acoustic "sediment sniffers" has greatly facilitated investigation beyond the shelf edge.

However, surficial observations of the sea floor may mislead the observer interested in events of longer-term significance. A case in point are the bed forms, such as ripple marks, seen and measured in submarine canyons, which are induced by bottom currents now demonstrated to be temporally variable in orientation, intensity, and origin (see Chapters 1, 2, and 4). Thus, a need exists to supplement direct and indirect observations of the sea floor with systems designed for long-term collecting of data on parameters such as bottom current velocities and directions, and suspended sediment concentrations. Systems of this type are already in wide use in shelf regions (see chapters in Swift et al. 1972; Stanley and Swift 1976).

For geologists, the element of time is paramount. Samples from the ocean floor (a total of about 460 at this writing) collected by the Deep-Sea Drilling Project provide one means of assessing the significance of this fourth dimension. Deep-tow geophysical packages, capable of fine resolution of sediment layers within the upper few tens of meters of the subbottom, offer an indirect means of evaluating the temporal as well as the spatial organization of deep marine sediment bodies (see Hess and Normark 1976, for a recent application of this technique to a modern deep sea fan).

Longer-term monitoring of seafloor or water column attributes requires of the sedimentologist the ability to utilize and interpret data that hitherto have lain within the province of the physical oceanographer. The most significant advances in marine sedimentology have been made through an awareness that this realm is a dynamic one and best rewards those with the ability to integrate the principles of fluid dynamics with those of sediment transport. We now recognize that mechanisms involved in deep marine environments may be assigned to suspension, traction and gravity-induced mass flow.

Progress in the study of suspensate sedimentation has been achieved largely through the use of optical techniques (Lisitzin 1972; Eittreim and Ewing 1972; Drake 1976; Pierce 1976). Such research led to the discovery of nepheloid layers as a significant element in outer margin sedimentation (Eittreim et al. 1976). Further growth in this productive field of investigation calls for sampling of suspensate-rich water bodies, concurrently with the measurement of physical and chemical water mass properties (see Chapter 3). There is evidence that the prolonged movement of large volumes of suspensate-rich waters, driven by thermohaline forces, contributes to the formation or modification of sediment wedges on the rises and in the deep ocean basins (cf. contour currents, Heezen and Hollister 1971). However, the role of nepheloid layers and geostrophic contour-following currents on marginal slopes and in fans and trenches remains obscure and unquantified.

The increasing evidence that virtually every part of the deep sea floor is influenced by moving water masses requires some reassessment of the role of tractional processes. The existence of large dune-like forms in environments generally considered to be dominated by downslope, gravity-controlled processes (Bouma and Treadwell 1975; Jacobi et al. 1975) suggests that tractional effects may be more widespread on deepsea fans (Hess and Normark 1976) than previously considered (see papers in Heezen 1977). The effects of flowing water at the seafloor are manifested by the molding of clay and the entrainment of sand, silt and clay particles, graphically portrayed by Heezen and Hollister (1971).

In addition to the thermohaline circulation systems, other possible sediment-affecting mechanisms such as internal waves and deep tidal flows increasingly are being recognized. Such deep perturbations, hitherto measured mainly in submarine canyons (see Chapter 1), almost certainly have counterparts on deepsea fans and in trenches. We need to know how constant are these driving forces in terms of direction and intensity. Are they permanent and how do they interface with gravity-induced flows, particularly those of low density? As the distinction between hemipelagic and mud turbidite flows (see Chapter 12) becomes better established, we should be able to evaluate the influence of persistently moving water masses on low-density flows proceeding downslope (see Stanley et al. 1971; Pierce 1976, Figure 5).

This growing realization of the sedimentological importance of the plexus of physical oceanographic parameters is matched by an awareness that gravity-generated processes also are highly varied. Deep marine sedimentology has progressed beyond the view that deep marine environments are characterized by a constant and gentle rain of sediment punctuated by rare incursions of turbidity flows. The transformation from slump to turbidity current, first postulated by Kuenen (1951), while still unobserved in the modern (at least by anyone who survived to tell!) remains a supposition. However, indirect evidence (see Chapter 7), theoretical considerations (Morgenstern 1967), and observations from the geological record (see Chapter 8) provide support for the existence of this phenomenon.

Other mechanisms exist within the spectrum of sediment gravity flow processes (Middleton and Hampton 1973; Carter 1975), and some of these are sufficiently distinct from "classical" turbidity currents to allow unequivocal identification of their deposits. Perhaps the debris flow process (Hampton 1972) provides the most readily recognizable sedimentary units both in ancient sequences (see Chapters 10 and 11) and in modern (Embley 1976) deposits. Less clear-cut is the distinction between the products of turbidity flows and mass or inertia flows (grain and fluidized flows, slurries, and so forth). Units that do not resemble the classic Bouma (1962) type turbidite conceivably may still have been transported by flows presenting the essential hydrodynamic attributes of turbidity currents. In such cases, departures from the ideal turbidite sequence may result from events occurring during the final stages of deposition (Walton 1967; Carter 1975). The possibilities for clarification offered by experiments and by careful examination of sequences assigned to this range of processes are far from being exhausted. Furthermore, we anticipate that monitoring of natural or artificially induced sediment gravity flows, particularly in the mid-canyon to upper fan valley regions and on delta front slopes, offers a new and potentially illuminating perspective on the problem of coarse sediment transfer to deep water.

As the study of geotechnical properties of the sea floor proceeds beyond the shelf edge, it becomes clear that biologic activity is an important factor in determining the stability of sediments. The obvious bioerosive effects produced in partially lithified canyon walls (see Chapter 6) are accompanied by more subtle changes such as those induced by organisms burrowing in soft sediment. Changes in pore water pressure caused by burrowers in metastable slope sequences may be sufficient to initiate failure and downslope movement. We can predict that similar processes may be effective in even deeper regions where high slopes prevail, such as trench or fan valley walls. The widespread occurrence of rock outcrops and their debris in modern slope and base-of-slope environments attests the prevalence of erosive agencies, in part biological.

Submarine weathering may supplement the erosive effects of organism and physical transport processes. With the exception of a few localized studies (e.g., Martin and Emery 1967) little effort has been directed towards the evaluation of submarine alteration of rock masses. Improved techniques of underwater drilling, such as that utilized in the Mid-Atlantic FAMOUS project (Davis et al. 1975), opens the door to further research in this area.

The *in situ* production of debris as a result of the processes enumerated above offers an alternative mechanism to those commonly invoked to explain the occurrence of large clasts in deep water environments—that is, eustatic changes in sea level and tectonic forces. This view of an internal source for the coarse (and fine) components of canyon-fan and trench complexes provides a new perspective for paleogeographic reconstructions of ancient basins (Chapter 11).

In this section we have touched on all the textural grades encountered beyond the shelf edge. Looking objectively, we clearly see that contemporary concepts regarding canyon, fan, and trench sedimentation derive largely from the study of sands and particularly turbidites. We must acknowledge that studies of this type of deepwater sedimentation received their impetus from the vision and enthusiasm of Philip H. Kuenen, to whom this book is dedicated, and after more than a quarter-century of concentration on this topic, we believe it to be timely that efforts are at last being directed towards understanding the origin of deepwater gravels (see Chapters 11 and 24 of this volume; Walker 1975; Helú et al. 1977) and towards elucidation of the fine fraction that constitutes the bulk of most deep marine sequences (see Chapters 3 and 12 of this volume; Rupke and Stanley 1974; Maldonado and Stanley 1976).

SUMMING UP

Fan: Originally a device for separating grain from chaff—
Webster's New World Dictionary, College Edition, 1966

In previous sections of this chapter, we have attempted to summarize, according to our doubtless idiosyncratic judgement, the present status and possible future development of research in canyon, fan, and trench sedimentation. The dynamic state of this field is evident from our consideration of the six major aspects outlined above, although the rate of growth appears to be uneven. Assuming that the pursuit of all of the topics cited will prove equally fruitful would be unduly optimistic. On the one hand, the application of economic pressures is likely to remain uneven, and on the other, our hypotheses and speculations will probably continue to outstrip the technological resources required to substantiate them. However, one advantage possessed by researchers in modern deeper waters over their brethren in the shelf realm is that international cooperation in this deep realm is less subject to the growing political constraints imposed by the extension of territorial interests.

If we can judge the maturity of a geological field in terms of the Lyellian principle of uniformitarianism, where the present provides the key to the past, we have to admit that the area of research with which we are concerned has not yet attained full maturity. The contributions in this volume are closely divided between ancient and modern categories and may be regarded as broadly representative of the present state of research. However, in terms of concepts and modeling, the study of ancient deposits clearly has provided the greater input to our understanding of the environmental trilogy dealt with here. While logistical obstacles have contributed to this bias, another factor has been the general reluctance, until recently, of sedimentologists to actively attack problems requiring rigorous involvement with the dynamics of sediment transport. Only when we can demonstrate that studies of existing systems are providing the bulk of the criteria required for interpretation of ancient sequences will the field attain conceptual maturity.

Until such time, we urge that students of deep marine sedimentation who wish to utilize the term *fan* bear in mind the original meaning of the word as defined by Webster and quoted at the head of this section!

ACKNOWLEDGMENTS

The authors express their appreciation to the Smithsonian Research Foundation (grant 71702120) and to the Royal Society (London) for financial support enabling us to undertake the editorial task that indirectly generated this contribution. The thoughts and hopes expressed herein derive in large measure from experiences and discussions shared with colleagues too numerous to name individually. We hope, nevertheless, that they will be assured of our gratitude.

REFERENCES

Aalto, K. R., 1976. Sedimentology of a mélange: Franciscan of Trinidad, California. *J. Sed. Petrol.*, 46: 913-29.

Almgren, A. A., 1978. Timing of Tertiary submarine canyons and marine cycles of deposition in the southern Sacramento Valley, California. In: D. J. Stanley and G. Kelling (eds.), *Sedimentation in Submarine Canyons, Fans, and Trenches.* Dowden, Hutchinson & Ross, Stroudsburg, Pa., chapter 19.

Bein, A., and Weiler, Y., 1976. The Cretaceous Talme Yafe Formation: a contour current shaped sedimentary prism of calcareous detritus at the continental margin of the Arabian Craton. *Sediment.*, 23: 511–32.

Belderson, R. H., Kenyon, N. H., Stride, A. H., and Stubbs, A. R., 1972. *Sonographs of the Sea Floor, a Pictorial Atlas.* Elsevier, Amsterdam, 185 pp.

Bloomer, R. R., 1977. Depositional environments of a reservoir sandstone in West-Central Texas. *Amer. Assoc. Petrol. Geol. Bull.*, 61: 344-59.

Bostrom, R. C., and Sherif, M. A., 1970. Disposal of waste material in tectonic sinks. *Nature*, 228: 154-56.

Bouma, A. H., 1962. *Sedimentology of Some Flysch Deposits.* Elsevier, Amsterdam, 168 pp.

———, and Treadwell, T. K., 1975. Deep sea dune-like features. *Mar. Geol.*, 19: M53-M59.

Burk, C. A. and Drake, C. L. (eds.), 1974. *The Geology of Continental Margins.* Springer-Verlag, New York, 1009 pp.

Cacchione, D. A., Rowe, G. T., and Malahoff, A., 1978. Submersible investigation of outer Hudson Submarine Canyon. In: D. J. Stanley and G. Kelling (eds.), *Sedimentation in Submarine Canyons, Fans, and Trenches.* Dowden, Hutchinson & Ross, Stroudsburg, Pa., chapter 4.

Carter, R. W., 1975. A discussion and classification of subaqueous mass transport with particular application to grain-flow, slurry-flow, and fluxoturbidites. *Earth Sci. Reviews*, 11: 146-77.

Carter, R. M., Hicks, M. D., Norris, R. J., and Turnbull, I. M., 1978. Sedimentation patterns in an ancient arc-trench-ocean basin complex, Carboniferous to Jurassic Rangitata Orogen, New Zealand. In: D. J. Stanley and G. Kelling (eds.), *Sedimentation in Submarine Canyons, Fans, and Trenches.* Dowden, Hutchinson & Ross, Stroudsburg, Pa., chapter 23.

Conolly, J. R., and Ewing, M., 1967. Sedimentation in the Puerto Rico Trench. *J. Sed. Petrol.*, 37: 44-59.

Cossey, S. P., and Ehrlich, R., 1978. Growth fault-controlled submarine carbonate debris flow and turbidite deposits from the Jurassic of Northern Tunisia: Possible canyon fill sequences. In: D. J. Stanley and G. Kelling (eds.), *Sedimentation in Submarine Canyons, Fans, and Trenches.* Dowden, Hutchinson & Ross, Stroudsburg, Pa., chapter 10.

Curray, J. C., and Moore, D. G., 1971. Growth of the Bengal deep-sea fan and denudation in the Himalayas. *Geol. Soc. Amer. Bull.*, 82: 563-72.

Davies, D. K., 1972. Mineralogy, petrography and derivation of sands and silts of the continental slope, rise and abyssal plain of the Gulf of Mexico. *J. Sed. Petrol.*, 42: 59-65.

Davies, W., and Cave, R., 1976. Folding and cleavage determined during sedimentation. *Sedim. Geol.*, 15: 89-133.

Davis, R. E., Williams, D. L., and von Herzen, R. P., 1975. ARPA rock drill report. *Woods Hole Oceanogr. Inst. Rept.*, pp. 75-28.

Dickinson, W. R., (ed.). 1974. *Tectonics and Sedimentation.* Soc. Econ. Paleont. Mineral., Sp. Publ. 22, 204 pp.

Dott, R. H., Jr., and Shaver, R. H., (eds.), 1974. *Modern and Ancient Geosynclinal Sedimentation.* Soc. Econ. Paleont. Mineral., Sp. Publ. 19, 380 pp.

Drake, D. E., 1976. Suspended sediment transport and mud deposition on continental shelves. In: D. J. Stanley and D. J. P. Swift (eds.), *Marine Sediment Transport and Environment Management.* Wiley, New York, pp. 127–58.

———, Hatcher, P.G., and Keller, G. H., 1978. Suspended particulate matter and mud deposition in upper Hudson Submarine Canyon. In: D. J. Stanley and G. Kelling (eds.), *Sedimentation in Submarine Canyons, Fans, and Trenches.* Dowden, Hutchinson & Ross, Stroudsburg, Pa., chapter 3.

Drooger, C. W. (ed.), 1973. *Messinian Events in the Mediterranean.* North-Holland, Amsterdam, 272 pp.

Dzulynski, S., Ksiazkiewicz, M., and Kuenen, Ph. H., 1959. Turbidites in flysch of the Polish Carpathian Mountains. *Geol. Soc. Amer. Bull.,* 70: 1089-118.

Eittreim, S., and Ewing, M., 1972. Suspended particulate matter in the deep waters of the North American Basin. In: A. L. Gordon (ed.), *Studies in Physical Oceanography,* 2. Gordon and Breach, New York, pp. 123-67.

——, Thorndike, E. M., and Sullivan, L., 1976. Turbidity distribution in the Atlantic Ocean. *Deep-Sea Res.,* 23: 1115-27.

Embley, R., 1976. New evidence for occurrence of debris flow deposits in the deep sea. *Geology,* 4: 371-74.

Emery, K. O., 1970. Continental margins of the world. *The Geology of the East Atlantic Continental Margin,* 1. General and Economic Papers, ICSU/SCOR Working Party 31 Symposium Cambridge, Rept. No. 70/13, pp. 7-29.

Fischer, R. L., 1974. Pacific-type continental margins. In: C. A. Burk and C. L. Drake (eds.), *The Geology of Continental Margins.* Springer-Verlag, New York, pp. 25-41.

Gorsline, D. S., and Emery, K. O., 1959. Turbidity-current deposits in San Pedro and Santa Monica Basins off southern California. *Geol. Soc. Amer. Bull.,* 70: 279-90.

Gross, M. G. (ed.), 1976. Middle Atlantic continental shelf and the New York Bight. *Spec. Symp. 2, Amer. Soc. Limn. Oceanogr.,* Lawrence, Kansas, pp. 1-441.

Hampton, M. A., 1972. The role of subaqueous debris flow in generating turbidity currents. *J. Sed. Petrol.,* 42: 775-93.

Heezen, B. C. (ed.), 1977. Influence of abyssal circulation on sedimentary accumulations in space and time. *Mar. Geol.,* Spec. Issue 23, pp. 1-215.

——, and Ewing, M., 1952. Turbidity currents and submarine slumps, and the Grand Banks earthquake. *Amer. J. Sci.,* 250: 849-73.

——, and Hollister, C. D., 1971. *The Face of the Deep.* Oxford University Press, New York, 659 pp.

——, and Rawson, M., 1977. Visual observations of the sea floor subduction line in the Middle-America Trench. *Science,* 196: 423-426.

——, Tharp, M. and Ewing, M., 1959. *The Floors of the Oceans. I: The North Atlantic.* Geol. Soc. Amer., Spec. Paper 65, 122 pp.

Helú, P. C., Verdugo, V. R., and Bárcenas, P. R., 1977. Origin and distribution of Tertiary conglomerates, Veracruz Basin, México. *Amer. Assoc. Petrol. Geol. Bull.,* 61: 207-26.

Hess, G. R., and Normark, W. R., 1976. Holocene sedimentation history of the major fan valleys of Monterey Fan. *Mar. Geol.,* 22: 233-51.

Hess, H. H., 1938. Gravity anomalies and island arc structure with particular reference to the West Indies. *Proc. Amer. Phil. Soc.,* 79: 71-96.

Hsü, K. J., 1970. The meaning of the word flysch–a short historical search. In: J. Lajoie (ed.), *Flysch Sedimentology in North America.* Geol. Assoc. Canada, Spec. Paper 7, pp. 1-11.

——, 1974. Mélanges and their distinction from olistostromes. In: R. H. Dott, Jr., and R. H. Shaver (eds.), *Modern and Ancient Geosynclinal Sedimentation.* Soc. Econ. Paleont. Mineral., Sp. Publ. 19, pp. 321-33.

Jacobi, R. D., Rabinowitz, P. D. and Embley, R. W., 1975. Sediment waves on the Moroccan continental rise. *Mar. Geol.,* 19: M61-M67.

Kay, M., 1951. *North American Geosynclines.* Geol. Soc. Amer. Mem. 48, 143 pp.

Keller, G. H., and Shepard, R. P., 1978. Currents and sedimentary processes in submarine canyons off the northeast United States. In: D. J. Stanley and G. Kelling (eds.), *Sedimentation in Submarine Canyons, Fans, and Trenches.* Dowden, Hutchinson & Ross, Stroudsburg, Pa., chapter 2.

Kelling, G., 1964. The turbidite concept in Britain. In: A. H. Bouma and A. Brouwer (eds.), *Turbidites.* Dev. Sediment., 3. Elsevier, Amsterdam, pp. 75-92.

——, and Holroyd, J., 1978. Clast size, shape, and composition in some ancient and modern fan gravels. In: D. J. Stanley and G. Kelling (eds.), *Sedimentation in Submarine Canyons, Fans, and Trenches.* Dowden, Hutchinson & Ross, Stroudsburg, Pa., chapter 11.

——, and Woollands, M. A., 1969. The stratigraphy and sedimentation of the Llandoverian rocks of the Rhayader district. In: A. Wood (ed.), *The Pre-Cambrian and Lower Paleozoic Rocks of Wales.* University of Wales Press, Cardiff, pp. 255-82.

Kepferle, R. C., 1978. Prodelta turbidite fan apron in Borden Formation (Mississippian), Kentucky and Indiana. In: D. J. Stanley and G. Kelling (eds.), *Sedimentation in Submarine Canyons, Fans, and Trenches.* Dowden, Hutchinson & Ross, Stroudsburg, Pa., chapter 16.

Kuenen, Ph.H., 1950. *Marine Geology.* Wiley, New York, 568 pp.

——, 1951. Properties of turbidity currents of high density. In: R. D. Russell (ed.), *Turbidity Currents and the Transportation of Coarse Sediments to Deep Water,* Soc. Econ. Paleont. Mineral., Sp. Publ. 2, pp. 14-33.

Lajoie, J. (ed.), 1970. *Flysch Sedimentology in North America.* Geol. Assoc. Canad., Spec. Paper 7, 272 pp.

Lisitzin, A. P., 1972. *Sedimentation in the World Ocean.* Soc. Econ. Paleont. Mineral., Sp. Publ. 17, 218 pp.

McCabe, P. J., 1978. The Kinderscoutian Delta (Carboniferous) of Northern England: A slope influenced by density currents. In: D. J. Stanley and G. Kelling (eds.), *Sedimentation in Submarine Canyons, Fans, and Trenches.* Dowden, Hutchinson & Ross, Stroudsburg, Pa., chapter 9.

Maldonado, A., and Stanley, D. J., 1976. Late Quaternary sedimentation and stratigraphy in the Strait of Sicily. *Smithsonian Contrib. Earth Sci.,* 16, 73 pp.

——, and Stanley, D. J., 1978. Nile Cone depositional processes and patterns in the late Quaternary. In: D. J. Stanley and G. Kelling (eds.), *Sedimentation in Submarine Canyons, Fans, and Trenches.* Dowden, Hutchinson & Ross, Stroudsburg, Pa., chapter 17.

Marshall, N. F., 1978. A large storm-induced sediment slump reopens an unknown Scripps Submarine Canyon tributary. In: D. J. Stanley and G. Kelling (eds.), *Sedimentation in Submarine Canyons, Fans, and Trenches.* Dowden, Hutchinson & Ross, Stroudsburg, Pa., chapter 7.

Martin, B. D., and Emery, K. O., 1967. Geology of Monterey Canyon, California. *Amer. Assoc. Petrol. Geol. Bull.,* 51: 2281-304.

Menard, H. W., 1955. Deep-sea channels, topography and sedimentation. *Amer. Assoc. Petrol. Geol. Bull., 39: 236-55.*

——, 1960. Possible pre-Pleistocene deep-sea fans off central California. *Geol. Soc. Amer. Bull.,* 71: 1271-278.

——, 1964. *Marine Geology of the Pacific.* McGraw-Hill, New York. 271 pp.

Middleton, G. V., and Bouma, A. H. (eds.), 1973. *Turbidites and Deep Water Sedimentation.* Soc. Econ. Paleont. Mineral., Pacific Section, Short Course, Anaheim, 157 pp.

——, and Hampton, M. A., 1973. Mechanics of flow and deposition. In: G. V. Middleton and A. H. Bouma (eds.), *Turbidites and Deep-Water Sedimentation.* Soc. Econ. Paleont. Mineral., Pacific Section, Short Course, Anaheim, pp. 1-38.

Miller, R. L., (ed.), 1965. *Papers in Marine Geology:* Shepard Commemorative Volume. Macmillan, New York, 531 pp.

Moore, B. R., and Clarke, M. K., 1970. The significance of a turbidite sequence in the Borden Formation (Mississippian) of eastern Kentucky and southern Ohio. In: J. Lajoie (ed.), *Flysch Sedimentology in North America.* Geol. Assoc. Canada, Spec. Paper 7, pp. 211-18.

Moore, D. G., 1969. *Reflection Profiling Studies of the California Continental Borderland: Structure and Quaternary Turbidite Basins.* Geol. Soc. Amer. Spec. Paper 107, 142 pp.

Morgenstern, N., 1967. Submarine slumping and the initiation of turbidity currents. In: A. F. Richards (ed.), *Marine Geotechnique.* University of Illinois Press, Urbana, pp. 189-220.

Murray, J. and Hjort, J., 1912. *The Depths of the Ocean.* Macmillan, London, 821 pp.

Mutti, E., 1974. Examples of ancient deep-sea fan deposits from circum-Mediterranean geosynclines. In: R. H. Dott, Jr., and R. H. Shaver (eds.), *Modern and Ancient Geosynclinal Sedimentation.* Soc. Econ. Paleont. Mineral., Sp. Publ. 19, pp. 92-105.

_____, Nilsen, T. H., and Ricci Lucchi, F., 1978. Outer fan depositional lobes of the Laga Formation (Upper Miocene and Lower Pliocene), East-Central Italy. In: D. J. Stanley and G. Kelling (eds.), *Sedimentation in Submarine Canyons, Fans, and Trenches.* Dowden, Hutchinson & Ross, Stroudsburg, Pa., chapter 15.

_____, and Ricci Lucchi, F., 1972. Le torbiditi dell'Appennino settentrionale: introduzione all'analisi di facies. *Mem. Soc. Geol. Italiana,* 11: 161-99.

Nelson, C. H., and Kulm, V., 1973. Submarine fans and channels. In: G. V. Middleton and A. H. Bouma (eds.), *Turbidites and Deep-Water Sedimentation.* Soc. Econ. Paleont. Mineral., Pacific Section, Short Course, Anaheim, pp. 39-78.

_____, Normark, W. R., Bouma, A. H., and Carlson, P. R., 1978. Thin-bedded turbidites in modern submarine canyons and fans. In: D. J. Stanley and G. Kelling (eds.), *Sedimentation in Submarine Canyons, Fans, and Trenches.* Dowden, Hutchinson & Ross, Stroudsburg, Pa., chapter 13.

Normark, W. R., 1970. Growth patterns of deep-sea fans. *Amer. Assoc. Petrol. Geol. Bull.,* 54: 2170-95.

Parker, J. R., 1975. Lower Tertiary sand development in the central North Sea. In: A. W. Woodland (ed.), *Petroleum and the Continental Shelf of North-West Europe.* I: Geology. Wiley, New York, pp. 447-52.

Payne, M. W., 1976. Basinal sandstone facies, Delaware Basin, West Texas and Southeast New Mexico. *Amer. Assoc. Petrol. Geol. Bull.,* 60: 517-27.

Pescatore, T., 1978. The Irpinids: A model of tectonically controlled fan and base-of-slope sedimentation in Southern Italy. In: D. J. Stanley and G. Kelling (eds.), *Sedimentation in Submarine Canyons, Fans, and Trenches.* Dowden, Hutchinson & Ross, Stroudsburg, Pa., chapter 22.

Pettijohn, F. P., 1957. *Sedimentary Rocks.* (2nd ed.) Harper & Brothers, New York 718 pp.

Pierce, J. W., 1976. Suspended sediment transport at the shelf break and over the outer margin. In: D. J. Stanley and D. J. P. Swift (eds.), *Marine Sediment Transport and Environmental Management.* Wiley, New York, pp. 437-58.

Picard, J. and Dietz, R. S., 1961. *Seven Miles Down.* Putnam, New York, 249 pp.

Piper, D. J. W., 1978. Turbidite muds and silts on deepsea fans and abyssal plains. In: D. J. Stanley and G. Kelling (eds.), *Sedimentation in Submarine Canyons, Fans, and Trenches.* Dowden, Hutchinson & Ross, Stroudsburg, Pa., chapter 12.

Revelle, R., Bramlette, M., Arrhenius, G. and Goldberg, E. D., 1955. Pelagic sediments of the Pacific. In: A. Poldevaart (ed.), *Crust of the Earth.* Geol. Soc. Amer., Spec. Paper 62, pp. 221-36.

Ricci Lucchi, F., 1975. Depositional cycles in two turbidite formations of northern Apennines (Italy). *J. Sed. Petrol.,* 45: 3-43.

Ross, D. A., Uchupi E., Summerhayes, C. P., Koelsch, D. E., and El Shazly, E. M., 1978. Sedimentation and structure of the Nile Cone and Levant Platform area. In: D. J. Stanley and G. Kelling (eds.), *Sedimentation in Submarine Canyons, Fans, and Trenches.* Dowden, Hutchinson & Ross, Stroudsburg, Pa., chapter 18.

Rupke, N. A., and Stanley, D. J., 1974. Distinctive properties of turbiditic and hemipelagic mud layers in the Algéro-Balearic Basin, Western Mediterranean Sea. *Smithsonian Contr. Earth Sci.,* 13, 40 pp.

Ryan, W. B. F., Hsü, K. J., et al., 1973. *Initial Reports of the Deep Sea Drilling Project,* 13. U. S. Govt. Print. Off., Washington, D. C., 1447 pp.

Saxena, R. S. (ed.), 1976. *Sedimentary Environments and Hydrocarbons.* New Orleans Geol. Soc., Amer. Assoc. Petrol. Geol., Short Course Notes, New Orleans.

Scholl, D. W., and Marlow, M. S., 1974. Sedimentary sequence in modern Pacific trenches and the deformed circum-Pacific eugeosyncline. In: R. H. Dott, Jr., and R. H. Shaver (eds.), *Modern and Ancient Geosynclinal Sedimentation.* Soc. Econ. Paleont. Mineral., Sp. Publ. 19, pp. 193-211.

Schweller, W. J., and Kulm, L. D., 1978. Depositional patterns and channelized sedimentation in active eastern Pacific trenches. In: D. J. Stanley and G. Kelling (eds.), *Sedimentation in Submarine Canyons, Fans, and Trenches.* Dowden, Hutchinson & Ross, Stroudsburg, Pa., chapter 21.

Scott, R. M., and Birdsall, B. C., 1978. Physical and biogenic characteristics of sediments from Hueneme Submarine Canyon, California coast. In: D. J. Stanley and G. Kelling (eds.), *Sedimentation in Submarine Canyons, Fans, and Trenches.* Dowden, Hutchinson & Ross, Stroudsburg, Pa., chapter 5.

Selley, R. C., 1975. Subsurface environmental analysis of North Sea sediments *Amer. Assoc. Petrol. Geol. Bull.,* 60: 184-95.

Sestini, G. (ed.), 1970. Development of the Northern Apennines Geosyncline. *Sedim. Geol. Spec. Issue* 4: 203-644.

Shepard, F. P., 1948. *Submarine Geology.* (1st ed.) Harper & Bros., New York, 338 pp.

_____, 1960. Mississippi Delta: Marginal environments, sediments, and growth. In: F. P. Shepard, F. B. Phleger, and T. H. van Andel (eds.), *Recent Sediments, Northwest Gulf of Mexico.* Amer. Assoc. Petrol. Geol., Tulsa, pp. 56-81.

_____, 1963. Submarine Canyons. In: M. N. Hill (ed.), *The Sea, 3.* Wiley-Interscience, New York, pp. 480-506.

_____, and Dill, R. F., 1966. *Submarine Canyons and other Sea Valleys.* Rand McNally, Chicago, 381 pp.

_____, Dill, R. F., and Heezen, B. C., 1968. Diapiric intrusions in foreset slope sediments off Magadalena Delta, Colombia. *Amer. Assoc. Petrol. Geol. Bull.,* 52: 2197-207.

_____, and Emery, K. O., 1941. Submarine topography off the California coast: Canyons and tectonic interpretations. *Geol. Soc. Amer. Sp. Paper* 31, 171 pp.

_____, and Emery, K. O., 1973. Congo submarine canyon and fan valley. *Amer. Assoc. Petrol. Geol. Bull.,* 57: 1679-91.

_____, and Marshall, N. F., 1978. Currents in submarine canyons and other sea valleys. In: D. J. Stanley and G. Kelling (eds.), *Sedimentation in Submarine Canyons, Fans, and Trenches.* Dowden, Hutchinson & Ross, Stroudsburg, Pa., chapter 1.

Silver, E. A., 1972. Subduction zones: Not relevant to present-day problems of waste disposal. *Nature,* 239: 330-31.

Stanley, D. J., 1961. *Etudes sédimentologiques des grès d'Annot et de leurs équivalents latéraux.* Inst. Franc. Pétrole Ref. 6821, Société des Editions Technip, Paris, 158 pp.

——, Got, H., Kenyon, N. H., Monaco, A., and Weiler, Y., 1976. Catalonian, Eastern Betic and Balearic margins: Structural types and geologically recent foundering of the Western Mediterranean Basin. *Smithsonian Contr. Earth Sc.,* 20, 67 pp.

——, Palmer, D. H., and Dill, R. F., 1978. Coarse sediment transport by mass flow and turbidity current processes in Annot Sandstone canyon-fan valley systems. In: D. J. Stanley and G. Kelling (eds.), *Sedimentation in Submarine Canyons, Fans, and Trenches.* Dowden, Hutchinson & Ross, Stroudsburg, Pa., chapter 8.

——, Sheng, H., and Pedraza, C. P., 1971. Lower continental rise east of Middle Atlantic States: predominant dispersal perpendicular to isobaths. *Geol. Soc. Amer. Bull.,* 82: 1831-40.

——, and Swift, D. J. P. (eds.), 1976. *Marine Sediment Transport and Environmental Management.* Wiley, New York, 602 pp.

Stille, H., 1955. Recent deformations of the Earth's crust in the light of those of earlier epochs. In: A. Poldervaart (ed.), *Crust of the Earth.* Geol. Soc. Amer., Spec. Paper 62, pp. 171-92.

Swift, D. J. P., Duane, D. B., and Pilkey, O. H. (eds.), 1972. *Shelf Sediment Transport: Process and Pattern.* Dowden, Hutchinson & Ross, Stroudsburg, Pa., 656 pp.

——, Stanley, D. J., and Curray, J. R., 1971. Relict sediments on continental shelves: a reconsideration. *J. Geol.,* 79: 322-46.

Trask, P. D. (ed.), 1939. *Recent Marine Sediments.* Dover Publications, New York, 736 pp.

van Vliet, A., 1978. Early tertiary deepwater fans of Guipuzcoa, Northern Spain. In: D. J. Stanley and G. Kelling (eds.), *Sedimentation in Submarine Canyons, Fans, and Trenches.* Dowden, Hutchinson & Ross, Stroudsburg, Pa., chapter 14.

Vedros, S. G., and Visher, G. S., 1978. The Red Oak Sandstone: A hydrocarbon-producing submarine fan deposit. In: D. J. Stanley and G. Kelling (eds.), *Sedimentation in Submarine Canyons, Fans, and Trenches.* Dowden, Hutchinson & Ross, Stroudsburg, Pa., chapter 20.

Vening Meinesz, F. A., 1948. *Gravity expeditions at sea, 1923-1938,* 4. Pub. Neth. Geod. Comm., Waltman, Delft, 225 pp.

von Rad, U., 1968. Comparison of sedimentation in the Bavarian Flysch (Cretaceous) and recent San Diego Trough (California). *J. Sed. Pet.,* 38: 1120-1154.

Walker, R. G., 1966. Shale Grit and Grindslow Shales: Transition from turbidite to shallow water sediments in the Upper Carboniferous of northern England. *J. Sed. Petrol.,* 36: 90-114.

——, 1967. Turbidite sedimentary structures and their relationship to proximal and distal depositional environments. *J. Sed. Petrol.* 37: 25-43.

——, 1975. Generalized facies models for resedimented conglomerates of turbidite association. *Geol. Soc. Amer. Bull.,* 86: 737-48.

Walton, E. K., 1967. The sequence of internal structures in turbidites. *Scott. J. Geol.,* 3: 306-17.

Warme, J. E., Slater, R. A., and Cooper, R. A., 1978. Bioerosion in submarine canyons. In: D. J. Stanley and G. Kelling (eds.), *Sedimentation in Submarine Canyons, Fans, and Trenches.* Dowden, Hutchinson & Ross, Stroudsburg, Pa., chapter 6.

Whitaker, J. H. McD. (ed.), 1976. *Submarine Canyons and Deep-Sea Fans, Modern and Ancient.* Dowden, Hutchinson & Ross, Stroudsburg, Pa., 460 pp.

Wilde, P., Normark, W. R., and Chase, T. E., 1976. Petroleum potential of continental rise off Central California - Summary. *Amer. Assoc. Petrol. Geol. Mem.* 25: 33-317.

Winn, R. D., Jr., and Dott, R. H., Jr., 1978. Submarine-fan turbidites and resedimented conglomerates in a Mesozoic arc-rear marginal basin in Southern South America. In: D. J. Stanley and G. Kelling (eds.), *Sedimentation in Submarine Canyons, Fans, and Trenches.* Dowden, Hutchinson & Ross, Stroudsburg, Pa., chapter 24.

Index